4-10-96

D1170751

WORK DESIGN

INDUSTRIAL ERGONOMICS

FOURTH EDITION

STEPHAN KONZ
KANSAS STATE UNIVERSITY

Publishing Horizons, Inc.
Scottsdale, Arizona

Editor	Nils Anderson
Developmental Editor	Gay L. Pauley
Production Manager	A. Colette Kelly
Production Editor	Carma Paden
Cover Design	Kevin Kall
Typesetting	Ash Street Typecrafters

Copyright © 1990, 1995 by Publishing Horizons, Inc.
Publishing Horizons, Inc.
An affiliate of Gorsuch Scarisbrick, Publishers
8233 Via Paseo del Norte, Suite F-400
Scottsdale, Arizona 85258

10 9 8 7 6 5 4 3 2 1

ISBN 0-942280-65-2

Printed in the United States of America.

Library of Congress Cataloging-in-Publication Data

Konz, Stephan A.
    Work design: industrial ergonomics / Stephan Konz—4th ed., p. cm.
    Includes bibliographical references and index.
    ISBN 0-942280-65-2
    1. Work design. I. Title.
T60.8.K66 1995
620.8—dc20   94-23944
                                                            CIP

CONTENTS

1 | INDUSTRIAL SOCIETY

OVERVIEW

The developments that make entire nations (rather than just a few individuals) rich are the pivotal developments of history. These developments have occurred both in the larger social structure and within the area of economic development (standard of living).

The 6 characteristics of industrial society (specialization of labor, energy from machines, standardization and interchangeable parts, production from machines, mass production and mass consumption, and the assembly line) have become part of our culture.

CHAPTER CONTENTS

1 Historical Background
2 Improved Living Standards
3 Industrial Society

KEY CONCEPTS

change
Industrial Revolution
industrial society
standard of living

1 HISTORICAL BACKGROUND

In the long tape of history, why are the last few inches so much better? Starting about 1750 in England, about 1800 in the United States, about 1825 in northwestern Europe, and later in the rest of the world, a profound change began—the **Industrial Revolution**—the change from an agriculture- to manufacturing-based society. The key word is **change.** Prior to that time there had been good times and bad times but the overall standard of living was no better in 1700 than it had been in 1000 or in Roman times. Roman houses in England had bathrooms; it wasn't until the time of Queen Victoria that bathrooms again were built in private houses. The rising wind of change has continued until today it is a firestorm. Slow and steady progress no longer is acceptable as people expect miracles from the great flowing rivers of technology.

2 IMPROVED LIVING STANDARDS

The following aspects of our technological society do not have any particular rank order but are all interrelated.

2.1 Increased Knowledge

The nature of knowledge is that it is cumulative and that it diffuses. Occasionally, the reservoir has stagnated, but the general trend is upward. Many innovations require developments in related fields. For example, the steam engine required development of a machine which could bore an accurate cylinder; both the steam engine and the boring machine required a coal technology for fuel and metal.

2.2 Diffusion of Knowledge

Literacy and education are essential. Due to the invention of printing (China and Korea in A.D. 800, Europe in 1400), knowledge can be mass produced at reasonable cost. This in turn has permitted the development of libraries throughout the land rather than in just a few major cities. The system of widely distributed professional journals ensures that specific knowledge in, say, industrial engineering, biochemistry, or electronics becomes available within a month of publication to specialists in that field in Johannesburg, Bombay, Eindhoven, and Manhattan, Kansas. Figure 33.4 shows how the median years of schooling in the United States have risen dramatically in this century. By 1970, over 30,000 Ph.D. degrees were granted each year in the United States. Television's impact on education of the population began within the last few generations.

2.3 Freedom from War

Throughout history, man has torn down prosperity through war, revolution, and military expenses. It is not a coincidence that the Industrial Revolution began in the United Kingdom and the United States. There was political stability; neither country was invaded; wars were fought on foreign soil (with the exception of the U.S. Civil War), and military spending was a low percentage of gross national product. Promising civilizations in the Arab world and China had been stopped by the Mongol invasions in the 13th century. During the period 1770 to 1945, France and Germany had "eaten each other up," but since then they have prospered. Since 1950, Germany and Japan have developed rapidly with low military expenditures, while the United States has acted as policeman to the world. In the last 30 years, the United States has had one of the lowest growth rates of the developed countries.

2.4 Capital

It takes money to make money. In England, the capital came out of the hides of the poor and the colonies; stories of the poverty of the working class at that time are well known (for example, Dickens or Marx). But this capital did permit "getting over the hump" so that future generations could have a better life. In the United States, the country itself was virgin; the natural resources had not been picked over for tens of generations. (The number of Native Americans in 1770 has been estimated as less than 1,000,000 for the entire area now covered by the United States.) In addition, as a country welcoming emigrants, the United States did not have the expense of raising those workers to adulthood.

Capital (deferred consumption) is needed for education, machine tools, buildings, bridges, research, and so forth, not only at the start of industrialization but also on a continuing basis. Low capital availability means few advances in the **standard of living** (the minimum of necessities, comforts, or luxuries held essential to maintaining a person or group). If a firm or country "eats its seed corn" (i.e., does not save), its long-run prosperity will be hindered.

2.5 A Development Orientation

A necessary condition of development is that a country must want to be developed. There must not be an entrenched dominant social class eager to maintain its power and the status quo. Democracy is related to economic development, as the developed countries tend to be democracies and the less-developed countries tend to be nondemocratic. A virtue of private enterprises is that they become independent sources of decision making (diffuse decision making). No single individual or authority can unilaterally veto an exploratory undertaking.

Historically, England had a merchant society eager for trade. The United States had few established social groups trying to avoid change. Japan, on the other hand, remained undeveloped—by choice—until the 1850s. Even today there are many countries illustrating Churchill's saying, "The inherent vice of capitalism is the unequal sharing of blessings while the inherent virtue of socialism is the equal sharing of miseries."

2.6 A Large Market

Many developments require a large customer base. The number of customers can be limited by technology and by politics.

In Roman times, land transport was very expensive. They had not yet invented horseshoes or, more important, the horse collar; when a horse pulled a load, the horse strangled. Oxen could only go about 10 miles per day. Thus, factories were scarce because any economy of scale from manufacturing was cancelled by the cost of getting goods to market (Hodge, 1990). Most transport of goods was by water. Low-cost land transport came with the train and the truck. (In 1860, there were 35,000 miles of railroad track in the United States.)

Countries traditionally have had barriers to foreign commerce. In the 1800s, Britain thrived not only due to its sea transportation but also due to its captive market (the British Empire). After World War II, Europe began to reduce internal trade barriers. Since then, there have been a number of other regional political groupings.

The largest market is the entire world. Active participation in international trade is closely associated with economic growth. Some explanations are economies of scale, keener competitive incentives, and economies of specialization.

3 INDUSTRIAL SOCIETY

Separate from the larger social context, **industrial society** has six key concepts:

1. specialization of labor
2. energy from machines
3. standardization and interchangeable parts
4. production from machines
5. mass production and mass consumption
6. the assembly line

3.1 Specialization of Labor

Specialization is not new. In the 1254 *Livre de Metiers (Book of Trades)* concerning the 101 Paris guilds, there were separate guilds in the leather industry for skinners, tanners, cobblers, harnessmakers, saddlers, and makers of fine leather goods; in carpentry there were guilds for chestmakers, cabinetmakers, boatbuilders, wheelwrights, coopers, and twiners. What is new is the degree of specialization throughout society. When we think of specialization we think of an industrial worker whose sole function is to add three screws and three nuts to each assembly on the line. But other jobs also have become specialized. In the United States, most farmers no longer raise their own food but specialize in one or two crops; a rancher who raises beef may buy the beef for his own table from the grocery store. Teachers formerly covered all grades and subjects in the "one room school"; now, they specialize by grade and subject. Physicians specialize in radiology, obstetrics, dermatology, and brain surgery. There are electrical, industrial, chemical, mechanical, ceramic, agricultural, and many other kinds of engineers. Even cleaning of blackboard erasers has become specialized. At Kansas State University, one person cleans all erasers on campus once a week.

3.2 Energy from Machines

Before the Industrial Revolution, power came from muscles (human or animal), wind, or falling water. (There was considerable use of water power in medieval Europe.) A horse, for example, can carry 120 kg, pull a cart of 500 kg on a rough road and 900 kg on a paved road, and pull 40,000 kg in a canal barge. In 1776, the steam engine began the liberation of mankind from physical effort while the Declaration of Independence began the liberation of mankind from totalitarianism—it was a very good year. It was Watt's steam engine, Faraday and Henry's electric motor, and Otto's internal combustion engine which permitted the enormous magnification of physical effort. Use of energy is probably the best single index of a society's standard of living.

The computer has accomplished the same magnification of mental effort. The telephone has magnified individual ability to communicate, while TV has magnified our ability to communicate to the masses (i.e., mass media).

3.3 Standardization and Interchangeable Parts

With galleys "as alike to another as swallows' nests," in the 16th century the 3,000-man Venetian shipyard could ready 100 galleys for battle in 60 days. The concept did not catch on in manufacturing due to the lack of accurate machine tools at that time.

Around 1800, Maudslay in England made some special machines to make standardized ship components but the real development of interchangeable parts had to await accurate machine tools. Interchangeable parts were put into use in the United States during the 1820s to 1850s in a few industries (gun manufacture, textiles, clocks) and spread gradually. Standardization tends to reduce the number of choices available to consumers—Ford's Model T was available only in black. Clever engineering can make many of the nonstyle parts usable on many different models and so reduce production costs drastically. However, sales departments constantly push for additional models, so there is a never-ending battle. An interesting example of the two strategies in practice is the Japanese approach of a few cars with a few models and few options vs. the Detroit approach of many models with many options.

3.4 Production from Machines
Craftsmen produce a few units/hour and their parts are not interchangeable, so the parts are expensive. By using machines (which can have capabilities of power, repeatability, and accuracy that human muscles, eyes, and reflexes cannot match) many units of interchangeable parts can be produced per hour, and so

the price is low. The continued development of the electric motor ("muscles") and computers ("brains") has resulted in the question, "How many slaves (machines) will each human supervise?"

3.5 Mass Production and Mass Consumption
These really are chicken and egg—you can't have one without the other, although group technology is a technique which tries to achieve the benefits of mass production with low sales volumes. Without mass production, cost is high; so prices are high; so sales are low. A large poor population is not sufficient because a customer must pay for the product. That is why high productivity is important. With high productivity, high wages can be paid and workers can be good customers, enabling firms to sell their products and pay high wages, and so forth.

3.6 The Assembly Line
In construction, you move the workers and the machines to the product; in manufacturing, you move the product past stationary machines and workers. When the sequence of operations is standardized, the machines can become specialized and the flow can be arranged into an assembly line.

REVIEW QUESTIONS

1. In 10–20 words for each concept, discuss the 6 key concepts of industrial society.
2. Is specialization of labor a development of the Industrial Revolution?
3. When did the Industrial Revolution begin?
4. What two events occurred in 1776?

5. What is "the inherent vice of capitalism"? What is the "inherent virtue of socialism"?
6. Discuss the relation between economic development and democracy.
7. Make a list of the slaves that you supervise.

REFERENCE

Hodge, A. A Roman factory. *Scientific American*, 106–11, November 1990.

2 | INDIVIDUAL CONTRIBUTORS

OVERVIEW

The "democracy" of ancient Greece was based on slaves to do the work. Now we have machine slaves rather than human slaves. Machine power comes from internal and external combustion, often distributed and applied locally by electricity. We compute and communicate with machines. We have organizational concepts such as interchangeable parts, assembly lines, scientific study of work, and high pay for workers.

Our lives have been profoundly influenced by the technical culture although most historians write only of the other culture—that of poets and politicians, the nonproducing verbalizers rather than the nonverbalizing producers.

CHAPTER CONTENTS

1 The Concept of Two Cultures
2 Vignettes

KEY CONCEPTS

accurate machine tools	Faraday and Henry	motion study
assembly line	Ford	Otto
Colt	Gilbreth	personnel selection
de Forest	Hollerith	prime movers
Edison	incentive wages	scientific study of work
electric illumination	interchangeable parts	Taylor
electric motor	internal combustion engine	Taylorism
electronics	machine computation	two cultures
ergonomics	mass consumption	Watt
factory	Maudslay	Whitney

1 THE CONCEPT OF TWO CULTURES

C. P. Snow wrote of the **"two cultures"**: one of scientists and engineers and the other of poets, artists, and writers. This chapter is written so that engineers not only can understand their culture but also appreciate the contribution their culture has made to society. As Galbraith commented: "Insist on the priority of dams, ditches, and fertilizer plants—it is these that feed poets."

Emerson said, "There is properly no history; only biography." However, the nontechnical historian ignores the people who have really changed our lives while paying extraordinary attention to the trivial behavior of politicians. For example, Barbara Tuchman's (1967) acclaimed history, *The Proud Tower: A Portrait of the World 1890-1914,* covers Europe and the United States during this period without *once* mentioning that during this time the automobile was developed, the radio was invented, people flew, and the telephone and electric illumination became common. Ford, Marconi, the Wright brothers, Bell, and Edison are not mentioned once. Another example is *The Oxford History of the American People* (Morrison, 1965), which uses 1,150 pages to cover America from prehistoric times to 1963. Whitney is covered in 26 words on the cotton gin and slavery; Colt is described in 10 words as the inventor of the "equalizer"; Bell is dismissed in 8 words as "Alexander Graham Bell, the inventor of the telephone"; and the sole mention of Edison is: "the American notion of a scientist still remained, as before the [Civil] War, a practical inventor such as Thomas Edison." Henry Ford and the auto industry are covered in 3 pages. There is no mention whatsoever of Taylor, Hollerith, Gilbreth, or de Forest. In contrast, the author Oliver Wendell Holmes is mentioned on 9 different pages.

Make your own list of contributors. You might consider the following in addition to those described in this chapter: Richard Arkwright, Alexander Graham Bell, John Bardeen, William Shockley, Walter Brattain, Henry Gantt, Gugliemo Marconi, Walter Shewhart, Louis Pasteur, Johann Gütenburg, Orville and Wilbur Wright, Edwin Land, Frank Whittle, Wilhelm Roentgen, Leo Baekeland, Elias Howe, Christopher Sholes, Cyrus McCormick, Vladimir Zworykin, Elmer Sperry, Henry Bessemer, Ronald Fisher, and George Stevenson. Who can deny the importance of (in a random order) low-cost steel, railroads, the reaper, automatic controls, the transistor, mechanical spinning, analysis of variance, the typewriter, "plastics," the sewing machine, the airplane, instant photography, the jet engine, printing (European inventor), the germ theory of disease, the telephone, statistical

quality control, radio, TV, production scheduling, and X-rays? Asimov (1964) lists over 1,000 contributors (and some of the above names are not included), so there is plenty of room for discussion.

The following vignettes give a brief view of some of the movers and shakers that made the world what it is.

> The reasonable man adapts to the world.
>
> The unreasonable man tries to adapt the world to himself.
>
> Therefore all progress comes from unreasonable men.
>
> G. B. Shaw

2 VIGNETTES

James Watt: Key Concept—A New Prime Mover, the Steam Engine

Before **Watt** there were four basic sources of power (**prime movers**): human, animal, wind, and water. Watt's steam engine permitted the power source to be located anywhere and to be of unlimited magnitude. It permitted implementation of the **factory** system (where machines are concentrated in one central location), cheap and fast transportation (steam-powered ships and trains), and urbanization as people forsook the rural idyll of the poet's imagination and the commuter's dream for a new life of working and living in the factory and city.

Watt was an instrument maker for the University of Glasgow and worked for several years on Newcomen steam engines. In 1764, he conceived the condenser. In the Newcomen engine the steam cylinder itself was cooled with water; then the metal of the cylinder had to be brought up to temperature again for the next stroke. In Watt's design, this inefficient cooling and heating was eliminated so fuel efficiency was doubled. The condenser was patented in 1769 but Watt ran out of money, and there was little progress until Matthew Boulton supported Watt in 1774. The first engine was installed in Wilkinson's foundry in 1776. Wilkinson's invention of an accurate boring mill, upon which Watt's cylinders were machined, had made possible Watt's engine. (Wilkinson's boring mill could make a cylinder that "erred from a true circle not more than the thickness of an old shilling." His mill could bore a 50-inch diameter; a worn shilling was .050 inches thick.)

Before Watt retired, rich and famous, in 1800 he made two additional improvements. First he made the engine double acting, which made rotary power practical as well as improving efficiency. Then he modified the ball governor used in flour milling by adding the critical concept of feedback so that the speed of the engine was self-regulating rather than just automatic. (Watt also invented steam heat when he heated his office in 1784.) Mining engineers in Cornwall, led by Trevithick, improved the energy efficiency of steam engines by a factor of six between 1810 and 1840; meanwhile, Carnot, in France, published his work on the ideal heat engine. Stevenson invented the railroad in 1814.

Using present-day concepts, application was slow. The carding mill in the New Salem village of Abraham Lincoln (built in 1832) was powered by oxen; a census of steam engines in France in 1833 reported 947 in use; the Secretary of the Treasury reported to Congress in 1838 that there were 3,010 steam engines in the United States (800 on steamboats, 350 in locomotives, and 1,860 in factories and public works). However, the factory (Latin *facio,* to make) system was in place in England by 1830. A factory involves: (1) adoption of machinery, (2) power from machines, (3) greater use of capital, and (4) collection of scattered workers into a regulated group.

Henry Maudslay: Key Concept— Accurate Machine Tools

It has already been mentioned that Watt's steam engine could not be manufactured until Wilkinson had a boring mill that could produce reasonably concentric cylinders. In a similar manner, the concept of interchangeable parts is based on the premise of each part being alike (from **accurate machine tools**); in practice, each part will not be alike unless the machines that make the parts are accurate. In 1807, Maudslay had 43 machines at Portsmouth make wooden pulleys for the British navy; each machine did one step. This was an early example of the specialized factory.

Maudslay is famous due to his improvement of the lathe (described by Farey in 1810 as "the most perfect of its kind") and his influence on the next generation of machine-tool builders. Clement, Roberts, Whitworth, and Nasmyth all worked at one time for Maudslay.

Clement improved lathe and planer design and worked on Babbage's mechanical computers. Roberts built a metal-planing machine in 1817, improved lathe gearing, and built many drilling machines. Whitworth introduced a standard screw thread and by manufacturing and selling standard gauges made his influence felt worldwide. Nasmyth invented a special-purpose milling machine, the shaper, and the steam forge hammer.

In Nasmyth's words:

Illustrating his often repeated maxim that "there is a right way and a wrong way of doing everything," Maudslay would take the shortest and most direct cuts to accomplish his objects. The grand result of thoughtful practice is what we call experience: it is the power or facility of seeing clearly, before you begin, what to avoid and what to select.

His "innate love of truth and accuracy" led him to develop a bench micrometer which he called "the Lord Chancellor" after the one from which there is no appeal.

Eli Whitney: Key Concept—Interchangeable Parts

Whitney was born in 1765 and spent his youth on the family farm in Massachusetts. The Revolutionary War caused shortages and Eli manufactured nails at the age of 14. At 23 he entered Yale. Graduating at 27 and wanting to study law, he intended to take a job as a tutor in the South. He visited a southern plantation and within two weeks had invented the cotton gin. Unfortunately for Eli, his factory burned down just as he was going into production. The gin was simple in design so any mechanic could make one— and they all did. Eli never made much money from the cotton gin. However, the cotton gin made slavery profitable—an "environmental impact" never dreamed of by Eli.

He decided his primary problem had been financing, so for his next project he turned to the federal government. In 1798 he signed a contract for 10,000 muskets for $13.40 apiece (normal price was $9.40) all to be delivered in 2 years. The key concept was that of interchangeable parts.

An inventive French mechanic, Honore Blanc, had developed a system of **interchangeable parts** for gunmaking. (Moveable type probably is the first example of completely standardized and interchangeable parts.) Thomas Jefferson had visited his shop in the 1780s, seen parts "gaged and made by machinery" and had tried to get him to move to America. Jefferson was president from 1801 to 1809, which was fortunate for Eli as it took him 10 years to complete the 10,000 muskets. (Secretary of State Thomas Jefferson reviewed patent applications in the evening; it was he who granted Eli Whitney's cotton gin patent.) Although Whitney emphasized interchangeable parts and even, in 1801, demonstrated a musket on which he could fit any of several locks, all of the 10,000 muskets had identifying marks on each part, something which truly interchangeable parts do not need. The Whitney muskets in the Smithsonian Museum have parts that cannot be interchanged. Simeon North (pistols) in 1799 and John Hall (rifles in 1824) made truly interchangeable parts. Whitney, although never really achieving his objective, had good public relations and so receives the credit for the "American system of manufacture." Yet, his cotton gin probably caused the Civil War; that should be fame enough for anyone.

Michael Faraday and Joseph Henry: Key Concepts—the Dynamo, the Source of Electricity; the Motor, a New Prime Mover

Michael **Faraday,** one of a blacksmith's 10 children, did not attend school. Fortunately he was apprenticed to a bookbinder where he not only bound books but looked inside them. In 1812 he attended a public lecture by Davy. Michael took careful notes and made colored illustrations and sent them to Davy, asking to be his assistant. Davy hired him in 1813. By 1825, Faraday was director of the laboratory.

In 1823, innocent of mathematics but the "prince of experimenters," he liquefied gases under pressure, in 1825 he discovered benzene, and in 1831 he had his greatest discovery, the dynamo and the electric motor. When Queen Victoria heard of the invention of the dynamo she asked "What good is it?" Faraday replied, "Madam, someday you will tax it!"

Joseph **Henry** came from a poor family. He left school at 10 and was apprenticed as a watchmaker. At the age of 16 he found a book, *Lectures on Experimental Philosophy,* in an abandoned church, read it, and became fired with the desire for education. He returned to school and by 1826 was teaching math and science.

In 1829, he made a vastly improved electromagnet which lifted 750 lbs; in 1835 he invented the electrical relay (in effect, the telegraph); in 1846 he became head of the Smithsonian Museum; and during the Civil War he founded the National Academy of Sciences. At the Smithsonian he began the policy of "publishing original research in a series of volumes and giving a copy to every major library on earth." Technology includes social concepts such as public libraries, technical societies, and technical journals as well as physical devices such as electric motors. If technology is an engine, then knowledge is its fuel; the fuel is becoming richer. Henry had a habit of not patenting his ideas as he thought that discoveries of science should be for the benefit of humanity. He discovered the principle of induction in 1831 but put his work aside at the end of August to be completed the following summer. Faraday published his results in November 1831.

The **electric motor** has three major advantages over the steam engine: It can be made any size (especially smaller), it can be started and stopped quickly, and it can be powered at a distance by use of wires. Somewhat surprisingly, motors were not applied until about 1880, but then their growth was rapid.

Samuel Colt: Key Concept—Assembly Line

Sam **Colt** designed a repeating pistol: "the six shooter," "Colt's Patent Pacifier," "The Difference."

He opened a factory in Paterson, New Jersey, in 1835 that used the **assembly line** for production: In Texas, where shooting was serious business, they liked it; the Texas Rangers ordered 100. The army thought it was "too complicated." His factory closed in 1842.

In 1847, the Mexican War began. General Zachary Taylor ordered 1,000 Colt 44s, and the weapon that tamed the Plains was back in production.

> The first workman would receive two or three . . . important parts and would affix them together and pass them on to the next who would add a part and pass the growing article to another who would do the same . . . until the complete arm is put together.

The National Bureau of Standards reports that the conveyor belt for assembly was not used until 1908.

Nikolaus Otto: Key Concept—the Internal Combustion Engine, a New Prime Mover

Reuleaux wrote in 1875 of the need for a small engine due to a high capital cost of steam engines: "How to make power independent of capital? . . . Engineers must provide small engines with low running costs. . . . These little engines are the true power units of the people."

In steam engines, combustion takes place outside the engine; in 1860 Jean Lenoir built the first engine where the combustion was internal. It was powered with illuminating gas and had a very poor efficiency, but it was a start.

Then in 1862, Nikolaus **Otto** (with his partner Eugen Langen in a relationship much like Watt and Boulton) began work on internal combustion engines. In 1876 (100 years after the steam engine) he produced the silent Otto, the first 4-cycle engine.

Improvements by fellow Germans Karl Benz, Gottlieb Daimler, William Mayback, and Rudolf Diesel came before 1900. The advantages of the **internal combustion engine** over steam engines were low capital cost, quick starts, and high power/weight. The advantage over the electric motor was the elimination of the tether—the wire. The age of the automobile and the airplane began.

Thomas Edison: Key Concept—Electric Illumination

Edison represents the classic tale of the self-made man: the poor boy who, without schooling or influence, made his way to fame and fortune by intelligence and hard work. Enrolled at birth in the school of hard knocks, later with 1,093 patents to his name, he was the most productive inventor in the history of the United States—probably in the history of the human race. He also is very quotable:

> There is no substitute for hard work. Genius is 99% perspiration and 1% inspiration. [To a job applicant] Well, we don't pay anything and we work all the time. We will make electric light so cheap that only the rich will be able to burn candles.

Early inventions of an improved telegraph and the stock market ticker permitted Edison (age 29) to found Menlo Park (the first industrial research laboratory in the world—in itself one of Edison's many inventions) in 1876. He hoped to produce a new invention every 10 days! In fact, during one four-year stretch before he became involved in finance, he obtained 300 patents, or one every five days.

In 1876 he improved the telephone and made it practical; in 1877 he invented the phonograph; in 1878 he announced he would tackle the problem of producing light by electricity.

Edison, who scorned theory, used the research method since then known as the Edison method—a patient trial of all possible alternatives. In 1879, he produced the first practical light bulb.

To apply the concept of **electric illumination,** he had to invent a host of auxiliary inventions—devices for sealing the bulbs, screw-in sockets, light switches, electric meters, safety devices—and then found the first electric utility, Consolidated Edison, which opened in 1882. He also founded what became General Electric.

Consolidated Edison began the concept of the electrical utility. At first, electrical power was sold only for lights. But in the 1890s, electricity was used for streetcars. Steam power hung on in manufacturing, and it wasn't until 1919 that over half of manufacturing power came from electric motors. Today it seems inconceivable to live without the electrical power grid.

His one contribution to science, the Edison effect, which led to the vacuum tube and electronics, he ignored.

Herman Hollerith: Key Concept—Machine Computation

After receiving a degree in mining engineering in 1880, **Hollerith** worked on statistics for the census office. Marking tallies with dip-pen and ink was just too slow. He needed a method of **machine computation** to automate the process. At first he tried edge-marked cards. Then he tried holes in paper tape that would have to be wound just to find the small amount of information that probably would be at the end of the roll. There is no evidence that he knew of the punched paper rolls used by French weavers since the mid-1700s (player pianos would use this concept in 1900), or weaver Jacquard's punched card (adapted from the roll), or even the punched card concept of Charles Babbage's analytical engine—a device that was never finished.

One day Herman took a trip to St. Louis. At this time train robbers were posing as passengers, and the government asked the railroads to keep track of everyone aboard. The conductors punched the ticket in specific places to indicate specific body characteristics—brown hair, blue eyes, medium weight—a punch-photograph.

Herman adopted the idea into the Hollerith card, the size of an 1890 dollar bill so it could fit existing file drawers. He developed a keyboard punching machine, rented his machines, and astounded the world with the speed of his census machine. IBM had been born.

Frederick Taylor: Key Concept—Scientific Study of Work

Frederick **Taylor,** the son of a wealthy Philadelphia family, attended Phillips Exeter prep school. Although he passed the Harvard entrance exams, his eyesight was impaired, so at 18 he became an apprentice machinist and patternmaker. Starting work at 22 during the recession of 1878 as an ordinary laborer at the Midvale Steel Co., he was successively time clerk, journeyman, lathe operator, gang boss, and foreman of the machine shop before being appointed chief engineer in 1887. He studied engineering at night and received a B. S. in mechanical engineering from Stevens Institute in 1883. The combination of practical experience and theory was to prove fruitful.

Taylor asked the question: "What is the best way to do this job?"

He did not accept opinions; he wanted facts—evidence. The questioning scientific approach (hypothesis, experiment, evaluation of data to prove or disprove the hypothesis) is well known today. It was known then and applied to chemistry and physics but not to the design of everyday jobs. This is Taylor's primary contribution—the application of the principles of science to improving jobs (the **scientific study of work**).

A famous example of this approach was his study of shoveling. In 1898, Taylor worked for Bethlehem Steel. In those days before mechanized material handling, when you wanted material moved you shoveled it by hand. In this one plant, 400 to 600 men spent most of their time shoveling. Various materials were shoveled each day by each man; each man furnished his own shovel. There was no training in shoveling techniques; pay was $1.15 a day.

One thing Taylor noticed was that with a constant-volume shovel the load was only 3.5 lbs when shoveling rice coal but 38 lbs when shoveling ore. The question was: "What is the best size shovel?" For the experiment, he had material shoveled with large shovels (i.e., heavy loads); then cut a little off the end of the shovel and had the same material shoveled; cut off a little more the next day, and so on. For subjects, he used two good, experienced shovelers (i.e., he replicated his data). Taylor used as his criterion the amount shoveled per day. The results indicated that maximum material was shoveled per day when the load on the shovel was 21.5 lbs (for distances up to four feet and heights less than five feet). For more on shoveling, see Box 18.3.

To apply this knowledge, Taylor had a toolroom established and special shovels purchased. The foreman was required to notify the toolroom of the work his gang would do that day. A large scoop was provided for shoveling ashes, a middle-sized shovel for coal, and a small shovel for ore.

Taylor instituted another concept that was quite radical for that day. He did not believe that all the benefits from increased productivity should be retained by the organization; the worker also should benefit. Therefore he specified a standard tonnage to be shoveled for each type of material. The worker was trained in the proper work method, given the proper tools, and put to work. When he achieved standard, he received a 60% bonus above the day wage rate. If he could not achieve standard even after training he was put on a different job. The concept of pay-by-results (**incentive wages**) and **personnel selection** (select the best people for each job) are commonplace today; they were radical innovations then.

After these methods were applied to the yard at Bethlehem, the same amount of work was done with 140 men; material handling cost for the company (including cost of the study, toolroom, and bonuses) was reduced from $.08/ton to $.04. Employee wages, of course, were 60% higher.

Taylor's scientific management is a good example that technology includes techniques as well as devices.

The concept that the job is to be designed by experts (engineers) and the worker's duty is just to follow instructions is called **Taylorism.** It has resulted in dramatic improvements in productivity and the standard of living for many cultures for approximately 100 years. However, the educational level of the workforce has risen enough in some countries so that greater participation of the workers in job design has become feasible. We now have "intelligence" distributed throughout the workforce, not just located in supervisors and staff. See Chapter 33 for some specific techniques.

Henry Ford: Key Concepts—High Pay for Workers, Low-Cost Auto, Mass Consumption

Just as Whitney is incorrectly credited with interchangeable parts, **Ford** is incorrectly credited with the auto **assembly line,** but the specialization of labor combined with use of conveyors had been used earlier in slaughterhouses (a "disassembly" line) (see Figure 2.1). The auto, however, was the product in the public eye. Ransom Olds applied the assembly line concept to building Oldsmobiles in 1899, 10 years before Ford installed an assembly line. By 1904, Oldsmobile production had reached 5,000/yr. Ford, however, dominated auto production. In 1924 Model Ts accounted for one-half of the world's motor vehicles. Ford's assembly line cut throughput time/car from 13 hours to 1. Formerly one man took 20 minutes to assemble a magneto by himself; using a subassembly line with 29 operators, time was cut to 13 man-minutes/unit.

High pay for workers and a low-cost auto were implemented with Ford's decision in 1914 to pay his workers $5 for an 8-hour day when they had been getting $2.50 for a 9-hour day. In addition he set up a $30,000,000 profit-sharing fund. How could he double wages while reducing costs? Through productivity. (High wages without accompanying productivity mean either losses or inflation). The concept of **mass consumption** is the key to our society. When his employees received low wages, they were not able to purchase cars; cars were for rich people (see Table 2.1). As long as cars were only for rich people, the total output of cars was small. And, looking at the problem as a businessman, Ford saw that even though his profit/car was satisfactory, his total profit was limited due to the small number of cars he could sell because of their high price.

To maximize his total profits (not because he was a social visionary), Ford set up his assembly line, made a standardized product ("any color you want as long as it's black") at high volume (and thus low cost), and paid his workers well (permitted by the high volume, which was permitted by the high sales, which were permitted by the high wages and low costs, which were permitted by the high volume which was permitted by the high sales which were permitted . . .). The natural result, as Horatio Alger would say, was that Henry Ford became the richest man in the world.

Yet, when all is said and done, the "main effect" of mass consumption of automobiles may really be less important than the "side effect" of personal

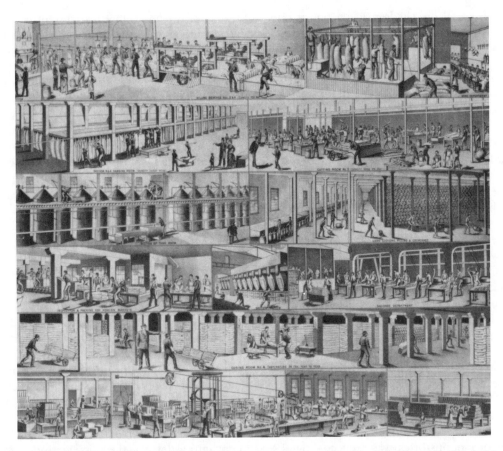

FIGURE 2.1 Assembly lines did not begin with the automobile, as is proven by the lithograph "Interior View of a Modern First Class Pork Packing & Canning Establishment of the United States of America" (Courtesy of the Chicago Historical Society), published in 1880. Titles under the respective pictures are: The Office; Killing Benches Nos. 3 & 4 (capacity 700 pr. hour); Fire Department (at dinner); Section No. 4 Hanging Room (capacity 10,000 hogs); Cutting Room No. 3 (capacity 5,000 pr. day); Section of Tank Room; Lard Coolers, Filling & Cooperage; Inspection & Packing for Foreign Markets; Boiler Room; Sausage Department; Curing Room No. 4 Temperature 38 Fah. Year to Year; Polishing; Canning Department Filling by Machinery, Meat Untouched by Hand, Soldering; Labelling. A question might be why it took so long to apply assembly line technology to other industries.

mobility obtained as a result of the automobile. In the United States it has increased our radius of daily travel from the limits of the streetcar to approximately 50 miles; this in turn has changed population distributions in both cities and small towns. Although factories were located in central cities at one time in order to obtain employees, effectively no factories have been built in central cities in the United States since 1940. Trade and distribution jobs began moving out of the central cities in the 1950s; during the 1960s, 9 of the 10 largest American cities lost population. It is difficult to determine in advance the main effects of developments; it is exceedingly difficult to anticipate the side effects and interactions occurring after a time lag.

Frank and Lillian Gilbreth: Key Concept—Motion Study

Frank **Gilbreth** was a super-organized man. An example of his organizing (publicized in *Cheaper by the Dozen*) was his posting of French grammar on the wall of his bathroom. While sitting on the toilet, his children were expected to learn French! He also

TABLE 2.1 Prices of Fords vs. wages/hour. Courtesy of Ford Archives.

Year	Type of Model T	Price ($)	Wage/Hour of Unskilled Labor in Ford Factory	Normal Hour/ Day	Hours of Work Required to Buy Car
1908	Roadster	825	.19	10	4,340
1913	Runabout	500	.26	10	1,923
1918	Runabout	434	.50	8	870
1923	Runabout	265	.75-.85	8	331

had the tonsils of all his children taken out at one time; he recorded the surgeon's hand motions with a movie camera. (Unfortunately his assistant forgot to put film in the camera.) But now you know what kind of man he was. His business was construction. A key construction job was laying brick. Frank, who had learned the bricklayer's trade as a teenager, applied his analytical skills to the study of bricklaying. He did this by studying the job in great detail—**motion study**.

In the existing method, bricks were dumped in a heap on the scaffold; the bricklayer bent over and picked up a brick. He then inspected it on all sides to select the best side for the wall face.

In Gilbreth's method, the bricks were inspected when they were unloaded from the freight car. They were placed with the best side up in packets of 90 lbs. The packets of oriented bricks were then placed on the scaffold. The scaffold was redesigned so that it could be raised or lowered easily, thereby reducing the distance the bricklayer had to reach. The mortar box and the packets were placed on the scaffold so that while the bricklayer picked up a brick with one hand, he scooped his trowel full of mortar with the other hand. The mortar was made of a standardized consistency so that the motion of tapping the brick into place with the trowel was no longer required. The number of motions/brick was reduced from 18 to 4.5. In a test of the new method, bricklayers laid 350 bricks/hour while the previous record for this type of construction had been 120 bricks/hour.

The reason Gilbreth's results attracted such attention was the task studied—bricklaying. Men had laid brick since antiquity. If there was any skill in which no more change could be anticipated, it seemed that bricklaying, with its 3,000 years of experience, certainly would be it. Yet one man, by systematic study, had made a 300% improvement over the experience of 3,000 years! A powerful example.

When Frank died his wife, Lillian, continued his work on motion study. (She is on the 40-cent stamp.) The many small elements of jobs have been named therbligs—Gilbreth spelled backward (almost). See Table 29.1.

See Box 2.1 about the rise of ergonomics.

Lee de Forest: Key Concept— Three-Electrode Tube (triode)

The telegraph was a great boon to speedier communication, but it required a wire connecting the sender and receiver. Marconi developed the wireless telegraph, the sending of dots and dashes without intervening wires. Edison, who had been a telegrapher himself, had discovered the Edison effect—electrons flowed across a gap between two electrodes inside a vacuum tube. Fleming developed the concept into a device—the two-electrode tube or rectifier. In 1906, **de Forest** added the third electrode (the grid). A varying weak signal on the grid could be converted into a varying strong signal between the

---

BOX 2.1 *The rise of ergonomics*

Taylor's and Gilbreth's successors were active in the management division of American Society of Mechanical Engineers and the Society for Advancement of Management. The first Ph.D. in industrial engineering in the United States was given to Ralph Barnes by Cornell University in 1933. Barnes' book *Motion and Time Study*, first published in 1936, had its seventh edition 44 years later in 1980. The American Institute of Industrial Engineers was founded in 1948; it is now the Institute of Industrial Engineers (Nadler, 1992).

During World War II, research was conducted to maximize human performance in military applications. In 1948 the Ergonomics Research Society (now Ergonomics Society) was founded in the United Kingdom.

Murrell developed the name **ergonomics** from the Greek *erg* (work) and *nomos* (laws, study of).

Over the years, this focus on "work" has expanded to cover all aspects of the human–machine interaction, not just the industrial workplace.

The International Ergonomics Society (IEA), the umbrella society of the various national societies, was founded in 1959. Table 2.2 shows the membership of the IEA-affiliated ergonomics society in each country in 1994.

Note that not all people interested in ergonomics necessarily are members of the affiliated society. For example, in the United States there are people interested in traffic ergonomics who belong to the Society of Automotive Engineers; in agricultural ergonomics who belong to the American Society of Agricultural Engineers; in safety who belong to the National Safety Council, and so forth.

TABLE 2.2 Ergonomics societies and recent membership; if there are multiple ergonomics societies in a country, only the IEA-affiliated society membership is listed.

COUNTRY/REGION	FOUNDING DATE	RECENT MEMBERSHIP Date	RECENT MEMBERSHIP Number	POPULATION (M)	MEMBERS PER M
Australia*	1964	1994	537	17.4	31.0
Austria*	1976	1990	42	7.5	6.0
Belgium*	1986	1991	140	9.9	14.0
Brazil*	1983	1988	244	143.3	2.0
Canada*	1968	1994	518	28.3	18.0
Czechoslovakia* (former)				15.7	
China*	1989	1989	300	1,151.0	0.3
Denmark**		1994	846	5.1	165.0
Finland**	1985	1994	90	5.0	18.0
France*	1963	1987	531	55.2	10.0
Germany*	1958	1991	700	76.6	9.0
Hungary*	1987	1988	90	4.6	20.0
India	1987	1991	100	866.0	0.1
Indonesia***	1988	1988	120	176.8	0.7
Israel*	1982	1992	120	4.3	28.0
Italy*	1961	1993	250	57.2	4.0
Japan*	1964	1993	1,864	124.0	15.0
Korea (South)*	1982	1988	250	43.9	6.0
Netherlands*	1962	1994	631	15.0	42.0
New Zealand*	1986	1994	114	3.3	35.0
Norway**		1994	155	4.2	37.0
Poland*	1977	1992	298	37.5	8.0
Russia*	1989	1994	150	150.0	1.0
Singapore***	1988	1990	36	2.6	14.0
South Africa	1984	1994	40	39.4	1.0
Southeast Asia*	1984	1990	86	250.0	3.0
Spain*	1988	1992	187	39.6	5.0
Sweden**		1994	298	8.5	35.0
United Kingdom*	1949	1992	1,030	56.5	18.0
United States*	1957	1993	5,102	248.8	21.0
Yugoslavia (former)	1973	1989	50	23.2	2.0
Affiliated Societies					
European Society for Dental Ergonomics		1994	37		
Human Ergology Society	1970	1994	230		
TOTAL			15,320		

*Member of IEA.
**Part of a regional umbrella group, the Nordic Ergonomics Society (founded in 1969), which is a member of IEA.
***Indonesia and Singapore are part of a regional umbrella group, the South East Asia Ergonomics Society.

Source: "Ergonomics societies and recent membership." *Ergonomics News,* Vol. 36, No. 6, 1993. Copyright © 1994 by Taylor and Francis, London. Reprinted with permission.

cathode and anode. Signals could now be continuous (rather than just discrete) and could be amplified. In 1912, he added the concept of a feedback circuit. Radio and the age of **electronics** began.

Lee, who received a Ph.D. from Yale in 1899 for his study of electromagnetic waves, never received much popular acclaim. Seniors at Yale voted him "nerviest and homeliest man in the class." Although he eventually had 300 patents, even his development of the feedback circuit was challenged by Armstrong. In 1934, after 14 years of litigation, the score was 6 court verdicts for de Forest and 6 for Armstrong. The United States Supreme Court made it 7-6 with its decision that de Forest had invented the feedback circuit 2 months before Armstrong.

Armstrong did receive credit for inventing FM radio in 1934, although the last lawsuit wasn't settled until 1967. By analogy to waves of water, electronic noise affects wave height (amplitude) more than the spacing between waves (frequency); Armstrong developed frequency modulated (FM) radio vs. the amplitude modulation (AM) radio. Scientific discovery and technological invention (like poetry) benefits from metaphor.

REVIEW QUESTIONS

1. What are the "two cultures"?
2. How are accurate machine tools and interchangeable parts related?
3. Prime movers are essential to an industrial society. Why was the steam engine important? Why was the electric motor important? Why was the internal combustion engine important?
4. Describe Taylor's shovel experiment using the words *task, subjects, controlled variable, criterion, results,* and *application.*
5. Who were the Gilbreths and why were they important?
6. *Ergonomics* is based on what two words?

REFERENCES

Asimov, I. *Biographical Encyclopedia of Science and Technology*. Garden City, N.Y.: Doubleday, 1964.

Ergonomics societies and recent membership. *Ergonomic News*, Vol. 36, No. 6, 1993.

Morrison, S. *The Oxford History of the American People*. New York: Oxford University Press, 1965.

Nadler, G. The role and scope of industrial engineering. In *Handbook of Industrial Engineering,* 2nd ed. Salvendy, G. (ed). New York: Wiley, 1992.

Tuchman, B. *The Proud Tower: A Portrait of the World 1890-1914*. New York: Bantam, 1967.

3 Work Smart Not Hard

OVERVIEW

The rich nation is the exception; techniques which make not only individuals but entire nations rich are the pivots of history. Our standard of living, which can be measured in leisure, health care, wine, or weapons as well as consumer goods, depends on our productivity. Productivity comes from efficient technology (combination of scientific, engineering, and managerial techniques) applied to land, materials, machines, and labor. The key to productivity is to WORK SMART, not to WORK HARD.

CHAPTER CONTENTS

1 Productivity and Living Standards
2 Productivity
3 Total Time for a Job or Operation
4 Work Smart Not Hard

KEY CONCEPTS

bad old days
components of productivity
pivotal events of history
work smart vs. work hard

1 PRODUCTIVITY AND LIVING STANDARDS

1.1 Bad Old Days

People have worked hard throughout history. People who extol the good old days as they protest technology fail to realize that the "good old days" were horribly **bad old days.** They visualize themselves in Athens talking to Socrates rather than as a slave in the silver mines; as a knight on a horse rather than as a starving serf; as D'Artagnan dueling with Cardinal Richelieu's men rather than as wielding a hoe. Millet's painting, "The Man with the Hoe," depicting agricultural labor in France during the 1830s, inspired Edwin Markham to write "The Man with the Hoe."

> Bowed by the weight of centuries he leans
> Upon his hoe and gazes on the ground
> The emptiness of ages in his face
> And on his back the burden of the world.
>
> Is this the Thing the Lord God made and gave
> To have dominion over sea and land,
> To trace the stars and search the heavens for power,
> To feel the passion of Eternity?
>
> What gulfs between him and the seraphim!
> Slave of the wheel of labor, what to him
> Are Plato and the swing of Pleiades?
> What the long reaches of the peaks of song,
> The rift of dawn, the reddening of the rose?

They left agriculture for the factory to improve their standard of living. And yet even this improved standard of living was low compared to today. Most people today consider Bob Cratchit (Scrooge's clerk in *A Christmas Carol* by Dickens) to be poor. Yet in his England, the country with the highest standard of living in the world in the 1840s, he was a well-paid worker. Cratchit could read, write, and calculate—certainly valuable traits since compulsory schooling did not begin in England until 1870. His wage was 15 shillings/week (30 pounds/yr); the average wage in the United Kingdom was 30 pounds (roughly twice the mean income/capita in the United States). Indoor laborers made 5 shillings/week, and weavers made 13 shillings/60-h week. Note that Cratchit lived in a four-room house and that his wife and 5 of 6 children did not work. The Factory Act of 1802 in the United Kingdom forbade parents to hire out their children for factory labor if they were less than 9 years old and limited child labor to 12 h/day.

The move to the city had drawbacks, as in 1900 the British government reduced the minimum height for soldiers to 60 inches from the 63 inches set in 1883 due to the lack of nourishment received by the ordinary citizen during this time. Yet progress continued. The height of Dutch Army inductees was 168 cm in 1850, 170 in 1900, and 175 in 1950 (Tanner, 1968). In summary, remember Sophie Tucker's famous words "I've been rich and I've been poor; believe me, rich is better."

1.2 Benefits from Productivity

Table 3.1 gives a concise view of the change, at 70-year intervals, in the standard of living in the United States since the first census in 1790. To help predict 2001, data are given for the halfway point between 1930 and 2000. What do you predict for 30 years from now? Figure 3.1 gives the output/work-hour from 1890 to 1992. Table 3.2 reports the changes in manufacturing productivity for some countries since 1964. Advances in labor productivity were smaller in the 1980s than in the 1970s or 1960s. There are substantial differences among the countries with the relative rankings changing with the decades.

People have climbed laboriously over hundreds of obstacles to successively higher and higher ledges of income. The rich nation is the exception; the techniques that made entire nations rich are the **pivotal events of history.** Was it done by working harder? No. By working more efficiently? Yes. We now are more productive—producing more output for the same input.

The benefits from increased productivity are not always taken in increased material goods; they can be spent on health care, wine, weapons, or increased leisure. Of course, with more productivity, there is more to share; "a rising tide lifts all boats." The choices have interactions. Education may be considered a goal in itself; yet the correlation coefficient between educational level and gross national product/capita in 75 countries is .89 (Harbison, 1963).

Before 1914 the 60-h week was typical for most industrial workers in France, Germany, England, Canada, and the United States. By 1922 the 48-h week was in general practice in industry throughout Europe, Australia, New Zealand, and Latin America. In 1926 Ford introduced the 40-h week in the United States. By 1948 the 40-h week was the norm in the United States, Australia, New Zealand, and the U.S.S.R. In 1953 most Western European countries and Japan worked 45–48 h/week; in 1963 it was 44–46 h/week; in 1973 it was 42–43 h/week. The decline has continued as the normal work week has become less than 40 hours in many countries (for example, 39 in France and 38.5 in West Germany). In 1984, manufacturing workers worked an average of 2,180 h/yr in Japan, 1,941 in the United Kingdom, 1,934 in the United States, 1,652 in West Germany, and 1,649 in France. In 1992 average annual hours/worker were 2,007 in Japan, 1,857 in the United States, 1,646 in France, and 1,519 in Germany.

Although working hours/year have declined over the years, in recent years they may be going up in the United States, especially for professional and managerial employees (Schor, 1991). Schor believes the reasons are that people prefer more income to more leisure and that the "system" now rewards employers for employing fewer people for longer hours rather than more people for shorter hours.

2 PRODUCTIVITY

2.1 Components of Productivity
As mentioned before, productivity is the ratio between input and output. One definition of output gives four essentials (**components of productivity**): labor, materials, energy, and information. The special character of a technical society is that the materials, energy, and

TABLE 3.1   A concise view of the change in the standard of living of the average person in the United States since 1790. The change was due to improved productivity.

Index	1790	1860	1930	1970-72
Population	4,000,000	31,000,000	123,000,000	213,000,000
Housing	Single-family log cabin.	Frame houses. 93% of dwellings single family.	Frame houses. 67% of dwellings single family. Multistory of brick with steel frame. 3.8 population/dwelling unit; 4.8 rooms/dwelling unit (median).	Frame houses. Apartments, condominiums. 63% own their own home. 3.1 population/dwelling unit; 5.0 rooms/dwelling unit.
House furnishings and equipment	Homemade furniture. No running water. Privy. Fireplace heat. Illumination by candle.	Factory-made furniture. Hot and cold running water in homes of rich. Stove (wood, coal). Illumination by kerosene and gas lamps. Matches.	Factory-made furniture. _Urban_ _Rural_ Electricity 96% 31% Indoor toilet 92 11 Running water 92 18 Mech. refrig. 56 15 Central heat 58 10 6,000,000 pianos	_% of Dwelling Units_ Flush toilet + H & C water + bath or shower 94% Running water 98 Telephone 94 TV 99+ Air conditioning 45 Home freezer 32
Food and drink	Local supply, little variety. Preserve by salt, pickling, and smoking.	Regional supply. Preserve also with ice. .2 cans of food/yr/capita. 178 lbs red meat/yr/capita (carcass weight).	Regional and world supply. Mech. refrig. in transport and home. 54 cans of food/yr/capita. 141 lbs red meat/yr/capita.	Regional and world supply. Fresh at all seasons, frozen, freeze dried, convenience foods. 101 cans of food/yr/capita. 192 lbs red meat/yr/capita (49 fowl, 11 fish). Food cost = 16% of disposable income.
Clothing	Made at home. Linen, wool, leather.	Some factory clothing, especially men's. Cotton and wool.	Factory-made for men and women. Rayon and silk plus cotton and wool.	Factory-made for all. Synthetic fabrics. Fabric treatments (permapress, soil and water resistant).
Health	Live expectancy at birth = 36. Doctors trained as apprentices. First hospital (1750). First medical school (1765). No public responsibility for health or sanitation.	Life expectancy at birth = 40. 149 hospitals; .0009 beds/capita (1873). 40 medical schools. 175 doctors, 18 dentists, 26 nurses/ 100,000 pop. First comprehensive water and sewer system (Chicago).	Life expectancy at birth = 60. 6,150 hospitals; .009 beds/capita. 125 doctors/100,000 pop.	Life expectancy at birth = 71. 7,100 hospitals; .0075 beds/capita. 174 doctors, 57 dentists, 353 nurses/100,000 pop. 36% of pop. have major medical insurance. Medicare.
Transportation and communication	Postal expense = $.01 yr/capita. No public roads.	Roads by local government. 35,000 miles of RR track.	227 pieces of mail/yr/capita. 200,000 miles of RR track. 53% of families own a car (1937). Some commercial air transport.	409 pieces of mail/yr/capita. .6 telephones/capita. 204,000 miles of RR track. 3,700,000 miles of highway. 83% of families own 1 or more cars (.47 cars and .1 truck/capita). 150,000,000 air passengers/yr. Man on moon.

(continued)

TABLE 3.1 continued

Work and leisure	Child labor; work until death; dawn to dark; 6-day week; no vacations.	65-70 h/wk for city and factory workers; no vacations. Circus, vaudeville, nonprofessional baseball. Newspaper circulation = .05/day/capita.	40 h/wk or less for 50% of wage earners. Begin work at 16-18; retire without pay before death; some vacations. Unions 15% of workers; 30% white collar (1940). 90,000,000/wk movie attendance. Radio. 6000 golf courses. Baseball and football pro sports.	56% of people over 16 hold at least one job. 40 h/wk or less average for entire work force. Retire at 65 (Social Security and company benefits). 6 or more paid holidays for 97% of work force; 15 days paid vacation after 15 years for 93%. Unions 23% of workers; 50% white collar. Newspaper circulation = .29/day/capita. 6600 radio, 707 TV stations; 1463 symphony orchestras, 713 opera companies, 763 museums, 11,000 golf courses. 100% of homes have TV. 3.7 visits/yr/capita to federal parks or recreation areas.
Education	No public education	Some tax-supported libraries. 80% literacy. Average schooling = 434 days. First land grant university (Kansas State University).	Education (including secondary and college) government responsibility. Free education for 12 grades. Median school completed by pop. over 25 = 8.0 yr. 96% literacy. School expense = 3.1% of GNP. 6200 public libraries.	Median school completed by population over 25 = 12.2 years. Literacy = 99%. School expense = 8.0% of GNP. 7109 public libraries.
Government	New York City government expense/yr/capita = $1.87.	150 public and private water supply systems. New York City government expense/yr/capita = $10.52.	New York City government expense/yr/capita = $189 (1935). Taxes = 10% of GNP. Government (all levels) workers = 9.7% of all workers.	Federal $729, state and local $731 expense/yr/capita (nationwide). New York City government expense/yr/capita = $1207. Taxes = 31% of GNP. Public responsibility for unemployment and recreation. Government (all levels) workers = 19.3% of all workers (1975).

information replace labor. There are four classic factors: land, materials, machines, and labor. (Sometimes materials and machines are called capital.) Overriding these four factors is a fifth factor, technology (the combination of scientific, engineering, and managerial techniques.)

Improved productivity from *land* might involve using better seed to grow 10% more corn/acre or better trees which will mature in 20 years instead of 25. Fertilizer or insecticides may increase crop yield. Output/unit of land increases. (For the industrialized countries, the farm population has shrunk to less than 5% of the population due to the high productivity of farmers.)

Improved productivity from *materials* might be use of a collector container to catch the drips from barrels of viscous chemicals so that 99.7% of a container's contents are used instead of 98.5%, or use of a noncorrosive material to extend the life of a bridge or truck. Insulation reduces fuel oil need. Output/

unit of material increases. Shields on the top of semi-trailer cabs reduce the coefficient of drag and thus save energy. (Other examples of drag reduction are air bubbles on airplane wing tips and the below-the-water bulb in the front of tankers.)

Improved productivity from *machines* might be scheduling a truck to haul materials both going and coming rather than returning empty, or using a ceramic cutting tool in a lathe so a higher speed can be used, or using a word processor to type a letter. Increasingly, machines are used to process information, not just materials, as we enter the "information economy." And electronic communication (phone, fax, E-mail) can replace transportation ("snailmail"). A machine's physical life might be 50,000 h. Use of 1 shift (2,000 h/yr) would let the machine be used 25 years. Two shifts (4,000 h/yr) would improve machine utilization and decrease the risk of obsolescence. (Because the problems of shift work are social rather than physiological, shift work may not be

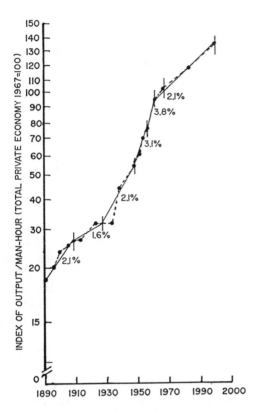

FIGURE 3.1 Labor productivity has increased for many years. From 1973–1982, the average rate was .63%/year; from 1983–1992, it was 1.13%/year. What do you predict for 1993–2002?

decreasing input more than the output decrease. Productivity should be recorded in nonmonetary units so comparisons are not distorted by inflation. Example indices are vouchers/week from the accounts payable office, number of student credit hours/teacher, and area cleaned/day by the janitors.

It also will be emphasized that productivity is a mixture of the factors of land, materials, machines, and labor. Later-developing nations have an advantage in that they can selectively accept ideas from an ever-larger store of transnational knowledge. This knowledge is not just physical hardware but also social knowledge (Quality Circles, interlibrary loans, double-entry bookkeeping, agricultural extension agents). The popular press often writes as if only reduction in labor costs is meaningful and ignores improved productivity for the other factors. Historically, the developed countries have substituted cheap energy and materials for labor (see Figure 3.2). The rise in the price of oil since 1973 has resulted in substitution of labor and capital for energy. In the United

utilized if public transport or recreation or shopping are closed during part of the day.) Output/unit of machine time increases.

Improved productivity from *labor* might be improvement in the work methods of a nurse to permit attending to more patients, or a simplified form so that a clerk could calculate more vouchers/h. Use of a fixture to hold parts can permit assembly with two hands instead of one. Output/unit of time increases.

Although the examples assumed the same input with an increase in output, improved productivity also can occur by decreasing input for the same output, by increasing output faster than input, or by

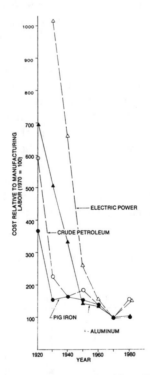

FIGURE 3.2 Relative costs of energy, materials, and labor change over time. The manufacturing labor wage in 1970 of $3.36 is the base of 100. Other costs are indexed on it. For example, the price of crude petroleum was $3.18/barrel in 1970, so the energy ratio is $3.36/$3.18 = 1.06. In 1960 the price of oil/labor was $2.88/$2.26 = $1.27. Then $1.06 × $1.27 = $1.35. Until 1973, the relative cost of energy vs. labor declined rapidly and the relative cost of materials vs. labor declined slowly. Thus engineers tended to substitute energy for labor. In the period 1973–1982, energy costs rose vs. labor, and so engineers focused on saving energy. Since 1982, energy costs have declined again relative to labor (as well as in absolute terms). What do you predict for the future?

TABLE 3.2 Average annual changes in manufacturing productivity.

Country	1964–73	1973–79	1979–86
U.K.	3.8	0.7	3.5
Canada	4.3	2.5	3.0
Japan	9.8	4.0	2.7
France	5.4	3.0	2.5
Italy	5.5	2.5	2.4
FR Germany	3.9	3.3	2.3
United States	3.4	3.5	2.3

States labor costs now dominate, while in the less-developed countries a shortage of foreign exchange or surplus of labor makes material cost or machine cost dominant.

2.2 Uses of Productivity

Table 3.3 shows how some countries have put more emphasis on leisure than others. This table is based on the nominal hours worked by a "male manufacturing worker" assuming full employment, zero illness, zero absenteeism, and zero overtime. They are representative values rather than official statistics. (In the United States in 1972, average hours were 40.8 in manufacturing but 34.5 in wholesale trade, 34.1 in services, and 40.4 in transportation.) Using the United States as an example, a "typical" male industrial worker would enter the labor force at age 18 and retire at 63—giving 45 years of work. A "typical" work week is 5 days but he receives 12 working days vacation (less when younger, more when older) and 9 paid holidays, so he works 239 days/yr. The "typical" work day of 8 hours (not including lunch) minus two "coffee" breaks of 10 min each gives 7.67 h of work/day. The total of 83,000 is affected by absenteeism, strikes, illness, unemployment (voluntary or involuntary), overtime, second jobs, more or less schooling, and early or late retirement. The 83,000 must support not only the worker's needs for food, furniture, and frivolity but also support the children, aged, blind, and other nonproducers (either directly or through taxes). In this book we will discuss "how to increase the size of the cake" rather than "how to cut the cake"; that is, how to multiply, not how to divide. (It seems the job of engineers is to increase the cake size and the job of politicians is to divide it.)

The output from the 83,000 h can be increased by working efficiently (**work smart**) or by working with more effort (**work hard**). Work smart is the desirable alternative because (1) there is more potential for improvement through reducing the excess work than through making the worker work harder, and (2) people don't like to work hard and therefore resist efforts made to make them do so.

3 TOTAL TIME FOR A JOB OR OPERATION

Figure 3.3 shows how capital substitution for labor can make labor noncompetitive. Figure 3.2 shows that substitution of energy and materials for labor has been worthwhile as they have become relatively cheaper than labor. In 1970, the manufacturing wage/h ($3.36) and the price of a barrel of oil ($3.18/barrel) were almost equal. Although there certainly will be fluctuations in the ratio of a barrel of oil to manufacturing labor in the future, the long-run tendency will be to increase the relative cost of oil and thus oil-based energy costs. Therefore, energy from other sources (such as coal) and conservation should become relatively more important. In some countries there is a great surplus of labor and thus the relative price of labor is low. In some countries foreign exchange is limited, which may cause energy or materials to be relatively expensive. Capital may be relatively expensive or cheap in various countries. Capital is relatively cheap in Japan, so Japan has tended to substitute capital for labor more rapidly than other countries. In addition, some industries are labor-intensive, some skill-intensive, and some knowledge-intensive. Thus engineers in different countries and organizations will have different objectives.

TABLE 3.3 Nominal working hours per lifetime for a "typical" male manufacturing worker in various countries in 1988. In general, hours/lifetime have declined since 1988.*

	France	U.K.	Sweden	Norway	India	U.S.	Japan	Taiwan	Indonesia	Hong Kong	China
Leave work force	60	60	62	67	60	63	60	60	58	60	60
Enter labor force	18	16	18	18	18	18	18	18	16	18	18
Working Years	42	44	44	49	42	45	42	42	42	42	42
Days/week	5	5	5	5	6	5	5.5	5.5	6	6	6
Nominal days/year	260	260	260	260	312	260	286	286	312	312	312
Vacation	36	15	25	20	30	12	14	14	14	7	0
Holidays	8	8	10	10	10	9	9	22	12	10	7
Working Days/Year	216	237	225	230	272	239	263	250	286	295	305
Hours (omit lunch)	7.80	7.5	8	7.5	7.5	8	8	8	8	8	8
Breaks	.33	.5	.33	.5	.33	.33	.5	0	.33	0	0
Hours/Day	7.47	7.0	7.67	7.0	7.17	7.67	7.5	8	7.67	8	8
Hours/Lifetime (000)	68	73	76	79	82	83	83	84	92	99	102

*Full employment, zero illness, zero absenteeism, and zero overtime are assumed as well as typical retirement. Numbers are not official but are judgments of students from that country studying in the United States.

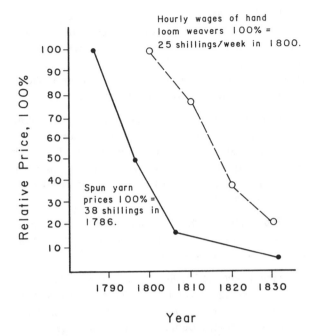

FIGURE 3.3 Substitution of capital (i.e., machinery) for labor can make wages noncompetitive. Cheyney (1908) showed how the introduction of machinery to spin yarn made it noneconomic to spin by hand. The introduction of mechanical weaving had a similar effect on the wages of hand loom weavers. Cheyney, Edward. *Industrial and Social History of England*, New York: Macmillan, p. 189; 1908.

3.1 Who Gets Benefits?

Any program to raise productivity by reducing time/unit must consider the workers' fears (1) that they will work themselves out of a job and (2) that employers will get all the benefits of the higher productivity. Policies to ensure adequate employment and satisfactory distribution of the benefits of the productivity are not merely desirable parts of productivity programs—they are the foundation. (The good of the whole conceals the cost to the few.)

Reducing time/unit has costs: (1) erosion of individual skills and experience, (2) the need for some workers to change their jobs and perhaps their place of residence, and (3) inability of some individuals to make the changes required. Thus workers, individually and collectively, need to have the costs of improved productivity not fall too heavily on any one person. One of the key aspects of Japan's high annual productivity increases is that the male Japanese worker is hired until age 55 and thus does not resist technological change. In the United States, employers attempt to use normal turnover and expansion of sales to cushion layoffs; severance pay and unemployment pay are a backup system.

People also differ on who should get the benefits of higher productivity: the workers through higher wages, the society through lower prices, or the person who risks his capital through greater profits.

Wars have been fought over this issue. When political rhetoric is brushed aside, the answer is that benefits must be split among the three. Naturally there is always discussion on the amount of the split. Samuel Gompers expressed labor's opinion concisely when asked what labor wanted: "More."

3.2 Extra Work Content

Time/unit can be considered to be composed of (1) basic work content and (2) extra work content. (For simplicity, the following examples will emphasize manufacturing, although the same concept can be applied to retailing, transportation, health services, and so forth. For simplicity, productivity is assumed to depend only on the time required for the person or machine.)

3.2.1 Poor Product Design

Five types of poor product design are as follows.

Improper design For example, don't design a product to be composed of weldments when a casting is more economical. (In some cases, weldments are better than castings.) Don't design to use slot-head screws instead of Phillips head screws. Design for easy maintenance. Coat steel used in corrosive atmospheres. If truckloads are weight-limited, use aluminum on the trucks instead of steel so more payload can be carried. Reduce cost of shipping also by using stackable containers and thus utilizing the cube.

Nonstandardization Use standard materials, not special materials. Use standard parts, not special parts. For example, can a standard washer, bracket, screw, bolt, and so forth, be used instead of a special? Lack of standardization splits the production volume between the parts, increases paper work, and makes supply of spare parts more expensive and difficult.

Incorrect quality standards The quality specified can be too low or too high. An example of low quality might be a plastic part instead of a metal part, or a container which allows the product to be damaged. An example of too high a quality is using precision threads when standard threads are sufficient. Overdesign ("goldplating") occurs as each engineer designs for the "worst case" rather than "real world." It is nice to have good quality, but not everyone can afford a Mercedes.

Material wastage A stamping might be designed which fits poorly on a coil-fed press, allowing material to be wasted. An auto might be designed with an engine with poor fuel economy. In the office a form might be designed which requires a large sheet of paper when a small piece of paper would do.

Energy wastage   A standard motor may be used in place of a high-efficiency motor. A conveyor motor may not be turned off during the lunch break. The lights may be left on in a room that is vacant. A semi-tractor may not have a shield on top of the cab to streamline the air flow.

3.2.2 Poor Methods   Four poor methods are the following.

Poor macro method   The wrong technique might be used. For example, a person might play "telephone tag" instead of using a telephone with an answering machine, a fax, or E-mail. A cellular phone in a vehicle permits executives, maintenance people, and sales people to call from a vehicle instead of having to stop and look for a public phone. Further examples might be data being entered by hand instead of by scanner and bar code; a factory not being organized for "lean production" and therefore having many uncorrected problems; a machine cutting one item at a time instead of 60 at a time; or a product being sent by rail when using a truck would be better (or vice versa).

Poor micro method   For example, a waitress might have to enter orders on a form with handwriting instead of circling preprinted information, or the wrong type or size of screwdriver, pliers, or other tool may be used.

Poor arrangement   Examples are machines arranged in a job-shop layout when a flow-line arrangement is better (or vice versa), supplies located in a crib when storage at the machine is better (or vice versa), or a bin at a workstation not conveniently located.

Poor equipment use   The equipment may not be used properly if the worker is not properly trained. Training should include not only the primary person assigned to the task but all people who do the task. The thrust has been to multiskilled workers so they can do all tasks (cross-training).

3.2.3 Poor Management   Five types of poor management are as follows.

Too many product models   Sales groups are always pushing for a wide variety of models to fulfill the needs of any possible customer. Unfortunately, this tends to fragment the market and results in very low volumes for each product and thus high cost/unit. If many different models must be produced for marketing reasons, insist that nonstyle components such as screws, washers, motors, brackets, clips, and so forth, be standardized so they, at least, can have reasonable production quantities and ease of spare parts supply.

Poorly designed product   The product may not be designed to withstand the stresses of normal use. The customer will be unhappy and will tend to refuse to buy more products from that organization, thus in the long run killing the organization. Rework, repair, and warranty expenses will be high.

Poor production scheduling   Poor scheduling can increase setup time, cause missed customer shipment dates, and require overtime and layoffs. A unit train hauling coal shows how scheduling a shipment all at once instead of in small amounts can improve transportation efficiency.

Poor maintenance   Equipment breakdowns annoy everyone. In addition to loss of production, product quality may suffer.

Poor safety and health   Safety pays. In addition to the moral responsibility of organizations to have safe working conditions, safety can be justified just from the economic benefits. (See Chapter 25.)

3.2.4 Poor Workers   Not all the excessive work is due to management. Some is due to workers. Workers can fail to start on time, quit early, and stretch breaktimes; they can be absent and thus cause extra work for their coworkers. They can cause poor quality—although most quality problems are due to poor product design, poor tools, and/or poor procedures, that is, from poor management.

In most operations or jobs, the effect of the worker in causing extra work is relatively minor compared to the extra work caused by poor design, work methods, or management. Obtaining better productivity (the benefits to be shared among the workers, the society, and the providers of capital) thus is a shared responsibility. The importance of better productivity and the method to achieve it can be demonstrated with a pegboard.

4 WORK SMART NOT HARD

Figure 3.4 is a photograph of the pegboard task: "work smart" in front, "work hard" at the right. The task is to put 30 pegs into the 30 holes. Figure 3.5 gives a left-hand/right-hand chart of the method used by the instructor in demonstrating condition A: blunt end of peg into nonchamfered hole. The students should time the instructor doing one assembly to the instructions "Work at a pace you can maintain for eight hours; assume you are paid by the hour." For reference, when 51 students assembled 10 boards, their mean time for 10 boards for condition A was 1.02 min.

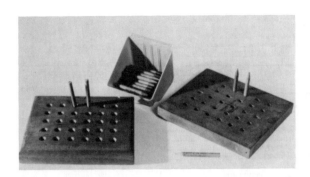

FIGURE 3.4 "Work smart not hard" can be demonstrated with assembly of pegs into a pegboard.

Condition B demonstrates WORK FAST. Use the method for condition A but with instructions "Work at a pace you can maintain for eight hours; assume you are paid by the piece." Mean time/assembly for 10 assemblies for the 51 students was .81 min—a reduction of 21%.

Condition C demonstrates WORK SMART—that is, reduce the excess work content. For *product design* changes, use chamfered holes instead of non-chamfered holes (turn board over) and insert the pointed end of the peg instead of the blunt end. For *manufacturing methods* changes, use two parts bins and two hands instead of one and fill the center holes first to avoid moving the hands over a barrier. Preorient the pegs in the bins. (Preorientation involves a cost but cost would be low if it could be done by the previous operator or by a vibratory feeder.) Figure 3.6 gives a left-hand/right-hand chart for condition C. Working at the pace of condition A, the mean time/assembly for 10 assemblies was .47 min—a reduction of 54%.

The effect of better management can be demonstrated by long production runs. To save demonstration time, use the learning curve rate of 94% calculated by Youde, who timed 300 consecutive assemblies of the pegboard (Youde, 1947). The 94% means that every time output is doubled, the new time is 94% of the previous time. Thus if time for 10 is .47, then time for 20 would be .94 (.47) = .4418; time for 40 would be .94 (.4418) = .4153; time for 80 would be .94 (.4153) = .3908, and so on.

The potential improvement for better product design, manufacturing methods, and management is unlimited. Convincing workers to work harder is difficult. Sweat, as they know, is inversely related to wealth.

ACT BREAKDOWN

SKETCH

Study File No. _____ Date _____
Oper. Name-Equip. Description _____
  Pegboard Assembly—Method A & B

Tools Used _____ 1 bin for pegs

Part Description   pegboard with non-
beveled holes   (30)-3/8 D 5/16 D peg

Part No. _____
Routing Hrly. Cap. _____

Study Hrly. Cap. _____
Analysis By _____ SK

Step No.	LEFT HAND DESCRIPTION	OBJECT	ACT	ACT	OBJECT	RIGHT HAND DESCRIPTION	
1			W	G	peg	in bin	Repeat
2			W	P	peg	in hole—column 6	5 times
3			W	G	peg	in bin	Repeat
4			W	P	peg	in hole—column 5	5 times
5			W	G	peg	in bin	Repeat
6			W	P	peg	in hole—column 4	5 times
7			W	G	peg	in bin	Repeat
8			W	P	peg	in hole—column 3	5 times
9			W	G	peg	in bin	Repeat
10			W	P	peg	in hole—column 2	5 times
11			W	G	peg	in bin	Repeat
12			W	G	peg	in hole—column 1	5 times
13							
22							
23							
24							
25							
26							

03080-219 - 12-66 -- IE-74-1   (OVER)

FIGURE 3.5 Detailed analysis of the pegboard assembly shows that the right-hand activity includes only the acts of get and place while the left hand has only inactivities of wait. (Acts are get, place, and dispose; inactivities are hold, wait, process, and drift.) The analyst "worked smart" by using G for get, P for place, and W for wait. The analysis is typed for clarity—normally it would be handwritten.

ACT BREAKDOWN

SKETCH

Study File No. _____ Date_____
Oper. Name-Equip. Description_____
    Pegboard Assembly—Method C

Tools Used ____ 2 bins for pegs

Part Description   pegboard with beveled holes
    (30)-3/8 D 5/16 D peg
Part No. _____
Routing Hrly. Cap. _____

Study Hrly. Cap. _____
Analysis By ____ SK

Step No.	LEFT HAND DESCRIPTION		OBJECT	ACT	ACT	OBJECT	RIGHT HAND DESCRIPTION	
1	Repeat	in bin	peg	G	G	peg	in bin	Repeat
2	5 times	in hole—column 3	peg	P	P	peg	in hole—column 4	5 times
3	Repeat	in bin	peg	G	G	peg	in bin	Repeat
4	5 times	in hole—column 2	peg	P	P	peg	in hole—column 5	5 times
5	Repeat	in bin	peg	P	P	peg	in bin	Repeat
6	5 times	in hole—column 1	peg	G	G	peg	in hole—column 1	Repeat
7								
24								
25								
26								

03060-219 - 12-66 -- IE-74-1   (OVER)

FIGURE 3.6   Reduction of excess work content by "working smart" demonstrates the benefits of using your brain instead of your muscles.

The British posted a guard on the cliffs of Dover in 1812 to watch for an invasion by Napoleon. The job was abolished in 1935. They should have worked smart, not hard.

REVIEW QUESTIONS

1. Give a few examples showing that the good old days weren't very good.

2. Briefly discuss the change in hours of work over the last century. What do you predict the typical hours of work/yr will be when you are 40? Give assumptions.

3. Briefly discuss the concept of increasing the size of the cake vs. dividing the cake.

4. Discuss replacing transportation with communication.

5. How much do you think manufacturing labor cost/h will increase over the next 25 years? Cost of oil/barrel? Cost of electricity?

6. What did Samuel Gompers say when asked what labor wanted?

7. Give the two reasons why it is more important to work smart than to work hard.

REFERENCES

Cheyney, E. *Industrial and Social History of England,* 189. New York: Macmillan, 1908.

Harbison, F. Education for development. *Scientific American,* Vol. 209, 140–47, September 1963.

Schor, J. *The Overworked American.* New York: Basic Books, 1991.

Tanner, J. Earlier maturation in man. *Scientific American,* Vol. 218, 20–27, January 1968.

Youde, L. A study of the training time for two repetitive operations. Master's thesis, State University of Iowa, 1947.

CHAPTER

4 ENGINEERING DESIGN

OVERVIEW

Engineering design differs from the scientific method. Remember the 5 steps of engineering design with the acronym DAMES: (1) Define the problem broadly, (2) Analyze in detail, (3) Make search, (4) Evaluate alternatives, and (5) Specify and sell solution. Concentrate your valuable design time on the "mighty few" rather than the "insignificant many"—that is, fight giants. Determine the benefits of your proposal with a benefit/cost analysis. After production begins, use EVOP to further improve the process.

CHAPTER CONTENTS

1 Scientific Method
2 Engineering Design
3 Important Problem First (Pareto)
4 Cost Allocation
5 Return on Investment
6 Evolutionary Operation of Processes (EVOP)

KEY CONCEPTS

annual savings	life of the application	satisfiers
benchmarking	one-time costs	scientific method
cost centers	optimum solution	standard cost
DAMES	overhead (burden)	
data vs. theory	Pareto distribution	
direct labor	passive vs. active observation	
Evolutionary Operation of	prerelease review	
Processes (EVOP)	return on investment	

1 SCIENTIFIC METHOD

If asked to list important inventions, the ordinary citizen tends to list devices such as the wheel, transistors, and the electric motor. Just as important, however, and perhaps even more important, are *concepts*. Examples of important concepts are technical societies, public libraries, the scientific method, and fast food (the restaurant as a factory).

Mowrer (1960) stated the importance of research:

> In plants change occurs almost entirely by means of the evolutionary mechanism; but in animals there is the capacity for another kind of "evolution" of change; namely learning. Learning by actual doing involves hazards. A still higher level of advantage accrues to organisms which can explore their environments not only in terms of actual performance but also perceptually.

The **scientific method** (see Table 4.1) is an efficient method of doing research (learning by doing). It becomes even more useful if combined with another concept, education (perceptual exploration of the environment), so that students need not "reinvent the wheel."

We can use a formula to predict when a beam will break; we do not need to build the beam. We can use a model to predict the best production schedule; we do not need actually to use all possible schedules. We can predict the amount of material handling required for a proposed plant; we do not need to build all possible plants. Nothing is so practical as a good theory. But, if we are going to trust the output of the formula (or model/theory) and not physically evaluate alternatives, it is essential that the formula (model/theory) is valid for the conditions of use.

Table 4.1 gives the 5 steps of scientific method as well as an example application to the optimum height for a keyboard. The essence of the scientific method is the feedback of **data vs. theory.** The ancient Greeks were good at the first three steps, but they considered it beneath their dignity to see if their theories checked with the reality. Mere data gathering without an underlying theory also is not very useful. The powerful combination is theory plus data with a comparison of the data vs. theory until the error is acceptable: a negative-feedback control circuit.

2 ENGINEERING DESIGN

Engineering design, although related to the scientific method, differs from it. Remember the five steps of engineering design with the acronym **DAMES**, where D = Define the problem, A = Analyze, M = Make search, E = Evaluate alternatives, and S = Specify and sell solution. See Table 4.2.

Box 4.1 gives an overview of the steps in releasing a part for manufacturing.

2.1 Define the Problem Broadly
Usually the designer is not given the problem but is instead confronted with the existing solution. The current solution is not the problem but just one solution among many possible solutions. The broad, detail-free statement of the problem should include the number of replications, the criteria, and the schedule. Using the example of Table 4.2, the replications are

---

TABLE 4.1 The scientific method has five steps. Step five, the critical step, compares data vs. the prediction.

Step	Example
1. Clearly state the problem you are trying to solve.	What is the optimum height at which a keyboard should be placed?
2. Construct a hypothesis or model.	Speed and accuracy of keying depend upon fatigue.
3. Apply analysis to the model. From the analysis, predict what will happen in various conditions.	Fatigue is caused primarily by supporting the arm weight; the higher the hand is held, the greater the torque which must be resisted and the greater the fatigue.
	Since the keystroker is assumed to be seated, the optimum keyboard position will be in the keystroker's lap (i.e., at the lowest feasible position).
4. Design and perform an experiment with the real situation. Compare vs. the model.	Have a number of individuals key the same material at a number of heights. Record their speed, accuracy, and preferred work heights.
5. Compare data from the real situation vs. the model's predictions:	See Guideline 2 of Chapter 15.
a. If the difference between the prediction and the data (called error) is acceptable, accept the model.	
b. If the error is too large, revise experiment or model until error is acceptable.	

TABLE 4.2 The five steps of the engineering design procedure can be remembered by the acronym DAMES (define, analyze, make search, evaluate, specify, and sell).

Step	Comments	Example
Define the problem broadly.	Make statement broad and detail-free. Give criteria, number of replications, schedule.	Design, within 5 days, a workstation for assembly of 10,000/yr of unit Y with reasonable quality and low mfg. cost.
Analyze in detail.	Identify limits (constraints, restrictions). Include variability in components and users. Make machine adjust to person, not converse.	Obtain specifications of components and assembly. Obtain skills and availability of personnel. Get restrictions in fabrication and assembly techniques and sequence. Obtain more details on cost accounting, scheduling, and trade-offs of criteria.
Make search of solution space.	Don't be limited by imagined constraints. Try for optimum solution, not feasible solution. Have more than one solution.	Seek a variety of assembly sequences, layouts, fixtures, units/h, hand tools, etc.
Evaluate alternatives.	Trade off multiple criteria. Calculate benefit/cost.	Alt. A: installed cost $1000; cost/unit = $1.10. Alt. B: installed cost $1200; cost/unit = $1.03.
Specify and sell solution.	Specify solution in detail. Sell solution. Accept a partial solution rather than nothing. Follow up to see that design is implemented and that design reduces the problem.	Recommend Alt. B. Install Alt. B1, a modification of B suggested by the supervisor.

BOX 4.1 *Releasing a part for manufacturing*

Good practice follows five steps for new parts or assemblies:

1. *Prerelease review* Before design engineering releases a part number to manufacturing, representatives from manufacturing should have a formal conference with design engineering. (A **prerelease review,** intended to reduce manufacturing problems, is a special case of a "useability review," which has the goal of reducing customer problems.) Typical representation would be industrial engineering, tool engineering, inspection, quality assurance, scheduling, and purchasing. The participation of purchasing is very important. The group challenges the details on the drawings. For example, can a specially designed item be replaced by a low-cost consumer item? Can standard threads replace precision threads? Can existing part numbers replace new part numbers? Should component tolerances be increased (to help component manufacturing) or be decreased (to help assembly)? Designers still may be willing to consider changes at this stage but "set their

feet in concrete" once the drawing is officially released for production. See Chapter 6 for value engineering, another cost avoidance technique.

2. *Outline production system* The manufacturing engineering group needs to decide on the general production concept. What machines will be used? Approximately what batch size? Job shop or production line?

3. *Detail* Now specific tools and fixtures need to be designed. What feed and speeds will be used? Will the operator use one or two hands? Will the operator sit or stand? What will the layout be? Will the conveyor feed from the left or the right?

4. *Installation* Concepts now need to be turned into reality. Translating drawings into reality often requires adjustments.

5. *Time standards* Time/unit needs to be established for cost accounting, scheduling, determining acceptable day's work, etc. See Chapter 26.

"10,000/yr," the criteria are "reasonable quality and low manufacturing cost," and the schedule is "within 5 days." Putting in too much detail makes you start with defending your concept rather than opening minds (yours and your clients') to new possibilities. At this stage, the number of replications should be quite approximate (within $\pm$ 500%). The importance of giving criteria is that there usually are multiple criteria (cost, quality, simplicity, etc.) rather than just one criterion. (Nadler, on the other hand, recommends that you start with an ideal—extreme—solution and then "back off.") The schedule identifies priorities and allocation of resources that can be used both in the design process and the replication of the products from the design.

An important distinction between science and engineering is that the scientist wants a precise answer while the engineer is willing to settle for a practical answer. Consider the following challenge. A girl sits on one end of a bench; a boy sits on the other end; the distance between them is X. In the first minute, they decrease the distance between them by 50%, in the second minute by a further 50%, in the third minute by a further 50%, etc. Will they ever meet? The scientist ponders and says "Never!" while the engineer smiles and says "Close enough for practical purposes!"

2.2 Analyze in Detail

Amplify step 1 (defining the problem) with more detail on replications, criteria, and schedule:

- What are the needs of the design users (productivity, style, comfort, accuracy, esthetics, etc.)?
- What should the design achieve?
- What are the design limits (also called constraints and restrictions)?
- What are the characteristics of the population using the design? For example, if designing an office typing workstation, the users would be adults within certain ranges (age from 18 to 65, weight of 50 to 100 kg, etc.).

Since people vary, designers can follow two alternatives. (1) Make the design with fixed characteristics and make the people adjust to the device. One example would be a fixed-dimension chair and selecting people to fit the chair; another would be a machine-paced assembly line and forcing each worker to work at the speed of the master cam. (2) Fit the task to the worker. The design adjusts to varying characteristics of the users. One example would be a chair which adjusts to individuals of different dimensions; another would be a human-paced assembly line with buffers so that all workers work at their own paces.

2.3 Make Search of Solution Space

At this stage the engineer designs a number of alternatives; one of the key distinctions between science and engineering is that in science there is only one solution while in engineering there are a number of solutions. Benchmarking is a technique of obtaining some good alternatives. See Box 4.2. Although the solution space is cut down by economic, political, esthetic, and other constraints, there is more than one feasible solution (see Figure 33.2). However, of the many feasible solutions (solutions which work), the engineer should try to get the best one—the **optimum solution.** The best will be a trade-off of the various criteria, which also change from time to time, so the designer must be careful not to eliminate alternative designs too early. Another problem is the tendency of designers to be **satisfiers** rather than optimizers. That is, designers tend to stop designing as soon as they have one feasible solution, when they have satisfied the problem. For example, when designing an assembly line, the engineer may stop as soon as there is a solution; when laying out the factory, the designer may stop as soon as one satisfactory layout is made. In order to get an optimum design, there must be a number of alternatives to select from; alternatives also suggest further alternatives, so stopping too soon can severely limit the solution quality and acceptance.

2.4 Evaluate Alternatives

A scientist tends to look for the single formula which describes one criterion of a situation; the engineer must trade off multiple criteria, usually without any satisfactory trade-off values. For example, one design of an assembly line may need .11 min/unit while another design may need .10 min/unit; however, the first design may give more job satisfaction to the workers. Which assembly line design should be used? How do you quantify job satisfaction? Even if you can put a numerical value on it, how many "satisfaction units" equal a 10% increase in assembly labor cost?

A simple ranking of alternatives is sufficient in some cases (good, better, best). More precise is a numerical ranking, using a single criterion with an equal interval scale. (Method A requires 1.1 min/unit while method B requires 1.0 min/unit; design A requires 50 m^2 of floor space while design B requires 40 m^2 of floor space.) Most managers, however, prefer a comparison combining all the various costs and benefits in terms of money. Even if the various costs and benefits can be put in such terms, they may be "different kinds" of money. For example, you must add operating costs (such as labor cost/unit), capital costs (machine purchase costs), maintenance costs (machine lubrication costs), product quality costs (reduced product failures in the field after three

---

BOX 4.2 *Benchmarking*

If you stand on someone's shoulders, you can see farther than that person can. The key concept of **benchmarking** is to learn from the experience of others and then apply the knowledge of someone else's product or process to your own product or process (Main, 1992).

Product Ford provides an example of benchmarking using a product. To build a better car, Ford compiled a list of 400 features that customers said were important, looked at how its competitors did each of these features, and then picked the one to beat. For example, Ford benchmarked door handles vs. Chevy Lumina, halogen headlamps vs. Honda Accord, fuel economy (Lumina), front bench seats (Lumina), easy-to-change tail lamp bulbs (Nissan Maxima), express window control (Maxima), tilt wheel (Accord), remote radio controls (Pontiac Grand Prix), and so forth.

Process Firms often compare designs of their own processes at different plants of their own firm, but the big improvements come from seeing how strangers do it. A key point here is that the stranger may be in a different country or industry. For example, Ford benchmarked handling of accounts payable vs. Mazda in Japan; Xerox benchmarked its order picking vs. L. L. Bean. Some guidelines for benchmarking follow:

- *Select a specific target.* After defining the problem, study your own process in detail before evaluating the approach of others.
- *Have people who make the decisions see the alternatives.* Senior decision makers often refuse to believe the teams and say "We can't be that bad."
- *Exchange information.* Be prepared to allow others to benchmark you also.
- *Avoid legal problems.* Benchmarking is not industrial espionage. Focus on existing products and processes, especially from noncompetitors.
- *Respect confidentiality.* Firms usually don't want their names used or data going to competitors.

---

years of use), environmental costs (CO concentration in the work area reduced from 40 ppm to 30 ppm), and so on. See Table 7.6 for one approach.

2.5 Specify and Sell Solutions

Scientists can state their conclusions in terms that only Ph.D.s can understand but engineers must communicate with the ordinary individual, so abstract theories must be translated into "nuts and bolts." Then this individual can have the audacity to say "No!" to your beautiful design! A humbling experience. The improvement occurring in a situation is a function of the quality of the design times the acceptance of the design. If "they" don't "buy" it, nothing happens. Engineers therefore accept modifications in their beautiful designs to gain acceptance; striving for the "whole loaf or nothing" usually gets "nothing." Then, after the design has been accepted, follow up to see that it is put into practice. (Many a good idea has been accepted with beaming smiles by people who have a firm resolve to let the implementation die due to apathy.) Then, when the design has been implemented, see whether it has reduced the original problem— are you part of the solution or part of the problem?

Although I have presented some of the difficulties of engineering design, engineering is a very satisfying profession. Herbert Hoover expressed it well:

It is a great profession. There is the fascination of watching a figment of the imagination emerge through the aid of science to a plan on paper. Then it moves to realization in stone or metal or energy. Then it brings jobs and homes to men. Then it elevates the standards of living and adds to the comforts of life. That is the engineer's high privilege.

The great ability of the engineer compared to men of other professions is that his works are out in the open where all can see them. His acts, step by step, are in hard substance. He cannot bury his mistakes in the grave like the doctors. He cannot argue them into the thin air or blame the judge like the lawyers. He cannot, like the architects, cover his failures with vines and trees. He cannot, like the politicians, screen his shortcomings by blaming his opponents and hope the people will forget. The engineer simply cannot deny he did it. If his works do not work, he is damned. . . .

On the other hand, unlike the doctor, his is not a life among the weak. Unlike the soldier, destruction is not his purpose. Unlike the lawyer, quarrels are not his daily bread. To the engineer falls the job of clothing the bare bones of science with life, comfort, and hope. No doubt as years go by the people forget which engineer did it, even if they ever knew. Or some politician puts his name on it. Or they credit it to some promoter who used other people's money. . . . But the engineer himself looks back at the unending stream of goodness which flows from his successes with satisfactions that few professions may know. And the verdict of his fellow professionals is all the accolade he wants.

3 IMPORTANT PROBLEM FIRST (PARETO)

Engineering time is a valuable resource; don't waste time on unimportant problems. Allocate design time to the important problems; neglect the minor problems. To check quickly whether a project is worth considering, calculate (1) savings/yr if material cost is cut 10% and (2) savings/yr if labor cost is cut 10%.

The concept of the Pareto distribution may help you find the important problems. (See Figures 4.1 and 4.2). (Lorenz also used curves to demonstrate the concentration of wealth, but Vilfredo Pareto's name is now associated with the concept of "the insignificant many and the mighty few.") Cause (x axis) and effect (y axis) are not related linearly; the Pareto curve (also called the ABC curve) can be approximated by a log-normal distribution. The key

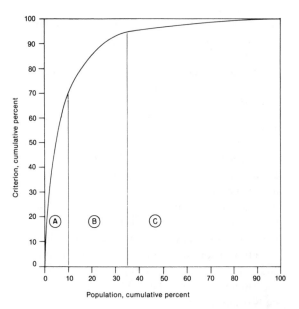

FIGURE 4.1 Fight giants. Many populations have a **Pareto distribution** in which a small proportion of the population has a large proportion of the criterion. Inventories, for example, can be classified by the ABC system. "A" items may compose 10% of the part numbers but 70% of the inventory cost. "B" items may compose 25% of the part numbers and 25% of the cost; so the total of A + B items composes 35% of the part numbers but 95% of the cost. "C" items thus compose the remaining 65% of the part numbers but only 5% of the cost. The concept is to concentrate your effort on "A" items so you don't use gold cannons to kill fleas.

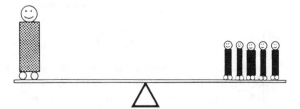

FIGURE 4.2 "The insignificant many and the mighty few" is another name for the Pareto distribution.

concept is that the bulk of the problem (opportunity) is concentrated in a few items, for example:

A few

products	produce	most of the direct labor dollars.
products	have	most of the storage requirements.
products	produce	most of the profit.
operations	produce	most of the quality problems.
machines	use	most of the energy.
operations	produce	most of the repetitive trauma problems.
time studies	cover	most of the direct labor hours.
routes	have	most of the customers.
days	have	most of the outgoing or incoming orders.
half-hours/day	have	the largest number of truck arrivals/half-hour.
salespersons	sell	most of the product.
farmers	grow	most of the food.
cities	have	most of the population.
engineers	have	most of the patents.
professors	have	most of the publications.
individuals	commit	most of the crimes.
individuals	have	most of the money.
individuals	eat	most of the food at a party.
individuals	drink	most of the beer.

Pareto diagrams are just a special form of histograms (frequency counts); the key difference is that the categories are put in sequence of largest first (instead of a random order) and the cumulative curve is plotted. Table 4.3 shows an example table of operations arranged in order of largest annual cost first. This then is translated into a cumulative figure such as in Figure 4.1. For examples that show the bars as well as the cumulative curve, see Figures 33.5, 33.6, and 33.8.

Farmers have always known that most of the butterfat is in the cream, not in the skim milk. Therefore, using the Pareto concept, if your design

TABLE 4.3 List categories with the most important one first and the least important last. Then plot the cumulative percent as in Figure 4.1.

Item Number	Operation	Work-Hours per year	Labor Cost/H	Labor Cost/Yr	Percent of Total	Cumulative Percent
1	Hand polish	65,000	8.75	568,750	15.7	15.7
2	Form grind	52,000	9.30	483,600	13.4	29.1
3	Drill and bore	56,000	8.25	462,000	12.8	41.9
906	Insert fittings	120	8.00	960	.0	99.9
907	Apply nameplates	60	8.50	510	.0	100.0
				3,600,000	100.0%	100.0%

concerns crime, it should concentrate on the few individuals who cause most of the crimes; if your design is to affect food consumption at a party, it should concentrate on the few individuals who eat the most; if it is to improve quality, it should concentrate on the few components which cause most of the problem, and so on. While engineers fight giants, who takes care of the midgets? Rather than ignore them, have Quality Circles and other employee participation groups work on them.

In order to maximize your design productivity, work on several projects at the same time rather than spending all your time on one project and then going on to the second project (i.e., parallel projects, not series projects). Not only will your time be spent more effectively, but the idea quality also will be better.

4 COST ALLOCATION

Organizations allocate costs into various categories to aid decision making. Table 4.4 gives a hypothetical cost breakdown of an electric fan. Different organizations, industries, and countries will have different categories and different ratios of costs in the categories depending on the product, industry, amount of competition, and so forth. A monopoly, for example, may have a higher profit/unit.

Direct materials and **direct labor** (also called "touch" labor, because they touch the product) are the most easily allocated costs. Of course, there may be difficulties, such as when multiple products use the same material or the same worker is assigned to work on many different products. For managerial control, firms usually keep raw material costs

TABLE 4.4 Hypothetical cost breakdown of an electric fan. Although the percent of each cost varies widely by product and industry, a rule of thumb is direct labor cost = 10% of manufacturing cost; operating indirect labor = 100–150% of direct labor; clerical/technical indirect labor = 100–250% of direct labor cost. These same ratios apply to the vendor who furnishes purchased components.

Category	Example		Cost/Fan ($)
Raw materials (need more work)	Steel for fan blades		.80
Purchased parts and assemblies (used as is)	Motor, bearings, knobs		4.00
Direct labor	Time to stamp out blades, paint blades, assemble knob to shaft		1.00
		Prime Cost	5.80
Indirect labor burden	Fork truck drivers, industrial engineers		4.00
Indirect materials burden	Toilet paper, light bulbs		.50
Capital burden	Depreciation for machines, conveyors, building		.72
		Manufacturing Cost	11.02
Selling cost	Warranties, sales salaries		2.00
		Total Cost at Factory	13.02
Profit	Interest, risk		.98
		Factory Selling Price	14.00
Distribution cost	Shipping, distribution chain expenses and profit		6.00
		Consumer Price	20.00

separate from purchased components costs; their total is the direct material cost. *Direct material + direct labor = prime cost.*

Overhead (burden) is the next level of costs to be allocated. It can be divided into indirect labor, indirect materials, and capital costs. Indirect labor includes the salaries and wages of the clerks, engineers, technicians, supervisors, inspectors, and others. Indirect materials burden includes the cost of the electricity, water, and other utilities and the cost of various supplies such as paper clips, pencils, lubricating oil, degreasing compounds used before painting, grinding wheels, and toilet paper. Capital burden includes the cost of the fork trucks, conveyors, machine tools, the building, property taxes, and so forth. Progressive firms now are allocating portions of overhead costs to local organizational units (**cost centers**) to encourage the reduction of waste. For example, a department could be charged $200/yr/desk for utilities, $100/yr for PC computer maintenance, $100/person/yr for janitorial support, or $20,000 for every lost-time back injury. *Prime cost + factory burden = manufacturing cost,* the cost of getting an item to the factory door.

In most circumstances customers are not standing outside your door in the rain clamoring to buy your product. To sell a product requires catalogues, product warranties, and so forth, as well as wages for those concerned with this work. *Manufacturing cost + selling cost = total cost at the factory.*

Next is the cost of capital (interest) and the cost of risk of the capital. They are usually totaled and called *profit.* For example, income taxes on the firm will be built into the cost structure and thus really are passed on to the customer; corporate income taxes really are a concealed sales tax. In the long run these expected costs and risks must be built into the product price in both capitalist and socialist organizations. If these costs are not included in the product price, the organization loses money, which must come either from the owner's assets or the taxpayer's assets. *Total cost + profit = selling price at the factory.*

Since in normal circumstances the customer does not come to the door, there are additional costs for distribution of the product. They include shipping and expenses of the retailing organization. *Factory price + distribution cost = consumer price.*

Within the factory, the engineer probably will deal mostly with the manufacturing cost of $11.02. Actually, fans probably will not cost $11.02 each day because on some days more scrap is made, on some days more or less labor is used, and so forth. The $11.02 is not the real cost but is a **standard cost.** Standard costs make assumptions about such things as the standard amount of material, a standard

price/unit of material, a standard amount of labor, a standard price/unit of labor, a standard amount of scrap, a standard cost of each type of defect, a standard use of energy/unit, a standard amount of overhead/unit. Unfortunately it is difficult to keep standard costs accurate over time as conditions change. For example, standard labor cost may be based on prorating the setup cost over 10,000 units/yr. However, over the years, production may have changed to 5,000/yr or 15,000/yr, causing an error in the standard setup hours/unit and thus an error in the standard cost. Another common problem is that the standard labor hours/unit (say painting time of 10 s/unit) may not be the actual time (painting time may be 9 s or 12 s). When making a cost analysis, try to use actual costs instead of standard costs.

When jobs are designed, a common criterion is the cost/unit, that is, we judge workstation A to be better than workstation B because labor cost/unit is lower for workstation A. Labor cost/unit should never be the sole criterion because material and overhead costs also are important. The following section discusses some of the important money costs to consider; Chapter 9 discusses the problem of criteria in more detail.

5 RETURN ON INVESTMENT

Do the benefits of a design outweigh the costs? Oxenburgh (1991) has written an excellent book giving 61 case studies of the economic benefits of ergonomics improvements. Making the benefit/cost comparison (determining the **return on investment**) requires three steps:

1. determining what is changed due to the design (e.g., product quality is better)
2. putting the changes into monetary units (e.g., improved quality is worth $.02/unit)
3. calculating the total benefits vs. costs (e.g., all the changes totaled give benefits of $4,700/yr; all the costs total $1,400/yr)

Most errors in decision making are due to poor data for steps 1 and 2; engineers and accountants tend to "overkill" step 3 with exotic formulas and complex calculations to four decimal points when they have errors of ± 50% in steps 1 and 2. Spend approximately 90% of your time in steps 1 and 2 and 10% in step 3; you will make better decisions than the person who reverses the ratio. The secret is valid data. Leave out judgments and opinions unless supported by data and quantified into dollars.

Figure 4.3 gives a cost analysis which emphasizes steps 1 and 2. It is oriented to metal-working

FIGURE 4.3 Cost analysis forms reduce omitting relevant data. The data shown are for a proposed specialized screwdriver to be used for automobile tune-ups. Costs can be given for two alternatives; for example, for direct labor, the cost is $.044 for the existing method and $.033 for the best manual proposal. The costs also can be given as a change from the reference value; for example, for pain and suffering, the existing method is the reference and the best manual proposal will save $.010/unit.

Project ___Screwdriver___

Part name _____ Part number _____ Used in dept. ___Tune-up___

Volume: ___800___ pcs/yr _____ pcs/day _____ pcs/h _____ h/pc

Labor cost/h ___$8.00___ Engineer ___SK___

A. Life of Application _____ yrs

B. Annual Cost on the Controlling Operation

	Existing Method	Best Manual Proposal	Best Mechanized Proposal
Direct labor, $/unit	.044	.033	
Relief labor, $/unit			
Downtime, $/unit			
Maintenance, $/unit			
Direct material, $/unit	reference	0	
Indirect material, $/unit	.007	.009	
Perishable tools, $/unit			
Tool regrind (repair), $/unit			
Utilities, $/unit			
Inspection, $/unit	reference	0	
Product quality, $/unit	reference	−.010	
Rework and scrap, $/unit	.002	.001	
Absenteeism cost, $/unit			
Safety and health, $/unit			
Turnover of workers, $/unit			
Other (specify) pain and suffering $/unit	reference	−.010	
TOTAL, $/unit	.053	.023	
TOTAL, $/year			
Line 1 Savings/year (Col. 2 or 3 − Col. 1)		$24.00	

C. One-Time Cost, $

Equipment, $	reference	0	
Jigs, fixtures for equipment, $	−	−	
Installation, $	−	−	
Operator retraining, $	reference	.67	
Engineering, $	reference	48.00	
Line 2 TOTAL one-time cost, $	reference	$51.67	

D. Benefit/Cost Calculations

Line 3 Total savings during application life = $120.00 (Line 1 × yrs)
Line 4 One-time cost = $51.67 (Line 2)
Line 5 Net savings = $68.33 (Line 3 − Line 4)
Line 6 Net savings/year = $13.67 (Line 5 / application yrs)
Line 7 Return on investment before taxes, % = 26% (Line 6 × 100 / Line 4)

manufacturing, but you can adapt it to your own industry. The example will evaluate use of a special-purpose screwdriver for use in making automobile tune-ups in a garage.

Key information required is: (1) project life, (2) savings/yr, and (3) one-time cost.

5.1 Life of the Application

The top of the form starts with basic information such as project, part name, part number, and where used. The first item is **life of the application,** which can be limited either by the life of the equipment (say a lathe would be worn out in 5 years) or by life of the product (say a fixture made obsolete by an anticipated change in product design in 3 years). Good practice is to use a project life of 20 years or less. It is very hard to predict the future; a project life of 30 or 40 years is highly unlikely. Assume the screwdrivers have an application life of 5 years.

The volume/yr is important. Be careful to calculate pcs/h and pcs/day by dividing by the proper number of days the product is made/yr; that is, most products are *not* made continuously, and so the rate/h usually cannot be multiplied by 2,000 hours/yr to obtain annual output. In many cases volume/h may change over the life of the application. Estimate volume/yr for each year; for a simple level of analysis (such as with this form) use the average; for a more precise analysis, use each year's estimate to calculate the benefits/cost for each year (see an engineering economics text such as White, Agee, and Case [1989] for the solution techniques). In our example, assume 800 tune-ups/yr with an average time of 1.0 h/tune-up.

An important question is the labor cost/h. Most cost reductions are justified on labor savings; the question is what is the proper labor-cost rate to use. A worker may be paid $6/h. Because of fringe benefits (insurance, pension, Social Security, holidays, vacations, etc.), the cost to the organization will be higher. (In 1983, fringe benefits averaged 37% of payroll costs in the United States.) Thus use cost of (for example) $8/h instead of $6/h. The burden in the factory may be allocated in proportion to direct labor cost (e.g., at 300% of direct labor); so cost/h may be given as $8 + 3(8) = $32/h. Use the direct labor cost ($8/h) rather than direct labor plus burden ($32/h). If labor cost is reduced, there is no reason to believe burden cost will be reduced; in fact, often burden expense increases (e.g., for more electrical power). Even for a constant burden cost, a lower amount of direct labor will mean that the burden rate/direct labor-hour will increase. Be very suspicious of cost-reduction proposals which use cost rates including burden. Burden changes can be listed separately. These imprecise cost estimates may be important in finely balanced decisions. In our case, assume labor cost of $8/h.

Next, record information for three alternatives: the existing solution, the best manual proposal, and the best mechanized proposal. The reason for requiring the best manual proposal is that engineers love machines and devices and thus often have a bias toward solutions which involve machines and devices. In our example, only simple handtools seem feasible, so the mechanized alternative will not be considered.

Two additional types of information are needed: annual savings and one-time costs.

5.2 Annual Savings

Annual savings, the second key item, is determined by calculating the savings/unit and then multiplying by annual volume. (Most items are not produced continuously, so it is not valid to multiply the daily rate by 240 days/yr.) In some situations, it may be easier to calculate annual savings for each subcategory directly:

- Direct labor is the cost of labor exerted specifically on this particular operation. An existing screwdriver for auto tune-ups may require 20 s/tune-up vs. 15 s/tune-up for a proposed screwdriver. In the example, existing cost would be (20/3,600) (8) = $.044/tune-up, while proposed cost is $.033.

- Relief cost is the cost for substitute labor on an assembly line (e.g., seven workers may work at six stations). In this auto tune-up example, there are no relief costs.

- Downtime is the cost of idle equipment or workers at this or other workstations. On tightly linked jobs, downtime becomes very important since downtimes add. In the screwdriver example we will assume that other workers are not tightly linked to this job and downtime is zero for both alternatives.

- Maintenance is the cost of equipment maintenance. Assume zero maintenance for both screwdrivers.

- Direct material is material used in the product. Assume $12 for materials with either tool (no cost difference).

- Indirect materials cover miscellaneous supplies and materials. Assume that the existing tool causes one stripped setscrew/50 tune-ups but the proposed tool, because it permits more torque, will probably have one stripped setscrew/40 tune-ups. Assume a setscrew costs $.10 plus $.25 for the labor required to go get the replacement screw from the stock room. Thus the existing tool cost is $.35/50 = $.007/tune-up while the proposed cost is $.009.

- Perishable tools (tool bits, grinding wheels, etc.) are used up by the process. Assume zero for both screwdriver alternatives.

- Tool regrind (repair) is the repairing, resharpening, or reworking of tools. Assume neither screwdriver will need repair.

- Utility costs include electricity, water, heat, and light. Assume no change in utility costs occurs with either screwdriver.

- Inspection costs include inspection both by the worker and by separate inspectors. Assume inspection cost of $1/tune-up regardless of the screwdriver used.

- Product quality is the improvement (degradation) in the product as expressed in warranties, lost customers, good will, law suits, and so forth. A product can be defective due to (1) a design defect (improper design, perhaps because the engineer did not anticipate how the product would be used) or (2) a manufacturing defect (item not made to specification). Legal costs are difficult to predict, but just defending a firm can cost $100,000; if you lose, it may cost much more. Assume that the proposed tool will give a very slightly better tune-up; estimated value of $.01/tune-up.

- Rework and scrap is the cost of the product quality before it leaves the department. These costs are notoriously underestimated, because people do not want to call attention to their errors. (See Figure 5.6.) Assume the existing tool requires rework in 1/70 while the proposed tool will require rework for 1/140; the rework time is estimated as 60 s. Then existing cost is $(60/3,600)(1/70)(8) = \$.002$/tune-up while the proposed is $.001.

- Absences can be planned (holidays, vacation) or unplanned. Unplanned absenteeism can be voluntary or involuntary. Involuntary absenteeism typically comes from sickness, accidents, and cumulative trauma. Payment (plus fringe benefits) goes to the absent person as well as to the replacement worker, and so costs are approximately doubled. Estimate the absenteeism rate for each alternative and the number of hours/yr absent. Multiply by the cost of an absent hour. Assume equal absences for both screwdrivers.

- Safety and health costs include medical costs and paperwork costs. Many doctors charge hundreds or even thousands of dollars/hour, and therefore costs become extraordinary. Estimate the cost of a safety/health problem and multiply it by the probability of its occurring for the various alternatives. Costs for a firm can be estimated from workers' compensation records (medical costs and disability costs) plus costs incurred that are not covered by workers' compensation. Probability of occurrence can be estimated from the Occupational Safety and Health Administration (OSHA) 200 logs (which record all on-the-job illnesses and injuries). A job might have an incidence of 8/200,000 hours. (The 200,000 is based on 100 people working 2,000 hours/yr.) If you estimate that incidence would be cut to 6, then there would be a savings of 2/200,000 h or 1 case/ 100,000 h. Assume equal health and safety costs for both screwdrivers.

- Turnover of workers should include (1) acquisition costs such as recruiting, selecting, hiring, and induction of employees; (2) development costs such as orientation, on- and off-the-job training, and loss in productivity due to incomplete training; and (3) separation costs such as severance pay and vacant position costs (Andersson, 1992). Assume equal turnover for both screwdrivers.

- Other costs are any additional costs you find relevant. Assume, in this case, that the proposed screwdriver will cause less fatigue, less muscle pain, and fewer cracked knuckles; you value this as $.01/tune-up.

The total cost considered for the existing screwdriver then is $.044 + .007 + .002 = \$.053$/tune-up while for the proposed tool it is $.033 + .009 - .01 + .001 - .01 = \$.023$/tune-up. Thus, savings/tune-up is $.03/tune-up; annual savings are $.03 \times 800 = \$24$/yr.

5.3 One-Time Costs

Next consider the **one-time costs,** the third key item, which include the following:

- Equipment cost is the purchased cost (including tax and delivery) of the equipment to your receiving dock. In this example, assume the existing tool costs $1 while the proposed tool costs $3. However, the existing tool will last the application life and its cost already has been paid, so consider its cost as zero. For a more expensive tool it may be worth considering its decline in value over the application life; that is, the difference between what you could sell it for now and what you could sell it for at the end of the application life.

- Jigs and fixtures often are required for equipment use. Assume as zero in the screwdriver example.

- Installation costs refers to the costs of getting the equipment from your dock installed and working. Typical expenses are for millwrights, electricians, and plumbers. Assume as zero in this example.

- Operator retraining refers to the loss in output while the operators adjust to the new procedure, tool, or device. Assume retraining for the proposed tool will require about 5 min; thus its one-time costs would be (5/60) (8) = $.67.

- Engineering costs (often forgotten in cost analyses) include time to investigate the new method, determine alternatives, calculate costs, sell recommendations, and install the new method, if accepted. For our example, assume it took the engineer 3 h to read about the proposed tool, talk to people, make this estimate, etc. Assume the engineer's wages are 200% of the mechanic's wages/h so engineering cost is 3 × $16/h = $48.

Thus, total one-time costs for the existing alternative are $0; one-time costs for the proposal are $3 + $.67 + $48 = $51.67.

5.4 Benefit/Cost Calculations

Total gross savings/yr, during the application life of 5 years, give total gross savings of $24 × 5 = $120. Subtracting the one-time expenses ($51.67) gives net benefits of $120 − $51.67 = $68.33. Returning to a yearly basis, this is $68.33/5 = $13.67. The return on investment is 13.67 × 100/68.33 = 26%.

In some cases it is difficult to estimate the amount of savings from a project. A useful technique is to calculate the savings for several alternatives, such as a 1%, a 2%, and a 5% improvement in productivity. The question then becomes not the exact improvement but whether the improvement is likely to be greater than the minimum required. For example, for a person costing $10/h (including fringes), the annual cost is $20,000 (assuming a 2,000 h work year). A 1% change would be $200, a 2% = $400, and a 5% = $1,000. Then a new chair costing $100 and improving output 1% (5 min/day) would pay for itself in .5 years. If it improved output .1%, then it would pay for itself in 5 years (i.e., a 20% return on investment.) The question management then needs to answer (assuming 20% is satisfactory) is "Do you think an improved chair will improve output at least 30 s/day"?

Organizations usually require new projects to have a proposed return on investment greater than they presently are making. Table 4.5 gives some example returns on investments for firms on an overall basis. The return needed for individual projects depends on the organization. This "expected mini-

TABLE 4.5   Return on investment and return on sales of selected organizations in 1992. Each year *Fortune* reports the results for the 500 largest U. S. industrial corporations, the 500 global industrial corporations, the 500 largest U. S. service corporations, and the 500 global service organizations.

NAME	TYPE	SALES RANK	SALES	ASSETS	NET INCOME	PERCENT RETURN ON Sales	Assets
			(000,000,000 Omitted)				
General Motors	U. S. Industry	1	132.8	191.0	(23.5)	(18)	(12)
Northrop	U. S. Industry	100	5.5	3.2	.12	2	4
Dean Foods	U. S. Industry	200	2.3	.86	.06	3	7
Rohr	U. S. Industry	300	1.3	1.4	.001	0	0
Gitano	U. S. Industry	400	.8	.25	(.24)	(29)	(95)
Block Drug	U. S. Industry	500	.6	.65	.057	10	9
General Motors (U. S.)	GLO Industry	1	132.8	191.0	(23.5)	(18)	(12)
Int. Paper (U. S.)	GLO Industry	100	13.6	16.4	.086	1	1
Alcan (Canada)	GLO Industry	200	7.6	10.1	(.112)	(1)	(1)
Dana (U. S.)	GLO Industry	300	5.0	4.3	(.382)	(8)	(9)
Kyocera (Japan)	GLO Industry	400	3.5	5.4	.191	5	4
Orckla (Norway)	GLO Industry	500	2.7	2.4	.033	1	1
AT&T	Diversified Service	1	65.1	57.2	3.8	6	7
Citicorp	Banking	1		213.7	.722		0
Fed. Nat. Mrtg.	Diversified Finance	1	14.5	180.9	1.62	11	1
Ahmanson	Savings	1		48.1	.204		0
Prudential	Life Insurance	1		154.7	.542		0
Sears	Retailer	1	59.1	83.5	(3.93)	(7)	(5)
United Parcel	Transportation	1	16.5	9.0	.516	3	6
GTE	Utilities	1	19.9	42.1	(.754)	(4)	(2)
Itochu (Japan)	Diversified Service	1	156.3	62.0	.030	0	0
Dai-Ichi Kangyo (Japan)	Bank	1		493.4	.372		0
Axa (France)	Diversified Finance	1		189.9	.401		0
Abbey Nat. (U. K.)	Savings	1		108.6	.559		1
Nippon Life (Japan)	Life Insurance	1		280.9	3.47		1
East Japan RR (Japan)	Transportation	1	18.8	61.6	.454	2	1
Elect. de France (France)	Utilities	1	39.4	117.6	.349	0	0

mum" rate might be modified in some situations. For example:

Category 1: laboratories, safety equipment, recreation facilities, etc., where no direct return is available, so return on investment is not used

Category 2: new projects with expectation of growth

Category 3: old products with a short life that may require greater return on investment

6 EVOLUTIONARY OPERATION OF PROCESSES (EVOP)

When designing, there are major decisions and minor decisions to be made, that is, major and minor in their effect on the outcome. Through trial and error, experimentation, luck, prototypes, pilot projects, and so forth, the engineer designs a process. Eventually there comes a stage in which the job "goes into production." There are still possibilities for improvement, but each individual change has only a small improvement potential. Yet your firm has a goal of "continuous improvement." Individual changes are not considered worthy of spending additional engineering resources investigating them and delaying production.

For example, consider painting part 123 in Department A. Considering that we could vary paint-thinner ratio, distance of part 123 from the paint nozzle, the nozzle diameter, the air pressure, the paint temperature, the temperature of part 123, and so forth, what is the optimum value of each of these variables? Or consider drilling holes in part 345 in Department C. Considering that we could vary drill rake angle, drill material, drill rpm, drill feed rate, coolant type, coolant volume, and so forth, what is the optimum value of each of these variables? To experiment would seem to require too much engineering time and expense.

You could use **passive observation** to examine the process through control charts. That is, you could examine control charts and hope to see a pattern which would allow you to guess the important patterns. But the process is not just what occurs; it is subject to modification. That is, you could actively investigate possible variables—that is, do an experiment **(active observation).**

Box and Hunter (1959) put forth the simple yet powerful concept that a process produces two things: (1) items for sale and (2) information about the process. That is, while we are painting part 123, we also are generating information about the effects of the various variables affecting the painting. In other words, we have been running an experiment—we just didn't realize it and collect and analyze the data!

Box and Hunter proposed that, to minimize the engineering expense and to study the process in its "production" version, the "experiment" be run by the plant operation personnel (the painter or drill press operator) on the production machines, thus eliminating the cost of the experimenter and the cost of the lab.

In most experiments, the experimental design attempts to maximize the amount of information obtained from the experiment, usually by minimizing the effects of "noise" (variability due to the process or the measurement of the process). This can be visualized as "cutting the weeds" (see Figure 4.4). The normal experiment tries to "cut the weeds" so the "crop" can be observed. However, there is an alternate strategy. Repeat the experiment over and over until the signal appears through the noise. Using the crop analogy, that means that if you watch long enough, eventually the crop rises above the weeds. **Evolutionary Operation of Processes** (EVOP) follows this second strategy, using the assumption that experimentation costs nothing since the process is producing product for sale.

Box and Hunter said that if the experiment is to be run by "shop" personnel (i.e., no college, no statistical training) and the primary objective of producing products for sale is not to be endangered, then the experiments must be simple and cautious so as to cause no scrap. The technique is called Evolutionary Operation of Processes since it follows the two essentials of evolution: (1) small variations and (2) selection of favorable variants.

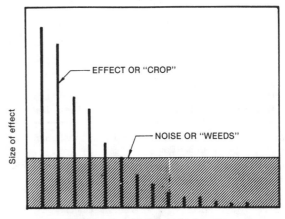

FIGURE 4.4 Traditional experimental design reduces the noise level to detect the effect—"cutting the weeds to detect the crop." EVOP uses another approach, continued experimentation until the signal shows through—"letting the crop grow above the weeds."

Table 4.6 gives some possible applications of EVOP. In most applications, the criterion seems to be the *yield* of the process. EVOP usually is used to maximize the use of material or energy or quality rather than to minimize the cost of labor. The very simple example in Table 4.6 is from Konz (1965). See Box and Draper (1969) for an excellent text on EVOP which not only gives administrative procedure for EVOP to systematically apply trial and error but also gives simple and elegant statistical techniques. EVOP is really another version of the operations research problem of "hill climbing"; EVOP, however, deals with a "noisy" signal.

Assume that a stainless steel bushing is being machined. We will use as the criterion the machining cost/bushing. Since we are looking for a minimum, look for the "bottom of the valley" rather than the "top of the mountain." The present feed and speed were selected from a handbook so we probably are close to but not at the bottom. For this specific bushing with its specific type of stainless steel and required type of cut, made on this specific machine, with a specific brand of cutting fluid at a specific volume, etc., etc., what is the optimum value of each variable?

An EVOP committee (composed of, perhaps, a manufacturing engineer, a quality control engineer, the supervisor, and the lathe operator) decides to vary feed and speed. At present, feed = .140 mm/rev and speed = 30 m/min; cost = $.25/unit. They set up a "search pattern" of four perimeter points around the center point (see Figure 4.5). The operator machined bushings for one day at each point, keeping track of output and tool life. The .16 and 31 point looked good as cost was $.22/unit. Should they shift to this point and run a new search; should they continue the search using the existing pattern; or should they try a different variable, such as volume

of cutting fluid? (In poker terms, should they bet, stay, or throw in the hand?) In our example, they decided to continue at the same five points for another week. After three weeks, they shifted to .14 and 32 as a center point and started a new search pattern. In general, it is good practice to overlap search areas as it helps to prevent "falling off a cliff." Eventually, after some months, the data shown in Figure 4.6 were available. They decided, as the new standard for feed and speed on this bushing, to machine the bushing with a feed of .16 mm/rev and a speed of 31 m/min for an anticipated savings of $.03/piece. EVOP savings, though worthwhile, rarely are spectacular.

In statistical terms, what has been described is a "full factorial design"; that is, data are gathered at each point. A fractional factorial design (data at only certain selected points) can be used for preliminary (screening) work. An example would be to determine which of eight potential variables is important. The reduction of data points can be very substantial, but only main effects (not interactions) are considered and someone quite knowledgeable in statistics would be needed to set up and analyze the data.

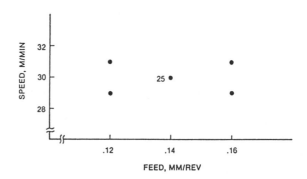

FIGURE 4.5 Focus of a search pattern is an initial center point (at .14 mm/rev and 30 m/min) with four perimeter points. Initial cost = $.25/unit.

TABLE 4.6 Example applications of EVOP.

Application	Criterion	Some Variables
Turning	Machining cost Surface finish	Feed, speed, tool geometry
Welding	Weld strength	Cooling rate, amps, rod type
Painting	Scrap rate	Paint-thinner ratio, gun distance
Casting	Yield	Pouring temperature, additive percentages, sand additive percentages
Chemical processes	Yield	Time, temperature, percentages of constituents and catalysts

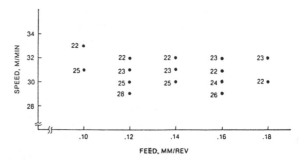

FIGURE 4.6 Optimum value of feed and speed is selected as feed of .16 mm/rev and speed of 31 m/min for an estimated savings of $.03/unit.

REVIEW QUESTIONS

1. Give the five steps of engineering design (DAMES).
2. Give three industrial applications of "fighting giants."
3. What is an OSHA 200 log?
4. What is a typical ratio of operating indirect labor to direct labor? of technical and clerical indirect labor to direct labor?
5. Discuss benchmarking an office operation.
6. What is a typical return on investment for a one-year certificate of deposit in a bank (i.e., a risk-free investment)? What is a typical return on investment of U. S. manufacturing industry (i.e., a risky investment)?
7. Why should cost reductions include an analysis for "best manual proposal" as well as "best mechanized proposal"?
8. Is EVOP an active or passive observation of the process?
9. The EVOP concept is that a process produces what two things?

REFERENCES

Andersson, E. Economic evaluation of ergonomic solutions: Part I—Guidelines for the practitioner. *Int. J. of Industrial Ergonomics*, Vol. 10, 161-71, 1992.

Box, G. and Draper, N. *Evolutionary Operation: A Method of Increasing Industrial Productivity.* New York: Wiley & Sons, 1969.

Box, G. and Hunter, J. Condensed calculations for evolutionary operations programs. *Technometrics*, Vol. 1, 77-95, 1959.

Konz, S. Selecting feed and speed under factory conditions. *Tool and Manufacturing Engineer*, Vol. 55 [7], 31-33, 1965.

Main, J. How to steal the best ideas around. *Fortune*, 102-6, October 19, 1992.

Mowrer, O. *Learning Theory and the Symbolic Process.* New York: Wiley & Sons, 1960.

Oxenburgh, M. *Increasing Productivity and Profit Through Health and Safety.* Chicago: CCH International, 1991.

White, J., Agee, M., and Case, K. *Principles of Engineering Economic Analysis*, 3rd ed. New York: Wiley, 1989.

5 | SEARCH

OVERVIEW

Searching for the best job design is divided into three categories. The first category is unstructured search, typified by brainstorming. Both the second and third categories are structured search. The second category is checklists. The third category uses the acronym SEARCH, where S = Simplify, E = Eliminate, A = Alter sequence, R = Requirements, C = Combine, and H = How often.

CHAPTER CONTENTS

1 Unstructured Search
2 Structured Search: Checklists
3 Structured Search: SEARCH

KEY CONCEPTS

alter sequence
brainstorming
checklists
combine operations
critical examination form
eliminate unnecessary work
exception principle
group technology

groupware
how often
initial vs. continuing costs
nominal group technique
prerelease review
red, yellow, and green
    operations
requirements

self-service
simplify operations
six honest serving men
unstructured search

As discussed in Chapter 4, engineering design requires:

- definition of the problem
- analysis
- search for a solution
- evaluation of alternatives
- specification and selling of the solution

This chapter discusses the "search for a solution." Edison, who had more inventions than anyone else in the history of the world, expressed it well: "Genius is 1% inspiration and 99% perspiration." What are some of the techniques for searching?

1   UNSTRUCTURED SEARCH

When trying to solve a problem, we may focus too narrowly. That is, possible solutions are rejected in our own minds due to assumptions, which, if considered carefully, turn out not to be limiting. What is desirable is obtaining many possible solutions; reducing them later is easy compared to the problem of getting a "large solution space." Creativity is considered to be an "orchid," which will be "blighted" by the "frost" of criticism. In **unstructured search,** ideas are to be encouraged and no criticism is allowed, because this may discourage contributions. Avoid "groupthink," the drive for consensus that suppresses disagreement and prevents the appraisal of alternatives.

Three techniques will be described to encourage inspirations: brainstorming, the nominal group technique, and groupware. All three techniques are used with groups of people.

1.1   Brainstorming   Consider the problem given in the process chart and flow diagram of Figures 7.7 and 7.8. Joe College was quite inefficient in supplying himself with beer. What are some other ways of obtaining beer?

In the **brainstorming** technique, the discussion leader states the problem clearly and then asks for oral suggestions from the group. The leader displays the idea publicly (say, on a blackboard) using different words if possible. (Paraphrasing the ideas is a useful technique to be sure the leader understands the idea.) The public display of the idea encourages the person contributing and also may serve as a "springboard" to other ideas by other people. No criticism is permitted of any idea. Ideas are not identified with their contributors—this reduces social pressure to accept or reject an idea because it came from someone with high or low status. The leader, knowing the tendency of some individuals to be extroverts, should discourage one or two individuals from

dominating the session by recognizing others in the group first; in some cases it may be necessary to recognize shy individuals purposefully and say "What do you think, Joe?" Ten minutes is sufficient for most brainstorming sessions.

Table 5.1 gives some brainstorming ideas for the beer drinking problem. After the brainstorming is complete, the group then begins evaluating the ideas. Another technique for encouraging participation is to take ideas from people in sequence.

1.2   Nominal Group Technique   There is some evidence that group participation *inhibits* creative thinking, however, because some people dominate and some just sit there. Another technique is the **nominal group technique,** called that because the participants are only nominally a group. They act more as individuals than as a group. This technique uses the following steps:

1. The leader states the problem.
2. In small groups, the leader gets suggestions from each person in turn. Participants cannot make a suggestion until their turn comes. In larger groups, each person makes suggestions on a sheet of paper. The idea is to force everyone to contribute. No criticism is allowed.
3. The leader publicly displays the ideas. If there are many, it may be worthwhile to consolidate them to a smaller number.
4. The merit (importance) of the ideas is evaluated by selecting the best ideas rather than eliminating the worst. It is suggested that participants vote on the square root of the number of alternatives displayed. That is, if there are 9 alternatives, vote for 3. The best would have a vote of 3; the second, 2; and the third, 1.
5. The leader publicly tallies the number of votes for each alternative. For example, alternative 1 got 8 points, alternative 2 got 10 points,

TABLE 5.1   Brainstorming suggestions to improve beer drinking (see Figures 7.7, 7.8).

Idea
Knock hole in wall to shorten distance
Switch position of TV and sofa
Move refrigerator
Drink from can
Move wastebasket
Put magnetic opener on refrigerator
Use pull-top can so no opener is needed
Have wife bring the beer
Move refrigerator next to sofa
Don't bother to inspect glass
Get two beers at a time
Keep beers in a cooler next to sofa

and so forth. Select the alternatives with the highest points (perhaps three or four) for further discussion.

1.3 Groupware

Groupware, a recent development in computer networks, allows multiple computers to simultaneously communicate with each other. Technically, what everyone does is communicate to central "bulletin boards," one for each topic. The bulletin board can hold not only comments but even databases. Thus, communication is not one to one but many to many. In this approach, the nominal group need not even be at the same physical location. Because the responses are anonymous, there is no social pressure restraining comments. Some remarks can therefore be quite blunt: "The company policy is stupid!!!" Such behavior goes by the picturesque label of "flaming."

Since nearly everyone in the company has access to the bulletin board, the information on the board is much more widely available than traditional sources of information. And because the information that supports decisions is now widely available to all levels of the organization, decisions (both good and bad) are fair game for comments by many people. This trend has profound consequences for organizations which emphasize authority over democracy.

Although groupware can speed communication, it also increases the mass of information available to everyone and thus causes information overload. At one time, people responded to every letter and phone call; this no longer is feasible when a person may receive hundreds of messages every day.

2 STRUCTURED SEARCH: CHECKLISTS

Some people feel that rather than depending on inspiration they will follow a more systematic approach. Why not follow "tried and true" techniques used by people who have worked on this type of problem previously? The approach is to use **checklists,** or lists of items that can be easily referred to.

Perhaps the most literary checklist is by Rudyard Kipling:

> I have **six honest serving men.**
> They taught me all I know.
> Their names are who and what and when
> And why and where and how.

Victor Morales expanded this into the **critical examination form** shown in Figure 5.1. The boxes ask suggestive questions about purpose, place, sequence, person, and means. The first column describes the existing situation; the second and third try to bring out disadvantages of the present method and advantages of alternatives; the fourth column asks for a decision on a part of the problem. This general purpose form tries to force the users to list alternatives.

Beardsley (1980) gives another technique to use the "5 W's and an H." Taking 6 pieces of paper, write WHO, WHAT, WHY, WHERE, WHEN, and HOW on the top of a sheet. Then, on the sheet with the WHO, write down all questions concerning who and the problem; on the sheet with the WHAT, write down all questions concerning what and the problem; and so forth. WHO questions might be: WHO should do it? WHO should bring the material? WHO should supervise? WHO should inspect? Don't worry about the sequence of questions. At this stage you are trying to generate ideas.

The second stage is to answer one of the questions on one of the sheets, such as Joe Roberts should bring the material. Now write JOE ROBERTS on the top of a sheet of paper and make up 6 questions concerning Joe Roberts starting with WHO, WHAT, WHY, WHERE, WHEN, and HOW. Examples might be: WHO is Joe Roberts? WHAT exactly should Joe bring? WHY did I select Joe? WHERE should Joe bring it? WHEN should Joe bring it? HOW should Joe bring it? Then go on to the next question on the first set of sheets, answer it and ask WHO, WHAT, WHY, WHERE, WHEN, and HOW questions.

Stage 3 is to review all the questions from stages 1 and 2 for ideas to solve the problem.

Specific checklists exist for specific situations. Table 5.2 shows the checklist developed by Paul van Wely and the Philips Ergonomics group.

Van Wely (1970), while at Kansas State as a visiting professor, developed a series of principles, rather than just a checklist (see Table 5.3). Stevenson and Baidya (1987) developed Table 5.4 as guidelines for reducing repetitive strain injuries (also known as cumulative trauma).

3 STRUCTURED SEARCH: SEARCH

Since others could make a checklist, so can I (and so can you). My six-item list can be remembered by the acronym *SEARCH,* where:

S = Simplify operations

E = Eliminate unnecessary work and material

A = Alter sequence

R = Requirements

C = Combine operations

H = How often

Situation Studied _____   Date _____

Why Studied _____   Study by _____

Where _____   _____ minutes/occurrence

How Often Repeated per Year _____   _____ minutes/year

	1	2	3	4
Purpose	What is achieved?	What would happen if it weren't done?	What could be done and still meet requirements?	What should be done?
Place	Where is it done?	Disadvantages of doing it there:	Where else could it be done? Advantages of doing it elsewhere:	Where should it be done?
Sequence	When is it done? After:	Disadvantages of doing it then:	Advantages of doing it sooner: Advantages of doing it later:	When should it be done?
Person	Who does it?	Why that person?	List two others who could do it.	Who should do it?
Means	What equipment and methods are used? Equipment: Method:	Disadvantages of equipment: Method:	How else could it be done? Advantages:	How should it be done?

FIGURE 5.1   Kipling's "six honest men" as expanded by Morales into a critical examination form.

Eliminate really should be done first although it is second in the acronym.

3.1 Eliminate Unnecessary Work and Material

Eliminate unnecessary work will be discussed under four subcategories: (1) eliminate unneeded work, (2) eliminate work where costs are greater than benefits, (3) use the exception principle, and (4) use self-service. Small cost reductions are the drops which combine into the ocean of productivity.

3.1.1 Eliminate unneeded work

One example of eliminating unneeded work would be that, when duplicating class handouts, the teacher uses both sides of a sheet of paper. This not only reduces the amount of paper used but also reduces the page assembly task (eliminating assembly altogether for a two-page handout). A second example is direct deposit of checks to a bank. This "win–win" approach not only is a winner for the payer, because paperwork is simplified, but is a winner for the payee because the inconvenience of check cashing is eliminated, there is no theft problem, and the payee has access to the funds even if not at a specific location when the check is issued. A third example is the book purchasing procedure now followed by Kansas State University. The original policy was to purchase all books requested by faculty members plus books judged important by the head librarian. The result was that the library was processing thousands of orders for books each year. It was discovered that approximately three years after a technical book was issued by a major publisher, there was over a 95% chance that the library would have ordered it. The present policy is to have a standing order with major publishers for 100% of books in certain categories. The publisher sends the book when published (thus getting it to the library 1 to 36 months sooner) and sends a monthly bill for all books sent that month. The library gets more books sooner, and at a substantial savings in paperwork cost.

Another example is pruning of mailing lists and internal distribution lists. Once/yr send a letter saying that unless the person completes the form and returns it, he will be taken off the list. Another technique is to make the continuation form page 2 or 3 of the report so you can see who really reads the report. Check your list by address so you aren't sending the same report to multiple people at the same address or one person for whom you have multiple spellings of the name. Another example is the use of left-justification of date and signature on a business letter instead of requiring an additional tabulation to the right side of the page. The typing of the typist's three initials on a letter also can be eliminated in most offices as either the typist is known or a single initial is sufficient.

Of course, errors cause much work. Perhaps a deburring operation can be eliminated by initially using sharper tools or tools with a different rake angle. See Chapter 12 for error reduction techniques.

3.1.2 Eliminate work where costs are greater than benefits
Consider eliminating backup and "just-in-case" procedures. Often much expense is incurred to protect against a rare event. For example, when I worked for a midwestern firm, the depart-ment making springs found one batch defective due to faulty material. I instituted a policy of testing each coil of material received. At the time, it seemed a good policy. With hindsight, I now realize the persistent cost of the inspection far outweighs the cost of the rare defective material. Another example of excessive zeal is the expense-account paperwork required by some firms. Few expenses are reduced, but the time spent processing paperwork is large. Another example is staffing of a service facility such as a tool crib. A question is what would happen if there was no attendant and self-service was used. The reply might be "they would steal us blind," "people wouldn't return items," or the like. The key question is whether the staffing cost exceeds the benefits of the reduced thefts and improved neatness. Perhaps a compromise is possible in which the facility is open only part of a shift or is restricted to key employees or supervisors. Most retail businesses have decreased staff and allow self-service, although restricting access to high-value items. Some restaurants have even eliminated serving you a portioned container of beverage (soft drink, coffee); instead they allow free refills, feeling the additional goodwill and decreased serving costs offset the increased beverage cost.

TABLE 5.2   Ergonomics checklist (Kellerman et al., 1963).

A.   **DIMENSIONS**

1. Has a tall man enough room?
2. Can a petite woman reach everything?
3. Is the work within normal reach of arms and legs?
4. Can the worker sit on a good chair? (height, seat, back)
5. Is an armrest necessary, and (if so) is it a good one? (Location shape, position, material)
6. Is a footrest required, and (if so) is it a good one? (Height dimensions, shape, slope)
7. Is it possible to vary the working posture?
8. Is there sufficient space for knees and feet?
9. Is the distance between eyes and work correct?
10. Is the work plane correct for standing work?

B.   **FORCES**

1. Is static work avoided as far as possible?
2. Are vises, jigs, conveyor belts, etc., used wherever possible?
3. Where protracted loading of a muscle is unavoidable, is the muscular strength required less than 10% of the maximum?
4. Are technical sources of power employed where necessary?
5. Has the number of groups of muscles employed been reduced to the minimum with the aid of counter-support?
6. Are torques around the axis of the body avoided as far as possible?
7. Is the direction of motion as correct as possible in relation to the amount of force required?
8. Are loads lifted and carried correctly, and are they not too heavy?

C.   **NOISE**

1. Is noise reduced to the minimum by means of technical measures at the source?
2. Are the sources of noise so insulated they hinder as few people as possible?
3. Is reflected noise reduced to the minimum by provisions on walls and ceilings?
4. Are the noisiest apparatuses located as far as possible from the ears?
5. Can sound signals and verbal instructions be readily distinguished from the ambient noise?

D.   **LIGHTING**

1. Is the lighting adequate for the nature of the work?
2. Has the fact that old lamps emit less light than new ones been taken into account?
3. Has glare on account of naked light sources, windows, or their reflection in shiny surfaces been avoided?
4. Has too great a brightness contrast between work position and surrounding been avoided?
5. Is reading of meters not impeded by reflection of light sources?

(continued)

TABLE 5.2   continued

E.   CLIMATE

1. Are the surroundings where the work is done not too warm or too cold for the nature of the work?

2. Have effective measures to prevent a high radiation temperature been taken?

3. Can the humidity of the air be kept within acceptable limits?

4. Are processes calling for a high degree of humidity insulated to the greatest possible extent?

5. Are draughts avoided and, at the same time, is there adequate ventilation?

F.   INFORMATION

1. Is the quantity of information received by the worker adequate and yet confined to essentials?

2. Does the information arrive in good time and is it perceived through the right sense?

3. Is the information clear and unambiguous?

4. Is urgent information given via the ears?

5. Particularly in the case of lengthy inspection tasks, is seeing replaced by hearing?

6. Can sound signals having different meanings be readily distinguished from each other?

7. Can prealignment, assembly, and setting be carried out rapidly and efficiently by touch?

8. Can the positions of components, control knobs, control buttons, and tools be perceived by touch?

G.   DISPLAYS

1. Can the meter be read quickly and correctly, according to a measure immediately suitable for use and with the required degree of accuracy?

2. Is the desired degree of accuracy really necessary?

3. Is the type of meter selected efficient?

4. Is the scale graduated properly and as simply as possible?

5. Are the letters, figures, and graduate marks clearly visible at the required range?

6. Is the pointer simple and distinct, and does it pass close to the scale without concealing the figures?

7. Are reading errors caused by parallax avoided?

8. Is a warning given, if the meter fails?

H.   CONTROLS

1. Have the knobs, wheels, grips, and pedals been adapted to the specific requirements of fingers, hands, and feet? (Location, dimensions, shape, direction of motion, counterpressure)

2. Have the controls been located logically, in the sequence of operation, and can they be easily recognized by shape, dimensions, marking, and color?

3. Are pedals avoided for standing work and confined to two for sedentary tasks?

4. Are all the selected controls efficient?

I.   PANELS

1. Is positioning of the controls in relation to the displays logical?

2. Is the relation between the direction of motion of the control and that of the deflection of the pointer or the reaction of the apparatus (e.g., On-Off) logical?

3. Has the panel the correct shape and dimensions in connection with sitting posture, grasping range, and direction of vision?

4. Are the meters, knobs, and buttons that are most important and most frequently employed appropriately located?

5. Are the meters so grouped and positioned in relation to each other that they can be read quickly and faultlessly?

6. Do the scale divisions and subdivisions of different meters correspond as far as possible?

7. Have larger panels been made easier to scan through separation of groups of meters?

8. If possible, is the process represented on the panel in the form of a diagram?

9. Are warning lamps clear and have they been located in the central part of the field of vision?

J.   PACED WORK

1. Are at least three vacant positions available at any moment?

2. Is the component to be mounted supplied ready-aligned?

3. Are the assembly positions equipped with ample guides for the feed movements? (Holes, stops, etc.)

4. Are faulty components removed on the basis of preliminary inspection?

5. Should the answers to questions 2, 3, and 4 be negative, are there at least six vacant positions?

a. Might it be useful to make a mockup of the machine or work setup?

b. Might it be useful to discuss the work setup with the operator concerned, the supervisor, or the departmental manager?

c. Might it be useful to discuss the work setup with your liaison for ergonomic matters?

3.1.3 Use the exception principle

The idea of the **exception principle** is to take action only in "exceptional" circumstances. As a first example, consider a reserved parking stall with a strange car in it. Normally the police would give a ticket. Using the exception principle, the police give a ticket only if the stall owner complains. This reduces police work and allows the owner to lend the stall to someone else if the owner is not using it that day. Membership cards from most societies are sent to members at the same time as the dues bill is sent. This saves the organization a great deal of paperwork.

3.1.4 Use self-service

Eliminating may involve having someone else do the task at no cost to you; this is known as **self-service.** Filling stations have eliminated attendants pumping your gas, retail stores have eliminated clerks picking out your

TABLE 5.3   The "eleven commandments" (van Wely, 1970).

I   AIM AT MOVEMENT
Use muscles, but *not* for holding or fixation.
Movement reduces monotony.

II   USE OPTIMUM MOVEMENT SPEEDS
Too-quick or too-slow movements are fatiguing and
inefficient; try to find the specific optimum speed
of each movement.

III   USE MOVEMENTS AROUND THE MIDDLE POSITIONS
OF THE JOINTS
Long duration or frequent use of extreme positions of a
joint, especially under load, are harmful and have
poor mechanical advantage; yet occasional extreme
positions, while not loaded, are desirable.

IV   AVOID OVERLOADING OF MUSCLES
Keep dynamic forces to less than 30% of the maximum
force that the muscle can exert; up to 50% is OK
for up to five minutes.
Keep static muscular load to less than 15% of the
maximum force that the muscle can exert.

V   AVOID TWISTED OR CONTORTED POSTURES
Do not use pedals in standing work.
Use arm supports only when the upper arms cannot
relax and be approximately perpendicular to the
floor.
Bending the head backward causes glare and sore necks.

VI   VARY THE POSTURE
Any fixation causes problems in muscles, joints, skin, and
blood circulation. Do not use pedals in microscope
work.

VII   ALTERNATE SITTING, STANDING, AND WALKING
Continuous sitting for more than one hour and
continuous standing for more than ½ hour are, in
the long run, too fatiguing

Standing for more than 1 hour a day is fatiguing and
causes physical abnormalities. Concrete floors are
very fatiguing; elastic supports such as wood,
rubber, or carpet are better.

VIII   USE ADJUSTABLE CHAIRS
For sitting longer than ½ hour continuously or longer
than 3 hours a day, use adjustable chairs and
footrests (if necessary). When adjusting,
remember:
a. Seat so that elbows are at about the height of the
working plane
b. Footrest so that no pressure occurs in the back
of the knees
c. Backrest so that the lower back is supported and
that the working plane be adjustable; a platform
for the short person is a simple solution.

IX   MAKE THE LARGE MAN FIT AND GIVE HIM ENOUGH
SPACE: LET THE SMALL WOMAN EASILY REACH
For standing it is essential that the working plane be
adjustable; a platform for the short person is a
simple solution.
Keep the working plane within 50 mm of elbow height.
Note: The working plane height is usually above
the table height.

X   TRAIN IN CORRECT USE OF EQUIPMENT: FOLLOW UP
People must be instructed and trained in good working
postures; sitting, standing, and especially lifting are
often done in the wrong way.

XI   LOAD PEOPLE OPTIMALLY
Neither maximum nor minimum are optimum.
An optimum load (physical and mental) gives better
performance, more comfort, less absenteeism, and
less harm.

---

merchandise, manufacturers have eliminated product assembly (the item now comes ready to assemble using simple tools). Another good use of others is to have them sort for you. The post office does this with local and out-of-town mail boxes; organizations can do this with multiple mail bins such as "company—this building," "company—other buildings," and "U. S. mail." Sorting of trash can make recycling worthwhile. For example, put a separate container for aluminum cans next to other trash containers so people will sort aluminum cans from other trash. (You may have to use a lid with a can-sized hole to reduce the insertion of paper trash by absent-minded people.)

Ford Hospital in Detroit treated about 250 chemotherapy cases/year with cisplatin. But cisplatin had a side effect of ravaging the kidneys. The hospital routinely admitted the patients one day early to "tank up" on water. Now they have the patients tank up at home (saving one day's hospital stay); the patient keeps a diary of liquid intake for the nurse to check upon admittal.

3.2 Simplify Operations

There are many examples of how to **simplify operations.** When giving directions (hotel, car rental, receptionist), use a map the individual can take instead of just verbal instructions.

There are a wide variety of jigs and fixture designs which can be used to simplify machining operations. When ejecting parts from a machine, use drop delivery instead of precise placement of the parts. Even better, if possible, is to use automatic ejection—either mechanically or with compressed air.

Shift from counting to sampling and statistics. Register keys in fast food restaurants can be keyed to specific items rather than price. Not only does this reduce errors but the information can be communicated automatically to the kitchen.

Figure 5.3 shows an office form used to substitute a checkmark for short standard phrases. Writing a letter takes considerable time. Can you write your answer on the other person's inquiry, make a photocopy for your files, and mail the original back? Some firms even have interoffice memo forms with an attached carbon copy to encourage such behavior. Many firms have form letters which are individually typed but the computer stops at the appropriate place for insertion of specific details such as name

TABLE 5.4   Guidelines to reduce repetitive strain (cumulative trauma) injuries. (Adapted from Stevenson and Baidya, 1987).

Guideline	Comment
1. Minimize the number and angular range of wrist extension movements in a repetitive cycle.	See Figure 11.9. Note that sharp bending of the wrist often accompanies tossing motions. Avoid prolonged holding of the wrist in an extended position.
2. Minimize ulnar deviation of the wrist, particularly for longer portions of the work cycle or with force applications.	Consider bent-handle tools and tilted work surfaces. Give careful attention to keyboard heights so the home row is even with the elbow.
3. Minimize radial deviation of the wrist.	Can occur with poorly chosen handtools, especially with poor workstations and postures.
4. Avoid highly flexed wrists, particularly when combined with finger flexion.	Grip strength decreases rapidly when wrist flexion increases.
5. Keep hand and arm movements at the mid-position of their range.	See Figure 11.9.
6. Don't repetitively rapidly decelerate the wrist.	Repeated shock-loading has a large cumulative effect, for example, hammering, tugging wires tight.
7. Avoid excessive finger and hand extension.	The grip becomes weaker as it becomes wider. Avoid thumb triggers and thumb buttons. "Scissors grip" motions (wide extension combined with resistance) are bad.
8. Avoid grasps and hand movements of "heavy" objects with a pronated (palm down) hand.	Check picking of objects from a conveyor or pallet. The effect is much worse if the object center of gravity is beyond the hand. Can sliding replace lifting?
9. Avoid excessive squeezing force.	Females have only 50–60% of male grip strength. For cutting with handtools, use motors instead of human force. See the "4th commandment" in Table 5.3.
10. Decrease gripping force and duration.	Replace pinch grips with power grips. Duration can be reduced by using tools with each hand alternately.
11. Minimize stress from tool handles.	Handles can be too short, too sharp, or too grooved (fit only a few hands). See Chapter 18.
12. Work at elbow height.	Reduce static loading or holding up the upper arm and forearm. Hand blood supply is reduced when the hands are above the heart.

Note: See Figure 5.2 for terminology.

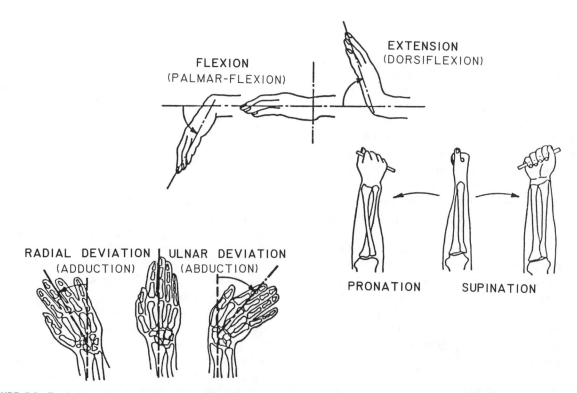

FIGURE 5.2   Terminology for hand movements and positions (Stevenson and Baidya, 1987). Radial/ulnar deviation and flexion/extension occur in the wrist joint. Pronation/supination is a function of the radius rotating around the ulna in the forearm.

OFFICE TRANSMITTAL SLIP

Kansas State University

Department of Industrial Engineering

Manhattan, Kansas

To:

_____

_____

_____

_____

_____

_____

( ) See Me
( ) Type
( ) File
( ) Signature
( ) Refer
( ) Necessary Action
( ) Comment
( ) Information
( ) Note, Sign, Return
( ) Per Conversation
( ) _____

Comment:

Date: _____ Signature: _____

FIGURE 5.3 Routing memo forms substitute a quick checkmark for writing phrases such as "please see me" or "for your signature."

and address of a particular person. Dictating equipment permits a person to "talk" a letter rather than write it; word processors permit easy modification of the output if necessary. Perhaps writing is not necessary at all and a phone call, with its advantages of speed and two-way interaction, can be used in place of written communication. Long distance phone calls using a wide area telephone system (WATS) typically cost about 30 cents/min for a phone call anywhere in the United States. Certainly this compares favorably with the managerial and secretarial cost of typing a letter.

Simplify your pollution and waste disposal problems by generating less waste. For example, in metal finishing operations (electroplating, anodizing, conversion coating, chemical etching, chemical milling) the products are rinsed after processing. The rinse water becomes contaminated with "drag-in." Reduce drag-in by allowing products to drain sufficiently before entering the rinse tank (5 to 10 s over the processing tank is sufficient). Rotate the product to aid dripping from concave surfaces and blind holes. A related example is reclaiming of cutting oil in machining operations by giving the chips time to drain. Reducing the variety of cutting oils and coolants not only simplifies inventory control, it also simplifies recycling of the oils and coolants. Some firms color-code chip bins—one for each type of oil or coolant.

3.3 Alter Sequence

To **alter sequence,** one can (1) simplify other operations, (2) reduce idle/delay time, and (3) reduce material handling costs.

3.3.1 Simplify other operations

Cleaning and deburring operations sometimes may be omitted if the operation sequence is changed; machining before hardening is easier than machining after hardening. Distance between holes can be controlled more easily if holes are drilled after assembly than if they were drilled in the components.

Formerly, boxes went down two lines, one automatic and one manual. On the automatic line, the boxes were weighed automatically and the weight was printed on the box with a jet printer. However, on the manual line, the boxes had to be pulled from the line, put on the scale, and the weight handwritten on the box. This physically demanding task was eliminated by moving the scale upstream to just after case sealing and *before* the lines split. Then all the boxes could be weighed and marked automatically.

Figure 5.4 shows how modifying *when* something is done can reduce physical strain on the operator.

For an office example, when photocopying multiple pages turn the original stack upside down and copy the last page first. Then the copied pages will be in the proper order; the original stack can be turned over in one motion.

Some time is more valuable. Customers at a hotel usually are in more of a hurry when they leave than when they check in. Use of a charge card upon checking in reduces checkout time.

3.3.2 Reduce idle/delay time

Travel agents now can issue you a boarding pass for a flight when you pick up your ticket instead of you obtaining one at the airport.

"Do it now" is a customer service concept that saves money and gives faster customer service. The concept is that when a customer (who may be an employee of your own firm) calls on the phone, the customer receives the answer during the same phone call. Implementing this concept requires (1) empowering employees to make a decision (instead of referring it to others) and (2) good information resources (usually on-line access to computer databases).

Another concept is "inside machine time." Many machines are semiautomatic. Once loaded, they process automatically, or loading/unloading is automatic and processing is manual. During the automatic portion, other operations can be done. For example, if machine 1 is automatic processing, spend the time loading machine 2. If a copy machine has a document feeder, then do some filing while it is copying. Semiautomatic operations can be combined.

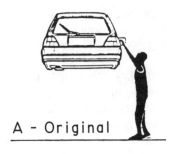

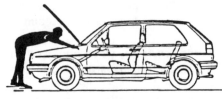

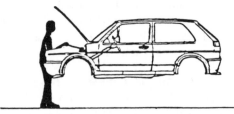

FIGURE 5.4 Modifying assembly sequence reduced strain for Volkswagen workers (Echard, Smolenski, and Zamiska, 1987). In Operation A, originally the operator suffered cervical strain (collected 8 months compensation) while installing a gas fuel neck 2 m above the floor. In the improved operation, the operation was done at another portion of the line so that the fuel neck was 1.7 m above the floor. In Operation B, originally the brake booster was installed after the engine, radiator, and bumper were installed. In the improved operation, installation of the brake booster was done before they were installed.

For example, at fast food restaurants, soft drink dispensers are semiautomatic (fill automatically on a trigger), allowing the employee to do other things (such as get lids) while it fills. Have service personnel process two lines of customers sequentially to minimize delay while the customer goes away and a new one arrives. Examples are bank drive-ins and cafeteria checkout.

Simultaneous vs. sequential is another concept. For example, can the operator reposition a part in the fingers while moving it or must the operations be done sequentially? See Table 29.13.

Central vs. local storage is another example. If items are stored locally there is less delay in obtaining them. However, central storage probably gives better control, less theft, and reduced inventory since storage is at one location instead of many. For example, New York City formerly had central storage of parts for sanitation department trucks. By decentralizing storage, they reduced delays in repairing trucks and thus truck downtime. However, the total stored increased (20 local stores, each with 5 units vs. 1 central store with 50 units); there also were more places for theft.

For restaurant examples, consider the sequence of ordering a pitcher of beer, having it drawn, and then paying your money. If you pay the bartender while the pitcher is being filled, you can get back to your table sooner! Fast food restaurants often have you pay your bill while the meal is being assembled or cooked. When serving breakfast, a good waitress will bring coffee to the table on the initial trip. A standard sequence also simplifies training and reduces errors. At McDonald's, for example, the standard sequence is drinks, sandwiches and pies, french fries, and ice cream.

3.3.3 Reduce material handling costs

Formerly, the U. S. mail was sorted by hand in each town. Now the mail from the smaller towns is sent unsorted (unless deposited in "local" boxes) to larger centers so that it can be sorted by machine. Then the sorted mail for the town is sent back to the local post office. This approach actually increases material handling costs but at the advantage of reduced processing costs.

Grouping similar parts into parts families and scheduling "relatives" together is known as **group technology.** It not only reduces material handling and paperwork but obtains the benefits of similar tooling and manufacturing skills.

Material handling costs can be reduced by modifying the sequence of when items are moved, using a bus system instead of a taxi system. The "taxi" goes from point to point quickly, but at high cost. If speed is not critical, the "bus," although it travels a longer distance, may be cheaper.

In the home, you can save steps if you don't bring wastebaskets to the trash barrel in the garage. Take an empty trash bag to the wastebasket and empty each wastebasket into the bag in turn.

3.4 Requirements

Two aspects of **requirements** must be determined: (1) quality (capability) costs and (2) initial vs. continuing costs.

3.4.1 Quality (capability) costs

Figure 5.5 shows the general shape of the cost vs. quality (capability) curve; beyond a certain point additional quality (precision, capability) has a very high cost. The question then becomes, is the quality used or specified the appropriate quality? (Remember the old joke that a tolerance is the smallest number that a design engineer can think of.)

It is important to have manufacturing people review design tolerances with the design engineer *before* the unit is released for production. Engineers tend to be reasonable at the **prerelease review,** but once the drawing becomes official everything becomes "set in concrete." Material costs can be reduced as well as manufacturing costs; for example, Emerson Electric saved $485,000/yr when they replaced some steel items in electric baseboard heaters with aluminum items. Thus it is good to have a purchasing department representative in the prerelease review meeting.

Many requirement costs are concealed because they do not show up in the standard cost system. Figure 5.6 shows that the ideal flow of items through a factory is considered to be a river but more realistically is a river with "eddies." These eddies are rework. Most rework is done "under the table." For example, if you told your boss that you had 10% scrap today, the boss would yell at you, rant, rave, etc. So you don't report it; you just fix the defectives, quietly, with no official paperwork.

Indirect materials, supplies, and utility costs often receive less attention than they should. One firm found they were gold plating electronic parts too thickly. They found out about it because people are conscious of gold. But I experienced a wastage of

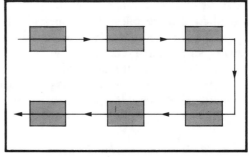

Idealized production flow

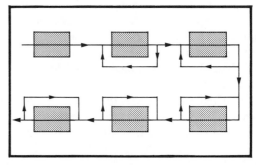

Real production flow

FIGURE 5.6 Ideal flow of product through a factory is shown above; a more realistic flow is shown below. Rework will cause "eddies in the river," substantially increasing the number of items processed at each station.

brightener in a motor plant. The supervisor had commented to the operator that the parts coming out of plating didn't seem very bright. So the operator put three times the required amount of brightener in the tank. The parts became bright, but at quite a cost, since the brightener was very expensive. DuPont found that a process was using 20,000 lb/h of steam although the requirement was 14,000 lb/h. They replaced the instructions "turn the valve about three turns" with a meter. An electric motor manufacturer used varnish on the motor coils. When they weighed the varnish on the coils and compared it with the varnish purchased, there was a 10% shortage. It turned out that the varnish was purchased in 55-gallon drums. When emptying the drums, they had turned them right side up when there was considerable product left inside. The solution was a stand which permitted the varnish to drip out of the drum over a period of hours.

Another requirements issue is the cost of components vs. the cost of assembly. The standard procedure for over 100 years has been interchangeable parts—that is, high-precision components to make assembly easy. But in some cases, selective assembly may give lower total costs. In selective assembly, the component is made to a lower tolerance and then sorted into sizes, such as small, medium, and large.

FIGURE 5.5 Cost rises very rapidly as quality (or product capability) increases. Do you really need that much quality?

Thus instead of each bushing fitting on each shaft, you would have three piles of bushings and three piles of shafts. You assemble small bushings on the small shafts, and so forth. The savings in component cost may be more than the additional sorting and more difficult assembly cost. Note, however, that if parts are not interchangeable, spare components cannot be sold, only spare assemblies.

3.4.2 Initial vs. continuing costs
A common tendency is to focus on **initial costs** and ignore **continuing costs** (operating and maintenance costs). Many maintenance, operating, and utility costs are not even known. However, the optimum decision is to minimize life-cycle costs (total of purchase, operating, utility, and maintenance costs over the item life) rather than the conventional minimizing of initial purchase cost.

A personal example was the bracket on my car's air cleaner. The bracket was too thin but it reduced the material cost of the bracket to the manufacturer by a fraction of a penny. However, it broke every 10,000 miles—forcing me to buy new brackets until, in desperation, I welded a washer to it to make it strong enough. Yet the auto companies have found that people are reluctant to pay attention to any cost of a car except first cost. Do you know what the maintenance cost or operating cost of your car will be over a 5-year period? Most people don't even think about it.

For military aircraft, the government has found that the maintenance cost over the plane's life is 30 times the initial cost. As the oil filter company ad says, you can pay me now (for the filter) or pay me later (for the engine repairs).

In fish farming, when high-initial-cost cupronickel cages replaced nylon nets (which must be cleaned often due to algae fouling), there were large maintenance savings. An electronics firm found using titanium plating baskets for acid dipping was better than steel baskets due to the much longer life of the titanium. However, it was a "hard sell" due to the high initial cost of the titanium baskets.

3.5   Combine Operations
Often, **combine operations** can be thought of as an argument for general purpose vs. special purpose. Sometimes the general purpose is best and sometimes the special purpose.

For materials, consider waxing your car. You can buy two separate compounds to clean and wax or one special-purpose compound which will allow cleaning and waxing in one operation. Farmers can combine seed, fertilizer, and insecticide in one pass through the field. Figure 5.7 shows a drilling example. Instead of drilling the hole in one operation and

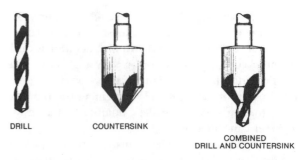

FIGURE 5.7   Special-purpose equipment (a combined drill and countersink) makes it possible to drill and countersink in one operation instead of two; the engineer must trade off the extra capital cost vs. the reduced operating cost.

countersinking in a second operation, the hole is drilled and countersunk in one operation. A pencil and eraser is a combined tool used in sequence rather than simultaneously. Another example is the cogeneration of electricity and heat. Formerly, many firms generated heat for their industrial processes or space heating but bought electricity from the power company. A number of years ago, the government changed the rules and required the utilities to buy all the electricity you want to sell to them at the highest price they pay for their own generation of power. Thus many firms have found it good to generate electricity for the utility while generating heat for themselves. Another energy saver is to preheat intake air from exhaust air.

The approach to job design was once to have a very specific job description—a specialist. One person was a carpenter, another an electrician A, another an electrician B, another a welder, etc. Although this allows a logical and organized workforce it also tends to be quite inefficient as it is difficult to find enough work to keep each and every specialization busy for a full day, day after day after day. One new automotive plant has only four job titles: operator, inspector, maintenance, and clerical.

Even the hooks on conveyors can be made single purpose or multiple purpose. The more expensive multiple-purpose hooks permit carrying multiple identical or even different items.

As an example of special-purpose equipment, McDonald's grills hamburgers on both sides simultaneously.

Advances in communications (faxes, E-mail, computer networks, express mail) now permit one central department to furnish service to a variety of local facilities—resulting in many economies of scale. For example, do you need a travel agent for each plant site? When there is sufficient volume (from many sites), agents often give discounts. Can you combine accounting functions of various local facilities? Other types of services? Note that local supervisors often resist consolidations because they measure their own status by the number of people they supervise.

3.6 How Often

A formal way of considering **how often** is to use economic lot size calculations (see Chapter 7). For "make" items, the trade-off is setup cost vs. inventory (capital, spoilage) cost; for "buy" items, the trade-off is purchasing cost vs. inventory cost. For example, when preparing a meal should a large portion be prepared and part frozen for use later? Technologies change the answer over time. Freezers and microwave ovens have decreased the storage problem and thus encouraged cooking larger batches. In manufacturing, computer-controlled machine tools, cell layouts, and group technology have encouraged smaller batches due to their lower setup costs. Inventories also have been divided into A, B, and C categories with close attention being paid to the inventories of the expensive A items. The inexpensive C items (nuts, bolts, washers, screws, etc.) have relatively little inventory cost but a large out-of-stock cost; thus a 6- to 12-month supply is ordered for them to keep purchasing costs low.

The question of what frequency also can be applied to sales. For example, assume it costs $100 to call personally on a potential customer. Then, if profit is 5% of sales, the sales/call should be $2,000. If sales/call are lower, the sales effort should be redirected—perhaps direct mail or phone contact could be used.

Should information be sent one-day Federal Express, two-day Federal Express, or U. S. mail? The how often question is especially relevant to maintenance, service, and inspection activities. Should a machine be oiled once an hour, a week, or a month? Should failed fluorescent lamps be replaced immediately or once a week or relamped only in groups? Should you pick up the mail once an hour, twice a day, or four times/day? Should the cash be taken to the bank once a day, twice a day, or what? Should the solution ph be tested once an hour, once a shift, or once a week? Should all suppliers receive equal attention or should some get more inspection than others? One way of "skimming the cream" (concentrating your resources where they will do the most good) is to label operators or operations or vendors as "red," "yellow," and "green" (**red, yellow, and green operations**). A green vendor is OK and inspection may even be omitted; a yellow vendor is caution and needs inspection; a red vendor is trouble and perhaps needs 100% inspection of all characteristics of all items.

REVIEW QUESTIONS

1. Give two unstructured search techniques.
2. Describe use of the nominal group technique.
3. Discuss groupware.
4. Give Kipling's six honest serving men (the five Ws and an H) as a poem.
5. Give the short phrases which SEARCH stands for.
6. Give an example of how "inside machine time" can be used.
7. Briefly discuss the prerelease review meeting concept.
8. Show flow of product through a factory as river with eddies. Why don't people report rework?
9. Briefly discuss selective assembly.
10. Why would vendors be labeled red, yellow, and green?

REFERENCES

Beardsley, J. Ingredients of successful Quality Circles. *Transactions of 2nd Annual International Association of Quality Circles*, 139–45, 1980.

Echard, M., Smolenski, S., and Zamiska, M. Ergonomic considerations: Engineering controls at Volkswagen of America. In *Ergonomic Interventions to Prevent Musculoskeletal Injuries in Industry*, 117–31. Chelsea, Mich.: Lewis Publishers 1987.

Kellerman, F., van Wely, P., and Willems, P. *Vademecum: Ergonomics in Industry*. Eindhoven, Netherlands: Philips Technical Library, 1963.

Stevenson, M. and Baidya, K. Some guidelines on repetitive work design to reduce the dangers of tenosynovitis. In *Readings in RSI*, Stevenson, M. (ed.). Kensington, NSW: New South Wales University Press, 1987.

van Wely, P. Design and disease. *Applied Ergonomics*, Vol. 1, 262–69, 1970.

6 VALUE ENGINEERING

OVERVIEW

Value engineering emphasizes reduction of excess work (cost) caused by poor design of products or procedures. The key concept is to highlight the cost of a function in relation to the function's value. From this, nonessentials are stripped away.

CHAPTER CONTENTS

1 Concept
2 Techniques
3 Sample Applications

KEY CONCEPTS

functional vs. prestige value
Pareto voting
Q-sort
value engineering

1 CONCEPT

Excess work can be caused by poor product design (see Chapter 4). The goal of **value engineering** (also called value analysis), which is applied to procedures (software) as well as products (hardware), is to reduce the excess cost of a design. The fundamentals of the systematic approach of value engineering were originated by Miles of General Electric in 1947. The basic concept is that many existing designs can be improved substantially, because the original design has excess costs. One common reason is that no design engineer can be expert in all phases of engineering, manufacturing, and purchasing; thus, the design may have potential for improvement. The excess cost may be due to a change in material or labor prices since the original design (either absolute or relative; see Figure 3.2), changing technology, different applications than originally envisioned, and lack of time to make a good original design. In contrast to this "second guessing" of existing designs, many firms now use value engineering techniques during the original design process. After all, why not do it right the first time? That's when change costs are smaller and benefits the greatest. (See Box 4.1 for comments on prerelease review meetings.)

2 TECHNIQUES

Table 6.1 gives the six steps of value engineering. First, select which product or procedure to study. The Pareto distribution (mighty few and insignificant many) helps focus the problem. If a product costs $5/unit and is used 3 times/yr, then its annual cost is $15.00. If you spend $500 of engineering time to reduce unit cost to $4, then you have spent $500 to save $3/yr. However, if you save 10% on another product which had annual costs of $50,000, then you spent $500 to save $5,000—a better buy! Note that $5,000/yr can come from different combinations: a $5,000 unit made once/yr, a $10 unit made 500 times/yr, or even a $.10 unit made 50,000 times/yr.

TABLE 6.1   Six steps of value engineering (Mudge, 1971).

1. Select the problem.
2. Get information.
3. Define and evaluate functions.
4. Create solutions.
5. Evaluate solutions.
6. Recommend a solution.

Note similarities to Table 4.1.

Within this range of potentially profitable projects, select products which are similar to other products (cousins), because the cousins may suggest alternate approaches and also may be additional application areas. Some projects come from noticing something on a stroll through the plant; chance favors the prepared mind.

Pareto voting is a formalized screening procedure. Since 80% of the value should be in about 20% of the items, voters select 20% of the items they consider important. Each item is voted on as "top 20%" or "not top 20%." Votes from all voters are tallied, and those receiving the most votes are investigated further.

Another screening technique (De Marle and Shillito, 1992) is the **Q-sort.** Each item is listed on an individual card. The cards will be sorted 3 times. Sort 1 is 2 piles of "low value" and "high value." Sort 2 splits the 2 piles into 4 piles of "very high," "high," "low," and "very low." Sort 3 splits the "high" and "low" into 3 piles of "high," "medium," and "low." The result is 5 piles of "very high," "high," "medium," "low," and "very low."

Second, get information from marketing, engineering, manufacturing, and purchasing. The diverse challenges facing the value engineering project usually cause projects to be done by interdisciplinary teams as no individual has such broad expertise. Because replacing special company components with standard purchasing components is a common solution, the team should have a purchasing member.

In addition to marketing and engineering data, get cost and constraint information. Cost information includes labor costs (setup time, time/unit, cost/h), material costs, packaging costs, tooling costs, and equipment costs. Get details of costs. For example, an electrical device which runs during a time of peak demand may cost four times as much per kWh as a device which is used off-peak. Constraints can be divided into customer and internal; yet another division is real vs. imaginary. Specifications often are figments of the imagination (e.g., based on a historical situation no longer relevant); that is, the engineer has been self-constrained.

Third, define and evaluate the function. Distinguish between value and function. Consider two tie clips: one a paper clip and one a diamond pin. Both accomplish the same function, but one has a value of $.01 and the other a value of $1,000. The functional value is the same but the personal (prestige, esteem) value is different (**functional vs. prestige value).** The engineer must consider not only function characteristics but also prestige characteristics. That is, not only what a product or service can do but also what will make it sell.

Mudge (1971) gives four rules for defining function:

1. Express functions in one noun and one verb.
2. Give work functions in action verbs and measurable nouns.
3. Give sell functions in passive verbs and nonmeasurable nouns.
4. Divide functions for each component into the primary (basic) one and secondary functions.

Table 6.2 gives examples of "good" nouns and verbs as well as some poor ones. Words such as *component, parts, nut, bolts,* and *bracket* are poor because they are too vague and don't identify the function. In units with 4 to 6 components, each component should have all its functions listed (*bushing:* provide support, provide adjustment, provide location, transmit force, provide connection; *key:* transmit force, provide location; *lock washer:* reduce friction, transmit force, provide location). In units with more components, use the Pareto principle and analyze just the "giants."

Then, for each component, identify which of the many functions is the basic function and which are the secondary ones.

Then consolidate and evaluate the functions on a form such as Figure 6.1. List all of the functions that appear for the components on the top portion of the form. Then using the "key letter" from the top portion of the form, compare function A (provide connection) with function B (transmit force). A *major* difference in importance (obvious decision) is assigned a 3; a *medium* difference in importance (short time needed to make decision) is assigned a 2; and a *minor* difference (considerable thought needed to decide which is more important) is assigned a 1. If A is more important and the difference is medium, write A2 at the intersection of the A row and B column. Now compare function A with function C. If A is more important than C and the difference is medium, write A2 at the intersection of the A row and C column. Continue until all the comparisons are made. Then total the "points" for each letter on the top portion of the form. You now have established the basic function (the function with highest weight). The remainder are secondary functions. The purpose of defining and evaluating the design's functions is to focus your thinking for the next stages—creating solutions and evaluating solutions. Perhaps it will "provoke you to the point of thought." Your objective is to compare costs of a function with value of a function—to cause a "creative discontent."

Fourth, create solutions. Solutions can be created either through unstructured search (such as the nominal group technique) or structured search (such as checklists). See Chapter 5 for examples.

Fifth, evaluate solutions. In practice, the fourth and fifth steps form a loop which is repeated a number of times until a number of solutions are considered. Perhaps the most important thing is to have

TABLE 6.2   Describing work functions with action verbs and measurable nouns is recommended by Mudge (1971). Sell functions should be described in passive verbs and nonmeasurable nouns.

WORK FUNCTIONS				
Action Verbs		**Measurable Nouns**		
amplify	interrupt	access	fluid	message
apply	make	circuit	force	oxidation
change	modulate	contamination	friction	protection
collect	prevent	current	heat	radiation
conduct	protect	damage	insulation	repair
control	provide	density	light	voltage
create	rectify	energy	liquid	weight
emit	reduce	flow		
enclose	remove		**Undesirable Nouns**	
establish	repel			
filter	retain	article	device	table
hold	shield	component	part	wire
impede	support			
induce	transmit			
insulate				

SELL FUNCTIONS			
Passive Verbs		**Nonmeasurable Nouns**	
decrease	appearance	exchange	prestige
improve	beauty	features	style
increase	convenience	form	symmetry
optimize	effect		

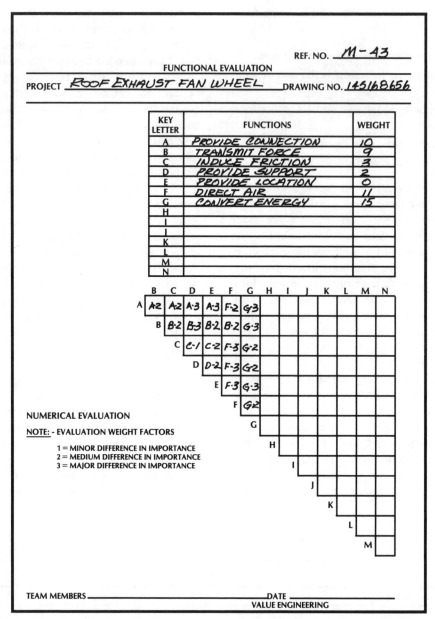

FIGURE 6.1 Functional evaluation is done in three steps: (1) List functions (without weights) on top part of form. (2) Tabulate differences in importance of the various functions (using key letter from upper part of form to determine row on lower part). (3) Total the weights for each function letter in lower part and enter as "weight" in upper part.

multiple solutions from which to pick. Do not stop with the first solution that works. Don't forget the selling functions.

Sixth, recommend a solution. Get all the information on the project together for the presentation to the decision maker. Determine exactly what needs to be done to implement the change (such as stop existing production in 43 days; change drawings 821, 822, and 917; rewrite service spares manual on pages 18 and 19; change standard cost for item, etc.). Make the proposal to the decision maker. Then, if accepted, get the paperwork started. (If the proposal is not accepted, can you settle for "half a loaf" rather than none?) Follow up to make sure the change is implemented. Audit the actual savings vs. your proposal so future calculations can be made more accurately.

3 SAMPLE APPLICATIONS

Men trust rather to their eyes than to their ears. The effect of precepts is, therefore, slow and tedious, while that of examples is summary and effectual.

Seneca

- **Test pumps.** Originally vacuum pumps were used to hold semiconductor wafers in place during testing. Fish tank pumps were modified to reverse their air flow at a savings of $100 per pump. A standard VE technique is to replace a special item with a low-cost consumer item.

- **Shampoo bottles.** Shampoo bottles originally were packaged in a clipboard container; changing to a six-pack holder (such as used in the beverage industry) saved over $100,000 the first year.

- **Coil stock.** Formerly, vendors sheared stock to exact size. Material purchased in coil stock and sheared in-house annually saved $133,000 in material and $46,000 in labor.

- **"O" Rings.** A chemical company used a special "O" ring at the rate of 2,000/month. It saved $17,000/yr by ordering standard rings in lots of 50,000 (i.e., a 2-year supply).

- **Nameplate.** The company nameplate used a nonstandard aluminum alloy. A standard alloy saved $9,500/yr.

- **Bushing and gasket material.** The specified material was red although the standard material is black. Switching to standard black saved $10,000/yr.

- **Molded name.** The company name was molded into a porcelain lamp receptacle for advertising. This made it a special item for the vendor. However, the only time the name was seen was when the lamp was changed by the maintenance man. Removing the company name gained a 10% price reduction from the vendor.

- **Brass oil-level gauge.** A special brass gauge was used on some compressors. A standard gauge, used on other company products, saved $500/yr.

- **Special nut.** A special nut was made in-house at $13.30 per hundred at the rate of 375,000/yr because suppliers did not make specials. When purchasing asked for a quote anyway, the vendor said they needed a minimum order of 25,000 for specials. When told of the 375,000/yr, the vendor said, "It's just become a standard!" Cost savings were $7/100.

- **Motor guard.** A guard for a totally enclosed electric motor on a candy machine cost $20. It turned out its real function was to prevent people touching the motor. Complaints had been received on the original machines that the motor ran hot. The guard was replaced with a $.25 adhesive nameplate which read, "This motor normally runs hot."

- **Test paperwork.** An extensive testing program (350,000 pages of test data with 4,500,000 entries) was challenged. It was discovered that the contract specified testing only during the research and development phase but it was not necessary during production. Savings on the contract were $720,000.

- **Rotary switch.** The original design had 167 parts and required 21 h/assembly. Value analysis generated 36 useful ideas. The revised unit had 10 parts and required .1 h/assembly.

- **Carton.** A side-opening carton replaced a top-opening carton; savings were 8%. Then the supplier suggested changing the carton material from gloss white to buff white; this saved 40%. Both changes were then applied to a variety of other cartons.

- **Milled contacts.** Copper contacts were hand-filed to remove the burrs from milling; cost was $1.58/unit. A forged contact at $.43/unit required no filing and also carried 20% more current.

- **Magnolia duct spacers.** Magnolia wood (which does not contaminate transformer oils) was being used for spacers. A value analyst took three years to experiment, test, and modify the material to chipboard; savings were $50,000/yr.

- **Packaging tape.** An airframe manufacturer used acetate fiber tape in packaging. Gummed paper provided the same function and saved $3,000/yr.

- **Shipping box.** An electric company packaged reels of punched tape in $1 plastic boxes. The new box (for $.15) is a plastic sandwich box from a dime store.

- **"Dead sharp" gaskets.** "Dead sharp" was considered to require lathe cutting by the vendor. When the high price was challenged, punching was considered satisfactory by the buyer; the price was reduced 20%.

- **Metal cabinet.** The cabinet was stiffened by spotwelding stiffeners. The spotwelding indentations were expensive to conceal. The stiffeners were deleted by making the cover thicker and bonding two attachment brackets. The savings on the contract were $10,000.

- **Dam culvert.** A culvert on Smithland Dam on the Ohio river was narrowed to 14 feet from 16 feet; savings were $695,000.

- **Two-stage contract.** The Bankhead lock needed replacement. When blasting for the new lock, contractors were afraid of damaging the existing lock; bids were high. The value engineering proposal was to divide the job into two stages. Stage 1 (for acquiring information) used highly skilled blasting crews and experimentation and data recording in blasting a small hole. Results were given to contractors for bid on stage 2—the big hole. The savings of $3,300,000 received a Presidential Management Improvement award.

- **Underground cables.** Contractor suggested modifying material enclosing cables for the Air Force. The Government Services Administration (GSA) and the contractor split the $100,000 savings. (The GSA, to encourage value engineering, splits savings in initial construction costs 50-50 with the contractor.)

REVIEW QUESTIONS

1. What is the goal of value engineering?
2. Functions are divided into which two categories?
3. Discuss Pareto voting.
4. List Mudge's four rules for defining function.
5. Discuss the distinction between functional value and prestige (esteem, sales) value. Give examples.

REFERENCES

De Marle, D. and Shillito, M. Value engineering. In *Handbook of Industrial Engineering*, 2nd ed. Salvendy, G. (ed.), Chapter 14. New York: Wiley, 1992.

Mudge, A. *Value Engineering*. New York: McGraw-Hill, 1971.

OPERATIONS ANALYSIS

OVERVIEW

Engineers can explore physically (i.e., trial and error in a physical situation). However, the engineer also can explore perceptually, using formulas and mathematical models. This chapter presents some tools of the trade for job design (see Table 7.14 for a summary). Be sure not to make the final decisions based solely on the calculations. The techniques are aids to decision making, not the decisions themselves. The techniques give a first draft; only rarely is the first draft the final decision.

CHAPTER CONTENTS

1 Location of One Item
2 Systematic Layout of Multiple Items
3 Balancing Flow Lines
4 Flow Diagrams and Process Charts
5 Multi-Activity Charts
6 Fish Diagrams
7 Decision Structure Tables
8 Breakeven Charts
9 Subjective Opinions
10 Economic Lot Size

KEY CONCEPTS

activity relationship diagram
adjective scales
balance delay percentage
balancing flow lines
body discomfort chart
Borg vote
breakeven charts
cost vs. distance
cycle time
decision structure tables
double tooling

economic lot size
element sharing
fish diagrams
flow diagrams
inventory cost
kitting
locating one item
multi-activity chart
precedence diagram
process charts
protocol

rating of perceived exertion
   (RPE)
relative vs. absolute rating
rework
satisfiers/optimizers
semantic differential
setup cost
subjective
Systematic Layout Procedure
   (SLP)

This chapter presents analysis techniques for the "big picture" (analysis of several operations or tasks) and then the "little picture" (analysis of an individual job). For the "miniature picture" (analysis of specific motions), see Chapters 15 and 29. For more on location of one item, systematic layout, and line balancing, see Konz (1994).

1 LOCATION OF ONE ITEM

1.1 Problem
Locating one item in a network of customers is a fairly common problem (see Table 7.1). The item can be a person, a machine, or even a building. The network of customers can be people, machines, or buildings. The criterion minimized can be distance moved by people or product, the amount of energy lost, or even time to reach a customer.

Typically, a user is interested in finding the location which minimizes the weighted distances moved—a minisum problem. (An example would be to locate a copy machine in an office.) Another possible objective is to minimize the maximum distance—a minimax or "worst-case" problem. (An example minimax problem would be to locate an ambulance so that everyone could be reached in no more than 15 minutes.)

There is extensive analytical literature on "planar single-facility location problems." See Chapter 4 in Francis, McGinnis, and White (1992) for an excellent discussion of the analytical techniques.

The following material, however, is not elegant math but "brute force" calculations. The reason is that with computers and even hand calculators, the engineer can solve all reasonable alternatives in a short time, perhaps 20 minutes. Then the engineer uses the material-handling cost calculations to gain insight into the location problem and, using this as one criterion (other criteria might be capital cost, maintenance cost, etc.), makes a recommended solution.

1.2 Solution
In the following example (Konz, 1970), consider (1) the item to be located as a machine tool, (2) the network of customers (circles in Figure 7.1) as other machine tools with which the machine tool will exchange product, and (3) the criterion to be minimized as distance moved by product.

In most real problems, there are only a few possible places to put the item; the remaining space is already filled with other machines, building columns, aisles, and so forth. In the example problem, the first solution to the problem will assume that there are only two feasible locations (A and B) for the new item. They are the rectangles in Figure 7.1.

Travel between a customer and A or B can be (1) straight line (e.g., conveyors), (2) rectangular (e.g., fork trucks down an aisle), or (3) measured on a map (e.g., fork trucks using one-way aisles, conveyors following aisles or connecting several machines, conveyors following nondirect paths). In real problems, travel may be a mixture of the three types.

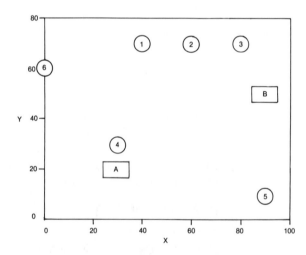

FIGURE 7.1 Should the customers (in circles) be served from location A or B? Table 7.2 shows the importance of each customer.

TABLE 7.1 Examples of locating an item in an existing network of customers with various criteria to be minimized.

New Item	Network of Customers	Criterion Minimized
Machine tool	Machine shop	Movement of product
Tool crib	Machine shop	Walking of operators
Time clock	Factory	Walking of operators
Inspection bench	Factory	Movement of product or inspectors
Copy machine	Office	Movement of secretaries
Warehouse or store	Market	Distribution cost
Factory	Warehouses	Distribution cost
Electric substation	Motors	Power loss
Civil defense siren	City	Distance to population
IIE meeting place	Locations of IIE members	Distance traveled
Fire station	City	Time to fire

Some customers are more important than others. Thus, the distance must be weighted. In a factory, a common index would be pallets moved/month. If the problem is location of a fire station, the weight of a customer might depend on the fire risk of the customer or the number of people occupying the site.

The operating cost of locating the item at a specific feasible location is:

$$MVCOST = WGTK\ (DIST) \qquad (7.1)$$

where $MVCOST$ = index of movement cost for a feasible location

$WGTK$ = weight (importance) of the Kth customer of N customers

$DIST$ = distance moved

$$MVCOST = \sum_{K=1}^{N} \ (|X_{i,j} - X_k| + |Y_{i,j} - Y_k|)$$
$$\text{(for rectangular)} \quad (7.2a)$$

$$MVCOST = \sum_{K=1}^{N} \ \sqrt{(X_{i,j} - X_k)^2 + (Y_{i,j} - Y_k)^2}$$
$$\text{(for straight line)} \quad (7.2b)$$

For the two locations given in Table 7.2, Table 7.3 shows the $MVCOST$. Movement cost at B is about 67,954/53,581 = 126% of A.

Assume you wish to know the cost at other locations than A and B for the above problem. By calculating costs at a number of points, a contour map can be drawn. This indicates that the best location is $X = 42$ and $Y = 40$ with a value of 32,000. Thus Site A is 6,000 from the minimum and Site B is 13,000 from the minimum.

The example, however, made the gross simplification that movement cost per unit distance is constant. Figure 7.2 shows a more realistic relationship of **cost vs. distance**, where most of the cost is loading and unloading (starting and stopping) or paperwork, and where cost of moving, when "acceleration and deceleration" are omitted, is very low. More realistically:

$$DIST = L_k + C_k \ \ (|X_{i,j} - X_k| + |Y_{i,j} - Y_k|)$$
$$\text{(for rectangular)} \qquad (7.3a)$$

$$DIST = L_k + C_k \ \ \sqrt{(X_{i,j} - X_k)^2 + (Y_{i,j} - Y_k)^2}$$
$$\text{(for straight line)} \qquad (7.3b)$$

where L_k = load + unload cost (including paperwork) per trip between the Kth customer and the feasible location

C_k = cost/unit distance (excluding L_k)

Assume for customers 1, 2, and 3 that $L_k = \$.50$/trip and $C_k = \$.001$/m; for customers 4, 5, and 6, $L_k = \$1$/trip and $C_k = \$.002$/m. Then the cost for

TABLE 7.2   Customers 1 to 6 can be served either from location A ($X = 30$, $Y = 20$) or from location B ($X = 90$, $Y = 50$). Which location minimizes movement cost?

Customer	Coordinate X	Coordinate Y	Weight or Importance	Movement Type
1	40	70	156	Straight line
2	60	70	179	Straight line
3	80	70	143	Straight line
4	30	30	296	Rectangular
5	90	10	94	Rectangular
6	0	60	225	Rectangular

TABLE 7.3   Cost of locating a new machine at locations A or B. Since $WGTK$ was pallets/month and $DIST$ was in meters, $MVCOST$ = meter-pallets/month.

Customer	Weight, Pallets/ Month	Site A Distance, Meters	Site A Cost, M-Pallets/ Month	Site B Distance	Site B Cost, M-Pallets/ Month
1	156	51	7,956	54	8,424
2	179	58	10,382	36	6,444
3	143	71	10,153	22	3,146
4	296	10	2,960	80	23,680
5	94	70	6,580	40	3,760
6	225	70	15,750	100	22,500
			53,781		67,954

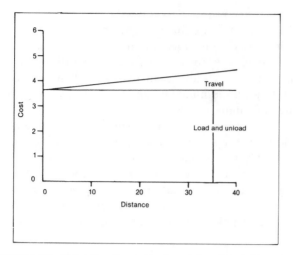

FIGURE 7.2   Material handling cost is almost independent of distance. See Guideline 2 of Chapter 14 for additional comments.

alternative A = $854 + $79.07 = $933.07 while the cost of B = $854 + $117.89 = $971.89. Thus B has a movement cost of 104% of A. When making the decision where to locate the new item, use not only the movement cost but also installation cost, capital cost, and maintenance cost. Note that the product *(WGTK) (DIST)* (that is, the $854) is independent of the feasible location; it just adds a constant value to each alternative.

Cost need not be expressed in terms of money. Consider locating a fire station where the *customers* are parts of the town and the *weights* are expected number of trips in a 10-yr period. Then *load* might be 1 min to respond to a call, *travel* is 1.5 min/km, and *unload* might be 1 min; the criterion is to minimize mean time/call.

Note also that the distance cost may rise by a power of 2—the *inverse square* law—for such problems as location of a siren or a light.

2 SYSTEMATIC LAYOUT OF MULTIPLE ITEMS

In contrast to the previous section on location of one item in a network of customers, this section discusses arrangement of the entire facility.

Systematic Layout Procedure (SLP) was developed by Richard Muther and is based on his extensive consulting work in plant layout (Muther, 1973; Muther and Haganaes, 1969; Muther and Hales, 1980). SLP deals with arrangement of entire facilities and can be used at the "block" (department) level or the "detail" (machine) level. When used to arrange displays within a panel, it may be called "link analysis." For more on control and display arrangement, see Chapters 19 and 20.

The following is a concise, simplified version of Muther's approach.

Step 1. The relationship chart (see Table 7.4) is the first step. Divide the facility into convenient activity areas (office, lathes, drill press, etc.). For more than 15 areas, analyze in two sections (e.g., layout of assembly departments and layout of component departments). For "Closeness Desired Between Areas," assign a letter grade: A = absolutely necessary, B = important, C = average, D = unimportant, and E = not desirable to be close. (Muther's technique use six levels instead of five.) Letters are used rather than numbers because numbers imply more precision to the judgment than is available. Avoid too many A relationships. About 10% As, 15% Bs, 25% Cs, and 50% Ds is a good goal. Support A, B, and E relationships with a "Reason for Closeness." Reasons will depend upon the problem, but common reasons are: 1 = product movement, 2 = supervisory closeness, 3 = personnel movement, 4 = tool or equipment movement, 5 = noise and vibration.

TABLE 7.4   Step 1 of systematic layout is to identify the desired closeness between areas with a letter grade. Give reasons for letters A, B, and E. Thus B/2 for drill press–lathes indicates a B importance for reason 2.

Area Number	Area Name	Office 1	Lathes 2	Drill Press 3	Punch Press 4	Plating 5	Shipping 6	Die Storage 7
1	Office							
2	Lathes	D						
3	Drill Press	D	B/2					
4	Punch Press	E/5	D	B/2				
5	Plating	D	C	D	D			
6	Shipping	C	D	D	C	B/2		
7	Die Storage	D	D	D	A/4	D	B/4	

In addition, you may have design concepts such as:

- truck docks on rear perimeter of the building
- executives on top floor of multi-floor building
- windows (view) for those with status
- cafeteria and toilets centrally located
- U-shaped flow (if shipping and receiving are one department or adjacent)
- storage kept together
- utilities in a spine (lower piping costs)

Step 2. Assign floor space to each activity area, along with physical features and restrictions (see Table 7.5). Remember Moore's corollary to Parkinson's law: "Inventories expand into whatever space is available, regardless of the need to maintain the inventory" (Moore, 1980). If the layout is a group of machines within an area, add space to the space for the machine alone. Consider space for the operator, for maintenance access, for movement of parts of the machine, and for local storage of product and supplies.

Step 3. Make an **activity relationship diagram** (see Figure 7.3). First, list all the A relationships from the relationship chart, then the Bs, Cs, Ds, and Es. Then make a diagram with just the As. Then add the Bs, keeping in mind the E restrictions. (For Es, walls and other barriers permit physical closeness while reducing visual and auditory distraction.) Then add the Cs; then the Ds.

Step 4. Make a scaled layout of at least two trials from Step 3 using the areas and restrictions of Step 2 (see Figure 7.4). First, use pieces of stiff paper for each department. Sketches tend to get "set into concrete" too soon. An alternative is a Computer-Aided Design (CAD) system in which areas can be rotated, moved, and so on. Don't forget that both areas and shapes of the departments can be adjusted. Some areas may be fixed in a specific location, for example, the shipping dock or the punch press department.

The reason for at least two layouts is that engineers are **satisfiers** rather than **optimizers**—they tend to stop designing as soon as they have a

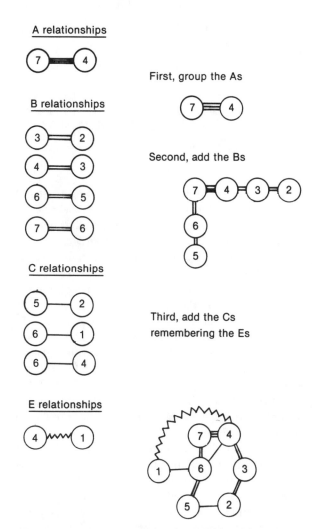

FIGURE 7.3 Step 3, the activity relationship diagram, first groups the As, Bs, Cs, and Es (left side of figure). (It may help to think of the lines as rubber bands pulling areas together. More bands make them closer together. The wavy line for the Es is a "spring" keeping them apart.) Then group all the As (right side of figure), add the Bs to the As (remembering the Es), and then add the remainder. Ds are not used.

TABLE 7.5 Step 2 of simplified SLP is to specify the amount of space (including about 20–30% for aisles) for each area. Give physical features and restrictions.

	Area Name	Desired m²	Restrictions
1	Office	50	Air conditioning
2	Lathes	40	Minimum of 10 m long
3	Drill press	40	
4	Punch press	50	Foundation
5	Plating	30	Water supply, fumes, wastes
6	Shipping	20	Outside wall
7	Die Storage	50	Crane
		280	

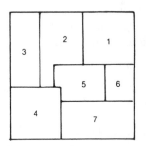 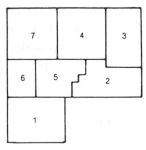

FIGURE 7.4 Step 4, scaled layout, makes several alternative arrangements. It may be desirable to slightly modify some of the desired areas or total plant area from Step 2 (to keep the building shape regular). A square (or nearly square) building shape minimizes wall perimeter and thus construction and energy costs.

solution that works instead of continuing to search for the best solution. (When the decision maker is satisfied to the point that it is not worth further effort to find something better, it is called *satisficing*.)

Note that E relationships do not have to be satisfied with distance. Walls and other barriers permit physical closeness while preventing the passage of noise, fumes, and other distractions.

Note also that A relationships do not have to be satisfied with closeness. For example, if the reason for the A is communication, the communication medium may be telephones, computer lines, faxes, video, or even pneumatic tubes; a distance of 50 or 500 ft is irrelevant. If the A is for product movement on a conveyor, again a distance of 100 or 500 ft is relatively insignificant.

Step 5. Evaluate the alternatives (see Table 7.6). The relevant criteria and their weights will change from situation to situation. Grade each layout (A = Excellent = 4; B = Good = 3; C = Average = 2; D = Fair = 1; and F = Bad = 0) and calculate the layout's "grade point" (grade × weight). If there is an existing layout, include it as one alternative. Defining the best as 100%, calculate the percent for the alternatives. Have the affected people sign off on the evaluation form. Then go back and select features from the alternatives to get an improved set of designs.

Step 6. Detail the layout (make a working drawing, using a 1:50 scale or possibly a 1:100 scale). At

this step, replace the department boundary lines with exterior walls (with doors), interior walls (with doors), and no walls (just department boundaries). Refine the estimate of the number of machines and operators (through detailed analysis of run and setup time, production schedules, consideration of alternative staffing patterns, use of simulation, etc.). Determine material handling and aisles. Locate machines and operators. Detail utilities (electricity, water, compressed air, and communications) and service areas (offices, toilets, breakrooms, toolrooms, nurse's office, etc.).

Present the alternatives to management for approval, and, after their modifications, install the final design.

3 BALANCING FLOW LINES

3.1 Standard Balancing Technique

The first question is whether you wish to balance the line. As is pointed out in the buffer design section (Guideline 3 in Chapter 14), it is feasible, and often desirable, to use an unbalanced line.

The line balance problem has three givens: (1) a table of work elements with their associated times (see Table 7.7), (2) a **precedence diagram** showing the element precedence relationships (see Figure 7.5), and (3) required units/minute from the line. To be determined are (1) the number of stations, (2) the number of workers at each station, and (3) the elements to be done at each station. The purpose of **balancing flow lines** is to minimize total idle time.

First, what is the total number to be made and in how long a time? For example, 20,000 units could be made in 1,000 h at the rate of 20/h, 500 h at 40/h, 250 h at 80/h, or in many other combinations. Continuous production is only one of the many alternatives. Hansen and Taylor (1982), for example, discuss the conditions when a periodic shutdown is best. Assume we wish to make the 20,000 units in 1,000 h at the rate of 20/h. Since we are dealing with a balanced line, each station will take 1,000 h/20,000 = .05 h/unit; **cycle time** = 0.5 h.

TABLE 7.6 Step 5 of simplified systematic layout planning is to evaluate the alternatives. Criteria and weights depend upon the specific management's goals.

Criterion	Weight	Present		Alternative 1		Alternative 2	
Minimum investment	6	A	24	B	18	A	24
Ease of supervision	10	C	20	C	20	B	30
Ease of operation	8	C	16	C	16	C	16
Ease of expansion and contraction	2	C	4	C	4	B	6
Total points			64		58		76
Relative merit (best = 100%)			84%		76%		100%

TABLE 7.7 Elements and work times for assembly line balancing problem. Each element time is assumed constant. In practice, each element time is a distribution.

Element	Work Time/ Unit, h
1	.0333
2	.0167
3	.0117
4	.0167
5	.0250
6	.0167
7	.0200
8	.0067
9	.0333
10	.0017
	.1818

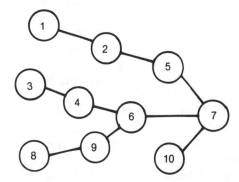

FIGURE 7.5 Precedence diagrams show the sequence required for assembly. The lines between the circles are not drawn to scale; that is, elements 4 and 9 both must be completed before 6, but 9 could be done before or after 4. Precedence must be observed; thus elements 3, 4, and 9 could not be assigned to one station and elements 8 and 6 to another. However, 8, 9, and 10 could be done at one station.

Second, guess an approximate number of stations by dividing total work time by cycle time: .1818 h/.05 h/station = 3.63 stations. Then use four stations with one operator at each.

Third, make a trial solution as in Table 7.8 and Figure 7.6. Identify each station with a cross-hatched area. Remember not to violate precedence. For example, elements 1 and 5 cannot be done at one station and element 2 at another. Then calculate the idle percentage (**balance delay percentage**): .0182/(4 × .05) = 9.1% in our case.

As an example of flexible thinking, consider Table 7.8. Here stations 1 and 2 are combined into one superstation; the elemental time now totals .0950. Since there are two operators, the time available is 1.000 and the idle time is .0050 at the station and .0025 for each of the operators. So far, there is no improvement over the solution of Table 7.8. However, note that there is *idle time at each station*. Therefore, the amount can be reduced at all stations until one has zero idle time. Thus, the line cycle time can be reduced to .0475.

Table 7.9 shows that the new idle time is .0083 h. Expressed in percentage terms, the idle time now is .0083/(4 × .0475) = 4.4% instead of the 9.1% of Table 7.8.

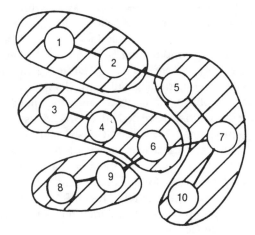

FIGURE 7.6 Solution of Table 7.8 is shown graphically.

TABLE 7.8 Trial solution for assembly line balance problem.

Station	Element	Element Time, h/unit	Station Element Time, h/unit	Station Idle Time, h/unit	Line Idle Time, h/unit
1	1	.0333			
	2	.0167	.0500	0	0
2	8	.0067			
	9	.0333	.0400	.0100	.0100
3	3	.0117			
	4	.0167			
	6	.0167	.0451	.0049	.0149
4	5	.0250			
	7	.0200			
	10	.0017	.0467	.0033	.0182

Idle percent = .0182/(4 × .05) = 9.1%

TABLE 7.9   Trial solution for assembly-line balance problem with two operators at one station and cycle time of .0475 h.

Station	No. of Operators	Element	Element Time (h)	Work Time (h)	Idle Time (h)	Cumulative Idle Time (h)
1	2	1	.0333			
		2	.0167			
		3	.0167			
		4	.0167			
		6	.0166	.0950	.0000	.0000
2	1	8	.0067			
		9	.0333	.0400	.0075	.0075
3	1	5	.0250			
		7	.0200			
		10	.0017	.0467	.0008	.0083

With larger lines, the problem complexity grows rapidly, and grouping the elements into zones (which either prevent or require certain elements to be done at the same station) is one attempt at simplification. Another source of complexity is the changes in product volume; that is, 20/h in May, 24/h in June, 26/h in July, and so forth. Another complication is that actual lines often are multiple-product lines. First we assemble a four-door Pontiac station wagon with a V6 engine, then a two-door Oldsmobile hardtop with a V6 engine, then a Pontiac four-door sedan with a V6 engine and air conditioning, and so on. In addition, there are mix changes (i.e., in March we produce 90% with air conditioning and in April 95%).

The problem complexity and the need for repeated solutions have led to efforts to use computer programs. However, Ghosh and Gagnon (1989), in their extensive survey of the literature, report that very few firms (about 5%) actually used computer programs for line balancing. Part of the problem is that user-friendly programs are not available. Another problem is that computers act on input as immutable facts.

3.2 Modifications to Standard Technique

If the computer solution gives some idle time at each station, then the cycle time can be decreased until the time at one station (the bottleneck station) is zero. Mariotti (1970) discusses some useful modifications.

First, consider **element sharing.** That is, operators/station need not equal 1.0. One possibility is more than one operator/station. Some examples are 2 operators/station (operators/station = 2), 3 operators/2 stations (operators/station = 1.5), and 4 operators/3 stations (operators/station = 1.33). This permits a cycle time that is less than an element time. For example, with 2 operators/station, each would do every other unit. (Often, combining two single stations with 1 operator each into one "super" station with 2 operators dramatically improves the balance.)

You also can have less than 1 operator/station by having operators walk between stations or having work done off-line (i.e., use buffers). Also it is possible in some situations to have operators from 2 adjacent stations share one or two elements. Station D does elements 16 and 17 on one-half or one-third of the units and Station E does elements 16 and 17 on the remaining units. Elements shared do not have to be adjacent stations if precedence requirements are not violated.

Second, remember that cycle times are not fixed. At the start, we assume a cycle time of .05 h (i.e., the line runs 1,000 h or 1,000/8 = 125 days). It may be more efficient to have a cycle time of .048 h (i.e., the line runs 20,000 × .048 = 960 h = 960/8 = 120 days). In addition to this balance cost, consider setup cost and inventory carrying cost. That is, what is the best combination of balance costs, setup costs, and inventory costs? The computer programs can do a quick check on many different alternatives.

Third, remember that elements often can be redefined. One possibility is to take former elements 16, 17, and 18 and eliminate element 17 by splitting it between elements 16 and 18. Or elements 16 and 17 might be combined and then split into 16a and 17a so that although the total time is the same, the relative allocation to 16 and 17 changes. Still another possibility is to split an element even further. For example, element 9 might be "pick up screwdriver, drive 20 screws, release screwdriver," with a time of .0167 h. For balancing purposes, it may be desirable to have element 9a be "pick up screwdriver, drive 15 screws, release screwdriver" with a time of .0125 h, and element 9b be "pick up screwdriver, drive 5 screws, release screwdriver" with a time of .0075 h. You have added extra work to the tasks (an extra pickup and release of the screwdriver) but may be able to cut the time of the bottleneck station and thus the time. Be careful about this technique since, when conditions change, the reason for the extra work may be forgotten and the unnecessary work retained without reason.

Fourth, interchange elements from the assembly task and the subassembly tasks. For example, a nameplate might be added at a subassembly station instead of the assembly station. An adjustment might be done on the assembly instead of the subassembly.

For more on line balancing, see Konz (1994).

4 FLOW DIAGRAMS AND PROCESS CHARTS

In contrast to the systematic layout and the location of one new item procedures, **flow diagrams,** and their associated **process charts,** are a technique for organizing and structuring a problem rather than a solution technique. Although a good engineer should be able to notice what is important in a problem and make corrections without the aid of a flow diagram and process chart, somehow solutions become more obvious with them.

Figure 7.7 shows a single-object process chart (the single object can be a person or an object). Figure 7.8 shows a flow diagram of Joe College, who is making a complex job of drinking beer in his apartment.

Figure 7.9 gives the five standard symbols for process charts: operations are circles, transportations or moves are arrows, inspections are squares, storages are triangles (upside down piles?), and delays (unplanned storage) are capital Ds. A circle inside a square is a combined operation and inspection. Some people put a number inside each symbol (identifying operation 1, 2, 3, or inspection 1, 2, 3) while others don't. Some people darken the circle for "do" operations but not for "get ready" and "put away" operations. Since a process chart is primarily a concise communication tool to yourself, take your choice. Since inspection without change was a square, inspection with change (an operation) is a circle inside a square. (A recent development in communicating and recording work methods is videotaping the operation.)

At the end of the analysis, summarize the number of operations, moves, inspections, storages, delays, and the total distance moved. Estimate times for storages and delays. Usually it is not worthwhile to determine or record operation or inspection times since this between-operations analysis usually is not concerned with within-operations methods. Process charts, because they give a "bird's-eye view" of the operation, often serve as methods documentation for others; you may wish to make a polished copy for them.

FIGURE 7.7   Single-object process charts follow a single object—in this example, an object. (Also see Figure 7.11, an assembly process chart.)

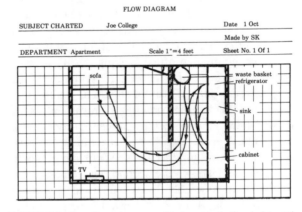

FLOW DIAGRAM

SUBJECT CHARTED	Joe College		Date 1 Oct
			Made by SK
DEPARTMENT Apartment		Scale 1"=4 feet	Sheet No. 1 Of 1

FIGURE 7.8   Flow diagrams are a graphical communication device used with process charts.

FIGURE 7.9   Five standard symbols for process charts are a circle (operation), arrow (move), square (inspection), triangle with point down (storage), and D (delay). A circle in a square is a combined operation and inspection.

Expense account paperwork was simplified at Intel by cutting 25 steps to 14 with the aid of a process chart (Main, 1981). Steps 5 and 7 were eliminated by having the accounts payable clerk, instead of the cash receipts clerk, collect refunds or unused traveler's checks. Steps 8, 10, and 11 were eliminated because another department also did these steps. Step 14 was eliminated as costing more than it saved. Step 19 proved unnecessary. The delays (steps 2, 6, 9, and 18) were cut. Expense accounts now are processed in days instead of weeks. Expense accounts less than $100 just require a petty cash voucher and a visit to the cashier. When the job was analyzed, the process chart was arranged horizontally and displayed on large horizontal strips of paper in a conference room. Each step then was questioned in detail by a group.

In hospitals, flow charts have helped reduce length of patient stays. For example, large doses of narcotics were easy to administer but such doses impaired patient digestion and therefore delayed discharge; now they give smaller doses of narcotics and discharge sooner. Some hospitals now give physicians an additional fee to make an additional bedside visit late in the day; as a result, many patients are discharged late in the day instead of the next morning. Detailed analysis of flow charts can yield an improved standard procedure—a **protocol.**

Draw—not to drafting standards—the flow diagram to scale on cross-section paper. It serves as a communication aid and shows overall relationships.

There are three types of flow diagrams: single object, assembly, and action-decision. Figures 7.7 and 7.8 showed the single-object type. Either an operator or an object is followed (i.e., follow the beer drinker or the beer container). Examples of following a person are vacuuming an office, making a bed, stocking shelves, changing a tire, unloading a semitrailer, handling material in the sandblast room, waiting on tables, and loading/unloading a dishwasher. Examples of following an object are purchase order preparation, check processing, and machining casting. Figure 7.10 shows how to handle scrap and **rework.**

Figure 7.11 shows the second type, an assembly process chart. Assembly flow diagrams tend to point out the problems of disorganized storage. Other examples are making a pizza, switch assembly, potting plants, making a whiskey sour, blood sample analysis, and reloading shotgun shells. There also are disassembly charts (beef packing-house, portioning pie). See Figure 7.12.

Figure 7.13 shows an action-decision flow diagram for a checkout operator in England while using a laser scanner. Note that in England the operator sits and the customer not only pays for the bag but also bags the groceries.

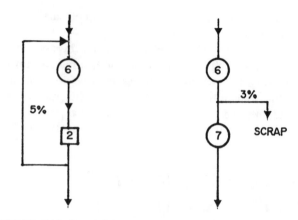

FIGURE 7.10   Rework (left figure) and scrap (right figure) should be included in flow process charts. Rework and scrap often are more difficult to handle than good product. Most items have some rework—even if it isn't officially reported. Material removal scrap occurs in machining and press operations; scrap units occur everywhere.

The purpose of flow diagrams and process charts is to focus your thinking. Use them with critical examination forms, Kipling's six honest men, and checklists (see Chapter 5).

5 MULTI-ACTIVITY CHARTS

Figure 7.14 shows a **multi-activity chart.** In different forms it has different names. If the columns represent people, it may be called a gang chart; if some columns are people and some machines, it may be called a man–machine chart. If one column represents the left hand and one the right, it may be called a left-hand right-hand chart. See Figure 7.15.

There can be two or more columns. The time axis (drawn to a convenient scale) can be seconds, minutes, or hours. The purpose of a multi-activity chart is improved utilization of a *column.* Improved utilization can mean less idle time, rebalanced idle time, or less idle time of an expensive component. See Guideline 4 of Chapter 13 for additional comments on utilization.

Some time is considered to be free. For example, "inside machine time" describes the situation in which an operator is given tasks while the machine is operating. Since the operator would be at the machine anyway, these tasks don't cost anything. But this assumes one operator for one machine. It is possible, however, to have more than one machine/operator. For example, in **double tooling,** there are two sets of tools/machine. One possibility is using two sets of tools on an indexing fixture to reduce idle time while waiting for a fixture to rotate. Another possibility is to have double fixtures to permit loading and unloading while the machine is processing a part on the other fixture. Or one person

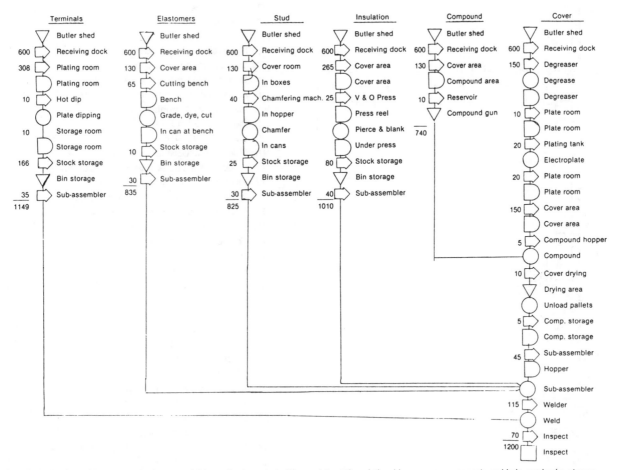

FIGURE 7.11 Assembly process charts are useful for methods analysis. They point out the relationships among components and help emphasize storage problems. Each column is an item with the assembly on the right. Figure 7.7 is a single-object flow chart.

can service two machines (i.e., .5 operator/machine). It also is possible to have .67 or .75 operator/machine. Do this by assigning 2 people/3 machines or 3/4. That is, the workers work as a team, not as individuals. In addition to improving productivity under standard conditions, the multiple cross-trained operators improve productivity when there is absenteeism. For task analysis, the chart shows time conflicts.

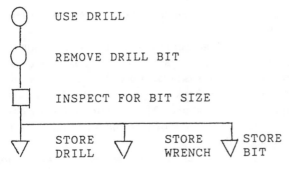

FIGURE 7.12 Disassembly process charts are useful in some situations.

Kitting is a general strategy of gathering components before assembly to minimize search and select operations and to ensure there are no missing parts. Subassemblies also may be useful. For example, McDonald's Big Mac has a special sauce. It is just a combination of relish and mayonnaise, premixed to ensure consistency.

To minimize operations and transportations, McDonald's makes and transports six hamburgers at a time from the grill to the serving area.

For each column, give cycles/yr, cost/min, and percent idle. Make the idle time distinctive by cross-hatching, shading, or coloring red.

A disadvantage of the multi-activity chart is that it requires a standardized situation. Nonstandardized situations are very difficult to show—for them, use computer simulation.

Example charts and columns might be: make lead molds (operator, machine, cooler); milling casting (operator, machine); cash checks (cashier, customer 1, customer 2); and serve meals (customer, waitress, cook).

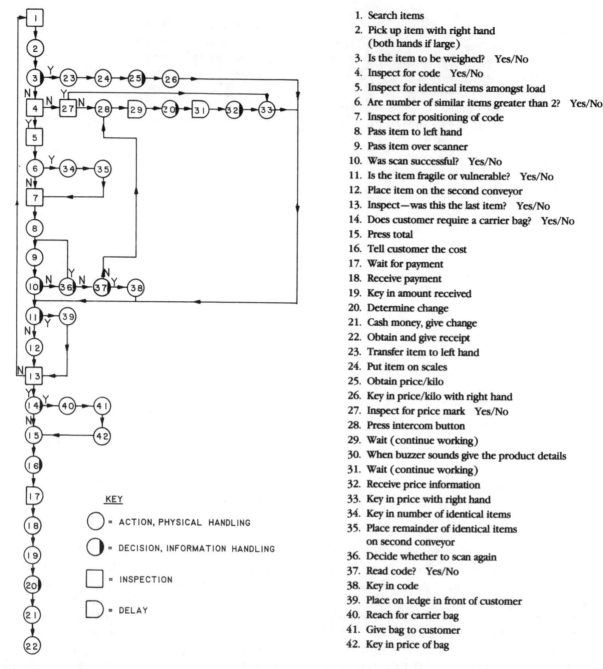

1. Search items
2. Pick up item with right hand (both hands if large)
3. Is the item to be weighed?   Yes/No
4. Inspect for code   Yes/No
5. Inspect for identical items amongst load
6. Are number of similar items greater than 2?   Yes/No
7. Inspect for positioning of code
8. Pass item to left hand
9. Pass item over scanner
10. Was scan successful?   Yes/No
11. Is the item fragile or vulnerable?   Yes/No
12. Place item on the second conveyor
13. Inspect—was this the last item?   Yes/No
14. Does customer require a carrier bag?   Yes/No
15. Press total
16. Tell customer the cost
17. Wait for payment
18. Receive payment
19. Key in amount received
20. Determine change
21. Cash money, give change
22. Obtain and give receipt
23. Transfer item to left hand
24. Put item on scales
25. Obtain price/kilo
26. Key in price/kilo with right hand
27. Inspect for price mark   Yes/No
28. Press intercom button
29. Wait (continue working)
30. When buzzer sounds give the product details
31. Wait (continue working)
32. Receive price information
33. Key in price with right hand
34. Key in number of identical items
35. Place remainder of identical items on second conveyor
36. Decide whether to scan again
37. Read code?   Yes/No
38. Key in code
39. Place on ledge in front of customer
40. Reach for carrier bag
41. Give bag to customer
42. Key in price of bag

KEY

◯ = ACTION, PHYSICAL HANDLING

◐ = DECISION, INFORMATION HANDLING

▢ = INSPECTION

◻ = DELAY

FIGURE 7.13   Action-decision flow diagram for a checkout operator in England using a laser scanner (Wilson and Grey, 1984).

6 FISH DIAGRAMS

Fish diagrams (see Figure 7.16), the cause side of cause-and-effect diagrams, are also known as Ishikawa diagrams due to their development in 1953 by Professor Ishikawa while on a quality control project for Kawasaki Steel. They can be considered a multidimensional list. Fish diagrams are widely used in Japanese Quality Circle meetings—meetings of about 10 production workers who meet on their own time to try to improve their jobs; by 1979 there

were 6,000,000 people in 600,000 groups in Japan (Aoki, 1979).

Start with a "fishhead"—a specific definition of a specific problem. Then put in a "backbone" and other bones. A good diagram will have three or more levels of bones (backbone, major bones, minor bones on the major bones). The diagram gives an easily understood overview of a problem and tends to trigger suggestions.

Figure 7.16 has improved grinding efficiency for the "head" and excessive grinding, working environ-

MULTIPLE ACTIVITY CHART

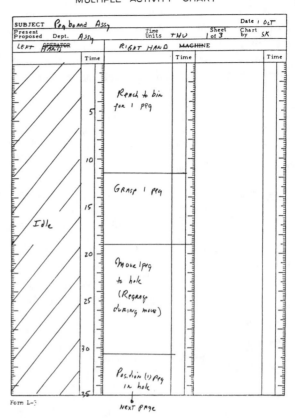

TIME CHART

METHOD: Proposed (1 operator 2 machines) CHARTED BY: Helen Gough DATE CHARTED: 19 Oct '76

OPERATION: Machining Gear Blanks

FIGURE 7.14 Multi-activity charts have many forms. The basic concept is to have two or more columns using a common, *scaled*, time axis. The chart may show idle time on a particular column, allowing activities to be moved between columns. It can show whether one operator can run two machines or three operators run four machines or two operators run three machines, when the operators are in a cell layout.

FIGURE 7.15 Left-hand right-hand charts are a type of multi-activity chart. For detailed hand movement analysis, Figure 29.1 shows a more common format.

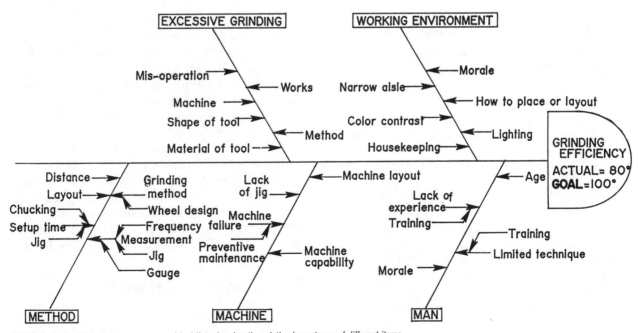

FIGURE 7.16 Fish diagrams are a graphical list, showing the relative importance of different items.

ment, method, machine, and man as the "major bones." Figure 7.17 was used at Bridgestone to reduce the variability of the viscosity of the splicing cement used in radial tires. Figure 7.18 shows the variability of the four operators before the Quality Circle studied the problem and after. For other

examples, consider defective berry boxes as the head, and equipment, material, workers, box design, and delays as the major bones; making a better pie shell (flour, fat, shaping shell, tools used, water, mixing ingredients, and baking); drilling a well (drill bit grinding, working environment, method, machine, operator); bartending serving efficiency (method, materials, operator, environment, and cash register); improved physiology lab teaching (skeletons, manuals, rats, cadavers, exams, instructor, assistants, visual aids, and handouts); barber shop utilization (sales promotion, environment, work policies, method, and man); and stabilizing the rice price in Indonesia (market monitoring, harvest forecasting, stock reporting, procurement operation, storage system, sales operation, and stock movement).

7 DECISION STRUCTURE TABLES

Decision structure tables are a version of *if* statements in computer programs; they also are known as protocols or *contingency* tables (see Tables 32.4, 32.6, and 32.7 for examples). They unambiguously describe complex, multirule, multivariable decision systems.

The discussion in Chapter 32 emphasizes making better quality decisions due to (1) better decision analysis (higher quality personnel make the decision, using complex analysis techniques if necessary) and (2) less time pressure at the time the decision is made. They are a "game plan" worked out in advance, not in the "heat of the battle." The educated guesses have been made and tested. However, in addition, decision structure tables are a good tool for methods analysis since they make the analyst consider all possibilities and force thoroughness. They also are good training aids.

A common example of a two-way decision structure table is a schedule with days as columns and hours, people, or machines as rows. A computerized version of a decision structure table is a spreadsheet. A spreadsheet is especially valuable in showing the results of "what if" decisions that have many effects.

Some example applications are spot welding, bowling prices, firewood prices, library checkout procedure, refund policy, spare parts prices, hiring policy, food scoop equivalents, bowling position for spares, CO_2 level for furnace, and flight scheduling.

FIGURE 7.17 Fish diagrams used at Bridgestone Tire to reduce cement viscosity variance (Cole, 1979). The four main categories were raw materials, operating methods, equipment, and human. The problem turned out to be an improper standard method as variability was greater for those who followed the standard procedure. See Figure 7.18.

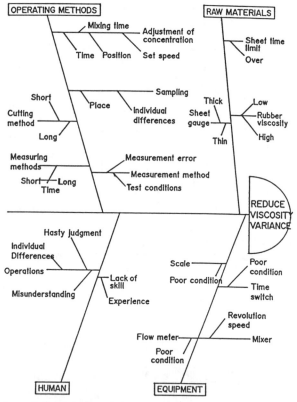

Source: Robert Cole, *Work, Mobility, and Participation: A Comparative Study of American and Japanese Industry.* Copyright © 1979 The Regents of the University of California. Reprinted with permission.

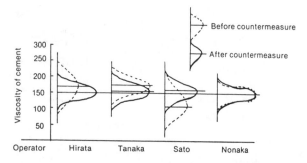

FIGURE 7.18 Distribution curves of four operators before and after the Quality Circle project (Cole, 1979). Reducing variability is an excellent quality improvement technique as the effects of various changes are seen much more clearly for processes with little variability. For another process improvement technique, see EVOP in Chapter 4.

8 BREAKEVEN CHARTS

A basic design problem is whether to use general-purpose equipment (low capital costs and high operating costs) or special-purpose equipment (high capital costs and low operating costs). At some production quantity (pieces, kg, etc.), the costs of the two methods are equal—the *breakeven point.*

The following example concerns making a part on an engine lathe vs. a turret lathe. Table 7.10 gives the information. The fixed cost, in this example, is setup cost + tooling cost. For the engine lathe, fixed cost is $11.05; for the turret lathe, it is $111.00. Initially, the engine lathe has an advantage of $111.00 − $11.05 = $99.95. For variable cost, consider both labor cost/unit and machine cost. The engine lathe variable cost is .2 h ($10.50/h) = $2.10 while the turret lathe is .1(11) = $1.10. The turret lathe picks up an advantage of $2.10 − $1.10 = $1.00 for every unit.

Figure 7.19 shows the cost equations graphically. The breakeven point, the point at which costs of both alternatives are equal, is:

$$\$111 + \$1.10 \ N = \$11.05 + \$2.10 \ N$$
$$N = 100 \text{ pieces}$$

That is, the engine lathe's initial advantage of $99.95 is overcome by unit 100. Below 100 units, theory says use the engine lathe; above 100 units use the turret lathe. In practice the problem is not so simple since errors or incorrect assumptions often are made.

First, consider tooling costs. If you already have the tooling, then the cost is zero for any additional project. On the other hand, if proposed tooling can be used by more than one project, spread the cost over all the projects.

Second, remember operating efficiency. That is, the standard time/unit from cost accounting may not be the same as the actual time. If, for example, the lathe department has been operating at 120% of standard, then both setup and operating times may be

1/1.20 = 83% of standard. Accurate operating times especially are important because a small error in labor costs has a large effect on the breakeven point.

Third, remember the effect of learning curves on labor time. See Chapter 28 for more on learning curves. The basic point is that time/unit varies with production quantity. Time is not constant. That is, instead of saying the engine lathe operator could make one unit every .2 h regardless of the quantity made, time/unit may be .20 h at 50 units; .18 at 100

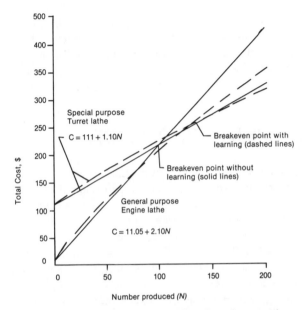

FIGURE 7.19 Breakeven charts show the effect of quantity on cost A vs. cost B or cost vs. income. This example shows that the cost of using the engine lathe is lower than the cost of using the turret lathe—but only up to 100 units. The dashed lines show the effect of learning. Assume a 90% curve and standard at 50 units for the engine lathe; a 95% curve and standard at 100 units for the turret lathe (see Chapter 28 for more on learning). Using learning increases the breakeven quantity from 100 units to 135.

TABLE 7.10 Information for breakeven example.

Item	Engine Lathe		Turret Lathe	
Machine cost, $/h				
Capital (depreciation)	.50		1.00	
Other burden	4.00	4.50	4.00	5.00
Labor cost, $/h		6.00		6.00
Machine + labor, $/h		10.50		11.00
Fixed costs				
Setup cost, h	.1		1	
Setup cost, $		1.05		11.00
Tooling costs, $		10.00		100.00
Total, $		11.05		111.00
Variable costs				
Manufacturing time, h	.2		.1	
Mfg. variable cost, $/piece		2.10		1.10

units; and .16 at 200 units. Conversely, at less than 50 units, time might be .22 at 25 units and .24 at 13 units. The alternative with a lower labor content probably will have less learning. Figure 7.19 shows that the breakeven point may be shifted very much if learning is considered.

Fourth, gamble if the odds are with you. For example, for an order of 80, the "best" alternative is the engine lathe. However, the advantage (using straight-line assumptions) is only $20. Thus, use the turret lathe if there is the possibility of repeat orders or other jobs on which the turret lathe tooling could be used.

It also is assumed that (1) both machines are available for work and (2) there is other work to keep both machines busy. In summary, breakeven charts are useful analysis tools, but be sure to check all the assumptions before making the decision.

9 SUBJECTIVE OPINIONS

When evaluating alternatives, the criterion can be performance (e.g., time, errors), physiological (e.g., heart rate, electromyogram [EMG]), or **subjective** (vote, opinion). In addition, data is history; to speak of the future, you need subjective opinion.

Rather than just asking someone's opinion, it is better to shape the question so that the answer you get is more precise and quantified (psychophysical scaling). See Hill et al. (1992) and Muckler and Seven (1992) for some comments on subjective evaluations. Four techniques are the Borg vote, body discomfort map, absolute adjective scales, and relative scales.

9.1 Borg Vote

The Borg **rating of perceived exertion (RPE)** vote (Borg, 1990) is described in more detail in Cardiovascular Response in Chapter 11. The basic concept of the **Borg vote** is to estimate heart rate of 30- to 50-year-old people by asking them a structured question. The technique is based on the psychophysical power law:

$$R = a + c (S-b)^n$$

where: R = response intensity

  a = constant for threshold (starting point)

  c = constant for measure
      (proportionality constant)

  S = stimulus intensity

  b = constant for threshold (starting point)

  n = power exponent

Although the Borg RPE scale was developed to estimate heart rate, it has become quite popular as a general index of stress. For example, Hagen (1993) demonstrated a .09 m longer lifting hook used by forestry workers was better by assessing heart rate, oxygen consumption, and force plate measurements as well as perceived exertion. The perceived exertion was compatible with the more complex measurements.

For another example, Freivalds and Eklund (1993) used multiple criteria to evaluate powered nutrunners. The perceived exertion votes were good estimators of the peak torque of the nutrunners. That is, rather than measure the torque of the nutrunner or the EMG of the forearm muscles, an engineer can have operators use alternative tools and give their opinion of their perceived exertion on the tools; the tools then can be selected from the perceived exertion vote.

9.2 Body Discomfort Chart

In the **body discomfort chart,** a person quantifies discomfort with a number for various body parts. Figure 7.20 combines Corlett and Bishop's (1976) concept of the body discomfort map with Borg's category ratio (CR-10) scale.

9.3 Adjective Scales (Absolute)

In the **adjective scales** measure, a person provides an opinion about a product or environment by ranking it on a scale between two opposite adjectives such as hot–cold, comfortable–uncomfortable, good–bad, and so forth (see Figure 7.21). Typically, the two adjectives peg either end of a nine-point scale (although sometimes a seven-point scale is used). By using a number of adjective pairs, it is possible to consider multiple characteristics of a problem. Have the positive end of the scale on the same side of the list for all adjective pairs. Placing the positive and negative ends randomly on both sides confuses people and gives less reliable results (Konz, Bennett, and Miller, 1986). The scores for the various pairs can be added arithmetically to get an overall evaluation. A more precise technique is to use a statistical technique called "factor analysis," which weights the contribution of each pair to the total (Rohles and Milliken, 1981). (Rohles and Laviana [1985] recommend retaining eigenvalues greater or equal to one and using a varimax rotation, then retaining descriptors which have loadings greater than .7 or less than -.7.)

The assumption underlying adjective scales is that it is possible to pick two adjectives which are opposites. However, people's mental image of what is opposite may not correspond to the two words chosen for the scale (Konz et al., 1986). One way to minimize this problem is to use only one end of the scale instead of both ends (Rohles, Woods, and Morey, 1989). For example, an office could be rated in the categories of air movement, aisle space,

FIGURE 7.20 Body discomfort is quantified with a number at various locations (Corlett and Bishop, 1976). In this version, additional information is desired for the foot, but sketches could be added for the hand, arm, etc. The rating scale is based on Borg's (1990) category ratio scale (CR-10) recommendations of 10 for "extremely strong" and .5 for "extremely weak." People are permitted to use decimals or go below .5 and above 10. A 10 is defined as the strongest value the person has experienced, but since the person can imagine an intensity even stronger, the absolute maximum could be 12, 13, or even higher.

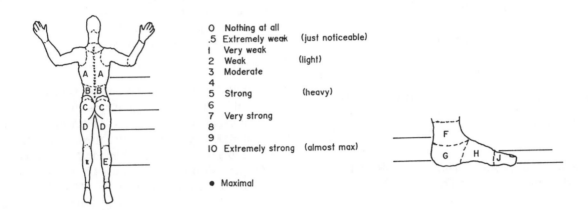

Source: Figure from N. Corlett and R. A. Bishop, "A Technique for Assessing Postural Discomfort," *Ergonomics,* Vol. 19, pp. 175-182. Copyright © 1976 by Taylor & Francis, London. Reprinted with permission. Table from G. Borg, "Psychophysical Scaling with Applications in Physical Work and the Perception of Exertion," *Scand. J. Work Environ. Health,* Vol. 16, Suppl. 1, pp. 55–58. Copyright © 1990 by Scand. J. Work Environ. Health, Helsinki, Finland. Reprinted with permission.

amount of dust, and so forth. The person would evaluate each with an acceptability vote:

6 = very acceptable

5 = acceptable

4 = somewhat acceptable

3 = somewhat unacceptable

2 = unacceptable

1 = very unacceptable

9.4 Relative (Paired) Votes
In Section 9.3, the person voted (using multiple criteria) for alternative A or B. For example, the person would vote concerning handtool A and then concerning handtool B; this is absolute rating. However, it is well known that a relative rating is more sensitive. That is, the person looks at the two items side by side and picks the preferred item. But relative rating does not allow determination of the strength of the preference. Considering the problem of **relative vs. absolute**

rating, Saaty (1980) developed a procedure which (1) allows the preference strength to be indicated also and (2) allows the resulting data to be compared statistically. Figure 7.22 shows the relative ranking scale. The person selects which member of the pair is preferred and the degree of preference. The data now have two values: preference and amount.

Then the data are reduced (Liu et al., 1990; Mitta, 1993) to one value, an eigenvector. Assume three lighting conditions were to be evaluated. Each person compares the pairs of conditions, indicating which condition is preferred and by how much. An example showing the result for one subject is given in Table 7.11.

Next the data are normalized and a single value *(w)* calculated (see Table 7.12). This single value then can be entered into standard statistical analyses such as analysis of variance.

Note that since the alternatives have to be paired for evaluation, the number of pairings rises rapidly

satisfied	___ : ___ : ___ : ___ : ___ : ___ : ___ : ___ : ___	dissatisfied
comfortable	___ : ___ : ___ : ___ : ___ : ___ : ___ : ___ : ___	uncomfortable
pleasant	___ : ___ : ___ : ___ : ___ : ___ : ___ : ___ : ___	unpleasant
acceptable	___ : ___ : ___ : ___ : ___ : ___ : ___ : ___ : ___	unacceptable

FIGURE 7.21 **Semantic differential** is the technical name for pairs of adjectives. Multiple pairs aid in evaluation of multiple dimensions of the situation.

10 —
9 — absolutely better
8 —
7 — significantly better
6 —
5 — much better
4 —
3 — somewhat better
2 —
1 — equal to

FIGURE 7.22 Relative ranking scale is used by (1) first selecting which member of the pair is preferred and then (2) indicating the degree of preference. Decimal votes (e.g., 3.5) are permitted.

with number of alternatives. Thus, if 4 tools are to be compared, there must be 6 pairings.

10 ECONOMIC LOT SIZE

When designing a job, a key question is how many identical units will be made at a time before shifting to a different model or product. (Most jobs involve working on *batches*—first batch A, then B, then C, and so on. A job is unusual if the same product is

TABLE 7.11 Pairwise comparison matrix for one subject. The matrix is filled in three steps. Step 1 is to put a 1 on the diagonal. Step 2 is to put the preference score at the intersection of the preferred condition. For example, assume condition 1 is preferred over condition 2 and the preference score is 2.5. (A score of 2.5 can occur either from the person using a decimal when voting or from averaging the results of several trials.) The third step is to put the reciprocals at the corresponding condition. For example, 1/2.5 = .4.

LIGHTING CONDITION		LIGHTING CONDITION	
	1	**2**	**3**
1	1.0	2.5	3.5
2	.4	1.0	3.5
3	.29	.29	1.0
Sum	1.69	3.79	8.0

TABLE 7.12 Normalized preferences for the subject in Table 7.10. The columns are normalized by dividing each entry by the column total. For example, 1/1.69 = .592. Then the eigenvector (*w*) is calculated by determining the row mean. The 1.001 grand total is approximately equal to 1. Condition 1 (score .563) is best.

LIGHTING CONDITION	LIGHTING CONDITION			ROW MEAN (*w*)
	1	**2**	**3**	
1	.592	.660	.437	.563
2	.237	.264	.437	.313
3	.172	.077	.125	.125
				1.001

made throughout the entire year.) The **economic lot size** problem is to balance one-time costs (paperwork and setup) vs. continuing costs (inventory); it can be formulated either within an organization (how many to make) or outside an organization (i.e., purchasing or how many to buy).

Table 7.13 presents data for an example problem. The solution will give the size of the *equal-size* lots.

10.1 Setup Costs
First, consider **setup cost.** If the annual requirement of 10,000 is made in 1 lot, then total setup cost/yr = (1 lot) ($50/lot) = $50. If 2 lots of 5,000 each are made, then total cost = $100; if 4 lots of 2,500 each are made, then total cost = $200, and so on. These annual costs also can be expressed in cost/unit; that is $50/10,000 = $.005/unit; (50 × 2)/10,000 = $.010/unit; (50 × 4)/10,000 = $.020/unit.

Using symbols:

$$STCTPU = \frac{\text{Annual setup cost, \$/yr}}{\text{Total units/yr}} =$$

$$\frac{S(N)}{A} = \frac{S(N)}{Q(N)} = S/Q \qquad (7.4)$$

where *STCTPU* = Setup cost, $/unit

S = Setup cost, $/lot

N = Number of lots/yr

A = Annual requirements, units/yr

Q = Quantity (number) of units/lot

Figure 7.23 shows setup costs/unit getting larger and larger as the number/lot gets smaller. Note the steeply rising curve for "small" lot sizes. If setup

TABLE 7.13 Information for economic lot size example.

Item		Value
Annual sales, units		10,000
One-time costs, $/lot		
Paperwork	20	
Tool setup	<u>30</u>	50
Continuing costs, $/unit		
Material cost, $/unit	.10	
Labor cost, $/unit	.20	
Manufacturing overhead, $/unit	<u>.50</u>	
Direct cost, $/unit	.80	
Inventory cost, % of inventory value/yr		
Obsolescence, deterioration	1	
Storage	1	
Capital	<u>18</u>	20
Inventory cost, $/yr/unit .20 ($.80/unit)		.16

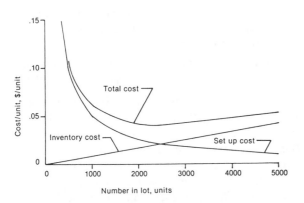

FIGURE 7.23   Total costs are composed of setup costs + inventory costs. We want to have a lot size such that total costs are as close as possible to the minimum.

were the only cost, we would make large lots. Recent emphasis on just-in-time manufacturing has emphasized reduction of setup cost so smaller lots can be run.

10.2   Inventory Costs

There is, however, another cost—inventory. **Inventory cost,** a continuing cost, depends on (1) the number in inventory and (2) the cost of keeping a unit in inventory.

The number in the inventory depends on the quantity we put in inventory each time we put some in, on the timing of when we put units in, and on the sales (withdrawal) pattern. Figure 7.24 shows a simple version of reality. Starting with an empty bin, we put Q items into stock. The stock level becomes $0 + Q = Q$. Then sales are absolutely equal every day, so the stock level declines in a straight line toward zero. Just at the second that the last item is sold, a new shipment arrives and the stock level goes back up to Q. The average amount in inventory is the beginning amount plus the ending amount divided by 2 or $(Q + 0)/2 = Q/2$. Geometrically we substitute a rectangle with height $Q/2$ for a triangle with height Q—both shapes have the same area. For a single lot of 10,000, the average inventory is $10,000/2 = 5,000$; for two lots of 5,000, the average inventory is $5,000/2 = 2,500$; for four lots of 2,500, the average inventory is $2,500/2 = 1,250$.

Cost of keeping an item in inventory depends on C, the cost or value of an item, and p, the annual inventory cost.

Cost of the item, C, is the direct cost (i.e., material, labor, and overhead).

Annual inventory cost, p, is expressed as a percent of inventory cost per year. It consists of obsolescence/deterioration/spoilage cost, storage cost, and capital cost. For obsolescence, calculate the annual cost for that specific product (e.g., $10,000) and then divide by the value of inventory (e.g., $1,000,000) to get the annual percent (e.g., 1% per year). Note that even if a product goes out of production, components can be used for service orders. For storage cost, calculate, for all products, the annual cost of heat, light, warehouse employees, and other costs (e.g., $100,000) and divide by the value of all inventory (e.g., $10,000,000) to get the annual percent (e.g., 1%). For capital cost, use the cost of alternative uses of capital in your organization, that is, your organization's expected return on investment. A typical value is 15–25%.

Thus, inventory cost is (Amount in inventory) (Inventory cost/unit) or, using symbols:

$$INCTPL = (Q/2)(Cp) \qquad (7.5)$$

where $INCTPL$ = Inventory cost per lot, $

$Q/2$ = Average inventory, units

C = Cost per unit, $/unit

p = Annual inventory cost, %/100

Converting this to inventory cost/unit:

$$INCTPU = INCTPL/A = QCp/2A \qquad (7.6)$$

where $INCTPU$ = Inventory cost/unit, $/unit

A = Annual requirements, units

Inventory cost/unit versus Q rises as a straight line (see Figure 7.23). Then total cost = Setup cost + Inventory cost.

$$TOCTPU = S/Q + QCp/2A \qquad (7.7)$$

where $TOCTPU$ = Total cost per unit, $/unit

Generally, we are interested in the minimum of this total curve, $QMIN$, or, more formally, where the slope equals zero. From calculus, this means setting the first derivative equal to zero and solving for Q:

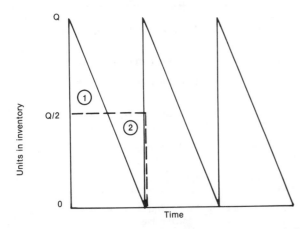

FIGURE 7.24   Inventory, in the most simple economic lot size formula, is represented by a triangle with height Q. Constant sales eventually result in zero inventory, at which instant a new shipment of Q arrives. The idealized triangle is replaced with an equivalent area rectangle (dashed lines) as triangle 1 = triangle 2.

$$Q/d \text{ cost} = 0 = -SQ^{-2} + Cp/2A$$

$$QMIN = \sqrt{\frac{2AS}{Cp}} \qquad (7.8)$$

Graphically, for a two-curve situation *where one curve is a straight line*, the minimum is at the point where the two component curves are equal. Or, from equation (7.8):

$$QMIN = \sqrt{\frac{(2)(10,000)(50)}{(.8)(.2)}} = 2,500 \text{ units}$$

In practice the problem is not so simple, since errors or incorrect assumptions often are made.

1. If, due to varying sales or production capability requirements, you will not produce *equal-size* lots, then equation (7.8) is not applicable; it might give a $QMIN$ too high or too low.

2. Remember to base tooling decisions over the total units produced, not just Q, the quantity released for this order.

3. Obsolescence often is overestimated. For example, if a product has a five-year life, obsolescence is not 20%—it is zero until at least the end of the model production and probably longer if spare parts are considered. Overestimating obsolescence tends to make predicted $QMIN$ smaller than actual $QMIN$.

4. Consider your sales pattern. If you ship most of your production shortly after you produce it, your average inventory is not $Q/2$ but $Q/3$, $Q/4$, or even $Q/10$. Overestimating the number in storage or the time in storage tends to make predicted $QMIN$ smaller than actual $QMIN$.

5. Equation (7.8) does not consider at least 4 different costs: overage, quality, learning, and inflation.

Overage costs consider the extra units needed in a lot; to get 20 completed units, you start with 25 at the first operation to allow for setup units and scrap. But to get 100 completed units, you also start with 5 extra, and 5/100 is a smaller percentage than 5/20. Since the cost of overage is not in equation (7.8), the predicted $QMIN$ is smaller than actual $QMIN$.

Quality is assumed to be the same for both small and large lots. In practice, once a process is set up correctly, errors are relatively few. Many setups mean more opportunities for errors. Thus, predicted $QMIN$ is lower than actual $QMIN$.

Learning is assumed to be zero; that is, time/unit was the same for lot sizes of 10,100 or 1,000. In practice, time/unit is lower for larger lots. Thus, predicted $QMIN$ is smaller than actual $QMIN$.

Inflation does not exist in equation (7.8). That is, future costs are assumed to equal present costs. Unlikely! Thus, the predicted $QMIN$ is lower than actual $QMIN$.

Finally, note the shape of the total cost curve in Figure 7.23. It is saucer shaped, not cup shaped. That is, departure from predicted $QMIN$ becomes expensive only with large departures from $QMIN$, and especially departures to small lot sizes.

Table 7.14 summarizes the analysis techniques.

TABLE 7.14 A summary of the analysis techniques given in this chapter.

Analysis Technique	Purpose of Technique	Typical Criterion
BETWEEN OPERATIONS Location of one new item	Locate one new item in an existing facility.	Minimize material handling costs.
Arrangement of entire facility	Rearrange all items in a facility.	Minimize material handling cost.
Assembly line balancing	Assign tasks to stations on an assembly line.	Minimize balance delay time.
Flow diagrams and process charts	Get overview of a task.	Eliminate unnecessary operations, movements, and inspections.
Multi-activity charts	Study coordination problems: between hands, people, and machines.	Reduce delay time.
WITHIN AN OPERATION Fish diagrams	Organize cost-reduction analysis.	Reduce costs; improve quality.
Decision structure tables	Have better quality decisions by workers; better communication of procedures.	Improve quality of procedure implementation.
Breakeven charts	Decide which process to use.	Minimize manufacturing cost.
Subjective opinions	Decide which procedure to use.	Choose among alternatives.
Economic lot size	Decide how many to make per order.	Minimize total of setup cost + inventory cost.

REVIEW QUESTIONS

1. Make an analysis of the location of a restaurant for meetings of the student chapter of the Institute of Industrial Engineers.

2. What are the three types of flow diagrams?

3. In a flow diagram and process chart, what are the standard symbols for an operation, transportation, inspection, delay, and storage?

4. What is the purpose of a multi-activity chart?

5. What is "inside machine time" used for?

6. What is "kitting"?

7. How is a body discomfort sketch used?

8. Give an example of a semantic differential question.

9. The economic lot size problem is to balance which two types of costs?

REFERENCES

Aoki, J. Japanese productivity: What's behind it? *Modern Machine Shop*, Vol. 52 [4], 117-25, 1979.

Borg, G. Psychophysical scaling with applications in physical work and the perception of exertion. *Scandinavian J. Work Environ. Health*, Vol. 16 (Supplement 1), 55-58, 1990.

Cole, R. *Work, Mobility, and Participation: A Comparative Study of American and Japanese Industry*. Berkeley: University of California Press, 1979.

Corlett, N. and Bishop, R. A technique for assessing postural discomfort. *Ergonomics*, Vol. 19, 175-82, 1976.

Francis, R., McGinnis, L., and White, J. *Facility Layout and Location: An Analytical Approach*. Englewood Cliffs, N. J.: Prentice-Hall, 1992.

Freivalds, A. and Eklund, J. Reaction torques and operator stress while using powered nutrunners. *Applied Ergonomics*, Vol. 24 [3], 158-64, 1993.

Ghosh, S. and Gagnon, R. A comprehensive literature review and analysis of the design, balancing and scheduling of assembly systems. *Int. J. of Production Research*, Vol. 27 [4], 637-70, 1989.

Hagen, K. Longer lifting hooks for forestry workers: An evaluation of ergonomic effects. *Int. J. of Ind. Ergonomics*, Vol. 12, 165-75, 1993.

Hansen, D. and Taylor, S. Optimal production strategies for identical production lines with minimum operable production rates. *IIE Transactions*, Vol. 14 [4], 288-95, December 1982.

Hill, S., Iavecchia, H., Byers, J., Bittner, A., Zaklad, A., and Christ, R. Comparison of four subjective workload rating scales. *Human Factors*, Vol. 34 [4], 429-39, 1992.

Konz, S. *Facility Design: Manufacturing Engineering*, 2nd ed. Scottsdale, Ariz.: Publishing Horizons, 1994.

Konz, S. Where does one more machine go? *Industrial Engineering*, Vol. 2 [5], 18-21, 1970.

Konz, S., Bennett, C., and Miller, B. Task lighting at a VDT workstation. *Proceedings of the Human Factors Society*, 192-96, 1986.

Liu, T., Narayanan, S., Subramanian, V., and Konz, S. Relative vs. absolute rating. *Proceedings of the Human Factors Society*, 1229-32, 1990.

Main, J. How to battle your own bureaucracy. *Fortune*, Vol. 103 [13], 54-58, June 29, 1981.

Mariotti, J. Four approaches to manual assembly line balancing. *Industrial Engineering*, Vol. 2 [6], 35-40, 1970.

Mitta, D. An application of the analytic hierarchy process: A rank-ordering of computer interfaces. *Human Factors*, Vol. 35 [1], 141-57, 1993.

Moore, J. Computer methods in facilities layout. *Industrial Engineering*, Vol. 12 [9], 82-93, September 1980.

Muckler, F. and Seven, S. Selecting performance measures: "Objective" vs. "subjective" measurement. *Human Factors*, Vol. 34 [4], 441-55, 1992.

Muther, R. *Systematic Layout Planning (SLP)*, 2nd ed. Boston: Cahners Books, 1973.

Muther, R. and Haganaes, K. *Systematic Handling Analysis (SHA)*. Kansas City, Mo.: Management and Industrial Research Publications, 1969.

Muther, R. and Hales, L. *Systematic Planning of Industrial Facilities (SPIF)*, Vols. I and II. Kansas City, Mo.: Management and Industrial Research Publications, 1980.

Rohles, F. and Laviana, J. Quantifying the subjective evaluation of occupied space. *Proceedings of the Human Factors Society*, 706-10, 1985.

Rohles, F. and Milliken, G. A scaling procedure for environmental research. *Proceedings of the Human Factors Society*, 472-76, 1981.

Rohles, F., Woods, J., and Morey, P. Indoor environment acceptability: The development of a rating scale. *ASHRAE Transactions*, Vol. 95, Part 1, 23-27, 1989.

Saaty, T. *The Analytic Hierarchy Process*. New York: McGraw-Hill, 1980.

Wilson, J. and Grey, S. Reach requirements and job attitudes at laser-scanner checkout stations. *Ergonomics*, Vol. 1, 27 [12], 1247-66, 1984.

8 OCCURRENCE SAMPLING

OVERVIEW

Occurrence sampling is a technique of gathering and analyzing data to aid decision making. After deciding how much accuracy and confidence are desired, a representative sample is gathered. The results then can provide information for decisions.

CHAPTER CONTENTS

1 Problem
2. Required Number of Observations
3 Representative Sample
4 Data Gathering
5 Data Analysis
6 Applications of Occurrence Sampling

KEY CONCEPTS

confidence	random sample
diminishing returns	randomization with restrictions
discrete/continuous sampling	relative/absolute accuracy
influence	reuse of data
p chart	stratification
periodicity	time standards

1 PROBLEM

Assume that in your organization material is moved with a fork truck. There seem to be long delays. Upon what do you bias your opinion? The question is asked, "Do we need another truck or is the present one idle too much?" This leads to a more specific question: "What is the present utilization of the truck?" It is important to remember that the purpose of occurrence sampling is to obtain information in order to make decisions; the purpose is not to demonstrate a knowledge of statistical theory.

We might have someone follow the truck for a specific time period—say, 20 working days—and record the following type of information for each day:

0700 Went to shipping dock
0706 Parked outside supervisor's office
0711 Left office with orders
0712 Entered first freight car with load

This would be a continuous time study. It gives a complete picture of the situation while it is studied, assuming the past is the same as the future. The problem is the expense of the study.

In order to cut the expense of the study, the truck might be observed for only 5 days or 1 day or even half a day. The expense has been cut, but we now have the problem of a representative sample—perhaps the Monday morning on which we made the study is not representative of the entire month. We could reduce the problem by following the truck for 2 h on Monday morning and 2 h on Tuesday morning. We still might worry about idleness in the afternoon. However, we could observe for 1 h on Monday morning, 1 h on Monday afternoon, 1 h on Tuesday morning, and 1 h on Tuesday afternoon. We still might worry about before and after breaks; about Wednesday, Thursday, and Friday; about the first week of the month vs. the second week, and so forth.

The end result is a sample composed of a large number of very short intervals: **discrete sampling** instead of **continuous sampling.** All time studies are samples from a population. The conventional time study (a "movie") is a continuous sample of n cycles (assumed underlying statistical distribution is normal). Occurrence sampling (a series of "snapshots") is a technique in which there are gaps of occurrences between the sample readings (assumed underlying statistical distribution is binomial). Although the statistical calculations are valid even if there are no gaps between the events sampled, continuous recording of occurrences does present questions as to whether the sample represents the population.

Occurrence sampling of times was first used by Tippett in the British textile industry in the early 1930s (Pape, 1992); it was introduced to the United States about 1940 under the name "ratio delay" since it is often used to study the ratios of various delays. It also is known as *work sampling* since the times sampled often are times of people working. More correct, however, is *occurrence sampling* since what are sampled are occurrences of various types of events. These events can be delays (such as in equipment utilization), work–task ratios (such as the proportion of material handling labor that is direct labor or the proportion of time spent on the telephone), or other ratios (such as the proportion of loads that are damaged). Work- and ratio-delay are poor adjectives since no connotation of work or delay is needed for the use of the technique.

Let us return to the fork truck problem. If we observe the fork truck many times (say, 1,000), then we will be quite confident in the information. If we observe the fork truck a few times (e.g., 10), our confidence will be much less. But there is a large difference in the cost of obtaining the information—the historic conflict between information and cost of obtaining information. You must make a trade-off. A small sample gives a low cost of information but a high risk that the sample is not representative of the population. A large sample gives a high cost of information but a low risk that the sample is not representative of the population. (See Figure 8.1.)

The sampling problem then becomes:

1. obtaining a sample whose size gives the desired trade-off between cost and risk

2. obtaining a sample representative of the population

The sample size problem will be discussed first.

2 REQUIRED NUMBER OF OBSERVATIONS

There is a general principle of statistics that the information obtained from a sample is a function of the

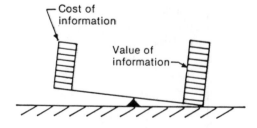

FIGURE 8.1 Trading off information (which increases with $\sqrt{n}$) vs. cost of information (which increases with n) is a common engineering problem.

square root of sample size, *n*. (It comes from the standard deviation of the error having that square root.) That is, if we get 2 units of information from a sample size of 4, then to get a total of 4 units of information, we need a sample size of 16; to get 8 units of information we need a sample size of 64.

Cost of obtaining information, however, generally increases directly with the sample size, *n*, rather than the square root of *n*.

In ordinary words, it is the law of **diminishing returns.** As *n* increases, there comes a point beyond which additional information is not worth the additional cost. Assume we can decide whether or not to buy a new fork truck if we know that the present truck is idle between 6% and 8% of the time. Then additional samples telling us that the idle percent is between 7.06 and 7.07% are not worth the additional cost.

How many observations should you make? It depends. It depends on:

A = Desired absolute accuracy

p = Proportion of occurrence

c = Confidence level desired

First, the standard deviation of a proportion, σ_p, is:

$$\sigma_p = \sqrt{\frac{p\,(1-p)}{n}}$$

where *p* = Proportion (decimal) of occurrence

n = Number of observations

Second, since sample sizes generally will be large (over 30), the normal distribution is assumed. See Figure 8.2.

Third, make the distinction between **relative** and **absolute accuracy.** This simple distinction, if not made, will cause much grief. If the mean proportion of occurrence, $\bar{p}$ = 40%, then 10% relative accuracy is (.1) (.4) = 4% absolute accuracy; 20% relative accuracy is 8% absolute accuracy; and 30% relative accuracy is 12% absolute accuracy. We need to use absolute accuracy in the calculations, but most people think in terms of relative accuracy. Be sure management understands the distinction when they make a statement such as "Make the study accurate to within 10%." The confusion arises because both the criterion and accuracy are in percent. If 10% accuracy on 25 lb is requested, then the 10% relative accuracy is not confused with the 2.5 lb absolute accuracy.

Fourth, how confident do you wish to be in your conclusions? Quite confident, very confident, absolutely confident, etc., are not precise enough, so confidence must be expressed in numbers—70% confident, 90% confident, 99.9% confident. What

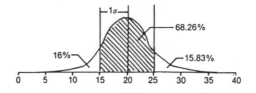
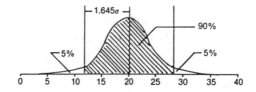
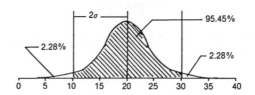
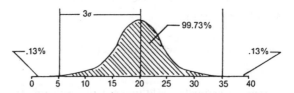

FIGURE 8.2 Normal distributions are symmetrical. The proportion of the total distribution included by a number of standard deviations does not depend on the value of the mean (see Table 8.1). Table 20.5 (the normal distribution) gives the area included for numbers of standard deviations other than above. (Note: It gives area from $-\infty$ to *z*, not area within ± *z*.)

TABLE 8.1 Confidence level for *z* values of equation $A = z\sigma_p$.

z, Number of Standard Deviations	Corresponding Confidence Level, %
±1.0	68
±1.64	90
±1.96	95
±2.0	95.45
±3.0	99.73

confidence expresses is the *long-run* probability that the sample mean is within the accuracy limits.

The probability calculations are similar to flipping a "true" coin; 8 of 10 is quite likely; 80 of 100 is very unlikely. It is the *number* of occurrences, not the percent, that is critical.

The formula to show the relationship between *p*, *n*, desired accuracy, and desired confidence level is:

$$A = z\,\sigma_p$$

or $A = z \sqrt{\dfrac{p(1-p)}{n}}$

or $n = \dfrac{z^2 p(1-p)}{A^2}$

$n = \dfrac{z^2}{s^2} \dfrac{(1-p)}{p}$

where p = Mean proportion occurrence, decimal
s = Relative accuracy desired, decimal
$A = sp$ = Absolute accuracy desired, decimal
z = Number of standard deviations for confidence level desired (see Table 8.1)
n = Number of observations

To determine the number of observations required:

1. Make a sketch, giving mean and A (see Figure 8.3).
2. Determine value of z (see Table 8.1).
3. Solve equation.

This may be clearer with some examples.

Example 1

Given   Estimated idle percent of fork truck = 40%
Relative accuracy desired = ±10%
Confidence level desired = 68%

To Find   How many observations, n, are required?
Solution   (.1) (.4) = .04 = 4% (see Figure 8.3)
68% confidence equals ±1 standard deviation (from Table 8.1)
Therefore:

$.04 = 1\sigma_p$

$.04 = 1 \sqrt{\dfrac{(.4)(.6)}{n}}$

$.04 = \sqrt{\dfrac{(.4)(.6)}{n}}$

$.0016 = \dfrac{.2400}{n}$

$n = .2400/.0016 = 150$

Statement   If the sample average of 150 observations is 40%, I can say with 68% confidence that the long-run idle percentage is between 36% and 44% with 40% being the most likely estimate, if the situation does not change.

Example 2

Given   Estimated idle percent of fork truck = 40%
Relative accuracy desired = ±10%
Confidence level desired = 95%

To Find   How many observations, n, are required?
Solution   (.1) (.4) = .04 = 4% (see Figure 8.3)
95% confidence equals ±1.96 standard deviations
Therefore:

$.04 = 1.96\sigma_p$

$.04 = 1.96 \sqrt{\dfrac{(.4)(.6)}{n}}$

$.0016 = 3.8416 \left(\dfrac{.24}{n} \right)$

$n = .922/.0016 = 576$

Statement   If the sample average of 576 observations is 40%, I can say with 95% confidence that the long-run idle percentage is between 36% and 44% with 40% being the most likely estimate, if the situation does not change.

Example 3

Given   Estimated percent of people wearing sweaters = 40%
Relative accuracy desired = ±20%
Confidence level desired = 68%

To Find   How many observations, n, are required?
Solution   (.2) (.4) = .08 (see Figure 8.4)

$.08 = 1\sigma_p$

$.08 = \sqrt{\dfrac{(.4)(.6)}{n}}$

$.0064 = \dfrac{.2400}{n}$

$n = .2400/.0064 = 37.5$

Statement   If the sample average of 38 observations is 40%, I can say with 68% confidence

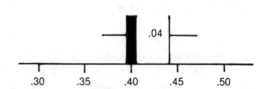

FIGURE 8.3   One-side sketch shows Examples 1 and 2.

that the long-run percentage of people wearing sweaters is between 32% and 48% with 40% being the most likely estimate, if the situation does not change.

Example 4

Given   Estimate percent of overcast days = 25%
Relative accuracy desired = ±5%
Confidence level desired = 95%

To Find   How many observations, n, are required?

Solution $(.05)(.25) = .0125 = 1.25\%$
(see Figure 8.5)

$$.0125 = 1.96 \sqrt{\frac{(.25)(.75)}{n}}$$

$$n = 4,618$$

In Example 4, we required "considerable" accuracy (5%) for a "small" target (1.25%). As proof that it is easier to aim at a big target, consider Example 5.

Example 5

Given   Estimated percent of days that are not overcast = 75%
Relative accuracy desired = ±5%
Confidence level desired = 95%

To Find   How many observations, n, are required?

Solution $(.05)(.75) = .0375$ (see Figure 8.5)

$$.0375 = 1.96 \sqrt{\frac{(.25)(.75)}{n}}$$

$$n = (1.96)^2(.25)(.75)/(.9375)^2 = 512$$

Table 8.2 shows the relation between p and n for a 5% value of *relative* accuracy and 95.45% confidence level. The table displays digitally the importance of not shooting at a small target. If we require ±5% relative accuracy about an occurrence with a mean of 1%, then we need 158,400 "shots" to hit the target (which has a width from .95% to 1.05%).

Examples 4 and 5 pointed out the importance of a precise statement of the problem. Contrast, from the table, the number of observations required for an occurrence with $p = .1$ ($n = 14,400$) vs. a nonoccurrence of $p = .9$ ($n = 175$).

Example 6 works the problem after gathering the data, with accuracy being the unknown and z being specified since p and n are by then known. You could specify A and calculate z if you wish.

Example 6

Given   In a completed study:

Machines idle	1,400
Machines working	2,600
Total observations	4,000

To Find   If I must be 95.45% confident of the answer, what is

  a. The absolute accuracy level?

  b. The relative accuracy level of percent idle?

Solution $p = 1,400/4,000 = .35 = 35\%$

$$A = 2 \sqrt{\frac{(.35)(.65)}{4,000}}$$

$$= \pm.015 = \pm 1.5\% \text{ absolute accuracy}$$

Relative accuracy = .015/.35

           = ±4.3% on machines idle

           = .015/.65 = ±2.3% on

               machines working

FIGURE 8.4   Example 3 is shown in a one-side sketch.

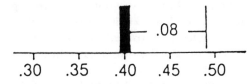

FIGURE 8.5   Examples 4 and 5 are shown in a one-side sketch.

TABLE 8.2   Relationship between p and n when a requirement of 5% relative accuracy and 95% confidence is imposed regardless of p.

p (Decimal)	5% of p	Resulting Range about p (Target)	Required Number of Observations
.01	.0005	.0095 to .0195	158,400
.02	.0010	.0190 to .0210	78,400
.03	.0015	.0285 to .0315	51,700
.04	.0020	.0380 to .0420	38,400
.05	.0025	.0475 to .0525	30,400
.10	.0050	.0950 to .1050	14,400
.15	.0075	.1425 to .1575	9,070
.20	.0100	.1900 to .2100	6,400
.30	.0150	.2850 to .3150	3,730
.40	.0200	.3800 to .4200	2,400
.50	.0250	.4750 to .5250	1,600
.60	.0300	.5700 to .6300	1,070
.70	.0350	.6650 to .7350	685
.80	.0400	.7600 to .8400	400
.90	.0450	.8550 to .9450	175
.95	.0475	.9025 to .9975	85

Statement   From the sample of 4,000 observations, I am 95.45% confident that the long-run idle percent is between 33.5% and 36.5% with 35% being the most likely estimate, if the situation does not change.

The above procedure is based on the sequence of plan study, do study, analyze data. If you are willing to analyze the data as you gather them (i.e., sequential sampling), the sample size can be reduced in some situations.

If there are several events to be studied (e.g., on a machine, you are interested in percent setup time, percent idle time, and percent rework time), you will get different Ns for the different percents. You will have to decide which percent and N to use—perhaps making several calculations and then compromising some of the initial desires on confidence and accuracy.

3 REPRESENTATIVE SAMPLE

There are three problems in obtaining a representative sample: stratification, influence, and periodicity.

3.1 Stratification

If you are taking a sample on political opinion, it is desirable to keep data from different strata (Kansas vs. New York, old vs. young, rich vs. poor, male vs. female, city vs. rural) separate (**stratification**). In the same way, when sampling for work design purposes, it is desirable to stratify your sample. Sample from morning and afternoon, first and second shift, Monday and Thursday, and so on. Data can always be combined.

The same item of information can belong simultaneously to several strata. A political poll could identify the same person as belonging to the male stratum, the Kansas stratum, the college graduate stratum, and others. By proper planning and by identification of the data, it is possible to "reuse" the data; that is, predict votes for males, for Kansas, for college graduates, and so forth. Data on the fork truck could belong to several strata: the morning stratum, the first shift stratum, etc.

If the sample is not properly stratified, you will fail to obtain potentially important information; more important, your sample will not faithfully represent the universe, and the risk of making incorrect assumptions about the population is greatly increased. The sampled population should be a good representation of the target population—the population to which your decision will apply.

An example might help. In 1936 the *Literary Digest* took a poll to see whether Landon or Roosevelt would win. They took a random sample of the people listed in the phone book. The poll predicted a Landon victory! Roosevelt won big, because people who weren't able to afford a phone voted heavily for Roosevelt. The sample had been unbiased but it also had not been representative.

An anecdote of George Bernard Shaw points up the importance of assumptions when you have mixtures. Beautiful woman to Shaw: "Let's get married and have children with my beauty and your brains." Shaw replied, "What if they had my beauty and your brains?"

Stratified samples have a higher efficiency (a smaller number of observations is required for a given risk) than nonstratified samples. More precisely, if the percent of occurrence is constant in all strata, then efficiency is the same for stratified and nonstratified sampling (Davidson, 1960). The higher efficiency is due to the lower variance (error) when strata are used. Stratification might not help, but it can't hurt.

Using the fork truck example, assume you took samples of 100 on Monday, Tuesday, Wednesday, and Thursday. Assume further that the truck was idle 10% of the time on Monday, 30% on Tuesday, 50% on Wednesday, and 30% on Thursday. Overall, $\bar{p} = (10 + 30 + 50 + 30)/400 = .3$. The variance is $(.3)(.7)/400 = .000\ 525$ and standard deviation $= .023$. Using a confidence band of ± 2 standard deviations, you would estimate the idle time as 30% ± 5%.

If you had kept your data stratified, you could predict idle times of .1, .3, .5, and .3. Variances would be .000 900, .002 100, .002 500, and .002 100; standard deviations would be .030, .046, .050, and .046. Your estimate of idle time then would be 10% ± 6% on Monday, 30% ± 9.2% on Tuesday, 50% ± 10% on Wednesday, and 30% ± 9.2% on Thursday.

3.2 Influence

The second problem of a representative sample is **influence**. The event being sampled must not change its behavior because it is being sampled. In many situations, changing behavior is not a problem. If you are observing the color of passing automobiles, the auto will not change from green to red because you looked at it. When recording free-throw percentage in a basketball game, the player will not make or miss the shot because you are keeping score in the stands.

But, if you always sample only the top of the barrel, bad apples seem to avoid the top. If Harry Nairdowell can see you coming on your sampling round, he may start working just before you record your observation. Even if Harry can't see you coming but can anticipate your presence since you always come at a specific time—you are periodic—the occurrence is not likely to be representative of the population.

3.3 Periodicity

Both the situation studied and the sample of the situation can be periodic or nonperiodic. **Periodicity** describes a situation in which a given behavior occurs regularly. For convenience we will call nonperiodic situations random even though they don't satisfy a mathematical definition of randomness. For engineering purposes they are sufficiently random.

The worst combination is a periodic situation and periodic sample with the same period length and initial point. For example, assume that Bill Kukenburger, the fork truck driver, every hour on the hour, drives by the desk on the shipping dock where Lisa Nimtz works and checks the time on his watch vs. the clock above her desk. If Sam Helal, your time study technician, observed Bill only every hour on the hour, his study would show that Bill did nothing but set his watch. If Sam observed Bill only at 14 minutes after the hour, he would never know the importance Bill attaches to having his watch set correctly.

Fortunately, although some machine cycles such as cam-controlled machines are completely periodic, most situations are not. But the possibility of a periodic situation, although rare, should be considered.

The most common technique of avoiding problems of influence and periodicity of the situation is to make the sample have no pattern. The easiest way to make a sample without a pattern is to make the sample random.

3.4 Sampling Procedure

Three sampling possibilities are random samples, random samples with restrictions, and periodic samples.

3.4.1 Random samples

Random samples are samples without a pattern. Assume that it has been calculated (see Section 8.2) that we need 10 observations on the first shift on Monday. That shift runs from 7:00 A.M. to 3:30 P.M. with 10-min coffee breaks at 9:00 and 1:30 and a 30-min lunch at 11:30. That is 480 minutes plus 30 minutes for lunch or 510 total. Label the minutes of the shift from 1 to 509. We need a representative sample from 1 to 119, no sample from 120 to 129, a sample from 130 to 269, no sample from 270 to 299, a sample from 300 to 389, no sample from 390 to 399, and a sample from 400 to 509.

Moder (1980) recommends against randomizing over the entire day but first stratifying by hours and then randomizing within the hours. That is, label the hours from 1 to 8 and randomly select 10 numbers from a random number table such as Table 8.3. Although the table is printed in blocks of three, the sequence is without pattern, and so the digits can be used in blocks of two, three, four, nine, or whatever.

To be precise, close your eyes and touch the table with a pencil; this selects the first number. Then if the first digit is odd, work up in the table; if it is even, work down. If you are not a purist, start in the upper left-hand corner and work down. The sequence will be random, but it also will be the same sequence every day.

Let's assume your pencil touched the ninth number from the top of column 6. The number is 413. The digit 4 indicates hour 4. Since the first digit, 4, was even, work down in the table. The next number is 1, then 6, 8, 6, 1, 9 (which will be skipped since

TABLE 8.3 Random numbers. All digits have equal probabilities. The sequence of digits has no pattern no matter which direction you move in the table (up, down, right, or left).

055	946	090	448	484	262	866	709	215	965
377	581	299	769	989	571	093	274	080	345
237	314	819	383	771	826	432	461	290	888
426	456	446	502	940	674	067	984	296	147
058	314	689	338	028	326	355	013	649	130
604	693	293	677	885	237	010	607	790	854
328	936	541	717	374	919	214	734	912	564
798	775	751	834	129	780	432	725	086	256
451	370	364	974	131	413	863	702	462	622
206	720	296	942	836	168	233	219	872	571
679	552	230	488	685	679	177	806	287	646
865	692	160	848	614	807	929	802	832	944
667	018	105	282	789	660	445	003	735	862
551	514	984	310	208	101	432	620	094	792
235	587	038	871	121	942	074	328	623	632
414	337	184	222	776	380	271	105	779	582
093	586	647	215	391	907	499	906	809	678
902	721	537	183	856	687	118	632	834	231
989	222	232	477	170	171	712	650	011	654
742	979	974	710	082	326	884	474	392	281
118	501	436	502	856	956	883	429	643	548

there is no hour 9), 3, 9 (skip), 6, 1, and 3. Now you know you will take 3 samples in hour 1, 2 in hour 3, 1 in hour 4, 3 in hour 6, and 1 in hour 8.

Next decide on the minute at which you will start your round of observation in each hour. Starting at the first time could be 41, then 16, skip 67, skip 80, skip 66, 10, and so on. The result is: 1 + 41 = 7:41, 6 + 16 = 12:16, etc. If the interval between rounds is not sufficient for you to complete the round before a new round is scheduled to start, toss out the starting time and select another random time.

Programs have been published by Weingast (1981), Whitehouse and Washburn (1981), and Taylor (1987). (Another program is in the DESIGNER program given to the instructors using this book.)

In addition to randomization of times, it also is feasible to randomize the direction of approach to a situation. For example, sometimes you come from the left, sometimes right. It also is feasible to have different people do the observing at different times. These additional randomizations can be done by computer or simple coin flips but the results probably should be recorded on a tour log such as Figure 8.6 so everything can be organized.

Simple randomization over the complete sampling period really is not very practical; in addition, it is inefficient (Moder, 1980).

3.4.2 Randomization with restrictions
Stratified random sampling has three advantages over non-stratified random: (1) It is more efficient. (2) It is easier to calculate the observation schedule than with a purely random sample. Using the fork truck example, it is easier to determine 20 sampling times on each of 10 days than to determine 200 sampling times randomized over the 10 days. (3) It is easier to

Date	Time	Entry point	Observer
8/16	8:07	A	J
8/16	8:33	C	J
8/16	8:51	C	J
8/16	9:43	A	J
8/16	9:46	B	J
8/16	12:26	A	M
8/16	12:56	A	M

FIGURE 8.6 Tour logs help keep all the times, entry points, observers, and other details organized.

modify a stratified plan. For example, if on day 3 you find that you have a meeting on the morning of day 4 and thus can't use the morning samples, you can change the schedule just for day 4. If randomization was over the entire 10 days, the schedule couldn't be changed since some of the times had already passed. If the plan is stratified, it also is easy to add or drop a day.

Randomization with restrictions can be used to encourage a representative sample. These restrictions are more important for small (e.g., n less than 30) sample sizes. For example, you might require that half of the observations occur before and after the midshift break or that 20% occur on each of the five days of the week. (By chance, random times could occur disproportionally more or less in a specific time slot.)

You also may wish to have a specified minimum time between observations if you are making multiple observations at a point. For example, if you are making observations of 10 different machines at each sampling time, you may wish that times between samples be at least 5 min so you can complete the observations on the 10 machines.

You might restrict the sampling schedule to match your schedule. For example, you may have a department meeting every Thursday from 9 to 10 and thus schedule no sampling during that time. This gives a small systematic bias of your sample vs. the population. Engineering judgment must be used in these situations. The goal is not a statistically perfect sample, but it is obtaining reasonably valid information with reasonable cost of information gathering allowing valid decisions to be made.

A random number (and thus time) may occur twice during the sampling period. This is valid. Just count that observation as two when it is done. Technically this is a situation with the minimum interval between samples equal to zero. The other extreme is where the interval between samples becomes quite large.

For maximum statistical efficiency, Moder and Kahn (1980) recommend a procedure, restricted random sampling, which is based on having a specified minimum interval between observation rounds— that is, no long "uncovered" gaps in sampling are permitted.

3.4.3 Periodic samples
We use the random sample when there are problems of periodicity or influence. Yet random samples are a lot of bother to calculate and tend to disrupt the observer's day by breaking it into segments of various lengths. Observers would like to use periodic samples.

The typical assumption is that the people being observed get tired of modifying their behavior if observed for a long enough time, so periodic samples can be used. An assumption! The periodicity of the situation and its phase relationship with the periodicity of the sample are also matters for engineering judgment. If you know the periods and their phases accurately, there is little need for the occurrence sampling study. Periodic samples also can be used when the thing being studied cannot be influenced by being observed.

Nonetheless, if there are a very large number of observations, say 1,000, then periodic sampling may be satisfactory. My personal recommendation is to avoid periodic sampling since it adds an additional possible error to the data. Your basic desire is to get clean data rather than "probably" clean data.

4 DATA GATHERING

4.1 Man or Machine?
The first question is whether the data will be gathered by a person or a device. Historically, the answer has been to use a person as devices did not have the capability to do the sampling. However, this has changed and devices now, in some cases, can do the sampling.

One possibility is to mount a TV camera to survey an area and connect the camera to a tape recorder which has a clock displayed in the scene. An observer then can watch the tape at leisure. If there is a desire to save tape, the tape can be used repeatedly or the camera turned on only occasionally.

Another possibility is to connect an event recorder to a machine. That is, it could record the times that the machine was running or the time the machine was stopped. Such devices now are attached routinely to long-haul trucks and continuously record speed and time the engine was running. In this way, the firm can identify how many times the driver exceeded the speed limit, how long the driver has stopped for breaks, or other behaviors. Another example is having a word processor operator monitored by a computer. The computer can record micro events (such as keystrokes/h) or macro events (files accessed/day). Clearly, in either situation, detailed information is available about the operator's performance. Strictly speaking, this electronic surveillance really is not sampling but continuous observation. The people being observed, however, may resent being "spied upon" by a camera. In addition, the camera field of view is relatively fixed, making it difficult to get a good view of events unless they are in a relatively small, fixed location. Carayon (1993) discusses worker stress caused by electronic monitoring.

Assuming a person does the study, the next question is *who*. The answer may depend on the study length. If the study is relatively short term (a week or so), it may be worthwhile to hire casual labor full time (at minimum wage) and study the situation relatively intensely. Of course, there will be some training time (perhaps a day) for the novice to become familiar with the task and the situation being studied. If the study is long term (a month or more, depending on the situation studied, which varies considerably), then using people already on the payroll (such as the supervisor or an engineer) for a portion of the day may be better. The assumption is that they can fit a given number of observations/day (perhaps 10–15) in among their other duties. Of course, their wage cost will be considerably higher than minimum wage, but their training time should be minimal.

4.2 Reuse Data?
For many situations, a simple cumulative data form, such as Figure 8.7, is satisfactory. The observer simply puts a tally mark in the appropriate place. At the end of the study, the total and percentages are calculated. Raw data and calculations and final answer are all on the same sheet.

The big disadvantage of cumulative recording is that the data are only used once. **Reuse of data,** in contrast, allows the data to be examined in terms of multiple criteria. Consider a political survey in which a 50-year-old male, living in Manhattan, Kansas, and who is a registered Republican, agreed with a particular issue. A 25-year-old female, living in Topeka, Kansas, and a registered Democrat, disagreed with a particular issue. Using cumulative tallies, this would be one for and one against. But the information is much more useful if the percent for and against can be identified by age, by town, by political preference, and other characteristics. In the same way, it is better to keep a fork truck observation so that it can be identified as occurring on Monday, in Dept. X, with driver Y, at time XXX, and so forth. That is, the data can be "reused" many times.

Formerly it was recognized that computers could do the analysis, but computers were relatively scarce and programming was difficult. Presently, standard programs such as SAS (Statistical Analysis System) are widely available for both mainframes and personal computers. See Box 8.1 for an example SAS program for use on a mainframe.

5 DATA ANALYSIS

If you have stratified your sample, you can use two different data analysis techniques. If you have no strata (subgroups), then you cannot use these two techniques.

Period	TALLY Working	Idle Driver absent	Idle Driver present	TOTAL Work	Idle (ab)	Idle (pr)	PERCENT Work	Idle (ab)	Idle (pr)
1	1111	111	1	4	3	1	50	37	12
2	1111 111			8			100		
3	1111 111			8			100		
4	1111 111			8			100		
5	1111 111			8			100		
6	1111 111			8			100		
7	1111 111			8			100		
8	1111 111			8			100		
9	1111 111			8			100		
10	1111		1	4	3	1	50	37	12
TOTAL				72	6	2	90.0	7.5	2.5

Situation _____ Clark Fork Truck _____ Observer _____ SK _____ Date _____ 8/17/89 _____

FIGURE 8.7 Combining data simplifies analysis but only allows one use of each entry.

5.1 Strata Comparison

It may be desirable to compare the occurrences between two different strata to see if there is a statistically significant difference. The following procedure assumes that the total observations are the same for both strata. See Natrella (1963) also for tables for unequal sample sizes. Allen and Corn (1978) present a chi-square procedure which can be used for unequal sample sizes.

Put the data in the following format:

	Class I	Class II	
Sample from strata A	r_a	S_a	r = number of occurrences
Sample from strata B	r_b	S_b	s = number of nonoccurrences
			$n = r + s$ = number of observations

For $n_a = n_b$

Example 1

Given $n_a = n_b = 16$

Strata	Times Idle	Times Working
Monday	9	7
Tuesday	1	15

To Find Is there a difference between Monday's and Tuesday's data?

Procedure	Number

Solution
1. Pick the smallest of the four numbers. 1
2. Select the other number in the same column. 9
3. Calculate "observed contrast." $9 - 1 = 8$
4. From Table 8.4, enter at sample size row, go across to the smallest number of step 1 and 2, or, if it is not given, the end of the row. Then go up to "minimum contrast"; 16 to 1 to 6 need contrast of at least 6.
5. Compare observed vs. minimum. $8 > 6$

Therefore, Tuesday's data are different from Monday's ($\alpha < .05$)

BOX 8.1   *SAS program for occurrence sampling*

See Figure 8.8 for an example output. In the actual program, data were entered for different days and a series of tables were generated for each machine. In addition, the data were resorted and tables printed by time of day vs. activity and by shift vs. activity as well as by product vs. activity. Note how the analyst can enter a code such as M21 and the computer will print May 21. Note how the analyst can enter a time and the computer will categorize it by hour.

Code	Comment
DATA MAY 21;	Data set is called May 21.
INPUT DATE$ 1-4 SHIFT 5-6 PRODUCT 7-9 TIME 10-14 TIMES$ 15-24 STATIONS$	Input of date, shift, etc., in the specified columns.
25-29 OPERATORS$ 30-34 ACTIVITIES 35-39 COMMENTS$ 40-65;	Items with $ are alpha-numeric.
IF TIME GE 0900 AND TIME LE 0959 THEN BLOCK = "0900-0959";	Blocks readings into hourly groups.
PROC FORMAT; VALUE $D M21 = "MAY 21";	Expands input of date M21 into May 21 on printed output.
VALUE $A TPL = "PALLET TOT:LVR"	Expands activity input.

DATA GE:
    SET MAY 21;
    IF STATION = "GE";
	Data set for east grinder is called GE.

PROC SORT DATA = GE OUT=GENEW; BY DATE PRODUCT;	Sort data set GE, create a new data set GENEW. Sort the new set by date, then product.

PROC FREQ DATA = GE;
    TABLES PRODUCT * ACTIVITY;
    FORMAT PRODUCT $P.;
    FORMAT ACTIVITY $A.;
    FORMAT DATE $D.;
    TITLE "GRINDER EAST";
	Do frequency analysis on data set GE with a table of product × activity. Title it Grinder East.

Shift				Activity		
Frequency Percent Row Percent Column Percent	Cleaning	Feeding Lids	Idle	Remove Lid Pack	Work Misc.	TOTAL
1	1 3.03 5.00 100.00	11 33.33 55.00 64.71	5 15.15 25.00 50.00	2 6.06 10.00 66.67	1 3.03 5.00 50.00	20 60.61
2	0 0.00 0.00 0.00	6 18.18 46.15 35.29	5 15.15 38.46 50.00	1 3.03 7.69 33.33	1 3.03 7.69 50.00	13 39.39
TOTAL	1 3.03	17 51.52	10 30.30	3 9.09	2 6.06	33 100.00

FIGURE 8.8   SAS output gives four numbers within each block. The top number is the number of the event occurring, the second is the percent of the overall total, the third is the percent of the row total, and the fourth is the percent of the column total.

TABLE 8.4 Relationships of sample size, smallest number (in Table), and "minimum contrast" for tests between two percentages. Binomial distribution used. Significance level of .05 for 2 tails; .025 for 1 tail.

Minimum Contrast for Statistical Significance

$n_a = n_b$

Sample Size	4	5	6	7	8	9	10	11	12	13	14	15	16	17	18
4-5	0														
6		0													
7-9		0-1													
10-11		0	1-2												
12-13		0	1-3												
14		0	1-2	3											
15-16		0	1	2-4											
17-19		0	1	2-5											
20		0	1	2-5											
30			0	1-2	6										
40			0	1-2	3-5	6-10	8-15	11-19							
50			0	1	3-4	5-7	7-10	9-13	14-24						
60			0	1	2-3	4-6	6-8	9-12	13-18	19-28					
70			0	1	2-3	4-5	6-8	8-11	12-15	16-23	24-33				
80			0	1	2-3	4-5	6-7	8-10	11-14	15-20	21-31	32-37			
90			0	1	2-3	4-5	6-7	8-10	11-13	14-18	19-25	26-42			
100			0	1	2-3	4	5-7	7-9	10-12	13-15	16-19	20-25	26-32	33-41	42-66
150			0	1	2	3-4	5-6	7-8	9-11	12-14	15-18	19-22	23-27	28-33	34-41
200			0	1	2	3-4	5-6	7-8	9-10	11-13	14-17	18-20	21-24	25-29	30-35
300			0	1	2	3-4	5-6	6-8	9-10	11-13	14-16	17-20	21-24	25-28	29-33
400			0	1	2	3-4	5	6-8	9-10	11-13	14-16	17-19	20-23	24-27	28-32
500				1	2	3-4	5	6-8	9-10	11-13	14-16	17-19	20-23	24-27	28-32

Sample Size	19	20	21	22	23	24	25	26	27	28	29	30	31	32
200	42-51	52-65	66-89	57-66	67-68	79-95	96-137	88-100	101-117	118-141	142-185	129-147	148-172	173-234
300	36-41	42-48	49-56	52-58	59-67	68-76	77-87	80-89	90-100	101-113	114-128			
400	34-38	39-44	45-51	49-55	56-62	63-70	71-79							
500	33-37	38-42	43-48											

Example 2

Given $n_a = n_b = 75$

Strata	Times Phone Busy	Times Phone Not Busy
1st Shift	45	30
2nd Shift	15	60

To Find   Is there a difference between the data for the first shift and second shift?

Procedure	Number

Solution 1. Pick the smallest of the four numbers.   15

2. Select the other number in the same column.   45

3. Calculate "observed contrast." $45 - 15 = 30$

4. From Table 8.4, enter at sample size row, go across to the smallest number of step 1 and 2, or, if it is not given, the end of the row. Then go up to "minimum contrast." There is no row for 75 so use 70 and 80 and interpolate, 70 to 15 to 12, 80 to 15 to 12. Thus minimum contrast is 12.

5. Compare observed vs. minimum. $30 > 12$
Therefore, first shift differs from second shift ($\alpha < .05$)

5.2 Control Charts
This test evaluates the effect of time (sequence) for your various strata (subgroups).

A special form of analysis of variance, called a control chart, was developed by Walter Shewhart during the 1930s. It has been used extensively in quality assurance work. A typical application in quality assurance would be monitoring a quality characteristic (say percent of parts with scratched paint). The percent of parts with scratched paint depends on many influences (operator, machine, type of paint, etc.), so you might think that an analysis of variance is desirable. However, in practice, there is a very strong relationship to sequence or time. That is,

if you can identify *when* something occurred, you can do the detective work to find out *where* and *why* it happened. The control chart helps identify *when* the process changes.

The same situation applies to occurrence sampling. Although attribute control charts and measurement control charts are used in quality assurance, occurrence sampling records the data in a discrete manner (yes–no, working–not working), so we use an attribute control chart. In fact we use only one of the various types of attribute control charts—the percentage chart or **p chart.**

The first step in the construction of a *p* chart is to plot the center line of the chart, $\bar{p}$. The average percentage is calculated by totaling the number of occurrences divided by the total number of observations.

The second step is to estimate the variability from the average that could occur by chance. The spread of the distribution is quantified by the standard deviation of *p*, symbolized σ_p. (Note that the *n* used for calculating the spread uses the sample size, not the total *n* of the study.) If all the individual readings were "slid along a wire parallel to the center line," they would form a histogram (see Figure 8.9).

The histogram describes the variability that could occur by chance. Variability beyond the histogram is assumed to occur not by chance but from a cause.

How far does the histogram spread? In theory, from negative infinity to positive infinity. However, the odds get quite small once we get out a ways from the center. If we go out $\pm 3\sigma_p$ in either direction from the mean percentage, there are very few chance occurrences; if we go out $\pm 2\sigma_p$ there are few; if we go out $\pm 1.5\sigma_p$ there are some, and so on.

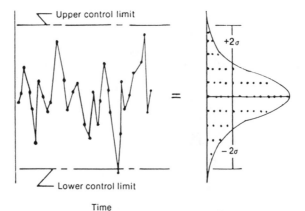

FIGURE 8.9 Control charts are a specialized form of analysis of variance which show the effect of time. Without the time scale it is a histogram (figure on right). The control limits indicate how much variability is likely to occur by chance. If a point falls outside $\pm 2\sigma$ limits, the odds are about 5% that nothing is different from the typical situation and about 95% that something has changed.

First, let's define the "very few," "few," etc. Although it's not statistically correct to use the normal as an approximation to the binomial for small sample sizes, the normal is commonly used because it is simple to use and gives answers that are close enough. Limits including $\pm 3\sigma_p$ include 99.7% of the chance occurrences; limits including $\pm 2\sigma_p$ include 95.45% (usually rounded to 95%); limits including $\pm 1.645\sigma_p$ include 90%. The trade-off is looking for trouble where none exists vs. not looking for trouble when trouble exists. Quality assurance limits commonly are $\pm 3\sigma_p$ although the chemical industry uses $\pm 2\sigma_p$ since the consequences of a change in the process are severe in the chemical industry. If a point is beyond the $3\sigma_p$ limits, the odds are 3/1,000 that the point is beyond the limit due to chance and 997/1,000 that the sample differs from the population.

For occurrence sampling, my recommendation is $\pm 2\sigma_p$ or even $\pm 1.645\sigma_p$ limits. If a point is beyond $2\sigma_p$ limits, the odds are 1/20 that chance has occurred and 19/20 that the sample differs from the population. The odds are still very good for $\pm 1.645\sigma_p$ limits: 1/10 for chance and 9/10 for cause.

Figure 8.10 shows an example chart. In constructing the chart, use all the observations to calculate the average, $\bar{p}$. *But*, the standard deviation for the limits uses the sample size n. That is, for a plot of 10 days, each an average of 50 occurrences, use $n = 500$ for the calculation of $\bar{p}$ but $n = 50$ for the calculation of σ_p. If n varies from sample to sample (e.g., 50 on sample 1, 52 on sample 2, 48 on sample 3, etc.), in theory the control limits should be calculated differently for each sample size. In practice, just

use the average sample size. Note also that the limits cannot exceed 0% or 100%. Thus if $\bar{p} = 60\%$ and $2\sigma_p = 70\%$, the upper limit would be 100% and the lower limit 0%.

All the discussion has emphasized the use of control *limits*—that is, the process is out of control (i.e., changed) only if a point is beyond the limit. However, in addition, an analyst can use *runs*. For example, eight events in a row that are all above the average (but not beyond the limit) have the same statistical significance as one point beyond the limit.

6 APPLICATIONS OF OCCURRENCE SAMPLING

The most common applications are to provide information for management decisions. Examples might be work vs. idle time of inspectors, nurses, doctors, teachers, supervisors, setup operators, or material handling equipment. In many studies, the categories of activity and of the items observed are not split into 2 or 3 but 10 to 20. That is, you might observe 10 different nurses and divide their activities into 18 different categories rather than observe just 1 nurse and divide the nurse's activity into 2 categories.

A slightly different application is to set **time standards,** which requires a count of the units produced during the time period as well as the occurrence sample.

Assume you had the data for a secretary shown in Table 8.5. From the sample, percent of time times 40 h gives the third column, the estimated hours

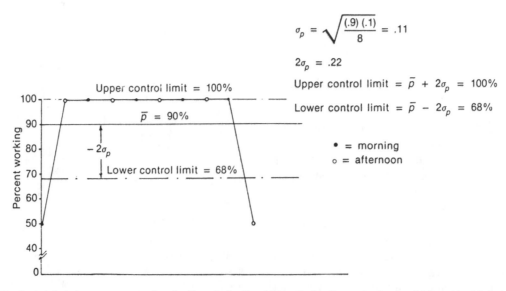

$$\sigma_p = \sqrt{\frac{(.9)\,(.1)}{8}} = .11$$

$2\sigma_p = .22$

Upper control limit $= \bar{p} + 2\sigma_p = 100\%$

Lower control limit $= \bar{p} - 2\sigma_p = 68\%$

• = morning
o = afternoon

FIGURE 8.10   Control charts for occurrence sampling should use limits either 1.645 or 2σ from the mean rather than the 3 used in statistical quality control, as the cost of a type 1 error (looking for a problem when none exists) is much lower in occurrence sampling. Use all the observations in calculating the mean but only the sample size for the limits.

TABLE 8.5 Example data showing how occurrence sampling data can be used to determine time standards.

Activity	Percent of Time Spent on Activity	Estimated Hours/Week on Activity	Output for Activity	Estimated Hours/Unit
Phone calls	12	4.8	147	.033
Letters	28	11.2	48	.233
Form A	3	1.2	121	.010
Form B	3	1.2	12	.100
Other	54	21.6		
		40.0		

spent on the activity. Then the fifth column, estimated h/unit, can be calculated by dividing by output. Note that standards need not be set on the entire list of activities during the week. Note that the worker's work method is assumed to be good; that is, we are setting time standards, not improving productivity.

An advantage of time standards from occurrence sampling is that the study can cover a large number of cycles of varied work (e.g., maintenance work, rework, office work). A disadvantage is that no performance rating was used. Performance rating of a crude nature can be done if the observer, when making the observation (allocating the activity to phone

call, typing letters, typing forms, etc.), simultaneously estimates a performance rating (90%, 120%, etc.). It should be recognized that the accuracy of the resulting standard will be improved only slightly by using this type of performance rating since instantaneous rating necessarily must be crude. Kinack (1967) recommends using conventional time study with performance rating for the important activities and using occurrence sampling for the minor activities; the rating from the important activities (e.g, 90%) is assumed to be valid also for the occurrence sampling activities.

See Box 8.2 for a summary of the steps in making an occurrence sample.

BOX 8.2 *Steps to make an occurrence sample*

Steps in making an occurrence sampling study:

Step	Example

1. State the problem:
  a. Give application of data.
  b. Determine categories.
	Determine idle percent for fork trucks in Dept. 8.
	Aid decision whether to buy new fork truck.
	Truck will be considered to be:
	  ▪ working
	  ▪ idle—driver absent
	  ▪ idle—driver present

2. Calculate sample size required.	Estimated percentages are:
	  ▪ working 75%
	  ▪ idle—driver absent 10%
	  ▪ idle—driver present 15%
	At this stage estimates are just the best guess.
	$A = 5\%$ (plus and minus) on all three
	$c = $ confidence $= 68\%$ on all three
	Required n are 75, 36, and 51, respectively, so use $n = 75$.

3. Plan to obtain a representative sample.	Take study during one entire week; half before lunch, half after. Divide week into 10 periods of 4 hours. For simplicity, take 8 observations per period for

(continued)

BOX 8.2   *Continued*

80 total observations. Make up random times for 10 periods. Walk into observation area by one of the three entrances—specific entrance chosen randomly.

Make up data form.

4. Make study.

5. Analyze results:

   a. Strata

   a. See Figure 8.10. $\bar{p}$ = .90; A = .034.

   Does not seem to be a morning vs. afternoon effect even though all idleness does occur on only two of the five days. The observed contrast is 4; required contrast is 5.

   b. Time

   b. See Figure 8.10. It seems there is a problem at the beginning and end of the week as points are beyond the control limits. The 2σ = 95% assumption is not valid for small samples; our n was 8, not the 30 needed for the assumption of normality.

   A check shows production control was unorganized at beginning of the week due to a late computer printout; not enough work was scheduled on Friday due to lack of knowledge of what would be made the following week.

6. Make conclusion.

   I am 68% confident that the long-run working percent is between 87% and 93%, with 90% the most likely estimate, if nothing changes.

7. Make decision.

   Make two decisions:
   a. Get production control organized.
   b. Don't buy new fork truck.

REVIEW QUESTIONS

1. What is the purpose of occurrence sampling?

2. Does information from a sample increase proportionally to sample size?

3. The number of observations to take in occurrence sampling depends upon what three things?

4. Briefly describe the three problems of obtaining a representative sample.

5. Discuss how a time standard could be set using occurrence sampling.

REFERENCES

Allen, D. and Corn, R. Comparing productive activities at different times. *Industrial Engineering*, Vol. 13 [3], 40-43, 1978.

Carayon, P. Effects of electronic performance monitoring on job design and worker stress: Review of the literature and conceptual model. *Human Factors*, Vol. 35 [3], 385-95, 1993.

Davidson, H. Work sampling—Eleven fallacies. *J. of Industrial Engineering*, Vol. 11 [1], 367-71, 1960.

Kinack, R. Activity evaluation technique—A quick and

easy procedure for developing time standards. *J. of Industrial Engineering*, Vol. 18, May 1967.

Moder, J. Selection of work sampling observation times: Part I—Stratified sampling. *AIIE Transactions*, Vol. 12 [1], 23–30, March 1980.

Moder, J. and Kahn, H. Selection of work sampling observation times: Part II—Restricted random sampling. *AIIE Transactions*, Vol. 12 [1], 32–37, March 1980.

Natrella, M. *Experimental Statistics*, Chapter 8, Handbook 91. Washington, DC: Supt. of Documents, 1963.

Pape, E. Work sampling. In *Handbook of Industrial Engineering*, 2nd ed. G. Salvendy (ed.), Chapter 64. New York: Wiley, 1992.

Taylor, D. Using Lotus 1-2-3 to generate random numbers and record observations of a system. *IE*, 46–52, 96, August 1987.

Weingast, J. Random samples. *American Industrial Hygiene Association J.*, Vol. 42 [1], A-15, 1981.

Whitehouse, G., and Washburn, D. Work sampling observation generator. *Industrial Engineering*, Vol. 13 [23], 16–18, 1981.

9 CRITERIA

OVERVIEW

Work design requires trade-off of multiple criteria. The criteria are vague and no explicit trade-off equations are available. Six design criteria are given.

CHAPTER CONTENTS

1 Organizations
2 Employees
3 Work Design Criteria

KEY CONCEPTS

criteria of job design
ego wants
enlargement/enrichment
foundations of job design
goals
limits (restraints)

machines as slaves
Maslow's stairs
physical wants
security wants
self-actualization wants
social wants

1 ORGANIZATIONS

1.1 Goals   Organization **goals** include survival and growth, among others. *Survival* is the first rule of any organism, whether bacterium, insect, man, corporation, or state. Carrying the analogy further, the organism must have sufficient nutrition; to an organization, money is the food. A surplus of income over outgo (called profits in capitalist countries and net favorable balance in socialist countries) is necessary in the long run. In the short run, a "diet" will not kill the patient, if the diet is not too severe or prolonged. Normally organizations set prices at a level which more than covers costs. If the price charged is higher than the market can bear or than the government permits, then either costs must be reduced or starvation begins. One exception is transfusions of public funds; this, in effect, has the taxpayer (a nonuser of the goods or services) pay some of the cost of the goods or services.

Growth is the second rule of an organism. Biological organisms, however, mature, stop growth, and die. Social organisms, such as auto manufacturers, universities, hospitals, and governments, are composed of "replaceable parts" and attempt to grow, grow, grow. Isn't a university with 15,000 students better than one with 7,000? Isn't a firm with 20,000 employees better than one with 10,000? Isn't a hospital with 500 employees better than one with 250? Isn't a bureau with 1,000 employees better than one with 500? Although some may not feel that bigger is better, more employees do give more power, prestige, status, and income to higher managers of an organization than do smaller numbers. Since the higher managers set the organization's goals, their rewards are what count. Thus, number of employees is a very important managerial criterion. Although not even admitted to exist as a criterion, it is a rare manager who would not prefer to supervise 2,000 rather than 1,000.

In the United States, a publicly proclaimed goal is a larger net income; that is, profits of $1,000,000 are considered better than profits of $200,000. A more sophisticated criterion is net income/assets; that is, return on investment of 10% is better than return of 5%. See Table 4.5 for returns on investment of some firms. Note that the return is not the same in all countries.

1.2 Limits   There are often **limits** (external **restraints**) imposed by society (usually governments and unions) but sometimes by public opinion or religious/moral values. Many decisions can be vetoed by the public due to concerns about pollution or safety, for example. These restraints can be quite different in developed and developing countries; they can be quite different in Judeo-Christian, Moslem, and Buddhist countries; and they can be quite different in capitalist and socialist countries.

Regulations (and enforcement) can be quite different concerning threshold limit values for toxic compounds. Environmental regulations can vary greatly, as do wage and employment regulations.

Japan provides an example of a religious/moral limit. The major firms (although not the small firms) have a "lifetime" employment policy for men (although not for women). Since labor is considered a fixed cost, there is tremendous pressure to continue producing product no matter what the price at which it can be sold. For example, as long as a car can be sold for a price exceeding the cost of purchased components, it will be cheaper to make it than not make it. Finding a local market may be a problem, so they export. Thus, prices in Japan often are considerably higher for products made in Japan than for the same product exported to other countries. Japanese firms often have a low return on investment; often it is lower than the rate a Japanese firm could make by investing money in a bank—a risk-free investment. That is, maintaining employment is a higher priority to Japanese firms than profits.

2 EMPLOYEES

The organizational goals are affected by the goals of the employees. These goals will be divided into physiological and psychological–social.

2.1 Physiological   Employees naturally want to work with the least stress. Chapter 11 discusses cardiovascular stress and biomechanics. The following chapters provide more information relating to physiology: Chapter 14, relationships among workstations; Chapter 15, workstation design; Chapter 17, manual material handling; Chapter 18, handtools; Chapter 19, controls; and Chapter 20, displays. The environment is covered in the following chapters: Chapter 21, eye and illumination; Chapter 22, ear and noise; Chapter 23, climate; Chapter 24, toxicology; and Chapter 25, safety.

2.2 Psychological–Social   **Maslow's stairs** in Figure 9.1 give perspective to the problem. Maslow said that there is a hierarchy of individual wants. **Physical wants,** the lowest level, concern basics such as the want for food, shelter, and health. Once these physical wants have been satisfied, the second level of wants, **security wants,** becomes important.

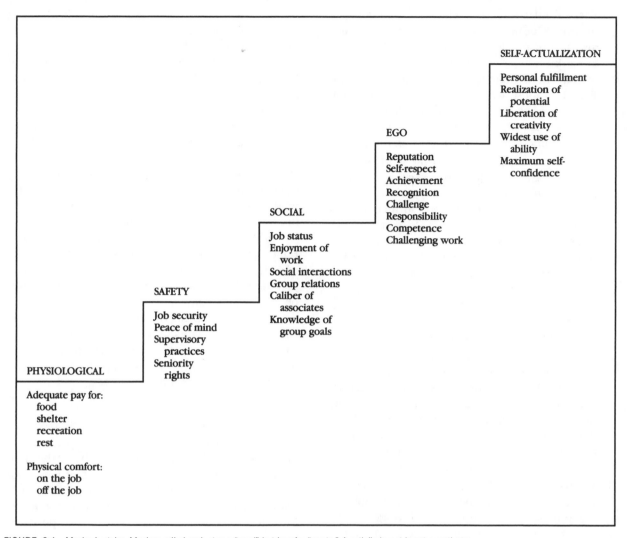

FIGURE 9.1   Maslow's stairs. Maslow called each step a "need" but I prefer "wants." A satisfied want is not a motivator.

In job terms, security wants might be having seniority on a job or having a supervisor who doesn't threaten you. The third level, **social wants,** becomes important after the second level is satisfied. Work examples of social wants are a job with status, having a job that you enjoy, having a job with friendly coworkers, or working in a physical location in which you can talk with fellow workers while working. The fourth level, **ego wants,** concerns challenge and achievement. Does the job challenge you? Do you have a feeling of contribution or are you "just a number"? The fifth level, **self-actualization wants**, concerns personal fulfillment and realization of potential. For example, is the organization "serving mankind" or are you merely making common items such as soap, autos, or chairs? Satisfying the fifth level may call for a "missionary" type of endeavor such as ecology, religion, or its secular equivalent, politics. Some find it in teaching, music, or in running their own businesses.

One question is how relevant these wants are to job design. Which of these wants, if any, are to be satisfied by the job and which outside of the job? As was pointed out in Table 3.3, males tend to work about 80,000 h during their lifetimes. In general, from the earnings of these 80,000 h, they support a family as well as provide for themselves and their wives during retirement. (In the United States, increasing participation of women in the employed labor force probably has added 20,000 to 30,000 h/family.)

Employees may prefer income from a monotonous, boring, dirty, dead-end job if they can own a sailboat and a Mercedes. Others may prefer less income and drive a used VW but enjoy the status of a white collar job.

Should an engineer be concerned with satisfying higher wants on the job? Some points for discussion are as follows.

1. There are dead-end jobs which must be done by someone. Idealists, perhaps with sounder hearts than heads, proclaim they would not be willing to work on an assembly line; this is true due to self-selection. Their opinions do not seem to be shared by those who actually work on assembly lines, probably because assembly line work is well paid and the workers are willing to trade satisfaction off the job for satisfaction on the job.

2. More and more, societies over the world guarantee the basics of life (food, shelter, clothing) whether an individual works or not. In the United States, in 1991, 14% of all income was government transfer payments. In addition, net income of unemployed (due to tax policies and benefits available to those not working) often is 70% to 90% of working income after taxes. This has led to *high*-seniority workers demanding that they be laid off first. Thus, if a job is dirty, boring, monotonous, dead-end, *and low paid*, finding employees is more and more difficult.

3. With increasing affluence and progressive tax structures, additional income becomes less meaningful. How motivating is additional income to a person whose primary problem is, on a Paris vacation, whether to stay at a two-star or three-star hotel? When one auto worker was asked why he worked only four of five days, he replied, "I can't make it on only three!"

4. A satisfied want does not motivate. With job enrichment it may be possible to design jobs to motivate workers; the motivated workers may work harder. However, unions in general oppose the concept of motivation. They favor working smart, not working hard.

5. Need there be a trade-off between fulfilling jobs and productive jobs? Can we "have our cake and eat it too"?

2.3 Satisfying Higher Wants
Higher wants will be discussed in sequence.

2.3.1 Security wants
In different cultures, managements have different approaches to job security. Large Japanese firms, for example, hire males at age 18 to work without layoffs until they are 55 (nenkō policy); their labor is considered a fixed cost. Since pensions don't start until 60, many hire back into their jobs at 55 as "temporary" workers (at 2/3 pay). Production fluctuations thus do not affect the permanent workers but do affect women, temporary workers, and subcontractors. The workers, in return, shift jobs freely (wages depend upon age, not the specific job), make methods and quality analyses during off-job hours to improve their jobs, and often

return early from vacations. A side effect is that since most costs except material are fixed costs, Japanese firms continue production almost regardless of the sales volume or price/unit. Japanese managers say they cannot afford to discharge their most valuable asset in temporary downturns since they must compete vs. cheap Asian labor from other countries.

In the United States, Procter and Gamble, for example, guarantees 2,000 h of work/yr for all permanent employees (those considered satisfactory after a 90-day probation period). The auto industry, however, guarantees income instead of work. If the worker is laid off, "supplemental employment benefits" pay up to 85% of wages for a year.

In the area of supervisory practices and seniority rights, unions have had a traditional concern. More recently, U. S. governmental agencies have exerted considerable pressure for less discrimination based on race, creed, color, sex, or age. Although organizations have complained about the extraordinary amount of paperwork now required, and although some arbitrary and discriminatory practices still occur occasionally, supervisory practices have become a relatively minor problem.

2.3.2 Social wants
In some Western societies, employees are gaining more say in the running of the business. Most dramatic is the "Mitbestimmung" of West Germany, the concept of joint command, of two hands on the tiller. Employees of firms with more than 2,000 employees elect almost 50% of the members of the Aufsichsrat, the supervisory board that hires and fires the senior executives who make up the Vorstand, the executives who run the firm. Workers also elect a Betriebsrat (works council); it, rather than the union, is the focus of shop-floor grievances, negotiates with management on changes in production organization, and has certain veto rights over hiring and firing. Other northern European countries have a form of Mitbestimmung with a smaller proportion of employee representatives. In England, where there is a strong us vs. them feeling, unions and management both reject Mitbestimmung as it requires cooperation rather than confrontation. Japanese organizations have a decentralized decision-making procedure with many different individuals "fixing their seal" on a proposal. There is a strong attempt to modify a proposal until there is 100% agreement.

In the United States there has been little formal sharing of authority between employees and management but considerable informal sharing due to the changing nature of work. In 1991, technical, sales, and administrative support composed 30.9% of the U. S. workforce, and managerial and professional specialty composed 26.5%—a total of 57.4%. Precision production, craft, and repair composed 11.3%, and operators, fabricators, and laborers composed 14.7%—a total of 26%. Managements traditionally have shared more decision making with white collar employees than with blue collar employees. Changing skill levels are shifting the shape of the workforce from a pyramid to a diamond.

At a superficial level, job titles can be changed to enhance job status. Changing janitor to building superintendent and garbage collector to sanitary engineer may be known to you already. In Texas, however, the Dallas School district calls school buses "motorized attendance modules" and their drivers "instructional facilitators." The Texas Secretary of State office advertised for an "administrative technician": Applicants "must have a demonstrated capacity to answer the phone when it rings without having to be told." When individuals design their own jobs, when does individual style become incompatible with order and efficiency?

Chapter 14 discusses many of the engineering aspects of job design: Guideline 1 discusses specialization; Guideline 3, assembly lines; and Guideline 6, social interaction.

2.3.3 Ego wants

Work can be made more challenging either by adding variety or by adding responsibility. Adding more variety has been called job **enlargement.** Examples would be to have a secretary run the duplicating machine as well as type, or to have an assembly line worker insert screws and tighten the screws rather than either insert or tighten. Adding more responsibility has been termed job **enrichment**: examples would be to have a secretary compose a letter as well as type it or to have an assembly line worker assemble and inspect rather than assemble or inspect.

Do workers want job enlargement and enrichment? Job enrichment is contrary to the traditional union approach of standard jobs, working conditions, and pay for all. Advocates of job enrichment often state that the workers will be more motivated and thus more productive. This "work hard not smart" philosophy is irritating to unions.

Job enlargement (less specialization, more variety) is not as controversial. Less specialization gives management more flexibility in job assignments and reduces union jurisdiction problems. Job rotation also can reduce the effects of cumulative trauma. Krawczyk et al. (1993) showed that spreading the work among different tasks, and thus different body parts, reduced the strain.

Salvendy (1978) commented that some people like to work at enlarged jobs, others prefer simplified jobs, and still others do not like work of any form!

Bennett (1973) divided jobs into four categories: physical (carry, lift), procedural (operate, follow procedures), social (talk, answer), and cognitive (decide, answer). Job enrichment seems most interesting to social and cognitive workers. Physical and procedural workers tend to be uninterested in more challenge and responsibility on the job; they tend to focus their lives on their off-work hours rather than the minority of hours they spend on the job. It may be just as well because, due to extensive education (see Figure 33.4), truly challenging work for everyone may be an impossible dream.

2.3.4 Self-actualization

In relatively few organizations do workers "throw their whole selves in." Japanese industry may be an exception. In 1974, the Matsushita Electric workers' song "Love, Light and a Dream" replaced the previous song "For the Building of a New Japan," which had been written in 1946 when rehabilitation of the war-devastated economy was the national goal.

The song is sung at the morning meeting, which was begun in 1933. Each section (10–20 people) has its own meeting, which lasts 10–15 min. On each day, one employee, in turn, goes up to the platform and reads the company creed and 7 objectives which all the others recite. Then he or she makes a 3–5 min speech on any subject (hobbies, family, job, etc.). Anyone else who wants to talk can then do so. The meeting is closed by singing the company song. The purpose is to (1) start the day with a refreshed recognition of the company mission, (2) have everyone learn to speak thoughts in public, (3) improve communication among the group, and (4) improve the group's knowledge of each other. American firms had company songs, but they are no longer sung.

Another technique is the Quality Circle movement discussed in Chapter 33.

Love, Light, and a Dream

A bright heart overflowing
With life linked together,
MATSUSHITA DENKI.
Time goes by but as it moves along
Each day brings a new spring.
Let us bind together
A world of blooming flowers
And a verdant land!
In Love, Light, and a Dream.

We trust our strength together in harmony
Finding happiness.
MATSUSHITA DENKI.
Animating joy everywhere,
A world of dedication,
Let us fulfill our hopes,
Shining hopes,

Of a radiant dawning.
With Love, Light, and a Dream.

Lyrics by Shoji Miyazawa
Music by Kozaburo Hirai

3 WORK DESIGN CRITERIA

3.1 Foundations Eight **foundations of job design** (underlying trends) will set the stage (Konz, 1985; Konz, 1987).

- *Foundation 1: People vary*. Variation in people in many dimensions (height, strength, training, etc.) is not new. However, more and more data have become available to designers to quantify this variation for different populations. See Chapter 10.

- *Foundation 2: People are more educated*. As shown in Figure 33.4, median years of schooling at age 25 in the United States is now over 12, although just 50 years ago it was 8.2. The same trend has occurred worldwide. Recently, management has begun to recognize that the workforce is educated and to encourage worker participation. See Box 9.1.

- *Foundation 3: People want a say*. Democracy has increased not only at the political level but also at the industrial level—at least in the developed countries. See Chapter 33 for a discussion of "small group activities."

- *Foundation 4: The world is becoming smaller*. Not physically smaller, of course, but changes in communication and transportation have had that effect. Consider telephone and computer networks, jet travel, interstate highway systems, and so forth. Multinational firms now manufacture on an international basis, not a national or regional basis.

- *Foundation 5: Machines are becoming more capable*. Machines have become more capable in power, control, and memory. In the United States, labor costs have more than doubled since 1970 while computer costs ($/instruction/s) are about 1% of 1970. Thus, considering the ratio of costs of computer/labor in 1970 as 1.0, the ratio now is less than .005. This helps explain why computers are finding more and more applications.

- *Foundation 6: Safety and health are more important*. The annual occupational death rate in the United States was 11.6/100,000 workers in 1933; in 1991 it was 3.9. The death rate for all accidents was 72.4/100,000 people in 1933; in 1991 it was 34.9. The American Industrial Hygiene Association had 160 members in 1940, and in 1994 it had over 12,400. The Human Factors and Ergonomics Society was founded in 1957; in 1994 it had over 5,300 members. The worldwide increase in the number and membership of ergonomics societies reflects increased concern for the worker.

- *Foundation 7: Job specialization is changing*. In developed countries, specialization in cognitive and social jobs (engineers, supervisors, teachers, etc.) has increased. Specialization in physical and procedural jobs (assembly line and clerical work) has decreased due to computerization and the movement of high-labor-content jobs to developing countries.

- *Foundation 8: Jobs are more interrelated*. With increasing technology, "No man is an island, entire of himself." More and more, "everything touches" and very few jobs stand alone. See Box 26.2. Figure 9.2 shows a conceptual work system.

BOX 9.1 *Productivity vs. stress reduction*

Frederick Taylor, working around the turn of the century, advocated "scientific management" to improve productivity. He improved handtools (such as shovels), gave better training, advocated rest breaks to reduce fatigue, and made other innovations. Taylor basically considered people as machines and aimed to improve the output of those machines. He did recognize the importance of motivation and advocated incentive wages. Taylor lived in a time when the educational level of workers as well as the level of technology were low.

However, the level of education has risen (see Figure 33.4) and the level of technology has advanced. Using a biological analogy, technology has advanced in "muscles" (motors, prime movers), in "nerves" (wired and wireless communication), and in "brains" (computers). The organizational design and management (ODAM) literature now emphasizes the individual, stress (physical and mental) reduction, social democracy, and quality of life. The ODAM writers tend to think of ergonomics as micro-solutions and their emphasis as macro-solutions. For more on macro-ergonomics, see Chapter 13.

FIGURE 9.2 Individuals are the focus in the work system advocated by Smith and Sainfort (1989). The individual is affected by environmental factors (physical and social), by task factors (e.g., repetitiveness, pacing), by technology (equipment), and by the organization (training, scheduling, roles, etc.). The individual also has personal influences, such as health, skills, motives, and so forth. Everything affects everything else.

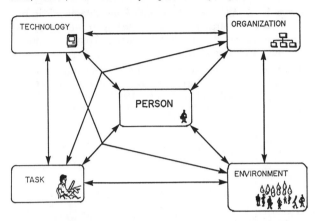

Source: M. Smith and P. A. Sainfort, A Balance Theory of Job Design for Stress Reduction, *Int. J. of Industrial Ergonomics,* Vol. 4, pp. 67–79. Copyright © 1989 by Elsevier Science, Amsterdam, Netherlands. Reprinted by permission.

The eight foundations lead to six ergonomic design criteria.

3.2   Criteria   There are six **criteria of job design:**

- *Criterion 1: Safety is first.* No job design is acceptable which endangers the worker's safety or health. However, life does not have infinite value. Managements must take reasonable precautions. Naturally the definition of "reasonable" is debatable. As pointed out above, we are putting increased emphasis on safety. After designing for safety, design for performance, then worker comfort, and finally consider higher wants (Bennett, 1972).

- *Criterion 2: Make the machine user-friendly.* The machine is to adjust to the worker, not the converse. If the system does not function well, redesign the machine or procedure rather than blame the operator.

- *Criterion 3: Reduce the percent excluded by the design.* Permit any person to use the machine or procedure. Gender, age, strength, or other physical attributes should not prevent people from participating in work or leisure activities. See Chapters 10 and 15.

- *Criterion 4: Design jobs to be cognitive and social.* Physical and procedural work now can be done by machines. Manufacturing jobs in the developed countries will take a smaller percent of the workforce. At present, over 50% of all jobs in the developed countries are in offices.

- *Criterion 5: Emphasize communication.* As shown in Figure 19.1, we communicate to machines with controls and receive information from machines by displays. We also need to improve communication among people to increase output and, perhaps equally important, reduce errors.

- *Criterion 6: Use machines to extend human performance.* The choice is not worker *or* machine, it is which machine to use. Workers reduce their own effort by using **machines as slaves.** Small machines (word processor, electric drill) tend to have costs (total of capital, maintenance, and power) of less than $.25/h. Even large machines (automobiles, lathes) tend to cost only $1–2/h. Labor costs, however, run $5–20/h. Thus, the real question is how many "slaves" (machines) the human will supervise and how to design the system to use the output of the slaves.

REVIEW QUESTIONS

1. List the five steps of Maslow's stairs; give examples for each step.
2. Why do unions dislike job enlargement and job enrichment?
3. What is the difference between job enlargement and job enrichment? Why should any organization use either?
4. Eight foundations of job design were listed. Give six.
5. Give the six work design criteria.

REFERENCES

Bennett, C. The human factors of work. *Human Factors*, Vol. 15 [3], 281–87, 1973.

Bennett, C. Designing for human what? *Bulletin of the Human Factors Society*, 3, February 1972.

Konz, S. Foundations of ergonomic job design. In *Industrial Ergonomics*, Alexander, D., and Pulat, M. (eds.), Chapter 4. Atlanta: Institute of Industrial Engineers, 1985.

Konz, S. Ergonomic job design. *International J. of Industrial Ergonomics*, 307–15, 1987.

Krawczyk, S., Armstrong, T., and Snook, S. Psychophysiological assessment of simulated assembly line work: Combinations of transferring and screw driving tasks. *Proceedings of the Human Factors and Ergonomics Society*, 803–07, 1993.

Salvendy, G. An industrial engineering dilemma: Simplified vs. enlarged jobs. *Proceedings of 4th Int. Conference on Production Research* (Tokyo). London: Taylor and Francis, 1978.

Smith, M. and Sainfort, P. A balance theory of job design for stress reduction. *International J. of Industrial Ergonomics*, Vol. 4, 67–79, 1989.

OVERVIEW

Anthropometric data permit the designer to design to fit the individual. Design to include as large a proportion of the population as possible—that is, minimize the number of people excluded by your design.

CHAPTER CONTENTS

1 Fit the Job to the Person
2 Population Values
3 Statistical Calculations

KEY CONCEPTS

anthropometry
DuBois surface area
ergonomics
exclude few

population percentile
principle of similitude
selection/job modification
surface area/volume

1 FIT THE JOB TO THE PERSON

1.1 Variability

Anthropometry, from the Greek *anthropos* (man) and *metrein* (to measure) explains how people vary. (**Ergonomics** also comes from the Greek: *erg* [work] and *nomos* [study of].)

This chapter will quantify human variability—put it into numbers. As Lord Kelvin said, "When you can measure what you are speaking about and express it in numbers, you know something about it; but when you cannot express it in numbers, your knowledge is of a meager and unsatisfactory kind."

People vary in many dimensions, for example, initiative, needs, dexterity, intelligence, visual acuity, imagination, determination, upper back strength, age, and leg length.

1.2 Selection vs. Job Modification

Assume a heavy box is to be moved from point A to point B. Realizing that people vary, there are two basic strategies.

One alternative is to use **selection.** That is, from the population of workers, select a strong person. This alternative can be labeled *fit the person to the job.* The other alternative is **job modification** so that almost everyone can do it. This alternative can be labeled *fit the job to the person.* In general, follow the second alternative, as the key to our improved living standard has been to challenge the environment and make the environment adjust to people rather than people adjust to the environment.

Given the decision to modify the job, one question is how much modification is necessary. That is, how much weight can employees lift? Or in other jobs, how far can they reach? How much space do they need?

1.3 Exclude as Few as Possible

Designers should design to **exclude few** so that as many people as possible can do the job. Not only does this minimize the number of people excluded but also this design strategy tends to make the job easier—benefiting everyone, not just the person who might have been excluded.

The percent of people to exclude depends on the trade-off between the seriousness of exclusion vs. the cost of inclusion. The more serious the exclusion, the smaller the percent that are excluded. For example, if a button is outside the normal reach distance but the operator can reach it by stretching, it is not good but it may still be allowed. However, if a control requires more strength to operate than the operator can exert, that is not allowable.

The cost of including a larger percent of the population may be negligible or it may be high. In most applications a taller door (so people don't hit their heads) adds little expense; however, if the door is in a warship and weakens the structure, the cost may be high. In a fighter airplane, a taller, heavier pilot requires a larger cockpit which increases the cross section, which may incur severe performance penalties. Thus most military organizations have a height restriction on fighter pilots. Or consider the weight of a tote pan to be used in a factory. If the weight is so low that a very high percentage of the population can move it easily, more tote pans/day must be moved. That is, 50 pans each weighing 10 kg are moved instead of 40 pans each weighing 12.5 kg. The engineer will have to balance the benefits of less weight/pan vs. more pans to move.

An excluded **population percentile** can be the upper, lower, or both upper and lower portions of the population. For example, a door might exclude the tallest .1% of the population. For manual dexterity, a test might exclude those with the least dexterity. For a factory job, a firm might exclude those who score low on an intelligence test (who can't learn how to do the job) *and* those who score high (who may become bored and quit).

Note that if the design is for the mean (50th percentile) instead of, for example, the 1st or 99th percentile, then many are eliminated. This is most obvious for reach distances. For example, if a pedal is designed so that the average person can just reach it, then 50% can't reach it.

2 POPULATION VALUES

2.1 Dimensions

Table 10.1 gives some useful physical dimensions of the nude U. S. adult civilian population (Kroemer et al., 1994). The data were gathered in 1988 using U. S. Army personnel (Gordon et al., 1989). (See Chapter 18 for some hand dimensions.)

Marras and Kim's (1993) data on Midwestern U. S. industrial workers (384 males, 124 females) showed that the army data were a good representation of the industrial workforce except for weight and abdominal dimensions. Mean male weight was 182.3 lb (vs. 173.0 for the army); mean female weight was 139.2 lb (vs. 136.7). For adjustment of nude data, they recommend:

- Shoe height adds 1 in. (2.5 cm) for males and .6 in. (1.5 cm) for females.

- Clothing increases torso breadths by .3 in. (.8 cm).

TABLE 10.1 Body dimensions (cm) of nude U. S. adult civilians (Kroemer et al., 1994).

	5th		50th		95th		Standard Deviation	
	Fem.	**Male**	**Fem.**	**Male**	**Fem.**	**Male**	**Fem.**	**Male**
HEIGHTS (Above Floor)								
Stature (height)	152.78	164.69	162.94	175.58	173.73	186.65	6.36	6.68
Eye height	141.52	152.82	151.61	163.39	162.13	174.29	6.25	6.57
Shoulder (acromial) height	124.09	134.16	133.36	144.25	143.20	154.56	5.79	6.20
Elbow height	92.63	99.52	99.79	107.25	107.40	115.28	4.48	4.81
Wrist height	72.79	77.79	79.03	84.65	85.51	91.52	3.86	4.15
Crotch height	70.02	76.44	77.14	83.72	84.58	91.64	4.41	4.62
HEIGHTS (Above Seat)								
Height (sitting)	79.53	85.45	85.20	91.39	91.02	97.16	2.49	3.56
Eye height (sitting)	68.46	73.50	73.87	79.20	79.43	84.80	3.32	3.42
Shoulder (acromial) height (sitting)	50.91	54.85	55.55	59.78	60.36	64.63	2.86	2.96
Elbow height (sitting)	17.57	18.41	22.05	23.06	26.44	27.37	2.68	2.72
Thigh height (sitting)	14.04	14.86	15.89	16.82	18.02	18.99	1.21	1.26
Knee height (sitting)	47.40	51.44	51.54	55.88	56.02	60.57	2.63	2.79
Popliteal height (sitting)[*]	35.13	39.46	38.94	43.41	42.94	47.63	2.37	2.49
DEPTHS								
Forward (thumbtip) reach	67.67	73.92	73.46	80.08	79.67	86.70	3.64	3.92
Buttock-knee distance (sitting)	54.21	56.90	58.89	61.64	63.98	66.74	2.96	2.99
Buttock-popliteal distance (sitting)[**]	44.00	45.81	48.17	50.04	52.77	54.55	2.66	2.66
Elbow-fingertip distance	40.62	44.79	44.29	48.40	48.25	52.42	2.34	2.33
Chest depth	20.86	20.96	23.94	24.32	27.78	28.04	2.11	2.15
BREADTHS								
Forearm-forearm breadth	41.47	47.74	46.85	54.61	52.84	62.06	3.47	4.36
Hip breadth (sitting)	34.25	32.87	38.45	36.68	43.22	41.16	2.72	2.52
HEAD DIMENSIONS								
Head circumference	52.25	54.27	54.62	56.77	57.05	59.35	1.46	1.54
Head breadth	13.66	14.31	14.44	15.17	15.27	16.08	0.49	0.54
Interpupillary breadth	5.66	5.88	6.23	6.47	6.85	7.10	0.36	0.37
FOOT DIMENSIONS								
Foot length	22.44	24.88	24.44	26.97	26.46	29.20	1.22	1.31
Foot breadth	8.16	9.23	8.97	10.06	9.78	10.95	0.49	0.53
Lateral malleolus height	5.23	5.84	6.06	6.71	6.97	7.64	0.53	0.55
HAND DIMENSIONS								
Circumference, metacarpal	17.25	19.85	18.62	21.38	20.03	23.03	0.85	0.97
Hand length	16.50	17.87	18.05	19.38	19.69	21.06	0.97	0.98
Hand breadth, metacarpal	7.34	8.36	7.94	9.04	8.56	9.76	0.38	0.42
Thumb breadth, interphalangeal	1.86	2.19	2.07	2.41	2.29	2.65	0.13	0.14
WEIGHT (Kg)	39.2	57.7	62.01	78.49	84.8	99.3	13.8	12.6

[*]Underside of the thigh.
[**]Rear of the calf.

- Clothing increases torso circumferences by .6 in. (1.5 cm).
- Shoes add 2.0 lb (.9 kg) of weight.
- Clothing (except shoes) adds 1.0 lb (.45 kg) of weight.

Military populations tend to be selected (young, healthy, fit) and so are biased estimators of the civilian population. Americans are clearly, on average, taller than Japanese and Chinese. (See Juergens et al. (1990) for 19 dimensions on 143 populations.) That is, design for an American population may not be appropriate for a non-American population. However, the U. S. population certainly is not homogeneous. In addition to the male-female difference, there are ethnic differences (people of Swedish descent are taller than those of Mexican descent), racial differences (blacks are taller than Asians), occupational differences (farmers are stronger than bookkeepers), and so forth. Naturally, adults differ from children. In addition, older (e.g., over 50) people tend to be weaker than those age 30; after 30, people even begin to shrink in height (due to changes in spinal disc thickness). In addition, clothing will increase many of the dimensions, especially with winter outdoor clothing.

A large part of the variation in human stature is in the length of the legs; the torso is relatively constant in height (White, 1975). There is a desire to predict various body dimensions from a person's height. Unfortunately most body dimensions have coefficients of determination vs. height of less than 50%—that is, the statistical relationship is poor.

Note that because a person is tall does not mean that the person has high intelligence. A short person does not necessarily have low manual dexterity. That is, because a person is average in one characteristic does not mean the person is average in another characteristic—even if all the characteristics are dimensions. For example, a person who is at the 50th percentile in height may be 40% in reach distance and 60% in weight.

Between species, however, there is a general relationship between size and shape—the **principle of similitude.** See Box 10.1 for the principle of similitude.

2.2 Strengths

Strengths for a specific muscle group vary greatly. Expect the coefficient of variation (std. deviation/mean) to be 50% or more. Strengths are greatly affected by the limb (arm vs. leg), by direction exerted, and, for arms, by whether it is the preferred hand. Heeboll-Neilson (1964) reported that the average difference in muscular strength between symmetrical muscle groups was 5–11%.

Table 11.6 gives arm strengths. Tables 18.1 and 18.8 give handgrip strengths and Table 18.2 gives finger strengths. Figures 19.2 and 19.11 give leg strengths. Figure 11.11 shows the effect of age.

Summarizing: (1) The leg is approximately 3 times stronger than the arm. (2) Direction is very important, with arm force at the nonoptimum angles being 50–80% of the force at the optimum angle. (3) The nonpreferred arm averages 60–150% of the strength of the preferred arm, depending on the angle and direction. (4) There seems to be no appreciable difference between the strength of the left and right legs.

2.3 Other Characteristics

Weight and center of mass of various parts of the body can be predicted from Table 15.2. Manual dexterity of the preferred and nonpreferred hand is discussed in Chapter 18.

Body surface area from the DuBois formulas (based on 5 subjects) is:

$$DBSA = .007\ 184\ (HT)^{.725}\ (WT)^{.425}$$

where

$DBSA$ = **DuBois surface area,** m^2
HT = Height, cm
WT = Weight, kg

Mitchell et al. (1971) recommend a corrected formula (based on an improved measurement technique and 16 subjects):

$$SA = .208 + .945\ DBSA\ \text{or}$$
$$= .208 + .006\ 789\ (HT)^{.725}\ (WT)^{.425}$$

where

SA = surface area, m^2
$DBSA$ = DuBois surface area, m^2

Van Graan (1969) reported that the total body area can be apportioned into two arms and hands = 18.1% (hands 5.1% and arms 13.0%), the two legs and feet = 35.9% (legs 29.8%, feet 6.1%), the trunk = 37.5%, and head and neck = 8.5%. Some people remember the division by the "rule of 9s": head and neck = 9%, each hand-arm = 9%, each leg-foot = 18%, and trunk = 36%. Mignano and Konz (1994) reported one clenched fist is 1.6% of body surface area, one hand open with fingers joined is 2.7%, and one hand open with fingers spread is 3.1%.

2.4 Anthropometric Sources

For anthropometric techniques, see Kroemer et al. (1994). For data, see Gordon et al. (1989).

BOX 10.1 *Principle of similitude*

Haldane (1928) called it "On Being the Right Size." The basic concept is that for every mass, there is an optimum shape. A change in size will require a change in shape. It really is the study of the ratio of **surface area/volume.**

Consider a man 2 m tall. If you made him a giant of 20 m, then his weight would increase by 10 (for height) × 10 (for width) × 10 (for thickness) = 1,000. However, his leg bone cross section only increases by 10 × 10 = 100, so with every step, the stress on the leg is 10 times what it would be in a normal human. So when he runs, he breaks his leg! The same concept limits trees to a maximum height of approximately 100 m, although the stressor for trees is the wind.

Consider the giant grasshoppers often found in grade D movies. Grasshoppers breathe through their skins. Giant grasshoppers would increase mass by the cube and oxygen intake by the square and so would die of oxygen starvation.

Kliebers Rule is that the metabolic rate = k (body weight)$^{3/4}$.

Roberts (1975) reported that people living in cold climates tend to be "spherical," which minimizes their surface area-to-volume ratio. However, in the tropics you want to be more like a radiator than a boiler, so long arms and legs are good. Elephants increase their surface area through "fins"— commonly known as ears. Gloves (i.e., separate fingers) lose more heat than mittens, due to their greater surface area/volume.

From a biomechanics viewpoint, big people not only have larger muscles (increasing by the cube) but also a longer moment arm (length of arm or leg), and so they can exert much more force or torque than small people.

For buildings occupied by people, the sphere is not generally used, even though it minimizes surface area in relation to volume. But the sphere is popular for storage of liquids and gases because it minimizes material cost and energy exchange with the environment. A popular compromise is a cylinder (e.g., for storage of pressurized gas). A dome minimizes material use for volume enclosed and strongly resists external pressure (although weaker for internal pressure); thus, it finds applications in military bunkers and igloos.

The cube is an efficient enclosure for cartons and boxes.

The shape used for industrial buildings is the square or rectangle with a wall height of about 15 feet (low bay) or about 30 feet (high bay). For heating and ventilating ducts, for the same air flow, round ducts have less perimeter than rectangular ducts; this results in less friction (i.e., smaller fans and thus lower energy costs) and heat transfer to the environment.

See Konz (1994) for a more extensive discussion of surface area/volume and of perimeter/area.

3 STATISTICAL CALCULATIONS

Probably the most useful single descriptor is the population mean. Although many anthropometric characteristics are not precisely fitted by a normal distribution, the normal distribution will give answers that are close enough, and so the normal distribution is used. Assume that the normal distribution allows us to further state that the mean is equivalent to the 50th percentile. The normal distribution has the further useful characteristic of being symmetrical.

The absolute variability of a population is given by the standard deviation. The relative variability of a population is given by the coefficient of variation, which is the standard deviation/mean. For example, in Table 10.1 the mean height of females is 162.94 cm and the standard deviation is 6.36 cm. The coefficient of variation is 6.36/162.94 = 3.9%.

See Figure 10.1 for how to sketch the normal distribution of a characteristic.

Assume you wished to calculate your percentile height. If you are male and have a height of 170 cm, then you are 175.58 minus 170.0 = 5.58 cm below the mean. Next you will have to convert from cm to standard units. The conversion is 1 standard deviation = 6.68 cm so you are 5.58/6.68 = .84 standard units below average in height. From a table of the normal distribution, this is the 20th percentile. That is, 20 percent of the population is shorter than you and 80 percent of the population is taller than you.

Assume you wished to calculate the 99% of the male population (that is, you are going to exclude the largest 1% of the population). Then from a normal table, 99% is 2.33 standard deviations above the mean. Converting from standard units to cm makes the distance above the mean equal to 2.33 (6.68) = 15.6 cm. Adding this to 175.58 gives 191.2 cm. If you wished to calculate the 1st percentile, subtract 15.6 from 175.58 to get 159.98 cm.

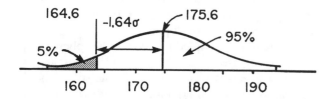

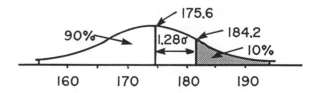

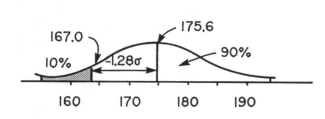

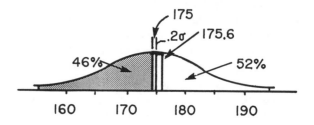

FIGURE 10.1   Sketch normal distribution of male stature height. From Table 10.1, the mean is 176.6 cm with std. deviation = 6.7 cm. Using the normal curve (such as Table 20.5):

(1) Put the mean at 175.6.

(2) Draw the curve concave downward between the mean and +/−1 standard deviation; that is, 175.6 − 6.7 = 168.9 and 175.6 + 6.7 = 182.3. These are the inflection points of the curve.

(3) Draw the curve upward outside this range.

(4) Have the curve approach the axis at the mean +/−3 standard deviations; that is, at 175.6 − 20.1 = 155.5 and 175.6 + 20.1 = 195.7.

For the vertical scale, if the height of the mean is 1.0, then the height of the curve at +/−1 standard deviation = .58, at +/−2 standard deviations = .12 and at +/−3 standard deviations = .01.

The 5th percentile height is 175.6 − 1.64 (6.7) = 164.6. The 10th percentile height is 175.6 − 1.28 (6.7) = 167.0. The 90th percentile height is 175.6 + 1.28 (6.7) = 182.2.

If you wish to determine what percentile is below 175 cm, this is 175 − 175.6 = −0.6 cm/6.7 cm/std. dev. = .09 std. dev. below the mean. From a table of the normal distribution, 46% of the U. S. male population is below 175 cm.

REVIEW QUESTIONS

1. If you design a footpedal for the distance an average person can reach, what percent of the population will not be able to reach the pedal?

2. Contrast the two opposing strategies of selection and job design.

3. Discuss what percentage of the employees you would exclude from a manual material handling job.

4. If King Kong (the giant gorilla) had really lived, why would he have had to be hollow?

5. Design a tote pan which 95% of the employees can lift. Assume the mean is 20 kg and one standard deviation is 5 kg.

REFERENCES

Gordon, C., Churchill, T., Clauser, C., Bradtmiller, B., McConville, J., Tebbets, I., and Walker, R. *1988 Anthropometric Survey of U. S. Army Personnel.* Natick, Mass.: U. S. Army Natick Research, Development and Engineering Center, 1989.

Haldane, J. *Possible Worlds.* Harper, 1928. Also reprinted in Newman, J., (ed.), *The World of Mathematics*, 952–57. New York: Simon and Schuster, 1956.

Heeboll-Neilsen, K. Muscular asymmetry in normal young men. *Communication 18*, Danish National Association for Infantile Paralysis, Copenhagen, Denmark, 1964.

Juergens, H., Aune, I., and Pieper, U. *International Data on Anthropometry.* Geneva: Int. Labour Organization, 1990.

Konz, S. *Facility Design: Manufacturing Engineering*, 2nd ed. Scottsdale, Ariz.: Publishing Horizons, 106–08, 1994.

Kroemer, K., Kroemer, H., and Kroemer-Elbert, K. *Ergonomics.* Englewood Cliffs, N. J.: Prentice-Hall, 1994.

Marras, W. and Kim, J. Anthropometry of industrial populations. *Ergonomics*, Vol. 36 [4], 371–78, 1993.

Mignano, B. and Konz, S. Surface area of the human hand. *Advances in Industrial Ergonomics and Safety VI.* London: Taylor and Francis, 1994.

Mitchell, D., Strydom, N., van Graan, C., and van der Walt, W. Human surface area: Comparison of the DuBois formula with direct photometric measurement. *Pflugers Archieves*, Vol. 325, 188–90, 1971.

Roberts, D. Population differences in dimensions. In *Ethnic Variables in Human Factors Engineering*, Chapanis, A. (ed.), Chapter 2. Baltimore, Md.: Johns Hopkins University Press, 1975.

van Graan, C. The determination of body surface area. *South African Medical J.*, Vol. 3, 952–59, August 1969.

White, R. Anthropometric measurements on selected populations of the world. In *Ethnic Variables in Human Factors Engineering*, Chapanis, A. (ed.), Chapter 3. Baltimore, Md.: Johns Hopkins University Press, 1975.

11 WORK PHYSIOLOGY AND BIOMECHANICS

IN PART 3 SCIENTIFIC BACKGROUND

OVERVIEW

First metabolism and the cardiovascular system are discussed. The responses of the system (heart rate, stroke volume, artery–vein differential, blood distribution) indicate the amount of strain on the body. The body also can go into oxygen debt. Gender, age, and training affect the responses. The skeletomuscular system also is affected by gender, age, and training.

CHAPTER CONTENTS

1 Metabolism
2 Cardiovascular System
3 Skeletomuscular System

KEY CONCEPTS

acclimatization
activity metabolism
aerobic vs. anaerobic
age effects
arterial–venous oxygen
    differential
basal metabolism
blood pressure
Borg vote
cardiovascular system
digestive metabolism

electromyogram (EMG)
gender effects
heart rate cost of work
heart rate measurement
isometric
job rotation
maximal voluntary contraction
    (MVC)
maximum oxygen uptake
mental load
oxygen debt

perceived exertion
percent fat
proportion of capability
pulmonary/systemic
    circulation
respiratory quotient
responses to exercise
skeletomuscular levers
stroke volume

1 METABOLISM

Metabolic rate can be estimated from tables and formulas and determined from measurements.

1.1 Tables and Formulas

Metabolic rate is divided into three parts: basal, activity, and digestion.

$$TOTMET = BSLMET + ACTMET + DIGMET$$

where

TOTMET = Total metabolic rate

BSLMET = Basal metabolic rate

ACTMET = Activity metabolic rate

DIGMET = Digestion metabolic rate

      (also called specific dynamic action)

Metabolic rate can be expressed in many different units (see Table 11.1). The standard power unit is watts, abbreviated W. The energy content of food is given in kcal (commonly spoken of as calories rather than the technically correct kilocalories).

1.1.1 Basal

Basal metabolism maintains body temperature, body functions, and blood circulation. Basal metabolic rate is determined as follows:

$$BSLMET = BSMET (WT)$$

where

BSLMET = Basal metabolic rate, W

  BSMET = 1.28 W/kg for males

         = 1.16 W/kg for females

     WT = Body weight, kg

The reason BSMET is lower for females is that females tend to have a higher percent of fat than males and fat has a limited metabolism. See Box 11.1.

A more complex formula also considers the effect of a person's age from 5 to 70 (Konz, 1984).

TABLE 11.1 Energy conversions.

Multiply Units by 1	To Equal Units	Multiply Units by 1	To Equal Units
Joule	1.0 Nm		
Joule	238.9 kcal		
Watt	1.0 J/s		
Watt	.85885 kcal/h	kcal/h	1.163 W
Watt	3.413 BTU/h	BTU/h	.293 W
Watt	.00134 HP	HP	746.0 W
Watt	6.116 (kg-m)/min	(kg-m)/min	.1635 W
Watt/sq m	.01718 mets	met	58.2 W/sq m
Watt/sq m	10.764 W/sq ft	W/sq ft	.0929 W/sq m
Kcal	3.968 BTU	BTU	.252 kcal
		BTU	776.65 ft-lb

Children have a higher basal metabolic rate/kg because they have a higher surface area/volume (and thus require more heat to maintain body temperature) and because growth takes energy.

1.1.2 Activity

Activity metabolism provides the energy for activities. See Table 11.2 for some examples. Working in a contorted posture (kneeling, bent over) increases metabolic rate (Freivalds and Bise, 1991). See Box 11.2 for data on walking. For lifting, Garg (1976) gave:

$$LFTMET = .0109 \ BWLB + F \ (LOADF)$$

where

LFTMET = Lifting metabolism, kcal/min

  BWLB = Body weight, lb

     F = Frequency of lift, lifts/min

LOADF = Load factor

       = LFBW (BWLB) + LF (W) + GENF (W)

  LFBW = Lift factor:body weight = .0002 for arm lift, .0012 for stoop, and .0019 for squat

    LF = Lift factor:load = .0103 for arm lift, .0052 for stoop, and .0081 for squat

 GENF = Gender factor = − .0017 G for arm lift, .0028 G for stoop, and .0023 G for squat

     G = Gender (female = 0; male = 1)

When estimating metabolic rate from tables or formulas, remember that most people do not work continuously. They take microbreaks. The duration of each task element can be estimated from videotape or occurrence sampling.

1.1.3 Digestion

Digestion and transformation of food within the body (**digestive metabolism**) also is called specific dynamic action. For every gram of carbohydrate burned, we get 4 kcal of energy; for every gram of fat, 9; for every gram of protein, 4.3. For the typical carbohydrate/fat/protein mixture of the U. S. diet and disregarding time since the meal:

$$DIGMET = .1 \ (BSLMET + ACTMET)$$

If you wish to make DIGMET constant in a laboratory experiment, have the subject standardize what is eaten at meals and control the time between the meal and the experiment.

1.1.4 Calories required

Assume a 70 kg male spent 24 h in bed without eating. Basal metabolism would be 70 (1.28) = 89.6 W. To convert this to kcal/h, multiply by .86. Thus, this male would require 89.6 W (.86 kcal/W-h) (24 h) = 1,849 kcal/day. (A kcal is what we normally think of as a calorie, as in a piece of bread has 100 calories.)

BOX 11.1   *Percent fat*

Percent fat can be determined three ways: (1) by underwater weighing, (2) by using known body characteristics and a formula, and (3) by measuring unknown body characteristics and using a formula. Underwater weighing, although the most accurate, requires special facilities.

Known characteristics are height and weight. Cowgill (1957) gave a formula for density for *adult males;* Cowgill said his formula was as accurate as those using skinfold measurements due to errors in skinfold measurements made by novices:

$DENSTY = .161 + .8 (HT)^{.242}/(1,000\ WT)^{.1}$
where
$DENSTY$ = Density of the human body, g/mL
HT = Height, cm
WT = Weight, kg

If the body is divided into two "compartments" with a mean *DENSTY* of .9 for fat and 1.1 for the lean body (rest of the body), the typical female will have a *DENSTY* of 1.03 and male of 1.06; the population range should be 1.02 to 1.08.

The next step is to calculate percent fat (Brozek et al., 1963):

$PERFAT = ((4.570/DENSTY) - 4.142)\ 100$
where
$PERFAT$ = Percent of the body that is fat

Lohman (1981) also comments on skinfold measurement problems.

For the unknown characteristics, Wilmore and Behnke (1969) recommend, for adult males, using:

$PERFAT = 74\ WACIRC/WT - 4464/WT - 8.2$
where
$WACIRC$ = Waist circumference, cm

Zuti and Golding (1973), for physically active adult males, recommend:

$PERFAT = 8.707 + .489\ WACIRC + .449\ PCSKIN - 6.359\ RWDMAX$
where
$PCSKIN$ = Pectoral (front, just below armpit) skinfold, mm
$RWDMAX$ = Right wrist diameter (max), cm

Sloan et al. (1962), for adult females, recommend:

$DENSTY = 1.076 - .000\ 880\ ARSKIN - .000\ 810\ ICSKIN$
where
$ARSKIN$ = Arm (back) skinfold, mm
$ICSKIN$ = Iliac crest (top of hip bone) skinfold, mm

The U. S. Army (1987) has developed a technique to predict percent fat for males and females.

For males:

$PERFAT = -8.4 - 3.75\ (A - N) - .045\ (A - N)^2 + .417\ HT$
where
A = Abdomen (belly button) circumference, in.
N = Neck (below Adam's apple) circumference, in.
HT = Height (barefoot), in.

For females the calculations are more complex as six measurements are needed:

WT = Weight (clothed, no shoes), lb
HT = Height (barefoot), in.
HP = Hip circumference (max) (where buttocks protrude the most), in.
F = Forearm circumference (max), in.
N = Neck (below Adam's apple) circumference, in.
WR = Wrist circumference, in.

These are entered into the equation:

$PERFAT = WTF + HPF - HTF - FF - NF - WRF$
where
WTF = Weight factor = $86.1 + .633\ WT - .001\ (WT)^2$
HPF = Hip factor = $8.84 + .0054\ HP$
HTF = Height factor = $.02 + 1.31\ HT$
FF = Forearm factor = $18.4 + .196\ F$
NF = Neck factor = $1.35\ N$
WRF = Wrist factor = $2.34 + .025\ WR$

There are a number of body mass indices which combine weight (*W*, kg) and height (*H*, m). Two examples are Quetelet's index (W/H^2) and the ponderal index ($W^{.33}/H$); Smalley et al. (1990) indicate that the indices have poor predictability of percent fat.

For those people who want to lose weight, the secret is not to eat fat. When people eat extra carbohydrates (sugar or starch), the body tends to burn most of it. But the body burns extra fat sparingly and instead saves it away. Alcohol is especially a problem as it decreases lipid (fat) oxidation about 33% (Suter et al., 1992).

The amount of fat increases with age. Garn and Harper (1955) report that lean body weight of adult males changes little between ages 20 and 60. However, fat increases from 11.5 kg (15.7% of body weight) at age 20–30 to 17.3 kg (22.4% of body weight) at age 50–60.

About 3% of body weight is "essential" fat (cell membranes, nerve tissue, tissue in and around various organs). Women have sex-specific fat (primarily breasts and hips) of 9 to 12%. Thus, anything over 3% for men and 15% for women is storage fat.

TABLE 11.2 Activity cost for various activities. For total energy cost, add the basal metabolism and, if appropriate, the cost of digestion. See Box 11.2 for equations for walking, carrying, and running.

W/kg	Activity
.4	Crocheting, eating, reading aloud, sewing by machine, sitting quietly, writing
.6	Playing cards, standing relaxed, typing with electric typewriter
.7	Paring potatoes, standing office work, sewing with foot-driven machine, standing at attention, violin playing
.8	Dressing and undressing, knitting a sweater
.9	Singing in a loud voice
1.0	Driving a car, tailoring
1.2	Dishwashing, typing rapidly
1.4	Washing floors
1.5	Cello playing, light laundry
1.6	Riding a walking horse, sweeping bare floor with broom
1.7	Golfing, organ playing, painting furniture with brush
1.9	Sweeping with hand (push) carpet sweeper
2.7	Doing heavy carpentry
3.0	Cleaning windows
3.1	Cleaning with upright vacuum cleaner
3.5	Dancing the waltz
4.1	Ice skating
4.5	Weeding a garden
5.0	Horseback riding (trot)
5.1	Playing ping pong
5.8	Dancing rhumba, playing tennis
6.6	Sawing wood (hand saw)
7.9	Playing football
8.9	Fencing

Assuming activity metabolism was .7 W/kg for the 16 waking hours, .7 (70) (.86) (16) = 674 additional calories would be needed: a total of 2,523 kcal. Assuming 10% for digestion, he needs 1.1 (2,523) = 2,773 kcal/day.

When people eat more than they burn, they will gain weight. Diets emphasize reducing eating. But more exercise is another alternative. There are 3,500 kcal/pound, 7,700 kcal/kg. If you reduce burning by just 20 kcal/day, that is 7,300 kcal/year or 2.1 lb/year. In 10 years, you have gained 21 lbs! The point here is that small changes in activity level (as well as in diet) can have large cumulative effects.

As a side note, in 1932 Klieber (1961) determined the following formula for a variety of animals (dove, rat, pigeon, hen, dog, sheep, human, cow, and steer):

$$DAYMET = 70 \; WT^{3/4}$$

where

$DAYMET$ = Metabolism during a day, kcal
WT = Body weight, kg

The exponent could have been predicted to be 2/3 since surface area increases by 2 and volume increases by 3. Peters (1989) says the difference is due to the differing shape of the animals as the size increases.

1.2 Measured Metabolism   In a laboratory situation where accurate determination of metabolic rate is important, it may be measured indirectly by measuring the oxygen consumption (indirect calorimetry). It is akin to measuring the fuel consumption of your car by measuring the oxygen intake through the carburetor. While the person has a clip on the nose to prevent nose breathing, a tube (connected to an oxygen supply) is inserted in the mouth. The body's conversion efficiency of fuel to energy is estimated. A recent development is instantaneously reacting sensors placed in the air flow of the exhaust air, eliminating the need for the noseclip.

$$TOTMET = 60 \; ENERGY \; (OXUPTK)$$

where

$TOTMET$ = Total metabolism, W
$ENERGY$ = Energy equivalent of 1 liter of oxygen, W-h/liter
This depends upon the **respiratory quotient** (RQ), which in turn depends upon the proportion of fat vs. carbohydrate metabolized during the exercise.
= 5.36 for RQ = .83 (rest)
= 5.66 for RQ = .86 (exercise up to 60% of maximum rate)
= 6.40 for RQ = 1.0 (100% of max oxygen uptake)
$OXUPTK$ = Oxygen uptake, VO_2, liters of oxygen/min

The metabolic rate of a laboratory task may be controlled to be constant by having the subject walk on a treadmill, pedal a bicycle ergonometer, or walk up and down steps to a metronome beat.

A related measurement is pulmonary ventilation (see Box 11.3).

2 CARDIOVASCULAR SYSTEM

This section will describe the anatomy of the system, the responses to exercise, and the limits.

2.1 Anatomy   Figure 11.1 shows an "engineer's view" of the cardiovascular system. The heart has two pumps. The right side pumps to the pulmonary circulation and the left side to the systemic circulation. The blood has transport and storage functions.

BOX 11.2 *Walking*

Pandolf et al. (1976) give walking metabolism (without a load) as a function of terrain and velocity.

$$TOTMET = C (2.7 + 3.2 (v - .7)^{1.65})$$

where

WLKMET = Walking metabolism (total), W/kg of body weight

C = Terrain coefficient

= 1.0 for treadmill, blacktop road

= 1.1 for dirt road

= 1.2 for light brush

= 1.3 for hard-packed snow: C = 1.3 + .082 (foot depression, cm)

= 1.5 for heavy brush

= 1.8 for swamp

= 2.1 for sand

v = Velocity, m/s (for $v > .7$ m/s (2.5 km/h, 1.56 miles/h)

A person walking on a treadmill in still air swings the arms and legs, and effective air movement is therefore about .9 m/s (Nishi and Gagge, 1970)

Walking stooped takes more energy (Morrissey et al., 1981): 12% more for a 10% stoop (90% of stature height), 51% more for a 20% stoop, and 91% more for a 30% stoop.

The metabolic cost of walking a specific distance is minimized when $v = 1.13$ m/s (4.05 km/h; 2.5 miles/h). For many time studies, the concept of normal (100%) pace is walking 3 miles/h (4.84 km/h; 1.39 m/s).

The pace of Methods-Time Measurement (MTM) is 3.57 miles/h (1.65 m/s) at a stride of 34 inches (.86 m); this also can be expressed as .000 015 h/stride; it is based on studies of 125 people walking a variety of paces (Maynard et al., 1948). MTM uses .000 015 h/stride for climbing stairs. For carrying a load of up to 16 kg, MTM uses .000 015 h/stride but .76 m strides. For walking through obstructed areas, MTM uses .86 m strides but .000 017 h/stride.

Work-Factor uses a pace of 3.7 miles/h (1.71 m/s) at a stride of 30 inches (.76 m). Time for normal walking (h) = .000 204 + .000 17 (number of strides).

The length of stride (L) divided by stature height (h) varies linearly with velocity; $L/h = .67$ at $v = .8$ m/s and $L/h = .9$ at $v = 1.7$ m/s (Alexander, 1984). Thus if a shorter person is walking at the same velocity as a taller person, the shorter person will take more steps.

Pandolf et al. (1977) give the following equation for standing or walking very slowly:

$$WLKMET = 1.5\ WT + 2\ (WT + WTL)\ (WTL/WT)^2 + C\ (WT + WTL)\ (1.5\ v^2 + .35\ VG)$$

where

WLKMET = Metabolic rate (total) for walking slowly, W

WT = Body weight, kg

WTL = Weight of a load on the shoulders, kg

v = Velocity (speed) of walking, m/s ($v < 1$ m/s)

C = Coefficient for terrain (see above)

G = Grade, %

The first term (1.5 WT) is the metabolic cost of standing without a load and is 1.5 W/kg. The second term is the cost of load bearing while standing. Walking on the level is $C\ (WT + WTL)\ (1.5\ v^2)$. The cost for climbing a grade is $C(WT + WTL)\ (.35\ VG)$.

The metabolic cost of carrying (walking with a load) depends on the load location. Soule and Goldman (1980) reported that loads on the head used 1.2 times the energy of carrying a kg of your own body; in the hands, loads required 1.4 to 1.9 times as much; on the feet, loads required 4.2 to 6.3 times as much. Legg and Mahanty (1986) found it took 6.4 times as much energy to carry a kg on the feet as the back.

As a general guideline for carrying, minimize the load's moment arm—both in the frontal and transverse axes.

The cost of running (Van der Walt and Wyndham, 1973) is:

$$RUNMET = -142/WT + 11 + .04\ v^2$$

where

RUNMET = Running metabolism (total), W/kg

v = Velocity, km/h

WT = Weight, kg

The pulmonary circulation starts at the right atrium ("little room" in Latin) and goes through a valve to the right ventricle (the pump itself). (Systole is the contraction of the heart muscle (the aortic valve is open); diastole is relaxation of the heart muscle (the aortic valve is closed). When heart rate

BOX 11.3   *Pulmonary ventilation*

$PULVNT = (LAPLOX)(OXUPTK)$

where

$PULVNT$ = Pulmonary ventilation, liters of air/min

$LAPLOX$ = Liters of air/liter of oxygen

= 20-25 at rest and for work less than 15 W/min

= 30-35 during maximal work

Respiratory frequency is between 10 and 20/min at rest. Most adults have a maximum of 40-45, but some athletes can go to 60 (Astrand and Rodahl, 1986).

Vital capacity, an index of an individual's lung capacity, is the maximum that can be exhaled following a maximum inspiration. Shephard (1972) says it can be estimated (with a standard deviation of 10%) from:

$VITALC$ (males) $= 56.3$ $(HT) - 17.4$ $AGE - 4,210$

$VITALC$ (females) $= 54.5$ $(HT) - 10.5$ $AGE - 5,120$

where

$VITALC$ = Vital capacity (standing), ml

HT = Height, cm

AGE = Age, years

Vital capacity is reduced 5 to 10% by a shift from standing to lying down.

Transfer of gases within the lung can be expressed by the following equation (Shephard, 1972):

$Resistance = 1/COND = 1/ALVENT + 1/(SOLFAC)(CO)$

where

$COND$ = Conductance = U, liters of gas/min

$ALVENT$ = Alveolar ventilation, V_2, liters of gas/min

$SOLFAC$ = Solubility factor, liters of gas/liter of blood

CO = Cardiac output, liters of blood/min

For oxygen, maximum $ALVENT$ is about 80, $SOLFAC$ is about 1, and maximum CO is about 25. Thus:

Oxygen resistance $= 1/COND = 1/80 + 1/(1 \times 25)$
$= .0125 + .0400 =$
$.0525$ min/liter of O_2

That is, oxygen input to the blood depends more on cardiac output (the .0400) than on alveolar ventilation (the .0125).

For carbon dioxide, maximum $ALVENT$ still is about 80, maximum CO is 25, but $SOLFAC$ is 5. Thus:

CO_2 resistance $= 1/COND = 1/80 + 1/(5 \times 25)$
$= .0125 + .0080 = .0205$ min/liter of CO_2

That is, elimination of carbon dioxide from the blood depends more on alveolar ventilation (the .0125) than on cardiac output (the .0080).

---

increases, it is by reduction of the diastolic time.) When the blood returns from the lungs (with more oxygen and less carbon dioxide) it enters the systemic circulation.

After passing through the left atrium and ventricle, the blood enters the arteries. Oxygen is removed and carbon dioxide added. In addition, the blood picks up new fuel at the intestines, which may be removed at the muscles (if they are working). When the muscles work, they add waste products to the blood, which then transports the waste to the liver (biotransformation), the kidney (elimination of water-soluble compounds in the urine), and intestine (elimination in feces).

2.2 Response to Exercise

The cardiovascular system has five **responses to exercise:** (1) heart rate, (2) stroke volume, (3) artery-vein differential, (4) blood distribution, and (5) going into debt.

2.2.1 Heart rate

Heart rate (with some exceptions) is a good estimator of metabolic rate. The exceptions are as follows: (1) Heart rate is increased by emotions (the effect is relatively larger when the person is sedentary—i.e., heart rate due to exercise is low). (2) Heart rate is increased by vasodilation (unacclimatized people may have heat stress, whereas acclimatized people sweat instead of vasodilating). (3) Heart rate may not increase as much as predicted at very heavy metabolic rates because other cardiovascular system responses may be increased also.

Nonetheless, for light and medium work loads, heart rate is a good predictor of metabolic rate. A predictor equation (Andrews, 1969) is:

$INCHR = K + .12$ $INCMET$

where

$INCHR$ = Increase in heart rate, beats/min

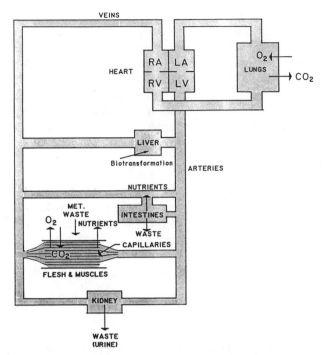

FIGURE 11.1 Engineer's view of the **cardiovascular system** shows a pulmonary circuit and a systemic circuit. The heart has two pumps.

In **pulmonary circulation**, the right ventricle sends blood to the lungs, where O_2 is added and CO_2 is removed. The blood returns to the left atrium.

In **systemic circulation**, the left ventricle sends blood out into the arteries. After passing through various organs and the capillaries, it returns in the veins. In the "marsh" of the capillaries, the blood gives up oxygen and nutrients to the flesh and obtains CO_2 and metabolic wastes. Nutrients are added to the blood from the intestines. Wastes are biotransformed in the liver and reinserted into the blood; the modified wastes are removed by the kidney (urine) and intestines (feces).

K = Constant

  = 2.3 for arm work (cranking)

  = −11.5 for leg work (walking) or arm + leg work (The 13.8 difference in the coefficient for the legs is due to venous pooling in the legs.)

$INCMET$ = Increase in metabolism, W

There are four ways to obtain a **heart rate measurement:**

1. *Light* can be used by shining a light on an artery in the earlobe. A photocell on the far side of the earlobe sees the light during "ebb tide" between beats. Another model works by reflected light. It is placed on the finger. Both earlobe and finger models are sensitive to body movement and so work best with a stationary person.

2. *Sound* is used by physicians with a stethoscope. (Women instinctively comfort their babies by carrying them on their left arm so the baby's head lies against the left chest and the baby can easily hear the mother's heart beating.)

3. *Palpation* is the detection, with the fingers, of the surge of blood that follows each beat. Common locations are the arteries in the wrist and neck. Count the number of beats in 10 or 15 s and multiply by 6 or 4. In working situations, this is done after the work stops—that is, measure recovery pulse.

4. *Electronics* is the most common and the only practical technique for a nonstationary person. Usually electrodes are attached to the chest and the EKG waveform is recorded.

   - To minimize muscle noise, place the electrodes above and below the heart in a triangle (sternum and outside each nipple about 5 cm). If the signal will be sent by radio, the third (ground) electrode is not needed. To improve the signal, wipe the skin with a solvent, add electrode paste, and avoid hairs.

   - Typically, the recording unit is about the size of a pack of cigarettes and is belt-mounted. Depending on the unit, the output can be displayed on the unit, can be stored on a small tape recorder, or can be sent by radio or by "hard wire."

Heart rate also can be estimated subjectively by asking for an individual's **perceived exertion.** See Table 11.3. The concept has been validated by many experimenters for many tasks. Ljunggren (1986) has reported it is not even necessary to have the individual's own perceived exertion—this exertion can be estimated by an observer.

TABLE 11.3 Borg's (1990) new Rating of Perceived Exertion (RPE) scale can be used to predict heart rate (**Borg vote**). Use the words for guidance but then vote with a number. Heart rate = 10 (Vote).

Vote	Subjective Description
6	—No exertion at all
7	—Extremely light
8	
9	—Very light
10	
11	—Light
12	
13	—Somewhat hard
14	
15	—Hard
16	
17	—Very hard
18	
19	—Extremely hard
20	—Maximal exertion

Source: G. Borg. "Psychophysical Scaling with Applications in Physical Work and the Perception of Exertion," *Scand. J. Work Environ. Health*, Vol. 16, Suppl. 1, pp. 55–58. Copyright © 1990 by Scand. J. Work Environ. Health, Helsinki, Finland. Reprinted with permission.

The standard prediction of maximum heart rate is:

$$HRMAX = 220 - AGE$$

where

$HRMAX$ = Maximum heart rate, beats/min

AGE = Age, years

Arstilla (1972) gives an alternate formula:

$$HRMAX = 200 - .66\,AGE$$

The standard deviation of the prediction is 10 beats/min. Thus, an average individual at age 30 would have a maximum heart rate of 190 (using the first formula) or 180 (using the second). Maximum heart rate is relatively unaffected by physical fitness. When exercising for cardiovascular fitness, initially exercise at 60 to 70% of $HRMAX$. An experienced exerciser can aim for 75 to 85% of $HRMAX$.

Figure 11.2 shows how the heart rate (primarily determined by aerobic oxygen supply) responds to constant-intensity exercise. **Heart rate cost of work** can be determined in three ways. The simplest is to subtract an individual's basal heart rate from the peak. For example, Joe's peak of 110 and basal of 70 give a task cost for him of 40 beats/min. Determining the basal is more complicated than it might seem. People subconsciously increase their heart rates just before exercising (Kozar, 1964). To get a good estimate of basal, measure it 5 or 10 minutes before exercise or after the work when heart rate has returned to basal—perhaps 5 min after light work and 10 to 15 min after moderate or heavy work. A problem with this first method is that it assumes that the peak represents the heart rate during the work—that the top of the curve is flat.

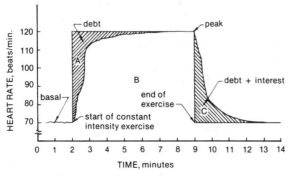

FIGURE 11.2   Aerobic response (and thus heart rate) lags the exercise onset. The deficit (area A) is replaced by anaerobic oxygen. **Aerobic** refers to reactions using oxygen from the lungs; **anaerobic** (without oxygen) reactions use oxygen stored as compounds in the blood. The anaerobic supply in the blood is composed of alactate, with an energy equivalent of about 1.9 liters of oxygen, and lactate, with an equivalent of about 3.1 liters. (Shephard, 1972, p. 398). During recovery (area C), the anaerobic oxygen in the blood is replaced but the replacement process itself uses oxygen ("interest") so area C is larger than area A. For task cost, use area B + C.

A second way is to consider the area under the curve (area B in Figure 11.2). However this assumes the heart rate instantaneously returns to basal. The best way is to consider work cost as area B plus C. For accurate work, the area under the curve should be calculated with a planimeter, but for most purposes the area can be estimated from a series of rectangles.

2.2.2 Stroke volume

The second method of adjusting oxygen supply to the body is **stroke volume**, the amount of blood pumped by the left ventricle, which can be predicted as:

$$SV = STROVB + .000\,050\,(TOTMET - 200)$$
$$TOTMET < 500$$

where

SV = Stroke volume, liters/beat

$STROVB$ = Basal male stroke volume, liters/beat (females = .9 male)

= $SI\,(DBSA)/1{,}000$

where

SI = Stroke index, mL/(beat-m²)

= $53.45 - .194\,AGE$

AGE = Age, years

$DBSA$ = DuBois surface area, m²

= $.007\,184\,HT^{.725}\,WT^{.425}$

HT = Height, cm

WT = Weight, kg

$TOTMET$ = Total metabolism, W. For $TOTMET$ over 500 W, add $.000\,025\,(TOTMET - 500)$ to the SV equations.

Stroke volume depends on body posture, exercise, and physical fitness. When lying down, stroke volume may be .12 liters/beat but for sitting and standing, a value of .08 is more typical. Exercise with the legs improves venous return; in fact, stroke volume may increase to about .40 liter. Exercise with the arms only tends to permit venous pooling in the legs and stroke volume changes little.

Maximum stroke volume is primarily a function of physical fitness. Maximum SV = .135 for excellent cardiovascular fitness, .120 for good, .100 for fair, .090 for poor, and .085 for very poor. See Table 11.4 for maximum oxygen uptake levels for these fitness levels. World-class athletes may be as high as .200 liters/beat, giving them a resting heart rate in the 50s. Stroke volume peaks at about 40% of maximum oxygen consumption.

Box 11.4 discusses blood pressure, and Box 11.5 discusses cardiac output.

TABLE 11.4 Maximum oxygen consumption VO_2MAX, mL/(kg-min) for U. S. males with various ages and degrees of cardiovascular fitness (Cooper, 1970).

CARDIO-VASCULAR FITNESS	AGE, YEARS			
	Under 30	30–39	40–49	50+
Very poor	<25.0	<25.0	<25.0	—
Poor	25.0–33.7	25.0–30.1	25.0–26.4	25.0
Fair	33.8–42.5	30.2–39.1	26.5–35.4	25.0–33.7
Good	42.6–51.5	39.2–48.0	35.5–45.5	33.8–43.0
Excellent	51.6+	48.1+	45.1+	43.1+

Source: The New Aerobics by Kenneth H. Cooper. Copyright © 1970 by Kenneth H. Cooper. Used by permission of Bantam Books, a division of Bantam Doubleday Dell Publishing Group, Inc.

2.2.3 A–V differential

The **arterial–venous oxygen differential** is the third method of adjusting oxygen supply to the body:

$$OXUPTK = CO \; (AVDIF)$$

where

$OXUPTK$ = Oxygen uptake, VO_2 at standard temperature (0^0 C) and pressure (760 torr) dry *(STPD)*, liters of oxygen/min

CO = Cardiac output, liters of blood/min

$AVDIF$ = Arterial-venous oxygen differential, liters of oxygen/liter of blood

While resting, the arterial oxygen content is 19 mL/100 mL of blood while the venous oxygen content is 15 mL/100 mL. That is, for every 100 mL of blood passing the muscles, the muscles get 4 mL of oxygen. However, in an emergency (fleeing from a tiger), the muscles can get 13 mL/100 mL of blood; that is, the veins drop to 6 mL. In highly trained athletes, the *AVDIF* can be 16. The coronary blood supply, even under normal circumstances, has an *AVDIF* of 17, and therefore, more oxygen for the heart must come from more blood, not an increase in the *AVDIF.*

2.2.4 Blood distribution

The fourth way of getting more oxygen to a muscle is through redistribution. During exercise, the capillary density increases from 200/mm² at rest to 600. Muscle blood flow can change from 2 mL/100 mL of tissue to 14. Blood stored in the lungs can change from 500 mL to 1,500. As exercise increases, the kidneys and intestines gradually use less blood and send their blood to the skin and muscles. If food is present in the stomach, cramps may result.

2.2.5 Debt

If you are "underdeposited," you go into debt. If the aerobic supply of oxygen (from the lungs) is not sufficient, the muscles then can draw upon the anaerobic oxygen stored in the blood **(oxygen debt).** However, the anaerobic supply is limited and it must be repaid—with interest. See Figure 11.2.

BOX 11.4 *Blood pressure*

Blood pressure (the pressure blood puts on the blood vessel walls) commonly is measured with a sphygmomanometer. The cuff is applied to the upper arm and while it is being tightened, one listens to the sound of the blood flowing in the artery below the cuff. After the cuff is tight, there is no sound because no blood is flowing in the artery. As the cuff is loosened, blood begins to flow turbulently and a noise is heard. This first sound is called systolic blood pressure. As the cuff is loosened further, the turbulence decreases until the flow becomes laminar and the sound disappears. The disappearance of the sound indicates diastolic blood pressure, the minimum or base pressure exerted by the blood on the artery walls.

Pulse pressure = Systolic pressure − diastolic pressure

Blood pressure can be estimated (Roozbahar et al., 1979) as:

$$SBP = 101.3 + .68 \; AGE$$
$$DBP = \;\; 63.7 + .36 \; AGE$$

where

SBP = Systolic blood pressure, mm Hg (mercury)

DBP = Diastolic blood pressure, mm Hg

AGE = Age, years

Life insurance studies indicate that blood pressures below 110/70 are optimal for a long life span. Resting diastolic pressure below 90 is satisfactory, from 90–100 is suspicious, and above 100 is poor.

BOX 11.5 *Cardiac output*

Cardiac output is the volume of blood/min from the left ventricle:

$$CO = HR \ (SV)$$

where
    CO = Cardiac output, liters/min
    HR = Heart rate, beats/min
    SV = Stroke volume, liters/beat

Cardiac output, for a resting young man, is about 5 liters/min; maximum for a normally sedentary young man is about 25 and for a world-class athlete about 35.

Cardiac output can be predicted from formulas for basal, activity, and skin blood flows.

Basal cardiac output is (Guyton, 1961):

$$COBASL = CI \ (DBSA)$$

where
    $COBASL$ = Basal cardiac output, liters/min
    CI = Cardiac index, liters/(min-m²)
       $= 4.29 - .029 \ AGE + .003 \ AGE^2$
          $(5 < AGE < 70)$
    AGE = Age, years
    $DBSA$ = DuBois surface area, m²

Activity cardiac output is (Astrand and Rodahl, 1986):

$$COACT = CLMW \ (TOTMET)$$

where
    $COACT$ = Activity cardiac output, liters/min
    $CLMW$ = Conversion from liters of blood to Watts
       $= .166$ for $TOTMET < 700$ W
       $= .114$ for $TOTMET > 700$ W
    $TOTMET$ = Total metabolism, W

Activity cardiac output increases primarily in the muscles, up to 18 times basal flow.

Skin cardiac output is (Stolwijk, 1970):

$$COSKIN = CSKIN \ (ISKINT) + CCORE \ (ICORET)$$

where
    $COSKIN$ = Skin cardiac output, liters/min
          $= .4$ for basal conditions
    $CSKIN$ = Skin coefficient, liters/C-min
    $ISKINT$ = Increase in skin temperature, C
    $CCORE$ = Core (hypothalamus) coefficient, liters/C-min
    $ICORET$ = Increase in core temperature, C

Rowell (1974) says, for tasks with $TOTMET < .2$ (VO_2MAX), the increase in $COSKIN$ comes from a decrease in $COBASL + COACT$ rather than an increase in CO. That is, for lighter work, the skin blood flow is just a redistribution of blood rather than an increase.

In a person who needs to lose heat, vasodilation brings hot central blood to the surface where heat is transferred to the environment by radiation and convection; $COSKIN$ can increase up to four times basal. This vasodilation circulation bypasses the muscles as the blood flows from the small arteries (arterioles) to the small veins (venules) through bypasses (arteriovenous anastomoses). This flow through a high resistance naturally increases the heart rate. Thus, the common observation is that heart rate increases in the heat.

However, when people are accustomed to the heat (heat **acclimatization**), the sweating ability is enhanced. Because they can sweat more and thus lose heat by evaporation, they do not need to lose as much heat by vasodilation. Since their blood does not use the high-resistance paths, heart rate is lower for acclimatized people than nonacclimatized people. Thus, although heat stress can affect heart rate, it does not affect the oxygen consumption.

2.3 Cardiovascular Limits

In determining cardiovascular limits, the first question is how to determine an individual's work capability. The second is what proportion of the capability should be used.

2.3.1 Capability

The capability of the cardiovascular system is determined through a test to determine the **maximum oxygen uptake, VO_2MAX,** mL/kg-min. VO_2MAX is a product of cardiac output and A-V differential. Table 11.4 shows categories of fitness for U. S. males. Females typically have VO_2MAX 15–30% below that of males (Vogel et al., 1986), so their critical values would be about 75% that of males. A male under 30 having a value of 33 mL/kg-min would be considered to be in fair shape. A female under 30 having a value of 33 would compare with $33/.75 = 44$ and be in good shape.

Wisner (1989) has many values of VO_2MAX for various countries. For developing countries, he emphasizes that the low values of aerobic power of many people are due to poor nutrition and parasitoses.

How is $VO_2 MAX$ determined? One possibility (a maximal test) is to put a person on a treadmill or bicycle ergometer and exercise the person until exhaustion. The $VO_2 MAX$ determined on a treadmill will be higher than on a bicycle, which will be higher than for lifting (Sharp et al., 1988). However, this requires trained staff and equipment. Another test is to record the distance a person can run or walk in 720 s:

$$VO_2 MAX = -10.3 + 35.3 \, DIST$$

where

$VO_2 MAX$ = Maximum oxygen uptake, mL/kg-min

   $DIST$ = Distance run in 720 s, miles

Bunc (1994) recommends recording the amount of time taken to run 2 km.

$$VO_2 MAX = 85.7 - 251.3 \, H \text{ (for males)}$$
$$= 61.9 - 124.2 \, H \text{ (for females)}$$

where

   H = Time to run 2 km, h

A submaximal test is the step test. There are two 5-min sessions of stepping up and down on a 40-cm step with 15 cycles/min for the first test and 25 cycles/min for the second. The heart rate is measured from the pulse from 30 to 60 s after work stops, from 90 to 120 s, and from 150 to 180 s (Tuxworth and Shahnawaz, 1977).

It should be mentioned that testing the capabilities of individuals for screening purposes has become controversial in the United States due to discrimination laws.

2.3.2 Proportion of capability

If $VO_2 MAX$ is 100% of capability, what **proportion of capability** is reasonable for work? Endurance work (lasting 4 to 8 h) can be carried out at a rate of 33–50% of a person's $VO_2 MAX$. The general concept is to avoid use of anaerobic metabolism. Rohmert (1973) said very fit individuals can work at 50% for 8 h/day for 6 days/week. Jorgensen (1985) recommends 50% for trained workers and 33% for untrained. If the task is primarily upper body work, the maximum watts should be about 30% less. See Kodak (1986) for a more extensive discussion. Mital et al. (1993) base their lifting guidelines on 21–23% of uphill treadmill aerobic capacity or 28–29% of bicycle aerobic capacity. Note that lifting has static components as well as dynamic.

For shorter or longer periods, see Figure 11.3. Kodak recommends a sustained maximum of 33% for an 8-h shift, 30.5% for a 10-h shift, and 28% for a 12-h shift (Kodak, 1986, Figures 11.1 and 15.1). Mital et al. (1993) reduce their 28–29% of bicycle aerobic capacity for 8 h to 23–24% for 12 h.

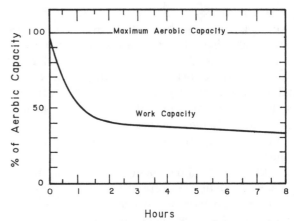

FIGURE 11.3   Work capacity declines with working time.

Assuming you wish to exclude only a small percentage of the population, the above are about 350 W, 5 kcal/min, and 100–120 heartbeats/min. Wisner (1989) states, "It is more or less universally admitted that heartbeat rates should not exceed 110 beats/min during the working day. During more intensive work periods, 130 beats/min should not be exceeded."

The fact that people can work hard does not mean that they should. High-metabolic-rate jobs should be considered prime candidates for mechanization because motors are a far more efficient power source than muscles. Consider fork lifts, balancers, manipulators, and horizontal transfer rather than vertical transfer. After all, although scientists may be interested in physiology, engineers should be interested in productivity.

Two administrative solutions are job rotation and part-time work. In **job rotation,** people shift jobs periodically within the day. In part-time work, the job is given to several people who work part of the day; for example, Joe works for 4 h in the morning and Pete for 4 h in the afternoon, or enough people are scheduled that the entire job is done in part of a shift (e.g., package handling for delivery services often is done within 2-3 h).

When setting work standards, industrial engineers usually use fatigue allowances. See Chapter 31.

2.4 Cardiovascular System: Gender, Age, and Training

2.4.1 Gender

For physical work, there are **gender effects** on capabilities. The average female has a $VO_2 MAX$ 15 to 30% below that of males (Vogel et al., 1986), which is due to a higher percent of body fat and a lower hemoglobin concentration in the blood.

Since fat tissue has little blood, blood volume for an adult male averages 75 mL/kg, whereas it is 65 for a female and 60 for a child (Astrand and Rodahl,

1986). For the same age and body weight, females have lung volumes about 10% lower than males. Females also have lower hemoglobin content than males (13.9 vs. 15.3), lower hematocrit, which is the relative amount of plasma and corpuscles (42 vs. 47), and lower arterial oxygen content (16.7 mL/100 mL vs. 19.2). Therefore, for submaximal work (oxygen uptake of 1.5 L/min), females need 9 liters of cardiac output to transport 1 liter of oxygen while men require only 8.

Although the above figures help explain the difference in athletic performance, cardiovascular differences between men and women should have little importance for most industrial work since most industrial tasks should not be designed to require maximum cardiovascular output.

2.4.2 Age

The body's physical performance peaks sometime around age 25. After that it is all "downhill" (**age effects**). See Figures 11.4 and 11.5.

For the specific index of $VO_2 MAX$, Dehn and Bruce (1972) reported, from three studies, that it declined 1.04, .94, and .93 mL/kg-min/year for males; in their own study, the decline was 1.32 for habitually inactive males and .65 for active males. Astrand et al. (1973) reported declines, over a 21-year period, for Swedish physical education instructors of .64 for males and .44 for females. Illmarinen (1992) says that

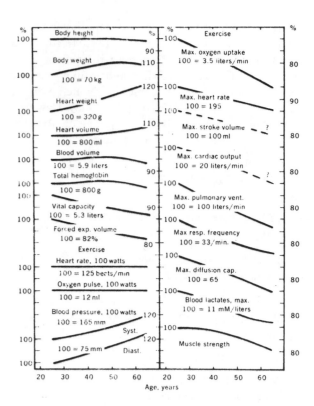

FIGURE 11.5 After 25 is "downhill" according to Astrand and Rodahl (1986). Heart rate, oxygen pulse, and blood pressure are reported for a total metabolism of 100 W (oxygen uptake about 1.5 liters/min).

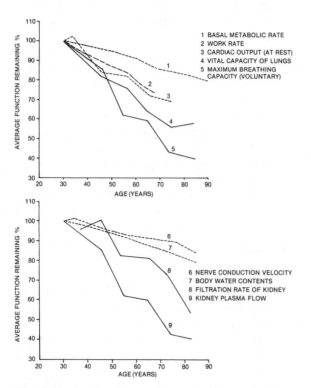

FIGURE 11.4 After 30 is "downhill" according to Shock (1962). In early life we have surplus capacity. As age increases, capacity eventually declines below requirements.

the decline in $VO_2 MAX$ after age 20-25 is 1-2%/year but that there are large individual variations. Bunc (1994) reports that for European subjects:

$$VO_2 MAXM = 57.7 - .404 \, AGE \text{ (for males)}$$

Std. dev. = 8.8 mL

$$VO_2 MAXM = 46.6 - .344 \, AGE \text{ (for females)}$$

Std dev. = 6.4 mL

where

$VO_2 MAXM$ = Mean maximum oxygen uptake, mL/kg-min

AGE = Age, years

Schacherer et al. (1992) combined gender, age, and fat to predict treadmill $VO_2 MAX$:

$$MALE \, VO_2 MAX = 66.734 - .315 \, AGE - .678 \, PFAT$$

$$FEMALE \, VO_2 MAX = 58.094 - .356 \, AGE - .494 \, PFAT$$

where

AGE = Age, years

$PFAT$ = Body fat, percent

Jackson et al. (1992) emphasize that most of the decline is due to physical activity level and percent body fat, not aging. The aging effect was just .27 mL/kg-min/year.

$$MALE\ VO_2\ MAX = 47.9 - .27\ AGE + 3.41\ SRPA$$
$$- .20\ PFAT - .09\ SRPA * PFAT$$

where

$SRPA$ = Self-report of physical activity

What does this decline in capability mean for jobs? Henschel (1970) has a good summary: "Cardiovascular capacity to perform light to moderate physically exhausting work is not grossly age-dependent up to age 65, although capacity for hard, exhausting work is strongly age-dependent, with maximum capacity between 20 and 25."

2.4.3 Training
Fitness has dimensions of cardiovascular fitness (endurance), muscle strength, and flexibility. Specific training techniques are more appropriately found in books on kinesiology or athletics. However, two general statements can be made: (1) For training the cardiovascular system for cardiac output, train with large muscle groups. (2) For training for strength of specific muscles, train the specific muscles. Some exercises strengthen weak muscles (work hardening), and other exercises stretch tight muscles and ligaments. If the exercise loads the muscles dynamically, relax and stretch them. If the exercise loads the muscles statically, the exercise should move them.

2.5 Response to Mental Work
Kalsbeek (1968) proposed heart rate variability (sinus arrhythmia) as an index of **mental load** (the load on the brain). Low variability went with high load. However, the indices indicated there is either mental load or no mental load, but not gradations of mental load. Typically, the researchers used the standard deviation of the interbeat interval as the index. Sharit and Salvendy (1982) recommended use of a "high pass filter" (the mean square successive difference) in place of the standard deviation because it reduced the effect of the respiratory rate on the heart rate. Atsumi et al. (1993) systematically investigated the high pass filter idea and determined it was best to use a sample of three successive beats.

$$RRV3 = \frac{Sum\ (RRI_i - RRI_m)^2}{3}$$

where

$RRV3$ = R to R variability of 3 beats (ms)^2

RRI_i = Individual RR interval, ms

RRI_m = Mean RR interval for 3 successive beats, ms

For an example calculation, see Table 11.5. The $RRV3$ is calculated on successive triads of heartbeats, adding a new value and dropping the oldest value. The rate of change for a person is calculated as:

$$CRRV3 = (WRRV3 - RRRV3)/RRRV3$$

where

$CRRV3$ = Change in $RRV3$, proportion

$WRRV3$ = $RRV3$ during work (ms)^2

$RRRV3$ = $RRV3$ during rest (ms)^2

Atsumi et al. (1993), who work for Toyota, used $RRV3$ to evaluate the stress caused by operating various alternatives for vehicle controls and displays during driving.

3 SKELETOMUSCULAR SYSTEM

3.1 Anatomy of Muscular Movement
Muscles and bones are arranged into three types of **skeletomuscular levers**. Note that there are two sets of muscles—the action muscles (protagonist) and the opposer (antagonist):

- First-class levers (see Figure 11.6) have the fulcrum in the middle, which is good for fine positional control. Common examples outside the body are the seesaw, scissors, and platform balance.
- Second-class levers (see Figure 11.7) have the fulcrum at one end with the force exerted through a longer lever arm than the resistance; the force has a mechanical advantage over the resistance. Common examples outside the body are wheelbarrows and refrigerator doors.
- Third-class levers (see Figure 11.8) also have a fulcrum at one end, but the force is exerted through a shorter lever arm than the resistance; resistance has a mechanical advantage over the force. An example outside the body is a forceps.

The amount of the mechanical advantage for a specific individual varies with the limb length and

TABLE 11.5 Example calculation of *RRV3*.

HEART RATE, BEATS/MIN	*RRI*, MS	DEVIATION, MS	DEVIATION SQUARED (MS)^2
91	659	−8	64
90	667	0	0
89	674	7	49
Mean	667		37.7

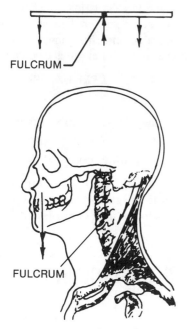

FIGURE 11.6   First-class levers have the fulcrum in the middle. A sports example is an oar in a rowboat.

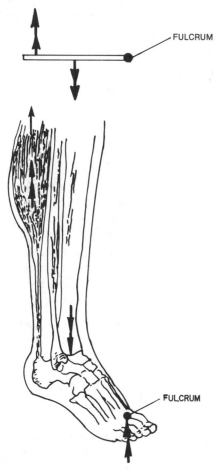

FIGURE 11.7   Second-class levers have the fulcrum at one end; the force has a mechanical advantage over the resistance (i.e., its moment arm is longer). A sports example is a pole in a pole vault when it is bent downwards just before the athlete springs off the ground.

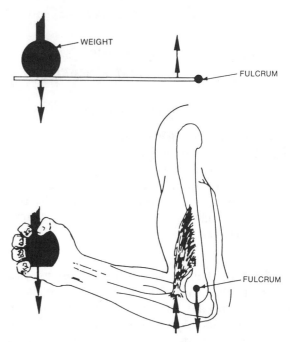

FIGURE 11.8   Third-class levers have a fulcrum at one end; the resistance has a mechanical advantage over the force (i.e., its moment arm is longer). A sports example is movement of an athlete's arm when throwing.

specific location of the attachment of the muscle to the bone. Many people have imperfect bodies. For example, a long neck may cause fatigue from supporting the weight of the head; unequal leg lengths (as little as 6 mm) greatly increases fatigue while standing.

Mechanical advantage also varies greatly by the direction and angle of exertion required by the task (Warwick et al., 1980). For an object straight ahead, their subjects (using both hands) could push with a force of 29.8 kg (100%), pull back with a force of 17.3 (58%), push right with 15.9 (53%), and push left with 17.0 (57%). They could lift up with 39.4 (132%) and press down with 34.7 (116%). See Chapter 17.

Ferguson and Mason (1988) studied force capability when using wrenches. They told designers: (1) at fastener heights below .5 m above the floor, don't require downward movement of the wrench; (2) at heights above .5 m, don't have horizontal movement of the wrench unless workers can brace themselves; and (3) at heights above 1 m, move the wrench up or down (not horizontally). Jorgensen et al. (1989) recommend that, when cutting meat with knives, cuts be made pulling toward the body and the knife edge should be down, not up.

Postures during maintenance often depart radically from standing erect (Haselgrave, Tracy, and Corlett, 1987). Using pushing horizontally as 100%, most forces while lying supine or working overhead while standing were 35 to 45%, although at some angles and positions, force capability increased to 140 to 240%.

Muscle strength of the arm vs. angle is shown in Figure 11.9 and Table 11.6. Figure 11.10 shows how sore elbow cases were influenced by the angle workers used when holding a screwdriver. That is, an improper work angle cannot only reduce strength but also can cause injuries. For additional comments on angles for workstations and handtools, see Chapters 15, 16, and 18.

Maximal voluntary contraction (MVC), the maximum sustained force of a muscle, declines exponentially with time. The asymptote at which a force could be held indefinitely formerly was given as 15%, but more recent data indicate it is below 10% (Rodahl, 1989, p. 236).

When there is a balance between muscle torque and external torque, the muscle length does not change. This is an **isometric** (static) muscular contraction; the person generates a single MVC. If there is a constant mass being moved, it is an *isoinertial* (dynamic) muscular contraction. If movement is constant, it is *isovelocity* (Kroemer et al., 1994).

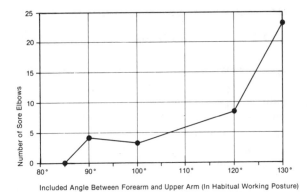

FIGURE 11.10   Sore elbow cases (238 patients from workers continuously using screwdrivers) were strongly related to the habitual angle (at work) between the forearm and the upper arm (Tichauer, 1973).

Formerly, most strength testing was isometric. However, if the strength testing is to predict lifting ability, isovelocity testing now is recommended due to its greater predictive ability as well as greater safety for the person being tested (Ayoub et al., 1986; Jiang et al., 1986; Mital et al., 1986a; Mital et al., 1986b).

Muscle activity can be evaluated with an electromyogram (see Box 11.6).

3.2 Skeletal System: Gender, Age, and Training Effects

3.2.1 Gender
The average female has less muscular strength than a male, although individual females may be stronger than individual males. The most simple approximation is that females have 2/3 of the strength of males. However, there are great differences by muscle groups. Laubach (1976) reported 56% for upper extremity strength, 72% for lower extremity strength, 64% for trunk strength, and 69% for dynamic strength. The differences are basically due to the smaller size (shorter lever arms) and muscle masses of women rather than gender in itself (Hosler and Morrow, 1982; Bishop et al., 1987).

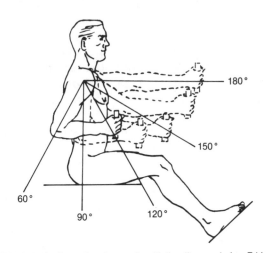

FIGURE 11.9   Strength varies greatly with the elbow angle (see Table 11.6).

TABLE 11.6   Forces (kg) exertable on a vertical handgrip with the right arm at various elbow angles (Damon et al., 1966, citing Hunsicker, 1955). Arm strength depends upon the angle, the direction, the arm, and the individual. See Figure 11.9. $N = 55$ college males. The mean coefficient of variation was 42%.

| | Elbow Angle, degrees | | | | | |
Movement	60	90	120	150	180	Left arm, % of Right
Pull	28.6	40.0	47.3	55.5	54.5	.94
Push	41.8	39.1	46.8	55.9	62.7	.92
Move right	19.1	16.8	15.5	15.0	15.5	1.30
Move left	23.6	22.7	24.1	24.5	22.7	.60
Up	22.3	25.5	27.3	25.5	19.5	.92
Down	23.2	24.1	26.4	21.4	18.6	.88

BOX 11.6 *Electromyography (EMG)*

For heavy work, the measurement of oxygen capability *(VO₂ MAX)* is of interest. However, for lighter, repetitive work, the activity of specific muscles may be of interest. The activity of muscles can be analyzed using an **electromyogram (EMG).** The EMG can provide information about the following (Marras, 1990):

- whether the muscle was in use (on/off)
- a relative activity level
- force generation (under special conditions)
- muscle fatigue

At the present state of technology, EMG usually is a laboratory tool rather than an industrial tool.

However, EMG analysis can lead to useful guidelines concerning muscular work, and these guidelines have direct industrial applications. In an interesting application, Parenmark et al. (1993) reported using EMG with biofeedback to train industrial workers so that the average muscular load on their arms was kept below 15–20% of their maximum voluntary contraction.

Mental tension, even in a physically inactive person, will cause a recordable EMG output, proportional to the degree of tension. Various EMG devices are sold which measure this EMG and feed it back to the individual (usually through an auditory tone) so that the person can try to relax (Rodahl, 1989, p. 240).

A larger hand can exert more grip force. But maximal muscle force is proportional to muscle cross-sectional area. Thus, when relating force to body dimensions, use the dimension squared. Franson and Winkel (1991) reported that about 35% of the gender difference in hand strength is due to hand size differences.

Note also that if rhythmic work is performed by the arms (such as most bench work), females have less arm muscle than males but also lighter bones and hands (Shephard et al., 1988).

3.2.2 Age

Figures 11.4 and 11.5 give some estimate of the decline of various functions of the body with age. Figure 11.11 gives estimates for muscle strength. Roozbahar et al. (1979) give

$$GS = 608 - 2.94\ AGE$$

where

GS = Grip strength of dominant hand, N

AGE = Age, years

Figure 11.12 shows the effect of disc degeneration of the back (Hult, 1954). However, Rhodes (1983), in a review of the relation between age and performance, reported that there were approximately an equal number of studies reporting that job performance increases with age, remains the same, and decreases. See also Waldman and Avolio (1986). Speed of movement decreases with age, and older people slow down most on more difficult movements (Brogmus, 1991).

3.2.3 Training

Basic gender differences in muscular strength are more marked after training (Hettinger, 1961) because, with the same kind of muscle training, strength increases faster and to a greater extent among men.

For industrial jobs, muscular training should target strength, endurance, and flexibility (Asfour et al., 1984). It is possible to substantially improve muscular strength and endurance with a short, intensive training program. To improve strength, the training load should be greater than 50% of the person's initial dynamic strength. Part of the increase in endurance time is due to the person learning to improve micromotions and neuromuscular coordination (Genaidy, 1991).

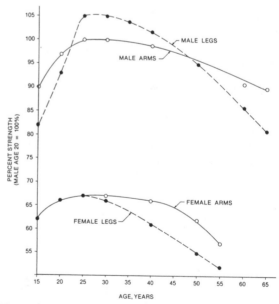

FIGURE 11.11 Isometric strength of arms and legs for men and women varies with age (Asmussen and Heeboll-Nielsen, 1962). Male strength at age 20–22 is set as 100%.

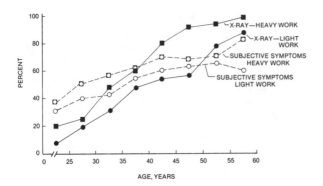

FIGURE 11.12 Lower back problems (lumbar insufficiency-lumbago-sciatica syndrome) were reported from workers in light and heavy industry (total $N = 1,200$) by Hult (1954). He used subjective symptoms of back pain and X-rays of disc degeneration.

REVIEW QUESTIONS

1. Why do females have a lower metabolic rate/kg than males?

2. Sketch a schematic of the cardiovascular system, showing pulmonary and systemic circulation. Identify input and output of the following to the blood: oxygen, carbon dioxide, nutrients, metabolic wastes.

3. Oxygen supply to a muscle is adjusted by what five techniques?

4. List four ways heart rate can be measured. What is the Borg vote concept?

5. What is the mean maximum heart rate for people 30 years old? What values include 95% of the population?

6. Define pulse pressure of the CV system.

7. Define cardiac output with a formula.

8. What is A-V differential?

9. After what age is it all "downhill"—at least physiologically?

10. What maximum level of VO_2 would you recommend for 8 h of work? What would be a typical heart rate for this?

11. Females, on average, are weaker than males. Therefore, they should not do physical work in industry. Discuss.

12. What can you use as an index of mental load?

13. What are electromyograms?

14. Sketch a 1st-, 2nd-, and 3rd-class lever for the human body.

REFERENCES

Alexander, R. Stride length and speed for adults, children and fossil hominids. *American J. of Physical Anthropology*, Vol. 63, 23–27, 1984.

Andrews, R. The relationship between measures of heart rate and measures of energy expenditure. *American Inst. of Industrial Engineers Transactions*, Vol. 1 [1], 2–10, 1969.

Arstilla, M. Pulse-conducted triangular exercise—ECG test. *Acta Medica Scandinavica*, Vol. 191, Supplement 529, 1972.

Asfour, S., Ayoub, M., and Mital, A. Effects of an endurance and strength training programme on lifting capability of males. *Ergonomics*, Vol. 27 [4], 435–42, 1984.

Asmussen, E. and Heeboll-Nielsen, K. Isometric muscle strength in relation to age in men and women. *Ergonomics*, Vol. 5, 167–69, 1962.

Astrand, I., Astrand, P., Hallbeck, I., and Kilbom, A. Reduction in maximal oxygen uptake with age. *J. of Applied Physiology*, Vol. 35, [5], 649–54, 1973.

Astrand, P. and Rodahl, K. *Textbook of Work Physiology*, 3rd ed. New York: McGraw-Hill, 1986.

Atsumi, B., Sugiura, S., and Kimura, K. Evaluation of mental work load in vehicle driving by analysis of heart rate variability. *Proceedings of the Human Factors and Ergonomics Society*, 574–77, 1993.

Ayoub, M., Selan, J., and Chen, H. Human strength as a predictor of lifting capacity. *Proceedings of the Human Factors Society*, 960–63, 1986.

Bishop, P., Cureton, K., and Collins, M. Sex difference in muscular strength in equally trained men and women. *Ergonomics*, Vol. 30 [4], 675–87, 1987.

Borg, G. Psychophysical scaling with applications in physical work and the perception of exertion. *Scand. J. Work Environ. Health*, Vol. 16 (Supplement 1), 55–58, 1990.

Brogmus, G. Effect of age and sex on speed and accuracy of hand movements and the refinements they suggest for Fitts' law. *Proceedings of the Human Factors Society*, 208–12, 1991.

Brozek, J., Grande, F., Anderson, T., and Keys, A. Densitometric analysis of body composition: Revision of some quantitative assumptions. *Annals of N. Y. Academy of Science*, Vol. 110, 113–40, 1963.

Bunc, V. A simple method for estimating aerobic fitness. *Ergonomics*, Vol. 37, [1], 159–65, 1994.

Cooper, K. *The New Aerobics*. New York: Bantam Books, 1970.

Cowgill, G. A formula for estimating the specific gravity of the human body. *American J. of Clinical Nutrition*, Vol. 5, 601-11, 1957.

Damon, A., Stoudt, H., and McFarland, R. *The Human Body in Equipment Design* 226-27. Cambridge, Mass.: Harvard University Press, 1966.

Dehn, M. and Bruce, R. Longitudinal variations in maximum oxygen intake with age and activity. *J. of Applied Physiology*, Vol. 33, 805-807, 1972.

Ferguson, C. and Mason, S. Design strategies for maximizing human force capability (HFC) when using spanners. *Int. J. of Industrial Ergonomics*, Vol. 2, 251-58, 1988.

Franson, C. and Winkel, J. Hand strength: The influence of grip span and grip type. *Ergonomics*, Vol. 34, [7], 881-92, 1991.

Freivalds, A. and Bise, C. Metabolic analysis of support personnel in low-seam coal-mines. *Int. J. of Industrial Ergonomics*, Vol. 8, 147-55, 1991.

Garg, A. *A Metabolic Rate Prediction Model for Manual Materials Handling*. Ph.D. dissertation, Ann Arbor: University of Michigan, 1976.

Garn, S. and Harper, R. Fat accumulation and weight gain in the adult male. *Human Biology*, Vol. 27 [1], 39-49, 1955.

Genaidy, A. A training programme to improve human physical capability for manual handling jobs. *Ergonomics*, Vol. 34 [1], 1-11, 1991.

Guyton, A. *Textbook of Medical Physiology*. Philadelphia: W. B. Saunders, 1961.

Haselgrave, C., Tracy, M., and Corlett, E. Industrial maintenance tasks involving overhead working. In *Contemporary Ergonomics 1987*, Megaw, E. (ed.). London: Taylor and Francis, 1987.

Henschel, A. Effect of age on work capacity. *American Industrial Hygiene Association J.*, Vol. 31 [4], 430-36, 1970.

Hettinger, T. *Physiology of Strength*. Springfield, Ill.: Charles Thomas, 1961.

Hosler, W. and Morrow, J. Arm and leg strength compared between young women and men after allowing for differences in body size and composition. *Ergonomics*, Vol. 25 [4], 309-13, 1982.

Hult, L. Cervical, dorsal and lumbar spinal syndromes. *Acta Orthopaedica Scandinavica* (Supplement 17), 1954.

Hunsicker, P. Arm strength at selected degrees of elbow flexion. *WADC Technical Report 54-548*. Wright Patterson AFB, Ohio, 1955.

Illmarinen, J. Job design for the aged with regard to decline in their maximal aerobic capacity: Part II—The scientific basis for the guide. *Int. J. of Industrial Ergonomics*, Vol. 10, 65-77, 1992.

Jackson, A., Beard, E., Wier, L., and Stuteville, J. Multivariate model for defining changes in maximal physical working capacity of men, ages 25 to 70 years. *Proceedings of the Human Factors Society*, 171-74, 1992.

Jiang, B., Smith, J., and Ayoub, M. Psychophysical modeling of manual materials-handling capacities using isoinertial strength variables. *Human Factors*, Vol. 28 [6], 691-702, 1986.

Jorgensen, K. Permissible loads based on energy expenditure measurements. *Ergonomics*, Vol. 28 [1], 365-69, 1985.

Jorgensen, M., Riley, M., Cochran, D., and Bishu, R. Maximum forces in simulated meat cutting tasks. *Proceedings of the Human Factors Society*, 641-45, 1989.

Kalsbeek, J. Measurement of mental work load and of acceptable load: Possible applications in industry. *International J. of Production Research*, Vol. 7 [1], 33-45, 1968.

Klieber, M. *The Fire of Life*. New York: Wiley, 1961.

Kodak, E. *Ergonomic Design for People at Work: Vol. 2*. New York: Van Nostrand-Reinholt, 1986.

Konz, S. *Work Design: Industrial Ergonomics*, 2nd ed., Scottsdale, Ariz.: Publishing Horizons, 1984.

Kozar, A. Anticipatory heart rate in rope climbing. *Ergonomics*, Vol. 7, 311-15, 1964.

Kroemer, K., Kroemer, H., and Kroemer-Elbert, K. *Ergonomics: How to Design for Ease & Efficiency*, Englewood Cliffs, N. J.: Prentice-Hall, 1994.

Laubach, L. Comparative muscular strength of men and women: A review of the literature. *Aviation, Space and Environmental Medicine*, Vol. 47 [5], 534-42, 1976.

Legg, S. and Mahanty, A. Energy cost of backpacking in heavy boots. *Ergonomics*, Vol. 29 [3], 433-38, 1986.

Ljunggren, G. Observer ratings of perceived exertion in relation to self ratings and heart rate. *Applied Ergonomics*, Vol. 17 [2], 117-25, 1986.

Lohman, T. Skinfolds and body density and their relation to body fatness: A review. *Human Biology*, Vol. 53 [2], 181-225, 1981.

Marras, W. Industrial electromyography. *Int. J. of Industrial Ergonomics*, Vol. 6, 89-93, 1990.

Maynard, H., Stegemerten, G., and Schwab, J. *Methods Time Measurement*. New York: McGraw-Hill, 1948.

Mital, A., Aghazadeh, F., and Karwowski, W. Relative importance of isometric and isokinetic lifting strengths in estimating maximum lifting capabilities. *J. of Safety Research*, Vol. 17, 65-71, 1986a.

Mital, A., Karwowski, W., Mazouz, A., and Osarh, E. Prediction of maximum acceptable weight of lift in the horizontal and vertical planes using simulated job dynamic strengths. *American Industrial Hygiene Association J.*, Vol. 47 [5], 288-92, 1986b.

Mital, A., Nicholson, A., and Ayoub, M. M. *A Guide to Manual Material Handling*. London: Taylor and Francis, 1993.

Morrisey, S., Ayoub, M., George, C., and Ramsey, J. Male and female responses to stoopwalking tasks. *25th Proceedings of the Human Factors Society*, 445-49, 1981.

Nishi, Y. and Gagge, A. Direct evaporation of convective heat transfer coefficient by naphthalene sublimation. *J. Applied Physiology*, Vol. 29, 830, 1970.

Pandolf, K., Haisman, M., and Goldman, R. Metabolic energy expenditure and terrain coefficients for walking on snow. *Ergonomics*, Vol. 19, 683-90, 1976.

Pandolf, K., Givoni, B., and Goldman, R. Predicting energy expenditure with loads while standing or walking very slowly. *Applied Physiology: Respirat. Environ.,* Vol. 43 [4], 577-81, 1977.

Parenmark, G., Malmkvist, A., and Ortengren, R. Ergonomic moves in an engineering industry: Effects on sick leave frequency, labor turnover and productivity. *Int. J. of Industrial Ergonomics,* Vol. 11, 291-300, 1993.

Peters, R. *The Ecological Implications of Body Size.* Cambridge: Cambridge University Press, 1989.

Rhodes, S. Age-related differences in work attitudes and behavior—A review and conceptual analysis. *Psychological Bulletin,* Vol. 93, 328-67, 1983.

Rodahl, K. *The Physiology of Work.* London: Taylor and Francis, 1989.

Rohmert, W. Problems in determining rest allowances: Part 2 — Determining rest allowances in different human tasks. *Applied Ergonomics,* Vol. 4 [2], 158-62, 1973.

Roozbahar, A., Bosker, G., and Richardson, M. A theoretical model to estimate some ergonomic parameters from age, height and weight. *Ergonomics,* Vol. 22 [1], 43-58, 1979.

Rowell, L. Human cardiovascular adjustments to exercise and thermal stress. *Physiological Reviews,* Vol. 54 [1], 75-159, 1974.

Schacherer, C., Rowe, A., and Jackson, A. Development of prediction models for physical work capacity: Practical and theoretical implications. *Proceedings of the Human Factors Society,* 674-78, 1992.

Sharit, J. and Salvendy, G. External and internal attentional environments II: Reconsideration of the relationship between sinus arrhythmia and information load. *Ergonomics,* Vol. 25 [2], 121-32, 1982.

Sharp, M., Harman, E., Vogel, J., Knaik, J., and Legg, S. Maximum aerobic capacity for repetitive lifting: Comparison of three standard exercise modes. *European J. of Applied Physiology,* Vol. 57, 753-60, 1988.

Shephard, R. *Alive Man.* Springfield, Ill.: C. T. Thomas, 1972.

Shephard, R., Vanderwalle, H., Bouhlel, E., and Monod, H. Sex differences of physical working capacity in normoxia and hypoxia. *Ergonomics,* Vol. 31 [8], 1177-92, 1988.

Shock, N. The physiology of aging. *Scientific American,* Vol. 206, 100-110, January 1962.

Sloan, A., Burt, J., and Blythe, C. Estimation of body fat in young women. *J. of Applied Physiology,* Vol. 17 [6], 967-70, 1962.

Smalley, K., Knerr, A., Kendrick, Z., Colliver, J., and Owen, O. Reassessment of body mass indices. *Am. J. of Clinical Nutrition,* Vol. 52, 405-08, 1990.

Soule, R. and Goldman, R. Energy cost of loads carried on the head, hands or feet. *J. of Applied Physiology,* Vol. 27, 687-90, November 1980.

Stolwijk, J. Mathematical model of thermoregulation. In *Physiological and Behavioral Temperature Regulation,* Hardy, J., Gagge, A., and Stolwijk, J. (eds.), Chapter 48. Springfield, Ill.: C. T. Thomas, 1970.

Suter, P., Schutz, Y., and Jequier, E. The effect of ethanol on fat storage in healthy subjects. *New England J. of Medicine,* Vol. 326, 983-87, 1992.

Tichauer, E. Ergonomic aspects of biomechanics. In *The Industrial Environment—Its Evaluation and Control,* Chapter 32. Washington, D. C.: Supt. of Documents, 1973.

Tuxworth, W. and Shahnawaz, H. The design and evaluation of a step test for the rapid prediction of physical work capacity in an unsophisticated industrial work force. *Ergonomics,* 20 [2], 181-91, 1977.

U. S. Army, *AR 600-9 (Section III Weight Control),* 13 Feb 1987.

Van der Walt, W. and Wyndham, C. An equation for prediction of energy expenditure of walking and running. *J. of Applied Physiology,* Vol. 34 [5], 559-63, 1973.

Vogel, J., Patton, J., Mello, R., and Daniels, W. An analysis of aerobic capacity in a large United States population. *J. of Applied Physiology,* Vol. 60, 549, 1986.

Waldman, D. and Avolio, B. A meta-analysis of age differences in job performance. *J. of Applied Psychology,* Vol. 71 [1], 33-38, 1986.

Warwick, D., Novak, G., Schultz, A., and Berkson, M. Maximum voluntary strengths of male adults in some lifting, pushing, and pulling activities. *Ergonomics,* Vol. 23 [1], 49-54, 1980.

Wilmore, J. and Behnke, A. An anthropometric estimation of body density and lean body weight in young men. *J. of Applied Physiology,* Vol. 7 [1], 25-31, 1969.

Wisner, A. Variety of physical characteristics in industrially developing countries—Ergonomic consequences. *Int. J. of Industrial Ergonomics,* Vol. 4, 117-38, 1989.

Zuti, W. and Golding, L. Equations for estimating percent fat and body density of active adult males. *Medicine and Science in Sports,* Vol. 5 [4], 262-66, 1973.

ERROR REDUCTION

OVERVIEW

Error reduction is a major goal of ergonomics. This chapter discusses many aspects of errors and error reduction. Ten guidelines for error reduction are given.

CHAPTER CONTENTS

1 Introduction
2 Error Analysis Techniques
3 Error Reduction Guidelines
   1 Get Enough Information
   2 Ensure Information Is Understood
   3 Have Proper Equipment/Procedures/Skill
   4 Don't Forget
5 Simplify the Task
6 Allow Enough Time
7 Have Sufficient Motivation/Attention
8 Give Immediate Feedback on Errors
9 Improve Error Detectability
10 Minimize Consequences of Errors

KEY CONCEPTS

closed-loop/open-loop system
detectability/criterion
ease of recovery
error-checking code
error costs
errors
fail-safe
information feedback
inspection
job aid

justifications for ergonomics
latent failures
lockout/tagout
maintenance budget
memory aids
mistakes
omission/commission
open and obvious
paired comparison
population stereotypes

protocols
redundancy
signal detection theory
signal/noise ratio
slips
social pressure
type 1/type 2
what you see is what you get

1 INTRODUCTION

The two primary **justifications for ergonomics** are reduction of physical stress on people and reduction of errors.

Errors are caused by poor design and management of equipment, procedures, and training, among other factors. Errors are therefore not beyond control ("acts of God") but are events that can be reduced. Although error reduction is discussed throughout the book, it is in many locations and thus is difficult to discuss as a specific topic. This chapter consolidates the information. The first section discusses error definitions, costs, and types.

1.1 Definitions
An error is defined as "an action other than desired takes place." See Table 12.1 for some error synonyms. We will be concerned with quick actions, not actions with long-term results such as selecting a poor handtool which gives cumulative trauma.

The general sequence is: normality—error—accident—minor loss (time, property, person)—major loss—catastrophe. An example of time loss would be the time to pick up a dropped part. An example of property loss would be damage to the dropped part. An example of person loss would be an injured toe from the dropped part. The goal is to maintain normality as much as possible.

If we consider accidents as due to unpredictable events and unpreventable events, then challenging the adjectives *unpredictable* and *unpreventable* can lead to reducing accidents. As discussed in Section 1.1 of Chapter 25, there is considerable similarity between minor and major accidents; thus, a study of minor accidents and even near accidents can be fruitful.

1.2 Costs and Cost Reduction
This section will consider error costs and strategies to reduce those costs.

1.2.1 Costs   Error costs (the negative results of mistakes) can range from a few seconds of time lost to multiple millions of dollars. The cost of human injury and death is especially difficult to quantify. Other obscure costs are costs of cleanup, fines, loss of market share, and legal costs. Context also can make a trivial error into a costly error. For example, what if a client's name is misspelled on a report and, because of this, the contract is not renewed? What if a potential vendor is late for a meeting and thus loses the contract because the customer considers the vendor unreliable. The vendor may not even realize an error was made!

Part of the reason it is difficult to quantify error costs is that errors and their costs often are concealed. People do not want to broadcast to the world (or at least to their supervisors) that they made an error. It is easier to conceal errors (at least to the formal, paperwork, on-the-record system). See Figure 5.6 for a short discussion of this effect on flow of product through a factory. Even governments conceal errors. For example, in 1970, the government of Iraq purchased about 96,000 tons of seed grain (Casey, 1993). It was treated with alkylmercury fungicide to inhibit spoilage. The grain was dyed red, and each bag displayed a skull and crossbones and a warning (in English) that the grain could not be consumed or fed to animals. But, in 1972, the crops failed in northern Iraq. The government distributed the grain without warnings that it was only for seeding and could not be consumed. About 60,000 Kurds suffered neurological damage; the government imposed a news blackout to the rest of the world to avoid embarrassment.

Our society does not perceive all costs equally. For example, society considers a death in a nuclear power plant as more important than a death in a coal mine. Occupational deaths (especially for large corporations) are considered more important than nonoccupational deaths. Therefore, error prevention receives special attention in large corporations and in

TABLE 12.1 Error synonyms.

ERROR	FLAW	FAULT	MISTAKE	LAPSE	FAILURE
failure	blemish	defect	error	breach	dud
fault	defect	deficiency	fumble	omission	flop
flaw	error	shortcoming	inaccuracy	slip	loser
lapse	fault	weakness	err	error	botch
mistake	imperfection	blame	blunder	delinquency	mess
delusion		wrong	miscalculation	mistake	muddle
fallacy		offense		peccadillo	default
misunderstanding		sin		decline	delinquency
		transgression			dereliction
					negligence
					snafu

certain industries (e.g., nuclear power generation, air transport).

1.2.2 Cost reduction

Generally, pay more attention to a problem if the error's potential cost ("potential energy") is higher. For example, the potential energy of 100 tons of poisoned grain is greater than 1 ton; an airplane with 500 people has more potential energy than one with 4 people; a procedure used by 1,000 people is more important than one used by 20; a procedure causing a system to "crash" is more important than one causing a 2 s delay; and so on.

In reducing errors, the analyst should be a scientist, not an advocate. The legalistic approach of blaming someone (instead of the engineering approach of solving the problem) does not address the root causes of the incident and, in fact, shields the incident from further scrutiny and treatment (Kirwan, 1992a). Punishment tends to lead to nonreporting of errors.

1.3 Types

There are many ways of categorizing errors. Errors can be divided into slips and mistakes. **Slips** are not deliberate, whereas **mistakes** result from conscious planning (i.e., the intended action is successful, but the outcome is not as anticipated). An example of a mistake would be if Jane got her hair caught in moving machinery while deliberately not wearing a hairnet. There is relatively little knowledge of why people deliberately violate established procedures and rules. However, because deliberate violations may occur, the potential impact of human error can increase dramatically (as well as the difficulty in preventing the error).

Errors can be of **omission** (passive, something not done) or **commission** (active, something done incorrectly). Commission can be further divided into errors of timing, sequence, selection, and magnitude (Proctor and van Zandt, 1993).

Errors can also be **type 1** (also called alpha risk and producer's risk) or **type 2** (also called beta risk and consumer's risk). See Figure 14.27. In type 1, a good item is rejected. In type 2, a bad item is accepted. A false alarm is a type 1 error. A failure of an alarm to function is a type 2 error.

Observed error = system error ± measurement error. The point here is that what is observed can be due to either the system or the measurement of the system.

Errors can be divided into perception, decision, and action. Did the person perceive the situation incorrectly? Was the perception correct but the decision wrong? Was the decision correct but the action insufficient?

An analysis of perception and decision, applied to inspection, is **signal detection theory** (SDT). See Box 12.1 and Figure 12.1. The SDT concept is that an inspector's performance is determined not only by the detectability of the target but also by the criterion used by the inspector.

Rasmussen et al. (1987) say behavior is at a skill (S), rule (R), or knowledge (K) level. Skill-based behavior (S) occurs in situations requiring highly practiced and essentially automatic behavior, with only minimal conscious control. The behavior can be inadequate or inappropriate. An example of inadequate behavior would be driving a fork truck along a familiar route and missing a turn. An example of inappropriate behavior would be driving the fork truck too fast. When the individual correctly understands the situation and forms a legitimate intention but an incorrect action is accidently triggered, the error is called a *slip*. Rule-based behavior (R) occurs when the person deals with a situation by a stored rule (procedure) of the following form: IF *situation* THEN *action*. The rules either are memorized or obtained from an aid. When the individual executes what is believed to be a correct action but the action is inadequate or inappropriate, the error is called a *mistake*. An example would be incorrect use of a map to decide which route to follow. Knowledge-based behavior (K) occurs when there are no useful rules to apply. The person must use problem-solving skills and knowledge of system characteristics. A K error is also called a *mistake*. An example would be driving through an area looking for an address not on a map and arriving at the wrong house. In general, for error reduction, eliminate applications of K behavior, then work on the R behavior.

If there is a conflict between a machine and a person, the common assumption is that the machine is wrong and the person is right. However, this can lead to catastrophes such as Three Mile Island and Chernobyl, in which the operators in both cases overrode the system. If your system design permits operator override, then communication to the operator and operator training need high priority.

The following material will emphasize technical approaches to error reduction. However, error reduction is managerial as well as technical. Consider rules and regulations (both governmental and organizational) and managerial approaches such as Total Quality Management and Quality Circles (see Chapter 33). If errors are to be reduced, error reduction needs to be an organizational goal.

2 ERROR ANALYSIS TECHNIQUES

Four error analysis techniques (checklists, decision structure tables, fish diagrams, and fault trees) are described in detail elsewhere in this book.

BOX 12.1   *Signal detection theory (SDT)*

An inspector's performance is determined by **detectability** (also called discriminability) of the error (the "target"). The easier it is to detect the target (due to good lighting, contrast, large target, etc.), the more likely it is the target will be detected. However, there is a second factor affecting inspection performance—**criterion.** What does the inspector consider to be a defect? How willing is the inspector to accept a bad item (type 2 error) or reject a good item (type 1 error)?

Assume an inspector is inspecting shafts for length. The shafts have a distribution about some mean length. If the inspector decides to reject all shafts less than 300 mm long, the criterion then is 300 mm. The inspector's decisions are shown in Figure 12.1. The curve on the left is for the bad shafts. For this curve, all decisions to the left of the criterion are correct rejections but all decisions to the right of the criterion are incorrect acceptances of the bad items (false alarms). The curve on the right is for the good shafts. For this curve, all decisions to the left of the criterion are incorrect rejections (misses) while decisions to the right of the criterion are good acceptances. A general goal is to minimize the inspector's mistakes (false alarms and misses).

However, if the criterion is moved to the left, false alarms increase (although misses decrease); if the criterion is moved to the right, false alarms decrease (although misses increase). Since hits and misses may have different costs, the solution is to minimize the sum of the costs of false alarms and misses.

Detectability influences the spread of the two curves. Assume a very good instrument was used. Then the spread of each curve would decrease, as would false alarms and misses.

Signal detection theory attempts to set the criterion at the point at which costs are minimized. Industrial applications of the theory are scarce. One practical problem is the difficulty in determining the curve shapes for each inspector. That is, different inspectors have different ratios of false alarms to misses; this ratio also may change during the day (because of vigilance decrements, information from the process, information from the supervisor, etc.). However, SDT does remind users that inspection decisions are a function of both the criterion and detectability.

FIGURE 12.1   Signal detection theory assumes an inspector's performance depends on (1) the detectability of the target and (2) the criterion used to make the decision (Sanders and McCormick, 1987).

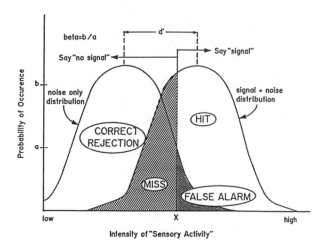

1. *Checklists.* Checklists range from detailed lists of specific questions to lists of general guidelines. See Structured Search in Chapter 5.

2. *Decision structure tables.* To minimize decision-making errors, use a decision structure table. See the discussion in Chapter 7 and Tables 32.4, 32.6, and 32.7.

3. *Fish diagrams.* You may wish to organize causes and effects of a situation. One approach is to make a list. However, the relationships often can be communicated more clearly in a graphical format than in a tabular format (list). Figures 7.16 and 7.17 show fish diagrams, which are graphical presentations of a list with subheadings.

4. *Fault trees.* Errors sometimes occur only when a specific chain of events occurs. If the chain can be broken, the error will not occur. The relationships of these events is called a fault tree. See Figure 25.2. Miller and Swain (1987) have a more extensive discussion of fault trees. See Kirwan (1992a, 1992b) for a detailed comparison of six different methods of human error identification.

3 ERROR REDUCTION GUIDELINES

Table 12.2 presents 10 general guidelines to reduce errors. The guidelines are organized in three categories: planning, execution, and allowing for error. Guidelines 1 through 5 consider planning.

GUIDELINE 1: GET ENOUGH INFORMATION

Getting enough information calls for (1) generating or collecting the relevant information and (2) ensuring that the user receives the information.

Generate Relevant Information

On receiving information, the first assumption made is that the information is correct. This is not always true, however. Organizations have a responsibility to their customers to provide correct information. Although customers usually are external to the organization, the same attitude should be used for fellow employees (internal customers).

If false information cannot be corrected, then the customer should be protected against costs resulting from receiving incorrect information. For example, if a phone number is no longer valid, a message informs the caller of this; if the number has changed, the new one is given.

Novices require a great deal of information; often the information given is not sufficient. Travelers often are novices in the local environment and so need considerable information concerning roads, airports, hotels, restaurants, and so forth. Road signage is especially important for novices.

TABLE 12.2 Error reduction guidelines.

Planning

1. Get enough information.
2. Ensure information is understood.
3. Have proper equipment/procedures/skill.
4. Don't forget.
5. Simplify the task.

Execution

6. Allow enough time.
7. Have sufficient motivation/attention.

Allow For Errors

8. Give immediate feedback on errors.
9. Improve error detectability.
10. Minimize consequences of error.

To perform many actions, a good view of the object helps. However, there may not be a good view because of failure to look, darkness, glare, lack of time, snow or rain, obstructions, or other factors. For example, when backing up a vehicle, you need to look behind the vehicle. But the field of view is not complete and it is difficult to judge distances. One solution is to park so backing up isn't necessary. If backing up is necessary, however, reduce its difficulty. For example, when backing up a semi-trailer, it is easier to see from the driver's side; therefore, design the shipping dock approach so that the driver can back up using the left-side mirror.

A warning can alert people to potential danger. For example, in baseball a warning track in the outfield alerts the fielder about a potential collision with the wall. A yellow band behind a needle on an instrument warns that the equipment response is in a caution zone. An auditory alarm sounds when a lift truck backs up.

Generating the relevant information may be difficult if the information is subjective. How do you evaluate potential employees so the best one is hired?

Ensure Information Reception

Be sure people have and know how to use information sources such as local contact persons, phone directories, information directories, and help services. With the increasing variety of communication techniques, you may need to be able to locate people by physical address, phone number, fax number, E-mail, or other method.

Sometimes the information is all there but people don't know how to find it. Just because an expert user, or even a casual user, knows how to obtain the information does not mean the novice knows how to find it. The designer may assume an expert is using a production device, but devices are used by maintenance workers, people on the first day of the job, and other untrained workers and their errors should be minimized also. Just because a phone has 25 buttons on it does not mean the person knows how to use them; just because a computer has help messages does not mean the user knows how to access them or, if read, understand them. Thus, a series of prerecorded messages needs to be backed up by a resource person who can help the user.

Learning how to use sources is best done through training, although trial (and error!) is a possibility.

GUIDELINE 2: ENSURE INFORMATION IS UNDERSTOOD

Many signs, labels, and warnings are poorly designed. Several techniques exist for improving communication.

Legibility of text can be improved by using the guidelines shown in Chapter 20. Auditory confusion among letters can be reduced by using Alpha, Bravo, Charley, and so forth (see Chapter 22). See Box 25.3 (PRUMAE) for comments on warnings.

Pictorial information supplemented with text is more effective than text only or pictures or pictograms only. See Chapter 32. Tufte (1983, 1990) has two excellent books demonstrating design of figures and tables. See Chapter 20 for information on figures, tables, and codes.

Information for use by the general public is more difficult to communicate than information for employees of your firm. The general public is a group of first-time users with very diverse characteristics (age, vision, language ability, etc.). Butler et al. (1993) showed that when giving directional information, using a series of signs was more effective than giving a map marked with "you are here" and letting people find their way. Andre and Koonce (1991) reported "you are here" maps were better when the map top was "track up" rather than north.

Words need to be translated and standardized (see Chapter 32). When giving instructions, don't say "second floor" when the elevator has only "upper level" and "lower level." That is, be sure that information uses the same words everywhere, not synonyms.

Be sure information is specific enough. "Meet at the south door of the office" is better than "meet at the office door."

Rather than assuming that the user has understood the information about a standard procedure, a better approach is to field test the procedure. For example, for assembly instructions for consumer products, videotape typical customers assembling the product. To test the effectiveness of inspection procedures, mark some defective units with paint visible under black light. Then include the defective units with regular production and see if the defects are caught. (This technique is to evaluate procedures, not to blame individuals. No blame or punishment should be given to individuals who do not catch the defects.)

When transmitting information to others, a good technique is to have them repeat the information back to you in their words to be sure they understood you. For example, a secretary could repeat back to the engineer, "Make three copies, sign your name, and mail." In technical terms the forward communication (you to them) has been fed back to you (**information feedback**) in the secondary communication (they to you). As the number of nodes increases, the need for feedback increases. For example, if you tell person A to tell person B to tell person C, the chance for error increases. As the message complexity increases, the chance for error also increases. If the message is transmitted without a permanent record (e.g., by conversation), the receiver must depend on memory, and the chance for error increases. If the information in the message is unexpected (e.g., it conflicts with standard procedure), the chance for error increases. Critical messages should therefore be written (so a reference is available for re-examination of the message) and require a confirmation from the user not only of the receipt of the message but of the user's understanding of its meaning.

GUIDELINE 3: HAVE PROPER EQUIPMENT/PROCEDURES/SKILL

This guideline discusses equipment, then procedures, then skill.

Equipment

Four aspects of equipment are design, amount, arrangement, and maintenance.

Design   With proper equipment, you can design out the potential error; then procedures and skill are not necessary. Properly designed equipment will tolerate foreseeable use and misuse. Consider an open manhole. People could be trained to walk carefully when near an open hole (skill). People could learn to avoid open manholes (procedure). But, if you put a cover on the hole (engineering), there no longer is the possible error of falling in! Engineering also has the virtue of being a permanent solution.

An agricultural example is manure pits on pig farms. Vapors from such pits can displace oxygen; so entry can cause death. Design the pit so there is no need to enter it (e.g., for maintenance), and put a cover on the pit so people don't fall in accidentally. (Also add warnings so people don't enter the pit deliberately.)

There are numerous design examples. A pedestal chair with 5 legs is more stable than one with 4 legs. The legs should not extend beyond the chair base or they will present a tripping hazard (see Guideline 4 in Chapter 15). Equipping an outlet with a ground-fault circuit interrupter minimizes the risk of electrocution. In sports, baseball players (especially amateurs) often are injured by sliding into bases. A breakaway base increases the margin for error because, if excessive force is applied to the base, it breaks away from its support (instead of being rigid and breaking the ankle). When designing mechanical or electrical connectors, make the connectors of different sizes and shapes so people cannot make the wrong connection. Design parts so they cannot be assembled in the wrong orientation.

Although the normal concept is to make equipment easy to use, in some cases the equipment should not be easy to use. A consumer products example is making prescription drug containers and disposable cigarette lighters difficult for children to use. In industry, you may wish to avoid casual users running or activating equipment. An example of this is **lockout/tagout,** which prevents the equipment from being activated while it is being maintained.

Some equipment needs guards. Be sure the guards are purchased and *remain* installed. (People often remove safety equipment.) If personal protection equipment should be used, be sure it is used. A common personnel policy is discharge for removal of guards or failure to use personal protection equipment. (Generally one warning is made and discharge occurs on the second offense.) If such policies are not enforced, it will be difficult to avoid accidents.

Equipment also can be designed to reduce the skill needed to avoid errors. For example, outside stairs might accumulate snow and ice. Instead of requiring care in walking, design the stairs with a tread of open expanded metal (see Figure 12.2) which minimizes snow/ice accumulation and also provides good friction.

Amount

Equipment can be duplicated. For example, you could have an umbrella at the office, in your car, and at home. Then if you forget to bring your umbrella, it is not a problem since an umbrella is already there. You can have duplicate pens or keys. Duplication is popular primarily for low-cost components. However, for critical situations, expensive equipment can be duplicated (e.g., triple redundancy in the Boeing 747 control system and double hulls in oil tankers). Even labor can be duplicated (copilot), or people can back up equipment or procedures. For example, a locksmith can open a door if you lose the key. If you forget to take enough money with you, perhaps you can use a credit card, write a check, get money from a cash machine, or borrow money.

Arrangement

The arrangement also is important. For example, components in parallel and components with a standby system improve system reliability, whereas components in series decrease system reliability. Availability can be improved by

FIGURE 12.2   Expanded metal stairs reduce slipping on snow and ice.

improving reliability or maintainability. For more on these topics, see Guideline 3 in Chapter 13.

Redundancy also is a powerful error-reducing tool because it provides a backup, safeguarding the system.

One use of redundancy is to repeat a procedure—checking the second answer versus the first. For example, add a column of figures twice. Have a test for cancer repeated. When doing machining or carpentry work, good practice is to make measurements twice. Bar codes include a check digit so that the sum of the information digits is confirmed by the check digit; if they don't agree, it is a "no read." To fire a missile, two people may both have to enter the same command. A device may transfer information two ways, for example, mechanically and electronically or over two separate paths. A message may be sent both by phone and by mail.

A second possibility for redundancy is to provide information in which some of the information confirms the rest. For example, when making appointments, it is good practice to give both the day and the date (Thursday, February 25th). The Thursday and the 25th are an **error-checking code** in that

Thursday confirms the 25th. When writing a check, you write the amount in words as well as writing the number. On a letter, a ZIP code is backed up by the name of the city and state.

Another redundancy variation is using a two-step procedure for critical actions. For example, to delete a computer file requires depressing two separate keys. To actuate critical controls may require initial actuation of an enabling control, or the control may have to be moved in two directions (e.g., over and then up).

Another redundancy technique is to duplicate equipment. For example, commercial jet aircraft have four engines but can fly on only one. You might have two cars, hoping that one will start. Duplicate your files on the computer in case they are erased (accident or virus).

Maintenance   The standard assumption is that the equipment is properly maintained and operates properly. Unfortunately, this is not always true. For example, at the Union Carbide plant at Bhopal, a cloud of escaping methyl isocyanate killed at least 2,500 people and severely injured tens of thousands more (although none of the staff of the chemical plant, who fled in time). The problem began when a maintenance crew, by mistake, added water to the methyl isocyanate, setting off a chemical reaction. The facility had a number of safeguards and emergency procedures, but they could not be implemented because pipe connections were missing, meters didn't work, and other problems. Management had made maintenance a low priority (Casey, 1993). It is clear that an important error reduction technique is to provide a sufficient **maintenance budget,** one that provides both funds for spare parts and funds for personnel.

Procedures

Procedures will be divided into computer procedures (software) and human procedures.

Computer procedures   Software errors are difficult to eradicate. For example, a programmer may not anticipate the "field" use. Failure of some Patriot missiles to intercept Scud missiles during the Gulf War has been attributed to accumulations of inaccuracies in the computer clock (Littlewood and Strigini, 1992). The designers anticipated that the Patriot system would be turned on and off, but the users just left the computer on.

Part of the problem of software is that response to change is not a "well-behaved" function (small changes in stimuli produce small changes in output).

Small changes in a program (changing one bit from a 0 to a 1) can cause a radical response.

Another software problem is that users may accept the computer's solution as "gospel" without realizing the assumptions built into the program (such as that input information is correct, that the programmer has considered all rare events, etc.)

Techniques for obtaining error-free software are beyond the scope of this text, but it is obviously an important problem.

Human procedures   Procedures describe the sequence of body motions necessary to accomplish a task. Skill (see the following section) is the eye/brain/hand coordination to do the body motion.

Designers often assume "everyone knows. . . ." In legal terms, it is **open and obvious.** A better approach is to assume "not everyone knows. . . ." The designer should make an explicit list of the things it can be assumed the user actually knows. Doing this will force realization of how little can be put on the list!

Standard procedures **(protocols)** should be developed. Then users should be trained in the protocol. The protocol can be considered a "technology transfer." Not all standard protocols are perfect, however. Figure 7.18 shows how errors were reduced at a Bridgestone plant after it was found the standard protocol gave worse results than a nonstandard protocol.

If people are going to use equipment, ensure that they are trained on that equipment. (It's not what we don't know that hurts us, it's what we think we know but don't!) When designing instruction manuals and operator manuals, have one manual for each model number and language. That is, do not force a person to try to find (to filter) the relevant information from a mass of irrelevant information. Dispose of obsolete information and equipment so people don't use the obsolete information or equipment by mistake.

Training may require a formal procedure of (1) writing down the training procedure and training aids, (2) designating specific people as trainers, and (3) recording that a specific person has been trained by a specific trainer on a specific date. On a larger scale, there may be special training, such as for airline pilots, electricians, nurses, and others. Training can be in vocational schools or in apprentice programs.

For assembly work, time is greater and errors tend to be greater if parts are missing and the assembly must be placed aside to be completed later. A good policy is not to start an assembly until all parts are present.

Skill

There are both physical and mental aspects of skill. The "skill" of the machine also can be called machine capability. Consider how much of the necessary skill should be in the machine and how much in the person. Properly selected tools may have the skill in the tool. For example, a proper box knife makes it very difficult to cut one's skin.

Designers normally assume a capable and fully trained operator. However, capability of even a fully trained person can be reduced by fatigue, prescription drugs (e.g., some antihistamines cause drowsiness), and recreational drugs (marijuana, alcohol). (Hahn and Price, 1994, discuss the effects of alcohol on job behavior.) In addition, some employees may not be fully trained. (This lack of full training can be due to management not providing the training or to the inability of the trainee to comprehend the standard training.) In addition, some employees may have physical or mental handicaps (and thus require eyeglasses, hearing aids, job aids, etc.).

Knowledge can be memorized or provided by a **job aid,** a source of constantly available information (e.g., assembly drawing). See Chapter 32. As computers have become lower in cost and capable of storing vast quantities of information, computerized job aids have become more feasible. There remains a problem of how to provide access to the specific information needed. See Chapter 32 on job instruction.

Training is a lifetime activity; that is, people need not only initial training but also refresher training. Refresher training may be critical because of expectancy. That is, when employees become very familiar with a task, they ignore some safety hazards and need to be reminded of the dangers.

Using the rule of practice makes perfect not only reduces errors but also reduces time. "A rookie mistake" is another way of saying that novices have many more errors than "old pros." To detect errors, therefore, check the rookies. Consider the technique of green, yellow, and red operators/operations discussed in Section 3.6 of Chapter 5.

Practice can be dress rehearsals (such as fire drills) or can be in simulators (for pilots). Nuclear power plant control rooms have emergency operating procedures for severe accident management. Practicing for various accident scenarios shows how to avoid the problems. However, in most jobs, practice is obtained by normal work on the job.

GUIDELINE 4: DON'T FORGET

Two approaches to avoid forgetting are to reduce the need to remember and to use memory aids.

Reduce the Need to Remember

Avoid verbal orders. The reason is that verbal orders leave no reference which can be consulted to refresh the memory. (If you do receive verbal information, create a database by writing it down so it can be referred to later if necessary.)

Make a list. Whenever you have something to do, put it on a list. Many electronic and paper organizers work on this concept. Of course, periodically you need to look at the list!

One way to reduce the need to remember is DO IT NOW, DO IT NOW, DO IT NOW.

People often forget where they put something. One solution is to put them in a standard place which is always checked (a place for everything and everything in its place). For example, if you want to take an item home from the office, when you are done with it, immediately place it in your briefcase so you don't forget to put it in later. Park in the same area of the parking lot so you can find your car easily. Keep all handtools on a pegboard with the tool shape in black to remind you what is missing. When you take something out of a file temporarily, write on the paper the name of the file so you know where to put it back, or leave the file folder partially pulled out.

Use Memory Aids

Memory aids are systems or devices to improve memory. People often depend on memory to recall information and, as is well known, memory is not perfect; thus, there is a need to make memory aids such as databases (files, books, notes, computer files, road maps) complete, convenient to use, and accessible. A phone number in the phone book is no help if you don't have the phone book! This explains the recent development of numerous organizers (both paper and electronic) to help people have easy access to information on appointments, addresses, phone numbers, and so forth. Forms aid remembering inasmuch as a blank spot on the form indicates something is missing.

Memory aids can have a downside. Consider a checklist. It minimizes forgetting but also stereotypes behavior (i.e., people ignore items not on the list).

Another memory aid is to reduce the amount to be remembered by remembering a pattern rather than specific items. (We can remember very complex patterns in music.) For example, schedule a meeting on the first Friday of each month. Then you only have to remember one day—the first Friday—rather than 12 individual dates. At McDonald's, the orders are filled in a standard sequence. Next time you are at McDonald's, observe and see if you can determine what the standard sequence is.

A calendar is also a memory aid. Writing appointments on a paper calendar works well, although there are many electronic alternatives now available. Will you remember to make the proper entry on the electronic calendar? Changes and corrections are relatively easy on the electronic versions, and the computer also can print out a daily "to do" list.

Giving appointment cards to people is a tried-and-true technique. They probably will lose the card, but at least they have something written down that they can transfer to their calendar. For best results, however, call them before the meeting and remind them!

GUIDELINE 5: SIMPLIFY THE TASK

Two ways to simplify tasks are to reduce the number of steps and to improve controls and displays.

Reduce the Number of Steps

An autodialer on the telephone can give one-button dialing in place of multiple-button dialing. It saves time as well as reduces errors. Other simplification examples are autologon procedures for computers and macros for word processing programs. When communicating, minimize the number of nodes in the communication network because each node is a potential source of error. For example, a verbal message is less likely to have errors if it goes through two people than if it goes through four.

Many retailers now have their local stores' computers communicate directly with various vendors' manufacturing computers. For example, a computer in a Wal-Mart store in Manhattan, Kansas, might communicate with a GE computer in Indiana, reporting that 40 units of product X were sold this week. The GE computer then would trigger replacement shipments and adjust manufacturing schedules. Although such direct communication originally was installed to reduce lead time, a significant benefit is the virtual elimination of errors in the "paperwork."

Taken to the extreme, reducing steps is automation—that is, eliminating the person as an operator and using the person only as a supervisor (of the machine). But we automate what we understand. What remains is complex, obscure, low probability, and not well understood. Thus, an argument can be made that the automation should handle the low-probability events since the cost of computer memory is minimal (although the engineering analysis cost of the low-probability events is high).

Improve Controls and Displays

We communicate to machines with controls. We and machines communicate to people with displays.

(Chapter 19 discusses how to reduce errors with controls; Chapter 20 discusses displays.) Controls and displays should not conflict with population stereotypes. See Table 19.8.

The designer may consider a task trivial and wonder why any instructions need to be given. However, novices may not even be able to do the task. Thus, instructions need to be field tested.

Confusion matrices show that people often confuse the characters zero and O, and 1 and l (zero/oh; one/el). For this reason, it is best to use all letter codes or all numeric codes. If a mixed numeric and alphabetic code is used, omit use of zero, oh, one, and el. Also do not use a dash and an underline—or at least distinguish between them.

In written communication, emphasize the important information by bold print, larger print, capitals, use of color, and so forth. Avoid complex words as they must be "translated." For example, is a flammable or a combustible liquid more dangerous? *Hint:* Flammable liquids have a flash point of $\leq$ 100 F; combustible ones have a flash point $>$ 100 F. *Second hint:* Gasoline is flammable; charcoal starter is combustible.

In many situations, the person faces data overload. The temptation is to simplify the task by having a computer filter the information presented. However, the filter may eliminate the relevant information. Consider having the type and level of filters under operator control so that the operator can decide how much information is filtered and how. Another possibility is changing the filter characteristics if certain types of responses go beyond predetermined limits.

See Box 12.2 for a discussion of inspection.

Guidelines 1 to 5 discussed error reduction through planning. Guidelines 6 and 7 consider execution aspects of error reduction.

GUIDELINE 6: ALLOW ENOUGH TIME

Although vigilance—too much time and too little to do—may be an inspection problem, the typical problem is too little time, resulting in stress and errors.

When people are under stress (fatigue, fear, personal problems, new on the job), they do not perform as well. (The Yerkes-Dodson law says that performance is an inverted U, with the low performance when the stress is either too low or too high.) One way to reduce the effects of the other stresses is to reduce the time stress. When under time stress, people take chances and shortcuts. This can lead to safety problems and to quality problems.

Some people may not plan ahead and allow enough time. If a report is due Wednesday and you

BOX 12.2 *Inspection*

The following borrows extensively from Drury (1992).

Inspection (viewing closely and critically) can be done manually, often using small tools, or by machines. For inspection by a fully automatic machine, a person will be involved in setup and calibration. Semiautomatic machines typically will examine an object and present a recommendation to a person for a decision. In all three cases, errors occur.

Tables 12.3 and 12.4 describe inspection outcomes. EFN is an index of system performance:

EFN = Effective fraction nonconforming (fraction of total input rejected by inspector)

$$= \text{False alarms} + \text{hits}$$
$$= (1 - p_1)(1 - DR) + p_2 (DR)$$
$$= (1 - p_1) - DR (1 - p_1 - p_2)$$

where

DR = Defect rate probability
p_1 = Probability of correct acceptance
p_2 = Probability of a hit

There are two basic techniques for repetitive inspection (see Guideline 4 in Chapter 14). The inspector can inspect all n items for characteristic A, then all n items for characteristic B, and so on. Or the inspector can inspect item 1 for characteristics A, B, . . ., then item 2 for each characteristic, and continue with the rest of the items. For minimum errors, use the first method. For example, a teacher should grade all exams for question 1, then all exams for question 2, and so on, because the opposite method (grading all questions of student 1, then all questions of student 2, etc.) is less consistent.

Table 12.5 lists the four basic inspection functions: present, search, decision, and action. Search and decision are the critical functions.

Search. Search time is affected by (among other things) field heterogeneity (number of types of nontargets), target uncertainty (number of types of targets), nontarget density, size differences (between target and nontargets), directed attention (cued or not cued on target location), peripheral position (angular separation from the fixation point), and meridian (angle of the axis of the eye) (Lee, Jung, and Chung, 1992). To maximize search performance, consider task, environmental, and personnel factors.

Task.
- Know exactly what to search for. Current information on the probability and importance of various types of defects improves performance. Good symptom descriptions aid mechanics and physicians.
- Know where to search. (Tell the inspector where defects are likely to be; for example, experienced radiologists know where to look on X-rays for cancer).
- Organize the items to be searched (as opposed to using a random pattern). For example, if you have to search a long list, organize the list into logical sections so you can skip areas not of interest. This is the procedure for computer menus.
- Note that, when inspecting on conveyor belts, inspectors tend to look only in a certain band (especially if they wear bifocals or trifocals); they ignore items straight down and on the far side of the belt. Belt orientation may be a factor also. Eastman Kodak (1983) says inspection on a moving belt is more accurate when the belt is perpendicular to the shoulders and moves toward the operator rather than parallel to the shoulders. However, Suresh and Konz (1991) found the belt should move parallel to the shoulders; they also recommended analyzing the scanning pattern to maximize viewing time, because if the inspection is done under time pressure, the last items inspected may have higher error rates.
- Make the object conspicuous in relation to its background (improve the contrast).
- Allow sufficient search time. (However, if the target occurs very rarely, there may be boredom and a vigilance problem.)
- Consider having machines do the search and then present candidate flaws to inspectors who decide which candidate flaws are faults.

Environment. Increase the visual size of the flaw (magnification, special lighting techniques, decreasing background clutter, etc.). Think of this as anticamouflage. See Section 2.5 of Chapter 21. Is the lighting adequate? Should ambient lighting be supplemented with portable (task) lights or even personal lights (flashlights)? For more on inspection lighting, see Section 4.2 of Chapter 21.

Personnel. A number of tests have been developed to identify good inspectors; however, due to

(continued)

BOX 12.2   *Continued*

legal reasons, it is difficult to implement these tests on industrial workers. The large differences in inspection performance among inspectors (as much as 10 to 1) are due not only to differences in search performance but also to differences in decision making. However, it should not be controversial to have inspectors pass periodic vision exams. See Section 1.1.2 on normal vision and Section 1.1.3 on eye problems in Chapter 21.

Decision Making. The inspector must compare the flaw vs. a standard. If the standard is only in the inspector's memory, it can vary (from the firm's standard, from inspector to inspector, or even from day to day for the same inspector). Therefore, use a job aid called a limit sample (see Chapter 32, Section 4) to change an absolute judgment to a more accurate comparative judgment. Note that it is very important to have an unequivocal objective standard. One of the practical problems of inspectors is that they have not received a precise definition of what is defective. For more on an inspector's criterion, see Box 12.1.

Training of inspectors should consider (1) what to perceive, (2) feedback on the effectiveness of

the search and decision, and (3) using the concept of learning the job step by step (progressive part training). An inspection manual should include:

- information on the cause and effects of each defect type
- conspicuousness and location of defects
- frequency of occurrence
- method for inspecting defects
- a clear representation of the standards

Kleiner and Drury (1993) give a detailed discussion of a specific training program they used for inspectors.

Note that social pressure can affect quality of decisions. Konz and Redding (1965) demonstrated many years ago that social pressure can be used either to enhance or to degrade decision-making accuracy; thus, it is a good idea to encourage procedures that use social pressure to enhance accuracy.

Complex decisions should be analyzed and the procedure presented in decision structure tables (see Chapter 7).

don't start until Tuesday night, you don't have any margin for delays, illnesses, and unexpected problems. Another example is driving a fork truck too fast and hitting a rack when turning.

A schedule change may force a reduction in the

allowed time. For example, an airplane may be late and the goal becomes reducing turnaround time from 30 minutes to 20 minutes. A production schedule may be changed and only 3 days are allowed before shipment instead of 5 days. In these cases, additional staff should be assigned to the task or errors will occur due to the time stress. The organization needs to have the flexibility to shift job assignments, which requires cross-trained employees.

See Chapter 31 for a discussion of fatigue allowances.

TABLE 12.3   Inspection outcomes and Type 1 and 2 errors.

	Innocent	Guilty
Acquit	OK	2
Convict	1	OK

TABLE 12.4   Inspection outcomes (Drury, 1992). Sinclair (1979) gives an extensive tabulation of data values of p_1 (values range from .90 to .99) and of p_2 (values range from .8 to .9).

Inspection Decision	TRUE STATE OF ITEM	
	Conforming	Nonconforming
Accept	Correct accept (p_1)	Miss ($1 - p_2$)
Reject	False alarm ($1 - p_1$)	Hit (p_2)

TABLE 12.5   Inspection functions (Drury, 1992).

TASK	DESCRIPTION
Present	Present items for inspection
Search	Search items to locate possible faults (called flaws)
Decision	Decide whether each flaw exceeds the standard (and is thus a fault)
Action	Take action to accept or reject the item

GUIDELINE 7: HAVE SUFFICIENT MOTIVATION/ATTENTION

Motivation

Motivation can be positive (helping performance) or negative (hindering performance).

Social pressure can help or hinder the performance of the decision maker (Konz and Redding, 1965). Social pressure often is confounded with rank. Many cases have been reported in which lower rank/status people did not want to challenge higher rank/status people and the higher ranked person's error was not corrected. For example, in 1923, 7 destroyers of the U. S. Navy ran aground in California because their group commander (1) required them to follow his ship at an interval of 13 s in a heavy fog and (2) ignored direction-finding signals and ran his ship (and the following 6 ships) onto the rocks (Casey, 1993). In a similar vein, the impact of the tanker *Torry Canyon* onto the rocks, spilling 31,000,000 gallons of oil, was due in part to lower ranking personnel not wanting to challenge the captain.

It is easy to recommend that decision makers consult others before making decisions, but this is difficult to implement—especially in hierarchical organizations. And, even if the decision maker consults, will others give their true opinion or just nod their heads and say yes?

People can be motivated by many things, not all of which are obvious. This is demonstrated when people deliberately disregard known rules. The crash of the Airbus 320 at the Mulhouse air show was due in part to "overconfidence of the pilot" who had turned off the automatic flight protection system (Casey, 1993). This accident, Three-Mile Island, and Chernobyl are all examples of operators shutting off automatic protection systems. When you are designing a system, will you permit the operator to shut off the safety features?

People can be motivated if given credit or publicity for their actions. See Figure 12.3 for an inspection example.

Attention

Attention lapses can range from the disastrous to the trivial. The *Herald of Free Enterprise* sank with 188 dead because, in part, the sailor whose job it was to close the bow doors was sleeping while on duty. Doctors in training (residents) often are scheduled to work 36-h shifts in a kind of rite of passage; unfortunately, they get tired and make mistakes. Although working very long hours causes errors, in general, people working the night shift do not have any more accidents/hour than people on the day shift. (A study

FIGURE 12.3   Publicly assigning responsibility can improve motivation; this tag came from a pair of shoes I bought. Another firm, Cooper Tire, calls this policy "The Tire with Two Names." Once a finished tire has been pulled from the press, the employee sticks a label inside with his or her name.

of miners in Minnesota, however, showed higher accident rates for people driving home after working a night shift.)

People may also lack attention due to substance abuse. It is difficult to test people's fitness for duty, especially on a continuous basis. That is, you can give drug tests which may (or may not) detect whether a person has previously taken drugs. But testing minute-by-minute alertness is difficult. One common technique is to have people work in a group so others can detect sleeping, heart attacks, drunkenness, or other problems. It also is possible to observe workers remotely through closed-circuit TV. The deadman's switch on a locomotive, if not held, stops the train. Some night security guards have to actuate a signal periodically to show they are still OK.

Attention lapses also can occur from distractions. Distractions can be from many causes (talking, listening to the radio, looking at people of the opposite sex, trying to do two things at once, etc.) and so, for critical decisions and actions, minimize distractions.

The following sections, Guidelines 8, 9, and 10, explore the concept of allowing for errors.

GUIDELINE 8: GIVE IMMEDIATE FEEDBACK ON ERRORS

Feedback distinguishes between a **closed-loop** and an **open-loop system.** See Figure 12.4; see also

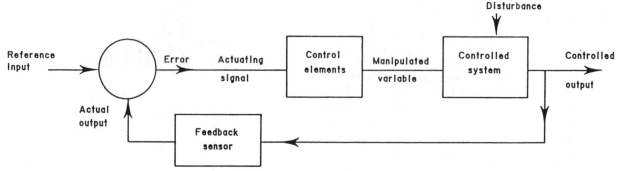

FIGURE 12.4   Feedback distinguishes a closed-loop system from an open-loop system.

Figure 23.8. Early error detection is aided by feedback or information. Control charts feed back the state of the process. Instruments feed back the status of equipment.

The two subdivisions of feedback are error detection and reducing delay.

Error Detection

The error can be detected by people and by machines. The person can be the person doing the action or someone else. (Feedback on manufacturing procedures is called inspection but feedback on office procedures tends to be called review.) See Box 12.2 for comments on inspection.

When a machine detects an error, it can display the information and wait for the human to take action through a control, or it can take action (either responding or not responding). If it takes action, it can actuate a display or not actuate a display. Displays can be audio, visual through instruments, or printed. Information also can come from failure to respond. Examples of failing to respond would be a door which didn't open when the wrong key was used and a computer program not responding when the wrong command was entered. The display also can be multimodal. For example, when a needle on an instrument goes into the red zone, an audio warning may sound (indicating there is a problem on some display).

Systems can correct errors without notifying the operator, but good practice is to notify the operator that an error has been corrected. Because what the machine considers to be an error may not really be an error, the machine may present the error as a *possible* error and await a human decision. An example of this approach is spelling-checking programs.

Another approach is to reduce the opportunities for error. The FAA has developed constraints on permitted computer input for aircraft inspectors (Layton and Johnson, 1993). For example, if an inspector selects Boeing as the aircraft make, only Boeing

models appear for selection in the model field. Furthermore, if Mythical Airways owns only Boeing 757 and 767 aircraft, only those models will appear on the screen for choice. When using the computer program *ERGO* which supplements this book, various values are checked for reasonableness. For example, if selecting metric or English, only an input of m or e is accepted. In calculating body fat, if weight does not have a reasonable relation to height, a message will appear: "Error—low input weight or high input height."

As discussed in Guideline 10, displays of errors are very useful to operators. Note that errors can be displayed to multiple people or locations.

Error messages should be specific and understandable. A message of "Error" flashing on the screen is not very useful. Only slightly more useful is "Type 3 error." Better would be "Invalid input—value too high"; better yet would be "Invalid input—value too high. Permissible range is 18 to 65." Do not combine types of errors in the message. For example, it would not be helpful if a fax machine gave a message "Number busy, please redial" when the other number was busy and also when you dialed a phone number instead of a fax number.

Alarms should not be too irritating or people will turn them off. In autos, the first seat belt alarm was an audio buzzer. It was so annoying that many people cut the alarm wires. Other alarms can be multilevel. For example, low fuel level in an auto could be signaled first by a steady warning light and then, as fuel gets even lower, by a flashing light and an occasional beep. Some cameras with timers will flash every second during the countdown until the last few seconds, and then the flash rate increases.

Reducing Delay

The longer the delay between the error and the detection of the error, the more difficult it is to detect the error. Early detection of an error tends to reduce the consequence of the error. (Failures with long delays are called **latent failures.**) A key feature of user-friendly word processing programs is **"what you see is what you get"**—abbreviated as WYSIWYG. That is, the screen shows whether you keyed an a or an s, capitals or lower case, bold or conventional. In WordPerfect 5.1, a superscript is indicated on the screen with one color and a subscript with another. A better clue is moving the character up or down (as is done in WordPerfect for Windows). The WYSIWYG concept minimizes the need for translation (reinterpreting information).

A short time between the error and its detection is one of the key features of the control charts used in statistical quality control. For example, if a defective part is detected before it leaves the department in which it is made, correction is easier than if the error is detected two weeks later when it is being assembled or two months later when the customer uses the equipment. Many cars have warning devices which make a sound if you remove the keys from the ignition while the headlights are on. In this case, you may forget (to turn off the lights), but you are able to make a correction before damage occurs.

GUIDELINE 9: IMPROVE ERROR DETECTABILITY

Considering an error as a "signal" and the background as "noise," detectability is improved when the **signal/noise ratio** is high. This improved contrast can be done by amplifying the signal or reducing the noise or both (see Box 12.1).

Amplify Signal

For an error to be detected, the body must perceive it (typically through the eye or ear) and then the brain must recognize the error. Some errors are not recognized as errors by the brain. A person may key "hourse" instead of "house." This obvious type of error can be detected with spelling-checking programs. But a person may think an error is correct, for example, writing "the plane truth" when it should be "the plain truth" or "the box weight is 40 kg" when it is 40 lbs. These errors are not detectable by a spelling checker since "plane" is an acceptable spelling, and it is the semantics that is incorrect. (For errors such as "the box weight are 40 lbs," there are grammar-checking programs which can check for tense, sentence construction, etc.) The use of the wrong unit (kg) is not detectable by spelling or grammar checkers. One solution for both the "plane" error and the "kg" error is checking by a person other than the person who initially did the work.

One way to enhance the signal is to match it. This **paired comparison** improves detection of differences. Inspectors, for example, often use a comparator which superimposes two images. Also, when someone gives you a message, you can match it by repeating it back to them (preferably with paraphrasing) to ensure that you understood it.

Alternatively, the size of the signal can be increased. The position of a switch is in itself a display, but not a very obvious one. One way to improve the display would be that when the switch is actuated, not only does the switch position indicate ON but also an indicator is turned on, creating a redundant display. Chapter 19 discusses control labeling.

The signal should not conflict with **population stereotypes** (expected relationships). For example, by convention, decaffeinated coffee is placed in containers with an orange or green stripe. Thus, do not place decaffeinated coffee in a nonstriped container or regular coffee in a decaffeinated container. Another example is putting caustic materials in food containers or near food containers. In Topeka, 12 people drank lye because it was stored in a container adjacent to a drink mix and the bartender picked up the wrong container (Casey, 1993). Do not keep chemicals in refrigerators or other locations where people expect to find food.

Location or time may affect ease/cost of error detectability. For example, on a new car, a low level of brake fluid can be detected quite easily by checking the level in each vehicle before shipment. Once the vehicle is delivered and driven, the driver would notice a warning indicator lighting and take the vehicle to the garage for service, but the cost of the error would be magnified manyfold by not detecting it at the factory.

Reduce Noise

In addition to increasing the signal, contrast can be improved by reducing the noise. Noise can have many dimensions. For example, improve the visibility of the target through various visual techniques, such as color coding (see Chapter 20). If the target is a sound, reduce background noise (see Chapter 22). See Guideline 7 of this chapter for comments on distractions.

GUIDELINE 10: MINIMIZE CONSEQUENCES OF ERRORS

With computer-based systems, it may be possible to indicate the consequences of a decision or action and ask for confirmation before execution. That is, important decisions or actions (1) have multiple steps and (2) are reversible. The more important the decision or action, the more steps (and people) should be necessary to implement it.

A goal is to design equipment and procedures so they are less sensitive to errors. Ideally, they should be **fail-safe,** so that errors are impossible. What if there is a power failure or a power surge? What happens if a bearing wears out? an instrument fails? a light bulb fails? the central air conditioning system fails? the address on a message is incorrect? an untrained person operates the equipment? the operator has a heart attack? the operator is drunk?

The well-designed system anticipates and deals with possible problems. For example, possible power failures imply emergency power systems; power surges imply circuit breakers; a bearing failure implies preventive maintenance and emergency shutdown procedures; instrument failure implies supplying redundant information (and, perhaps, more reliable instruments); light bulb failure implies multiple light sources and spare bulbs; central air conditioning failure implies windows that open, portable fans, and ability to move people to another location; incorrect addresses imply using return addresses; and so on.

Ease of recovery also is important. Designers should analyze the error recovery path to be sure that recovery doesn't cause additional problems. For example, if a mistake is made on a computer input, can the error be corrected easily? What if a box jams on a conveyor turn? What if a paycheck is lost?

Design the system to minimize the spread of the error throughout the system. If the computer fails, it should not affect the local area network. If the power plant fails, it should not cause the power grid to fail. If the tire has a blowout, it should not cause the auto to crash.

Many accidents and catastrophes result from combinations of failures. For example, the *Herald of Free Enterprise* sank because a worker was sleeping (and didn't close the bow doors), there was no instrument to show the captain that the doors were not closed, the sea was heavy, the ship was going relatively fast, and the bow was down because of the load.

For comments on equipment availability, reliability, and maintainability, see Guideline 3 of this chapter.

Error effects can be minimized by guards. Guards can be on equipment (auto seat belts, air bags, roof on lift truck, rubber handles to prevent burns and electrical shocks), in the environment (stair railings), on people (hard hats, gloves), or in procedures (putting a stop on a check you don't want cashed).

REVIEW QUESTIONS

1. What are the two primary justifications for ergonomics?

2. What is the point of the story about the poisoned grain in Iraq?

3. What is the difference between a slip and a mistake?

4. What is the difference between an error of omission and an error of commission?

5. The 10 error reduction guidelines are divided into what three general categories?

6. What is the point of the story about the Bhopal disaster?

7. If a critical message is to be transmitted through a loop of several people, what is a good procedure?

8. What is the point of the story about the Patriot missile failure?

9. Is a flammable or combustible liquid more dangerous? What point did discussion of this information make?

10. What does WYSIWYG stand for?

11. What did the operators at Chernobyl, Three Mile Island, and the Airbus 320 crash at Mulhouse all do?

12. Why did 12 people in Topeka drink lye?

REFERENCES

Andre, A. and Koonce, J. Spatial orientation and wayfinding in airport passenger terminals: Implications for environmental design. *Proceedings of the Human Factors Ergonomics Society,* 561–65, 1991.

Butler, D., Acquino, A., Hissong, A., and Scott, P. Wayfinding by newcomers in a complex building.

Human Factors, Vol. 35 [1], 159–73, 1993.

Casey, S. *Set Phasers on Stun.* Santa Barbara, Calif.: Aegean Publishing, 1993.

Drury, C. Inspection performance. In *Handbook of Industrial Engineering,* 2nd ed. Salvendy, G. (ed.), Chapter 88. New York: Wiley, 1992.

Eastman Kodak. *Ergonomic Design for People at Work.* Belmont, Calif.: Lifetime Learning Systems, 1983.

Hahn, H. and Price, D. Assessment of the relative effects of alcohol on different types of job behavior. *Ergonomics,* Vol. 37 [3], 435–48, 1994.

Kirwan, B. Human error identification in human reliability assessment. Part 1: Overview of approaches. *Applied Ergonomics,* Vol. 23 [5], 299–318, 1992a.

Kirwan, B. Human error identification in human reliability assessment. Part 2: Detailed comparison of techniques. *Applied Ergonomics,* Vol. 23 [6], 371–81, 1992b.

Kleiner, B. and Drury, C. Design and evaluation of an inspection training program. *Applied Ergonomics,* Vol. 24 [2], 75–82, 1993.

Konz, S. and Redding, S. The effect of social pressure on decision making. *Journal of Industrial Engineering,* Vol. 16 [6], 381–84, 1965.

Layton, C. and Johnson, W. Job performance aids for the flight standards service. *Proceedings of the Human Factors Ergonomics Society,* 26–29, 1993.

Lee, D., Jung, E., and Chung, M. Isoresponse time regions for the evaluation of visual search performance in ergonomic interface models. *Ergonomics,* Vol. 35 [3], 243–52, 1992.

Littlewood, B. and Strigini, L. The risks of software. *Scientific American,* 62–75, November 1992.

Miller, D. and Swain, A. Human error and human reliability. In *Handbook of Human Factors,* Salvendy, G. (ed.), Chapter 28. New York: Wiley, 1987.

Proctor, R. and van Zandt, T. *Human Factors in Simple and Complex Systems.* Needham Heights, Mass.: Allyn and Bacon, 1993.

Rasmussen, J., Duncan, K., and Leplat, J. *New Technology and Human Error.* New York: Wiley, 1987.

Sanders, M. and McCormick, E. *Human Factors in Engineering and Design,* 6th ed. New York: McGraw-Hill, 1987.

Sinclair, M. The use of performance measures on industrial examiners in inspection schemes. *Applied Ergonomics,* Vol. 10 [1], 17–25, 1979.

Suresh, A. and Konz, S. Movement and part orientation in paced visual inspection. In *Advances in Industrial Ergonomics and Safety III,* Das, B. (ed.), 597–600. London: Taylor and Francis, 1991.

Tufte, E. *The Visual Display of Quantitative Information.* Cheshire, Conn.: Graphics Press, 1983.

Tufte, E. *Envisioning Information.* Cheshire, Conn.: Graphics Press, 1990.

Macro Ergonomics

OVERVIEW

Jobs must be designed from the macro ergonomics level as well as the micro ergonomics level. Eight guidelines give some of the "big picture" concepts to complement the micro ergonomics guidelines given in the other chapters.

CHAPTER CONTENTS

1 Plan the Work, Then Work the Plan
2 Reward Results
3 Optimize System Availability
4 Minimize Idle Capacity
5 Consider Shiftwork
6 Use Filler Jobs or Filler People
7 Reduce Fatigue
8 Communicate Information

KEY CONCEPTS

active vs. passive rest
availability
carrots vs. sticks
circadian rhythm
core time
double tooling
downtime
duplicate components
filler job
filler people
financial and nonfinancial
    motivation
fixed costs and variable costs

fixed shifts
flex time
hours/year
idle low-cost components
job rotation
job sharing
life-cycle costs
maintainability
modularization
muscle and central fatigue
not one on one
off-peak
part-time workers

plan the work, work the plan
pools
preventative maintenance
reliability
revising schedules
rotating shifts
series vs. parallel arrangements
staggered work times
standby
temporary workers
working rest

GUIDELINE 1: PLAN THE WORK, THEN WORK THE PLAN

If a sailor does not know
to which port he is sailing,
no wind is favorable.

Seneca

Much blue collar work is repetitive with a countable output. The goals are set by others. White collar work, on the other hand, tends to be self-directed, with a great variety of tasks. Self-directed people need to determine their desired direction and then work toward the goal (**plan the work; work the plan**). For a book on strategy, see Merrill and Merrill (1987); for a book on tactics, see Kaufman and Corrigan (1987).

The two steps are (1) plan the work and (2) work the plan.

1.1 Plan the Work   There are four steps to planning the work.

1.1.1 List goals   First you need to set long-range, lifetime major goals. It is convenient to divide these into work-related and non–work-related goals. Examples of work-related goals are to get a 10% raise in salary, to be promoted, to transfer from a line to a staff job (or vice versa). Example nonwork goals are to get married, to buy a new car (house, take a vacation), to spend more time with your spouse (children, parents), to become more religious. It may help you to focus your thinking to ask, "If I were to die in a plane crash in six months, what would I do until then?"

1.1.2 Set goal priorities   Now prioritize your work and nonwork goals by assigning an A, B, or C to each goal.

1.1.3 Make a "to do" list   This list is of *activities* to reach the goal. For example, a nonwork goal might be to spend more time with your spouse. Activities might be to go to a movie, go out to dinner, or be home by 6:00, for example. A work goal might be to get a report done by the first of the month. Activities might be to make a rough draft of figures, set up a table on the spreadsheet, get cost figures from accounting, and so on. The to do list should be written. If you have accomplished no activities toward the goal within a week, either add activities or consider revising the goal.

1.1.4 Set activity priorities   Next to each activity, mark an A, B, or C. You now are ready to work your plan.

1.2 Work the Plan   There are three approaches to working the plan.

1.2.1 Start with As, not with Cs   The concept here is to start the work period working on your A goals. The reason is that in the day you really have little free time because most of it will be taken up with required duties (answer phone, go to meetings, eat lunch, drive to work, handle the mail); so work on your A goals first. Using the Pareto concept, only 20% of your activities will have a high payoff, so concentrate your efforts on the giants and put off the midgets as long as possible. Perhaps some of the tasks can be eliminated, delegated, or simplified. For example, instead of writing a letter, can it be dictated? Can a phone call replace a letter?

1.2.2 Do it now   Avoid procrastination. It may help to standardize the location of free time for your A activities. For example, lock the door of your office and don't answer phone calls for the first 30 min while working on your As. Spend Saturday afternoon working on your nonwork As.

1.2.3 Cut big activities into bits   Big goals may seem overwhelming, so they are put off. However, each journey of a thousand miles begins with a single step. Assume your A goal is to get a 10% raise and among the activities leading toward that goal is to have a monthly report done on time. However, you won't be able to complete it today. Get started on it and do as much as you can.

GUIDELINE 2: REWARD RESULTS

My object all sublime
I shall achieve in time
To make the punishment fit the crime
The punishment fit the crime.

The Mikado, Act II

Work smart, not hard is the primary message of this book. However, reasonable effort cannot be ignored. The challenge is to get people to work hard *and* smart.

2.1 Types of Motivation   Very briefly, motivation can be divided into positive and negative (**carrots vs. sticks**). Positive motivation can be internal (self-motivation) or external. External can be divided into **financial and nonfinancial motivation** (pay vs. other incentives).

2.2 Financial Rewards   Figure 13.1 shows four different financial reward plans.

2.2.1 Pay independent of output

Curve A, the horizontal line, shows that pay has no relation to performance. Who would use such a plan? Most organizations! Pay by the month (salary) is an example; workers are paid even if they don't come to work (within limits). Salary often is used when an individual's specific contribution is difficult to count. Examples are administrative and technical jobs such as deans, researchers, accountants, engineers. Sometimes there is a countable output, but it is irregular and of unknown quality. Examples might be patents or designs from an engineer or articles published or classes taught by a professor. Salary also is used for situations where there is an undefined relationship between effort expended and the value of the contribution. Examples might be classroom teaching or TV news announcing.

A variation on curve A is pay by the hour. The worker is not paid unless physically present at the workplace. Pay-by-the-hour plans range from pay for the entire day (once you arrive at work) to plans which pay to the nearest 10th of an hour for time clocked in. Although physical presence on the premises does not guarantee a useful contribution, it has better odds than not being there at all!

2.2.2 Pay independent but step function

Curve B shows pay independent of output but there is a delayed step-function response by the organization. For example, if a professor does a good job, there may be a larger raise at the end of the year. Motivation is increased and longer lasting when the size of the step function is larger, the relationship between the behavior and the reward is clearer, and the reward is automatic (i.e., does not depend upon someone's judgment).

For example, if someone works paid overtime or receives a bonus, pay this in a separate check. (Some people consider the regular paycheck is to be shared with the spouse but the bonus is theirs.) Pay it quickly—say within 24 h.

There has been a growing trend to give a special one-time payment to reward current performance—that is, in response to the question "What have you done for me lately?" (Kopelman, 1987). Five advantages are (1) it can be given for a specific contribution, (2) it can be substantial (not being spread out over a year), (3) it can be awarded soon after the specific contribution, (4) it does not make the extra pay permanent, and (5) it increases compensation flexibility.

Many firms give an annual bonus to all employees, depending upon the firm's success. This can range from a bonus of $100 to each worker at a retail store that achieved the goal of preventing shrinkage from theft to a bonus of $1,000 to all employees of Ford Motor Co. when company profits are high. Although the Ford employees like the money, there is very little relation between an individual's performance and the firm's profit.

2.2.3 Piecework with a guaranteed base

Curve C shows piecework with a guaranteed base. Just coming to work gives the guaranteed hourly rate. But for every 1% increase in output, there is a 1% increase in pay. Although the organization's direct labor cost per unit stays constant with greater output, total cost per unit declines because fringe costs (holidays, health costs) decline on a per unit basis. In addition, overhead (burden) per unit declines. Thus, the employee wins and the firm wins.

For more on incentives, see Box 26.2.

2.2.4 Piecework without a guarantee

Curve D pays you what you are worth. The self-employed (doctors, dentists) are in this situation as well as people on commissions (stockbrokers, insurance agents). Curve D is rare in the industrial workforce.

2.3 Nonfinancial Motivation

Pay is a great motivator, but it is not the only motivator. An interesting phenomenon is the "Pygmalion" effect. It is based on the play *Pygmalion,* later made into the movie *My Fair Lady.* It concerns Liza Doolittle, a poor flower girl adopted by an eccentric professor. When treated like a lady, she acted like a lady. The point is that if you expect your workers to do well,

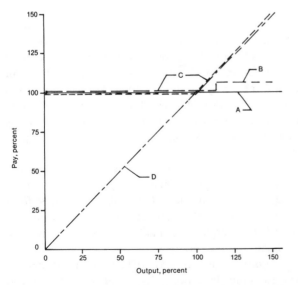

FIGURE 13.1 Curve A shows pay as being independent of output. Most people are paid using curve A. Curve B shows pay as being independent of output but with a step function after a delay (e.g., a 5% raise at the end of the year). Curve C shows incentive pay with a guaranteed base. Curve D shows incentive pay with no guarantee.

praise them, and trust them, they will do well and will deserve your trust. If you criticize people, don't trust them, and so forth, they will do poorly and not deserve your trust.

There are many nonfinancial motivators.

2.3.1 Carrots

"Carrots" generally work better than "sticks"—perhaps because negative rewards are difficult to implement fruitfully in the real world.

Some examples of carrots might be stars on the helmets of football players for good plays, a reserved parking place for perfect attendance, public recognition, praise. (Don't ignore the power of public praise and recognition; the military have found that soldiers will die for such "baubles.") Excellent organizations work hard at creating opportunities for showering pins, buttons, badges, and medals on employees. Titles are cheap; distribute them generously.

2.3.2 Sticks

Some "sticks" are verbal abuse by a supervisor (coaches do this a lot) and penalty points for tardiness and absenteeism. For example, late for work = .5 point, unexcused absence = 3 points, sick with a doctor's excuse = 2 points for first day only. All points are dropped to zero every four months. When 9 points are accumulated, you get an oral warning, 12 points is written warning, 14 points is termination.

GUIDELINE 3: OPTIMIZE SYSTEM AVAILABILITY

$$\text{Availability} = \frac{\text{Uptime}}{\text{Total Time}}$$

$$= \frac{\text{Uptime}}{\text{Uptime} + \text{Downtime}}$$

$$= \frac{MTBF}{MTBF + MTR}$$

where

$MTBF$ = Mean time between failures (reliability)

MTR = Mean time to repair (maintainability)

This guideline discusses improving the availability of a service channel while the following three guidelines discuss better matching of service channel availability and requirements.

In calculating optimum system availability, use life-cycle costing (rather than just initial cost).

Life-cycle cost = Initial capital cost
+ Operating cost over the
system life
+ Maintenance cost over the
system life
+ Downtime cost over the
system life

There are three strategies to improve availability: (1) increase **reliability** (uptime), (2) increase **maintainability** (decrease downtime), and (3) make loss of availability less costly.

3.1 Increase Reliability

Increase $MTBF$. One alternative is to increase the capability or decrease the load. An example of more capability is to use a 3 hp motor when only a 2 hp motor is needed. The additional capability makes failure less likely (although with additional capital and operating costs). A lower load would be using special controls to reduce starting problems as well as operating the motor in a less stressful (cooler, cleaner) environment.

Another possibility is to consider **series vs. parallel arrangements;** in a series arrangement, the system works if *every* component in a path works; in a parallel arrangement, the system works if *any* path works.

$$R_{\text{series}} = R_1 R_2$$
$$R_{\text{parallel}} = 1 - U_1 U_2$$
$$U_1 = 1 - R_1$$
$$U_2 = 1 - R_2$$

For example, if $R_1 = .9$ and $R_2 = .8$, then in a series system the system reliability is $.9(.8) = .72$. In a parallel system, the system reliability is $1 - .1(.2) = .98$.

For example, in a truck, two batteries could be installed in parallel; if one fails, the other does the job. In addition, life should be longer since there is less stress on the 2 batteries than 1. Redundancy in people generally is quite expensive and isn't used very much, although copilots in aircraft are one example. A critical item could be inspected by 2 independent people and accepted only if both pass it.

For series systems, reduce the number of components and/or steps because each is a possible source of failure. For example, when making a request for a hotel wake-up call, contact the operator directly rather than requesting that the desk clerk give the message to the operator.

Another possibility is to use a standby circuit. For example, in a boat if you run out of fuel you could switch to the standby tank. If the power screwdriver failed at the assembly station, you replace it with a new power screwdriver. Some slide projectors have standby bulbs built in so that if the main bulb fails, you switch in the standby bulb. If the air conditioner fails, you use the fan. In a car, if you have a flat tire, you switch to the standby tire. Standby people are quite common, ranging from the utility operator on a flow line to a relief pitcher in baseball.

Modularization is a variation of the standby concept. In place of the original unit being repaired, it is discarded and the standby unit replaces it. An example is small electric motors, which are no longer repaired, just replaced. Automobile spark plugs rarely are cleaned and reused any more; they are just discarded. Modules permit relatively quick switching and can be done by less skilled people. In addition, it may not be either technically feasible or economic to repair the unit locally.

Note that, in a standby system, there generally is some time and effort involved in switching and the system may not return to normal. The switching itself may have a failure rate—for example, you forgot the jack and so can't change the tire! Or the baseball manager may not make the switch to the relief pitcher in time!

Critical systems (such as electricity, fuel, and computers) need **standby** (backup) systems. At a minimum, if the system goes down it should be failsafe. That is, no damage or injury should result from the downtime.

3.2 Increase Maintainability   Decrease *MTR*.
The strategy is to cut downtime.

Downtime = Fault detection time

(time to find the device doesn't work)

+ Fault location time

(time to find the problem)

+ Logistics time

(time to get repair parts)

+ Repair time

(time to fix unit)

A standby unit has only fault detection time; it is working (perhaps not returning the system to 100% performance) while fault location, logistics, and repair time occur. For equipment, this requires good planning for spares. For people, this requires good planning for cross-training of personnel.

Fault location time can be reduced. For example, maintenance personnel can be given beepers and service vehicles can have radios; this reduces the time until help arrives. Many firms offer expert advice over the telephone to local maintenance personnel so the device doesn't need to be sent elsewhere for simple repairs. One example is advice, using 800 numbers, for computer software problems.

Logistics time also can be reduced by administrative procedures. For example, good records allow people to find the supplier for a replacement part quickly. Low-level authorization for ordering parts is quicker than requiring lots of paperwork and multiple signatures. Establishing relationships with vendors so that a telephone call is sufficient instead of a letter reduces the ordering time. Having a rotatable pool of spare components allows the item to be repaired at leisure while the replacement is immediately available. Stocking spares at a number of locations may help, although nationwide 24-hour delivery courier services have made distance less important.

Repair time is usually a small portion of downtime. It can be cut by **modularization,** in which the repair involves merely switching units. Put permanent lighting fixtures *inside* machines to aid the vision of maintenance workers. Maintenance personnel do more than just repair items. Table 13.1 gives some standard maintenance tasks.

3.3 Make Loss of Availability Less Costly
There are two possibilities: (1) preventative maintenance and (2) partial function.

Preventative maintenance is scheduled maintenance. You pick the time you don't need the availability and then do the maintenance. This probably will increase your maintenance expenses over a "if it ain't broke, don't fix it" philosophy.

Partial function may not be too much of a problem—if the partial function is temporary and the "partial" is closer to full capacity than to zero. For example, on an assembly line, two identical stations

TABLE 13.1 Categories of maintenance tasks (Ostrofsky, 1977). Relative to direct labor expense (which has been analyzed intensively for over half a century), maintenance has many opportunities for improved methods and cost reductions.

Task	Comment
Troubleshoot	The task which isolates a fault or failure to keep the desired level in the system.
Inspect	Observation or test to determine the condition or status of the system (or component of the system).
Calibrate (or adjust)	The task required to regulate or bring the condition or status of the system (or component of the system) to the desired level.
Remove	The task required to remove a desired portion of the system.
Repair	The task required to restore a given level of the system to operating condition.
Replace	The task required to replace the desired portion of the system, given that a removal has occurred.
Service	The replenishment of consumables needed to keep a given level of the system in operating condition.

could do a job in parallel. Then if one station fails (equipment failure, employee absence, or whatever), the remaining stations can get by. Firms often split orders between vendors rather than having one vendor as the sole source. The decision to go multiple source or sole source for equipment, vendors, operators, and others should be considered in detail.

GUIDELINE 4: MINIMIZE IDLE CAPACITY

Minimizing idle capacity has two divisions: **fixed costs and variable costs.**

4.1   Fixed Costs   This is divided into problem and solutions.

4.1.1   Problem   Annual cost of many machines and people varies little with output.

Equipment depreciation (i.e., capital cost), for example, usually depends more on equipment age than equipment hours of use. Think of used car prices; they depend more on the model year than on whether it has been driven 60,000 miles or 80,000 miles. The capital cost and property taxes of a factory or warehouse will not vary whether it is used 8, 16, or 24 h/day, or 5, 6, or 7 days/week.

An airline's cost for an airplane flight is about the same whether there are 50 or 150 passengers. A professor will receive the same pay whether there are 20 or 40 students/class; a hospital's cost will be approximately the same whether there are 70 or 100 patients; a hotel's cost will be approximately the same whether there are 300 or 500 guests. The reason is that direct labor cost is a small percentage of total cost.

Even if direct labor is a high proportion of job cost, the wages, especially salaries, vary little with output. A secretary will receive the same pay whether typing 1,000 letters/yr or 1,500 letters/yr; a cook receives the same pay whether cooking for 50 or 100; a bus driver receives the same pay whether there are 10 or 40 on the bus.

Even for hourly wages, costs are relatively fixed, because employers are reluctant to lay people off for minor ups and downs. Part of this is because of the economic cost of the layoff (unemployment taxes on the employer go up, disruption of the existing workforce, slowdowns by the existing workforce, etc.), and part is due to the natural desire of managers not to cause economic problems for their employees.

Thus a desirable social goal is to get more use of the existing capacity. This also has good economic benefits for the firm since the incremental cost of using the existing capacity is low.

4.1.2   Solutions   Four solutions are to operate more hours/yr, use pools, revise work schedules, and encourage off-peak use.

Operate more hours/yr   More **hours/yr** may spread a constant number of customers over more hours, thus requiring fewer facilities. For example, if a machine shop ran for two shifts instead of one, fewer machines would be required (for the same total daily output). Kansas State University runs classes from 7:30 A.M. to 5:30 P.M. (instead of 8 to 12 and 1 to 5) to get better use of classrooms, and thus requires fewer classrooms.

The longer hours may attract more customers. Service facilities often use this strategy. Examples are resorts open more than one season, fast food restaurants open for breakfast, and discount stores open in the evenings. Universities have evening classes for nontraditional students as well as short courses and intersessions. High school gymnasiums are opened to nonstudent users when students are not using them.

Use pools   Pools are based on the concept that peaks (or valleys) in one group don't coincide with peaks (or valleys) in another group. For example, assume the typing requirements of 10 individuals go to one typist and the requirements of 10 other individuals go to another typist. The pool concept is that typing from any of the 20 might go to either of the two typists. Motor pools are an example for vehicles. Other pool examples are technicians, computers, printers, copiers, and lift trucks. You trade off reduced buying of duplicate staff or equipment against loss of specialized services. Pools do not have to be within an organization. Rental cars are an example of a pool among organizations.

Revise schedules   It may be that you have control of when the output is due. If so, **revising schedules** by dividing or joining categories can minimize idle time. For example, divide the month or year into categories. The police department at Kansas State University formerly issued parking permits for everyone at the beginning of the academic year in August. A few years ago they began issuing permits for students in August and for faculty and staff in January. They replaced one very large peak with two smaller peaks.

This technique can be extended. The state of Kansas divides all automobile licenses into 12 months—depending on the last name; thus they divide the load into 12 groups. Many firms formerly sent out bills on the first of the month. Now they divide the month into 20 working days and divide their bills into 20 categories, depending on the customer's last name.

Your purchasing department can schedule shipments to the dock for all days of the week and all weeks of the month instead of Mondays and the first of the month.

Hospitals have found that it is not necessary to do all surgeries during the morning only; thus they can do more operations with fewer operating rooms and less staff.

Be sure to do maintenance and inventory work during slow periods rather than peaks. However, Kansas State University saves over $100,000/yr in lower lighting costs by having custodial work done during the day instead of the night; it also is easier to recruit custodial staff.

Encourage off-peak use   Even if you don't have full control of when the service is to be performed, you can encourage customers to be serviced **off-peak,** that is, in the valleys. One example is the "demand ratchet charge" of utilities in which electrical use in peak periods is very severely penalized; this results in the utility requiring less generating plant. Telephone companies charge less for long-distance calls during nonbusiness hours. Airlines and hotels charge less for use during slack times. Barbers charge less for haircuts during the week, universities charge less for tuition during summer school, courier services charge less for two-day delivery than one-day delivery. Thus, your sales or specials should be timed to fill in the valleys of demand rather than amplifying the peaks.

4.2 Variable Costs

4.2.1 Problem   Consider the relation among the resources. The general concept is to improve utilization among members of a team. The concept is that some team members are more expensive than others; thus it is more important to keep some parts of the team busy than other parts.

The term *team* is defined broadly. Team members include machines, tools, and facilities as well as people. For example, a lift truck driver and the lift truck will be considered to be a team.

In the United States, the expensive part of the team generally is human labor. Labor cost (including fringes such as holidays, vacations, social security, etc.) ranges from $4/h to over $20/h. Although about 1,850 h/yr is representative for most jobs, we will round it out to 2,000 h/yr. Thus, an operator being paid $8/h probably has a cost (assuming 30% fringes) of $10.40/h or $20,800/yr. A minimum wage operator would be about $4/h or $8,000/yr.

A machine's costs are for capital, utilities, and maintenance.

Capital cost usually is the most important. With a 10-year life, a $20,000 machine (such as a lift truck)

would be $2,000/yr or $1/h; a $2,000 machine (such as a small computer) would be $.10/h; a $1,000 machine (conveyors, a workbench with handtools) would be $.05/h.

Utility costs are primarily power, which primarily is electricity. At 7 cents/kWh, even a big motor (1 hp or .75 kW) costs only $.05/h.

Maintenance varies, but few machines require $1,000/yr for maintenance, that is, $.50/h. A value of $.25/h seems more representative.

Thus, if capital is $.25/h, utilities are .05, and maintenance is .25, this is only $.55/h. Few machines cost over $1/h. In sum, in the United States the important thing is to maximize use of the person, the expensive part of the system.

4.2.2 Solutions   Three possibilities are duplicate components, idle low-cost components, and not one on one.

Duplicate components   To maintain utilization of the high-cost components, have low-cost **duplicate components** (substitutes). For example, there might be a typewriter in the office which could be used if the word processor broke down. There might be more than one printer. On a production line, there might be a spare power handtool in case one failed. Since people are flexible, if the computer broke down it might be possible to shift the worker to another task, such as filing or duplicating. If one person in a group "fails" (e.g., by being absent), perhaps a substitute worker can be used (assuming cross-training).

Idle low-cost components   You may decide that part of the system, the low-cost part, will be idle and not worry too much about it. In **double tooling,** two duplicate tools are used (**idle low-cost components**)—generally on a semiautomatic machine. For example, while the part in fixture B is being machined, the operator unloads the part from fixture A and loads a new part. Then, while the part in fixture A is being machined, fixture B is unloaded and loaded. Another example is two Coke dispensers at McDonald's to reduce serving time. McDonald's is more concerned with minimizing time of the server and the customer than the utilization of inexpensive pop dispensers. An example with people is an executive and secretary. It is more important to reduce idle time of the executive than of the secretary. On a production line, the supervisor may keep 30 workers when there is work for only 29 so that illness and other absences don't stop the line.

Duplicate service lines for a single queue server permit faster customer service and reduce worker idle time. An example would be a bank drive-in teller servicing two car lanes. After the first customer is served, the teller can begin serving a customer in the other lane while the first lane is "unloaded" and then "loaded" with a new customer. The same concept can be used for cafeteria checkout and any other situation where customer movement takes appreciable time.

Not one on one   A common assumption is that there is one worker for one machine. But as shown above there can be double tooling—two for one. The numbers need not be integers, however. You can have fractions—2/3 worker/machine or 3/4 worker/machine. How can this be done? You can't cut a person into pieces! Part-time workers will be discussed in the next guideline. But even with full-time workers, 2/3 worker per machine can be achieved by assigning 2 workers to 3 machines, which results in 2/3 worker/machine or 3/2 machines/worker. The worker simply walks back and forth among the machines (many machines are semi-automatic or automatic and need only occasional tending). Often this is called a cell layout, and it illustrates **not one on one.**

Machines and people also can be shared between functions. For example, a lift truck driver (or vehicle) can work for Department A for 4 h, B for 2 h, and C for 2 h. A mechanic, secretary, or technician can be shared between Departments E and F. Although supervisors like to "own" people and equipment (it increases their status as well as making their job easier by giving them complete control), from the organization's viewpoint the system is suboptimal.

GUIDELINE 5: CONSIDER SHIFTWORK

5.1 Extent   Definitions of shiftwork vary by country, but the U. S. Bureau of Labor Statistics classifies people as being on shiftwork if they do not start work between 7 A.M. and 9 A.M. On that basis, in the 1970s about 27% of Americans had some sort of shiftwork (Kodak, 1986). The percentage seems to be increasing.

One reason for shiftwork is economic. Equipment is available 168 h/week. Using it for one shift of 8 h/day for 5 days/week uses the equipment and facilities only 40 h or 24% of capacity. Many manufacturing firms, therefore go to two shift operations—80 h/week. This policy is especially appropriate if capital is relatively expensive vs. labor or the equipment becomes obsolete quickly.

Another reason for shiftwork is "social need" for the service. Police and hospitals obviously need to be available 24 h/day, 7 days/week. But society has tended to demand that other services be available

more hours/week. Now service operations feel the need to be available for their customers for more than 40 h/week. Examples are restaurants, entertainment, retail stores, and transportation firms. (Incredible as it may seem, in the 1950s banks were open only from 9 to 3 Monday to Friday and 9 to 12 on Saturday (i.e., "banker's hours"); most merchants closed at 5:30 in the evening and were closed on Sunday.)

5.2 Shiftwork Problems   The primary economic problem is that people tend to dislike shiftwork. This presents problems in recruiting staff. Additional pay may be necessary in some cases, especially if the workers can find comparable work which does not require shiftwork. There does not appear to be any firm evidence that productivity is lower or safety is worse when shiftwork is used. (Many people have looked for such evidence for many years, so not finding it should be fairly good evidence that shiftwork in itself doesn't lower productivity or safety.)

Much concern also has been expressed over worker health during shiftwork. Numerous studies have been made and there does not seem to be any generalized health decrement with shiftwork. This may be due to self-selection of workers. Shiftworkers are a survivor group and tend to be noncomplainers (stoics). If there is a health problem, it may be gastrointestinal, because shift workers tend to have poor eating habits. See Table 13.2.

There is a sleeping problem if the shiftwork requires work during the night. Figure 13.2 shows the normal **circadian rhythm** (24-h cycle). Night workers (when they are working nights) sleep about 1 h less per 24-h period than day workers, and night workers' sleep has a lower quality. Thus, they accumulate a sleep deficit, which may be made up during weekends. See Table 13.2.

The primary employee problem with shiftwork is social. Although "social death" is exaggerated, most social and leisure activities do tend to occur during the evening and during weekends. Family life and contact with children and spouses also conflict with shiftwork.

5.3 Shiftwork Criteria   Schwarzenau et al. (1986) suggest:

- *Any shiftwork plan must be compatible with legal or union contractual requirements.* An example requirement is that at least 11 hours intervene between the start of two successive shifts.
- *A shift system should have only a few night shifts in succession.* Full entrainment of physiological functions to night work is difficult. Even permanent night workers have problems due to

TABLE 13.2   Tips for shiftworkers (primarily from Kodak, 1986).

Sleeping (after night shift)
- Plan your sleeping time. Don't alter it. Be sure others know your schedule. Train your children.
- Consider sleeping in two periods (5-6 h during day and 1-2 h before returning to work). This may permit a better fit with family/social activities.
- Have a light (not zero or heavy) meal before sleep. A glass of milk may help. Avoid caffeine (the body needs 4-5 h to halve the amount of caffeine in the bloodstream at any given time).
- Develop a good sleeping environment. Have a dark, quiet, cool (air-conditioned) room. Turn telephone volume to off. Use earplugs.
- If under emotional stress, relax before going to bed. One possibility is light exercise.

Digestion
- For B shift, eat your large meal before going to work. Eat lightly at other meals. Avoid large meals before going to sleep.
- For C shift, eat a number of small meals rather than one large one. Avoid gassy, greasy, or acidic foods during the shift.
- Avoid alcohol, especially before the C shift.
- Avoid caffeine.

readapting to day cycles during weekends, holidays, and vacations.

- *A shiftworker should not work more than 7 shifts in succession.* Normal work schedules have 5 shifts and then a weekend. Many good plans have 5 in succession. Thus, a maximum of 5 shifts in succession for shiftworkers seems more reasonable than 7.

- *Weekends are important.* Days off during the week are not as valuable as weekend days. At least some weekends should have 2 full successive days off.

- *When rotating shifts, chase the sun.* Shiftworkers can stay on the same shift (fixed shifts) or the workers on the various shifts can switch to other shifts periodically (rotating shifts). Jet lag experience has shown it is easier to go west than east.

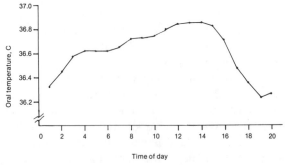

FIGURE 13.2   Oral temperature of 59 male ratings peaked around 4 P.M. (Colquhoun et al., 1968). The subjects slept from 2300 to 0630; performed light duties (primarily short tests) from 0800 to 1600; and had meals at 0700, 1200, and 1700. McFarland (1975) says the circadian rhythm recovers from "jet lag" at the rate of one recovery day per time zone crossed.

Similarly, shifts should rotate from day to evening to night (forward) rather than night to evening to day (backward).

- *Keep the schedule simple and predictable.* Workers want to be able to plan their personal lives. Work schedules must be understandable and must be publicly posted in advance so people can plan. I recommend 30 days in advance.

5.4 Possible Plans

Shift and rest periods will be coded:

A = A (day) shift, usually 7 or 8 A.M. to 3 or 4 P.M.
B = B (afternoon, evening) shift, usually 3 or 4 P.M. to 11–12.
C = C (night) shift, usually 11–12 to 7–8 A.M.
D = Day shift for 12 h, usually 6–7 A.M. to 6–7 P.M.
N = Night shift for 12 h, usually 6–7 P.M. to 6–7 A.M.
R = Rest day

The basic division is between five-day shifts and seven-day shifts.

5.4.1 Five-day shifts

Within 5-day shifts, most commonly the work is 16 or 24 h/day. For 16 h/day:

- fixed shift (A and B only)
- alternation of A and B shift (usually weekly); occasionally A and C shift

For 24 h/day (see Table 13.3):

- fixed shifts of A, B, C
- fixed C, weekly alternating of A and B
- rotating rapidly

Daytime supervision tends to prefer a rotating system as they get to know people on all three shifts. Weekly rotation is difficult for the workers. Rapid rotation has fewer physiological and social disadvantages but is not as simple. Note that the Wednesday C shift is worked Saturday morning. An alternative is to work the C shift only 32 h/week.

5.4.2 Seven-day shifts

Since the week has 168 h, each person works 42 h/week. In addition to the problems of the 5-day shift, there is the loss of weekends.

- Fixed 8-h shifts. For 2 days off/week, you need 12 crews. People working for 7 days continuously have problems and considerable interference with weekend activities. Table 17.4 in Kodak (1986) gives an example schedule.
- Weekly rotating 8-h shifts. Table 13.4 gives a plan for 4 crews, using a limit of 5 continuous shifts. Table 17.5 of Kodak (1986) gives some plans with a limit of 7 continuous shifts.

TABLE 13.3   Five-day, 24-h shift schedules with three 8-h shifts (Kodak, 1986).

Schedule	Crew	Week 1 MTWTF	SSu	Week 2 MTWTF	SSu	Week 3 MTWTF	SSu	Week 4 MTWTF	SSu
Fixed shifts	1	AAAAA	RR	AAAAA	RR	AAAAA	RR	AAAAA	RR
	2	BBBBB	RR	BBBBB	RR	BBBBB	RR	BBBBB	RR
	3	CCCCC	RR	CCCCC	RR	CCCCC	RR	CCCCC	RR
Fixed C, weekly alternating A and B	1	AAAAA	RR	BBBBB	RR	AAAAA	RR	BBBBB	RR
	2	BBBBB	RR	AAAAA	RR	BBBBB	RR	AAAAA	RR
	3	CCCCC	RR	CCCCC	RR	CCCCC	RR	CCCCC	RR
Weekly rotating, forward	1	AAAAA	RR	BBBBB	RR	CCCCC	RR	AAAAA	RR
	2	BBBBB	RR	CCCCC	RR	AAAAA	RR	BBBBB	RR
	3	CCCCC	RR	AAAAA	RR	BBBBB	RR	CCCCC	RR
Rapidly rotating, 2-3 for all three shifts	1	AABBB	RR	CCAAA	RR	BBRCC	CR	AABBB	RR
	2	CCAAA	RR	BBRCC	CR	AABBB	RR	CCAAA	RR
	3	BBRCC	CR	AABBB	RR	CCAAA	RR	BBRCC	CR

- Rapid rotating 8-h shifts. Table 13.5 gives a plan for four crews, using a 2-2-3 rotation and a 2-2-2 rotation. Although the 2-2-3 schedule has 2 rest days coincide with a weekend only 1 weekend in 4 (and the 2-2-2 even less), they usually are preferred over the fixed or weekly rotating schedules due to better opportunities for social interactions with families and friends.

- Fixed 12-h shifts. See Table 13.6. These "compressed" plans have the advantage that the worker comes to work only 7 days of every 14 instead of the 10 days of 14 for normal work. This permits less commuting and is used by

TABLE 13.4   Seven-day, 24-h shift schedule with 8-h shifts (Kodak, 1986). The schedule is shown for crew 1; crews 2, 3, and 4 start at weeks 6, 11, and 16.

		Week	Day of the Week MTWTF	SSu
Four crews, 5-day work week, 20 week cycle	Crew 1	1	CCCCC	RB
		2	BBBBR	RA
		3	AAAAR	RC
		4	CCCCR	BB
		5	BBBRR	AA
	Crew 2	6	AAARR	CC
		7	CCCRB	BB
		8	BBRAA	AA
		9	AARRC	CC
		10	CCRBB	BB
	Crew 3	11	BRRAA	AA
		12	ARRCC	CC
		13	CRBBB	BB
		14	RRAAA	AA
		15	RRCCC	CC
	Crew 4	16	RBBBB	BR
		17	RAAAA	AR
		18	RCCCC	CR
		19	BBBBB	RR
		20	AAAAA	RR
Return to week 1				

some to "moonlight" (take another job). People working 12-h days may have difficulty finding babysitters at home, they may be quite fatigued, and overtime is not feasible.

- Rotating 12-h shifts. See Table 13.6.

Schwarzenau et al. (1986) and Kodak (1986) give other examples of shift schedules beyond the ones given here, with extensive discussion.

Williamson et al. (1994) reported that computer operators who transferred to 12-h shifts from 8-h shifts had improved psychological health with no loss in productivity. It may be that short periods of long shifts (including the highly prized 4-day break) are easier to adjust to than long periods of shorter rotating shifts.

Monk and Folkard (1992) tell how to make shift-work tolerable.

GUIDELINE 6: USE FILLER JOBS OR FILLER PEOPLE

The problem is to match worker time to job requirement time. For example, a fast food restaurant may have a need for workers primarily at meal times; so what does it do with the worker during the middle of the afternoon? In a factory, a maintenance worker may be scheduled to work until 4 but finishes a job at 3:15. What should be done for the remaining 45 minutes? Many workers have idle time while a machine is "on automatic" and is either loading/unloading or operating. "Machine time" examples are word processing operators while the machine prints, numerical control operators while the machine processes parts, disc jockeys while the song is being played, a cook while the food is cooking.

TABLE 13.5   Seven-day, 24-h shift schedules with 8-h shifts, rapid rotation (Kodak, 1986).

Schedule	Crew	Week 1 MTWTF	SSu	Week 2 MTWTF	SSu	Week 3 MTWTF	SSu	Week 4 MTWTF	SSu
2-2-3 rotation	1	RCCAA	BB	BRRCC	AA	ABBRR	CC	CAABB	RR
four crews	2	BRRCC	AA	ABBRR	CC	CAABB	RR	RCCAA	BB
	3	ABBRR	CC	CAABB	RR	RCCAA	BB	ABBRR	CC
	4	CAABB	RR	RCCAA	BB	BRRCC	AA	ABBRR	CC

Schedule	Crew	Week	MTWTF	SSu	
2-2-2 rotation,	1	1	CCAAB	BR	
four crews,		2	RCCAA	BB	
8-week	2	3	RRCCA	AB	
schedule		4	BRRCC	AA	
	3	5	BBRRC	CA	
		6	ABBRR	CC	
	4	7	AABBR	RC	
		8	CAABB	RR	Return to week 1

TABLE 13.6   Seven-day, 24-h shift schedules with 12-h shifts (Kodak, 1986).

Schedule	Crew	Week 1 MTWTF	SSu	Week 2 MTWTF	SSu	Week 3 MTWTF	SSu	Week 4 MTWTF	SSu
EOWEO, every	1	DDRRD	DD	RRDDR	RR	DDRRD	DD	RRDDR	RR
other weekend	2	RRDDR	RR	DDRRD	DD	RRDDR	RR	DDRRD	DD
off, fixed shift	3	NNRRN	NN	RRNNR	RR	NNRRN	NN	RRNNR	RR
	4	RRNNR	RR	NNRRN	NN	RRNNR	RR	NNRRN	NN
EOWEO, every	1	DRRNN	RR	RDDRR	NN	NRRDD	RR	RNNRR	DD
other weekend	2	RDDRR	NN	NRRDD	RR	RNNRR	DD	DRRNN	RR
off, rotating	3	NRRDD	RR	RNNRR	DD	DRRNN	RR	RDDRR	NN
	4	RNNRR	DD	DRRNN	RR	RDDRR	NN	NRRDD	RR

The two general strategies to minimize idle time are to (1) adjust the work load but keep the workforce constant and (2) adjust the workforce but keep the work load constant.

6.1 Adjust Work Load

Use **filler jobs** to adjust the work load. It is easier to minimize idle time when the work idle time is in relatively large blocks. For example, a job with 25% idle time could have work 6 h and then be idle for the last 2 h of the shift. It also could have 225 cycles of 2 min—with each cycle having .5 min idle time. It is more difficult to find an additional job (filler job) for the idle time when the time is in short segments. Thus, try to concentrate the idle time.

It is important to have broadly defined job descriptions so that transferring workers from task to task is not a problem. See Chapter 34.

6.1.1 Short jobs

Longer jobs can be broken into shorter jobs. Consider the task of waxing a car—it might take 6 h. If you have just Saturday afternoon to work, you might be reluctant to start and leave unfinished work. Try to think of it as 4 jobs, each taking a shorter time (wax top, wax hood and fenders, wax door and sides, and wax rear); now work will not be "unfinished." Now you can do 2 jobs this Saturday, 1 Monday, and 1 Tuesday. Or consider a secretary who has an exam to type, duplicate, and assemble. If the entire job takes 40 min, the secretary may be reluctant to start it 20 min before quitting time. Think of it as three jobs: typing, duplicating, and assembling. Then do just the typing before quitting time.

Make a list of short, low-priority jobs that can be done in the idle time. Many routine maintenance and clerical tasks fit in this category. Opening the mail and returning phone calls are good filler jobs. Don't do these low-priority, "machine-time" tasks during "prime" time but reserve them for idle time, end-of-shift time, or while the machine is operating. Service personnel (receptionists, tool room clerks, etc.) often have considerable time between customers. A receptionist could type letters and answer the telephone. A tool room clerk could sharpen tools. For another example, police officers in Manhattan, Kansas, formerly reported back to the police station approximately 30 min before the end of the shift to dictate reports. Now each officer has a battery-operated tape recorder. Whenever a report is needed, the officer parks the car in a prominent place and dictates into the recorder. This not only completes the report while the topic is still fresh but also makes the officer visible and thus acts as a deterrent—increasing useful time/shift. One of the guidelines at McDonald's is "If there is time to lean, there is time to clean."

6.1.2 Scheduling

may increase or decrease idle time. For example, if a meeting starts 15 min after the start of work, most people will waste the 15 minutes. (However, if the meeting is of supervisors, it may be best to allow them 15 min to get their people organized for the day.) In addition, meetings tend to fill the time available. To shorten meetings, have an agenda and schedule the meeting partly on the participants' time rather than on the organization's time (i.e., schedule them to overlap lunch or quitting time). If possible, schedule meetings for a time of reduced work load (a less busy time of the day, week, or month).

Supervisors should assign more work to subordinates than subordinates have time to do. The reason is the variability of the estimated time and the variability of the actual time for each job. No one wants to come to the boss and say "I'm out of work." What they will do is "stretch" the job as much as possible.

6.2 Adjust Workforce

Use **filler people** to adjust the workforce. Three possible ways are to use staggered work times, temporary workers, and part-time workers.

6.2.1 Staggered work times

Make an analysis of the work requirements vs. time. The most efficient staffing pattern may have **staggered work times,** with varied starting and stopping times for different workers.

For example, in a restaurant, the dishwasher can start after the waiters; the waiters need not stay as long as the dishwasher. In a mail order business, people opening the mail should arrive before those filling the orders. Hospital emergency rooms have periods of peak demand that should have peak staffing and periods of low demand that should have low staffing. Staggered schedules spread the demand on production facilities (workstations) but also the demand on service facilities (parking lots, toilets, food service facilities, etc.).

When people do the same task, staggering their work hours permits keeping the business open for more hours at no additional cost. For example, if Mary goes to lunch from 11:30 to 12:00 and Betty goes from 12:00 to 12:30, someone can always be present to answer the phone. If Mary works from 7:30 to 4:30 and Betty works from 8:30 to 5:30, then customers can be served from 7:30 to 5:30 instead of 8 to 5. Staggered hours often are popular with employees also as it may simplify their personal life.

A greater modification is changing the work from the standard 8 h/day, 5-day week pattern. There are many alternatives. They are called "compressed" plans as they compress the number of days worked, although expanding the number of hours/day. One plan is 7 shifts of 12 h each during a 14-day period; another is 4 shifts of 10 h during each week. See Guideline 5 of this chapter for further discussion. Flexible schedules tend to be popular with employees because they give employees greater control of their personal lives. Firms with flexible schedules usually have lower absenteeism and tardiness.

Note that flexible schedules imply greater cross-training of workers since the business must be able to continue when a specific individual is not there. Some firms reduce this problem by defining certain time as **core time** (a time when everyone must be there) and **flex** (i.e., flexible) **time** (a time in which some but not all people are present).

Over the longer period of the year, most organizations have busy times and slack times. Require vacations to be taken during slack times, not busy times.

6.2.2 Temporary workers

Short-term workers **(temporary workers)** have a long history. Farmers have hired daily and seasonal labor for thousands of years. (Remember the Bible story of the workers in the vineyard. Some worked all day and some worked for just part of a day.) Construction firms traditionally have had a small core staff and many temporary people hired for a specific job. Many firms hire students and faculty for work during the summer. Starting about 50 years ago, special firms began furnishing temporary secretaries and office help. They furnished trained workers for the amount of time needed. Gradually this service has spread to other jobs such as warehouse labor.

Within the technical and executive areas, however, the concept was to consider all workers as core workers. (There was some business subcontracted out to consulting firms and freelancers, such as professors.)

More recently, however, the practice of hiring temporary executives, engineers, accountants, and other professionals has grown considerably. They tend to cost more than the permanent staff (even though they don't get as many fringe benefits). The big advantage for the organization is that it can staff for the bottom of the business cycle and have a no-layoff policy for its core staff. (The famous Japanese "lifetime employee policy" applies only to a core staff of males and excludes temporary workers and females.) The temporary executives and engineers can be removed from the payroll very quickly and with no problems or trauma to the organization. The temporary workers (freelancers) are like entrepreneurs. They have freedom and mobility and possibly high income but little security.

6.2.3 Part-time workers

People can be hired for less than 8 h/day, 5 days/week (**part-time workers**). There are a number of advantages to the organization:

- The organization may be able to better match work requirements and worker availability (e.g., part-time workers for McDonald's at mealtimes).

- The organization, for the same hourly rate, may be able to get a better quality part-time than full-time person. For example, a part-time college student may work for $5/h but wouldn't consider working full-time for $5/h.

- The part-time worker may work boring, tedious, or fatiguing work part time but would not do it full time.

- Labor costs will be lower. Fringe benefits costs probably will be lower. Pension costs should be lower since most part-time workers quit before retirement. In addition, part-time workers tend to be under 30, so their medical costs tend to be lower. The wage itself tends to be lower as the worker has little experience and thus is in the bottom of the wage bracket, not the top.

- Increasing or decreasing work hours often is quite easy with part-time workers. For example, you can ask them to work 20 h this week instead of 15 (33% increase) or tell them they will only work 15 h this week instead of 20. It is very difficult to cut hours of full-time people and an extra 8 h/week (20% change) is the usual maximum increase.

Job sharing, that is, two people each working part time to fill one position, also provides benefits to management: (1) It allows for an expanded range of skills and experience; (2) both employees may work extra hours during peak loads; (3) noncoverage of the job (due to vacations, illnesses, absenteeism) becomes minimal; and (4) if one person leaves, the other provides coverage until a new person is hired.

There are many part-time plans. Part-time can be (1) part time within a day or (2) full time within a day but part time within the week or month. For example, one Kansas firm has a monthly peak load. They hire people to work only the first two weeks of the month. People also can be hired to work only weekends or Wednesdays or whatever. See Nollen (1982) for more on alternatives to the 8-h, 5-day work schedule.

GUIDELINE 7: REDUCE FATIGUE

Fatigue can be defined as a reduction in the capacity to do work. See Chapter 31 for fatigue allowances.

7.1 Types of Fatigue

There are two categories: **muscle** (local) and **central fatigue.**

1. muscle fatigue results from

- physically demanding static work
- physically demanding dynamic work

2. central fatigue results from

- light-effort work
- perceptual/inspection work

7.1.1 Muscle

See Box 13.1 for some background on muscle fatigue.

Physically demanding static work Examples of static work are standing, holding your arm above your head (such as in painting or assembly), carrying a load in your arms (static work on the arms, dynamic work on the legs), and holding your hands above the heart while assembling (static work on the arms, dynamic work on the fingers). As can be seen from the examples, static work often is combined with dynamic work. Static work usually is more limiting than dynamic work. See Guideline 1 of Chapter 15 for more on static work.

Physically demanding dynamic work An example would be moving cases of product in a warehouse, such as from pallets to a conveyor.

Figure 11.3 shows recommended maximum aerobic-capacity levels—primarily based on whole-day energy expenditures. These limits are based on cardiovascular and nutritional considerations. For shorter periods, muscle fatigue becomes limiting. Figure 11.3 gives 85% of maximum aerobic capacity as a limit for 5 min, 70% for 20 min and 50% for 60 min. For short, intense work, the activity should be paced by the individual doing it, not by a machine or a predetermined rest schedule. The rest can be dynamic work at a lower metabolic rate, using other muscles.

BOX 13.1 *Muscle fatigue*

Muscles can be divided into two types (Chaffin and Andersson, 1984):

- Slow twitch (type I) are smaller, have a high capacity for aerobic metabolism (and thus are good for sustained or endurance activities), and have a long (e.g., 100 ms) rise time to peak tension. When stained with ATP-ase, they are dark. An example is the leg's soleus muscle—a postural muscle. Slow twitch fibers are richly surrounded by capillaries and have great potential to store and use oxygen.

- Fast twitch (type II) are larger, depend mostly on anaerobic metabolism (and thus are good for power), and have a short (e.g., 10 ms) rise time to peak tension. When stained with ATP-ase, they are light. An example is the eye muscle. Fast twitch fibers are common when there is a need for short bursts of high exertion levels.

Strength training increases the thickness of fibers, whereas endurance training increases the muscle's ability to store and use oxygen.

During muscle contraction, the muscle produces waste products and also requires nourishment and oxygen. During static contractions, the blood flow drops drastically, so little waste is eliminated or nourishment furnished. In dynamic work, the interruption in blood flow is intermittent instead of continuous. Thus, static work is more fatiguing than dynamic work.

Figure 13.3 shows muscle recovery time as a function of work intensity and duration. It points out the inefficiency of sustained heavy work. For example, assume a person worked for 1.0 min at 20% of MVC (maximum voluntary contraction). Then a rest of .25 min is needed—that is, the person is working 1.0/1.25 = 80% of the time. But if the work was for 2.0 min, then the required rest is 1.2 min—that is, the person is working 2.0/3.2 = 62% of the time. The penalty for sustained work is even higher for higher percentages of MVC. Figure 13.4 shows the typical endurance vs. strength for muscles.

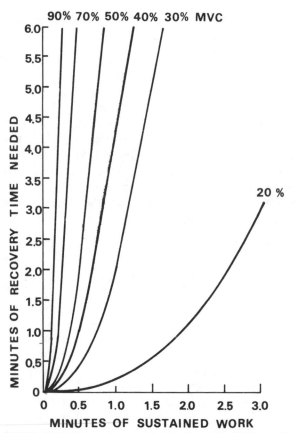

FIGURE 13.3   Recovery times as a function of work duration and work intensity (Rodgers, 1984). MVC = Maximum voluntary contraction.

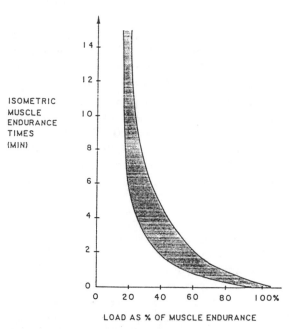

FIGURE 13.4   Small decreases in the load can make a big increase in the muscle endurance time (Chaffin and Andersson, 1984).

7.1.2 Central

The physiology of central fatigue is not as clear-cut, because there are strong motivational effects.

Light-effort work   Light-effort work primarily involves use of the fingers. Examples are assembly work, clerical work, and service counter work. There is a relatively low metabolic rate. There may be problems of boredom and monotony; in some cases the work may be highly repetitive (short cycle).

Perceptual/inspection work   The metabolic work here is minimal, but mental work is high. Some inspection work may have boredom and monotony components. See Chapter 21 for techniques of improving visual tasks and Chapter 20 for comments on displays. Data entry, VDT work, and assembly using a binocular microscope are examples combining perceptual work, light-effort work, and static work.

7.2 Fatigue Reduction

Fatigue is reduced by (1) preventing it in the first place and (2) curing it with rest.

7.2.1 Fatigue prevention

As pointed out in Box 13.1, preventing fatigue is better than curing it. Prevention involves reducing the stress level and reducing the duration of continuous exposure. (Although most of the knowledge on fatigue concerns muscle fatigue, it seems logical that central fatigue also should be reduced by having intensity and duration of stress reduced.)

There are many examples throughout this text of how to reduce the stress level.

7.2.2 Rest

Reducing duration of continuous exposure means increasing the rest time. However, rest time does not necessarily mean not working. That is, in most cases the fatigue can be overcome simply by shifting to another task—that is, **working rest.** In other words, a change is a rest.

Working rest   The benefits of working rest will be greater, the greater the shift in use of the stressed system. For example, if the eyes are stressed, there is greater benefit if the work becomes manipulative or lifting. There is some benefit (but not as much) from shifting to similar work (e.g., pack 12-ounce bags in cartons for an hour, then pack 8-ounce bags for an hour; pick cases from racks at ground level and racks 1 m above the floor instead of just from ground level; paint walls and then paint trim; saw and then nail). Define jobs so that workers get supplies, lay aside completed product, and maintain their machines rather than having one person assemble, another do material handling, and another do maintenance.

Smith (1981) reviewed 50 references on boredom and found that a change in task was as good as stopping work.

In addition to switching within one job, it also is common to switch tasks between workers—**job rotation.** A specific example is in a warehouse. An operator scans boxes coming down the conveyor and keys a number into the computer (at a rate of 1 box/s), indicating to which dock the box should be sent. After 2 h, the worker shifts responsibilities with a person unloading the conveyor. At a lamp plant, inspectors inspecting items on a belt conveyor shift jobs with other inspectors every hour. In a sausage plant, a crew of 6 divides the duties among the 6; they rotate jobs every 30 min. In a packaging operation, where there are 14 different lines of snack foods being packaged into cardboard boxes, workers shift every 60 min. In a warehouse, workers pick cases from pallets to a conveyor for 1 shift and then the next day unload cases from the conveyor. In another warehouse, order pickers alternate picking large orders and small orders. (The large orders are considered more difficult.) At the EPCOT center, the tour guide stays with a group through the entire tour rather than just showing a single attraction and passing the group on to another guide.

Job rotation, in addition to reducing fatigue, reduces the feeling of inequity among workers as everyone shares the good and bad jobs. As a third benefit, job rotation reduces the boredom and monotony of a job; this reduction of psychological fatigue through job enlargement depends on how varied the different jobs are.

Ramsey et al. (1974) reported that inspectors did better with short (0–24 min) duration sessions than with long (50–74 min) sessions.

Nonworking rest   It may be necessary to rest the entire body as well as specific portions. Many industries give a 10- or 15-min coffee break at the end of 2 h of work. Lunch provides another break.

At present there is not much knowledge on how nonworking time should be spent. For example, some Japanese firms have their workers do calisthenics to warm up before doing manipulative work. If this is really beneficial, why don't all firms do it in all countries?

What is the best way to recover from mental fatigue? Should the rest be **active vs. passive?** (Active means doing something else while passive means just relaxing.) For example, do you recover better from mental fatigue by moderate exercise, by light exercise, or by sitting? Should the exercise be automatic (jogging, swimming) or something requiring attention (calisthenics)? Should the environment be stimulating (such as conversation) or nonstimulating (take a nap)?

Benefits from a break decline exponentially with break duration. That is, the benefit from minutes 0–5 is greater than from minutes 6–10, which is greater than minutes 11–15. Thus, frequent short breaks are better than occasional long breaks.

If the individual is self-paced and works independently of others, it probably is better for the individual to self-select the time and duration of the break. However, there may be administrative difficulties since workers tend to stretch breaks.

GUIDELINE 8: COMMUNICATE INFORMATION

Figure 19.1 shows the flow of information in the human–machine system. The information flows from the person to the machine through controls. The information flows from the machine to the person through displays. In addition, in the larger system, information flows (it is hoped) among people. Long-term information can be considered job instruction and training. Short-term information can be considered "command and control."

8.1 Job Instruction/Training   A work method in the mind of the engineer must be transferred to the mind of the operator. The worker can memorize this information (be trained) or can look it up each time (have job aids). See Chapter 32 for more details.

8.2 Command and Control   The concept of command and control is of relatively short messages which trigger behavior patterns. See Chapter 20 for techniques for improving communication of alphanumeric information as well as information from instruments. Box 25.3 discusses the problem of warnings. The problem of noise and speech interference is discussed in Chapter 22.

REVIEW QUESTIONS

1. In Plan the Work there are four steps; in Work the Plan there are three. Give the seven steps.

2. Sketch the reward vs. output curve for a person on salary.

3. Give the formula for availability. What is availability if mean time between failure = 1,000 h and mean time to repair is 10 h?

4. Give an example of a standby system in baseball.

5. Four solution techniques are given for minimizing idle capacity. Give the four and an example of each.

6. Why should a queue server have two queues?

7. What is the primary problem with shiftwork?

8. If you want to have a short meeting, when should it be scheduled?

9. Give an example of a compressed work schedule.

10. Is static or dynamic work more limiting?

11. Give three examples of working rest.

12. Table 13.2 gives nine tips for shiftworkers. Give six of them.

REFERENCES

Chaffin, D. and Andersson, G. *Occupational Biomechanics.* New York: Wiley & Sons, 1984.

Colquhoun, W., Blake, M., and Edwards, R. Experimental studies of shift work, I: A comparison of rotating and stabilized 4-hour shifts. *Ergonomics,* Vol. 11 [5], 437–53, 1968.

Kaufman, P. and Corrigan, A. *How to Use Your Time Wisely.* Stamford, Conn.: Longmeadow Press, 1987.

Kodak, *Ergonomic Design for People at Work,* Vol. 2.

New York: Van Nostrand-Reinhold, 1986.

Kopelman, R. *Managing Productivity in Organizations.* New York: McGraw-Hill, 1987.

McFarland, R. Air travel across time zones. *American Scientist,* Vol. 63 [1], 23–30, 1975.

Merrill, A. and Merrill, R. *Connections: Quadrant II Time Management.* Salt Lake City: Publishers Press, 1987.

Monk, T. and Folkard, S. *Making Shiftwork Tolerable.*

London: Taylor and Francis, 1992.

Nollen, S. Work schedules. In *Handbook of Industrial Engineering,* Salvendy, G. (ed.), Chapter 11.8. New York: Wiley & Sons, 1982.

Ostrofsky, B. *Design, Planning and Development Methodology.* Englewood Cliffs, N. J.: Prentice-Hall, 1977.

Ramsey, J., Halcomb, C., and Mortagy, A. Self-determined work/rest cycles in hot environments. *International J. of Production Research,* Vol. 12 [5], 623–31, 1974.

Rodgers, S. *Working with Backache.* Fairport, N. Y.: Perinton Press, 1984.

Schwarzenau, P., Knauth, P., Kiesswetter, E., Brockmann, W., and Rutenfranz, J. Algorithms for the computerized construction of shift systems which meet ergonomic criteria. *Applied Ergonomics,* Vol. 17 [3], 169–76, 1986.

Smith, R. Boredom: A review. *Human Factors,* Vol. 23 [3], 329–40, 1981.

Williamson, A., Gower, C. and Clarke, B. Changing the hours of shiftwork: A comparison of 8 and 12 hour shift rosters in a group of computer operators. *Ergonomics,* Vol. 37 [2], 287–98, 1994.

14

ORGANIZATION OF WORKSTATIONS

OVERVIEW

This chapter discusses the organization of workstations. Six guidelines are given.

CHAPTER CONTENTS

1 Use Specialization Even Though It Sacrifices Versatility
2 Minimize Material Handling Cost
3 Decouple Tasks
4 Make Several Identical Items at the Same Time
5 Combine Operations and Functions
6 Vary Environmental Stimulation Inversely with Task Stimulation

KEY CONCEPTS

assembly line
balance-delay time
balanced line
bottleneck station
buffer
bus/taxi
capital costs
carrier
cellular manufacture
cycle time
decoupling
distance insensitive

elements
environmental stimulation
fixed cost/trip
flow lines
get and put-away times
get ready, do, put away
group technology
help your neighbor
just-in-time
more places than people
multifunction
one-worker line

on-line/off-line buffers
operation only line
order-picking line
paired station
rocks in the river
special-purpose equipment
station blockage
station starvation
task stimulation
transportation/communication
type 1/type 2
utility operator

GUIDELINE 1: USE SPECIALIZATION EVEN THOUGH IT SACRIFICES VERSATILITY

Specialization is a key to progress. Use special-purpose equipment, material, labor, and organization. Seek the simplicity of specialization; thereafter distrust it, but first seek it.

1.1 Equipment

Special-purpose equipment, designed for unique tasks, has the advantages of greater capability and lower production cost/unit but the disadvantages of slightly higher capital cost and less flexibility.

Special-purpose equipment often can perform functions that general-purpose equipment cannot. As the designer designs specialized equipment and the user uses it, design restrictions of general-purpose equipment are eliminated and major improvements often result. For example, an ordinary grinder is used to remove very little material and give a high surface finish. A special-purpose grinder may be able to remove large amounts of material (rough cuts) as well as leave a good finish so that the job may be done on one machine in one setup instead of on two machines. A general-purpose cash register will have numbers on the keys so any price product can be registered. A special-purpose register has special keys for each item (hamburger, small Pepsi) instead of prices; pressing the key actuates the proper price and perhaps may even display the complete order on a screen for the employee to use for order picking. When McDonald's cooks hamburgers, they use a special-purpose grill which cooks the hamburger on both sides simultaneously.

Lower production cost/unit results from the specialized nature of the machine's components. The components run at maximum speed, give minimum variability, require minimum labor time to operate, or have other unique features.

In theory, special-purpose equipment has fewer components since many of the components needed to make general-purpose equipment general purpose are not needed. Fewer components should mean a simpler machine and thus a lower capital cost. However, the number of copies of each special-purpose machine is small, so design and build costs must be spread over a few machines instead of many. Thus, special-purpose equipment has a higher capital cost than general-purpose equipment.

A penalty of special-purpose equipment is its lack of flexibility. It does one job very well but only one job. What if you don't have just one job?

1.2 Material

Specialized materials have the same types of advantages and disadvantages as specialized equipment with the most common trade-off being higher material cost vs. greater capability.

For example, when you use tool steel for dies instead of the cheaper low-carbon steel, you trade off greater capability and longer die life vs. a higher material cost. A titanium basket in the plating department trades longer life and less maintenance vs. higher initial material cost. A throw-away syringe in the hospital trades better sanitation and elimination of cleaning costs vs. higher initial cost/syringe. A rug in the office trades lower maintenance costs vs. higher capital costs (than tile).

1.3 Labor

Labor specialization affects both labor quality and quantity.

1.3.1 Quality

Quality of output of a specialist is potentially high due to the skill being in the tool and due to practice. The jack-of-all-trades is master of none.

When specialization is high, the specialist develops or purchases special-purpose machines and tools—thus the statement "the skill is in the tool."

With special-purpose tools, many hours spent at the same task, and a more restricted variety of skills required, quality should be better for the specialist. At least in theory a Ford carburetor tune-up operator should be more skilled at tune-ups of Ford carburetors than a carburetor tune-up operator (who works on all models of cars) or a tune-up operator (who works on all aspects of the field). The specialist may be much farther out on the learning curve; that is, the specialist may have tuned 7,500 Ford carburetors while the mechanic may have tuned 150. The brain surgeon may have operated on 1,000 brains while the general surgeon may have operated on 1.

1.3.2 Quantity

Quantity of output/time usually is higher for a specialist (that is, labor time/unit is lower) for the same reasons that quality is higher. With a restricted range of skills, training time is less for the specialist.

Since the individual is trained deeply rather than broadly, the general rationale has been that a less talented individual is required and so a lower rate of pay is justified. In most industries, the generalist (tool and die maker) is paid more than the specialist (turret lathe operator). Therefore, specialized labor costs less—both because of its greater productivity/unit and its lower wages/h.

From the individual viewpoint, specialized work may be repetitive and monotonous. It has been difficult to recruit workers for monotonous jobs if they have low pay. If the job has high pay, many workers can be found whether the job is monotonous or not.

1.4 Job Organization   Adjectives describing specialization are *rigidly structured, inflexible, disciplined,* and *machinelike.* The overwhelming characteristic is the need for high volume of a standardized product. Levitt mass-produced homes by breaking home construction into 26 steps and reversing the assembly line (the product stood still and the worker moved). If you are going to do nothing but brain surgery, you need many patients requiring brain surgery; if all you do is tune up Ford carburetors and it takes 30 min/carburetor, you need about 15 Fords/day to keep busy.

If, instead of time/unit of 30 min, time/unit = 1 min, then output/day is 450 (allowing for breaks), output/month is about 10,000, and output/yr is about 120,000. Can you sell 120,000 identical units/yr? For time/unit of .1 min (6 s), output is 1,200,000/yr. Can you sell that many?

Most firms don't have that amount of volume and can use specialization only as a desirable goal. One approach finding growing favor is **group technology,** which attempts to get the benefits of mass production from batch production. "Families" of parts are manufactured in "cells." The crux of the problem is to identify, from the large variety of components manufactured in a typical firm, the similar components (the families) so that common solutions can be found for similar problems. Then members of the family are scheduled close together (close in time) and produced close together (close in space). Benefits include many of the advantages typically cited as advantages for specialization, such as lower setup cost/unit, less paperwork cost, increased use of special-purpose fixtures, and so forth. However, the benefits primarily are in *avoided* costs, so managers find it difficult to spend money for savings that don't show up in the cost accounting system.

GUIDELINE 2: MINIMIZE MATERIAL HANDLING COST

Material handling does not add value—just cost. Reduce the cost by analysis of its components.

Cost of material handling can be broken down as follows:

- Material handling cost/yr = Capital cost + Operating cost
- Operating cost = (Number of trips/yr) (cost/trip)
- Cost/trip = Fixed cost/trip + (Variable cost/distance) (Distance/trip)

2.1 Capital Cost of Systems   The **capital costs** of material handling (return on investment and depreciation) do not vary appreciably with the amount of material moved. For example, you may purchase an electric fork truck for $20,000 and a recharging station for $3,000. Then $23,000 invested at 10% returns $2,300. This cost occurs whether you use the truck one h/month or 100 h/month; whether your downtime is 10% or 90%. In addition, depreciation depends more on age of equipment than usage. For example, resale value of the fork truck after 2 years might be $5,000 if you used the truck 1 h/day and $4,000 if you used it 8 h/day. A used conveyor probably will sell for the same amount regardless of use. Conversely, capital cost/unit moved can become very low if the equipment is kept busy.

Thus, if utilization is poor, the lowest total cost might be obtained for a system with high operating cost but low capital cost. If utilization is high, the high capital cost alternative may be best.

Eliminating peak loads by scheduling may eliminate need for some equipment. For example, schedule shipments for 5 days a week, not just Thursday and Friday; receive material from vendors five days/week. Move by priority—not just first come, first served.

2.2 Number of Trips/Year   The ultimate would be to make number of trips equal zero, eliminating not only operating cost but also capital cost. Question the need for the trip; it may not really be required. A repair or maintenance call might be eliminated if higher quality maintenance is used (e.g., use a component which requires service once every 120 days instead of 60 days). A sales call might be eliminated by using a letter or electronic communication. The opposite of **transportation** is **communication.**

Reduce the number of trips by scheduling and combining trips. For example, a trip from San Francisco to New York with a stop in Chicago is less expensive than two trips, one to Chicago and one to New York. The trade-off is among reduced travel cost, increased scheduling problems, and capacity/trip. The same type of trade-off must be made in other problems. Should the clerk take each item to the duplicating machine or accumulate a batch before going? Should the money be deposited in the bank once a week, once a day, or once an hour? Should the operator send material from the machine to the next station once/min, once/h, or once/day? Should the pallet hold 50 units and move once/day or hold 25 and move twice/day?

2.3 Fixed Cost/Trip   **Fixed cost/trip** has two components: (1) information transfer (mainly paperwork) and (2) start and stop.

Information transfer is a material handling cost that often is overlooked. Reduce these costs by using *line* production. In a job shop, even if the many paperwork forms are completed correctly, costs are substantial. It is not just the cost of filling out the forms; it is the cost of transporting the forms, filing the forms, transferring information from one form to another, and so on. Mistakes occur. Products can get misplaced, workers can run out of supplies, trips can be made to the wrong place. One of the major advantages of line assembly is the standardization of routing and scheduling—thereby reducing information transfer costs. Computer and electronic sensor technology, however, can reduce information transfer costs; so single-product assembly lines no longer are always desirable. For example, electronic sensors can "read" a box number as it moves along a conveyor, send the information to a "brain," which then consults its "memory," decides to send this box to station 14, and then moves its "arm" to put the box in station 14.

Start and stop (pick up and put down; load and unload; pack and unpack) is a substantial cost that does not vary with the distance moved. A large part of the cost of flying a commercial airplane is the take-off and landing cost, just as a large portion of a fork truck's cost and time is spent picking up and putting down the load. If load time = 1.0 min, travel time = .01 min/m, and unload time = 2.0 min, a 50 m trip costs 3.5 min and a 100 m trip costs 4.0. Twice as far doesn't cost twice as much (see Figure 7.2).

Much transportation (and communication) is **distance insensitive** (i.e., price varies little with distance). Over the last 100 years technology has increased this insensitivity. Water transport especially is insensitive to distance. Goods can be moved from Japan to New York for not much more than from Boston to New York. Land transport has been speeded by improved highways and air transport by improved airplanes. Improved electronics—both wired and wireless—speed communication of words and data. These increases in insensitivity affect plant design and location because products now come from a specialized factory in Dusseldorf or Milan or Chicago rather than a number of local plants, each with a variety of products with low production volumes. Increased mobility of workers (due to the automobile) and of products (due to trucks instead of railroads) permits decentralization of places of employment and, thus, decentralization of cities.

Within a factory, fork trucks and tractor-trailer trains have increased distance insensitivity. Formerly, factories were built vertically to minimize product movement distance. Offices (paperwork factories) are still in the early stage of development (multistory buildings with workstations—desks—close together).

2.4 Variable Cost/Distance

Costs/distance are a function of energy consumed and labor cost. Energy consumption does not make much difference in many cases since it is a small portion of the total cost. Electric fork trucks do tend to be cheaper to operate than diesel trucks, which are cheaper than gasoline trucks, which are cheaper than liquid petroleum (propane) trucks. Low resistance to motion also helps; so trains use less energy than trucks and ships are cheapest of all per km-ton.

Care must be taken that low energy costs are not overcome by high labor costs. Move more product/labor hour—the ultimate is infinite volume or zero labor hours. Large oil tankers use much less labor/barrel of oil transported than small tankers, as do large trucks vs. small trucks—more volume for the same labor. For distances within a plant of over 150 m, tractor-trailer trains may be more economical than fork trucks. If the route is standardized (and perhaps with less than 20 destinations), the train need not have a driver because sensors and computers can replace a driver in some applications—the same volume with less labor.

2.5 Distance/Trip

Reduce distances by efficient layout and arrangement. Trade off short trips for local supply vs. decreased inventories for central supply. For example, fewer micrometers are needed with a tool crib than if everyone is issued one, but distance moved to get the "mike" becomes 100 m, not 2 m. Micrometers have low capital costs, whereas fork trucks have high capital cost. Thus, giving each department its own fork truck reduces distance traveled but the extra capital cost can be very high. Sharing may work; that is, department A "owns" it in the morning and B in the afternoon.

All other things being equal, a short trip costs less than a long trip. Figure 14.1 shows a "bus route" around an area. A **bus** goes around the area on a standardized route. Material going from B to A must first go to C and D before coming to A.

Why use a bus instead of a **taxi** system (point-to-point service)? First, cost of movement tends to be relatively insensitive to distance due to high pick up and put down costs in relation to movement costs (see Figure 7.2). Second, if distance moved is important, usually it is total distance moved by the *carrier* (distance/loop × number of trips around loop) that is important rather than distance moved by the objects. Third, time for the physical movement of products tends to be relatively small in relation to the time in storage at each end of the move and thus not very important. Overhead conveyors often wind for long distances around the ceiling to act as a work-in-progress storage. In the international oil trade,

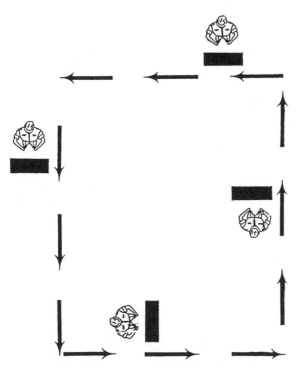

FIGURE 14.1   "Bus service" (such as from a power and free conveyor or an automatically guided vehicle) follows a standardized route while "taxi service" (such as from a lift truck) goes directly to the desired location. Bus service tends to be cheaper than taxis but it may take an hour or two for the delivery instead of the minutes of the taxi.

tankers often leave port without a specified destination. Their cargo is sold when they near Europe.

Emphasize minimizing total material handling cost—in itself a subset of the goal of reducing manufacturing cost—rather than minimizing distance/trip, variable cost/distance, and so forth.

GUIDELINE 3: DECOUPLE TASKS

This guideline is adapted from Konz (1994). See Box 14.1.

3.1 Introduction   Figure 14.2 shows three types of **flow lines.** In one extreme, the **operation only line,** a single component goes through a series of operations (item is processed or changed at stations); no additional components are added. There are operations and transportations. Examples are an engine block being machined on several machines, a steel rolling mill, a rotary index table with several "heads," and university class enrollment (visit various tables to sign up or pay for various activities).

In the other extreme, the **order-picking line,** items are accumulated together without any operation at the station. There are transportations but no operations. Examples are order picking in a warehouse and a customer obtaining food in a cafeteria.

BOX 14.1   *Just-in-time*

A technique called **just-in-time** currently is popular in management circles. Unfortunately, some managers have become enthusiasts of this technique without understanding why the Japanese found it so useful.

The simplistic view is to reduce work-in-process inventories to very small amounts to reduce inventory costs. In effect, this means small buffers. However, inventory cost reduction is not the reason for using just-in-time.

What is the cost of inventory—that is, buffers?

A rule of thumb is that annual inventory carrying cost is about 30% of the product price. (The 30% is primarily the cost of capital, although there is some cost for the building, heat, lights, etc.) Thus, a product costing $100 will cost .30 ($100) = $30 to store for 1 year. For simplicity, assume a factory works 250 days/year. This means it costs about 30%/250 production days = .125% of the product cost to store an item for 1 production day or (assuming an 8-h shift) .125%/8 = .0156%/production hour. A $100 buffer then would have an inventory cost of $100 × .125% = $.125 = 12.5

cents/production day or 1.57 cents/production hour. Unless there are truly enormous inventories between stations, therefore, there is little inventory savings available.

Why do the Japanese advocate use of just-in-time? Because it forces management to solve problems. Consider the flow of inventory through a facility as a river upon which you are using a boat. When the river level is reduced (inventory is reduced), **rocks in the river** (problems) appear which were hidden by the large inventory. Lowering the river forces the management to solve the problems (scheduling, quality, etc.) which formerly could be concealed by large inventories. After the problems are solved, the river is lowered further and more problems are solved. The solution of these hidden problems is the justification for just-in-time, not minor reductions in inventory costs. In theory, managers should have solved the problems without "holding their feet to the fire," but just-in-time seems a practical technique to get their attention and make them do what they should have done anyway.

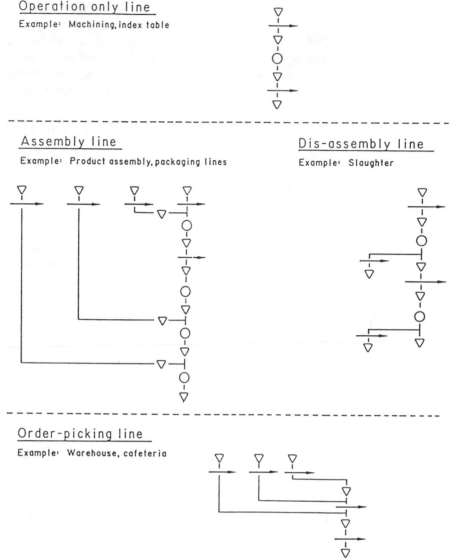

FIGURE 14.2   Flow lines can be operation only, assembly, or order picking.

The most common flow line, the **assembly line,** has both operations performed and items added at the station. There are both operations and transportations. Examples are product assembly (autos, TVs, clothing), packaging lines, chemical processing lines, and filling lines. A reverse version is the disassembly line, often used in food processing. Example disassembly lines are slaughter operations and grain mills.

Flow lines do not need to make a single product continuously. Figure 14.3 shows three alternatives: (1) the single product made continuously, (2) multiple products made sequentially in batches, and (3) multiple products made simultaneously. Alternative 1 might be product A made continuously. Alternative 2 might be product A made all day Monday, product B made all day Tuesday and Wednesday, product C made Thursday, and so on. Alternative 3

might be one of product A made at 8:00 A.M., two of product B made at 8:07 and 8:14, one of product C made at 8:17, one of product A made at 8:21, and so on.

The **elements** (work) of the task are divided among the line's stations. If the amount of work is equal at each station, it is a **balanced line;** if not equal, it is an unbalanced line. Depending upon the type of line, one or more elements can be done at each station. In addition, there is the possibility of the same element's being done at more than one station. A well-designed line will have:

- minimum idle time at the stations
- high quality (enough time at each station for operators to complete assigned work)
- minimum capital cost (for both equipment and work in process)

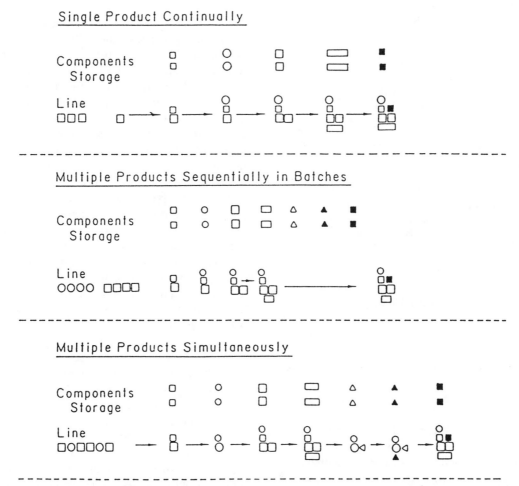

FIGURE 14.3 Flow lines can make a single product continuously, multiple products sequentially in batches, and multiple products simultaneously.

Although it may be obvious, (1) the transport between stations need not be by conveyor, (2) the transport between stations need not be at a fixed speed or time interval, and, very important, (3) there may be storages between the operations or transportations. There may be a storage before and after each operation and transportation. This storage is known technically as a **buffer,** a bank, or float. Its purpose is **decoupling** the line (isolating components of the line).

There are two primary reasons for decoupling: (1) line balancing and (2) shocks and disturbances.

3.2 Line Balancing
The line balancing challenge occurs because the mean times for stations A, B, C . . . are not equal.

Assume operation A took 50 s, B took 40 s, and C took 60 s. Then, assuming there was no buffer, the line would have to index at the speed of the slowest station (the **bottleneck station**), station C. At stations A and B there would be idle time, called **balance-delay time,** the difference between the **cycle time** (time/unit of the line) and the work time.

Thus, without buffers, the line speed must be set considering the speed of the slowest station.

Figure 14.4 shows another aspect of the problem of mean times—the variation in the ability of human operators. Typically performance among operators varies about two to one. The best can produce twice what the worst can. Assuming symmetry and putting the average operator at 100%, the range is from about 67% to 133%. If the line is set at the speed of the average operator, 50% can work faster and 50% cannot keep up. (There is a temptation for those who cannot keep up to reduce safety and quality.) Thus the speed at a station (assuming the typical balanced line where work content is equal at each station) cannot be set at the speed of an average (e.g., 1.0) operator; instead it is set at the speed of a slow operator. Referring to Figure 14.4, if a slow operator is defined as a 90% operator (i.e., 90% of the operators can do the job in the mean cycle time), this is 1.28σ below the mean. Considering the range of performance from 2/3 to 4/3 as 6σ, then $(4/3 - 2/3)/6 = 1/9$. Then $-1.28(1/9) = .142$. Then $1.0 - 0.142 = .858$ for the station speed. That is, if the station could be

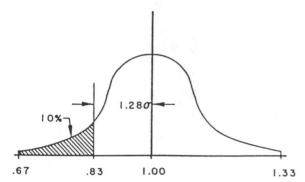

FIGURE 14.4 Variability of cycle times increases the inefficiency of flow lines. If the pace of individual decoupled workstations and stations arranged in a flow line are the same, then the flow line would have to be 17% slower to accommodate the slow operators (slow is defined as a pace 90% of operators can do).

set for an average operator instead of a slow operator, then it could be 17% faster (as [1.0 − .858]/.858 = 1.17). Therefore, without buffers, the line speed must be set considering both the speed of the slowest station and the mean time of the slow operator on the slow station.

But there is a third problem. Cycle times vary.

3.3 Shocks and Disturbances

Shocks and disturbances make the cycle lines vary. Figure 14.5 points out the problem of distribution of operation times. Assume operation C has a mean of 60 s and σ = 4 s. To include 95% of the times, then station C, and thus the line, would operate at a cycle time of 65.1 s.

The variability of cycle times can occur for many reasons.

First, consider a station with both an infinite supply of incoming components and perfect removal of the completed unit. There may be a temporary shift in the mean because the operator normally doing operation C is absent and a substitute is doing the job. If the temporary worker took 66 s instead of 60 s, then each station on the entire line also would take 66 s, if there were no buffer. Or the time of the regular operator may vary for normal reasons—tools becoming dull, short breakdowns of the machine, operator's stopping to light a cigarette or sneeze, talking to the supervisor, or dropping a part. (If the operator is a machine or robot, cycle times still can vary. For example, a machine may fail if the incoming part arrives upside down.) Figure 14.6 shows typical cycle time distributions for unpaced work (i.e., with sufficient buffers) and paced work (usually work without buffers). The distributions for unpaced work usually are positively skewed (a few long times with a lower absolute minimum time). Paced work has a higher mean time (i.e., pace is slower) and the curve is more symmetrical.

Second, consider the station as part of a line. The station's time can vary because of inadequate supply from the preceding station (**station starvation**) or because the following station is not yet ready to accept a unit (**station blockage**). In addition, there is the problem of what to do with a defective unit (it is scrap if it can't be fixed, rework if it can be fixed). Time problems include when to repair the item (now or later); space and transport problems include where to put the defective component or assembly.

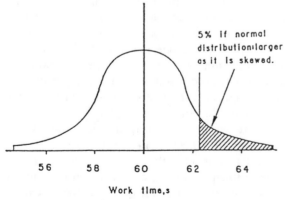

FIGURE 14.5 Distribution of times of a specific operator points out that only 50% of the times are done in the average time or less. If a paced line is set so that 95% of the cycles are completed before the line indexes, then the cycle time must be slower than average time. If average time = 60 s and σ = 2 s, then the line cycle time (based on this station and assuming normality) would be 60 + 1.64(2) = 63.3 s.

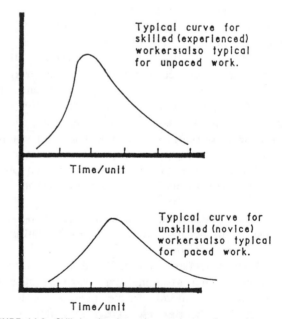

FIGURE 14.6 Skilled workers have a lower variance and more skew as well as a lower mean. Unpaced time distributions are similar to skilled time distributions; paced time distributions are similar to unskilled time distributions. See Figure 26.2 for distribution of times when output is restricted.

Thus, without buffers, the line speed must be set considering: (1) the speed of the slowest station, (2) the mean speed of the slowest operator on the slowest station, and (3) the slowest cycle time of the slowest operator on the slowest station.

Buffers give the flow line flexibility or tolerance. Figure 14.7 shows the flow line without buffers (the rigid system) as being similar to a train, while the flow line with buffers (the flexible system) is similar to a string of trucks. In the "train," each rail car must travel at the same speed as every other rail car; if one rail car breaks down, the entire train stops. With "trucks," each vehicle can go at its own speed; if one vehicle stops, the remaining vehicles can continue.

The total penalty of a line without buffers versus a line with buffers is difficult to determine because most lines have some buffers. See Box 14.1. However, a U. S. machine-paced auto assembly line (i.e., a line with few or no buffers) usually has a balance-delay time of 8 to 15%. U. S. auto companies also allow an extra relief of 22 min/480 min shift (about 5%) for a machine-paced line (total of 46 min per shift). Thus, U. S. auto companies have a built-in inefficiency of 13 to 20%, ignoring the problem of slowing the line down to the speed of the slowest station. Toyota has demonstrated that auto lines can be built with buffers. They have a buffer of 3 cars approximately every 10 stations. Each worker has a button that can stop the line if there is a problem such as bad quality. In U. S. auto plants there are no buffers and only a few people are authorized to stop the line (since if the line stops, everybody stops), so they rarely stop the line. For other industries, Kilbridge (1961) gave 5 to 10% as a typical balance-delay percent. Figure 14.8 shows that balance-delay percent usually declines for longer cycle times.

Buxey (1978) reported on a number of lines in Scotland. In case 1 there were 20 stations with a cycle time of 15 s; items/station (i.e., buffer) were 1.5; the balance loss was 20%; and there was an additional loss because cycle time was determined by adding 20% to the time of the work at the most difficult station. Approximately 3% of the items were not

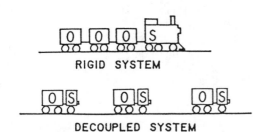

FIGURE 14.7 Flow lines without buffers can be symbolized as a train, with initial storage (S) followed by operations (O), that is, S-O-O-O. Flow lines with buffers can be symbolized as trucks, that is, S-O, S-O, S-O.

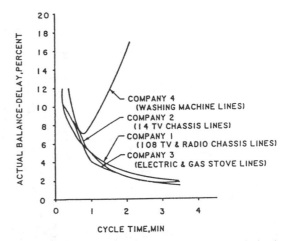

FIGURE 14.8 Balance-delay time usually declines when cycle time is longer (Kilbridge and Webster, 1961). Reprinted by permission of Kilbridge and Webster, *Management Science*, Vol. 2 [1], 1961. Copyright 1961. The Institute of Management Sciences.

completed when they finished the line and had to be finished at a repair station. In case 2 there were 13 stations with a cycle time of 10 s; items/station were 2; balance loss was 4%. In case 3 there were 15 stations with a cycle time of 1 h; items/station were 2. The balance loss was 6%, but in addition there was a utility operator, resulting in an additional cost of 1/15 = 6.7%.

These figures point out that although the fixed-pace line is a manager, it is an inefficient manager.

3.4 Buffer Design

Figure 14.9 is a schematic of a workstation. The total workstation is composed of an operator, machine, energy and information input, energy and information output, material input, material output, input product storage, and output product storage. Buffers increase the size of input storage and output storage.

There are two buffering techniques: (1) decoupling by changing product flow, and (2) decoupling by moving operators.

3.4.1 Decoupling by changing product flow

There are three subcategories: (1) buffers at or between the stations, (2) buffers due to carrier design, and (3) buffers off-line.

Buffers at or between stations   One possibility for a buffer at or between stations is a physical barrier on a conveyor. Figure 14.10 shows two common arrangements. In the upper figure, a piece of wood, a pipe, or piece of steel is placed across the conveyor. The pieces from the upstream workstation move along the conveyor until they hit the barrier and stop. The operator lifts them across the barrier, works on them, and puts them back on the conveyor

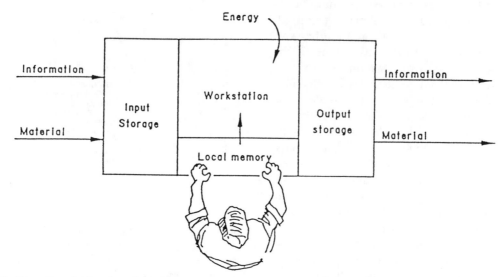

FIGURE 14.9 Schematic workstation shows that both information and material flow into input storage. Then they are transferred to processing, where they are transformed, using energy and local memory. Then the information and material go to output storage before being transferred to the next station.

downstream of the dam. Rotary tables (lower figure) can be used to increase the size of the reservoir upstream of the dam. Parts stay on the rotary table and go round and round until picked up.

Buffer capacity can be increased by increasing the time the item is available to the operator or increasing the space within the reach of the operator. Increase time by having the operator face upstream. Arm motions are easier forward than backward, and when the object can be seen, timing can be better. If the object must approach from the operator's rear, use a rear view mirror so the operator

need not turn around. Increase space within reach with a rotary table, as in Figure 14.10. A fixed-pace conveyor with items fixed to the conveyor tends to be a poor design. For a moving assembly, where access time is minimized, the components can be stored on a second conveyor positioned above the assembly conveyor or behind the operator and moving at the speed of the assembly conveyor. Figure 14.11 shows two techniques of putting more items within the operator's reach.

The buffer can be designed for line balancing purposes as well as for shocks and disturbances—that is, the line can be unbalanced. Figure 14.12 shows a schematic pair of stations.

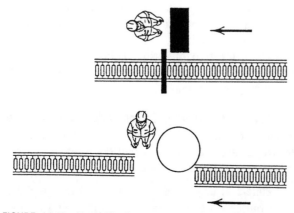

FIGURE 14.10 Physical barriers on a conveyor can dam the flow of parts, creating a reservoir or buffer. The lower figure shows a larger reservoir created by using a rotary table. If there is a "flood" and the reservoir capacity is insufficient, then the flow must be stopped upstream. This is called station blocking. If there are not enough items in the reservoir, then the downstream station must shut down. This is called station starvation. The variability of the number of items in a buffer is an index of the buffer effectiveness. That is, if there is no variance in the number in the buffer, there is no need for the buffer!

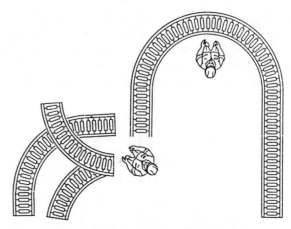

FIGURE 14.11 Increase storage space at a station by using the cube of the space or by making a U in the line to take advantage of the operator's ability to turn. Close spacing of items on a conveyor belt and slow belt speed is preferable to wide spacing with high speed (since more items are within reach), even though average rate of arrival is the same.

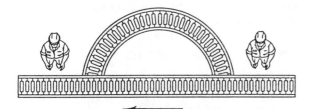

FIGURE 14.12   Buffers can be used for balancing as well as for shock absorbers. Assume the first station produces 60 units/h for 4 h—putting 30/h on the line and 30/h into the buffer (curved portion). Then the first station shuts down. For the second 4 h of the shift, the second station is fed from the buffer.

First we will discuss buffer input rate equal to buffer output rate. For example, assume line output is 10/h or 80/shift of 8 h, and that station A produces 20/h for 4 h and station B produces 10/h for 8 h. During the morning, A sends 10/h to B and puts 10/h into the bank. After 4 h, the bank has 40 units. Then B is fed from the bank at the rate of 10/h while operator A works elsewhere.

Suppose the buffer input rate is not equal to output rate. Assume, for example, that A has a rate of 20/h, B has a rate of 15/h, and desired line rate is 15/h. Then operate B at a rate of 15/h for 8 h. A produces 20/h but for only 6 h. During the 6 h, A sends 15/h to B and puts 5/h into the buffer. At the end of 6 h, the buffer holds 30 units. For the last 2 h of the shift, feed B from the buffer while A works elsewhere. Or, assume A has a rate of 20/h, B a rate of 15/h, and the desired line rate is 20/h. Then operate A for 8 h at 20/h and have B work 8(20/15) = 10.7 h/day. (It may be easier in some cases to have a partial shift or Saturday work than to work 2.7 hours of overtime each day.)

If a machine must be fed at a constant rate, it is more efficient for operators to load a magazine at their own speed and the machine to work from the magazine than for a machine to feed the machine directly. This applies even if the machine cycle time from the magazine is no faster than the mean operator feed time (Corlett, 1982). See also Figure 14.13.

The concept so far has been to maximize time a unit is available to a station. Sometimes there are items that do not need to stop at each station; then a technique is needed that bypasses unwanted stations. Figure 14.14 shows the general concept. Figure 14.15 shows this idea applied to cafeterias, where the product is the human customers and the stations are the salads, desserts, drinks, and so forth.

Buffers due to carrier design   Items often are moved between workstations by carriers; the **carrier** can be a pan, box, pallet, index table, hook, or cart. If the carrier may be removed from the line, the

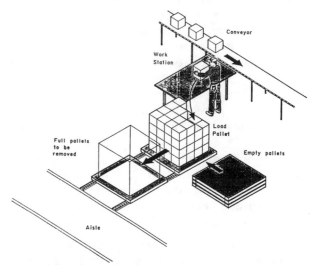

FIGURE 14.13   Pallets on tracks increase the output buffer by decoupling the operator from the material handlers. When the operator completes a pallet, it is pushed into the removal position, a new pallet put into position, and work is resumed. Without the track, the operator must stop work when the pallet is full until the material handler moves the pallet. This tends to result in overstaffing of material handlers.

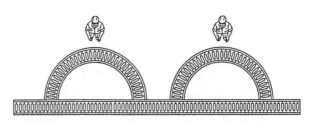

FIGURE 14.14   Bypassing stations not needed permits reduced work-in-process time.

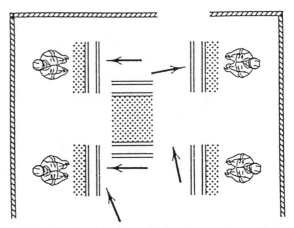

FIGURE 14.15   Scramble-system cafeterias assume customers will stop at only a fraction of the available service areas. Since the goal is to minimize the customer's waiting, the system permits customers to skip stations at which no service is desired.

effect can be that of a buffer. Transport between stations can be with wheeled carts pushed by hand. Heavy items (such as with tractor assembly or mounting jet engines to aircraft) can be pushed if on airfilm pallets. If power is desired, it can be from overhead (power and free conveyor) or below (a towline). See Figure 14.16.

Another possibility is to keep the carrier on a conveyor but make the path omnidirectional. See Figure 14.17.

The following gives the advantages and disadvantages of multiple items/carrier. (It may be possible to obtain many of the advantages of multiple items/carrier with one item/carrier by batch processing the one-item carriers.)

1. *Labor.* Consider pickup and putdown of tools. In a high-volume operation, say a station time of .1 min, the operator picks up the tools at the start of the shift and does not put them down until break time. Thus, **get and put-away times** are prorated over many units so the cost/unit is small. If the station time is longer, say 1.0 min, then the operator does a number of operations and uses a variety of tools, picking up and putting down each one. However, if the items come in a carrier with multiple

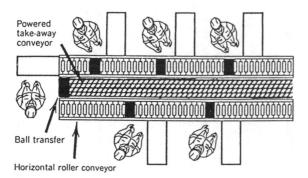

FIGURE 14.16 Removing a carrier from the line's path aids buffer creation. The carrier can be completely mobile (e.g., hand cart, air pallet) or normally connected to a power source but able to be disconnected (power and free conveyor, towline).

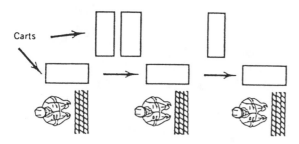

FIGURE 14.17 Multipath conveyors use ball transfers at intersections and horizontal gravity roller conveyors for accumulators. Take-away is by powered rollers. Quick modifications to the number and combinations of stations permit multiple products to be built simultaneously. It is a cart system but with conveyors.

items, the operator can do a number of items with only one get and one put-away.

$$TTIME = (GTIME + PATIME)/N$$

where

$TTIME$ = Transport time/unit

$GTIME$ = Get time/unit

$PATIME$ = Put-away time/unit

N = Number of units

Assume that $GTIME + PATIME = X$. Then, for $N = 1$, $TTIME = X$; for $N = 2$, $TTIME = .5X$; for $N = 4$, $TTIME = .25X$; for $N = 8$, $TTIME = .125X$. The point of diminishing returns comes fairly rapidly. In addition, as N increases, the reach and move distances in the workstation increase, so the "do" times also increase. Kilbridge (1961) estimated get time + put-away time as 13% of do time for electronics assembly done on worker-paced conveyors.

Time/cycle will have a smaller variance. Although a carrier with 4 units probably will have approximately 4 times the mean do time of carriers with 1 unit, the variance of the longer cycle time probably will be less than the variance of the 4 short times, because long- and short-unit processing times may cancel each other on the multiple-item cycle time.

Multiple items/carrier encourage use of both hands since 2 units can be worked on at the same time. When using multiple items/carrier, make N an even number.

Multiple items/carrier may restrict access to individual units on the carrier. For humans, this may just mean more time. For robots and machines, however, the restricted access may make the operation not feasible or the device very expensive.

2. *Material handling.* Multiple units/carrier give more units/foot of the line. For a specified distance, multiple units/carrier give more buffer. For a specified buffer size, the buffer will fit into a shorter distance for multiple units/carrier.

Multiple units/carrier require heavier carriers. The carrier will be harder to push, pull, and lift—although fewer carriers will be moved. Motors rather than muscles may be required for the heavier carriers.

Disposal of rejected units is more difficult for multiple units/carrier. If the defective unit remains on the carrier, take care that additional work is not done on it; be sure it does not get included with good units. The defect also fills up buffer space. If the defect is removed from the carrier, empty space moves down the line. In addition, a carrier and means of transport must be provided to move the rejects to the rework station or the scrap pile.

3. *Equipment costs.* Carrier cost/unit usually is less when there are multiple units/carrier. The cost/carrier is higher but fewer carriers are required. For example, 1 carrier holding 8 units might cost $200 while 1 carrier holding 1 unit might cost $40.

Automatic processing of units is more difficult with multiple items/carrier. One possibility is to use duplicate heads on one machine so that all units on the carrier are done simultaneously; another possibility is to do the units in sequence by indexing either the carrier or the head. However, unless this station is the bottleneck station, the extra heads or indexing equipment may be just extra capital expense, since one head may have had sufficient capacity. If a robot is used, it may have to be a more expensive robot.

Buffers off-line   **On-line buffers** (or those in-line) can handle minor disturbances. Major shocks (machine breakdowns, employee absenteeism, learners, etc.) and line balancing problems may benefit from **off-line buffers.** Off-line buffers can be remote either in time or in space.

Additional time to compensate for a disturbance can be obtained by overtime work, partial shifts, and working holidays on the existing equipment. For example, assume the buffer size between stations = 1 h. Then, for any shock of less than 1 h is (e.g., short machine breakdown, small difference in production between a learner and experienced worker), fill the buffer up again by having the worker work extra time at the end of the shift.

Additional space for an off-line buffer can be made several ways. Figure 14.18 shows an off-line buffer, a semicircular conveyor. In this situation the buffer acts as a standby circuit. A standby circuit analysis considers the device (in this case, the conveyor with units on it), the device that detects circuit failure (the operator), the switch that activates the standby system (the operator), and the reliability of

the standby device (how good the units on the semicircular conveyor are). Naturally the off-line buffer does not have to be conveyorized. It simply may be items stored on a cart or pallet, either close to the workstation or perhaps several hundred feet away.

Off-line buffers can include processing as well as storage. See Figure 14.19. For example, assume station B was able to produce only 15 units/h when the line rate was 20 units/h. Then a duplicate of station B could be built (5 or 500 ft. from the line) and operated the number of hours necessary to obtain the extra units. This technique is most useful when the capital cost of an additional workstation is relatively low. The extra workstation can be used for training purposes, or to use idle time of a worker on another job with a large machine-time component, or to use idle time of a worker whose primary job does not require 8 h/day.

So far we have decoupled by moving the product. There is another alternative (often overlooked): move the operator.

3.4.2   *Decoupling by moving operators*

There are four subcategories: (1) utility operator, (2) help your neighbor, (3) *n* operators float among *n* stations, and (4) *n* operators float among more than *n* stations.

Utility operator   Figure 14.20 shows the utility operator approach. In this concept, most of the operators work at specific workstations. However, one operator is a **utility operator,** or relief operator. The utility operator's assignment is to help the individual operators if they have trouble for a few minutes, if they want to go to the toilet, and so forth. In the U. S. auto industry, for example, a common situation is 1 relief worker for every 6 stations; that is, 7 people work at 6 stations with 1 always being off. The line doesn't stop for breaks. (However, if capacity is greater than demand, it may be worthwhile to

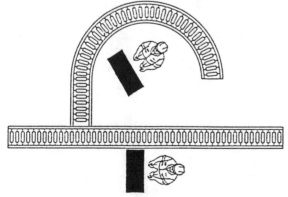

FIGURE 14.18   Off-line banks (semicircular portion) can function as standby components in the production system.

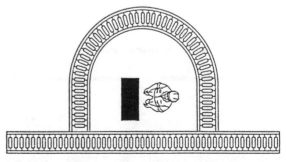

FIGURE 14.19   Standby systems can include processing as well as storage. The main advantages are better utilization of people and equipment and easier training. Disadvantages are increased material handling and scheduling problems.

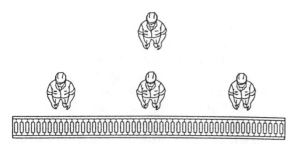

FIGURE 14.20   Utility (relief) operators move from station to station while the station operators all stay at their own stations. The utility operator must know all the jobs and so frequently becomes the trainer. In many cases, the utility operator is given minor management responsibility and is called a group leader or working supervisor.

eliminate the "tag relief" system and stop the line during breaks, saving money. Chrysler has used this strategy.) In other industries, the utility operator does not have formal times assigned to relieve specific operators but just helps out when and where needed.

The duties of this operator vary widely; additional work is needed for when there is no line worker taking a break. In these situations, the utility operator often is called a group leader (working supervisor); responsibilities include training new employees and making minor decisions (if the supervisor is absent) in addition to working at the various stations. Minor maintenance work or product rework are other duties.

Help your neighbor   Figure 14.21 shows the help-your-neighbor approach. Each operator, by management directive, is "your brother's keeper." That is, you help your neighbor not because you are a good person but because it is part of your job responsibility. If management does not formally require operators to help each other out, then those helping may feel that they are "suckers." If a fellow worker gets behind, this is not a reason for everyone else to take a break until the person catches up. One approach is to divide the work at a station into thirds. The middle third is the sole responsibility of the station operator,

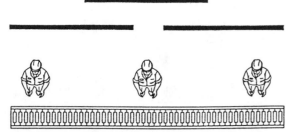

FIGURE 14.21   **Help your neighbor** means shared responsibility. The line shows responsibility extends both directions from each operator.

the first third is entitled to help from the upstream operator, and the last third is entitled to help from the downstream operator.

***n* operators float among *n* workstations**   Figure 14.22 shows the *n* operators float among *n* workstations approach. Stations are not given specific times/unit; instead, the operators are just told the total time for the entire unit. For example, instead of saying that each of the 5 stations had 1.0 min of work/unit, the operators would be told there is a total of 5 min of work. It is the operator's decision as to which operator does what job at what rate; the operators move upstream and downstream as they desire; and the operators decide when to switch jobs. This type of line management tends to be efficient since buffer problems are minimized (if a buffer = 0, the operator moves to another station); group pressure to produce is high; the line runs at the average speed of the group (rather than the speed of the slowest member); everyone tends to be able to do every job (so absences or illness are less of a problem); and, since the line need not be balanced, there is no balance-delay time.

***n* operators float among more than *n* stations**   Figures 14.23 and 14.24 show two examples of the *n* operators float over more than *n* stations approach.

FIGURE 14.22   *n* operators among *n* stations is the full-float design. Supervisors assign operators to the station as a whole instead of to individual workstations. Operators move from station to station as the need arises. Thus, the station need not be balanced.

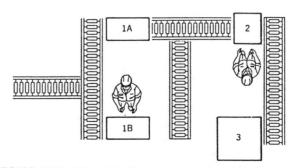

FIGURE 14.23   Paired station line for semiautomatic machines is a version of *n* operators among more than *n* stations. The two stations for an operator can be identical (see 1A and 1B) or different (see 2 and 3). The machine can either run automatically while the operator loads/unloads the other station, or can load/unload automatically while the operator processes. Thus operator utilization is high while machine utilization is low. Naturally there need not be a conveyor since material handling could be by cart or hand.

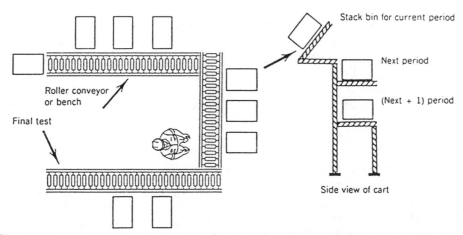

FIGURE 14.24 One-worker line uses a U shape to cut worker walking. Since final test was done upon completion of each unit, the worker got quick feedback on quality, so quality problems were reduced drastically (Gargano and Stewart, 1975). The wheeled carts hold a complete period (day or week, depending on the unit) supply of parts. The next two periods' supply is on the second and third shelf, which helps keep an inventory count and pinpoint supply problems.

The rationale of **more places than people** is the minimization of the total cost of the system. In the United States, labor costs tend to be higher than capital costs. Labor wages of only $5/h become labor costs of $6/h after fringe benefits are considered. At 2,000 h/yr, this means minimum labor cost is $12,000/yr. Many operators have an annual cost of over $25,000. If a workstation costs $5,000, lasts 3 years, and has a scrap value of $2,000, then annual cost is ($5,000 − $2,000)/3 = $1,000/yr. Thus it is far more important to keep the $12,000/yr operator busy than to keep the $1,000/yr workstation busy.

Figure 14.23 shows the **paired station** approach. Two stations are built (if they are identical, it is called double tooling); the operator goes back and forth between the two. This approach is quite useful when the operator has a large idle time due to machine time. This also makes the line more reliable, because if one machine fails, the other still can work, and although output would not be up to full potential, it would not fall to zero.

A second variation (Figure 14.24) is the **one-worker line.** The work required for a product may require more space for components, tools, and equipment than can be located conveniently at one workstation. It is not necessary to split the total task among several operators. Break the job down into several stations (for example, three). The operator works at station A, completing a number of units, perhaps 25. Then the 25 units *and the operator* move to station B, completing the required operations at station B. Then the 25 units and the operator move to station C. Since there is only one operator, there is no balance-delay time and the stations need not have equal cycle times (i.e., the line need not be balanced). Buffer sizes are not critical.

Another version is a single-product line with, for example, 12 workstations and 4 operators. See Figure 14.25. The workers work at a specific station and send products to the next station. Then they walk to another station (not necessarily adjacent) and begin work again. Different assignment policies can be used. For example, some operators can work at all 12 stations, and others (such as beginners) might be restricted to just a few.

Another concept is **cellular manufacture.** All the machines and equipment needed to produce a product are grouped in a cell. For example, three operators might run 10 machines. Operators switch among machines as needed. The goal is maximum use of the expensive cost component, human labor. Note that since the operators are surrounded by machines, they may be exposed to high noise levels.

Another version is multiple lines, each set up for a single product (say line A for product A, line B for

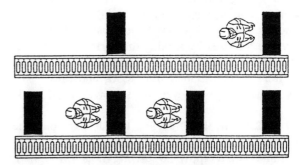

FIGURE 14.25 More places than people is a different way of saying *n* operators among more than *n* workstations. The rationale is to keep the operator busy and not worry about machine utilization. Note that if the conveyor is eliminated from the figure, *n* operators among more than *n* stations is similar to a job shop. Flow lines are concepts; material handling and layouts follow many patterns.

product B, line C for product C). The workers might work on line A on Monday, line B on Tuesday and Wednesday, line C on Thursday morning, and then line A for the rest of the week. A furnace manufacturer had 9 lines for assembly of 171 models. Teams of 2 roved from one line to another, doing complete assembly of a given model. Meanwhile the setup operator converted a line for the next batch of another model. Each team specialized in a range of models, and so they didn't have to know all 171 models. If there was a shortage of components for any line, workers simply shifted over to another line until the supply was sufficient. Production scheduling was very flexible. This flexibility is very useful when there is insufficient demand for just one product for an entire year. Although equipment is duplicated, setup and put-away costs are minimized; maintenance can be done during normal working hours instead of at premium pay hours.

The more places than people concept is very useful for short runs and operators who are learning.

Letting the worker walk requires walking time, which must be added to the work time. The MTM predetermined time system (see Chapter 29) allocates 5.3 TMU/foot (.2 s). Thus walking 10 ft requires 2 s. If 10 ft were required for every 20 units, then add 2/20 = 0.1 s/unit. From a physiological and comfort viewpoint, standing with occasional walking reduces venous pooling in the legs and is less fatiguing than standing without walking.

GUIDELINE 4: MAKE SEVERAL IDENTICAL ITEMS AT THE SAME TIME

Tasks can be broken into three stages: (1) **get ready,** (2) **do,** and (3) **put away.** Reduce cost/unit by prorating the get-ready and put-away stages over more units. Manufacturing similar parts in sequence (parts families) reduces setup time by minimizing the number of changes necessary. Items may differ only at a later stage: B is blue, C is green, and D is red; or B has 1 hole, C has 2, and D has 3. Decrease lead time and increase lot sizes by making the early operations as item A (unpainted part or part without holes). Then, for item B, pull item A from stock and finish. The assembly example analyzes physical vs. mental work.

4.1 Assembly Example   In this example, the pick-up and put-down of handtools is minimized. The example could apply to driving nuts on a bolt, putting pickles on hamburgers, or marking cans in a store, but the example in Table 14.1 is for soldering.

Trade off prorating the pick-up and put-down over more units vs. the increased distances moved during the *do* portion of the cycle. (See the comments about multiple items/carrier in Guideline 3.) Distances moved increase because of the larger workstation. In general, it takes less time to move a greater distance than to have additional reaches and grasps during the pick-up and additional move-asides and releases during the *put away*. As the size of the workstation increases, there is a tendency to begin moving the product instead of the worker; this mechanized handling and increased worker specialization leads to the assembly line.

4.2 Inspection Example   In Table 14.1, the inspector must inspect n items for m characteristics (see Figure 14.26). The *get ready* is the mental work of fixing in the mind the quality standard for the characteristic m. The *do* is the sensing of the object, comparing it with the mental standard, making a decision, and executing the decision. The *put away* is the mental transfer of the quality standard from working memory to long-term memory. The following example could be a potato inspector looking for mold, eyes, or cuts (symbolized by circles, squares, and triangles in Figure 14.26); it could be a machine shop inspector inspecting for surface finish, concentricity, and length; it could be a teacher looking for 3 key words in a student's examination; or it could be a typist looking for misspelled words, punctuation mistakes, and tense mistakes in a letter.

One item at a time requires less physical handling of items but considerable mental manipulation of characteristics. One characteristic at a time (less mental work but more physical work) ensures that a characteristic is not omitted, that the mental standard is more consistent, and that there is less halo effect of one characteristic on another characteristic. In the extreme it may cause boredom.

Konz and Osman (1977) had 24 women inspect numbers on slides. For one item at a time, they made 16% errors for defect A and 9% for defect B; for one characteristic at a time, they made 5% for defect A and 7% for defect B. Effectively all the errors were type 2 errors. See Figure 14.27 for an explanation of type 1 and 2 errors. Total inspection time was held constant. Su and Konz (1981) found that, if inspection is easy, there is no difference in accuracy between inspection for one or multiple characteristics at a time but a considerable time penalty for one characteristic at a time. If the inspection is difficult, there is a trade-off needed as one characteristic at a time gives fewer errors but an increase in time.

GUIDELINE 5: COMBINE OPERATIONS AND FUNCTIONS

Do several steps at the same time by using **multi-function** materials and equipment rather than

TABLE 14.1   Items can be assembled two ways, the one-item-at-a-time method or the multiple-unit method. Items can be inspected one item or one characteristic at a time.

ONE-ITEM-AT-A-TIME METHOD (ASSEMBLY)	TIME		MULTIPLE-UNIT METHOD	TIME	
Get soldering iron	3		Get soldering iron	3	
			Solder diode on unit 1		5
Solder diode on unit 1		5	Move iron to unit 2		1
Put iron away	2		Solder diode on unit 2		5
			Move iron to unit 3		1
Additional work		X	Solder diode on unit 3		5
			Put iron away	2	
Get soldering iron	3				
Solder diode on unit 2		5	Additional work		3X
Put iron away	2				
Additional work		X			
Get soldering iron	3				
Solder diode to unit 3		5			
Put iron away	2				
Additional work	X	X			
Soldering time/unit = 15/3 + 15/3			Soldering time/unit = 5/3 + 17/3		
= 5 + 5 = 10			= 1.7 + 5.7		
			= 7.4		

ONE-ITEM-AT-A-TIME METHOD (INSPECTION)	ONE-CHARACTERISTIC-AT-A-TIME METHOD
Recall standards for circles	Recall standard for circles
Inspect for circles on unit 1	Inspect for circles on unit 1
Dispose of standard for circles	Aside unit 1
	Get unit 2
Recall standard for triangles	Inspect for circles on unit 2
Inspect for triangles on unit 1	Aside unit 2
Dispose of standard for triangles	Get unit 3
	Inspect for circles on unit 3
Recall standard for squares	Aside unit 3
Inspect for squares on unit 1	Dispose of standard for circles
Dispose of standard for squares	Recall standard for squares
Etc.	Etc.

single-function materials and equipment. Material cost/unit will be lower. Labor cost/unit will be reduced. Capital cost/unit generally will be higher. Total cost will be lower since material and labor cost/unit generally are more important than capital cost/unit.

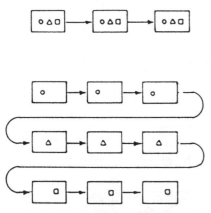

FIGURE 14.26   Two inspection strategies can be used. One item at a time (upper figure) is faster but one characteristic at a time (lower figure) probably gives better quality.

5.1 Multifunction Materials   Use a compound which waxes at the same time that it cleans. Use a water-pump lubricant which also has a rust inhibitor. Although the basic chemicals will cost the same whether sold together or separately, the manufacturing and distribution costs will be reduced since one container is used instead of two, material handling will be less for one container than two, stocking expense on the store shelf will be lower for one container, and advertising expense will be lower. Labor cost will be lower and quality may be better since, if the two compounds were sold separately, one might not be used. Henry Ford had supplies delivered in special wooden boxes; the boxes became part of the Model T floor. The ultimate multifunction container is the ice cream cone.

Paperwork can fill multiple functions. Figure 14.28 shows a dividend check. If the computer prints the address as well as the name, the check can be used with a window envelope so that reproducing the address on the envelope is not required. The check also serves as a change-of-address form. By writing the new address on the check, the user saves

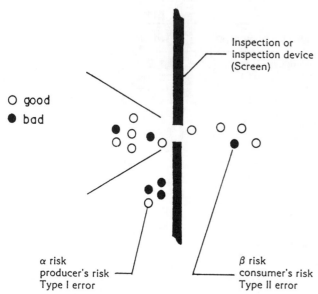

FIGURE 14.27 Inspection devices sort good and bad items into two categories. However, they are not perfect. The rejecting of good material as bad is called a **type 1** risk, α risk, or producer's risk. The accepting of bad material as good is called a **type 2** risk, β risk, or consumer's risk. Remember which is which by remembering that 1 comes before 2 and α comes before β; that is, 1 and α are on the left, and 2 and β are on the right.

not want to go to the meals, the appropriate numbers were crossed off with a grease pencil when they received the badge at registration.

A farmer can use a fertilizer which adds nitrogen and trace compounds and can plant two crops (each with a different growth time) at the same time. In the home, use a premixed spread of peanut butter and jelly to reduce labor cost. (Labor savings don't get much emphasis in most homes since production volume is low and labor cost is considered to be zero.) In the office, use carbon paper to make duplicate copies at the same time as the original is made.

5.2 Multifunction Tools   A multifunction tool is used to combine operations. Farmers can fertilize at the same time they plow. In the home, one compressor can cool both a refrigerator and a freezer. Use a popcorn popper that melts the butter as it pops the popcorn. In the factory, a special-purpose drill can drill and countersink at one stroke; a special-purpose drill press can drill several holes at one stroke just as a multiple die can punch several holes at one stroke; a lathe tool can form as well as cut off; a fork truck can lift and move. In the office, a ruler can be used as a straightedge as well as a measuring device.

GUIDELINE 6: VARY ENVIRONMENTAL STIMULATION INVERSELY WITH TASK STIMULATION

Tasks will be divided into low and high stimulation.

6.1 Low-Stimulation Tasks   Many industrial tasks are quite automatic for the operator and require little conscious attention. Some even require little physical movement. Some examples of this minimum of mental attention and minimum of physical movement are inspecting items on a moving conveyor,

the writing of a letter and a stamp; the company saves by reducing processing time (letters to the company would require several steps of internal processing before they got the proper location); and errors are reduced because the users do not need to transmit their names or account numbers since they are on the check. Use of the punched hole to signify change of address permits machine sorting of the returned checks. Another example of multiple function paperwork is a magnetically coded identification card which acts also as a key. Figure 14.29 shows a convention registration badge which also served as three meal tickets. For registrants who did

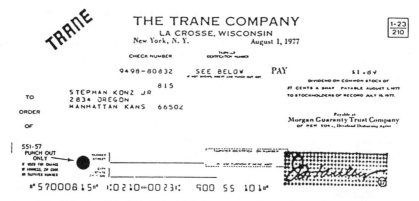

FIGURE 14.28 Paperwork can fill multiple roles. This dividend check transfers money, addresses the envelope, and serves as a change-of-address form.

STEPHAN A. KONZ (15)
(16)
Kansas State University (17)
Manhattan, Kansas

FIGURE 14.29 When registering, the convention attendees indicated whether they would attend the three meals. The appropriate number is marked out for any meal not paid for. The badge then serves as an up-to-three meal ticket as well as a badge and simplifies accounting problems.

monitoring an automatic pilot in an airplane, watching a chemical process indicator board, and monitoring a radar screen. If the person is seated rather than walking, if the environmental temperature, humidity, lighting, noise, and air velocity all are controlled at a constant level, the brain has minimum **task stimulation.** Performance (measured, for example, as percent of defective units noticed) declines as the brain "goes to sleep" for periods of 1 to 20 s—even longer in some cases! (See the vigilance literature for more on this topic.) Operators strongly dislike this type of work situation—solitary confinement is the most feared of punishments. Variety is not the spice of life, it is the very essence.

Cure by adding stimulation to either the task or the environment.

6.1.1 Add physical movement to the task

For example, have the night security guard walk around as well as monitor a TV screen. Eliminate automatic equipment and replace it with something requiring operator movement and attention. Have inspectors dispose of rejects as well as indicate that they are rejects. Machine-paced tasks result in very poor quality if the products can pass the station while the operator is "asleep." If machine pacing must be used (and usually it need not be), the operator should be required to make a conscious act to pass an item rather than make a conscious act to reject an item.

6.1.2 Add stimulation to the environment

The easiest **environmental stimulation** solution is to let operators talk to each other. Managers who do not permit their employees to talk to each other are the same ones who proclaim (while sitting in their chairs) that chairs will make employees lazy. Aside from this attitude that only people with ties should have chairs and be permitted to talk, there is little reason not to arrange workstations so that workers with low-stimulation tasks are permitted to talk. Figure 14.30 shows 7 schematic arrangements with

my judgment of their stimulation value. More stimulation occurs when people are face to face, they are close, there are no barriers, and noise is less. Reducing noise and encouraging visual and auditory contact also improves communication, especially useful for feedback on work quality.

Another common example of external stimulation is the use of background music, which should be stimulating but not too stimulating (no vocals, played only a fraction of the time, neither "soft strings" nor "big brass"). It should be a "soft fog."

Windows also furnish stimulation. Light normally is furnished by electric illumination and ventilation by fans; so keep the windows small to minimize energy losses. Many large windows have their upper portions covered with venetian blinds, curtains, or shades to reduce glare, indicating that they are too

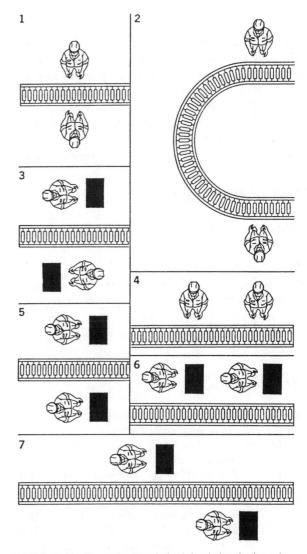

FIGURE 14.30 Vary environmental stimulation designs for the workplace. Vary the stimulation by adjusting orientation in relation to other people, by changing distance, and by using barriers such as equipment between stations. The lowest stimulation situation is 7.

large for their actual function. The important design feature is to have a view of the horizon with perhaps a 20° view of the ground.

Paint walls a variety of pastels, not just "industrial green." In large open office areas (e.g., holding over 10 people), paint each of the four walls a different color; use art work. Color-code stock racks, bins, pipes, chairs, and tables to establish "ownership" (territory) and identify function as well as add variety.

When people cannot see each other, increase visual contact with mirrors or even closed-circuit TV. Headset communicator radios keep people who are physically isolated (such as guards) from being socially isolated.

In general, the stimulation should not be expected to improve productivity over what a fresh person could do but just to reduce the decrement in performance.

6.2 High-Stimulation Tasks

Many of the nonroutine tasks which require concentration are office tasks. Over half of the U.S. working population works in an office. However, even engineering design, calculations, and detailed inspection rarely require complete freedom from environmental stimulation. Konz (1964) demonstrated that background music had no effect on output or errors for a repetitive clerical task (mark sensing numbers), output or errors of hand addition of numbers, or output on a creative mental task (anagrams). Workers may occasionally need freedom from high-information-content noise (conversation or vocal music); so it may be best to have two workstations—their usual one and one for concentration (a "think tank"). It could be created by just closing a door or by using an entirely separate enclosed area. Auditory privacy can be obtained by physical enclosure but also by masking the high-information-content noise with low-information-content noise such as air ventilation noise or background music. Evaluations of "office landscape" arrangements have indicated that auditory privacy, although necessary, is not sufficient. Visual privacy is needed also. The important thing about visual privacy seems to be to obscure the face. To that end, use a barrier from .6 to 1.6 m above the floor. In general, orient chairs at workstations so the worker's eyes do not meet the eyes of passersby.

Many employees are perfectly willing to be distracted and will talk at length about sports, sex, politics, or the weather. If work standards are vague (as in most offices), and if conversation interferes with their work or the work of others, discourage excess conversation with desk or machine orientation (side by side rather than face to face, not facing open doors) or with head-high barriers.

REVIEW QUESTIONS

1. How does labor specialization affect quantity of output, quality of output, and unit cost/h?

2. What is balance-delay time?

3. Discuss how the slowest operator slows down a balanced line.

4. Assuming the arrival rate is the same, why is close spacing and a slow conveyor belt speed preferable to wide spacing and a high belt speed?

5. Briefly describe the group technology concept.

6. Discuss the rationale of just-in-time.

7. What is the rationale of making several identical items at the same time? Discuss in relation to making hamburgers, citing get ready, do, and put away.

REFERENCES

Buxey, G. Incompletion costs versus labor efficiency on the fixed-item moving belt flow. *Int. J. of Production Research*, Vol. 16 [3], 233-47, 1978.

Corlett, N. Design of handtools, machines, and workplaces. In *Handbook of Industrial Engineering*, Salvendy, G. (ed.), New York: Wiley & Sons, 1982.

Gargano, H. and Stewart, F. Material handling system is key to efficient assembly operation. *Material Handling Engineering*, Vol. 30 [4], 53-55, April, 1975.

Kilbridge, M. Non-productive work as a factor in the economic division of labor. *J. of Industrial Engineering*, Vol. 12 [3], 155-59, 1961.

Kilbridge, M. and Webster, L. The balance delay problem. *Management Science*, Vol. 2 [1], 69-84, 1961.

Konz, S. The effect of background music on productivity of four tasks. Ph.D. dissertation, Urbana: University of Illinois, 1964.

Konz, S. *Facility Design: Manufacturing Engineering*. Scottsdale, Ariz.: Publishing Horizons, 1994.

Konz, S. and Osman, K. Team efficiencies of a paced visual inspection task. *J. of Human Ergology* (Japan), Vol. 6, 111-19, 1977.

Su, J. and Konz, S. Evaluation of three methods for inspection of multiple defects/item. *Proceedings of 25th Annual Meeting of the Human Factors Society*, Rochester, N. Y., 1981.

OVERVIEW

The previous chapter discussed organization of groups of workstations. Chapter 17 covers manual material handling at and between workstations. This chapter discusses the physical design of workstations through 14 workstation design guidelines.

CHAPTER CONTENTS

1 Avoid Static Loads and Fixed Work Postures
2 Reduce Cumulative Trauma Disorders
3 Set the Work Height at 50 mm Below the Elbow
4 Furnish Every Employee with an Adjustable Chair
5 Use the Feet as Well as the Hands
6 Use Gravity; Don't Oppose It
7 Conserve Momentum
8 Use Two-Hand Motions Rather Than One-Hand Motions
9 Use Parallel Motions for Eye Control of Two-Hand Motions
10 Use Rowing Motions for Two-Hand Motions
11 Pivot Motions About the Elbow
12 Use the Preferred Hand
13 Keep Arm Motions in the Normal Work Area
14 Let the Small Woman Reach; Let the Large Man Fit

KEY CONCEPTS

acceleration/deceleration
area of vision
bit
coefficient of friction
damaging wrist motions
dominant hand/eye
eye focus/eye travel
Fitts' law

gravity as a fixture
heel strike
human power
line of sight
movement angles
normal work area
optimum work height

physiological cost vs. organizational cost
seating posture variability
static load
three contact rule
venous pooling
weight penalty

GUIDELINE 1: AVOID STATIC LOADS AND FIXED WORK POSTURES

In the following discussion, the cardiovascular effects of static loading from posture will be discussed. The biomechanical effects of torques about the spine (from the body and from objects) are discussed in Chapter 17.

Static (isometric) **load** is bad for the blood supply of a specific muscle as well as the total body. Figure 15.1 shows how static work (muscle does not move) increases both systolic and diastolic blood pressure, while rhythmic (isotonic) work does not change diastolic pressure much and only slightly increases systolic pressure. In addition, metabolic wastes tend to accumulate in the muscles during isometric work due to the minimum blood flow. When a job has a static load, consider increasing the recovery time. See Box 16.1.

1.1 Standing
Three aspects of standing are venous pooling, shoes, and floors.

1.1.1 Venous pooling
Nonmovement of the legs during standing causes a problem. Standing without walking results in more discomfort than active standing (walking 2–4 min every 15 min) (Konz et al., 1988). The veins are the body's blood storage location. If the legs don't move, the blood from the heart tends to come down to the legs but not go back up; this is called **venous pooling.** Since the blood supply required by the muscles does not change, the heart tries to maintain a constant cardiac output (mL/min) by increasing the beats/min to compensate for the lower mL/beat. See Table 15.1. The work of the heart also is increased because the heart must do the entire pumping job. Normally the "milking action" of the leg muscles helps move the blood.

From a circulation viewpoint, walking is clearly better than standing. In addition to adding to the heart's load, venous pooling causes swelling of the legs (edema) and varicose veins.

1.1.2 Shoes
Opila et al. (1988) found that the line of gravity is forward of the spine (even L4-5), meaning that the body normally has a forward bending moment, counterbalanced by ligament forces and back muscle forces. If the heel is too high (cowboy boots, women's high heels), the center of gravity is moved forward, causing an additional torque while standing. Cowboy boots are designed to prevent the foot going through the stirrup. Women's high heels are designed to make the woman arch her back. If it was to increase her height a simple horizontal platform would do. Neither shoe is designed for standing. (In the 1970s there was even a fad in which shoes had no heel and thick soles; they made you lean forward to stand!)

The general goal is to have the hips parallel to the floor (i.e., to stand with weight equally on both feet). However, unequal leg lengths or uneven shoe wear may be a problem. A difference of as little as 6 mm in leg length can cause considerable stress on the back. Shoe heels may wear unequally, presenting another problem. Many people wear down a corner of the heel when walking. Then they are standing not on a flat surface but on a curved surface. A solution is to have a shoemaker build up the heel with heel plates. Figure 15.2 shows how a bar rail can be used to vary the work posture.

For most tasks, ankle support is not needed. Higher sides (e.g., boots) often are used in sports activities and walking over uneven terrain; the high sides also can protect the leg from vegetation and bites.

A good shoe will mold itself to the foot, giving support over a relatively large area and thus minimizing pressure on the foot. Wooden shoes (Dutch shoes), once broken in, give good area support. In modern shoes a cushioned inner sole does the same job more quickly.

The outer sole of the shoe also should be cushioned. This can be done by material (e.g., crepe) or by form (e.g., ripple soles), or a combination of material and form. In many jobs, resistance to slipping is important. In this case, consider deck shoes—that is, shoes with no arch. They increase contact area (more "rubber on the road") and have high **coefficient of friction** (see Figures 15.4 and 15.5) materials. See Table 17.5 for shoe and floor coefficients of friction. Ramsey and Senneck (1972) point out that rubber-soled boots give excellent grip on clean surfaces but poor grip on greasy or muddy surfaces. They recommend boots with tungsten carbide-tipped

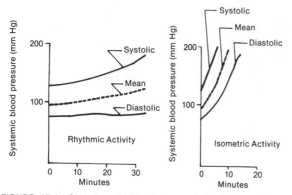

FIGURE 15.1 Static (isometric) loads increase both systolic and diastolic blood pressure (Shephard, 1972). During rhythmic activity, diastolic blood pressure increases very little and systolic increases slightly. For more on blood pressure, see Box 11.4.

TABLE 15.1   Cardiovascular effects of sitting and standing for 16 adults between 18 and 51 with mean age of 30 (Ward et al., 1966).

Criterion	Stand (a little walking)	Sit	Sit/Stand %
Cardiac output, L/min	5.1	6.4	125
Stroke volume, mL/beat	54.5	78.3	144
Mean arterial pressure, torr	107.0	87.9	82
Heart rate, beats/min	97.2	84.9	87
Total peripheral resistance, dynes/cm²-s.	1820.0	1207.0	66

studs. An intensive French study (Hopkins, 1966) recommended: (1) the heel should be eliminated (i.e., use deck shoes) and (2) the tread design should have many "braking edges," acting equally in the longitudinal and transverse directions. The best tread studied had small, flexible rectangular cleats with drainage canals which facilitated movement of liquid from under the sole.

When shoes are used for standing work, they need to be slightly oversized—1/2 to 1 size too large, as the feet will swell during the day. Thus, those shoes should be purchased after work, not on weekends or before work. A Buckingham palace guard commented, "Stood on parade four to five hours at a time. The trick is to keep the weight off your heels. That's why guards' boots bulge in front—lots of room to wiggle your toes." It is the rare person whose feet are both the same size. See Figure 15.3. How much do yours differ? Shoes with more holes for laces have more adjustability for foot size differ-ence—that is, shoes with four lacing holes adjust better than shoes with three lacing holes.

1.1.3   Floors

Hard floors cause standing fatigue in the feet, legs, and back. Metal gratings are the worst of all since they not only have no resilience but also have minimum surface area, thus acting as knives. A conventional ladder (with a rung or 3-inch step) is another example of a minimum surface area; if a ladder is used for standing (e.g., in an orchard) instead of just for moving between levels, then the step should be 12 inches deep so that it supports the entire foot. Concrete is fatiguing; plastic or cork tile is slightly better; wood is better; and best of all is carpet.

FIGURE 15.2   Bar rails can make industrial workers as well as tavern customers comfortable (Rodgers, 1984). Use bar rails along conveyor lines as well as individual workstations. The rail should have a nonsharp surface for the foot and allow either foot to be used. A small platform is another alternative. The changed height of the foot rotates the thigh bone forward and reduces stress on the lower back muscles.

FIGURE 15.3   Left feet, on the average, are the same size as right feet (Rys and Konz, 1989). However, for a specific individual, there can be considerable variability between the left and right foot.

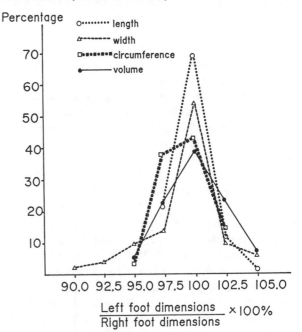

Source: M. Rys and S. Konz. "Adult Foot Dimensions," in *Advances in Industrial Engineering and Safety,* A. Mital (Ed.). Copyright © 1989 by Taylor & Francis, London. Reprinted with permission.

If the entire floor is not carpeted, individuals can modify their own work area by covering a local area with a mat, a rug, or a wooden platform (a favorite of machine-shop operators). Some people even use pieces of cardboard.

Rys and Konz (1988) demonstrated that carpet reduced standing heart rate by 5% (vs. standing on concrete). When selecting a mat, look for the following features: It should compress 3–4% under adult weight (Hinnen and Konz, 1994) and should have beveled edges to reduce tripping. The underside of the mat should be constructed so that the mat does not move on the floor. The size of the mat depends on how much movement the person makes—24 inches deep by 36 inches wide is a common size. In some industries, periodic cleaning and sanitizing of the mat are important.

Although carpet has higher capital cost than tile, carpet should have lower maintenance cost. A solid-color carpet shows spots and stains more than a multicolor (sculptured) carpet. Covering an unheated concrete floor with carpet will give about a 1% savings in heating cost. Cold floors with high thermal conductivity (metal, concrete) cool the feet, increasing vasoconstriction and restricting blood flow to the feet.

A light-colored floor surface (painted concrete, light-color carpet) increases reflected light, improving lighting distribution and thus improving visual acuity. See Chapter 21.

1.2 Slips and Falls   The section will be divided into problem, analysis, and solutions.

1.2.1 Problem   Falls are very low-probability events. However, the exposure (e.g., the number of steps taken by the population of the United States every day) is enormous. The annual death rate from falls in the United States is about 11,700; about 6,500 of these are in the home. Occupational exposures result in approximately 1,500 deaths and 300,000 injuries (Leamon, 1992). The injury typically is a musculoskeletal strain injury to the low back, ankle, or knee (Manning et al., 1988). In industrial fatalities, falls account for 12.5%, which is greater than the total for electrical current, fires, burns, and poisons of all types (Leamon, 1992). Construction workers, cleaning personnel, transportation workers, and restaurant serving personnel appear to be at particular risk (Chaffin et al., 1992). For some unknown reason, there is little public attention to these figures, and slips frequently are not recorded by accident recording procedures. Truly, slipping is a pedestrian activity!

1.2.2 Analysis   A fall can occur when the floor level changes or when there is no change in level.

The problem with a change in level is that often the change in level is not noticed; the result is an arrested foot (caught, trip, stumble) rather than a slip. Hammer (1991) reports that almost 50% of all tractor accidents occur when mounting or, especially, dismounting from tractors. For vehicle steps, about 80% of the reported accidents are while descending (Nicholson and David, 1985). Falls also can occur from scaffolds, roofs, furniture, and so forth. The problem is focused in job trades such as roofer, painter, maintenance (Helander, 1991). The major problem here is that the fall is of a greater distance; the body velocity and thus deceleration become greater.

With no change in level, the key event is the heel strike; this is the phase of the stride cycle in which a slip occurs most often. See Figures 15.4 and 15.5. Walkers strike the surface with a force of about 1.5 (body weight); a typical angle is 30°. The slipping coefficient of friction *(SCOF)* usually is less than .25 although faster velocity or pushing or pulling a cart can increase the values (Chaffin et al., 1992).

Some factors affecting the coefficient of friction *(COF)* are velocity of contacting surfaces, normal loads on contacting surfaces, and contaminants (Chaffin et al., 1992).

FIGURE 15.4 **Heel strike** is when most slips occur (Chaffin et al., 1992). The vertical foot force is normal force; the horizontal foot force is foot shear force. Since the two vertical forces are equal (i.e., the floor is stiff enough to hold the person's weight and inertia), the key is the relation of the horizontal forces. If the foot's shear force is less than the friction force, the foot will not slip.

The slipping coefficient of friction *(SCOF)* required to stop the foot from slipping is horizontal shear force/normal force.

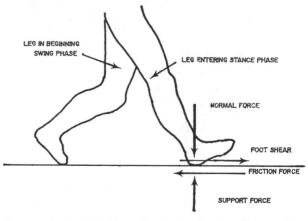

Source: D. Chaffin, J. Woldstad, and A. Trujillo, "Floor/Shoe Slip Resistance Measurement." *Am. Ind. Hygiene Assoc. J.,* Vol. 53, No. 5, pp. 283–289. Copyright © 1992 by Am. Ind. Hygiene Assoc., Fairfax, VA. Reprinted with permission.

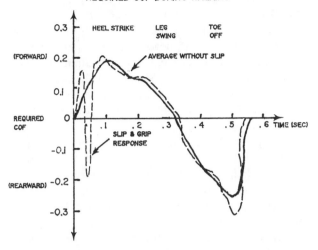

FIGURE 15.5 Slipping coefficient of friction (SCOF) required for walking without slipping is shown with the solid line (Chaffin et al., 1992). The dashed line shows a microslip and recovery during a step. Leamon (1992) defines a *microslip* as a slip less than 2 cm, a *slip* as 8–10 cm, and a *slide* as uncontrolled movement of the heel. Microslips occur very often and normally are not perceived by the person. A slip usually is perceived and the person typically jerks the upper body, moves the arms, etc. A slide involves loss of control and usually a fall.

Heel–floor closing velocities may exceed 50–100 cm/s, and so the situation is dynamic, not static. In addition, the ratio of static coefficient of friction to dynamic coefficient of friction is not constant for all materials, contaminants, and surface geometries. Therefore, static COF alone is not sufficient.

The normal load can be separated into two separate but additive components:

$$FF = FA + FD$$

where

FF = Friction force

FA = Force of adhesion

FD = Force of deformation

Adhesion is most affected by lubrication, which lowers the *COF*.

Contaminants provide lubrication. Lubrication depends on area of contacting surfaces (a larger area gives lower dynamic *COF*), roughness (rougher gives lower dynamic *COF*), velocity of surfaces (higher velocities give higher dynamic *COF*), vertical loads (larger loads give lower dynamic *COF*), and fluid viscosity (higher viscosity gives lower dynamic *COF*). See Table 17.5 for some coefficients of friction.

Deformation is affected by material stiffness, contact area roughness, and normal loads applied. Rougher floors have higher dynamic *COF*. Heel size, because it is so small, usually is not a consideration. When testing surface *COF*, it is important to use normal forces representative of people, not very light loads.

1.2.3 Solutions

Solutions can be engineering related or behavioral.

Engineering For stairs with only one or two steps, use handrails, color change of walls and floors, and lighting to emphasize their location. For more on stairs, see Templer (1992). Hammer (1991) gives detailed recommendations for tractor mounting/dismounting. One long-known but often violated rule is the **three-contact rule** (at each phase of mounting/dismounting, at least three limbs should maintain contact with the steps or handles at the same time).

Miller et al. (1991) investigated steps with metal grating (e.g., used for vehicle ingress or egress), noting that the predominant foot movement is vertical rather than horizontal and that the contact point is on the ball of the foot rather than the heel. For soles, they emphasize that the relative rank changes when the sole is covered with water or diesel. (Diesel is worse than water.) Oil-resistant rubber is always best; crepe's slip-resistance coefficient drops greatly when covered with water or diesel and crepe then becomes even worse than leather.

When the fall can be a long distance (scaffold, roof, etc.), special precautions need to be taken (railings, safety harnesses, safety nets, etc.). See OSHA 29 CFR 1926 subpart M (Fall Protection).

Because walking is an automatic activity that people normally don't think about, a small change in height (such as a one-step stairs) is very dangerous. Ideally, the stairs should be replaced with a ramp. If a step must be used, add a handrail and change the floor color at the step. A temporary small change in height (an obstruction causing a tripping hazard or a hole) calls for housekeeping and maintenance. A toeboard (4 inches high and with a clearance between the toeboard and the floor of less than 1/4 inch) prevents the foot slipping over the edge and also prevents objects being pushed over the edge.

Walking surfaces should be kept clean and dry (minimizing both dry and wet lubricants), and rough surfaces are good. Concrete surfaces can be roughened (with sandblasting or chemicals); steel surfaces can have gratings (Miller et al., 1991) or other treatments to increase their roughness. If abrasive strips are used, the gap between each tape should not exceed about 1 inch. Avoid oil-treated mops to remove dust. Interior tile floors near outside doors are a problem because people track in snow and rain. Carpet these areas.

Shoes should have a soft heel with a tread design to let lubricants escape. Typically, the person falls before the shoe sole contacts the floor so the heel is the key area.

A final engineering solution is to reduce the impact force (akin to padding auto interiors). One possibility is to use carpets, especially on stairs. Carpets not only reduce impact damage but have better *COF.* Maki and Fernie (1990) reported carpet (vs. hard surfaces) attenuated hip impact force 23% and hand impact force 6%. Another possibility is to provide grab rails (e.g., on stairs) so the fall is interrupted. Use a railing on platforms and around permanent holes. The railing (according to OSHA) should have a smooth top rail 42 inches above the floor and an intermediate rail at about 21 inches. It must be able to withstand a 200-lb force at any point on the top rail.

Behavior   People can and do walk safely on slippery surfaces (e.g., ice). However, it is important to know the surface is slippery. Unexpected or invisibly slippery surfaces (e.g., at night) are quite dangerous. Shorten the length of the stride (which reduces foot velocity, gives smaller foot shear forces, and keeps the body center of gravity between the feet). James (1983) gives $COF\text{-}S = 158\ (SL)$ where $COF\text{-}S$ is coefficient of friction for stability and SL is step length, cm. For example, for a typical step length of 60 cm, $COF\text{-}S = .38$, while for a step length of 30 cm the $COF\text{-}S$ drops to .19. Leaning forward helps keep the body center of gravity between the feet and, if you fall, it will be forward rather than backward. Walking flat-footed increases the heel contact area.

1.3 Sitting

At first thought, getting tired of sitting is surprising. However, upon reflection, remember that students get very "twitchy" whenever the lecture approaches 60 min. What if you had to sit continuously for 120 min as many workers do? Sitting will be discussed in more detail in Guideline 4.

1.4 Head and Neck

Various parts of the body may have a torque about the body center of gravity. Table 15.2 gives the weight of various body segments. Kaleps et al. (1984) give mass distribution properties of the head.

The head, for example, weighs about 7.3% of body weight so, if you weigh 90 kg, your head weighs about .073 (90) = 6.6 kg. In common terms, this is the same as a bowling ball!

For most people in most situations the head is easily supported on the neck. Tired neck syndrome tends to occur when the neck is long and thin and/or the head is tilted considerably forward or backward.

The head position will be affected by what is being observed; that is, the head might be tilted forward to reduce distance to an object to improve visibility (inspection, fine assembly, VDT work). The head position also will be affected by the presence and location of arm supports. For example, if there is a desk or table the person may lean forward, with the amount of the lean depending on the support height. If the person is seated, head position depends on the chair design (for example, the semireclining position in automobile seating). Kroemer and Hill (1986) suggest a typical forward head tilt when seated at a VDT of 10° to 15°. (See display location in Chapter 20 for additional comments.)

In general, keep the **line of sight** (see Figure 20.10) below the horizontal. (This not only reduces head–neck fatigue but also tends to reduce direct glare from lighting.) Chaffin (1973), using electromyography and subjective pain measurements, suggests a maximum forward inclination of 30°.

Sore necks occasionally occur in people with bifocals because they tilt the head backward to be able to see through the bottom of the bifocals. A solution is to use single-vision glasses at the workstation. Another solution technique is modification of the vertical relationships of the chair and workstation; that is, raise or lower the chair or table. Another technique is to change the location or orientation of the item being worked on. For example, Hamilton (1986) reported less neck tension for VDT operators when the source document was on a holder than when it was flat on the table.

1.5 Arms

If you weigh 90 kg, then one of your hands weighs about .0065 (90) = .6 kg, a hand plus a forearm about .0227 (90) = 2 kg, and an entire arm about 4.4 kg. If you hold a 25-g feather, you also are holding 4,400 g of bone and muscle. Complete elimination of the 25-g feather doesn't reduce the load much. Thus, avoid work where the upper arm is not vertical, that is, where the hands are above heart level (thereby moving the upper arm forward) or upper arm is abducted (upper arm angled from the side). See Figure 15.6. Sakakibara et al. (1987) comment that working with the hands elevated also may make the head tilt back, causing neck pain and possible vertigo.

(Position of the arm also substantially affects the flow of blood as well as arm and hand temperature. To minimize the flow of blood and drop arm temperature about 1 C, hold your hand above your head. To maximize the flow of blood from the warm central torso, hold your arm straight down or lie down and put your arm at your side.)

Weight of the arms can be supported by a worksurface or chair arms. Many worksurfaces have sharp edges, making padding desirable. (See Figure 15.14 later in this chapter for some commercially available designs for worksurface edges.) Supporting the arms also reduces tremor and thus permits more accurate work with the hands. Naturally there should be no

TABLE 15.2 Weight and center of mass for various body segments in adult males (Clauser et al., 1969).

BODY SEGMENT	WEIGHT OF SEGMENT/ TOTAL BODY WEIGHT		SE	LOCATION OF CENTER OF MASS AS RATIO OF SEGMENT SIZE
Head		7.28	.16	.46 (top of head/ht of head) .40 (back of head/head length)
Trunk		50.70	.57	.38 (suprastern/trunk length)
Hand	0.65		.02	.18 (meta 3/styl-meta 3 length) .56 (med aspect/hand breadth)
Forearm	1.61		.04	.39 (radiale/rad-styl length) .49 (ant aspect/ap at cm)
Forearm + hand	2.27		.06	.63 (radiale/rad styl length) .52 (ant aspect/ap at cm)
Upper arm	2.63		.06	.51 (acrom/acrom-rad length) .51 (ant aspect/ap at cm)
Total arm	4.90		.09	.41 (acromion/arm length)
Both arms and hands		9.80		
Foot	1.47		.03	.45 (heel/foot length) .54 (sole/sphyrion height)
Calf	4.35		.10	.37 (tibiale/calf length) .42 (ant aspect/ap at cm)
Calf + foot	5.82		.12	.47 (tibiale/tibiale height) .33 (ant aspect/ap at cm)
Thigh	10.27		.23	.37 (trochanterion/thigh length) .53 (ant aspect/ap at cm)
Total leg	16.10		.26	.38 (troc/trochanteric height) .63 (ant aspect/ap at cm)
Both legs and feet		32.20		
		99.98		

An improved estimate can be made for some segments using the following equations where X is total body weight, kg:

SEGMENT	EQUATION	STANDARD ERROR
Trunk	$.551 X - 2.837$	1.33
Head and trunk	$.580 X + .009$	1.36
Total arm	$.047 X + .132$	.23
Upper arm	$.030 X - .238$	.14
Thigh	$.120 X - 1.123$	.54
Foot	$.009 X + .369$	.06

sharp edges, corners, bolt heads, or other impediments in the work area to snag any body area (hands, hips, thighs, etc.). In addition, if the operator reaches into a cardboard box, the forearm can be protected by wearing a sock (with the toe cut off) on the forearm.

One reason the arms are elevated is that the work requires close visual attention and thus is elevated to move it closer to the eyes. An alternative is to magnify the object, which allows "friendly" positions for both the head and the hands. A 4X (multiplying size by 400%) lens mounted on a stand (or post-mounted with a swivel arm) with accompanying light will allow a field of view of approximately 25 × 75 mm with severe restrictions on the depth of field. A 2X lens (multiplying size by 200%) will have a field of view of approximately 100 × 500 mm and a reasonable depth of view. Both approaches tend to keep the head in a relatively fixed location. Obtain more freedom for the head by wearing spectacles with magnification lenses (e.g., 1.5X) or by taking a TV picture with a magnification lens and then having the operator view the screen. See Chapter 21 for more on vision.

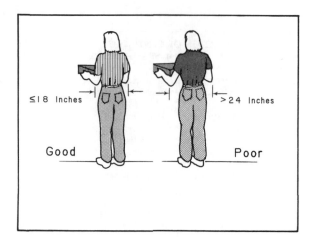

FIGURE 15.6 Minimize arm abduction by keeping the upper arm vertically downward. The elbows can be raised either by carrying something too wide or by having the hands too high (say at heart level).

GUIDELINE 2: REDUCE CUMULATIVE TRAUMA DISORDERS

This guideline is covered in detail in Chapter 16. Table 15.3 summarizes all the guidelines for hand–arm position and movement given in Chapters 15 and 16.

GUIDELINE 3: SET THE WORK HEIGHT AT 50 MM BELOW THE ELBOW

3.1 Optimum Height   For comments on the location of controls, see Chapter 19.

The **optimum work height** for manipulative hand–arm work (50 mm below the elbow; slightly

TABLE 15.3 Guidelines for hand–arm position and movement.

TEXT PAGE	GUIDELINE
	HEIGHT
200	Set the work height at 50 mm below the elbow.
	JOINT MOVEMENT
232	Don't bend your wrist.
235	Don't lift your elbow.
242	Don't reach behind your back.
	HAND MOTIONS
208	Use the feet as well as the hands.
209	Use gravity, don't oppose it.
209	Conserve momentum.
211	Use two-hand motions.
212	Use parallel motions for eye control.
213	Pivot motions about the elbow.
213	Use rowing motions.
214	Use the preferred hand.
214	Keep arm motions in the normal work area.
216	Let the small woman reach; let the large man fit.

below heart level) is based on both productivity (i.e., cost to the organization paying the employee) and physiological cost (i.e., cost to the individual to achieve a specified output).

Barnes (1940) stated the hand should be "allowed to work one to three inches lower than the elbow, the average height of the work surface. . . ." Although Barnes did not cite any evidence for his recommendation, all the following studies tend to confirm his judgment. The key points are as follows:

- Work height is defined in terms of *elbow height* rather than a fixed height from the floor. Since individual heights will vary, any fixed-height design must be wrong.
- Optimum height is slightly *below* the elbow. Some research indicates it is farther below than other research but the consensus is below.
- Optimum height from the elbow is the *same* for both sitting and standing. This means work height from the floor will differ for sitting and standing, unless the chair height is adjusted.
- Work height is not table height. Most items (mechanical assemblies, keyboards, hamburgers on a grill, etc.) have a thickness—typically 25 to 50 mm. Thus, if the optimum height for the hands is 50 mm below the elbow and the object being worked on is 50 mm thick, the table should be 100 mm below the elbow.

Konz (1967) surveyed the literature. There is a relatively flat performance response around the optimum. That is, although the optimum performance seems to be about 50 mm below the elbow, output does not decrease more than a couple of percent within a range from 125 mm below to 25 mm above the elbow; beyond this the penalty is greater.

Wiker et al. (1989) reported, from a laboratory study, increases in time of less than 8% when the hand was kept below the shoulder but increases of over 20% when above the shoulder. In industrial work, the penalties of nonoptimal positions should be even greater due to fatigue.

Physiological cost (vs. organizational cost i.e., cost to the worker rather than gain to the organization) also has been used as a criterion for many years; the effect of physiological cost now is known as cumulative trauma.

Within the range of knee to chest, there seems to be a greater penalty for working above the optimum than below. This may be related to upper arm abduction when working above the optimum. See Figure 11.10. As pointed out in Section 1.5 of this chapter, it is good policy to keep the hands below the heart level. However, if the operator is seated and needs to reach below the knee (e.g., to get supplies), this involves considerable twisting and bending, which is bad for the back. Consider storing such items some distance away so the operator must rise from the chair to obtain the supplies.

There is little information concerning the optimum work height when a downward force is required, such as a polishing or sanding operation. It probably is with the lower arm at about a 45° angle—a distance below the elbow of about 150 to 200 mm. Magnusson and Ortengren (1987) reported that butchers preferred a table height 17 to 22 cm below the elbow with a 5–10° tilt; they recommend varying the height during the day, because different heights stress different parts of the body.

For manual material handling, the optimum height is at the knuckle, with the arm vertical.

3.2 Solution Techniques

See Box 15.1 for comments on a specific type of workstation—the computer workstation. In general, the three approaches are to change machine height, adjust elbow height, and adjust work height on the machine.

3.2.1 Change machine height

Changing machine height may not be practical if a variety of people use the machine. However, many machines and people are one-on-one. That is, one person is the sole machine operator for a long time period—1, 2, or even 5 years. Figure 17.23 shows how conveyor heights can be adjusted depending on whether loading or unloading is performed. The workstation could be adjusted to individual worker heights. Many conveyors (as well as tables) have the pin-and-hole system on the legs; adjustment is simple. Many lightweight machines are mounted on a worksurface (computer keyboard, sewing machine); certainly spending one hour to adjust the bench height would be a small price to pay for freedom from back and shoulder pain.

Another alternative is a multiple-level table. A common example is a desk with a lower platform for the keyboard. Figure 15.7 shows a multiple-level table for VDTs. (See also Box 15.3.) In one welding shop I visited, welding tables were built with surfaces 500, 700, and 1,000 mm above the floor; when welders had thick pieces, they used the 500 and 700 mm surfaces. A variety of surfaces also permitted the worker to shift surfaces (even for the same item) to get a variety of postures.

The table height can be made completely adjustable by using a scissors lift. (See Figure 17.16.)

Wick and DeWeese (1993) describe modifying a packing workstation by adjusting the height of the workstation, putting a tilt on the workstation table, and putting the supplies in better positions. **Damaging wrist motions** (force to the wrist or fingers while the wrist is not in neutral position) were reduced from 3,000/h to 1,400, high-force pinch grips ($\geq$ 8 lbs) from 4/cycle to 0/cycle, and extreme motions of the shoulder, neck, and back from 44/cycle to 0/cycle; in addition, cycle time was reduced 12%.

A person unloading a pallet transported on a lift truck can leave the pallet on the truck forks while unloading and occasionally stop unloading to raise the pallet and thus reduce bending. One firm had approximately 20 packaging stations in which operators placed product into cardboard boxes. They had the operators rotate to the next station every hour so that they would use a variety of muscles and variety of postures during the day. Since the operators varied in height, they had a packing stand designed and built which was variable in height (10 heights) and could be changed in height within 10 seconds. It even had an adjustable tilt surface so the box could be tilted toward the operator at varying degrees to facilitate loading product into the box.

3.2.2 Adjust elbow height

Rather than move the machine, move the person. Usually this technique is applied to sitting operators in adjustable-height chairs. It also can be applied to standing operators by having short operators stand on a platform. In one plant, the output of the workstations was placed on a conveyor. The workstations were set at a level appropriate for an operator 68 inches tall. Maintenance built platforms that were approximately 36 inches wide and 72 inches long with a height of 3 and 6 inches; the operator had a choice whether to use no platform, the 3-inch platform, or the 6-inch platform.

In agricultural labor, it often is necessary to work at or below the level of the feet; examples are planting vegetables or picking from low-growing plants such as strawberries. Vos (1973) divides such jobs into stationary jobs and jobs requiring movement (as along a row of beans).

For stationary work, sitting is best, squatting is almost as good, and bending and kneeling are poor. If bending or kneeling is necessary, then supporting the body with one hand on the knee or ground will decrease net energy required by 25 to 55% over no arm support. If a tool is used, it should be usable by either hand, so the support hand can be alternated.

For movement, squatting always requires less energy than sitting on a 175-mm stool attached to a belt around the waist; picking performance was 6 to 7% higher with squatting also. Although bending required 1 kcal/min more than squatting when movement was 2 m/min, its curve "crossed over" at 4 m/min forward speed; at 10 m/min, bending required 1 kcal/min less than squatting. Since bending

BOX 15.1 *VDT workstations*

VDT stands for *video display terminal*, but another meaning is *very demanding task*. It is demanding physically due to the sedentary, minimum-activity posture for most of the body combined with highly repetitive motions by the fingers. It is demanding mentally due to the constant attention required. It is demanding visually due to the poor-quality character reproduction on screens.

VDT tasks include

- data entry (document-intensive because the operator routinely enters data from documents)

- data acquisition (screen-intensive because information from a form is matched with information from the screen)

- word processing (document- and screen-intensive because attention is focused on both places)

- interactive (occasional user of screen such as travel agent, CAD/CAM user, programmer)

VDT job organization   Traditionally, the office worker (white collar or pink collar) was treated differently (better) than blue collar workers. However, as the office has begun to have production aspects, individualized consideration has begun to disappear. VDT operators will likely have fewer complaints if they are treated as individuals rather than "faceless production workers." However, remember that VDT stands for very demanding task, and one reason for that is the high number of repetitions of keystrokes.

Assume a person keys words at the rate of 70 words/minute. Assuming that a typical word (including spaces and punctuation) has 7 strokes, then 70 words/min $\times$ 7 strokes/word = 490 strokes/min. Keying 400 minutes/shift (allowing some time for breaks and other work) then results in 490 $\times$ 400 = 196,000 strokes/day. In a month, this is 21 days $\times$ 196,000 = 4,116,000 strokes; in a year, this is 235 days $\times$ 196,000 = 46,060,000 strokes. In 10 years, this is 460,600,000 strokes! It takes about 22 years to reach 1 billion strokes. Yet organizations want a person to work for 40–45 years.

We would be very impressed with any machine which operated over 1 billion cycles. We need to reduce the stress on the "human machine." Stress can be reduced in two primary ways: reduce exposure and increase worker ability to endure the stress. See Chapter 16, especially Section 1.4.2

and Box 16.1.

VDT lighting   See the discussion on VDT lighting in Chapter 21.

VDT furniture   The workstation has three key items: the screen, the keyboard, and the document holder. The user has two key items: the eyes and the hands. How should these five items be arranged?

Start with a seated person; this provides a location for the eyes and hands. However, since size varies, (1) the hand location in relation to the floor will not be the same for everyone, (2) the eye location in relation to the floor will not be the same for everyone, and (3) the relation between the eyes and hands will not be the same for everyone. The above obvious statements are given because they lead to a key guideline: *Workstation furniture must be adjustable.*

Next locate the primary visual element ahead of the eye; this is the screen for screen-intensive tasks and the document for document-intensive tasks. The consensus is that this screen or document should be below the head, but the exact location is controversial since real people (as opposed to drawings in books) rarely sit upright. The screen (document) should be perpendicular to the line of sight (Brand and Judd, 1993). The distance from the eyes depends on the size and quality of the displayed material. Considering the many trade-offs, the consensus seems to be 20 to 30 inches (500–750 mm) from the eyes. Avoid distances of less than 20 inches (Jaschinski-Kruza, 1990).

Next locate the documents (screen). To minimize the need for visual accommodation, both the document and screen should be the same distance from the eyes. Since the head can move side to side easily (especially if the chair swivels), side by side is a common arrangement. If the document is used more, place the document ahead and the display at its right or left (operator's choice). If the display is used more, place the display straight ahead and the document on the operator's preferred-hand side, to avoid cross-body reaches when the operator is changing or marking the document. For additional screen location comments, see Section 4.1 in Chapter 20.

Locate the keyboard in relation to both the eyes and the hands. Although VDT operators spend considerable time looking at the keyboard

(continued)

BOX 15.1 *Continued*

(because of the extra function keys), location vs. the hands is more important than location vs. the eyes. The forearm should be approximately horizontal with the hands at or slightly above elbow height. The keyboard should be directly ahead of the person so no twisting is needed. Many operators like wrist rests (to rest the forearm while reading or waiting for the computer to finish a step). Sauter et al. (1987) point out that the wrist support should not present any pressure points to the wrist but should be gently contoured and padded.

Finally, the workstation should have some space for work-in-process storage of documents, supplies (discs, pens), and personal items (family pictures, purse). For VDTs used by engineers and executives, the input and output are less standardized, and so additional space is needed to hold drawings, reports, documents, and other material for interactive work with the computer.

To repeat, the key consideration is adjustability of the furniture. For an example workstation, see Figure 15.7.

---

causes local overload of the back muscles, bending should be used with caution.

3.2.3 Adjust work height on machine
Figure 15.8 shows how a work height can be adjusted with a multiple-level table. Barbers raise the height of children's heads by having them sit on a board across the chair's arms. Items also may have to be lowered. If container walls make the arms lift too high, lower the walls. For example, tip component boxes on their side, tip them on a 45° angle (on a support), or cut out the side of the box with a cardboard knife.

Perhaps a container has deep sides so it can hold lots of product. Consider a short-sided box filled more often. Also consider the orientation of the item in the container. Orienting the item vertically instead

of horizontally may decrease the number of very deep reaches and moves. In sacking groceries, have the bag on its side to reduce high reaches. Another alternative, of course, is to have a lower shelf built into the checkout stand so the bag can remain upright. (The Europeans have another approach to reducing the stress on the checkout operator. They have the operator sit and make the customer do the sacking!)

If the part/assembly is held in a fixture, design the fixture to permit both vertical and horizontal adjustment. For example, cars are now assembled in

DIFFERENT PEOPLE

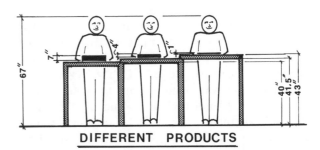

DIFFERENT PRODUCTS

FIGURE 15.7 Adjustability is the key to a well-designed VDT workstation. The keyboard and display are on separate tables. The display table can be adjusted in height, can be tilted (to reduce glare on the screen), and can be swiveled. The keyboard table can be moved up and down as well as in and out. The source document holder is on an adjustable arm. Note the rounded edges and corners on the furniture. Photo courtesy of IBM.

FIGURE 15.8 Multiple-level tables permit easy work height adjustment. In the top view, the same thickness part can be processed by different people using different portions of the table. In the lower view, the same person can process different thickness parts using different portions of the table.

rollover frames that allow the operator to work on the car while it is tilted at various angles.

GUIDELINE 4: FURNISH EVERY EMPLOYEE WITH AN ADJUSTABLE CHAIR

4.1   Justification   Think of the chair as a tool. Sitting reduces physiological load compared to standing. For example, dentists who work seated with the patient horizontal have fewer backaches. Table 15.1 shows the results from one of the many studies in the literature showing the benefits of sitting. For more on chairs, see Rodgers (1984), Eastman Kodak (1983), Chaffin and Andersson (1984), and Corlett et al. (1986).

The cost of an adjustable chair is very low. Chair cost is entirely a capital cost since operating cost and maintenance costs are zero. Assuming a cost of $200/chair, a life of 10 years, and one shift operation of 2,000 h/year, the cost per hour is $200/20,000 h or 1 cent/h. (Often the real comparison is between a chair without good ergonomics vs. one with ergonomic features, and the incremental cost of the good chair is $100 or less—that is, .5 cent/hour.)

The cost of labor (wages plus fringes) of the person sitting in the chair will vary with the job. Typically this will be at least $10/h, with $15 or $20 fairly common. Assuming a cost of $10/h, a .1% change in productivity equals 1 cent/h. The improved productivity is more likely to occur from the person working more minutes rather than more output/min. That is, a secretary who does not have to get up to relieve a sore back will not key more words/min of working but will work more minutes. In a 480-min work day, .1% is .5 min or 30 s. The question becomes whether a person will work 30 s more per day with a good chair than a poor chair. Naturally the breakeven time becomes lower if the chair cost is less than 1 cent/h or the wage cost is over $10/h.

When selecting specific chairs, it is recommended that the engineer select from various models using the features discussed below. Then, however, the *users* should try the alternatives in their specific jobs before purchasing. After all, they will be using the chairs, not you. Some people advocate a 5-min trial, some a 30-min trial, and some an all-day trial for each alternative. Then the alternatives should be lined up side-by-side before making the final decision.

Assume that there are three chairs in the "finals." Users will test the three in turn.

First they should adjust the chair to their preference. (Note that the initial setting biases the selected value. For example, the selected seat height will be lower when the initial setting is low than when the initial setting is high (Helander and Little, 1993).)

Then they should sit and work in the chair for at least two hours and vote on their body part discomfort (see Fig. 7.20) every 30 minutes. In addition, they should evaluate the chair on a chair feature checklist. See Figure 15.9 for an example checklist.

Finally, select one of the chairs, using the discomfort votes, checklist evaluation, price, and any other information as input. Since not all people may like the same chair, a decision will have to be made whether to standardize on one chair or let employees select their own. Although it is obvious, it needs to be noted that not all tasks done in chairs are the same, and different chairs will be selected for different situations.

One of the primary factors in selecting a chair is the importance of easy adjustability. The chair should have adjustable seat height, back height, back in–out location, and back tension. It may have even more adjustments, but those are the basics. If only one person will sit in the chair, then the adjustment need only be made once, and speed is not very important. If multiple people share the chair, then easy adjustment is important. People will have to be trained on how to make the adjustments and which setting to use.

FIGURE 15.9   Chair evaluation form (Drury and Coury, 1982).

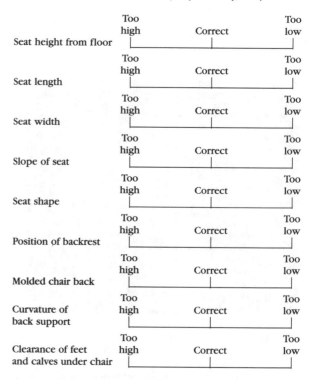

Source: This figure was first published in *Applied Ergonomics*, Vol. 13, 195–202, 1982, and is reproduced here with the permission of Butterworth-Heinemann Ltd., Oxford, UK.

4.2 Chair Design

Chair components can be divided into seats, backrests, armrests, and legs/pedestal. See Figure 15.10 and Table 15.4.

4.2.1 Seats

This will be divided into seat construction and seat dimensions.

Seat construction   When sitting, the weight is not supported by the entire buttocks but by two small areas called the ischial tuberosities (sitting bones). The blood vessels in the tissue overlying the tuberosities (and the tissue on the heels in the feet) are arranged in a special manner to reduce the

effects of pressure. Figure 15.11 shows the comprehensive force for normal sitting and for sitting cross-legged. The amount of load depends also on the sitting posture, because when sitting upright the center of mass of the body is directly above the tuberosities and the feet support about 25% of body weight. Leaning forward shifts more load to the feet and leaning backward reduces the load support of the feet (Chaffin and Andersson, 1984).

A well-designed chair is not contoured (form fitting), because contouring forces the body into one standardized position—keeping the pressure on the same area. A well-designed chair thus permits micro-postural changes. (When driving long distances, periodically make a small adjustment in the power seat tilt and height.)

Cushioning is desirable because it reduces pressure by increasing support area. The upholstery should give way about 25 mm. If it gives way too much, the body is not firmly supported and must be supported by the muscles. Some lunch counters encourage rapid eating (thus maximizing seats available) by using small-diameter hard stools with no backrests.

A curved front edge (waterfall front) is desirable to maximize the surface area contacting the underside of the thigh. Avoid upholstery beading in this area.

The material of the seat (and backrest) should be fabric since it breathes and reduces sliding of the body (better coefficient of friction). In some applications a stain-repellent finish may be desired. Avoid plastic since it causes perspiration problems (body

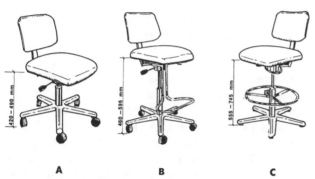

A   **B**   **C**

FIGURE 15.10 Industrial chairs, since they must be used for sitting 8 h/day, can have features selected for specific jobs. Models A, B, and C show different seat height ranges. Model A shows rug casters, B shows normal casters, and C has no casters. Model A has no footrest, while B has a footrest which moves with the seat and C has a footrest which doesn't move. There are a variety of mechanisms to adjust the height of the seat and the location (vertical and horizontal) of the backrest.

TABLE 15.4 Recommended dimensions for office chairs (Chaffin and Andersson, 1984). Dimensions are in cm, angles in degrees.

Feature	British Std. (BS 3079 & 3893)	European (CEN)	Different et al.	Danero & Zelnik	Grand-jean	German Std. (DIN)	Swedish Std. (SS)
Seat							
Height	43-51	39-54	35-52	36-51	38-53	42-54	39-51
Width (breadth)	41	40	41	43-48	40-45	40-45	42
Length (depth)	36-47	38-47	33-41	38-42	38-42	38-42	38-43
Slope angle	0-5	0-5	0-5	0-5	4-6	0-4	0-4
Backrest							
Top height	33				48-50	32	
Bottom height	20						
Center height		17-26	23-25	19-25	30	17-23	17-22
Height		10	15-23	10-20		22	22
Width (breadth)	30-36	36-40	33	25	32-36	36-40	36-40
Horizontal radius	31-46	40 min	31-46		40-50	40-70	40-60
Vertical radius	convex					70-140	convex
Backrest-seat angle	95-105		35-100	95-105			
Armrest							
Length	22	20	15-21			20-28	20
Width (breadth)	4	4	6-9				4
Height	16-23	21-25	18-25	20-25		21-25	21-25
Interarm rest	47-56	46-50	48-56	46-51		48-50	46

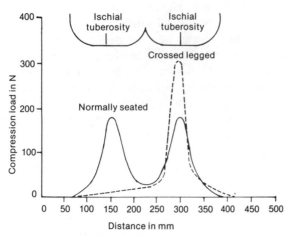

FIGURE 15.11   Sitting cross-legged increases pressure upon the ischial tuberosities (Seating in Industry, 1970). Padding the seat reduces the pressure. Avoid "bucket" seats (form fitting) as they don't permit micropostural changes and thus changing of the pressure point.

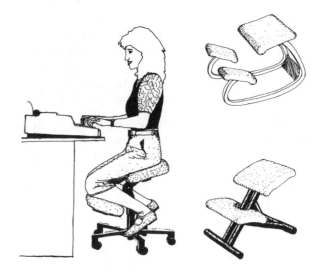

FIGURE 15.12   A thigh–back angle of 135° can be achieved with different design "kneelers" as well as with "saddle chairs."

heat can't dissipate, especially if wearing synthetic-material clothing such as nylon or acetate underpants, panty hose, or slips) and probably has a shorter life due to rips.

The typical chair seat has a backward slope of 1°–5°, which encourages use of the backrest. However, some people advocate having the seat slope forward (Mandal, 1982; Congleton et al., 1986; Bendix and Bloch, 1986). They cite evidence that the relaxed, minimum-strain position between the back and thigh is about 135°—a "sleeping" angle. There are several ways of supporting the body on a forward tilt seat so it doesn't slide forward. The chair can have a "saddle horn," the arms can rest on a tilted table, the support can come from the knees (see Figure 15.12), or the support can come from the feet. At the present time, the vast majority of chairs still have a backward slope; therefore, an option permitting the user to select the desired slope should be good.

Seat dimensions   See Table 15.4 for recommended dimensions.

The most critical dimension is the height. Measure it from the work height, not the floor. As pointed out in Guideline 3, the work should be slightly below elbow height for manipulative work although keyboard work possibly should be slightly higher. The distance between the top of the work and the top of the seat is occupied by three things: the work, the worksurface, and the thighs. Burandt and Grandjean (1963) recommend this distance to be 275 +/− 25 mm. Since the 95% thigh thickness for both U. S. males and females is 175 mm, it is important when purchasing worksurfaces to make them thin so there is maximum space for the thighs and

the work. Thus, do not have a drawer between the thighs and the work. In addition, since the knees occupy the space ahead of the chair, the machine must not have any panel projecting downward from the work, thus preventing the knees occupying the space. (This is emphasized because some machine designers don't seem to realize where the knees go when you sit down and insist on designing machines that cannot be operated from a seated position.)

Once the upper portion of the body is satisfied, turn your attention to the lower portion, that is, the distance from the seat to the floor. If the feet are not flat on the floor, they should be supported by an adjustable footrest (see Figure 15.13), a footrest on the chair pedestal, or a footrail on the workstation.

Since the operator dimensions and work thicknesses vary, seat height should be adjustable. A common problem is that the seat cannot be lowered far enough to permit sufficient thigh clearance or to permit short workers to put their feet on the floor. (This latter problem can be solved with a footrest.)

FIGURE 15.13   Footrests (as well as chairs) should be adjustable. Photo courtesy of Toledo Metal Furniture.

For seat width, the wider the better. Wide seats (benches) not only accommodate more of the population but also permit more varied postures. If armrests are used, they should be farther apart than the seat.

For seat depth, the most common problem is a seat that is too deep. The user either must sit forward (and lose the backrest support) or sit back (and have the legs dangle in the air).

4.2.2 Backrests

The prime function of the backrest is to take a load off the back—to rest the back. This works only if there is contact between the back and the backrest. The ideal backrest is adjustable both horizontally and vertically. In the best design, the backrest has a spring action so that the backrest "tracks" with the back and thus moves in and out as the back moves in and out. It should be relatively small if there is arm movement and elbow impact is to be avoided. The shape should be concave to give area support to the back, especially in the lumbar region. It should be stiffly upholstered and covered with fabric, not a nonpermeable material. See Table 15.4 for recommended dimensions. Note that seats without good backrests (e.g., most automobile seats, nonadjustable chairs) can be retrofitted with a small pillow placed between the back and the backrest.

4.2.3 Armrests

Armrests support the arms and thus take some load off the ischial tuberosities. However, they do restrict movement and may prevent a chair being brought close to a workstation. A possible solution is to use short armrests that support the elbows but do not restrict placing the chair next to the worksurface. A chair with only one arm may be an alternative. It is a common design error in lounge furniture to have armrests (sides) that are far too high. If the posture is relatively fixed (precise assembly), consider support pads on the table. See Figure 15.14. For keying work, consider wrist supports. A pad is placed on the front edge of the keyboard and the wrists are rested on it while not keying (for example, while proofreading or waiting for the computer to complete a search). Note that if repetitive keying is done while the arms are supported on armrests, then the elbows are held farther from the body than if there are no armrests; this, in turn, increases the radial and ulnar deviation at the wrist.

4.2.4 Legs/pedestal

In most circumstances the seat should swivel. (An exception is when operating a pedal.) If the task requires turning the torso and the operator is locked into position by the chair seat, there is twisting in the torso if the chair does

FIGURE 15.14 Arm supports prevent sharp edges of furniture from creating pressure points on the forearms. For operators working along a conveyor, use pressure pads to prevent pressure points on the thighs. Photo courtesy of Toledo Metal Furniture.

not swivel. In addition, a swivel seat permits micro adjustments in posture, may permit easier entrance and exit from the chair, and increases the operator's range of reach.

If the chair has a pedestal with 4 or 5 horizontal supports, the horizontal supports need to be designed to minimize tipping. Minimize the tip angle (the angle from the vertical at which the unloaded chair will tip completely). Generally, this requires 5 legs, each 72° apart (rather than 4 legs 90° apart); the feet should be set in a circle at least as big as the seat itself. Chaffin and Andersson (1984) recommend a minimum radius of 300 mm to prevent the chair from tipping over and a maximum radius of 350 mm to prevent the worker from tripping over the chair base.

If the chair is moved around the workstation, use casters. If the floor is carpeted, use rug casters (wider than conventional casters). If the chair is not moved, don't use casters, because they will force the operator to use leg muscles to keep the chair from moving. Another alternative is to mount the chair on rails to permit the operator to move along the workstation yet not be concerned with chair stability or precise control of chair movements.

4.3 Nonchair Alternatives

In some jobs, it is not possible for the operator to sit on a chair 100% of the time. In that case, permit the operator to sit for a smaller percent of the time. For example, some firms have pull-down benches mounted along the walls. During rest breaks, the operators can sit instead of stand. Or assume a service counter operator (e.g., bank teller, airline service counter) stands while serving the customer. If you can't get the management to consider a seated workstation, suggest pull-out seats for use when there are no customers.

Another alternative is a sit/stand stool where the person's legs are almost vertical. The post office uses such a stool for people manually sorting letters to bins. Another alternative used for keying operations is the Balens kneeler (see Figure 15.12). Kneelers generally fatigue the knees, however, but they can be used for short periods (e.g., 30 min) several times/day.

4.4   Seating Posture Variability   Recent evidence has shown that even with an "optimum" chair design, people still have back pain—if they sit in one posture all day. Physiologically, this seems to be connected with nutrition of the intervertebral disc. The disc is avascular and depends upon diffusion for its nutrition. Diffusion takes place when there are alternations of pressure on the disc (Grieco, 1986; Bendix, 1986). Prolonged sitting also causes discomfort in the buttocks (due to surface pressure) and increases foot volume. The person should therefore get up and walk around occasionally, i.e., have **seating posture variability**. One way to vary postures is to have the telephone placed so you stand while answering it.

Grieco (1986) recommends a "physiological pause" of 15 min for every 75 min of seated work. From an economic viewpoint, this need not be idle time but just different work. Design the job so people do not sit continuously. Possible techniques would be to have the word processor operator go get the mail and sort it; the executive take a letter to the secretary (if the executive needs movement) or the secretary take the letter from the executive (if the secretary needs movement); the inspector or operator get a supply of parts (instead of having them furnished by a material handler); a receptionist/secretary stand when dealing with a customer but sit when keying; or the secretary run the copy machine as well as key. Movement also gives the eyes a chance to rest.

In addition, people doing seated work should be told of the problems of disc nutrition and encouraged to be active in their nonwork activities.

GUIDELINE 5: USE THE FEET AS WELL AS THE HANDS

The foot moves more slowly than the hand and, due to the construction of the ankle vs. the wrist and the weight of the leg vs. the arm, it is not as dexterous.

An example of the speed of the arms vs. the strength of the legs is the English longbow vs. the Continental crossbow. The longbow was powered by the arm and "a first-rate English archer, who in a single minute was unable to draw and discharge his bow 12 times with a range of 240 yards and who in those 12 once missed his man was very lightly esteemed" (Heath, 1971). The crossbow, which had a greater range, was limited in its speed of fire since it was cocked with the "belt and claw" method. The bowstring was looped over a claw attached to the belt and the bow pushed with the feet. Fortunately for the English, the Chinese repeating crossbow, which was developed in the first century and which could fire 10 bolts in 15 s, was unknown in Europe. It was cocked with the arms.

For power generated from muscles (**human power** in contrast to power from machines), the 2 legs have approximately 3 times the power of the 2 arms; the arms are slightly more efficient per kg of muscle but the legs have much more muscle (Davies and Sargeant, 1974). With 2 arms a man can generate about 1/25 hp; with 2 legs about 1/10 hp. The 1/10 hp estimate is based on a long-term human work rate of 5 kcal/min and 20% efficiency, yielding a 1 kcal/min work output. (1 hp = 746 W = 10.7 kcal/min.)

Don't use pedals for standing work. The body would be supported unevenly. In addition, weight on the entire leg or even the entire body must be moved. Although the muscles can compensate for the unnecessary strain, the resulting unnecessary energy expenditure, pain, and fatigue are a reflection on the engineer's competence. In addition, since the operator is off balance, reaction time in an emergency is increased.

Pedals can be used for power and control.

5.1   Power   Power generation can be continuous (bicycle) or discrete (nonpowered automobile brake pedal).

Continuous power usually is generated with both limbs with a rotary pedal arrangement so each limb can rest for 50% of the cycle although output is continuous. Bicycle pedaling (at 20 to 25% efficiency for an experienced cyclist) is first in efficiency among traveling animals and machines (Wilson, 1973). For more on human power generation, see McCullagh (1977), Whitt and Wilson (1982), and Brooks et al. (1986).

Discrete power usually is applied by one leg since application time usually is less than 10 s and thus fatigue is not a problem. There does not seem to be any power advantage to using the right or left foot (Mortimer et al., 1970). Force using both feet is about 10% higher than that using a single foot, but since people will not always use both feet, the designer should not design for use of both feet.

Force capability depends on a number of factors, but individual capability and pedal location are very important. See the discussion in Chapter 19 and Figures 19.11 and 19.12.

5.2 Control   An example application of a foot pedal for continuous control is the auto accelerator. Examples of discrete (on–off) applications are a pedal-controlled punch press and an automobile foot switch for high–low beam. See Chapter 19.

In most industrial situations, the time it takes a person to move a foot to a control is not important because the movement can be done at the operator's leisure or can be done simultaneously with some other motion. However, in vehicles, control time is important. Time to move the foot from an accelerator pedal to a brake pedal is about .55 s; if the pedal acts as a rocker switch and your foot is already on the pedal, it takes about .28 s to move the foot from a toe-depressed position to a heel-depressed position (Konz et al., 1971). Using a conventional brake pedal with your foot resting on the brake, you can save about .27 s (about 1 car length at 60 mph) over moving your foot from the accelerator. However, with present pedal designs, poising the left foot on the brake pedal is tiring; so you should use left-foot braking only in heavy traffic. All the above times include signal detection time (time to detect a red light), decision time (time to decide to push pedal), and effector time (time for the nerve impulse to travel to the foot and the foot to act).

GUIDELINE 6: USE GRAVITY; DON'T OPPOSE IT

Consider the weight of the body and weight of the work.

6.1 Body   As pointed out in Guideline 1, holding 25 g of feathers also requires holding 4,500 g of arm. Therefore, make movements horizontal or downward—avoid lifting.

In certain circumstances, the weight of the human body can be used to increase the force on a lever or pedal. From a conservation of energy viewpoint this is just a substitution of potential energy for kinetic energy with no net gain. The key practical point, however, is that the potential energy can be applied over a period of time and the kinetic energy can be released all at once. An example is a wheel with a ratchet and a release.

6.2 Work   Gravity can be used to move material to the work. Examples are paint from a paintbrush, a welding bead from a welding rod, solder on a solder joint. Gravity also can be used to hold components before assembly **(gravity as a fixture).** For example, contrast driving a screw into the ceiling vs. doing it horizontally or vertically downward.

Use gravity in the feeding and disposal of components. Vibratory feeders (see Figure 15.15) use gravity to orient components mechanically before they are fed into machines. An additional advantage of the vibratory feeder is that the inexpensive orientation of parts for feeding permits parts ejected from the *preceding* operation to fall in a random orientation. If the component is fragile, cushion the fall of the ejected part with a chute. (A chute also permits horizontal transportation at zero labor cost.) For greater distances, use gravity wheel conveyors. For maximum use of the floor space while still using gravity, use spiral conveyors and chutes.

GUIDELINE 7: CONSERVE MOMENTUM

The goal of conserving momentum is to avoid unnecessary **acceleration** (increase in speed) and **deceleration** (decrease in speed) since they take both time and energy. For example, in the Work-Factor predetermined time system, a 45-cm reach with a change in direction is given 39% more time due to the change in direction.

7.1 Stirring and Polishing Motions   There are a number of operations in which the arm is in relatively continuous motion. An ancient example was the conversion of grain to flour with a mortar and pestle. This pounding, with its acceleration and deceleration of the hand, was replaced by grinding in a circular motion. Eventually water power replaced muscles for this job and now the millstone is powered by mechanical power.

Figure 15.16 shows how this same principle applies to hand-polishing operations, whether they

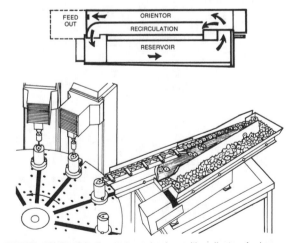

FIGURE 15.15 Orienting parts can be done with a vibratory feeder as well as by human fingers. Since orienting parts can be done by machine at low cost, less care is needed for part orientation during disposal from the previous operation.

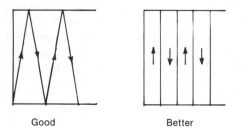

Good   Better   Best

FIGURE 15.16   Avoid abrupt decelerations in vacuum sweeper, mop, and hand polishing operations (Barnes, 1940).

are with a rag (in a factory or on your car) or whether the pressure is on the end of a pole or hose (broom, mop, vacuum cleaner). Writing vs. printing is another example. In a kitchen, stirring soup is the same situation. (If there is insufficient mixing of product with circular stirring in a circular container, use a rectangular container to furnish the turbulence rather than requiring a zigzag motion.) Bicycle pedals are a leg example of the principle of circular motion.

7.2   Disposal Motions

Acceleration/deceleration can be minimized in disposal motions. In sports, an important principle is to "follow through" to impart maximum velocity and to minimize deviations from the desired path. Abrupt deceleration gives poor performance. Figure 15.17 shows how abrupt deceleration can be avoided when loading/unloading a punchpress. Reduce precise disposal of completed units; instead, toss them aside. If the unit might be damaged by abrupt deceleration, cushion its fall with a resilient surface or by using chutes. If the part must be oriented for the following operation, orient the part with a vibratory feeder. In addition to reducing the strain on the body, there are considerable time

benefits. For example, using MTM technology (see Chapter 29), an 18-inch move to toss aside a part is an M18E and an RL1—a total of .63 s. A precise placement would require an M18B, a P1SE, and an RL1—a total of .89 s. This is an increase of 42%. In addition, the precise placement usually requires eye control, and so other motions cannot be done simultaneously.

7.3   Grasping Motions

Grasping motions can affect acceleration/deceleration. Figure 15.18 shows how the sharp edge of a bin can be modified with a rolled edge. If there is a rolled edge the operator does not need to slow down to prevent injury. Figure 15.19 shows how the thickness of the tabletop can affect the speed of grasping small objects (such as coins) from a flat surface. With a thick tabletop, the hand must come to a stop for a precise grasp. A thin tabletop permits the fingers to slide the object to the edge and then get a good grasp. A thin table with a small front lip improves the ease of the grasp even more.

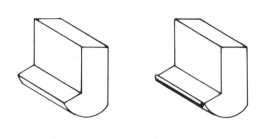

Bad   Good

FIGURE 15.18   Avoid knife edges on bins by using a rolled edge.

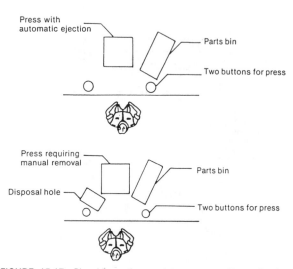

FIGURE 15.17   Place bins so the operator can grasp or dispose "on the fly." The top sketch shows the right hand inserting the part and automatic disposal. The lower sketch shows the right hand inserting and the left hand removing the part.

Bad   Better   Best

FIGURE 15.19   Grasping small objects (such as coins on a counter) is easier when the lip acts as a wedge under the object to raise it off the surface and aids grasping on both sides. The table lip also reduces the chance of objects rolling off the table and gives a more favorable hand orientation. Note the counter design at McDonald's.

7.4 Transport Motions It requires extra effort and extra time to transport weight in the hand (**weight penalty**). Figure 15.20 shows the recommendations given by 3 predetermined time systems as well as some experimental data. The amount of extra time is indicated by the slope of the lines; the intercept indicates varying definitions of a normal pace.

Our study was concerned with only the *control* effects of the additional weight; fatigue effects are additional. The increase in time per move is 3%/kg. The increase in physiological cost (as measured by the integrated output of the three axes of a force platform) was 6%/kg. Since the time/move increased 3%/kg, this is an increase of about $6 - 3 = 3$%/kg in the acceleration–deceleration forces. These values are compatible with Ayoub's (1966).

GUIDELINE 8: USE TWO-HAND MOTIONS RATHER THAN ONE-HAND MOTIONS

8.1 Human Power Andrews (1967) demonstrated that cranking, for loads up to 25 W, required 10% fewer watts when using 1 arm instead of 2 arms. Beyond 25 W, cranking with 2 arms is about 25% more efficient than with 1 arm. For exerting a static pull, 1-arm work required 42% more watts at a 5 kg load, 18% at a 10 kg load, and 127% more at a 15 kg load.

8.2 Manipulative Work In most uses of the arm, it is moved to some location because the hand is going to manipulate something. In this case also, 2 hands are better than 1. Using 2 hands takes more time and effort than using just 1 hand, but more is produced so cost/unit is lower. The reduced cost is

both individual cost (physiological cost/unit) and organizational cost (time/unit).

Nichols and Amrine (1959), using heart rate as the criterion, reported that for equal amounts of pieces moved, 1-hand motions had a smaller increase in heart rate than nonsimultaneous but symmetrical 2-hand motions. Salvendy and Pilitsis (1974) showed that kcal/min of 1-hand and 2-hand motions did not differ significantly, and thus, on a per-unit basis, cost was less when using 2 hands simultaneously.

Reduced time/unit for 2 hands vs. 1 was reported first by Barnes et al. (1940). However, speed of movement is confounded with accuracy of movement. Both speed and accuracy can be combined into one index, bits/second. See Box 15.2.

Konz, Jeans, and Rathore (1969) had 7 women move a stylus back and forth at 7 different angles (0°, 30°, 60°, 90°, 120°, 150°, and 180° where 0° = "3 o'clock") and 2 distances (225 and 400 mm). Trials of 18 s were performed with the right hand, left hand, and all combinations of angles with both hands. The average for the right (preferred) hand was 12.9 bits/s, the average for the left hand was 11.7, and for both hands working at the same time it was 21.2. Thus, using just the preferred hand gave $12.9/21.2 = 61$% of the potential output and using just the nonpreferred hand gave $11.7/21.2 = 55$% of the potential output.

If the person is assumed to be working at maximum output in all 3 conditions, why isn't the rate the same for all 3 conditions?

For the 2-hand condition, the maximum bits/s at any angle combination for the 225-mm distance was 23.0 and the minimum was 20.1, a difference of 2.9. At the 400-mm distance, the maximum was 22.4 and minimum was 19.3, a difference of 3.1. Thus, the maximum effect of the visual field is about 3 bits/s.

What is the effect of 2 hands vs. 1 hand? When the right hand moved at 225 mm, the mean was 12.9. With both hands the total was 21.4, an addition of $21.4 - 12.9 = 8.5$ bits/s. When the right hand moved at 400 mm, the mean was 13.0. With both hands the total was 20.8, an addition of $20.8 - 13.0 = 7.8$ bits/s. Increases for both hands over the nonpreferred hand averaged 9.4 bits/s. Therefore, the body can produce 8–9 bits/s more with 2 hands than with 1 hand. These values are compatible with Langolf et al. (1976), who reported fingers could process 38 bits/s, wrists 23, and arms only 10.

It seems the bottleneck in the output of the brain–eye–muscle system is neither the brain (command system) nor the eyes (tracking subsystem) but the muscles and nerves (effector and feedback subsystem). In other words, the limiting factor in hand–arm manipulative movements is not the ability of the brain to command or ability of the eyes to

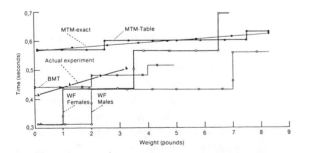

FIGURE 15.20 Time penalty for weight is not consistent by source. Experimental data (Konz and Rode, 1972) indicate that time should increase 3%/kg. The slope of the MTM exact method is considerably lower. Methods-Time-Measurement, Work-Factor, and Basic Motion Times use a step function for the weight penalty; they do not disclose how their recommendations were obtained.

BOX 15.2   *Information content of motions*

Shannon (1948) defined information in terms of a signal-to-noise ratio transmitted from a transmitter over a channel to a receiver. Since the ratio was logarithmic and log to the base 2 ($\log_2$) customarily has been used, the unit of information commonly has been called a **bit,** short for *bi*nary digi*t.* Fitts (1954) and Fitts and Peterson (1964) developed this concept into a formula for hand–arm movements. Fitts' concept was that any movement was limited by the amount of information to be processed. Fitts used distance of the move as the signal and width of the target as noise. **Fitts' law** is

$$I = \log_2 \frac{D}{W/2}$$

where

   I = Information/move, bits
   D = Distance of the move
   W = Width (in movement direction) of target

The validity of this formula has been confirmed by others, especially in a series of elegant investigations by Hoffman (Gan and Hoffman, 1988; Hoffman and Gan, 1988; and Hoffman, 1991a, 1991b). Drury (1975) and Hoffman (1991a) showed it applied to foot motions also.

Fitts' law can be solved to determine movement time:

$$MT = a + b\,I$$

where MT = Movement time, ms

   a = constant
   b = constant

Gan and Hoffman (1988) showed that Fitts' law for arm motions is only applicable for visually controlled motions—that is, when $ID > 3$. When $ID < 3$, the motion is ballistic and just amplitude (but not target width) is important. Hoffman (1991a) gives the following:

$$HMT = 56.6 + 4.43\,A^{.5}$$
$$ID < 3 \text{ bits (ballistic)}$$
$$- 56.9 + 59.2\,ID$$
$$3 < ID < 6 \text{ bits (visually controlled)}$$

$$FMT = 107.5 + 6.5\,A^{.5}$$
$$ID < 3 \text{ bits (ballistic)}$$
$$= -57.7 + 115\,ID - 34.6 \log_2 (W)$$
$$3 < ID < 6 \text{ bits (visually controlled)}$$

where HMT = Hand movement (discrete, not reciprocal) time, ms
   FMT = Foot movement (discrete, not reciprocal) time, ms
   A = Movement amplitude, m
   ID = Index of difficulty, bits
   W = Target width, mm

The foot takes about 1.7 times as long as the hand for ballistic movements and 2 times as long for visually controlled movements.

---

supervise but the ability of the nerves and muscles to carry out the orders. Remember this, as "the spirit is willing but the flesh is weak."

GUIDELINE 9: USE PARALLEL MOTIONS FOR EYE CONTROL OF TWO-HAND MOTIONS

Gilbreth (1911) first stated:

> When work is done with two hands simultaneously, it can be quickest and with least mental effort, particularly if the work is done by both hands in a similar manner, that is to say, when one hand makes the same motions to the right as the other does to the left.

Barnes (1940) was more concise:

> Motions of the arms should be made in opposite and symmetrical directions, and should be made simultaneously.

However Barnes also stated:

> Eye fixations should be as few and as close together as possible.

Which principle has precedence? Figure 15.21 poses the dilemma.

Konz (1983) discusses various studies. See also Hassan and Block (1967), Raouf and Arora (1980), and Raouf and Tsuchiya (1986). The conclusion, using both time/unit and physiological cost/unit, is to minimize the degree of spread rather than worry about the symmetry of the motions. The problem with spread is the cost of eye control. See Figure 15.22.

The various predetermined time systems enable us to estimate the cost of eye control. The two eye activities are a change in viewing distance (**eye**

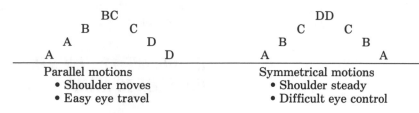

FIGURE 15.21 Parallel or symmetrical motions? When is each best?

focus) and a change in line of sight (**eye travel**). The normal **area of vision** is a circle with a 100 mm diameter at a viewing distance of 450 mm.

Eye focus is allocated .0044 minutes by MTM. The version of MTM used by General Motors gives .0020 min for eye focus, .0030 for eye reaction, and .0030 for eye interpretation. Work-Factor gives eye focus from .0025 to .0100, depending on the location of the initial and final focus location; .0050 is a typical time. In addition, Work-Factor allows extra time for eye inspect.

Eye travel is .009 *(W/D)* in the MTM system, where *W* is the distance between the points and *D* is the perpendicular distance to the line of travel. General Motors gives a flat .0050. In Work-Factor, it is called eye shift. For a shift of 0° to 5°, time = .0004, which increases to .0015 for a 40° shift; beyond 40° it is called a head or body turn. In Work-Factor, you use either eye focus or eye shift but not both. White et al. (1962) give .0012 min for a 10° shift, .0015 min for a 20° shift, and .0020 min for a 40° shift.

GUIDELINE 10: USE ROWING MOTIONS FOR TWO-HAND MOTIONS

When moving both hands, should the hands move in a rowing motion or an alternating motion?

Nichols and Armine (1959) reported that alternating motions had heart rates 1.5 beats/min higher than when the same motions were performed in a rowing manner. Konz, Jeans, and Rathore (1969) reported that force platform output (i.e., work cost) was 10% greater when alternating motions were used

in place of rowing motions. In both these studies the work output was controlled to be the same for alternating and rowing motions.

For both types of motions, the hands move in a relatively flat plane with considerable acceleration and deceleration at each end of the stroke. Alternation, however, has more movement of the shoulder and twisting of the torso. See also Guideline 9.

Note that for human generation of power (bicycle, winch), the handles or pedals are arranged so that the arms or legs alternate strokes while the path is circular to conserve momentum. Harrison (1970) demonstrated that maximum power output came from a device in which both pedals were at the same angular position on both sides of the hub instead of the 180° out of phase used for bicycles. A large flywheel was used to return the pedals for the power stroke. In addition, a "forced" motion, in which kinetic energy of the limbs was fed back into the mechanical system, had greater power output than a "free" motion, in which the kinetic energy was absorbed by the limbs.

GUIDELINE 11: PIVOT MOTIONS ABOUT THE ELBOW

The question is "For horizontal movements at a height, does direction of movement affect (1) movement speed, (2) movement accuracy, and (3) movement physiological cost?"

Yes. Yes. Yes.

Figure 15.23 shows the effect of various **movement angles** on motion time. The least time is with a pivot about the elbow (i.e., movement of only the forearm) and maximum time is a pivot about the shoulder (movement of both the forearm and upper arm). As a rule of thumb, use standard time for movements pivoted about the elbow (say 30°–60° for the right hand, 150°–180° for the left), give 5% additional time for straight-ahead moves (say 60°–120° for either hand), and give 15% additional time for cross-body moves (over 120° for the right hand, below 60° for the left).

Figure 15.24 shows the effect of various angles on accuracy. The somewhat surprising conclusion is

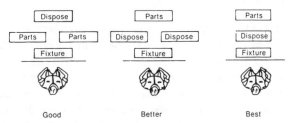

FIGURE 15.22 Proper location of disposal chute reduces eye control as well as conserving momentum.

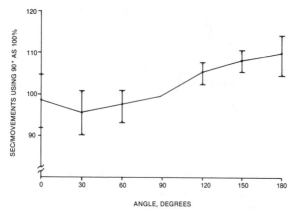

FIGURE 15.23 *Pivoting about the elbow takes less time than cross-body movements (Briggs, 1955; Konz, Jeans, and Rathore, 1969; Konz and Rode, 1972; Schmidtke and Stier, 1961). Output at 90° (straight ahead, with 0° as 3 o'clock) is defined as 100%.*

that cross-body movements (i.e., pivoted only on the shoulder) are more accurate than movements pivoted about the elbow.

Physiological cost of movements at various angles has been studied by a number of people, including Konz and Rode (1972). Physiological cost is lower for movements pivoted about the elbow. (Moving the total arm and hand requires moving 4.9% of body weight; moving just the forearm and hand requires moving 2.3% of body weight.) Thus, although cross-body movements are more accurate, they take longer and are more costly to the person. Avoid them. Encourage elbow pivoting motions by

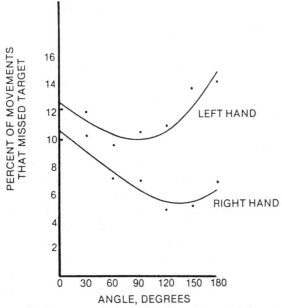

FIGURE 15.24 *Preferred hands are more accurate. In addition, maximum accuracy for a hand is better for cross-body movements than pivoting about the elbow (Konz, Jeans, and Rathore, 1969; Konz and Rode, 1972). 0° = 3 o'clock.*

putting bins ahead of the shoulder—not straight ahead of the nose.

GUIDELINE 12: USE THE PREFERRED HAND

The **dominant hand** (preferred hand) is about 10% faster for reach-type motions; it also is more accurate (Konz, Jeans, and Rathore, 1969). The preferred hand, because it is used more, gets more practice. The good news is that it is farther down the learning curve and thus faster; the bad news is that it is more exposed to cumulative trauma. The preferred hand is 5 to 10% stronger (Dickson et al., 1972; Kroemer, 1974).

About 10% of the population uses the left hand as the preferred hand.

In addition to a dominant hand, you have a **dominant eye** (your "shooting" eye). You can check which eye is the dominant one by holding out your arm. Align an object with your thumb. Close one eye. If the object moved, you closed your dominant eye. Fisher (1974) studied 300 people; 88% used the right hand as the writing hand (the writing hand is the best single indicator of which side of a person is dominant), 60% had the right eye as the dominant eye, and 54% had the right eye as the more acute eye. There was a nonsignificant relationship between the writing hand and eye dominance and acuity but a significant relationship between dominant eye and most acute eye.

In general, work should come into a workstation from the operator's preferred side and leave from the nonpreferred side. The reason is that reach and grasp are more difficult motions than dispose and release. Avoid transfer grasps. However, if the new item is obtained on the same side as the disposal, a body turn is eliminated.

GUIDELINE 13: KEEP ARM MOTIONS IN THE NORMAL WORK AREA

First we will discuss the work area in the horizontal plane at elbow height, then the reach distance above and below that plane.

13.1 Elbow Height   Maynard (1934) presented a dimensionless sketch of an inner and outer semicircle for the right and left hand. Asa (1942) gave the sketches some dimensions, based on measurements of 30 male students. Farley (1955) gave male and female dimensions based on average General Motors operators: Height (in shoes) was 1,750 mm for males and 1,500 mm for females.

Squires (1956) suggested the shape shown in Figure 15.25 because the elbows do not stay at a fixed point during movement but move in an arc as the forearm pivots. The coordinates of the arc *PQ* are given by the equation:

$$x = A \cos \theta + B \cos (65 + 73/90) \theta$$
$$y = A \sin \theta + B \sin (65 + 73/90) \theta$$

where

A = Distance *EC* = Elbow to shoulder projection distance

B = Distance *CP* = Distance from elbow to end of thumb

θ = Angle given at any instant by the radius which sweeps out the arc *DC*, degrees; *D* = Point at which θ = 0; *C* = Point at which θ = 90.

Konz and Goel (1969) determined values for the range of the U. S. population. They measured 40 men and 40 women, selected to be representative of the U. S. population in 1960. The 50th percentile male was 1,735 mm; the 50th percentile female was 1,508 mm. A = 112 mm for the 5th percentile male, 152 for the 50th, and 198 for the 95th; the corresponding values for females were 91, 145, and 188. B = 378 for 5th percentile males, 412 for 50th, and 457 for 95th; the corresponding values for females were 356, 376, and 414. AC = 211 mm for males and 194 for females; this was estimated as 50% of the mean elbow-to-elbow distance (*Weight, Height, and Selected Body Dimensions*, 1965). The angle of

abduction (upper arm vs. the horizontal) was 65°. The resulting values (reach to end of thumb with no "back assist") depict the **normal work area** and are plotted in Figures 15.26 and 15.27. See also Das and Grady (1983).

13.2 Reach at All Heights

Figure 15.28 is a general guideline for average U. S. workers (males and females combined). Reach distances are assumed to be identical for sitting and standing, although one of the practical advantages of standing work is a greater reach distance (as the torso moves forward). Remember that the arm is pivoted on the shoulder, not the nose, so it is easier to reach ahead of the shoulder than ahead of the nose. However, the reach distance decreases rapidly for reaches outside the shoulders because the elbow pivots (see Figure 15.25). Note that horizontal reach distance while standing decreases quickly for heights below the waist.

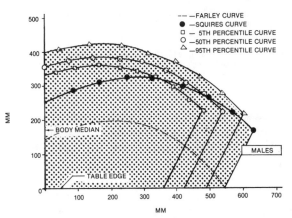

FIGURE 15.26 Normal male work area (right hand) for the 5th, 50th, and 95th percentile U. S. male (Konz and Goel, 1969). The left-hand area is mirrored.

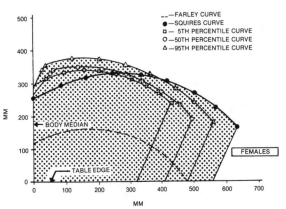

FIGURE 15.27 Normal female work area (right hand) for the 5th, 50th, and 95th percentile U. S. female (Konz and Goel, 1969). The left-hand area is mirrored.

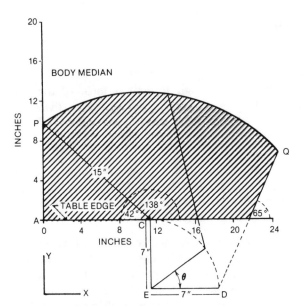

FIGURE 15.25 "Windshield wiper" pattern is shown for the right hand (Squires, 1956). *CD* is the projection of the right elbow during movement of the hand from point P to Q.

FIGURE 15.28 Approximate reach distances for average U. S. workers (male and female) (Putz-Anderson, 1988).

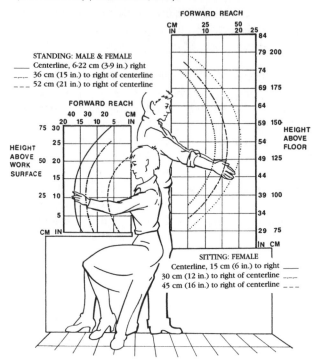

Source: V. Putz-Anderson, (ed.). *Cumulative Trauma Disorders.* Copyright © 1988 by Taylor and Francis, London. Reprinted with permission.

Reach distance into bins which are loaded from the rear can be decreased by having the bin bottom slant toward the operator. Have the entire front of the bin open (although a strip across the top may be needed for rigidity).

Note that just because it is possible to reach these distances does not mean it is desirable. Closer is better. See Figures 15.29 and 15.30. See also Figure 17.24.

Allow enough space for standing workers' feet. A space 150 mm × 150 mm × 500 mm wide (6 in. × 6 in. × 20 in.) is sufficient. See Figure 15.31. If there is rotating machinery near the toe, a guard gap 150 mm above the floor probably will be sufficiently small (Bottoms and Butterworth, 1990).

GUIDELINE 14: LET THE SMALL WOMAN REACH; LET THE LARGE MAN FIT

As was pointed out in Chapter 10, the designer designs to include a certain proportion of the population rather than for the population mean. If the designer selects a chair to fit under a table but uses the 50th percentile thigh dimension, 50% of the people will not fit under the table.

FIGURE 15.29 Five desirable design features are shown in the good layout (Rodgers, 1984). Can you find them? First, the operator has a 90° turn to the pallet instead of a 180° turn to the pallet. Second, there is a shorter reach to the cartons since the long axis of the pallet is parallel to the shoulders instead of perpendicular. Third, there is a shorter horizontal reach to the conveyor. Fourth, there is less vertical reach down to the pallet. Finally, there is storage under the work table.

Therefore, design to permit most of the user population to use the design. An alternative statement is "Exclude few." Another statement is "Design for the tall; accommodate the small."

14.1 User Population   Some population data are given in Chapter 10. The problem in selecting a specific population has become more difficult for the engineer. Some points to consider:

- Jobs now must be designed for both sexes. Formerly, a job could be considered a male job or a female job. Changing cultural values and laws have modified that, and the designer now must consider the range from small woman to large man instead of the range from small woman to large woman or small man to large man.

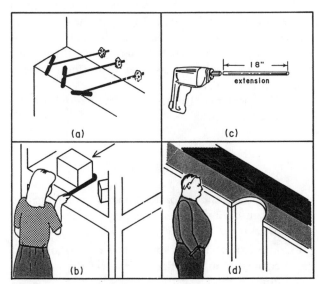

FIGURE 15.30 Four ways of reducing strain from forward reaches are given (Rodgers, 1984). Figure (a) shows how reaching for a control behind a barrier can be reduced by putting a long handle on the control. Figure (b) shows use of a hook to obtain cartons when order picking from the second-level rack. Figure (c) shows an extension to a tool. Figure (d) moves the man closer to the conveyor by putting a cutout in the table. If you don't want to cut a semicircle in the table, don't have a table at all between the operator and the conveyor and just have a table to the left and right of the operator.

FIGURE 15.31 Allow room (150 × 150 × 500 mm) for the feet (DeLaura and Konz, 1990).

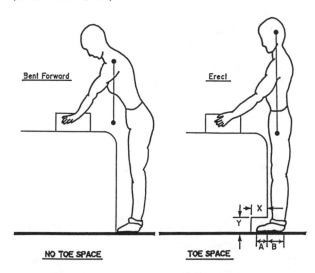

Source: D. DeLaura and S. Konz. "Toe Space" in *Advances in Industrial Ergonomics and Safety II,* B. Das (ed.). Copyright © 1990 by Taylor and Francis. Reprinted with permission.

- Multiperson use of equipment and workstations is becoming more common with greater shift-work and more part-time workers. That is, people of different dimensions will be using the equipment—either within the same shift (e.g., 10 different people use the duplicating machine) or over multiple shifts (e.g., 3 different police officers use the same vehicle over a 24-hour period). Thus, easy equipment adjustability becomes important.

- Civilian industrial population data are not the same as military population data. Military populations heavily emphasize youth and physical fitness and are primarily male. General population data for adults also include retired people as well as those who have never worked.

- International populations may have to be considered. Many countries have "guest workers." In addition, many firms are multinational with facilities in many countries and thus may use the same workstation in multiple countries.

14.2 Percent to Exclude

As indicated in Figure 15.32, people can be excluded from the job in three ways:

1. Exclude the lower percentile (e.g., short people or weak people).
2. Exclude the upper percentile (e.g., tall people).
3. Exclude both lower and upper percentile (e.g., people with low and high intelligence from an assembly-line job).

The proportion to exclude depends upon the seriousness of being designed out and the cost of including more people. For example, consider design of a tote box to be used on the assembly line. We might design the box to hold a weight that 90% of the population could lift. In other words, 10% would be excluded. This might give, for example, a total load of 20 kg. If you would like to exclude fewer, then the load would drop, perhaps to 18 kg. In this

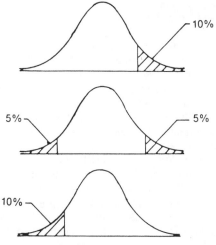

FIGURE 15.32 Excluding 10% can mean the upper percentile, some from each end, or the lower percentile.

case, excluding fewer gives a continuing lower productivity—a trade-off of including more people and making the movement easier for everyone vs. the cost of requiring more movements. Or consider the design of a door height. A height of 80 inches would require only a small percent of people to duck their heads. However, increasing the height to 86 inches would include effectively 100% of the population.

There would be no continuing cost and perhaps a small additional initial cost. However, the cost to the person of ducking the head is small also.

Design so that a small woman (Cathy Rigby, Sally Fields) can reach and a large man (John Wayne, Michael Jordan) can fit.

The most practical design technique is to make the machine adjustable.

REVIEW QUESTIONS

1. Sketch the response of blood pressure vs. time for static and dynamic work.
2. Give three results of venous pooling.
3. Give the characteristics of good floor mats.
4. Approximately what percent of industrial fatalities are from falls?
5. Give three techniques for reducing falls on one-step stairs.
6. What is the function of a toeboard?
7. For walking with slip-resistant shoes, is the sole or heel more important?
8. If your feet differ in size, is it better for shoes to have 4 lace holes or 2? Why?
9. What is arm abduction?
10. If a person keys 70 words/min, how many keystrokes is this in a working lifetime? Give assumptions.
11. Discuss whether you think easy-adjusting chairs are desirable.
12. When keying, should the arms rest on the wrist rest? Justify your answer.
13. What is the optimum work height for manipulative work?
14. What are the 3 basic solution techniques to obtain optimum work height? Give an example of each.
15. Approximately how much does it cost/hour for a chair used in an office or factory? Show calculations.
16. Should a chair be form-fitting or not? Why?
17. Should a chair seat swivel? Why?
18. How does sitting affect disc nutrition?
19. Approximately how much stronger are the legs than the arms?
20. What is a vibratory feeder?
21. What has McDonald's done to the lips of their counters to aid the order taker?
22. Is the bottleneck in manipulative motions in the brain, the eyes, or the muscles?
23. When using two hands, should the motions be rowing or alternating?
24. How can you determine your dominant eye?
25. People can be excluded from jobs from the lower percentile, upper percentile, or from both. Give an example of each of the three.

REFERENCES

Andrews, R. The relative energistic efficiencies of one-armed and two-armed work. *Human Factors,* Vol. 9 [6], 573–80, 1967.

Asa, M. A study of workplace layout. M. S. thesis. State University of Iowa, 1942.

Ayoub, M. Effect of weight and distance travelled on body members' acceleration and velocity for three-dimensional moves. *International J. of Production Research,* Vol. 5 [1], 3–21, 1966.

Barnes, R. *Motion and Time Study,* 2nd ed. New York: Wiley & Sons, 1940.

Barnes, R., Mundel, M., and MacKenzie, J. Studies on one- and two-handed work. Bulletin 21, *Studies in Engineering*. Iowa City: University of Iowa, March 1940.

Bendix, T. Chair and table adjustments for seated work. In *The Ergonomics of Working Postures,* Corlett, N., Wilson, J., and Manenica, I. (eds.). London: Taylor and Francis, 1986.

Bendix, T. and Bloch, I. How should a seated workplace with a tiltable chair be adjusted? *Applied Ergonomics,* Vol. 17 [2], 127–35, 1986.

Bottoms, D. and Butterworth, D. Foot reach under guard rails on agriculture machinery. *Applied Ergonomics,* Vol. 21 [3], 179–86, 1990.

Brand, J. and Judd, K. Angle of hard copy and text-editing performance. *Human Factors,* Vol. 35 [1], 57–69, 1993.

Briggs, J. A study in the design of work areas. Ph. D. dissertation. Lafayette, Indiana: Purdue University, 1955.

Brooks, A., Abbott, A., and Wilson, D. Human-powered

watercraft. *Scientific American,* Vol. 256 [12], 120-30, December 1986.

Burandt, U. and Grandjean, E. Sitting habits of office employees. *Ergonomics,* Vol. 6 [2], 217-28, 1963.

Chaffin, D. Localized muscle fatigue—Definition and measurement. *J. Occupational Medicine,* Vol. 15 [4], 346-54, 1973.

Chaffin, D. and Andersson, G. *Occupational Biomechanics.* New York: Wiley & Sons, 1984.

Chaffin, D., Woldstad, J., and Trujillo, A. Floor/shoe slip resistance measurement. *American Industrial Hygiene Association J.,* Vol. 53 [5], 283-89, 1992.

Clauser, C., McConnville, J., and Young, J. *Weight, Volume and Center of Mass of the Human Body, AMRL-TR-70.* Dayton, Ohio: Aerospace Medical Research Laboratory, 1969.

Congleton, J., Ayoub, M., and Smith, J. The design and evaluation of the neutral posture chair for surgeons. *Human Factors,* Vol. 17 [2], 127-35, 1986.

Corlett, N., Wilson, J., and Manenica, I. (eds.). *The Ergonomics of Working Postures.* London: Taylor and Francis, 1986.

Das, B. and Grady, R. The normal working area in the horizontal plane: A comparative analysis between Farley's and Squires' concepts. *Ergonomics,* Vol. 26 [5], 449-59, 1983.

Davies, C. and Sargeant, A. Physiological responses to standardized arm work. *Ergonomics,* Vol. 17 [1], 41-49, 1974.

DeLaura, D. and Konz, S. Toe space. In *Advances in Industrial Ergonomics and Safety II,* Das, B. (ed.). London: Taylor and Francis, 1990.

Dickson, A., Petrie, A., Nichole, F., and Calnan, J. A device for measuring the force of the digits of the hand. *Biomedical Engineering,* 270-73, July 1972.

Drury, C. Application of Fitts' law to foot pedal design. *Human Factors,* Vol. 17, 368-73, 1975.

Drury, C. and Coury, B. A methodology for chair evaluations. *Applied Ergonomics,* Vol. 13, 195-202, 1982.

Eastman, Kodak Co. *Ergonomic Design for People at Work.* Belmont, Calif.: Lifetime Learning, 1983.

Farley, R. Some principles of methods and motion study as used in development work. *General Motors Engineering J.,* Vol. 2 [6], 1955.

Fisher, G. Handedness, eye dominance and visual acuity. *Ergonomics News Letter,* 2, August 1974.

Fitts, P. The information capacity of the human motor system in controlling the tolerance of the movement. *J. of Experimental Psychology,* Vol. 47 [6], 381-91, 1954.

Fitts, P. and Peterson, J. Information capacity of discrete motor responses. *J. of Experimental Psychology,* Vol. 67, 103, 1964.

Gan, K-C., and Hoffman, E. Geometrical conditions for ballistic and visually controlled movements. *Ergonomics,* Vol. 31 [5], 829-39, 1988.

Gilbreth, F. *Motion Study.* New York: D. van Nostrand, 1911.

Grieco, A. Sitting posture: An old problem and a new one. *Ergonomics,* Vol. 29 [3], 345-62, 1986.

Hamilton, N. A mathematical model for the reduction of neck tension. *Proceedings of the Human Factors Ergonomics Society,* 1063-67, 1986.

Hammer, W. Safe access to farm tractors and trailers. *J. Agricultural Engineering Research,* Vol. 50, 219-37, 1991.

Harrison, J. Maximizing human power output by suitable selection of motion cycle and load. *Human Factors,* Vol. 12 [3], 315-29, 1970.

Hassan, M. and Block, S. A study of simultaneous positioning. *J. of Industrial Engineering,* Vol. 18 [12], 682-88, 1967.

Heath, L. *The Grey Goose Wing.* Greenwich, Conn.: New York Graphic Society, 1971.

Helander, M. Safety hazards and motivation for safe work in the construction industry. *Int. J. of Industrial Ergonomics,* Vol. 8, 205-23, 1991.

Helander, M. and Little, S. Preferred settings in chair adjustments. *Proceedings of the Human Factors and Ergonomics Society,* 448-52, 1993.

Hinnen, P. and Konz, S. Fatigue mats. In *Advances in Industrial Ergonomics and Safety VI,* Aghazadeh, F. (ed.). London: Taylor and Francis, 1994.

Hoffman, E. A comparison of hand and foot movement times. *Ergonomics,* Vol. 34 [4], 397-406, 1991a.

Hoffman, E. Capture of moving targets: A modification of Fitts' law. *Ergonomics,* Vol. 34 [2], 211-20, 1991b.

Hoffman, E. and Gan, K-C. Directional ballistic movement with transported mass. *Ergonomics,* Vol. 31 [5], 841-56, 1988.

Hopkins, S. Elusive factor in falls; the shoe sole. *National Safety News,* Vol. 94 [5], 34-39, November 1966.

James, D. Rubber and plastic in shoes and flooring. *Ergonomics,* Vol. 26 [1], 83-99, 1983.

Jaschinski-Kruza, W. On the preferred viewing distance to screen and document. *Ergonomics,* Vol. 33 [8], 1055-63, 1990.

Kaleps, I., Clauser, C., Young, J., Chandler, R., Zehner, G., and McConville, J. Investigation into the mass distribution properties of the human body and its segments. *Ergonomics,* Vol. 12, 1225-37, 1984.

Konz, S. Design of workstations. *J. of Industrial Engineering,* Vol. 18, 413-23, July 1967.

Konz, S. *Work Design: Industrial Ergonomics,* 2nd ed. Columbus, OH: Publishing Horizons, 1983.

Konz, S. and Goel, S. The shape of the normal work area in the horizontal plane. *AIIE Transactions,* Vol. 1 [4], 359-70, December 1969.

Konz, S., Jeans, C., and Rathore, R. Arm motions in the horizontal plane. *American Inst. of Industrial Engineers Transactions,* Vol. 1 [4], 359-70, December 1969.

Konz, S. and Rode, V. The control effect of small weights on hand-arm movements in the horizontal plane. *AIIE Transactions,* Vol. 2, 228-33, September 1972.

Konz, S., Rys, M., and Harris, C. The effect of standing barefoot on a hard tile surface. *Proceedings of 10th Int. Ergonomics Association.* London: Taylor and Francis, 1988.

Konz, S., Wadhera, N., Sathaye, S., and Chawla, S. Human factors considerations for a combined brake–accelerator pedal. *Ergonomics*, Vol. 14 [2], 279-92, 1971.

Kroemer, K. Horizontal push and pull forces. *Applied Ergonomics*, Vol. 5 [2], 94-102, 1974.

Kroemer, K. and Hill, S. Preferred line of sight angle. *Ergonomics*, Vol. 29 [9], 1129-34, 1986.

Langolf, G., Chaffin, D., and Foulke, J. An investigation of Fitts' law using a wide range of movement amplitudes. *J. of Motor Behavior*, Vol. 8 [2], 118-28, 1976.

Leamon, T. The reduction of slip and fall injuries: Part I—Guidelines for the practitioner and Part II—The scientific basis (knowledge base) for the guide. *Int. J. of Industrial Ergonomics*, Vol. 10, 23-34, 1992.

Magnusson, M. and Ortengren, R. Investigation of optimal table height and surface angle in meatcutting. *Applied Ergonomics*, Vol. 18 [2], 146-52, 1987.

Maki, B. and Fernie, G. Impact attenuation of floor coverings in simulated falling accidents. *Applied Ergonomics*, Vol. 21 [2], 107-14, 1990.

Mandal, A. The correct height of school furniture. *Human Factors*, Vol. 24, 257-69, 1982.

Manning, D., Ayers, I., Jones, C., Bruce, M., and Cohen, K. The incidence of underfoot accidents during 1985 in a working population of 10,000 Merseyside people. *J. Occupational Accidents*, Vol. 10, 121-30, 1988.

Maynard, H. Workplace layouts that save time, effort and money. *Iron Age*, Vol. 134, December 1934.

McCullagh, J. (ed.). *Pedal Power*. Emmaus, Penn.: Rodale Press, 1977.

Miller, J., Lehto, M., and Rhoades, T. Prediction of slip resistance in climbing systems. *Int. J. of Industrial Ergonomics*, Vol. 7, 287-301, 1991.

Mortimer, R., Segel, L., Dugoff, H., Campbell, J., Jorgenson, C., and Murphy, R. *Brake Force Requirement Study*. National Highway Safety Bureau Final Report FH-11-6952, April 1970.

Nichols, D. and Amrine, H. A physiological appraisal of selected principles of motion economy. *J. of Industrial Engineering*, Vol. 10, 373-8, September 1959.

Nicholson, A. and David, G. Slipping, tripping and falling accidents to delivery drivers. *Ergonomics*, Vol. 28 [7], 977-91, 1985.

Opila, K., Wagner, S., Schiowitz, S., and Chen, J. Postural alignment in barefoot and high-heeled stance. *Spine*, Vol. 13 [5], 542-47, 1988.

Putz-Anderson, V. (ed.). *Cumulative Trauma Disorders*. London: Taylor and Francis, 1988.

Ramsey, J. and Senneck, C. Anti-slip studs for safety footwear. *Applied Ergonomics*, Vol. 3 [4], 219-33, 1972.

Raouf, A. and Arora, S. Effect of informational load, index of difficulty, direction and plane angles of discrete moves in a combined manual and decision task. *Int. J. of Production Research*, Vol. 18 [1], 117-28, 1980.

Raouf, A. and Tsuchiya, K. A study of simultaneous hand motions on a horizontal plane for unequal task difficulty and unequal angles. In *Trends in Ergonomics III*, Karwowski, W. (ed.). Amsterdam, Elsevier, 1986.

Rodgers, S. *Working with Backache*. Fairport, N. Y.: Perinton Press, 1984.

Rys, M. and Konz, S. Standing work: Carpet vs concrete. *Proceedings of the Human Factors Society*, 522-26, 1988.

Rys, M. and Konz, S. Adult foot dimensions. In *Advances in Industrial Ergonomics and Safety I*, Mital, A. (ed.). London: Taylor and Francis, 1989.

Sakakibara, H., Miyao, M., Kondo, T., Akagawa, T., and Kobayashi, F. Relation between overhead work and complaints of pear and apple orchard workers. *Ergonomics*, Vol. 30 [5], 805-15, 1987.

Salvendy, G. and Pilitsis, J. Improvement in physiological performance as a function of practice. *International J. of Production Research*, Vol. 12 [4], 519-31, 1974.

Sauter, S., Chapman, L., Knutson, S., and Anderson, H. Case example of wrist trauma in keyboard use. *Applied Ergonomics*, Vol. 18 [3], 183-86, 1987.

Schmidtke, H. and Stier, F. An experimental evaluation of the validity of predetermined elemental time systems. *J. of Industrial Engineering*, Vol. 12, 182-204, 1961.

Seating in industry. *Applied Ergonomics*, Vol. 1 [3], 159-65, 1970.

Shannon, C. A mathematical theory of communication. *Bell System Technical J.*, Vol. 27, 379-423 and 623-55, 1948.

Shephard, R. *Alive Man*. Springfield, Ill.: C. T. Thomas, 1972.

Squires, P. The Shape of the Normal Work Area. Report 275, Navy Department, Bureau of Medicine and Surgery, Medical Research Laboratory. New London, Conn., 1956.

Templer, J. *The Staircase: Studies of Hazards, Falls and Safer Design*. Cambridge, Mass.: MIT Press, 1992.

Vos, H. Physical workload in different body postures while working near to or below ground level. *Ergonomics*, Vol. 16 [6], 817-28, 1973.

Ward, R., Danziger, F., Bonica, J., Allen, G., and Tolas, A. Cardiovascular effects of change of posture. *Aerospace Medicine* (now *Aviation Space and Environmental Medicine*), Vol. 37, 257-59, March 1966.

Weight, Height and Selected Body Dimensions of Adults, United States: 1960-62. Public Health Service Publication 1000-11-8. Washington, D. C., 1965.

White, C., Eason, R., and Bartlett, N. Latency and duration of eye movements in the horizontal plane. *J. of the Optical Society of America*, Vol. 52 [2], 210-13, 1962.

Whitt, F. and Wilson, D. *Bicycling Science*. Cambridge, Mass.: MIT Press, 1982.

Wick, J. and DeWeese, R. Validation of ergonomics improvements to a shipping workstation. *Proceedings of the Human Factors and Ergonomics Society*, 808-11, 1993.

Wiker, S., Langolf, G., and Chaffin, D. Arm posture and human movement capability. *Human Factors*, Vol. 31 [4], 421-41, 1989.

Wilson, S. Bicycle technology. *Scientific American*, Vol. 228, 81-91, March 1973.

16 CUMULATIVE TRAUMA

OVERVIEW

Cumulative trauma affects the muscles, joints, and ligaments over a period of months and years. The three major areas affected are the hand/wrist, the shoulder/neck, and the lower back. Reduce these problems with engineering and administrative approaches.

CHAPTER CONTENTS

1 General Comments

2 Hand/Wrist

3 Shoulder/Neck/Elbow

4 Back

KEY CONCEPTS

administrative solutions
automation
carpal tunnel syndrome (CTS)
cross-trained
cumulative trauma disorders
    (CTD)
damaging wrist motion
disc
duration: short, moderate, long
force
industrial athletes
job enlargement

job rotation
joint deviation
leg-length discrepancy
ligaments
mechanization
muscles
nerves
neutral position
one-sided work
pain/impairment/disability/
    compensation
psychosocial problems

recovery/work ratio
repair
repetitive
risk factors
safety
sit–stand workstation
tendons
30 s rule
toxicology
work glasses
working rest

1 GENERAL COMMENTS

1.1 Problem

During the years from 18 to 64, more people are disabled from musculoskeletal problems than any other category of disorder (Putz-Anderson, 1988). The rate has been rising; cumulative trauma disorders (CTDs) were 21% of work-related illnesses in 1982 and 56% in 1990 (Rigdon, 1992).

Safety concerns are for short-term (time frame of seconds) effects of physical agents on the body. An example is cutting off a finger in a punch press. See Chapter 25, Safety. **Toxicology** generally deals with long-term (time frame of years, decades) effects of chemicals on body organs. An example would be exposure to acetone for 10 years, causing damage to the central nervous system. See Chapter 24, Toxicology. **Cumulative trauma disorders,** this chapter, concern intermediate-term (months, years) effects of body activity upon the nerves, muscles, joints, and ligaments. An example would be back pain due to lifting.

Cumulative trauma disorders also are called occupational overuse syndrome or repetitive strain. The cumulative trauma gradually wears away at the body. Symptoms appear sooner for people with weak bodies and later for people with strong bodies but, with enough trauma, even the strong eventually fail. Thus, although it is good to have a strong body, the emphasis in this chapter is on reducing the trauma. That is, the discussion deals with the ergonomic approach of making the job fit the person instead of the selection approach of making the person fit the job. The ergonomic approach benefits everyone—although the need will not be as evident for those with strong bodies.

1.2 Anatomy

Ergonomists primarily are concerned with **muscles** that control movement of various bones. **Tendons** connect muscles to bones. **Ligaments** (strong ropelike fibers) connect bones together. **Nerves** supply the communication within the body. Motor nerves (efferent signals) bring messages from the brain to the muscles. Sensory nerves (afferent signals) bring messages from the body (muscles, pain transducers, pressure transducers, etc.) to the brain. Autonomic nerves control various functions, such as sweat production. Arteries supply nutrients, oxygen, and hormones to the muscles; veins remove waste products. The interchange takes place in the capillaries.

Figure 16.1 shows the bones of the hand and arm. See Figures 16.2 through 16.5 for sketches showing control of the wrist and fingers. Table 16.1 gives the functions of the forearm muscles.

FIGURE 16.1   Bones of the upper extremity (Putz-Anderson, 1988). Ligaments (strong ropelike fibers) connect bones together to form a joint. Joints make a compromise between stability and mobility. The shoulder joint emphasizes mobility while the ankle joint emphasizes stability. Sprains are a tearing of the ligament—within itself or from the bone. Where ligaments are subject to friction, a lubricating device called a *bursa* (a small, flat, fluid-filled sac lined with synovial membrane—the body has about 200 of them) shields the structure from rubbing against the bone. An inflamed bursa is called *bursitis.*

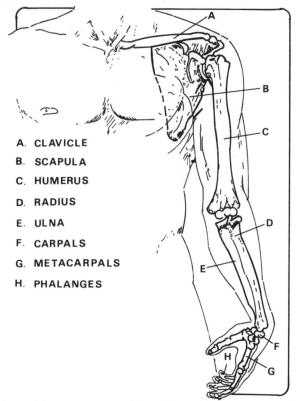

A. CLAVICLE
B. SCAPULA
C. HUMERUS
D. RADIUS
E. ULNA
F. CARPALS
G. METACARPALS
H. PHALANGES

Source: V. Putz-Anderson, (ed.). *Cumulative Trauma Disorders.* Copyright © 1988 by Taylor and Francis, London. Reprinted with permission.

See Figure 18.14 for parts of the fingers. For hand anthropometry, see Tables 18.4 and 18.5. See Figure 5.2 for hand movement terminology. See Section 2 of this chapter for comments on carpal tunnel syndrome.

Figure 16.6 shows how the bone and muscle are connected. Strains (tearing or severe stretching of muscle or tendon fibers) and sprains (tearing of ligaments) usually result from a single act rather than repetitive activity.

Figure 16.7 shows the shoulder.

Figure 16.8 shows an overview of the back. Figure 16.9 shows how the vertebrae and discs are held together with ligaments. Figure 16.10 shows how the rear ligament in the back is quite narrow in the lumbar region—the typical location for back injuries. Figure 16.11 is a detail of a spine segment. Figure 16.12 shows a detail of a **disc,** which is the material between vertebrae. Note the ringlike structure, which gives great strength. Figure 16.13 shows a side

FIGURE 16.2   Right palm view. Three flexor muscles flex the wrist on the palm side (Luttgens, et al., 1992). Of the 19 muscles of the fingers and thumb, the 10 entirely within the hand are called intrinsic muscles; the 9 located outside the hand (but with tendon attachments on the thumb or fingers) are called extrinsic.

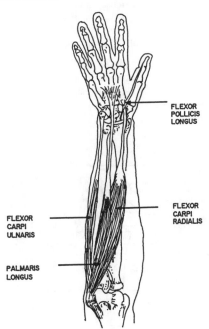

FIGURE 16.3   Back of right hand (anterior) view. Three extensor muscles extend the wrist on the back side (Luttgens et al., 1992). Radial flexion of the wrist (abduction) is produced by the flexor carpi radialis, extensor carpi radialis longus, and the abductor pollicis longus of the thumb. Ulnar flexion (adduction) is produced by the extensor and flexor carpi ulnaris.

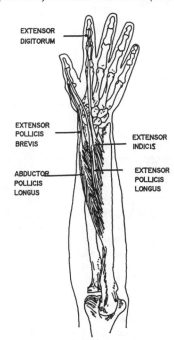

FIGURE 16.4   Right palm view. Flexing (closing) the fingers is done with the deep muscles of the forearm (Luttgens et al., 1992). Flexing (closing) muscles have over twice the strength of extensor (opening) muscles.

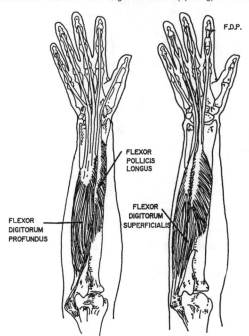

FIGURE 16.5   Back of right hand (anterior) view. Extending (opening) the fingers is done with two sets of muscles (Luttgens et al., 1992).

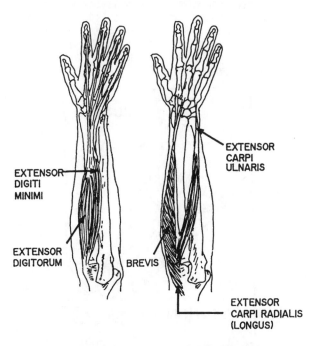

TABLE 16.1   Forearm muscle functions.

WRIST: Rotate the radius on the ulna:	WRIST: Extend the hand:	DIGITS: Extend the digits (except thumb):
▪ pronator teres ▪ pronator quadratus ▪ supinator	▪ extensor carpi radialis longus ▪ extensor carpi radialis brevis ▪ extensor carpi ulnaris	▪ extensor digitorum ▪ extensor indicis ▪ extensor digiti minimi
WRIST: Flex the hand:	DIGITS: Flex the digits:	DIGITS: Extend the thumb:
▪ flexor carpi radialis ▪ flexor carpi ulnaris ▪ palmaris longus	▪ flexor digitorum superficialis ▪ flexor digitorum profundus ▪ flexor pollicis longus	▪ adductor pollicis longus ▪ extensor pollicis brevis ▪ extensor pollicis longus

FIGURE 16.6   Tendons connect muscles to bones (Putz-Anderson, 1988). Synovial sheaths, containing a lubricant called synovial fluid, often surround the tough, ropelike tendon. A tendon in a sheath is like a wire in a soda straw. *Strain* is tearing apart of tendon fibers (akin to fraying of a rope). *Tendinitis* (also called tendonitis) is inflammation of the tendon. *Tenosynovitis* (also called tendosynovitis, tendovaginitis, and peritendinitis) results when the sheath produces excessive synovial fluid,making the sheath swollen and painful. Tendons have virtually no blood supply and thus are incapable of self-repair. An inflammatory reaction may begin in the damaged tendon area. The inflammatory tissue carries with it a blood supply for tendon repair but also a nerve supply which can cause severe pain (Rowe, 1985).

If the tendon sheath constricts (stenosis), it is called *stenosing tenosynovitis*. The most common example is DeQuervsain's disease, affecting the long abductor and short extensor tendons on the side of the wrist and base of the thumb. The person is unable to bridge a wide span and may have difficulties with a combination firm grip and forearm roll (wringing motion).

If the tendon becomes locked in a swollen sheath, the movement will be snappy or jerky. In a finger this is called *trigger finger*. This often occurs with extended use of sharp-edged tools.

A sheath that swells with fluid may form a ganglionic cyst (a bump).

The unsheathed tendons in the elbow can get tendinitis. The humerus bone of the upper arm splits into two condyles at the elbow; the epicondyle is the outside protrusion on each condyle. Golfer's elbow (also called medial epicondylitis) is on the inside (medial side) of the elbow and often occurs when the forearm is rotated while the wrist is bent. Tennis elbow (also called pitcher's elbow, bowler's elbow, and lateral epicondylitis) is on the outside (lateral side) of the elbow and often occurs when the arm is used for impact or jerky throwing motions.

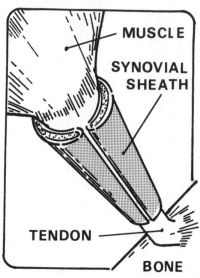

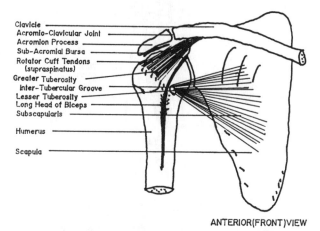

ANTERIOR (FRONT) VIEW

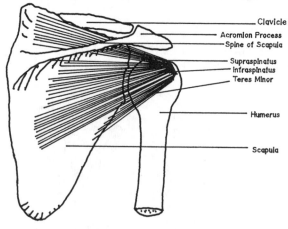

POSTERIOR (REAR) VIEW

FIGURE 16.7   Right shoulder views (Rowe, 1985) show that the shoulder (designed for mobility over stability) is stabilized almost exclusively by tendons rather than by the mechanical fit of opposing bones or snug ligaments.

Perhaps most important is the cojoined sheet of three tendons (supraspinatus, infraspinatus, and teres minor) collectively known as the rotator cuff.

Source: V. Putz-Anderson, (ed.). *Cumulative Trauma Disorders.* Copyright © 1988 by Taylor and Francis, London. Reprinted with permission.

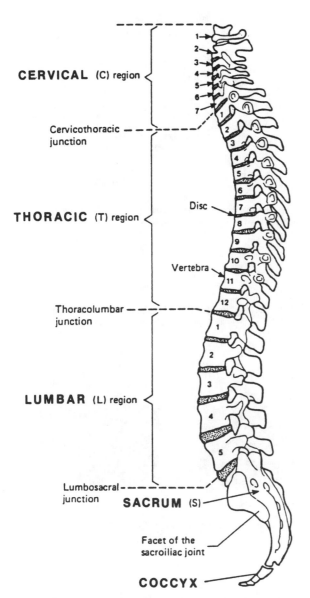

CERVICAL (C) region

Cervicothoracic junction

THORACIC (T) region

Disc

Vertebra

Thoracolumbar junction

LUMBAR (L) region

Lumbosacral junction

SACRUM (S)

Facet of the sacroiliac joint

COCCYX

FIGURE 16.8   The low back (Lumbar 4 and Lumbar 5, abbreviated L4 and L5) is the problem area. Note how these discs angle downward, instead of being horizontal. The weight of the torso therefore tends to push these discs forward. The disc pressure is higher when sitting than standing (50–100% greater), especially when sitting with a slumped back. The pressure can be reduced by using chairs with armrests (which can support some of the torso weight).

Due to cumulative compressive loading on the discs, body stature shrinks about 1.1% each day (Kramer and Gritz, 1980), but sleep restores the shrinkage. Krag et al. (1990) found that the change is relatively rapid: 26% of the 8-h loss occurred in the first hour upright and 41% of the 4-h recovery occurred in the first hour recumbent.

FIGURE 16.9   Ligaments hold the vertebrae and discs together (Ring, 1981).

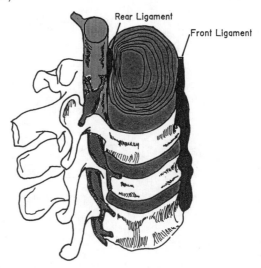

Rear Ligament

Front Ligament

Source: L. Ring, *Facts on Backs.* Copyright © 1981 by Institute Press, Loganville, GA. Reprinted with permission of L. Ring.

FIGURE 16.10   The rear ligament narrows as it descends (Ring, 1981). In the lumbar region, the ligament is only half its original width, exposing unprotected discs on either side.

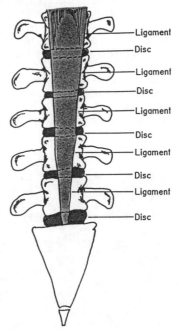

Ligament
Disc
Ligament
Disc
Ligament
Disc
Ligament
Disc
Ligament
Disc

Source: L. Ring, *Facts on Backs.* Copyright © 1981 by Institute Press, Loganville, GA. Reprinted with permission of L. Ring.

view including the nerve root. Figure 16.14 shows a top view of a prolapsed (slipped) disc squeezing a nerve.

Note that not all bodies are perfect. Figure 16.15 shows the effect of **leg-length discrepancy** (one leg longer than the other) for people with and without back pain.

Although the back is very complex, mathematical models of the spine have been developed. The initial models were two-dimensional, but three-dimensional models have been developed which predict lumbar compression as well as shear and torsional forces during various activities (Marras and Sommerich, 1991a, 1991b).

FIGURE 16.11 Spine segment detail shows how discs separate bony segments (Ring, 1981).

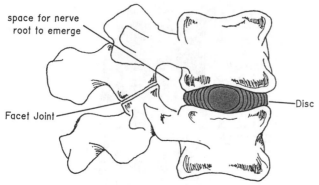

Source: L. Ring, *Facts on Backs.* Copyright © 1981 by Institute Press, Loganville, GA. Reprinted with permission of L. Ring.

FIGURE 16.12 Discs have a ringlike structure surrounding a sac of fluid (Ring, 1981). In the young adult, the discs are so strong that when violence is applied to the vertebral column (e.g., parachutist whose chute didn't open or someone who falls from a building) the bones usually give way first, providing the discs are healthy (Ring, 1981).

Discs fail because they wear out. "Normal daily activities ultimately exact their toll in the form of microscopic stretchings, tearings and ravelings of the casing fibers. The increments of microdamage become cumulative until, finally, the casing is no longer capable of containing the packing material. Usually slow leaks of the fluid of the gel packing occur from time to time and the disc begins to narrow and gradually go flat. Less commonly, the weakened casing may bulge or a sudden blowout may occur" (Rowe, 1985, p. 140).

The two load-bearing structures (the facets, excluding the synovial membrane, and the intervertebral discs) do not have nociceptors, the peripheral nerves concerned with the reception of pain. *Thus they can be injured without pain* (Garg, 1992), and low-back pain may be the culmination of a series of point-in-time painless injuries.

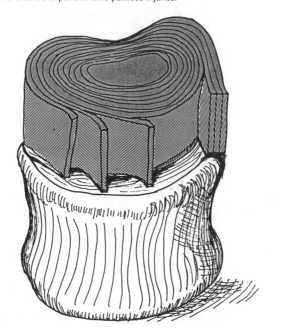

Source: L. Ring, *Facts on Backs.* Copyright © 1981 by Institute Press, Loganville, GA. Reprinted with permission of L. Ring.

FIGURE 16.13 Disc degeneration reduces the space available for the nerve to emerge, possibly causing pinching of the nerve (Ring, 1981).

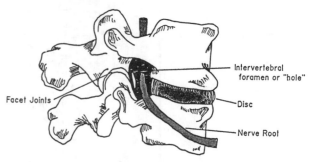

Source: L. Ring, *Facts on Backs.* Copyright © 1991 by Institute Press, Loganville, GA. Reprinted with permission of L. Ring.

FIGURE 16.14 Prolapsed (herniated, ruptured) discs pinch the nerve (Ring, 1981). The public incorrectly calls this a "slipped" disc.

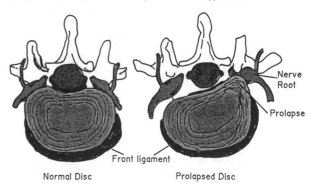

Source: L. Ring, *Facts on Backs.* Copyright © 1991 by Institute Press, Loganville, GA. Reprinted with permission of L. Ring.

FIGURE 16.15 Leg-length discrepancy (Contreras, Rys, and Konz, 1993) affects back pain.

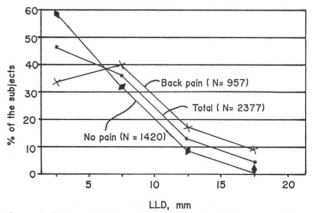

Source: R. Contreras, M. Rys, and S. Konz. "Leg Length Discrepancy," *The Ergonomics of Manual Work,* W. Marras, W. Karwowski, J. Smith, and L. Pacholski (eds). Copyright © 1993 Taylor and Francis, London. Reprinted with permission.

1.3 Risk Factors Cumulative trauma disorders are a function of many risk factors, both occupational and nonoccupational.

1.3.1 Risk factor formula

The primary occupational factors are

$$CTD = A\ (RD)^a\ B\ (JD)^b\ C\ (FO)^c$$

where CTD = Average risk of a cumulative trauma disorder for a specific part of the body

RD = Repetition/duration of joint use factor

$\quad = D\ (FH)^d\ E\ (HOURS)^e\ F\ (DAYS)^f$
$\quad\quad G\ (YEARS)^g$

FH = Frequency/hour

$HOURS$ = Hours/day

$DAYS$ = Days/year

$YEARS$ = Years/work lifetime

JD = Joint deviation factor

FO = Force resisted by muscle factor

A, B, C, D, E, F, G = Constants

a, b, c, d, e, f, g = Exponents

Naturally the constants and exponents would be different for different problems (wrist problems, neck problems, back problems, etc.) Unfortunately, at the present time, no one knows the values of any of the constants or exponents for *any* of the problems! We don't know whether the formula should be multiplicative or additive. In addition, the formula assumes no interactions. Yet interactions are known to exist. For example, it is known that force at extreme deviations is worse for the body than force at the neutral position. Regarding interaction between force and repetition, Silverstein et al. (1986) reported that highly repetitive jobs had a 2.8 odds ratio of injury compared to low repetition jobs. If the job was highly repetitive and required high force, the odds ratio jumped to 30. In addition, the formula would give average values for a population, not values for specific individuals.

Thus, at present, engineers should reduce the various risk factors (e.g., reduce force from 4 kg to 2 kg) while realizing that an exact prediction of the reduction of risk is not available. (Use of tobacco and cancer are an analogous situation. We are not able to predict the probability of lung cancer for a 22-year-old female, Jane Doe, who smokes 12 cigarettes/day of brand X. However, it is still worthwhile for her to reduce the number of cigarettes smoked.)

The three main occupational **risk factors** are repetition/duration, joint deviation, and force.

1.3.2 Repetition/duration

In general, a job is considered **repetitive** if the basic (fundamental) cycle time is less than 30 s (**30 s rule**). A total cycle may take longer. For example, total cycle time to pack a box with product may take 60 s. But if this requires placing 20 items in the box in 10 fundamental cycles of 4 s + 20 s to obtain and dispose of the box, then this is a repetitive task.

The 30 s rule deals primarily with hand/wrist motions; back or shoulder movements (i.e., more stressful movements) might be repetitive with a several minute interval.

However, if the task is only done for 15 minutes/shift, then there is relatively little risk of cumulative trauma because the total duration is short. **Short duration** will be defined as < 1 h/day, **moderate** as 1 to 2 h, and **long** as > 2 h.

In addition, within limits, the body is self-repairing. (**Repair** is used in the broad sense of furnishing fuel and oxygen and removing waste as well as repair of the cells themselves.) The more time the body has to repair itself the better. In addition, some repair can take place even within a motion (but not with static loading). For example, if the hand is moved in and out from a bin, the muscles for moving out can rest while the muscles for moving in are used and vice-versa. A complication of repair is that the body may not return to the original state but to a more proficient state (training).

If the work rate is not high, the repair can be done concurrently with the work (see Box 16.1) and the work can be done continuously all day.

During working rest, a joint can be repaired while it is idle and other parts of the body are working. For example, the left wrist joint could be repaired while the person used the right wrist. The fingers could rest from keying while the brain is making a decision. In addition, most jobs have "microbreaks" of 1–30 s due to the actions of other operators or the machine. In machine-paced work, the operator has less control over when the microbreak occurs and thus less control over when rest/recovery occurs.

In addition, the joint can be repaired during non-working rest (lunch breaks, coffee breaks, sleeping at home, etc.).

See Box 16.1 for comments on the recovery/work ratio. Note that repair and recovery is faster for muscles than for tendons, ligaments, and nerves.

Duration over the months and years is significant because the body will not always be able to repair itself completely and the insult to the flesh becomes cumulative. In addition, the repair process slows down with age. Older people heal more slowly than younger people.

Repetitions have less harmful effects if the joint (muscles, tendons, ligaments) is trained. Athletes go through preseason conditioning drills. Workers

---

BOX 16.1 *Recovery/work ratio*

Both the amount of recovery and the distribution are important.

Amount. Repair time can be calculated as a ratio of exposure time, or a **recovery/work ratio**. For example, in a 24-h day, if a specific joint on a person is used for 8 h, then there are 16 h available for recovery. That is, there is 16/8 = 2 h of recovery for 1 h of exposure. If the joint is used for only 4 h/day (say by using the other arm also), then there is 20/4 = 5 h of recovery for 1 h of exposure. Note that overtime can cause considerable CTD problems: 12 h of work gives 12/12 = 1 h of recovery for 1 h of exposure. Perhaps the extra work can be given to a part-time worker.

Weekends, holidays, and vacations increase the time available for recovery.

Guideline: Increase recovery time.

Distribution. The above assumes that each minute of repair/recovery time is equally effective (i.e., recovery rate vs. time is a horizontal line). But recovery rate declines exponentially with time. That is, the amount of recovery for minutes 5–10 is less than for minutes 0–5. Thus, for the same total length of break, many small breaks are better than occasional longer breaks. A single break of 15 minutes is not as effective as 3 breaks of 5 minutes. An 8-h break is not as effective as 4 breaks of 2 h. Thus it is better to rotate jobs within days rather than between days. For example, have Joe work on job A in the morning and have Pete work on it in the afternoon. This is better than Joe working for 8 h on job A on Monday and then on job B for 8 h on Tuesday. Waersted and Westgaard (1991) suggest that rotation probably should be after an hour or two rather than after four hours.

Guideline: Frequent small breaks are better than occasional long breaks.

---

should be considered **industrial athletes** and should also have conditioning (work hardening). This can be warmup exercises before work. People who have lost conditioning (vacation, layoff, injury) need to take precautions before going full speed.

The above discussion assumed the body was moving and the general thrust was to reduce the extent and frequency movement. However, CTD also can occur when the body moves too little. Static loading minimizes nutrient and waste exchange in the muscles, eventually causing pain. Thus CTD can come from too much or too little activity. The human body is meant for movement—but not too much!

1.3.3   Joint deviation
Ideally, the joint should operate at the **neutral position** (zero **joint deviation**). Since different joints have different ranges of motion, express the deviation in relative terms (percent of maximum deviation) as well as in absolute terms (degrees). Posture affects joint deviation, which then affects the internal force required to counteract the external force.

1.3.4 Force
Ideally, the internal force on the joint should be low. The observed **force** on the joint typically is an external force multiplied by some lever arm (i.e., we are really talking about a torque). Reduce not only the magnitude of the external force and its lever arm but also the length of time the force is applied.

Torque exertion capability varies little with normal postures, but extreme postures (e.g., lying on the stomach, leaning sideways from a ladder) cause large differences in capabilities (Mital and Channaveeraiah, 1988).

1.3.5   Other occupational risk factors
Cold exposure is a risk factor for CTD because it causes vasoconstriction, thereby reducing blood flow. Reduced blood flow means a reduced supply of nutrients and a slower repair process. See Section 4 of Chapter 23.

Vibration probably is a risk factor due to interference with blood flow as well as mechanical trauma to the body. Handtool vibration also increases grip forces. See Box 18.2.

1.3.6 Nonoccupational risk factors
Nonoccupational risk factors can be from trauma outside of work, such as sewing, musical instrument keying, sports (e.g., tennis elbow, golfer's elbow, lifting weights, bowling), or from a nonperfect body. The lack of perfection can be anatomical (e.g., small carpal tunnel, weak back muscles) or physiological (diabetes, insufficient hormones, use of oral contraceptives). The body also can be nonperfect due to an injury.

1.4 Solutions The ergonomic solutions that follow are divided into engineering and administrative. Administrative procedures should be considered temporary measures, to be superseded by permanent engineering solutions. Medical treatment after the injury is covered very briefly.

Some of the solutions may involve equipment such as adjustable workstations, wristrests, footrests, supports, handtools, and so forth. Two of the many vendors are AliMed (1-800-225-2610) and North Coast Medical (1-800-821-9319).

1.4.1 Engineering solutions

First, analyze the job with videotape. See Box 16.2.

One possible solution is **automation**—that is, eliminate the person. If there is no person, there is no possible injury. Use a robot to load boxes, use a palletizer to place cartons on a pallet, use bar coding to eliminate keying.

A second possibility is to reduce the number of cycles or reduce the difficulty of a cycle. Two approaches are mechanization and job enlargement.

Mechanization means that a machine does part of the job but there still is an operator present. For example, on a packaging operation, each box might need a label applied. The present method might be to use a pinch grip to peel a label off a sticky-back tape. A mechanized method might use a machine to peel the label off the tape and present it to the operator—eliminating the pinch grips of peeling the label.

Job enlargement (the opposite of job simplification) increases the total job content for each person. The same motions are done, but by a larger number of people. If the job takes 4 min instead of 2 min, the repetitive motions per person are reduced.

Minimizing joint deviation (from the neutral posture) is a third engineering approach. See examples below in Sections 2, 3, and 4.

Minimizing force duration and amount is a fourth engineering approach. See examples below in Sections 2, 3, and 4.

Although most cumulative trauma problems concern the hand/wrist, shoulder/neck, and back, there are some problems with the leg. For example, carpetlayers use a device actuated by the knee to stretch carpets. Carpetlayers represent .06% of the workforce but submit 6% of the knee injury claims. Normalizing the data by using bodyweight (*BW*), impact forces on the knee are 1.3 *BW* for walking, 3 *BW* for running, and 4 *BW* when using the knee-kicker (Village et al., 1993).

BOX 16.2 *Videotaping jobs*

Videotaping jobs is useful for methods analysis and training. (See Chapter 27 for comments on use of videotaping for time standards.) Although most cameras take satisfactory pictures in normal factory lighting, a low light capability is handy as is the ability to shoot a high number of frames/second.

The following list provides some shooting tips:

- Study a variety of operators (if possible). Operators usually will have small variations in technique; these can be detected on tape but are difficult to detect by direct observation. For example, how are items oriented? Is the wrist deviation the same? Is the sequence of steps modified?

- Plan the location of camera and subject ahead of time. Use a tripod. Multiple views are best. Consider some combination of a front view, a side view, a back view, floor level views, "stepladder" views (i.e., a partial plan view), overall views, and closeup views. Begin the scene with a full view (far shot, wide-angle view) and then zoom in as desired.

- Take lots of cycles for your stock tape. Each scene should have several cycles. You can always edit, but it is expensive to go back and shoot more tape.

- If later there might be a question of when the tape was made, a camera feature which continuously displays the date on the screen is desirable.

- Use audio as a notepad while filming, identifying what you are shooting and what seems interesting. If the tape is used later for a management presentation or training, you can dub in a voice reading a script.

- Have both an engineer and the operator analyze the stock tape. The VCR should have a freeze-frame and a single-frame advance. The operator can point out to the engineer why some things are done, perhaps with audio dubbing. The tape also is a way to show the operator some problems and how other operators do the task.

1.4.2 Administrative solutions

Administrative solutions seek to reduce exposure and increase worker ability to endure the stress. Examples of reducing exposure are job rotation and part-time workers; examples of increasing ability to endure the stress are exercise, stress reduction, and supports.

Job rotation   In **job rotation,** people rotate jobs periodically within the shift. The concept is of **working rest** — the specific part of the body rests but the person is still working. The concept works best when the alternate work is not similar. For example, a person picking boxes off a conveyor could rotate with a person tallying the paperwork. A person using a right-handed workstation could rotate to a left-handed workstation (this method assumes that not all workstations emphasize the use of one hand). A person keying could shift with a person answering phone inquiries. Less satisfactory would be a shift from packing large bags of snack foods to small bags of snack foods or shifting from keying reports to keying forms. As a practical matter, however, it is difficult to rotate jobs between different supervisors.

In addition to the stress-reduction benefits, job rotation has some managerial advantages in that people are **cross-trained** (able to do more than one job), which allows more flexible scheduling, and there is a perceived fairness because everyone shares good and bad jobs.

Job rotation requires broad job descriptions rather than narrow ones, for example, "office worker" rather than "secretary 1," "secretary 2," "word processing operator," and so forth.

During rest, recovery from fatigue declines exponentially. Therefore, frequent short breaks are better than occasional long breaks. The rotation should be within the day rather than between days. Rotation every hour is better than rotation every four hours. See Box 16.1.

Part-time workers   Part-time workers can be used when there is no other job with which to rotate. For example, keying could be done by a person for only 4 h/day by hiring people to work only 4 h/day. The assumption is that the part-time people are not stressed during off-work hours (from another job or hobby).

Other advantages of part-time workers, in addition to reducing cumulative trauma, are lower cost/hour (wage and fringes), better fit to fluctuating demand (a 25% increase in time requires each person to work 5 h/day vs. 10 h/day for a full-time worker), and (possibly) higher quality workers. Disadvantages of part-time workers are less time on the job (and thus less learning), possible moonlighting for other employers (and thus being tired when working for

you), and a high fixed cost per employee of hiring and training.

Exercise   In general, exercise can improve cardiovascular fitness (endurance), flexibility, and strength; for CTD prevention, the emphasis should be on flexibility and strength. Exercises should be tailored to the specific set of muscles, tendons, and ligaments that are stressed. Some exercises strengthen weak muscles and other exercises stretch tight muscles and ligaments. Strengthening (work hardening) is needed for people new to a specific task and for people returning to work (e.g., after vacations, injuries). Note that stress can be caused by static loads as well as dynamic loads. Exercises can be designed to improve the body for dynamic activity, to counteract the effect of static loading, and to reduce tension. If the work loads the muscles statically, the exercise should move them. If the work loads the muscles dynamically, relax and stretch them. *Pause Gymnastics* (Gore and Tasker, 1986) has an excellent set of exercises.

Stress reduction   Social factors both on and off the job can cause stress. Examples of on-the-job stressors are work load, deadlines, and interpersonal relationships. Examples of off-the-job stressors are domestic problems and financial problems. Stresses also can be due to personality (perfectionists, self-pushers, workaholics, etc.). Stresses can either aggravate localized CTDs such as carpal tunnel or cause diffuse muscle conditions (Wigley et al., 1992). Example diffuse symptoms are pain, weakness, numbness and tingling, and tissue swelling.

Some specific methods of treatment are relaxation (learning to let go), rest (especially microbreaks and working rest, in addition to rest at home), postural modification, exercises (warm-up before work, dynamic exercises to counteract static loads), and medicine to minimize sleep disturbances.

Supports   An appealing idea is to support the body. Four possibilities are available commercially: wrist braces, elbow braces, back braces (back belts, lifting belts), and lower arm supports. The wrist, elbow, and back braces are worn on the body; the lower arm support holds the lower arm and is suspended from above or attached to the chair with a swiveling mechanism.

These supports are a cumulative-trauma type of personal protective equipment. However, unlike safety types of personal protective equipment (hard hats, safety shoes, etc.), these supports may have an effect on the underlying anatomy. In general, it seems the anatomy is made weaker because the support was used. Use of these supports must therefore consider not only what happens when they are worn but also what happens when they are no longer worn.

The wrist splint is used to keep the wrist from bending. This is relatively noncontroversial off the job (especially at night). However, wearing the splint during work may not allow the muscles to recover as quickly. In addition, the splint may actually cause injury because the person moves against the resistance of the splint.

Back belts are popular but the scientific evidence tends to be against their use (Rys and Konz, 1993). They do seem to help reduce lifting stress. However, they may lead to a false sense of security (people think they are Superman when they wear a belt); they weaken the body so that injury occurs when they are not worn; and firms tend to use back belts (low-cost, temporary, personal protective devices) rather than implement permanent engineering changes.

Arm support did not provide worthwhile benefits during welding or light assembly (Jarvholm et al., 1991).

1.4.3 Medical/rehabilitation

Once a person has a CTD, medical personnel such as occupational therapists, physical therapists, and physicians enter the battle.

Physical treatments such as ice (to reduce inflammation), heat (to increase blood flow), exercise (such as strengthening and stretching), splints, and massage can be used. Consult medical personnel for specific recommendations.

Chemicals (drugs) can be used to reduce swelling, promote healing, and so forth. Consult medical personnel for specific recommendations.

If all else fails, surgery (see Figure 16.16) is a last resort. For example, for carpal tunnel syndrome, the ligament roofing the carpal tunnel is cut. Unfortunately the surgery (besides being expensive) does not always prevent the problem from recurring. Owen (1994) reports "80% of carpal tunnel syndrome surgery is highly successful." Naturally surgery can't help if the nerve has been permanently damaged.

2 HAND/WRIST

2.1 Problem   The CTD problem can occur in the tendons, the muscles, or the nerves. (See Figure 16.17.)

There is a specific problem with the hand and wrist. See Box 16.3. The hand is small and flexible because the muscles which power it are in the bulky forearm. The fingers are moved by tendons leading from the fingers through the wrist (*carpal* in Latin) tunnel to the muscles. These tendons (think of them

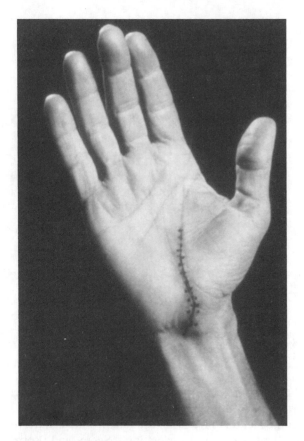

FIGURE 16.16 Surgery is a last resort. For CTS, the median nerve typically is compressed in the palm—about 2 cm from the wrist crease. Note that surgery does not have a 100% success rate.

as wires inside of a sheath) can fray on the bones; the result is called tenosynovitis.

The median nerve also goes through the carpal tunnel. If the median nerve (think of it as a cotton rope) is pinched at the carpal tunnel, the result is a compression neuropathy called **carpal tunnel syndrome (CTS).** (Note that the median nerve can be pinched at a number of locations in the arm and shoulder; thus not all pinched median nerves are CTS.) Pinching the median nerve causes numbness in the thumb and index finger. CTS tends to cause pain in the evening rather than during work.

The compressed nerve can be detected with different tests. Nonclinical tests are Phalen's test and Tinel's test. Phalen's test (wrist-flexion test) has the patient flex the wrist; if numbness or tingling occurs within 60 s, CTS is indicated. In Tinel's test (percussion test), the examiner lightly taps over the median nerve; CTS is indicated if tingling results. The most common clinical test is nerve conduction (a pinched nerve conducts more slowly; conduction latencies are longer). Johnson and Evans (1993) show that nerve conduction tests have large variances both among and within individuals and strongly question

FIGURE 16.17 Section of right wrist (looking toward fingers). Note that the shape of the median nerve conforms to the available tunnel space (Tanaka and McGlothlin, 1993).

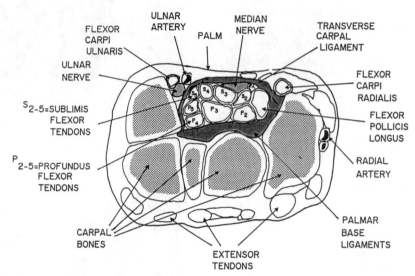

Source: S. Tanaka and J. McGlothlin, "A Conceptual Quantitative Model for Prevention of Work-Related Carpal Tunnel Syndrome (CTS)," *Int. J. of Industrial Ergonomics,* Vol. 11, pp. 181–193. Copyright © 1993 by Elsevier Science, Amsterdam, Netherlands. Reprinted with permission.

the criterion of 4.3 ms as indicating carpal tunnel syndrome. Another clinical test is vibrometry, based on the fact that fingers using a pinched median nerve are less sensitive to vibration (Neese and Konz, 1993).

When CTS occurs in both wrists it is called bilateral.

2.2 Risk Factors   Tanaka and McGlothlin (1993) proposed the following risk model for CTS which shows the complexity of the problem:

$$EL = k\ (a)F\ (b)\ R\ e^{cD}$$

where:

EL = Exposure limit

k, a, b, c = Constants and coefficients

e = Base of natural logarithm (2.71 . . .)

F = Force (internal) exerted by the finger–hand–wrist–arm complex

R = Repetition, cycles/min

D = Deviations of the wrist/max deviation, $0 \le D \le 4$.

= Flexion dev/Max flexion + External dev/Max external + Ulnar dev/Max ulnar dev + Radial dev/Max radial dev

Vibration is considered through the greater force it requires to hold a vibrating tool. At present a time factor (duration of exposure and rest/work relationship) is not included in the model. Psychosocial factors (stress) also are not included.

2.3 Solutions   Naturally, not all CTS cases are occupationally related. Some are related to an imperfect body (e.g., diabetes) or nonoccupational trauma (e.g., piano playing). In addition, it should be recognized that mental stress (both on and off the job) increases general tension and thus may increase the effects of cumulative trauma. Finally, a few malingerers may report symptoms to reap financial rewards (this is called "compensation neuroses").

Solutions will be divided into engineering and administrative.

2.3.1 Engineering   The three engineering solution approaches are repetitions/duration, joint deviation, and force.

Repetition/duration   Repetitions influence CTS more than force (Armstrong et al., 1985). A possible solution, therefore, is to reduce the lifetime use of the joint. For example, use a foot-operated control instead of a hand-operated control. Use a telephone (voice) instead of E-mail (keying). Split a 5,000 repetition/day job into 2,500 for the left wrist and 2,500 for the right instead of 5,000 on one wrist.

Joint deviation   The consensus is that joint deviation influences CTS. Schoenmarklin and Marras (1991) report that wrist CTDs are better predicted by wrist velocity and acceleration than wrist angle. Scheonmarklin and Marras (1993) give data on the dynamic capabilities of the wrist.

Guideline: Don't bend your wrist.

BOX 16.3 *Assumptions for work-related CTS (Tanaka and McGlothlin, 1993)*

1. The carpal tunnel (see Figure 16.17) is a tightly constrained space. Therefore, an increase in the content inside the tunnel, or pressure from outside, will raise intra-tunnel pressure.

2. If the pressure increase is sufficient, it will cause local venous congestion within the vascular plexus of the nerve, as well as ischemia in arterioles nourishing the nerve, leading to endoneurinal edema. In this initial stage of CTS, the patient feels numbness, tingling, and pain.

3. The edema increases the effect of the initial compression, creating a vicious circle. Fibroblasts within the nerve tissue lead to scarring.

4. Thus any disease (local or systemic) which increases intra-tunnel pressure can cause CTS.

5. Under normal conditions, when manual work is performed, synovial lubrication of the tendon sheaths is adequate for smooth gliding of the tendons. There is "reasonable and comfortable" usage.

6. In some work situations, workers may not be able to stay within the reasonable and comfortable range, and so the tendons become strained and tendon lubrication becomes inadequate.

7. Bending of the wrist reduces carpal tunnel size, increasing pressure.

8. Inadequate lubrication leads to increased friction and inflammation.

9. If the stress is reduced at this time, inflammation will subside and normal lubrication will be restored.

10. If the stress is not reduced in time, the inflammation may progress to swelling of the structures. (Tendons need a much longer time to restore normal lubrication than muscles to recover from fatigue; it also takes time for inflammation to subside.)

11. The swelling further increases pressure and the cycle continues.

12. Initially, various treatments to restore local circulation will give temporary relief. Examples are shaking the hand, administration of nonsteroidal medication or local steroid injection, or a wrist splint worn at night.

13. However, continued stress may lead to chronic tenosynovitis and permanent damage to the median nerve.

14. Carpal tunnel release surgery can be done, but its effectiveness depends on the degree of damage to the median nerve and its recovery potential.

Source: S. Tanaka and J. McGlothlin, "A Conceptual Quantitative Model for Prevention of Work-Related Carpal Tunnel Syndrome (CTS)," *Int. J. of Industrial Ergonomics*, Vol. 11, pp. 181–193. Copyright © 1993 by Elsevier Science, Amsterdam, Netherlands. Reprinted with permission.

A goal is to keep the wrist in the neutral (handshake) position. Do this by changing the job or by changing the tool.

Changing the job may simply require a change in worker posture. Standing tends to permit longer reaches than sitting. Perhaps the person's chair is too high or too low, thus resulting in an awkward wrist orientation. The work orientation may be adjusted by tilting the entire worksurface or by tilting a fixture or bin on the surface. For example, well-designed keyboards are adjustable in angle so the person can change them. Note that if a keyboard operator uses a chair with armrests and rests the elbows on the armrests while keying, this may put the wrists in a more deviated position (radial on left hand, ulnar on right). See Figure 5.2.

Changing the job may require a change in product design. For example, should a screw be slot, Phillips, or hex? How would these different types affect operator hand, elbow, shoulder, and neck orientations?

Changing the handtool angle is another possibility. (See Figures 18.21 through 18.24.) The concept is to bend the tool, not the wrist. Note that one perfect angle is unlikely. Much more likely is one best angle for each posture and direction of movement. For example, Fogelman et al. (1993) reported, for poultry deboning, that a −30° blade in a dagger grip was best for a table cut and a +30° blade held normally was best for a hanging cut. Jegerlehner (1991) reported that at Deere they reduced wrist deviations on the spindle lever of a tapper by using a ball handle on the ends of the levers.

Force Force influences CTS. Reduce the force duration and amount.

Marras (1992) says that high velocity and, in particular, acceleration increase cumulative trauma.

Drury and Wick (1984) define a **damaging wrist motion (DWM)** as a bent wrist involving a force. McCarty et al. (1991) give 1,000 DWM/h as an upper ergonomic limit. In addition, all pinch grips >8 lbs are considered dangerous. Figure 16.18 shows gripping thin objects with a cup grip rather than a pinch grip.

Can force be eliminated? Is a robot desirable? If a person performs the action, can a clamp or fixture replace the hand?

Can force be reduced? In handwriting, for example, a ballpoint pen requires about 180 g of force, pencils and fountain pens about 120 g, and felt tips about 100 g (Kao, 1979). For assembly, can the component designs be modified (e.g., use of a taper) so less force is required during assembly? Can component tolerances be modified so less force is required during assembly? Can previous processes be modified so less trimming is needed? Can the tool (pliers, scissors, clippers) or object be returned with a spring instead of requiring a muscle action? Can lubricants be used?

Can a small diameter grip on handtools be replaced with a larger diameter grip so less force is needed to prevent movement of the grip in the hand? Can the grip have a better coefficient of friction so less grip force is needed? Can the grip be designed to reduce the point pressure on the roof of the carpal tunnel? For example, a screwdriver handle could have a ball shape instead of a conventional long ellipse. See Chapter 18 for more comments on handtools. In the meatpacking industry, the worker originally held a power knife and cut the meat; in the revised method, the power knife was mounted in a fixture and the meat held with both hands and pulled through the knife. The knife also was designed so a sharp blade could be installed every hour.

Vibrating tools require more grip force than non-vibrating tools of the same size and weight (Armstrong et al., 1987). Therefore reduce vibration.

Gloves (see Box 18.1) decrease gripping force on an object; if you wish to exert the same amount of control of an object with and without gloves, you will have to exert more force with the gloves.

Note that decreased hand force requirements tend to increase accuracy of movement and thus quality.

2.3.2 Administrative solutions   See Section 1.42 and Box 16.1.

3 SHOULDER/NECK/ELBOW

3.1 Problem   Here the CTD problem seems to be primarily in the shoulder and the neck although the elbow is occasionally involved. CTD of the neck and shoulder also is called occupational cervio-brachial disorder (OCD) and upper limb disorder (ULD). Shoulder region pain ranks second only to low back and neck pain in clinical frequency (Sommerich et al., 1993). Disorders can be related to

1. tendons
   - rotator cuff tendinitis
   - calcific tendinitis
   - bicipital tendinitis
   - tendon tear
   - bursitis
2. muscular shoulder pain
3. nerve-related disorder (suprascapular nerve)
4. neurovascular disorder (thoracic outlet syndrome)

Job titles associated with low risk of shoulder/neck CTD are VDT operator, assembly operator (some jobs), and cash register operator; medium-risk job titles are garment worker, assembly operator (some), and packer; high-risk titles are welder and letter carrier (Winkel and Westgaard, 1992).

3.2 Solutions   See Box 16.4 for a procedure to prioritize upper-arm CTD investigations. See also Table 16.6 and Table 16.7. Solutions will be divided into engineering and administrative.

FIGURE 16.18   Avoid pinch grips and use cup grips (Tayyari and Emanuel, 1993).

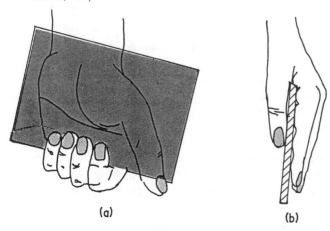

(a)   (b)

Carrying a thin item in a cupped hand (a) is less straining than holding it by a "pinch" grip between the thumb and fingers (b), which causes strain in the hand.

Source: F. Tayyari and J. Emanuel, "Carpal Tunnel Syndrome: An Ergonomics Approach to Its Prevention," *Int. J. of Industrial Ergonomics,* Vol. 11, pp. 173–179. Copyright © 1993 by Elsevier Science, Amsterdam, Netherlands. Reprinted with permission.

BOX 16.4 *Rapid upper limb assessment (RULA)*

RULA (McAtamney and Corlett, 1993) uses an additive model emphasizing posture to determine priorities for action.

Stage 1 (Observe job; score for posture, repetition, and force)

Step 1: Determine posture score (Figure 16.19) for Group A:

- upper arm
- lower arm
- wrist
- wrist twist

Step 2: Determine posture score (Figure 16.20) for Group B:

- neck
- trunk
- legs

Step 3: Determine frequency score (Table 16.2) for

- Group A
- Group B

Step 4: Determine the force (load) score (Table 16.3) for

- Group A
- Group B

Stage 2 (Calculate risk index)

Step 5: Determine the posture index (Table 16.4)

for Group A. Enter into appropriate box in form (Figure 16.21).

Step 6: Determine the posture index (Table 16.5) for Group B. Enter into appropriate box in form (Figure 16.21).

Step 7: Enter repetition and force scores into appropriate boxes for Group A and Group B.

Step 8: Calculate total score for Group A and Group B. Enter into appropriate boxes.

Step 9: Calculate grand score, using Figure 16.22.

Step 10: Take action.

Action

Level	Score	Comment
1	1-2	Posture is acceptable if not maintained or repeated for long periods.
2	3-4	Further investigation is needed; changes may be required.
3	5-6	Investigation and changes are required soon.
4	7	Investigation and changes are required immediately.

Source: This figure was first published in *Applied Ergonomics*, Vol. 24, No. 3, pp. 91–99, and is reproduced here with the permission of Butterworth-Heinemann Ltd., Oxford, UK.

3.2.1 Engineering solutions   Decrease repetition/duration, joint deviation, and force.

Repetition/duration   The first guideline (as with the hand and wrist) is to reduce the lifetime use of the joints.

To protect the shoulder, can manual material handling be eliminated (e.g., with a machine)? Can a clamp replace static holding?

A key design concept is to minimize **one-sided work.** That is, design the job so that both hands do the work rather than just the right or left hand. Avoiding one-sided work applies to assembly work as well as manual material handling.

Repeated movement of the neck is rare. If there is repeated neck movement from side to side, consider using a swivel chair seat.

Frequent use of the elbow means frequent movement of the lower arm, which requires considerable energy. The primary problems, however, occur when

frequent use is accompanied by extreme deviations or considerable force—as in throwing (lateral epicondylitis or tennis elbow) or rotating the forearm while the wrist is bent (medial epicondylitis or golfer's elbow).

Joint deviation   Reduce joint deviation from the neutral position. People have considerable variability in their techniques of doing any task. Therefore, when studying people in various jobs, don't assume everyone uses the same motions and has the same joint deviations.

To protect the shoulder, keep the upper arm vertical downward, not horizontal or even elevated. See Figure 11.10.

Guideline: Don't lift your elbow.

The work can be moved or the shoulder can be moved. For example, the work height can be reduced. See Figures 5.4 and 17.27. The point here is

FIGURE 16.19   Posture scores for Group A: upper arm, lower arm, wrist, and wrist twist (McAtamney and Corlett, 1993).

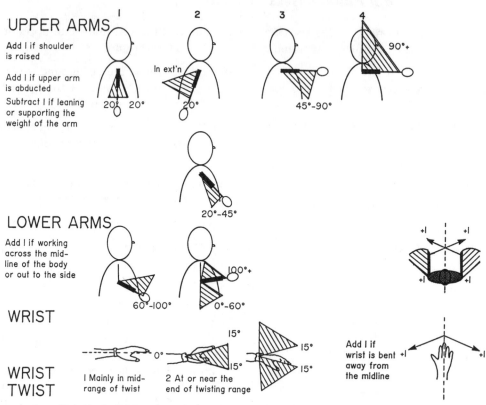

Source: This figure was first published in *Applied Ergonomics,* Vol. 24, No. 3, pp. 91–99, and is reproduced here with the permission of Butterworth-Heinemann Ltd., Oxford, UK.

FIGURE 16.20   Posture scores for Group B: neck, trunk, and legs (McAtamney and Corlett, 1993).

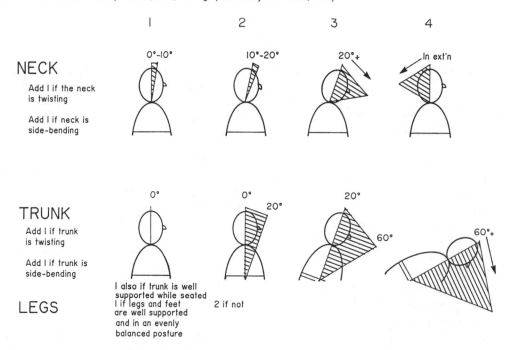

Source: This figure was first published in *Applied Ergonomics,* Vol. 24, No. 3, pp. 91–99, and is reproduced here with the permission of Butterworth-Heinemann Ltd., Oxford, UK.

TABLE 16.2 Frequency scores for RULA (McAtamney and Corlett, 1993).

SCORE	SITUATION
1	Activity is mainly static (held longer than 1 min). or Activity is repeated more than 4 times/min.

Source: This figure was first published in *Applied Ergonomics*, Vol. 24, No. 3, pp. 91–99, and is reproduced here with the permission of Butterworth-Heinemann Ltd., Oxford, UK.

TABLE 16.3 Force (load) scores for RULA (McAtamney and Corlett, 1993).

SCORE	SITUATION
0	No resistance or Less than 2 kg intermittent force or load
1	2–10 kg intermittent force or load
2	2–10 kg static load or 2–10 kg repeated force or load
3	≥ 10 kg static load ≥ 10 kg repeated force or load
4	Shock or forces with a rapid buildup

Source: This figure was first published in *Applied Ergonomics*, Vol. 24, No. 3, pp. 91–99, and is reproduced here with the permission of Butterworth-Heinemann Ltd., Oxford, UK.

TABLE 16.4 Group A (upper arm, lower arm, wrist, wrist twist) index (McAtamney and Corlett, 1993).

| | | WRIST POSTURE SCORE | | | | | | | |
| | | 1 | | 2 | | 3 | | 4 | |
Upper Arm	Lower Arm	W. Twist 1	2	W. Twist 1	2	W. Twist 1	2	W. Twist 1	2
1	1	1	2	2	2	2	3	3	3
	2	2	2	2	2	3	3	3	3
	3	2	3	3	3	3	3	4	4
2	1	2	3	3	3	3	4	4	4
	2	3	3	3	3	3	4	4	4
	3	3	4	4	4	4	4	5	5
3	1	3	3	4	4	4	4	5	5
	2	3	4	4	4	4	4	5	5
	3	4	4	4	4	4	5	5	5
4	1	4	4	4	4	4	5	5	5
	2	4	4	4	4	4	5	5	5
	3	4	4	4	5	5	5	6	6
5	1	5	5	5	5	5	6	6	7
	2	5	6	6	6	6	7	7	7
	3	6	6	6	7	7	7	7	8
6	1	7	7	7	7	7	8	8	9
	2	8	8	8	8	8	9	9	9
	3	9	9	9	9	9	9	9	9

Source: This figure was first published in *Applied Ergonomics*, Vol. 24, No. 3, pp. 91–99, and is reproduced here with the permission of Butterworth-Heinemann Ltd., Oxford, UK.

TABLE 16.5 Group B (neck, trunk, legs) index (McAtamney and Corlett, 1993).

| | TRUNK POSTURE SCORE | | | | | | | | | | | |
| | 1 | | 2 | | 3 | | 4 | | 5 | | 6 | |
NECK POSTURE SCORE	Legs 1	2	Legs 1	2	Legs 1	2	Legs 1	2	Legs 1	2	Legs 1	2
1	1	3	2	3	3	4	5	5	6	6	7	7
2	2	3	2	3	4	5	5	5	6	7	7	7
3	3	3	3	4	4	5	5	6	6	7	7	8
4	5	5	5	6	6	7	7	7	7	7	8	8
5	7	7	7	7	7	8	8	8	8	8	8	8
6	8	8	8	8	8	8	8	9	9	9	9	9

Source: This figure was first published in *Applied Ergonomics*, Vol. 24, No. 3, pp. 91–99, and is reproduced here with the permission of Butterworth-Heinemann Ltd., Oxford, UK.

to ask why the hands are elevated. Figure 19.13 shows how controls can be placed at a more convenient height. Try to place the hand position so the elbows are below the shoulders (tucked in) rather than abducted (moved away from the body centerline). Figure 15.6 shows how carrying a narrower load will permit the elbows to be tucked closer to the body. The work height also can be increased. An example is placing a spacer under the work. If people work with different size objects, consider adjustable height work surfaces. See Figure 15.7.

The height of the shoulder (relative to the work) can be increased if the work is too high. For an example, see Figure 5.4. In another example, a van was originally spray-painted by workers standing on the floor—requiring them to reach upward. The job was modified by having the workers stand on a platform, thus painting horizontally and downward. This had the additional benefit of less paint settling back onto the painter's face.

FIGURE 16.21   Scoring form for RULA (McAtamney and Corlett, 1993).

TASK: _____

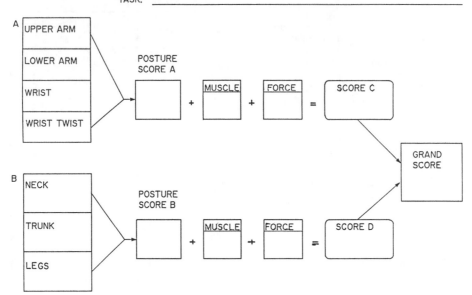

Source: This figure was first published in *Applied Ergonomics,* Vol. 24, No. 3, pp. 91–99, and is reproduced here with the permission of Butterworth-Heinemann Ltd., Oxford, UK.

A variation of moving the shoulder or the work is to use a **sit–stand workstation:** Sometimes the person sits and sometimes the person stands. One possibility is to have part of the job done sitting and part standing. For example, inspection and record keeping work can be done sitting whereas assembly is done standing. Another possibility is to have adjustable height worksurfaces and chairs so that the same task can be done sitting or standing, at the person's option. Note that standing postures may be better than sitting postures. For example, reach is better with standing due to better movement of the hips; in addition, the feet can move. Occasional standing also uses different body muscles than continuous sitting.

Even if the relative height of shoulder and product cannot be changed, can the orientation of the work be changed? Reorientation generally involves tilting the work toward the worker. For packing and assembly work, consider tipping a box on its side or at an angle to reduce elbow elevation during the move or reach. A slanting worksurface also can be used to improve elbow (and wrist) orientation. The work often is elevated to the eye level to improve vision. A better solution is to improve the lighting or to use magnification.

For information on the neck, see Figure 16.23. The neutral position is facing forward and slightly downward (10–15°). Neck flexion of over 20° is a risk factor for neck disorders in electronic assembly (Kilbom et al., 1986). Also see Table 16.8. Can the work surface be tilted?

People who use the telephone by cradling it between their head and shoulder should expect a sore neck. Two possible modifications are a speakerphone (which may require a private office due to the noise) or a head-mounted microphone and headphones. For VDT work, screens and documents

FIGURE 16.22   Grand score index for RULA (McAtamney and Corlett, 1993).

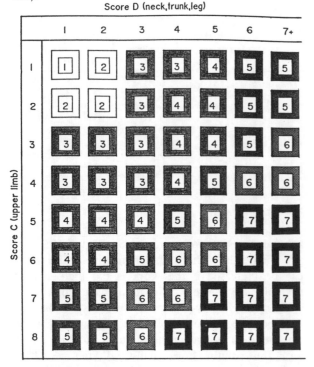

Source: This figure was first published in *Applied Ergonomics,* Vol. 24, No. 3, pp. 91–99, and is reproduced here with the permission of Butterworth-Heinemann Ltd., Oxford, UK.

TABLE 16.6 Engineering (E) and administrative (A) controls to reduce risk of shoulder CTDs (Adapted from Sommerich et al., 1993).

1. **REPETITION**
   Do not design tasks at either end of the repetition continuum (highly repetitive or static postures).
   - (E) Design sequential tasks, not short-cycle repetitive ones.
   - (A) Rotate tasks, especially if highly repetitive or static.
   - (E or A) Reduce speed of highly repetitive movements (less than 1 s/cycle).
   - (A) Minimize use of incentive pay plans for tasks with high-speed arm movements.
   - (A) Train workers to relax nonessential muscles.

2. **POSTURE**
   Minimize shoulder flexion and abduction.
   - (E) Keep work close to the body and about elbow height.
   - (E) If the arms must be abducted or flexed, minimize the time.
   - (E) Tool grip (inline or pistol) depends on posture. See Figure 18.21.
   - (E) Have adjustable chairs and worksurfaces. Have sit/stand options.

3. **FORCE**
   Minimize hand-held weight.
   - (E) Suspend tools.
   - (E) Hold parts with fixtures, not hands.
   - (E) Use mechanical aids for lifting and moving.
   - (E) Keep static load ≤ 20% of maximum strength.

   Minimize forceful or heavy work.
   - (E) Minimize direct load-bearing activities (time and weight).
   - (A) Consider job rotation, provided alternate job reduces shoulder loading.
   - (E) Use power tools instead of hand-powered tools. Maintain tools.
   - (E) Ensure parts can be assembled easily.

4. **REST**
   Provide rest breaks.
   - (E) Design breaks into the work. Breaks at the individual's option are best.
   - (A) If piece-rate incentives are used, require people to take their breaks.
   - (A) Teach people about advantages of micro and mini-breaks. Encourage use of breaks. See Box 16.1.

TABLE 16.7 Sample Checklist for upper extremity cumulative trauma disorders (Lifshitz and Armstrong, 1986). A "no" indicates a risk factor. Checklist tested only in an automobile final assembly plant. Also see Table 5.4.

NO	YES	RISK FACTORS
		Physical Stress
___	___	1. Can the job be done without hand/wrist contact with sharp edges?
___	___	2. Is the tool operating without vibration?
___	___	3. Are the worker's hands exposed to temperatures > 70 F (21 C)?
___	___	4. Can the job be done without using gloves?
		Force
___	___	1. Does the job require exerting less than 10 lbs (4.5 kg) of force?
___	___	2. Can the job be done without using a finger pinch grip?
		Posture
___	___	1. Can the job be done without wrist flexion or extension?
___	___	2. Can the tool be used without wrist flexion or extension?
___	___	3. Can the job be done without deviating the wrist from side to side?
___	___	4. Can the tool be used without deviating the wrist from side to side?
___	___	5. Can the worker be seated while performing the job?
___	___	6. Can the job be done without a clothes-wringing motion?
		Workstation Hardware
___	___	1. Can the worksurface orientation be adjusted?
___	___	2. Can the worksurface height be adjusted?
___	___	3. Can the tool location be adjusted?
		Repetitiveness
___	___	1. Is the cycle time longer than 30 s?
		Tool Design
___	___	1. Are the thumb and finger slightly overlapped in a closed grip?
___	___	2. Is the tool handle span between 2 and 2.75 inches (5 and 7 cm)?
___	___	3. Is the tool handle made from material other than metal?
___	___	4. Is the tool weight below 9 lbs (4 kg)? Note exceptions to the rule.
___	___	5. Is the tool suspended?

should be located so the gaze is down and ahead rather than horizontal and to the side. (See Chapter 20, Section 4.1.) Bifocals cause people to hold the head in unnatural postures when working at VDTs; therefore, workers should use single-vision glasses (**work glasses).**

To reduce strain on the elbow, keep both the upper and lower arm pointed downward if weight (of the arm or an object) must be supported. If the angle between the upper and lower arm is approximately 90° (as with keying or bench assembly), then the forearm and hand weight must be supported by muscles, causing fatigue. Typical solutions are arm pads (see Figure 15.14) or occasional rest breaks.

Force   Reduce the magnitude and duration of the force. (See Chapter 17 for comments on manual material handling.)

To protect the shoulder, reduce static load of tools or product with balancers. See Figures 17.17 and 17.22. Overhead welding is often a problem. If a tool action is horizontal, the ideal suspension is

FIGURE 16.23   Neck postures (Hidalgo et al., 1992). For flexion, neutral is $-15° ≤ α ≤ 15°$; moderate is $15° ≤ α ≤ 45°$; severe is $≥ 45°$. For extension, neutral is $β ≤ 15°$; moderate is $≥ 15°$. For rotation, neutral is $ɤ < 15°$. See also Table 16.8.

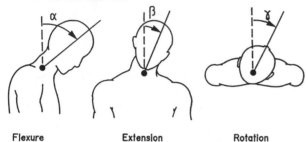

Flexure   Extension   Rotation

Source: J. Hidalgo, A. Genaidy, R. Huston, and J. Arantes, "Occupational Biomechanics of the Neck: A Review and Recommendations." *J. of Human Ergology,* Vol. 21, pp. 165–181. Copyright © 1992 by Center for Academic Publications, Tokyo. Reprinted with permission.

TABLE 16.8   Posture checklist (Keyserling et al., 1992) for neck, trunk, and legs. A zero indicates insignificant risk. An X indicates a potential risk. A star indicates a significant risk.

Job Studied	PERCENT TIME POSTURE USED IN JOB		
	Never	<1/3	>1/3
Neck			
1. Mild forward bending (>20°)	0	0	X
2 Severe forward bending (>45°)	0	X	*
3. Backward bending (>20°)	0	X	*
4. Twisting or lateral bending(>20°)	0	X	*
Trunk			
5. Mild forward bending (>20°)	0	X	*
6. Severe forward bending (>45°)	0	*	*
7. Backward bending (>20°)	0	X	*
8. Twisting or lateral bending(>20°)	0	X	*
General body/legs			
9. Standing stationary (no walking or leaning)	0	0	X
10. Standing, using footpedal	0	X	*
11. Knees bent or squatting	0	X	*
12. Kneeling	0	X	*
13. Lying on back or side	0	X	*

Total X = _____   Total * = _____

Comments:

Source: M. Keyserling, M. Brouwer, and B. A. Silverstein, "A Checklist for Evaluating Ergonomic Risk Factors Resulting from Awkward Postures of the Legs, Trunk, and Neck." *Int. J. Industrial Ergonomics,* Vol. 9, pp. 283–301. Copyright © 1992 by Elsevier Science, Amsterdam, Netherlands. Reprinted with permission.

above the tool center of gravity with the thrust axis at elbow height.

For work at shoulder height, can the opposite hand support the body? Can hands be alternated? Can the body be turned to give a better use of the muscles? Can feet or torso be repositioned to reduce stretching? Welders have many shoulder problems because they hold a relatively heavy object in a precise location (i.e., static load). Consider a balancer (see Figure 17.17) for the welding equipment.

When push or pull movement is needed, push or pull below the shoulder and above the hip, because in this range the muscles are strongest. When cutting with a knife, have the cutting edge down. Cutting away from or toward the body has twice the strength of cross-body motions.

Valves require a torque to open and close. Therefore, determine whether the moment arm can be increased so that the force is less.

Note that even if the hand is empty, an arm is about 4.9% of body weight (see Table 15.2). Consider supporting this weight. For example, can the arm rest on a support when using a VDT mouse?

Shoulder symptoms probably will differ for static and dynamic work (Torner et al., 1991).

Reduce dynamic loads by using the guidelines in Chapter 17, Manual Material Handling. Can motors furnish the power instead of muscles? Can two surfaces be the same height so sliding replaces lifting? Can a short section of gravity roller conveyor bridge the gap between the workstation and a powered takeaway conveyor—thus permitting sliding instead of lifting? Sometimes force requirements can be reduced by maintenance, such as using sharp equipment (drills, knives) and properly lubricating equipment (handtools, valves) or product (to reduce sticking).

For the neck, the load typically is static (e.g., microscope or VDT work). Waersted and Bjorklund (1991) point out that mental effort increases muscle tension, adding to the biomechanical force requirements. Helander et al. (1991) give suggestions for improving microscope work.

3.2.2 Administrative solutions   See Section 1.42 and Box 16.1.

4 BACK

4.1 Problem   CTD problems with the back primarily occur due to manual material handling, although some problems occur from body movement without a load. Back problems also can occur from lack of movement, such as from sitting.

Khalil (1991) gives some statistics:

- On any given day, 6,500,000 people in the United States are in bed with back pain.
- 75,000,000 Americans have back pain problems.
- In industrialized societies, 80% of working adults will develop back pain during their career.
- In the United States, only colds cause more physician visits than back pain. An estimated 50% of all chiropractor visits are due to low-back symptoms.

In 1988, estimated total compensable cost of low-back disorders in the United States was $15.3 billion. In 1988, 35% of all Liberty Mutual workers' compensation claim costs were for low-back disorders. About 1/3 of the costs was for medical treatment and 2/3 for lost wages. The cost distribution is quite skewed, with 25% of the cases accounting for 95% of the costs: average cost/claim = $7,400 while median cost = $360 (Snook, 1991).

Snook (1991) points out there is a hierarchy of low-back problems:

1. low-back **pain**
2. low-back **impairment** (reduced ability to perform various musculoskeletal activities)—about 11% of U. S. working population
3. low-back **disability** (time lost from the job or assignment to restricted duty)—about 2% of workers each year
4. low-back **compensation** (reimbursement for medical costs and lost wages)—in the early 1980s, .75 claims presented each year/100 workers with approximately a 10-1 range from the best to worst industry

4.2 Risk factors

Khalil (1991) gives risk factors for low-back pain (LBP):

- individual physical factors (weight, physique, age, gender, flexibility, etc.)
- psychological factors (depression, anxiety, job dissatisfaction)
- task demand factors (posture, speed, repetition, twisting, prolonged sitting or standing, etc.)
- environmental factors (workplace design, slippery floors, distractions, bulky containers, etc.)

For another review of risk factors, see Pope et al. (1991).

The following is a concise summary of Garg and Moore's (1992) review in which 19 risk factors are divided into personal and job categories.

4.2.1 Personal

Eleven personal factors are discussed briefly in the following.

Age Incidence is maximal between age 35 and 55 but the exact form of the risk is not fully understood.

Gender The incidence of low-back pain is equal among males and females. However, women (vs. males performing similar work) performing physically heavy jobs have higher incidences of low-back pain and have a larger percent of expensive back injury claims. However, about 75–80% of back injury claims are filed by males.

Anthropometry Body dimensions (stature, weight, build), in general, do not have a strong correlation with low-back pain. However, taller people seem to have more low-back pain.

Physical fitness and training Although the importance of physical fitness and training is generally accepted, there is little epidemiological evidence supporting its benefit to low-back injuries. However, regular exercise with large muscle groups (e.g., jogging, swimming, cycling) may have considerable benefit and is unlikely to do harm.

Lumbar mobility There is no evidence that reduced spinal mobility is a risk factor for low-back pain.

Strength Although mean isometric trunk muscle strength is higher for normal healthy people than for people with low-back pain, there is considerable overlap between those with and without back pain.

Medical history A previous history of back or sciatic pain is one of the most reliable predictive factors for subsequent work-related back problems. A person with a history of repeated low-back pain should not be given a job requiring heavy lifting.

Years of employment Although inexperience would seem a likely risk factor, epidemiological evidence is lacking.

Smoking Although many studies have found smoking a risk factor, some have not.

Psychosocial factors Patients with chronic and disabling low-back pain often have significant **psychosocial problems.** (Psychosocial problems include hysteria, anxiety, emotional instability, depression, job dissatisfaction, family problems, disabled relatives, and alcoholism.) It is not clear which comes first, the back disorder or the psychological problems. After successful back rehabilitation, most psychosocial problems tend to dissipate. Johansson et al. (1993) also emphasize the need for analysis of the psychosocial variables as well as the physical (ergonomic) work load.

Structural abnormalities Most abnormalities, at least as a single factor, do not seem to predispose to low-back pain. However, there are studies showing evidence for and against low-back pain for most abnormalities, so the evidence is mixed.

4.2.2 Job risk factors   The following lists eight job-related risk factors.

Heavy physical work   The consensus is that low-back pain and disc degeneration are more prevalent among heavy physical jobs.

Lifting   Low-back pain is clearly related to lifting. See Chapter 17, Manual Material Handling, for a discussion of the risk factors in lifting.

Bending, stretching, and reaching   When lifting is combined with torques about the spine, risk of low-back pain increases.

Twisting   Twisting when lifting increases risk of low-back pain.

Pushing and pulling   Pushing and pulling increase risk of low-back pain.

Prolonged sitting and standing   Static load (lack of movement) increases risk of low-back pain.

Whole-body vibration   Vibration (e.g., for truck drivers) is a low-back pain risk factor.

Accidents   Acute (sudden) events as well as cumulative trauma can cause low-back pain.

4.3 Solutions   Solutions depend on the two problems: (1) underuse of the back and (2) overuse of the back.

4.3.1 Underuse of the back   When people sit for prolonged periods, the intervertebral disc is not properly nourished. The disc receives its nutrition by diffusion, which occurs when there are alterations of pressure (i.e., from movement). The solution is to avoid prolonged sitting. In addition, people in sedentary jobs should be encouraged to be active in their nonwork activities.

The problems seem to be worse when pro-longed sitting is combined with whole-body vibration as, for example, with truck drivers (Kelsey et al., 1984; Griffin, 1990). Note that many vehicles vibrate at a frequency similar to the body's natural frequency (4 to 6 Hz) and so the motion is amplified (resonance) (Frymoyer et al., 1983). Even worse effects result when driving is combined with prolonged head rotation, for example, when driving an industrial truck in a sideways sitting position (Wikstrom, 1993). Truck drivers who also load and unload their vehicles have more musculoskeletal complaints than drivers who just drive (van der Beek et al., 1993).

Possible solutions might include reducing the vibration transmitted to the driver, good posture, and variability in posture (so the same muscles and ligaments are not continuously loaded). For additional comments, see Section 4.4 Seating Posture Variability in Chapter 15.

Prolonged bed rest actually can inhibit healing in the back, and the general recommendation now is a moderate level of activity rather than inactivity.

4.3.2 Overuse of the back   Examination of the occupational risk factors in Section 4.2.2 shows that manual material handling is the primary problem. An entire chapter, Manual Material Handling, covers the engineering/ergonomic aspects of reducing back stress.

Twisting (such as when reaching for a part behind you) also is a problem. Demonstrate this by moving your extended horizontal arm slowly from straight ahead to the side and then to the rear. Just after you pass "3 o'clock" you will begin to feel the strain. This strain occurs sooner when sitting than standing because sitting tends to lock your torso in place. Design workstation supply and disposal areas so operators don't have to twist to get supplies or dispose of completed units. This may even require placing items a little farther away so that the operators must get off their chairs or move their feet (rather than just twist). A swiveling chair tends to be better than a nonswiveling chair for assembling while seated.

Guideline: Don't reach behind your back.

REVIEW QUESTIONS

1. Sketch a tendon, tendon sheath, muscle, and bone.
2. What is the difference between tendinitis, tenosynovitis, stenosing tenosynovitis, and trigger finger?
3. What is the difference between golfer's elbow and tennis elbow?
4. Assume a person did a task lasting 1 minute, 450 times/shift, 5 days/week. Ignoring the weekend, what is the recovery/work ratio?
5. What are the three primary occupation risk factors for cumulative trauma disorders?
6. What amount of time does Konz assign to short, moderate, and long duration?
7. Briefly discuss nonoccupational risk factors for CTD.

8. What is the difference between automation and mechanization?

9. Outline the administrative approaches to reduction of CTD.

10. What part of the body is injured in carpal tunnel syndrome?

11. Give some examples of how wrist joint deviation can be reduced.

12. How does vibration affect grip force?

13. Why should one-sided work be minimized?

14. How do bifocals cause neck problems?

15. Comment on 7 of the 11 personal risk factors for low-back pain.

16. Comment on 5 of the 8 occupational risk factors for low-back pain.

17. How is disc nutrition related to movement?

18. What can be done about underutilization of the back?

REFERENCES

Armstrong, T., Fine, L., and Silverstein, B. *Occupational Risk Factors: Final Contract Report to NIOSH,* No. 22-82-2507. Cincinnati, 1985.

Armstrong, T., Fine, L., Goldstein, S., Lifshitz, Y., and Silverstein, B. Ergonomics considerations in hand and wrist tendinitis. *J. of Hand Surgery,* Vol. 12-A [5], Part 2, 830-837, 1987.

Contreras, R., Rys, M., and Konz, S. Leg length discrepancy. In *The Ergonomics of Manual Work,* Marras, W., Karwowski, W., Smith, J., and Pacholski, L. (eds.). London: Taylor and Francis, 1993.

Drury, C. and Wick, J. Ergonomic applications in the shoe industry. In *Proceedings of the 1984 Conference on Occupational Ergonomics,* Toronto, 489-93, 1984.

Fogelman, M., Freivalds, A., and Goldberg, J. An ergonomic evaluation of knives for two poultry cutting tasks. *Int. J. of Industrial Ergonomics,* Vol. 11, 257-65, 1993.

Frymoyer, J., Pope, M., Clements, J., Wilder, D., MacPherson, B., and Ashikaga, T. Risk factors in low-back pain. *J. Bone Joint Surgery,* Vol. 65A, 213-18, 1983.

Garg, A. Occupational biomechanics and low-back pain. In *Ergonomics: Low Back Pain, Carpal Tunnel Syndrome, and Upper Extremity Disorders in the Workplace*, Moore, J. and Garg, A. (eds.). Philadelphia: Hanley and Belfus, 1992.

Garg, A. and Moore, J. Epidemiology of low-back pain in industry. In *Ergonomics: Low Back Pain, Carpal Tunnel Syndrome, and Upper Extremity Disorders in the Workplace*, Moore, J. and Garg, A. (eds.). Philadelphia: Hanley and Belfus, 1992.

Gore, A. and Tasker, D. *Pause Gymnastics.* North Ryde NSW 2113, Australia: CCH, 1986.

Griffin, M. *Handbook of Human Vibration.* San Diego: Academic Press, 1990.

Helander, M., Grossmith, E., and Prabhu, P. Planning and implementation of microscope work. *Applied Ergonomics,* Vol. 22 [1], 36-42, 1991.

Hidalgo, J., Genaidy, A., Huston, R., and Arantes, J. Occupational biomechanics of the neck: A review and recommendations. *J. of Human Ergology,* Vol. 21, 165-81, 1992.

Jarvholm, U., Palmerud, G., Kadefors, R., and Herbets, P. The effect of arm support on supraspinatus muscle load during simulated assembly work and welding. *Ergonomics,* Vol. 34 [1], 57-66, 1991.

Jegerlehner, J. Ergonomic analyses of problem jobs using computer spreadsheets. In *Advances in Industrial Ergonomics and Safety III*, Karwowski, W. and Yates, J. (eds.). London: Taylor and Francis, 865-71, 1991.

Johansson, J., Kadefors, R., Rubenowitz, S., Klingenstierna, U., Lindstrom, I., Erngstrom, T., and Johansson, M. Musculoskeletal symptoms, ergonomic aspects and psychosocial factors in two different truck assembly concepts. *Int. J. of Industrial Ergonomics,* Vol. 12, 35-48, 1993.

Johnson, S. and Evans, B. Tracking median nerve conduction as a method of early detection of carpal tunnel syndrome. *Proceedings of the Human Factors and Ergonomics Society,* 759-63, 1993.

Kao, H. Differential effects of writing instruments on handwriting performance. *Acta Psychologia Taiwanica,* Vol. 21, 9-13, 1979.

Kelsey, J., Githens, P., O'Conner, T. et al. Acute prolapsed intervertebral disc: An epidemiologic study with special reference to driving automobiles and cigarette smoking. *Spine,* Vol. 9, 608-13, 1984.

Keyserling, M., Brouwer, M., and Silverstein, B. A checklist for evaluating ergonomic risk factors resulting from awkward postures of the legs, trunk, and neck. *Int. J. of Industrial Ergonomics,* Vol. 9, 283-301, 1992.

Khalil, T. Ergonomic issues in low back pain: Origin and magnitude of the problem. *Proceedings of the Human Factors Society,* 820-24, 1991.

Kilbom, A., Persson, J., and Jonsson, B. Risk factors for work-related disorders of the neck and shoulders— With special emphasis on working postures and movements. In *The Ergonomics of Working Postures,* Corlett, N., Wilson, J., and Manenica, I. (eds.). London: Taylor and Francis, 1986.

Krag, M., Cohen, M., Haugh, L., and Pope, M. Body height change during upright and recumbent posture. *Spine,* Vol. 15 [3], 202-7, 1990.

Kramer, J. and Gritz, A. Korper-langenaderunger durch druckabhangige Flussigkeitsverschiebung im zwischenwirbel Aabschuitt. *Z. Orthop.,* Vol. 118, 161-64, 1980.

Kroemer, K. Avoiding cumulative trauma disorders in shops and offices. *Am. Industrial Hygiene Association J.,* Vol. 53, [9], 596-604, 1992.

Lifshitz, S. and Armstrong, T. A design checklist for control and prediction of cumulative trauma disorders in hand intensive manual jobs. *Proceedings of the Human Factors Society,* 837–41, 1986.

Luttgens, K., Deutsch, H., and Hamilton, N. *Kinesiology: Scientific Basis of Human Motion,* 8th ed. Dubuque, IA: Brown and Benchmark, 1992.

Marras, W. Toward an understanding of dynamic variables in ergonomics. In *Ergonomics: Low Back Pain, Carpal Tunnel Syndrome and Upper Extremity Disorders in the Workplace,* Moore, J. and Garg, A. (eds.). Philadelphia: Hanley and Belfus, 1992.

Marras, W. and Sommerich, C. A three-dimensional motion model of loads on the lumbar spine: I. Model structure. *Human Factors,* Vol. 33 [2], 123–37, 1991a.

Marras, W. and Sommerich, C. A three-dimensional motion model of loads on the lumbar spine: II. Model validation. *Human Factors,* Vol. 33 [2], 139–49, 1991b.

McAtamney, L. and Corlett, E. N. RULA: A survey method for the investigation of work-related upper limb disorders. *Applied Ergonomics,* Vol. 24 [3], 91–99, 1993.

McCarty, M., Thayer, C., and Wick, J. Reducing risk of repetitive motion injuries in wafer processing: A case study. In *Advances in Industrial Ergonomics and Safety III,* Karwowski, W. and Yates, J. (eds.). London: Taylor and Francis, 83–86, 1991.

Mital, A. and Channaveeraiah, C. Peak volitional torques for wrenches and screwdrivers. *Int. J. of Industrial Ergonomics,* Vol. 3, 41–64, 1988.

Neese, R. and Konz, S. Vibrometry of industrial workers: A case study. *Int. J. Industrial Ergonomics,* Vol. 11 [4], 341–45, 1993.

Owen, R. Carpal tunnel syndrome: A products liability perspective. *Ergonomics,* Vol. 37 [3], 449–76, 1994.

Pope, M., Frymoyer, J., Anderson, G., et al. *Occupational Back Pain: Assessment, Treatment and Prevention.* St. Louis: Mosby Year Book, 1991.

Putz-Anderson, V. (ed.). *Cumulative Trauma Disorders.* London: Taylor and Francis, 1988.

Rigdon, J. The wrist watch. *Wall Street Journal,* September 28, 1992.

Ring, L. *Facts on Backs.* Loganville, Ga.: Institute Press, 1981.

Rowe, M. *Orthopaedic Problems at Work.* Fairport, N. Y.: Perinton Press, 1985.

Rys, M. and Konz, S. Lifting belts? In *The Ergonomics of Manual Work,* Marras, W., Karwowski, W., Smith, J., and Pacholski, L. (eds.). London: Taylor and Francis, 1993.

Schoenmarklin, R. and Marras, W. Quantification of wrist motion and cumulative trauma disorders in industry. *Proceedings of the Human Factors Society,* 838–42, 1991.

Schoenmarklin, R. and Marras, W. Dynamic capabilities of the wrist joint in industrial workers. *Int. J. of Industrial Ergonomics,* Vol. 11, 207–24, 1993.

Silverstein, B., Fine, L., and Armstrong, T. Hand–wrist cumulative trauma disorders in industry. *British J. Industrial Medicine,* Vol. 43, 779–84, 1986.

Snook, S. Low back disorders in industry. *Proceedings of the Human Factors Society,* 830–33, 1991.

Sommerich, C., McGlothlin, J., and Marras, W. Occupational risk factors associated with soft tissue disorders of the shoulder: A review of recent investigations of the literature. *Ergonomics,* Vol. 36 [6], 697–717, 1993.

Tanaka, S. and McGlothlin, J. A conceptual quantitative model for prevention of work-related carpal tunnel syndrome (CTS). *Int. J. of Industrial Ergonomics,* Vol. 11, 181–93, 1993.

Tayyari, F. and Emanuel, J. Carpal tunnel syndrome: An ergonomics approach to its prevention. *Int. J. of Industrial Ergonomics,* Vol. 11, 173–79, 1993.

Torner, M., Żetterberg, C., Anden, U., Hansson, T., and Lindell, V. Workload and musculoskeletal problems: A comparison between welders and office clerks (with reference also to fishermen). *Ergonomics,* Vol. 34 [9], 1179–96, 1991.

van der Beek, A., Frings-Dresen, M., van Dijk, F., Kemper, H., and Meijman, T. Loading and unloading by lorry drivers and musculoskeletal complaints. *Int. J. of Industrial Ergonomics,* Vol. 12, 13–23, 1993.

Village, J., Morrison, J., and Layland, A. Biomechanical comparison of carpet-stretching devices. *Ergonomics,* Vol. 36 [8], 899–909, 1993.

Waersted, M. and Bjorklund, R. Shoulder muscle tension introduced by two VDT-based tasks of different complexity. *Ergonomics,* Vol. 34 [2], 137–50, 1991.

Waersted, M. and Westgaard, R. Working hours as a risk factor in the development of musculoskeletal complaints. *Ergonomics,* Vol. 34 [3], 265–76, 1991.

Wigley, R., Turner, W., Blake, B., Darby, F., McInnes, R., and Harding, P. *Occupational Overuse Syndrome: Treatment and Rehabilitation.* Wellington, New Zealand: Department of Labor, 1992.

Wikstrom, B. Effects from twisted postures and whole-body vibration during driving. *Int. J. of Industrial Ergonomics,* Vol. 12, 61–75, 1993.

Winkel, J. and Westgaard, R. Occupational and individual risk factors for shoulder-neck complaints: Part I—Guidelines for the practitioner. *Int. J. of Industrial Ergonomics,* Vol. 10, 79–83, 1992.

Winkel, J. and Westgaard, R. Occupational and individual risk factors for shoulder-neck complaints: Part II—The scientific basis (literature review) for the guide. *Int. J. of Industrial Ergonomics,* Vol. 10, 85–104, 1992.

OVERVIEW

Manual material handling is a major safety problem. Force limits are given for pushing and pulling. Carrying recommendations are given for both distance and local movement. After explaining NIOSH guidelines, principles of material handling are discussed. If you can't make the job perfect, make it better.

CHAPTER CONTENTS

1 Background
2 Pushing and Pulling
3 Holding
4 Carrying
5 Lifting
6 Guidelines for Manual Material Handling

KEY CONCEPTS

acute/chronic
angle of asymmetry
approaches
balancers
bend the knees
biomechanical approach
horizontal location
J hook
job severity index
lifting duration

lifting frequency
lifting index
load constant
manipulators
one-hand lift
origin/destination
physiological approach
psychophysical approach
recommended weight limit (RWL)
recovery time

scissors lift
spinal torque
static load
tipping aids
turntables
vertical location
vertical travel distance
workstation positioners

1 BACKGROUND

1.1 Problem   The problem of manual material handling (MMH) is demonstrated by some example statistics:

- 27% of all industrial injuries associated with MMH
- 670,000 injuries/yr in the United States
- 60% of all money spent on industrial injuries
- 93,000,000 work days/yr on industrial injuries

The injuries generally are to the skeletomuscular system—muscles, ligaments, and especially the back. The trauma may be **acute** (result from a one-time event) but most is **chronic** (result from cumulative events over a time period). That is, it is cumulative strain or repetitive trauma. However, a specific event may have been the last straw on a previously weakened system and thus give the appearance that the problem is acute rather than chronic. Managements may like to think of MMH injuries as acute (and therefore an "act of God" and beyond their control) rather than chronic (and therefore subject to change).

1.2 Criteria   Various groups have given recommendations for the amount that a person can push, pull, carry, and lift. Their criteria will be grouped into three categories: biomechanical, physiological, and psychophysical.

1.2.1 Biomechanical   The **biomechanical approaches** emphasize the forces and torques of MMH and their effect on the parts of the body. In general, the back is the weak link—especially the low back with its L4–L5 disc and L5–S1 disc. See Figure 16.8. The emphasis is on measuring external forces and torques due to the task, measuring body dimensions and strengths, and then computing the effect on the back. Another approach (*Force Limits,* 1980) emphasizes intra-abdominal pressure. (Intra-abdominal pressure is obtained by having a person swallow a pressure transducer which then radios the pressure while the person does lifting.)

1.2.2 Physiological   The **physiological approaches** emphasize the energy requirements of the task and the effects on the cardiovascular system. Energy requirements have little impact on occasional lifting but can become important for repetitive lifting—for example, once/min. In general, the goal is to limit metabolic rate to less than 5 kcal/min.

1.2.3 Psychophysical   The **psychophysical approach** has people actually do various MMH tasks under controlled conditions. Based on the amount of MMH that the sample of workers does for the sample of time, predictions are made as to what the population of workers will do in real jobs. A virtue of the psychophysical approach is that it combines the biomechanical and the physiological stresses.

1.3 MMH Variables   The many variables in MMH will be divided into three broad categories: individual variables, technique variables, and task variables. The general thrust is to (1) increase the strength of the worker, (2) decrease the stress due to technique and task, or (3) increase strength *and* decrease stress.

1.3.1 Individual   Selection variables include gender, age, back muscle strength, arm strength, and intra-abdominal pressure. Although there is no doubt that it is safer to have a strong person do MMH than a weak person, there are many questions when it comes to selecting individuals. In addition, our society tends to frown on excluding people from jobs. The question becomes not "Who has the strongest back?" but "What can be done so a petite woman can do the job?"

1.3.2 Technique   Training variables include body posture, hand orientation, foot position, accelerations, and lifting training. Although there is no doubt that a trained, knowledgeable person will be less likely to be injured during MMH than a novice, there are many problems when it comes to specific techniques. The primary challenge is that people are not machines and they forget, make mistakes, never learn, and so on.

1.3.3 Task   Ergonomic variables include object, weight, ease of handling, initial and final height of object moved, angle of rotation, lift symmetry, clothing, and thermal environment. A strong advantage of modifying the task to improve MMH is that it is a permanent change as opposed to the temporary changes of personal selection and training. Although selection and training will not be ignored, this chapter will focus on ergonomics, that is, on job design.

2 PUSHING AND PULLING

After giving recommended force limits, some task modifications will be given.

2.1 Force Limits   Force limits will be divided into horizontal and vertical.

2.1.1 Horizontal push and pull   The amount of force an individual can exert is strongly influenced by the direction and height of the exertion. For example, Warwick et al. (1980) reported that male subjects, with the object straight ahead and using both hands, could push with a force of 29.8 kg (100%) but could pull back with a force of 17.3 kg (58%), push right with 15.9 (53%), and push left 17.0 (57%). They could lift up with 39.4 kg (132%) and press down with 34.7 (116%). Winters and Chapanis (1986) reported that almost all males could push with the thumb (when inserting a card into electronics) with 10 to 15 kg. They told the IBM designers to provide an insertion tool, to reduce insertion force, and to orient the cards at the favorable locations and positions. Although some forces are exerted parallel to the shoulders, most exertions are perpendicular to the shoulder line.

Perpendicular to shoulders   Table 17.1 gives recommended upper force limits (Eastman Kodak, 1986). *Force Limits in Manual Work* (1980) has a detailed set of tables for (1) standing, pushing with 1 hand; (2) standing, pushing and pulling with 2 hands; and (3) kneeling, pushing and pulling with 2 hands. Snook and Ciriello (1991) have detailed push and pull recommendations. Tables 17.2 through 17.4 give recommendations by Mital et al. (1993). In summary:

- Two hands are better than one.
- Force capability goes down as it is exerted more often.
- Females are weaker than males—especially in pushing.
- Push at waist level rather than shoulder or knee level.

- Pull at knee level rather than waist or shoulder level.

If rails are used, large loads can be pushed. Wyndham and Heyns (1967) reported that 70-kg male miners could push 900-kg cars at 3.7 km/h (with oxygen consumption of 2.1 L/min and mechanical efficiency of 15%). They reported that pushing a 700-kg car (i.e., loaded) at 2.4 km/h and returning it (weight 400 kg) empty at 4.8 km/h was most efficient.

Pulling a load in a rickshaw is at least 6 times more efficient than carrying the same load on the head. The rickshaw studied could have been improved by making it of light metals and using better tires and bearings, but that would have increased its capital cost (Datta et al., 1978).

For many handling tasks, the arms and shoulder muscles are the weakest link—a point sometimes overlooked by users of computerized biomechanical models, which tend to focus on the lower back. Arms and shoulders tend to be limiting when the activity is repetitive (and thus causes local muscle fatigue) or when the posture is poor, as when pushing with arms fully extended, pushing or pulling with one arm, or pushing or pulling above the shoulder or below the hip. In general, arm strength (both isometric and isokinetic) is greatest at .5 (reach distance), drops some at .75 (reach distance), and is lowest at 1.0 (reach distance) (Kumar, 1987).

Lack of a solid support (because of slippery floor or nothing to brace yourself against) also causes problems. Table 17.5 shows that some types of floors and shoe soles have different coefficients of friction. Use high-friction flooring, especially on ramps, stairs, and areas which might become wet.

TABLE 17.1   Upper force limits for horizontal pushing and pulling perpendicular to the shoulders (Eastman Kodak, 1986). The recommendations permit a large majority of workers to do the job.

POSTURE	LOCATION	MAXIMUM FORCE, Kg	EXAMPLE ACTIVITIES
Standing: whole body involved	Between waist and shoulder	23.0	Truck and cart handling Moving equipment on wheels or casters Sliding rolls on shafts
Standing: primarily arm and shoulder muscles, arms	Between waist and shoulder	11.2	Leaning over an obstacle to move an object Pushing an object at or above shoulder height
Kneeling	Between waist and shoulder	19.2	Removing or replacing a component from equipment (as in maintenance work) Handling in confined areas (as in tunnels or large conduits)
Seated	Between chest and shoulder	13.3	Operating a vertical lever (as floor shift on heavy equipment) Moving trays or product on and off conveyors

TABLE 17.2   Recommended maximum initial TWO-HANDED PUSHING forces (kg) for male (female) industrial workers at 2.1-m pushing distance (Mital et al., 1993). Mital et al. also have initial and sustained forces for other distances and frequencies. Values in parentheses are for females.

Handle Height (cm)	Population %	PUSHING FREQUENCY				
		10/min	5/min	1/min	1/5 min	1/8 h
144(135)	90	20(14)	22(15)	25(17)	26(20)	31(22)
	75	26(17)	29(18)	32(21)	34(24)	41(27)
	50	32(20)	36(22)	40(25)	42(29)	51(32)
	25	38(24)	43(25)	47(29)	50(33)	61(37)
	10	44(26)	49(28)	55(33)	58(38)	70(41)
95(89)	90	21(14)	24(15)	26(17)	28(20)	34(22)
	75	28(17)	31(18)	34(21)	36(24)	44(27)
	50	34(20)	38(22)	43(25)	45(29)	54(32)
	25	41(24)	46(25)	51(29)	54(33)	65(37)
	10	47(26)	53(28)	59(33)	62(38)	75(41)
64(57)	90	19(11)	22(12)	24(14)	25(16)	31(18)
	75	25(14)	28(15)	31(17)	33(19)	40(21)
	50	31(16)	35(17)	39(20)	41(23)	50(25)
	25	38(19)	42(20)	46(23)	49(27)	59(30)
	10	43(21)	48(23)	53(26)	57(30)	68(33)

Source: A. Mital, A. Nicholson, and M. M. Ayoub, *A Guide to Manual Materials Handling*. Copyright © 1993 by Taylor and Francis, London. Reprinted with permission.

TABLE 17.3   Recommended maximum initial TWO-HANDED PULLING forces (kg) for male (female) industrial workers at 2.1-m pulling distance (Mital et al., 1993). Mital et al. also have initial and sustained forces for other distances and frequencies. Values in parentheses are for females.

Handle Height (cm)	Population %	PULLING FREQUENCY				
		10/min	5/min	1/min	1/5 min	1/8 h
144(135)	90	14(13)	16(16)	18(17)	19(19)	23(22)
	75	17(16)	19(19)	22(20)	23(23)	28(26)
	50	20(19)	23(22)	26(24)	28(28)	33(31)
	25	24(21)	27(25)	31(28)	32(32)	39(35)
	10	26(24)	30(28)	34(31)	36(36)	44(39)
95(89)	90	19(14)	22(16)	25(18)	27(21)	32(23)
	75	23(16)	27(19)	31(21)	32(25)	39(27)
	50	28(19)	32(23)	36(25)	39(29)	47(32)
	25	33(22)	37(26)	42(29)	45(33)	54(37)
	10	37(25)	42(29)	48(32)	51(37)	61(41)
64(57)	90	22(15)	25(17)	28(19)	30(22)	36(24)
	75	27(17)	30(20)	34(22)	37(26)	44(28)
	50	32(20)	36(24)	41(26)	44(30)	53(33)
	25	37(23)	42(27)	48(30)	51(35)	61(38)
	10	42(26)	48(31)	54(34)	57(39)	69(43)

Source: A. Mital, A. Nicholson, and M. M. Ayoub, *A Guide to Manual Materials Handling*. Copyright © 1993 by Taylor and Francis, London. Reprinted with permission.

Parallel to shoulders   Operators at a workstation may move objects parallel to the shoulders (e.g., moving an object on or off a conveyor). Since this uses the weaker shoulder muscles, Eastman Kodak (1986, p. 391) recommends 50 to 60% of the forces in Table 17.1.

2.1.2   Vertical push and pull

Table 17.6 gives recommended upper force limits (Eastman Kodak, 1986). Specific individuals can exert more force; lifts also could be higher for occasional exertions. Of course, lower forces are better for the operator than the upper limits of the table.

2.2   Task Modifications

Before existing equipment is replaced, consider maintenance and lubrication of the bearings, wheels, and casters of the present equipment.

Avoid muscle-powered pushing or pulling for ramps, long distances, and high frequency moves.

TABLE 17.4 Maximum force (kg) for ONE-HAND PUSH AND PULL while standing (Mital et al., 1993).

CONDITION	GENDER	
	Male	Female
Push: one time	16	11
Push: repeated	11	7.5
Pull: one time	15	10
Pull: repeated	10	7

Source: A. Mital, A. Nicholson, and M. M. Ayoub, *A Guide to Manual Materials Handling.* Copyright © 1993 by Taylor and Francis, London. Reprinted with permission.

Figure 17.1 shows the effect of posture when pushing. Figure 17.2 shows the beneficial effect of drum carts (mechanical aids). Of course some aids are better than others. Figure 17.3 shows that, even without mechanical aids, momentum can help. Figure 17.4 shows some solutions to handling of rolls. As a preventative measure, take the wheels off heavy carts. This will prevent pushing them and require the workers to use mechanical equipment to move them. Figure 17.5 shows the benefits of large wheels. Put ball transfer tables and wheel conveyors on a slight incline to reduce required pushing forces, yet not so great an incline that the object moves when you don't push it. Although it seems obvious, defective wheels and casters should be replaced and good ones lubricated.

3 HOLDING

In most manual material handling, there is movement. Something is moved or lifted. However, there are some situations in which the worker holds the object without any movement. In some situations,

TABLE 17.5 Coefficients of friction of floors and shoes. Developed from Kroemer (1974). Use 0.3 for the coefficient of friction of metal–metal and 0.4 for wood–metal and wood–wood.

Coefficient of Friction	Floors	Floor		Shoes: Soles
		Clean	Soiled	
1.0	Soft rubber pad	.8	.6	Rubber-cork
.8	End grain wood	.75	.55	U.S. Army-U.S. Air Force std.
.7	Concrete, rough finish	.7	.5	Rubber-crepe
.65	Working decorative, dry	.6	.4	Neoprene
.5	Working decorative, soiled	.5	.3	Leather
.4	Steel			
				Shoes: Heels
		.7	.5	Neoprene
		.65	.45	Nylon

TABLE 17.6 Upper force limits for vertical pushing and pulling while standing (Eastman Kodak, 1986). The recommendations permit a large majority of workers to do the job. For seated downward pulls, use 85% of table values.

ACTION	LOCATION	MAXIMUM FORCE, KG	EXAMPLE ACTIVITIES
Pull down	Above head	55.1	Hook grip (safety shower handle, manual control)
		20.4	Power grip with less than 5-cm grip diameter (operating a chain hoist)
Pull down	Shoulder level	32.1	Hook grip, activating a control
			Threading operations (as in paper mfg. and stringing cable)
Pull up	25 cm above floor	32.1	Lifting an object with one hand
	Elbow height	15.1	Raising a lid or access port, cover, palm up
	Shoulder height	7.7	
Push down	Elbow height	29.3	Wrapping, packing, sealing cases
Push up (boosting)	Shoulder height	20.6	Raising a corner or end of an object (pipe, beam) Lifting an object to a high shelf

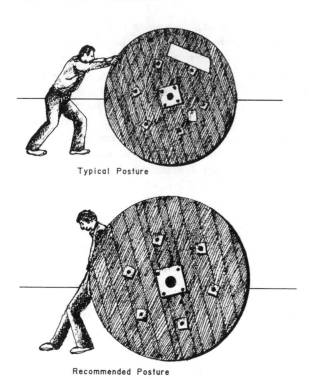

Typical Posture

Recommended Posture

FIGURE 17.1   Posture when pushing makes a difference. Ridd (1983) reported peak intra-abdominal pressure was 50% less when pushing with the back.

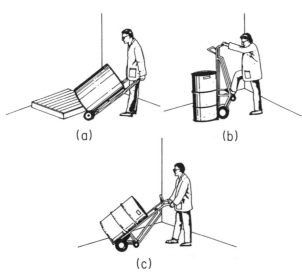

(a)

(b)

(c)

FIGURE 17.2   Drum carts reduce stress (Eastman Kodak, 1986). (a) shows a two-wheel cart; it should have a latch over the top rim during movement. (b) and (c) show a four-wheel cart, useful for heavier drums. In (b), the foot can be used to help tip the drum. Note that the hand positions are different in (b) and (c). There also is another handle, forward of the main handles, which is useful when tilting and maneuvering the drum for scales and pallets.

there may not even be any external object and what is supported is part of the body itself. This gives a **static load.** Guideline 1 in Chapter 15 discusses cardiovascular effects of static load.

Figure 17.6 shows a typical exponential trade-off between load and time (Chaffin and Andersson, 1984).

Figure 5.4 shows how automobile workstations were redesigned to reduce strain of holding part of the body. Figure 17.7 shows how a tipping aid can reduce strain from holding a drum.

4 CARRYING

Carrying can be carrying on the back and/or shoulders (long distance) and carrying with the hands (short distance).

4.1 Back and Shoulders (Long Distance)
Table 17.7 shows the cost of carrying heavy weights long distances; carrying objects in the hand (an asymmetric loading of small muscle groups) is the worst—approximately 50% more energy is required than for a load on the back. Figure 17.8 shows how to carry a serving tray. Transporting loads for longer distances (over 10 m) normally should be done with mechanical aids. These mechanical aids can be mechanically powered (lift trucks, conveyors) or human powered (carts). A good motto is "Everything on wheels." See Box 11.2 for a discussion of walking. Even within the local area, aids should be used for repetitive carrying (chutes, jib cranes, carts, wheel conveyors). Use the human as a "beast of burden" only as a last resort.

The key guidelines are the following:

1. *Minimize the moment arm of the load vs. the spine.*

2. *Carry large loads occasionally rather than light loads often.* However, this minimization of energy cost increases stress on the back and thus may lead to back pain.

Balogun et al. (1986) showed that a frontal yoke was worse than a transverse yoke because the trunk was bent to the side of the body on which the yoke was placed.

In addition to the transport cost, also consider the problems of loading and unloading. Teamwork is one possibility. A loads B who carries to C who unloads B; periodically A, B, and C switch jobs. The Korean A-frame has legs and the porter uses a walking stick. To rest, the porter squats, slips the frame off, and props it with the stick.

4.2 Hand Carrying (Short Distance)
See Table 17.8. Mital et al. (1993) also have tables for two-handed carrying. Hand carrying is more difficult to mechanize than long-distance carrying. However, consider the use of balancers, manipulators, and even robots to minimize humans moving product

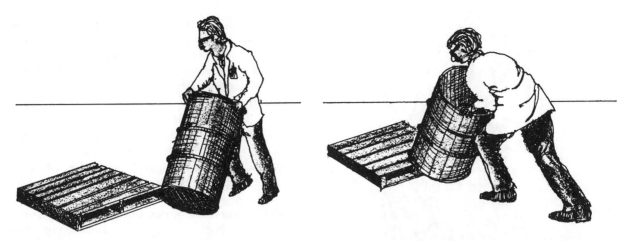

FIGURE 17.3   Chiming a drum (rotating a tipped drum) requires little effort; with skill the drum momentum can be used to move it onto a pallet (Eastman Kodak, 1986). A straight push, however, as shown on the right, requires considerable effort. A drum cart is another alternative.

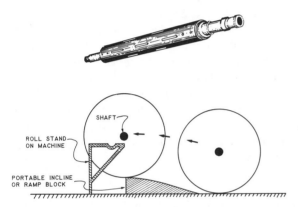

FIGURE 17.4   Rolls present problems. The lower figure shows a ramp block for setting a roll on a machine or roll stand. The roll is pushed along the floor to gain momentum to easily climb the incline. An air shaft (the upper figure) reduces problems of handling heavy center bars and roll shafts. The air shaft is hollow with internal rubber tubes that expand as opposed to lugs, leaves, or buttons to grip rolls or cores. Besides being lighter in weight than solid shafts, they are easier to insert and move. (Sketch courtesy of Liberty Mutual Insurance.) Shafts should be stored on a wheeled, A-frame shaft cart.

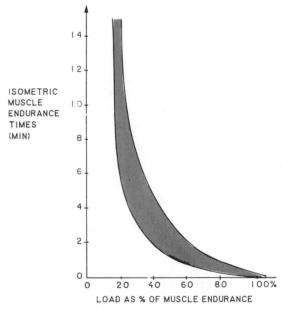

FIGURE 17.6   Exponential curve is a typical trade-off for time vs. load. The maximum long-term load is 15–20% of maximum one-time capacity.

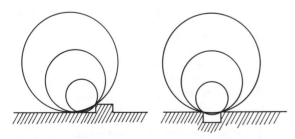

FIGURE 17.5   Pushing force depends on the ratio of a wheel radius to the height of an obstruction. A small radius gives a small lever arm, and so small wheels can get hung up on bumps, objects on the floor, and holes. Therefore buy carts with large wheels and then clean and maintain the floor (Courtesy of Liberty Mutual). In addition, remove door sills. If the gap between door and floor must be sealed, use a flap on the door.

around a workstation. Short-distance transport often is associated with twisting and bending, which stress the back, while the load stresses the muscles of the arms and shoulders.

Continuous carrying primarily is limited by the cardiovascular system, although there may be local muscle fatigue problems. Figure 17.9 shows an example of a simple carrying aid. Figure 17.10 shows recommended tray handholds. Figure 17.11 shows maximum weight to be carried in the hands. Figure 15.6 shows the abduction problem caused by a wide container.

FIGURE 17.7 **Tipping aids** permit a drum or carboy to be counterbalanced when pouring. The danger of spills is reduced and operator muscle stress is reduced (Eastman Kodak, 1986).

For very short distances (.5 to 2 m), Mueller, Vetter, and Blumel (1958) emphasized the metabolic penalty of moving objects from the floor. See Figure 17.12. Raising the initial location greatly reduced cost.

Burton (1986) pointed out that shopping bags with handles were not as desirable as bags without handles. Those without handles tended to be clutched to the body and thus closer to the body's center of gravity. They also tended to be placed on a table rather than the floor, reducing the stress of picking them up. Morrissey and Liou (1988) found that handles on boxes led to a significantly lower weight carried (8% less); boxes without handles tended to be clutched close to the body, allowing the forearms, upper arms, and body to support the box, not just the wrists and forearms as is required with handles. Figure 17.13 points out that people may strain their backs simply to avoid soiling their clothes.

Mital and Ilango (1983) demonstrated that carrying containers with liquids instead of lead shot resulted in a 12% reduction of load carried. The movement of the liquid was the problem.

Carrying objects up and down stairs is especially dangerous as (1) the hands are occupied holding the object and thus not free to grasp a handrail if there is a slip and (2) the object may impair vision. Therefore move loads between levels with hoists, platforms, and elevators, not manually on stairs (Chaffin, 1987).

Randle (1987) gave the following equation for predicting intermittent load carriage in the arms (walking continuously but carrying a load every other 30 s).

TABLE 17.7   Comparison of seven modes of carrying on a horizontal plane (Datta and Ramanathan, 1971). The 50-kg subjects carried 30-kg weights 1 km.

	Criteria		
		Incremental Heart	
Carrying Method	**Energy, kcal/min**	**Beats/Min**	**Comments**
Double pack 15 kg in two packs across shoulder—one in front and one in back	337	50	Load must be divided in two. Special harness. May not be suitable for repeated short trips without quick release harness.
Head Basket on head	348	64	Requires training. Body movement restricted. Very suitable for repeated short distance carrying. Arms free.
Rucksack High pack across the shoulder	368	62	Not good for repeated short-distance carrying.
Sherpa Load in a gunny sack supported on back by a strap across forehead	387	57	Requires practice. Arms can carry stick for hill climbing.
Rice bag Load in a gunny sack supported on back by holding two corners of sack with hands or hooks	414	60	No advantage over rucksack. Rather unsafe.
Yoke 15 kg is suspended by 3 ropes on each end of a bamboo strip across shoulder. Strip held with one or two hands	434	66	Uncomfortable without practice. Contorted posture.
Hands Two canvas bags with padded handles; 15 kg in each hand	486	81	Least efficient and most fatiguing. Even worse with only one hand.

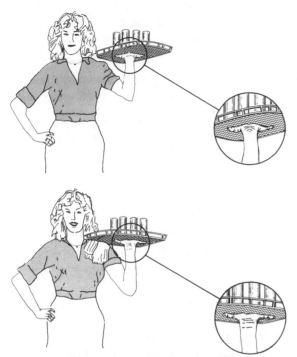

FIGURE 17.8   Serving trays stress the wrist. Reduce this stress by supporting one end on the shoulder; a towel makes a good pad. Note that the forearm is parallel to the axis of the hand to reduce twisting of the wrist under load. By wrapping the fingers around the tray edge, the load is supported by the forearm muscles instead of the finger muscles. The right hand remains free to pick up small items, open doors, carry a tray stand, and so forth. If there is enough room between tables, replace the tray with a cart.

TABLE 17.8   Recommended weight of carry (kg) for one-handed infrequent carrying (Mital et al., 1993). Reduce weight by 30% if carrying is performed frequently.

CARRYING DISTANCE, M	POPULATION PERCENTILE	MALES	FEMALES
30.5	90	6.5	5.5
	75	8.5	7.0
	50	11.0	8.0
	25	13.5	9.0
	10	15.5	10.5
61.0	90	6.0	5.5
	75	8.0	6.5
	50	10.0	7.5
	25	12.0	8.5
	10	14.0	9.5
91.5	90	6.0	5.0
	75	7.5	6.5
	50	9.0	7.0
	25	10.5	8.0
	10	12.0	9.0

Source: A. Mital, A. Nicholson, and M. M. Ayoub, *A Guide to Manual Materials Handling.* Copyright © 1993 by Taylor and Francis, London. Reprinted with permission.

FIGURE 17.9   **J hooks** reduce sheet-carrying problems (Rodgers, 1984). For large sheets, the J hook with a large grip contact area reduces finger strain caused by sharp edges of the sheet (if no hook is used). In addition, because the right hand can be placed in a comfortable position, the posture (and thus back stress) is better; walking and maneuvering with the sheet is easier than without the hook.

$$M = 1.25 \ (GV)^2 + .1$$
$$[47.33 \ G + 87.75 \ L + 21.96 \ W$$
$$+ 197.51 \ V]$$

where

M = Metabolic rate, kcal/h

G = Treadmill gradient, %

V = Walking velocity, km/h

L = Load weight, kg

W = Body weight, kg

5 LIFTING

5.1 Revised NIOSH Lifting Guidelines   In 1993 the National Institute of Occupational Safety and Health (NIOSH) revised its lifting guidelines (Waters et al., 1993, 1994).

The basic concept is to have a load constant of 51 lb (23 kg). This is the maximum that can be lifted or lowered. The **load constant** is multiplied by various factors (all equal to or less than one) to obtain the **recommended weight limit (RWL).** If a person lifts less than RWL, NIOSH considers the task acceptable; if the person lifts more than RWL, NIOSH considers the task not acceptable.

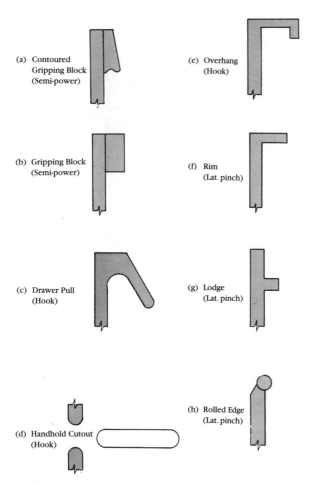

(a) Contoured Gripping Block (Semi-power)

(b) Gripping Block (Semi-power)

(c) Drawer Pull (Hook)

(d) Handhold Cutout (Hook)

(e) Overhang (Hook)

(f) Rim (Lat. pinch)

(g) Lodge (Lat. pinch)

(h) Rolled Edge (Lat. pinch)

FIGURE 17.10   Tray handholds are ranked from (a) to (h) (Eastman Kodak, 1986). The contoured gripping block (a) is superior to the straight gripping block (b) because it provides more grip security and stability when the weight is handled in high lifts; both permit an oblique (semipower) grip. Grips (c) and (d) permit a hook grip and work well for lifts below chest height; they present problems for lifts above shoulder height. Grips (e), (f), (g), and (h) put pressure on a localized part of the hand or only permit use of a lateral pinch grip.

The NIOSH biomechanical criterion is a 350 kg (3.4 kN) compressive force on L5-S1. The metabolic criterion is 9.5 kcal/min (max aerobic capacity of 50% female, age 40), multiplied by 70% (due to arm work) and by 50% (for 1 h), or 40% (for 2 h) or 33% (for 8 h). The NIOSH psychophysical criterion is based on the 75% female (99% male) for a 34-cm wide box, 76-cm vertical displacement, and lifting frequency of 4 lifts/min.

5.1.1 Basic formula   The NIOSH formula is:

$$RWL = LC \times HM \times VM \times DM \times FM \times AM \times CM$$

where

RWL = Recommended weight limit, lb
LC = Load constant = 51 lbs

HM = Horizontal multiplier, proportion
VM = Vertical multiplier, proportion
DM = Distance multiplier, proportion
FM = Frequency multiplier, proportion
AM = Asymmetry multiplier, proportion
CM = Coupling multiplier, proportion

For metric units, LC = 23 kg. Enter distances in cm.

The NIOSH formula **does not apply** if any of the following occur:

- lifting/lowering in which the feet move more than one or two steps
- lifting/lowering with one hand
- lifting/lowering for over 8 h
- lifting/lowering while seated or kneeling
- lifting/lowering in a constrained or restricted workspace
- lifting/lowering hot, cold, or contaminated objects
- lifting/lowering with unpredicted conditions (unexpected heavy loads, slips, falls)
- lifting/lowering of an unstable load (center of gravity varies significantly during the lift)
- lifting/lowering while carrying, pushing, or pulling
- lifting/lowering with wheelbarrows or shovels
- lifting that is high speed (speed of about 30 inches/second—a lift from floor to table height in < 1 s)
- unreasonable foot/floor interface (≤ .4 coefficient of friction between the sole and the floor)
- unfavorable environment (temperature significantly outside 66–79 F (19–26 C) range; relative humidity outside 35–50% range)

The required input information depends on whether significant control over the object is needed (1) at the origin of the movement only or (2) at origin and destination. Control at destination (precise placement) is implied by (a) a regrasp near the destination, (b) a momentary hold near destination, or (c) a need to carefully position or guide the load near the destination.

If control of the object is needed only at the origin, you need:

- initial horizontal location of the hands from the ankle midpoint
- initial vertical location of the hands
- initial angle of asymmetry of object center
- vertical travel distance between the lift origin and destination
- frequency of lifts per minute

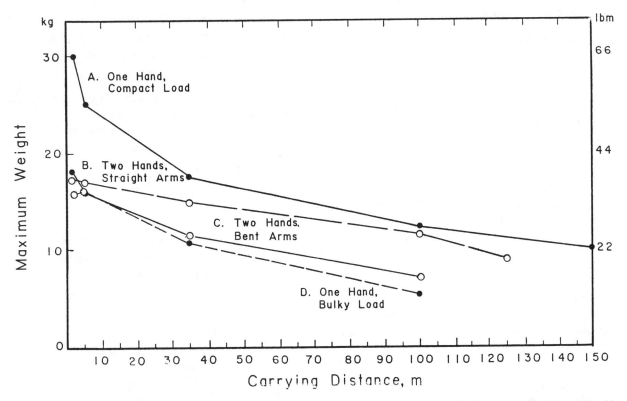

FIGURE 17.11   Maximum weight to be carried for four types of grips (Eastman Kodak, 1986). (A) is for one hand holding a compact load close to the side (e.g., carrying a briefcase). (B) is for two hands with the arms straight and the object rested against the lower abdomen during the carry. (C) is for two hands with the elbows bent and the object rested against the upper abdomen; the biceps in the arms are used to maintain elbow flexion. (D) is for one hand holding a bulky object at least 25 cm from the side to prevent it from banging against the legs during the carry. This statically loads the shoulder.

- lifting duration (h) and recovery time (h)
- hand-container coupling classification

If control is needed at both **origin and destination,** you need in addition:

- final horizontal location of the hands
- final vertical location of the hands
- final angle of asymmetry of object center

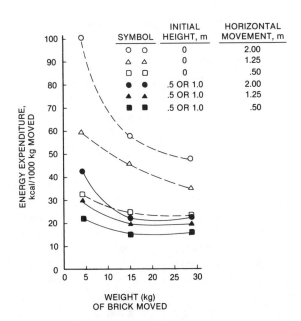

FIGURE 17.12   Minimize metabolic load by not picking up the object from the ground (Mueller, Vetter, and Blumel, 1958).

FIGURE 17.13   Protect your back by protecting your clothes.

5.1.2 Formulas for multipliers

Horizontal multiplier

$$HM = BIL/H$$

where

HM = Horizontal multiplier, proportion

BIL = Body interference limit = 10 for inches
= 25 for cm

H = Horizontal distance from the large knuckle at the end of the third finger to the ankle midpoint (midpoint of inner ankle bones)—this gives the object's **horizontal location.** The maximum value of H (functional reach limit) = 25 inches (63 cm).

If H is measured, it is measured from the projection of the hand midpoint (i.e., load center) to the midpoint of a line connecting the ankles. Note that the feet should not be aligned if a turn is involved.

If H is not measured, estimate H from:

*Vertical location of
hands from the floor*

< 10 inches $H = 10 + W/2$

≥ 10 inches $H = 8 + W/2$

where

W = Width of the container in the plane perpendicular to the shoulders

The value of HM declines as the hands move farther from the spine. If $H \leq BIL$, then the value of HM = 1.0.

Vertical multiplier

$$VM = 1 - VC\,|\,V - KH\,|$$

where

VM = Vertical multiplier, proportion

VC = Vertical constant = .0075 for inches
= .003 for cm

V = Initial vertical height of knuckles— this gives the object's **vertical location.** Note that this usually is several inches above the container bottom. The maximum value of V (vertical reach limit) is 70 inches (175 cm).

KH = Knuckle height (optimum height) of typical lifter (assumed stature height of 66 inches (165 cm))

= 30 inches (75 cm). When $V \leq 30$ inches, the lifting is considered as whole body; when $V > 30$ inches, the lift is considered upper body.

The value of VM declines from 1.0 for any height departing from optimum knuckle height of 30 inches. The concept is a 22.5% penalty for lifts from the floor or shoulder (60 inch; 150 cm).

Distance multiplier

$$DM = .82 + DC/D$$

where

DM = Distance multiplier, proportion

$.82$ = Multiplier at maximum hand height of 70 inches (175 cm)

DC = Distance constant = 1.8 for inches
= 4.5 for cm

D = Distance moved vertically (absolute value)—the **vertical travel distance,** inches or cm

If $D \leq 10$ inches (25 cm), the value of $DM = 1.0$. Maximum $D = 70 - V$ (for inches) and $175 - V$ (for cm).

Asymmetry multiplier

$$AM = 1 - .0032\,A$$

where

AM = Asymmetry multiplier, proportion

A = Angular deviation (also called **angle of symmetry**) (degrees) of the midpoint of the two hands (container center) from straight ahead (neutral body posture; sagittal plane). A can range from 0 to 135°; ignore direction as clockwise movement is considered equivalent to counterclockwise movement. The concept is a 30% penalty for a 90° angle.

Frequency multiplier   See Table 17.9. **Lifting frequency** can range from less than one in 5 minutes (.2 lifts/min) to 15 lifts/min. It is the mean number of lifts in a 15-minute period. If lifting is not continuous, count the number of lifts in 15 minutes and divide by 15. Any frequency less than .2 lifts/min is set equal to .2 lifts/min. The frequency multiplier varies depending on **lifting duration**/session and whether the initial vertical location of the hands is above or below typical knuckle height (30 inches or 75 cm).

TABLE 17.9   Frequency multiplier (FM).

FREQUENCY LIFT/MIN	≤ 1 HRS		> 1 BUT ≤ 2 HRS		> 2 BUT ≤ 8 HRS	
	V < 75 CM (30 IN)	V ≥ 75 CM (30 IN)	V < 75 CM (30 IN)	V ≥ 75 CM (30 IN)	V < 75 CM (30 IN)	V ≥ 75 CM (30 IN)
≤0.2	1.00	1.00	.95	.95	.85	.85
0.5	.97	.97	.92	.92	.81	.81
1	.94	.94	.88	.88	.75	.75
2	.91	.91	.84	.84	.65	.65
3	.88	.88	.79	.79	.55	.55
4	.84	.84	.72	.72	.45	.45
5	.80	.80	.60	.60	.35	.35
6	.75	.75	.50	.50	.27	.27
7	.70	.70	.42	.42	.22	.22
8	.60	.60	.35	.35	.18	.18
9	.52	.52	.30	.30	.00	.15
10	.45	.45	.26	.26	.00	.13
11	.41	.41	.00	.23	.00	.00
12	.37	.37	.00	.21	.00	.00
13	.00	.34	.00	.00	.00	.00
14	.00	.31	.00	.00	.00	.00
15	.00	.28	.00	.00	.00	.00
>15	.00	.00	.00	.00	.00	.00

Lifting duration/session in hours has 3 categories:

- short = .001 h to ≤ 1 h with recovery time of at least 1.2 (duration)
- moderate = > 1 h but ≤ 2 h with recovery time of at least .3 (duration)
- long = > 2 h but ≤ 8 h

During recovery time, the person is resting or has light work (such as sitting, standing, walking, monitoring). If a person does not meet the recovery criterion, omit the recovery time and add the work times together.

Coupling multiplier   See Table 17.10 and Figure 17.14. The coupling multiplier depends on the height (i.e., whether *V* is ≤ 30 inches) of the initial and final hand–container coupling and whether the coupling is good, fair, or poor. Table 17.11 defines optimal container design, handle design, and hand-hold cutout design. The following is an example problem and solution.

Example problem with solution. Assume a container with a weight of 15 lb was lifted and control was needed at both origin and destination. (Waters et al., 1994, has 10 detailed examples.)

TABLE 17.10   Coupling multiplier (CM).

COUPLINGS	V < 75 CM (30 IN)	V ≥ 75 CM (30 IN)
Good	1.00	1.00
Fair	0.95	1.00
Poor	0.90	0.90

Load weight = 15 lb
Initial horizontal location = 12 in.
Initial vertical location = 33 in.
Final horizontal location = 12 in.
Final vertical location = 22 in.
Initial angle of asymmetry = 5°
Final angle of asymmetry = 6°
Lift frequency = .5 lifts/min

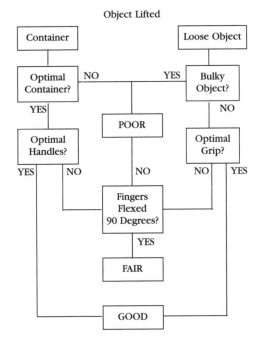

FIGURE 17.14   Decision tree for coupling quality (Waters et al., 1994).

TABLE 17.11 Hand-to-CONTAINER coupling classification and definitions of optimal container, handle, and cut-out.

1. good	—	optimal container design with optimal handles or optimal handhold cut-outs
	—	loose parts or irregular objects with comfortable grip (hand can easily be wrapped around object)
2. fair	—	optimal container design with non-optimal handles or non-optimal handhold cut-outs
	—	loose parts with no handles or handhold cut-outs
	—	irregular objects with a grip in which fingers can be flexed about 90°
3. poor	—	non-optimal design containers with no handles or handhold cut-outs
	—	loose parts or irregular objects that are bulky or hard to handle

	OPTIMAL	NON-OPTIMAL
CONTAINER design		
Frontal length	≤ 40 cm (16 in)	Non-optimal is failure to meet
Height	≤ 30 cm (12 in)	one or more optimal conditions
Surface	smooth, non-slip	
Edge	non-sharp	
Center of mass	symmetric	
Load	stable	
Gloves required	no	
HANDLE design		
Diameter	1.9 to 3.8 cm (0.75 − 1.5 in)	
Length	≥ 11.5 cm (4.5 in)	
Clearance	≥ 5 cm (2 in)	
Shape	cylindrical	
Surface	smooth, non-slip	
HANDHOLD CUT-OUT design		
Height	≥ 3.8 cm (1.5 in)	
Length	≥ 11.5 cm (4.5 in)	
Clearance	≥ 5 cm (2 in)	
Container thickness	≥ 1.1 cm (.43 in)	
Shape	semi-oval	
Surface	smooth, non-slip	

Lift duration/session = 3 h

Recovery time = 6 h

Then initial $HM = 10/12 = .83$

Final $HM = 10/12 = .83$

Initial $VM = 1 − .0075\,|\,33\text{-}30\,| = .98$

Final $VM = 1 − .0075\,|\,22\text{-}30\,| = .94$

Vertical distance multiplier $= DM = .82 + 1.8/11$
$= .98$

Frequency multiplier $= FM = .81$

Initial asymmetry multiplier $= AM = 1 − .0032\,(5)$
$= .98$

Final asymmetry multiplier $= AM = 1 − .0032\,(6)$
$= .98$

Initial coupling multiplier $= CM = 1.0$

Final coupling multiplier $= CM = .95$

Thus RWL at lift origin $= 51 \times .83 \times .98 \times .98$
$\times .81 \times .98 \times 1.0$
$= 32.6$ lb

RWL at destination $= 51 \times .83 \times .94 \times .98$
$\times .81 \times .98 \times .95$
$= 29.7$ lb

The 29.7 lb at the destination is limiting since it is lower.

The lifting index (LW/RWL) = 15/29.7 = .51. Since the lifting index should be less than 1.0, the situation is satisfactory.

If the situation is not satisfactory, see Table 17.12 for suggestions.

TABLE 17.12 Ergonomic design suggestions (Waters et al., 1994).

SITUATION	SUGGESTION
$HM < 1$	Bring load closer to body (remove horizontal barriers, reduce object size). Avoid lifts near the floor; if unavoidable, object should fit easily between the legs.
$VM < 1$	Raise (lower) the origin (destination) of lift. Avoid lifting near the floor or above the shoulders.
$DM < 1$	Reduce vertical distance between origin and destination.
$AM < 1$	Reduce twisting by reducing distance between origin and destination; an alternative is to *increase* the distance so the worker must step rather than twist the body.
$FM < 1$	Reduce frequency rate. Reduce duration. Provide longer recovery periods (working rest of light work).
$CM < 1$	Improve containers, handles, cutouts, and handholds for irregular objects.
RWL at destination $>$ RWL at origin	Eliminate need for control at destination. Change job or container.

The above describes analysis of a single task.

The Guideline also has a "multi-task" procedure. This is for the situation in which several tasks are done. An example would be alternately placing a box on a conveyor on the left and on the right; another example would be lifting the first, second, and third row of boxes off a pallet.

The multi-task procedure is:

1. Compute two versions of the recommended weight limit: frequency-independent (FIRWL) and single-task (STRWL).

2. Compute two versions of the **lifting index**: frequency-independent (FILI) and single-task (STLI).

3. Compute the composite lifting index (CLI).

1. Recommended weight limit

   A. Frequency-independent RWL (FIRWL). For each task, compute RWL but set the frequency multiplier to 1.

   B. Single-task RWL (STRWL). For each task, multiply FIRWL by its appropriate frequency multiplier. (Each STRWL is calculated as if each task were the only task being performed.)

2. Lifting index

   A. For each task, compute the frequency-independent lifting index (FILI). The formula is:

   $$FILI = L/FIRWL$$

where

   FILI = Frequency-independent lifting index
   L = Load of the task
   FIRWL = Frequency-independent RWL

   B. For each task, compute the single-task lifting index (STLI). The formula is:

   $$STLI = L/STRWL$$

where

   STLI = Single-task lifting index
   L = Load for that task. If the load varies, use the average of the various loads as this best represents the metabolic demand.
   STRWL = Single-task recommended weight limit

3. Composite lifting index

   A. Renumber the tasks in order of the STLI value (i.e., greatest physical stress first).

   B. Compute composite lifting index (CLI). The formula is:

   $$CLI = STLI_1 + ILI$$

where

   CLI = Composite lifting index
   $STLI_1$ = Largest (first) single-task RWL
   ILI = Sum of incremental lifting indices
   $= FILI_2 (1/FM_{1,2} - 1/FM_1) + FILI_3$
   $(1/FM_{1,2,3} - 1/FM_{1,2}) \ldots$
   $+ FILI_n (1/FM_{1,2,3\ldots n} - 1/FM_{1,2\ldots n-1})$

The subscripts are for the new task numbers. The FM values are from the frequency table.

Example Problem with Solution. Assume an operator moved three types of boxes from a 36-inch high conveyor to three pallets; the 10-lb boxes were moved 4/min (task 1), the 20-lb boxes were moved 2/min (task 2), and the 30-lb boxes were moved 1/min (task 3). The job is done for less than 1 h/day.

1. Recommended weight limit

   A. Assume the FIRWL (i.e., using a frequency multiplier of 1) for the three tasks was 15, 20, and 20 lbs.

   B. From Table 17.9, the appropriate multipliers are .84, .91, and .94. Then the single-task RWL (STRWL) is (15)(.84) = 12.6, (20)(.91) = 18.2, and (20)(.94) = 18.8 lbs.

2. Lifting index

   A. Compute FILI. For the three tasks, FILI is 10/15 = .67, 20/20 = 1, and 30/20 = 1.5.

   B. Compute STLI. The average weight for tasks 1, 2, and 3 are 10, 20, and 30 lbs, so the STLIs are 10/12.6 = .8, 20/18.2 = 1.1, and 30/18.8 = 1.6.

3. Composite lifting index

   A. Renumbering the tasks in order of stress (i.e., STLI value), the 30-lb weight task is numbered 1, the 20-lb task is numbered 2, and the 10-lb task is numbered 3.

   B. Compute CLI. From the new task 1, the initial lifting index is 1.6.

   Then add the FILI value from task 2, adjusted by the combined frequencies of tasks 1 and 2. The frequency of task 1 is 1/min so the FM value is .94. The combined frequencies of tasks 1 and 2 are 3 lifts/min so the FM value is .88. The additional FILI is 1.0 (1/.88 − 1/.94) = 1.0 (1.136 − 1.064) = .07.

   Then add the FILI value from task 3, adjusted by the combined frequencies of tasks 1, 2, and 3. The combined frequencies of tasks 1, 2, and 3 are 7 lifts/min so the FM value is .70. The additional FILI is .67 (1/.70 − 1/.88) = .67 (1.429 − 1.136) = .20.

   The total CLI is 1.6 + .07 + .20 = 1.9.

An IBM-compatible PC computer program to do the NIOSH lifting calculations is available on a 3.5-inch disk for $2 postpaid. Send $2 in cash or a check to Prof. Stephan Konz, Dept. of Industrial Engineering, Kansas State University, Manhattan, KS 66502.

5.2 Force Limits

The formula used (Mital et al., 1993) is:

$$FL = A \ (F) \ (DIST)$$

where

FL = Force limit, kg

A = Age factor

  = 1.0 for males < 40 years

  = .915 for males 41–50 years

  = .782 for males 51–60 years

F = Frequency factor

  = 1.0 for frequencies < 1/min

  = .7 for frequencies ≥ 1/min

$DIST$ = Distance factor

  = $50.15 + .332 \ V - .0066 \ V^2 -$
  $.000 \ 087 \ 7 \ V^3 - .647 \ H -$
  $.003 \ 72 \ VH + .000 \ 073 \ 5 \ V^2 H$

V = Vertical distance of the load from the shoulder, cm

H = Horizontal distance of the load from the shoulder, cm

Freivalds (1987) has an interesting comparison of the Force Limits guideline and the 1981 NIOSH guideline:

- Only males were considered by Force Limits, vs. males and females for NIOSH.

- Age is an adjustment for Force Limits but not for NIOSH.

- Frequency has only two categories for Force Limits but varies continuously for NIOSH; NIOSH also varies the factor depending on lifting duration.

- Force Limits is based on a maximum abdominal pressure of 90 mm Hg for the fifth percentile weight and fifth percentile height male. However, NIOSH used three criteria: (1) compressive force on the L5/S1 disc (350 kg for Action Limit and 650 kg for Maximum Permissible Limit), (2) metabolic rate (3.5 kcal/min for AL and 5.0 kcal/min for MPL), and (3) psychometric votes.

- The two guidelines do not agree very well—with the Force Limits guideline, on average, permitting a load of 1.8 times the NIOSH load.

5.3 MMH Guide

Mital et al. (1993) have a 114-page book with detailed recommendations. For example, their Table 4.2 gives 720 recommended weights of lift for male industrial workers for two-handed symmetrical lifting for 8 h. It has 4 variables:

- box size (75, 49, and 34 cm)

- frequency of lift (1/8 h, 1/30 min, 1/5 min, 1/min, 4/min, 8/min, 12/min and 16/min)

- lifts (floor to 80 cm, floor to 132 cm, floor to 183 cm, 80 to 132 cm, 80 to 183 cm, and 132 to 183 cm)

- population percentile (90, 75, 50, 25, 10)

There also are modifying values for work duration (1, 4, 8, 12 h); limited headroom (full upright, 95% upright, 90% upright, 85% upright, and 80% upright); asymmetrical lifting (0–30°, 30–60°, and above 90°); load asymmetry (sidewise shift in frontal plane of 0, 10, 20, 30 cm); coupling (comfortable handles, poor handles, no handles); load clearance at destination (unlimited to 30 mm, 15 mm, 3 mm); and heat stress (up to 27 C WBGT (Wet Bulb Globe Temperature), 32 C). Their values are based on a maximum load of 27.2 kg.

6   GUIDELINES FOR MANUAL MATERIAL HANDLING

The NIOSH guidelines, the Force Limits, and the MMH Guide follow the approach of recommending maximum weights for specific situations. This section, however, will follow a different approach and give design guidelines without any specific numerical recommendations.

Table 17.13 summarizes the excellent work of Ayoub and his colleagues (1987).

Table 17.14 gives 10 guidelines for occasional lifting. They are discussed in more detail below. They are divided into three groups: (1) select individual, (2) teach technique, and (3) design the job.

6.1   Select Strong People Based on Tests

A number of studies (Liles, 1986; Ayoub et al., 1987; Herrin et al., 1986) have shown the importance of the job severity index:

$$JSI = f(WEIGHT/CAPACITY)$$

where

JSI = Job severity index

$WEIGHT$ = Weight lifted or moved

$CAPACITY$ = Capacity of worker for that specific task

One approach is to select workers with large capacities. The point of this first guideline is that the selection needs to be based on specific tests, rather

TABLE 17.13 Manual material handling design guidelines; adapted from Ayoub et al. (1987).

Guideline	Examples
1. Eliminate heavy MMH	Use mechanical aids (hoists, lift tables, conveyors). Provide best work height (change height of work or height of worker).
2. Decrease stress	Reduce object weight (split load between two workers or containers, reduce container size, reduce container (tare) weight). Change type of MMH (lower instead of lifting, push not pull, pull not carry). Reduce distances (both horizontal and vertical) in reaching for object and in transporting it. Avoid twisting (both standing and seated work). Make object easier to handle (handles, balanced containers, reasonable width) Increase recovery time (lower frequency, job rotation).

than simply selecting large males. In addition, the capacity needs to be determined for the specific task, not all possible tasks. That is, a person might have good upper body strength but poor leg strength or vice versa.

Note that use of the **job severity index** forces analysis not only of the worker capacity but of the task requirements. Naturally it is very desirable to reduce the requirements of the task as this not only permits a larger proportion of the population to do the job but reduces the stress on those doing the job. There are a number of computer models now available to aid in the analysis (Karwowski et al., 1986; Chaffin and Evans, 1986).

X-rays are not recommended as a screening device due to the radiation exposure as well as their

TABLE 17.14 Guidelines for occasional lifting.

Guideline

Select Individual
1. Select strong people based on tests.

Teach Technique
2. Bend the knees.
3. Don't slip or jerk.
4. Don't twist during the move.

Design the Job
5. Use machines.
6. Move small weights often.
7. Put a compact load in a convenient container.
8. Get a good grip.
9. Keep the load close to the body.
10. Work at knuckle height.

poor predictive ability (Rowe, 1985; Chaffin and Andersson, 1984). If employees are going to be eliminated from lifting jobs, use a multiple-criteria screen including such variables as strengths of specific muscle groups, previous back injuries, uneven leg length, and percent body fat. For example, a large belly adds a constant stress to the back muscles. Uneven leg length puts an unbalanced load on the back.

Ayoub et al. (1987), for example, recommend measuring an individual's weight, arm strength, shoulder height, back strength, abdominal depth, and dynamic endurance. Various regression equations and tables then are used to predict capacity for a specific task. The goal is to use only people who have a capacity of at least 150% of requirements.

6.2 Bend the Knees Brown (1975) says, "The straight back, bent knees technique has been enthusiastically promoted for 40 years with no decrease in back injuries . . . which raises a question as to the benefit of lifting procedures advocated." Perhaps it's because no one uses the bent knees technique. One possible reason there has been a lack of worker acceptance of squat lifting is that when you **bend the knees** you lower the torso and thus require more energy. Garg et al. (1983) compared squat vs. free-style lifting (subjects could lift any way they wanted to) and found that, when free-style lifting was used, the subjects tended to pull the load toward the rear when lifting it, thus decreasing strain, vs. a straight vertical lift as typically assumed for a stoop lift. Andersson and Chaffin (1986) evaluated stoop lifting vs. four types of squat lifting. They concluded that the squat lifting techniques are best and emphasized (1) keeping the load close to the body, (2) using a straddle stance (feet not aligned) for bulky loads, and (3) keeping the back aligned (as when standing erect) throughout the lift. Figure 17.15 shows the **one-hand lift** (Lovested, 1980). During the stoop lift of items out of the container, the worker is trained to support the back muscles by holding one side of the container with the hand that is not grasping the object. However, don't use training as a substitute for design.

6.3 Don't Slip or Jerk The problem with slipping is a sudden unexpected load on the back. Reduce slipping by having a high coefficient of friction between the shoe and the floor. See Guideline 1 of Chapter 15. In summary, this involves workers having shoes with high coefficient of friction soles and large contact areas. It also involves high coefficient of friction floors—primarily a matter of good housekeeping.

Data gathered on a force platform (e.g., Konz and Bhasin, 1974) indicate that peak forces and

FIGURE 17.15   One hand supports the back while the other hand grasps.

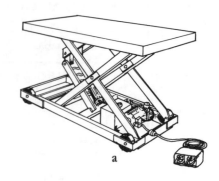

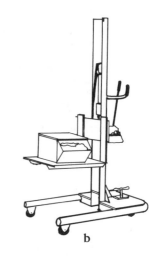

FIGURE 17.16   **Workstation positioners** reduce lifting, bending, and twisting. (A) shows a scissors lift. (B) shows a bin on an adjustable-height cart—useful for temporary workstations such as stocking of shelves. (C) shows self-leveling trucks (similar to cafeteria tray dispensers).

torques during a lift are very "spiked." The peak force often occurs during the initial lowering of the body before the object is even grasped! The accelerations and decelerations during the lifting and lowering should be fast enough that the body gets the benefits of momentum but not so fast as to cause injury. Professional weightlifters use "clean and jerk" movements but these movements are highly practiced in precisely standardized situations, not industrial lifts. Patterson et al. (1987) demonstrated that, when lifters know the weight of the object being lifted, they have less strain on their bodies because they use different movement patterns. They believe teaching the worker to recognize the type of load to be lifted is important. Marras et al. (1986) also pointed out that additional strain occurs because of unexpected loads.

6.4   Don't Twist During the Move   Although spine biomechanics are complex during symmetrical two-handed lifting (as in NIOSH 1981 guidelines), they are even more complex when twisting is involved. These variables are now being quantified (Mital et al., 1993), but much work needs to be done.

Twisting while bent over (such as when taking parts off the floor) probably is especially bad. A general recommendation is to tell the workers to move their feet instead of twisting but this is difficult to get people to follow.

Figure 17.16 shows how the task can be modified to reduce bending and twisting.

In general, the strategies of personnel selection (Guideline 6.1) and training (6.2, 6.3, and 6.4) have not been very fruitful in reducing back problems. This recalls the analogy of the open manhole. The selection approach is to select people who can walk around the open hole. The training approach is to train the people to walk around the open hole. The engineering approach is to put a cover on the hole. The engineering approach is recommended because it solves the problem permanently. The selection and training approaches do not guarantee a solution and require continual repeating as the employees either forget or change jobs.

6.5 Use Machines

Although this section concerns *manual* material handling, the first solution to consider is eliminating the manual handling by using machines. For example, instead of moving a power tool around the workstation, use a balancer (see Figure 17.17) or manipulator (see Figure 17.18). Figure 17.19 shows the use of a turntable on a scissors lift to mechanize palletizing. In automation, the operator would be replaced with a palletizer.

In some circumstances, robots (manipulators with "brains") are appropriate.

Common machines used to eliminate manual handling are hoists, lift trucks, and conveyors. When loading/unloading trucks, use portable and telescoping conveyors. Note that lift trucks have a wide variety of attachments (forks, clamps, rotators, rams, push–pull, etc.). In England, mail carriers use carts. See Figure 17.20.

A lever arm can be considered a machine. Figure 17.7 showed how a tipping aid reduced the load on the muscles. Figure 17.21 shows how moving the fulcrum of the ladder support reduced the load. Figure 17.22 shows how a simple lever system can help transfer loads between levels. Other examples of lever arms are pry bars, ramps, and screws.

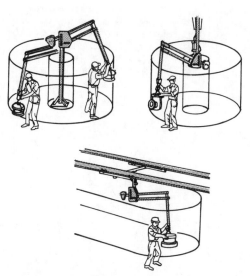

FIGURE 17.18   **Manipulators** (balancers with arms) can support tools or can be used to move product around the workstation.

6.6 Move Small Weights Often

From a viewpoint of reducing strain on the skeletomuscular system, small is beautiful. That is, small weights are better than big weights. What this means in practice is that weights must be reduced or loads moved more frequently.

The total weight to be moved is the weight of the product plus the weight of the container. Can a steel container be replaced with a plastic or fiberboard container?

For example, assume 1,000 recycled flat cardboard boxes (each weighing .6 kg) need to be moved

FIGURE 17.17   **Balancers** reduce tool weight from pounds to ounces. Note that the balancer is suspended from a jib crane to minimize horizontal forces as the tool is moved about the workplace. Ulin et al. (1993) showed that an increased weight tool increased the torque at the elbow and shoulder, decreasing the comfort of the neck and back as well as the arm.

FIGURE 17.19   **Turntables** can reduce stress when loading/unloading. The turntable is rotated after a few items are moved. The turntable can be mounted on the floor and a pallet placed on the turntable. If the turntable surface is rollers or ball casters or an air table, rotation can be manual. If the turntable is mounted on a **scissors lift** (as shown), horizontal transfer can replace lifting/lowering.

FIGURE 17.20 Mail is moved on carts in England, not in a shoulder bag as is typical in the United States. The precut pieces of twine are looped around the frame for easy access.

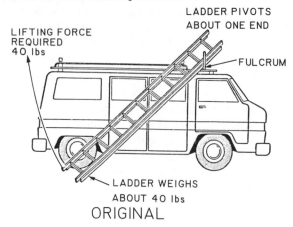

FIGURE 17.21 Positioning the fulcrum can help (Kantowitz and Sorkin, 1983). The original fulcrum was at the right, so when raising the left end of the ladder the entire weight had to be lifted above the head. When potential female employees couldn't do the job, the task was analyzed and the fulcrum moved (lower figure). Now the weight of the right portion of the ladder counterbalances the portion being raised. This helped not only the new female workers but also the existing male workers.

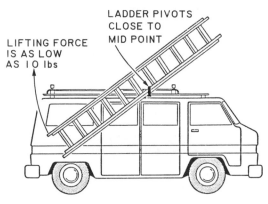

Source: B. Kantowitz and R. Sorkin, *Human Factors.* Copyright © 1983 by John Wiley & Sons. Reprinted by permission of John Wiley & Sons.

from a truck to a pallet. The knocked-down boxes can be tied in bundles of 25 at the supermarket or in bundles of 20. If there are 25 in a bundle, 40 bundles of 15 kg need to be moved. If there are 20 in a bundle, 50 bundles of 12 kg need to be moved. In general, keep loads (including container weight) less than 15 kg—especially if the load is awkward. Below 10 kg has relatively little strain and the number of moves rises quite rapidly.

Pay close attention to the container size that vendors use when shipping to you. Get the purchasing department away from their desks and out to the receiving dock! Reducing strain on your receiving personnel also may reduce strain on vendors' shipping personnel.

One design strategy is to make the load weight so high (say 80 kg) that people must use a machine to move it and don't even consider manual handling.

Naturally, one way to reduce a weight is to let gravity do the work; that is, lower loads rather than lift loads. See Figure 17.23.

Another way to reduce a weight is to lift with a partner rather than lift the entire weight yourself. Or it may be possible to lift one end of the object, move it, and then lift the other end and move it (as with furniture).

There are a variety of simple techniques to orient and reorient items on conveyors. Some are just a protruding bar or angled piece of metal. If you see a person twisting to orient an item on a conveyor, ask why the orientation isn't done by a device.

When sliding a carton or item into or from a workstation from a wheel or roller conveyor, have the movement downhill. For example, when loading a carton, have the conveyor at a small angle immediately after the assembly area so as to minimize the force required to slide the packed carton.

6.7 Put a Compact Load in a Convenient Container
Within a job design the key concept is to keep the load close to the spine. Tichauer (1973) proposed considering the **spinal torque:**

$$SPINET = OBJWT \ (OBMARM)$$

where

$SPINET$ = Spinal torque that body exerts in reaction, kg-m

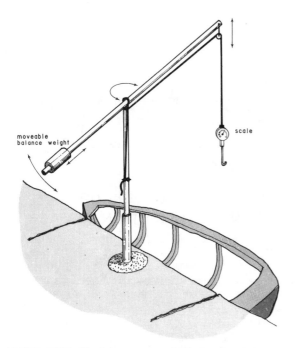

FIGURE 17.22 Simple lever arms can drastically reduce strain on people moving items between levels. Fishermen can lift loads out of the unstable boat and also weigh them.

$OBJWT$ = Object weight, kg

$OBMARM$ = Moment arm of the object, m

= $DISTO + DISTCG$

$DISTO$ = Distance between spine and object (closest portion), m

= .2. However, Damon et al. (1966) say mean male waist depth is .20 to .25 m so .3 might be a better value. See Figure 17.13. Note the effect of a protruding belly.

$DISTCG$ = Distance from closest portion of the object to its center of gravity = $L/2$, m

For example, if $L = .2$, the $DISTCG = .1$ and $SPINET = .3\ OBJWT.$

Thus, because of the bulk, it is more difficult to lift 25 kg of feathers than 25 kg of iron.

Konz and Coetzee (1977) reported that increasing a box's volume (up to a 30-cm cube) did not bother men but strongly bothered women.

6.8 Get a Good Grip   For tray grips, see Figure 17.10. Cardboard boxes with gripping holes are poor because (1) the hand can't rotate as the object is lifted from the knee to above the waist and (2) the cardboard surface area is small and tends to put too much pressure on the hands. For bags, Smith (1981) reported that people were willing to lift 10% less for a 70% full bag than a 90% full bag; they were willing to lift 10% more when the bag had handles on both ends. Of course most objects don't have handles. Course and Drury (1982) reported that the best grip for boxes without handles was an opposition hand-hold—one hand on the upper-outer corner and the other on the opposite lower-inner corner. Although it causes a higher load on the lower hand than both hands on the bottom, it gives more stability.

6.9 Keep the Load Close to the Body   This is a follow up of the *SPINET* equation above. Actually the equation should include an additional term for the body in addition to the term for the load. That is,

$SPINET = OBJWT\ (OBMARM) + UBWT\ (BMARM)$

where

$UBWT$ = Upper body weight, kg. See Table 15.2.

$BMARM$ = Upper body moment arm, m

For example, a 70-kg person has a head of .073 (70) = 5.1 kg, two arms and hands of .098 (70) = 6.9 kg, and a torso of .51 (70) = 35.7 kg. The key problem is not during the carrying of the load but the reaching for and disposal of the load. Let's assume, during a .3-m reach, the center of gravity of the head is .15 m forward, the arms are .2 m forward,

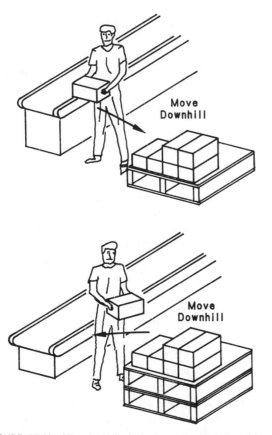

FIGURE 17.23 Move downhill whether from a conveyor to a pallet or a pallet to a conveyor. When loading a conveyor from the pallet, adjust the conveyor to be low; raise the pallet by having the lift truck driver place it on a platform (such as an empty pallet) or a platform. When loading a pallet from a conveyor, have the conveyor higher and the pallet on the floor.

and the torso is .1 m forward. Then the head torque is .76 kg, the arm torque is 1.38 kg, and the torso is 3.57 kg—a total of 5.71 kg-m, even with a zero torque from the object! If an object is only .1 m closer (e.g., by having the item reached for come down the operator side of the conveyor instead of the center of the conveyor), then .05 (5.1) + .1 (6.9) + 0 = .94 kg-m instead of 5.7 kg-m!

Another way to keep the load close to the body centerline is to consider the foot position when turning with the load. Rather than having both feet aligned, have the foot opposite the direction of turn ahead and the foot on the side of the turn to the rear (Konz and Bhasin, 1974).

Orient pallets and bins to reduce reach distance; see Figure 17.24.

When loading a pallet with cases, consider use of a wooden backboard on two sides of the pallet. This permits the operator to toss the case rather than leaning over with the heavy case to place it precisely on the pallet. The same concept of a backboard applies when tossing cartons onto a conveyor; if the rails are too short, the operator needs to exert too much care. (On the other hand, where cartons are being picked from a conveyor, have no rails so sliding can replace lifting. See Figure 17.25.)

Poulsen and Jorgensen (1971) gave the following formula concerning back pain from working stooped (that is, with no external load):

$$MISBCK = .4 \ (WT) \ (\sin \alpha) \ / \ LTSLPR$$

where

$MISBCK$ = Maximum isometric back muscle strength needed to avoid back pain while working standing stooped at angle α, kg

WT = Body weight, kg

α = Acute angle between the back and the

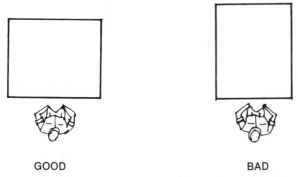

GOOD   BAD

FIGURE 17.24 Orient pallets and bins to minimize reach distances. The long axis of the pallet or bin (not the short axis) should be parallel to the shoulders. Another option is to unload half of the pallet and then rotate the pallet 180°.

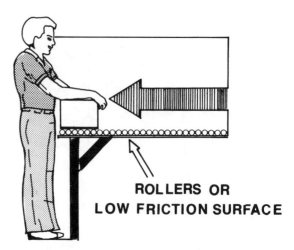

ROLLERS OR LOW FRICTION SURFACE

FIGURE 17.25 Slide, don't lift, to reduce strain. Eliminate rails on the conveyor where items are being removed whether the removal is from the end of the conveyor or the side. *Leave room for the feet.*

line of gravity, degrees

$LTSLPR$ = Maximum long-term static load proportion of the maximum isometric strength. A value of 15% is commonly used.

Thus if a worker bent over at an angle of 20° and $LTSLPR = .15$, then $MISBCK = .4 \ WT \ (.342)/.15 = .91 \ WT$. That is, this individual would have back pain unless the individual's back strength was at least 91% of the individual's body weight. The solution generally is to raise the height of the object being worked on.

6.10 Work at Knuckle Height   Of course, the first thought should be to avoid lifting altogether. Drag it, pull it, push it, but don't lift it. Wheeled luggage and trash containers are examples. If a lift is necessary, however, the work behind the NIOSH guideline emphasizes the desirability of material handling at knuckle height. The other side of the coin is don't put a load on the floor or work above your shoulder.

6.10.1 Knuckle height   See Guideline 3 in Chapter 15 for techniques of adjusting height. Also see Figure 17.16.

For example, when loading cartons into a semi from a conveyor, have the conveyor height adjustable. When loading cartons into a semi from a pallet, leave the pallet on the lift truck's forks. Periodically adjust fork height. If you don't want to tie up the lift truck, have the pallet placed on a wheeled lift table.

6.10.2 Don't put a load on the floor   There is a large penalty for lifting from the floor. (In addition to the stress of lifting a load from the floor, there is the stress of putting it on the floor.) Avoid this penalty by not putting an object on the floor in the first place. Figure 17.26 shows how laundry workers reduced their strain by having drop-delivered items fall onto a pallet instead of the floor.

Lovested (1980) reported that many problems are associated with workers obtaining parts by reaching into deep wood, cardboard, and wire containers placed on the floor. Some solutions included using boxes which had one-half of one side that could be folded down, cutting away the side of cardboard boxes, placing boxes on their sides (on a table or stand), using shorter boxes (19 in. high vs. former 33 in. high), and placing the short box on a platform. The stress on the back also can be reduced by picking up multiple items with each reach, thus reducing the number of times the operator has to bend over.

6.10.3 Don't lift above the shoulder

The final location of the load is as important as the initial location. High lifting is bad from two viewpoints. (1) Muscular strength is relatively poor above the shoulder because the relatively small shoulder and arm muscle groups take over for the leg and back muscles. (2) It is dangerous to remove an object from a high shelf because the object may be dropped and may be damaged (also possibly damaging you!). See Figure 17.27.

If you have to stand on tiptoe, the stress, of course, is even worse on the arms and legs. In addition it stresses the feet and legs. Tiptoe standing (i.e., over-reaching) decreases maximum weight of lift 10 to 15% over reaching as high as possible with the feet flat (Mital and Aghazadeh, 1987).

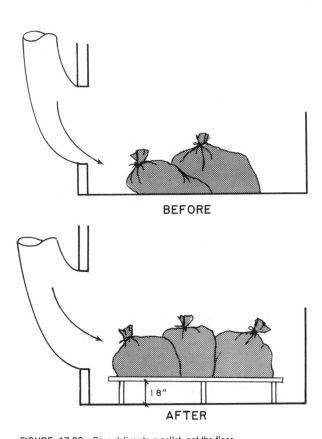

FIGURE 17.26 Drop deliver to a pallet, not the floor.

FIGURE 17.27 Avoid high shelves. Adding the high shelves on this cart also added to the weight that had to be pushed and made the load unstable.

REVIEW QUESTIONS

1. The NIOSH lifting guidelines are based on minimizing stress where in the body? The Force Limits guidelines are based on minimizing stress where in the body?

2. Is arm strength greatest at reach distance, .75 (reach distance), or .5 (reach distance)?

3. How does wearing an apron protect you against back injury?

4. The NIOSH guideline assumes (a) one-hand lift using handles, (b) one-hand lift without handles,

(c) two-hand lift with or without handles, or (d) two-hand lift with handles.

5. The NIOSH lifting guideline has a load constant of 23 kg and six multipliers. Name the six multipliers.

6. The NIOSH guidelines are for (a) men and women, with no age adjustment, (b) men and women, with a 10% reduction for those over 50, (c) men only, with no age adjustment, or (d) men, with a 10% reduction for those over age 50.

7. Briefly discuss the job severity index.

8. Discuss the open manhole analogy. Which approach is best?

9. How should a 40 × 48 × 9-inch pallet be oriented to minimize strain when loading or unloading boxes from the pallet?

10. What is the concept of the one-hand lift?

11. What could be carried with a J hook?

12. Sketch a lever system to move loads between levels.

REFERENCES

Andersson, G. and Chaffin, D. A biomechanical evaluation of five lifting techniques. *Applied Ergonomics,* Vol. 17, 2–8, 1986.

Ayoub, M. M., Selan, J., and Jiang, B. Manual material handling. In *Handbook of Human Factors,* Salvendy, G. (ed.), Chapter 7.2. New York: Wiley & Sons, 1987.

Balogun, J., Robertson, R., Goss, F., Edwards, M., Cox, R., and Metz, K. Metabolic and perceptual responses while carrying external loads on the head and by yoke. *Ergonomics,* Vol. 29 [12], 1623–35, 1986.

Brown, J. Factors contributing to the development of low back pain in industrial workers. *American Industrial Hygiene Association J.,* Vol. 36 [1], 26–31, 1975.

Burton, A. Spinal strain from shopping bags with and without handles. *Applied Ergonomics,* Vol. 17 [1], 19–23, 1986.

Chaffin, D. Manual materials handling and the biomechanical basis for prevention of low-back pain in industry—An overview. *Am. Ind. Hygiene Assoc. J.,* Vol. 48 [12], 989–96, 1987.

Chaffin, D. and Andersson, G. *Occupational Biomechanics,* New York: Wiley & Sons, 1984.

Chaffin, D. and Evans, S. Computerized biomechanical models in manual work design. *Proceedings of the Human Factors Society,* 96–100, 1986.

Course, B. and Drury, C. Optimum handle positions in a box holding task. *Ergonomics,* Vol. 25, 645–62, 1982.

Damon, A., Stoudt, H., and McFarland, R. *The Human Body in Equipment Design,* 226–27. Cambridge, Mass.: Harvard University Press 1966.

Datta, S., Chatterjee, B., and Roy, B. The energy cost of rickshaw pulling. *Ergonomics,* Vol. 21 [11], 879–86, 1978.

Datta, S. and Ramanathan, N. Ergonomic comparison of seven modes of carrying loads on the horizontal plane. *Ergonomics,* Vol. 14 [2], 269–78, 1971.

Eastman Kodak Ergonomics Group, Rodgers, S. (ed.). *Ergonomic Design for People at Work: Vol. 2.* New York: Van Nostrand-Reinhold, 1986.

Force Limits in Manual Work. Guildford, Surrey England: IPC Science and Technology Press, 1980.

Freivalds, A. Comparison of United States (NIOSH Lifting Guidelines) and European (ECSC Force Limits) recommendations for manual work limits. *AIHA Journal,* Vol. 48 [8], 698–702, 1987.

Garg, A., Sharma, D., Chaffin, D., and Schmidler, J. Biomechanical stresses as related to motion trajectory of lifting. *Human Factors,* Vol. 25 [5], 527–39, 1983.

Herrin, G., Jaraidi, M., and Anderson, C. Prediction of overexertion injuries using biomechanical and psychophysical models. *American Industrial Hygiene Association J.,* Vol. 47 [6], 322–30, 1986.

Kantowitz, B. and Sorkin, R. *Human Factors.* New York: Wiley & Sons, 1983, pp. 642–43.

Karwowski, W., Mulholland, N., and Ward, T. A knowledge-based expert system for the analysis of risk of overexertion in manual lifting. *Proceedings of the Human Factors Society,* 101–105, 1986.

Konz, S. and Bhasin, R. Foot position during lifting. *American Industrial Hygiene Association J.,* Vol. 35 [12], 785–92, 1974.

Konz, S. and Coetzee, K. Prediction of lifting difficulty from individual and task variables. *Proceedings of the 4th Int. Congress on Production Research.* Tokyo, 1977.

Kroemer, K. Horizontal push and pull forces. *Applied Ergonomics,* Vol. 5 [2], 94–102, 1974.

Kumar, S. Arm strength at different reach distances. In *Trends in Ergonomics/Human Factors IV,* Asfour, S. (ed.). Amsterdam: Elsevier, 1987.

Liles, D. The application of the job severity index to job design for the control of manual materials-handling injury. *Ergonomics,* Vol. 29 [1], 65–76, 1986.

Lovested, G. Reducing warehouse material handling strains. *Proceedings of 24th Annual Meeting of the Human Factors Society,* 653–54, 1980.

Marras, W., Lavender, S., and Rangarajulu, S. The effects of expectation on trunk loading. *Proceedings of the Human Factors Society,* 91–95, 1986.

Mital, A. and Aghazadeh, F. Psychophysical lifting capabilities for overreach heights. *Ergonomics,* Vol. 30 [6], 901–909, 1987.

Mital, A. and Ilango, M. Load characteristics and manual carrying capabilities. *Proceedings of the Human Factors Society,* 274–78, 1983.

Mital, A., Nicholson, A., and Ayoub, M. M. *A Guide to Manual Materials Handling.* London: Taylor and

Francis, 1993.

Morrissey, S. and Liou, Y.-H. Maximum acceptable weights in load carriage. *Ergonomics,* Vol. 31 [2], 217–26, 1988.

Mueller, E., Vetter, K., and Blumel, E. Transport by muscle power over short distances. *Ergonomics,* Vol. 1, 222–25, 1958.

Patterson, P., Congleton, J., Koppa, R., and Huchinson, R. The effects of load knowledge on stresses at the lower back during lifting. *Ergonomics,* Vol. 30 [3], 539–49, 1987.

Poulsen, E. and Jorgensen, K. Back muscle strength, lifting and stooped working postures. *Applied Ergonomics,* Vol. 2 [3], 133–37, 1971.

Randle, I. Predicting the metabolic cost of intermittent load carriage in the arms. In *Contemporary Ergonomics 1987,* McGaw, E. (ed.). London: Taylor and Francis, 1987.

Ridd, J. A practical methodology for the investigation of materials handling problems. In *Ergonomics of Workstation Design,* Kvalseth, T. (ed.). London: Butterworths, 1983.

Rodgers, S. *Working with Backache.* Fairport, N. Y.: Perinton Press, 1984.

Rowe, M. *Orthopaedic Problems at Work.* Fairport, N. Y.: Perinton Press, 1985.

Smith, J. A manual material handling study of bag lifting. Paper at annual meeting of American Industrial Hygiene Association, Portland, Oregon, 1981.

Snook, S. and Ciriello, V. The design of manual handling tasks: Revised tables of maximum acceptable weights and forces. *Ergonomics,* Vol. 34 [9], 1197–1213, 1991.

Tichauer, E. Ergonomic aspects of biomechanics. In *The Industrial Environment—Its Evaluation and Control,* Chapter 32. Washington, D. C., Supt. of Documents, 1973.

Ulin, S., Armstrong, T., Snook, S., and Keyserling, M. Examination of the effect of tool mass and work postures on perceived exertion for a screw driving task. *Int. J. of Ind. Ergonomics,* Vol. 12, 105–15, 1993.

Warwick, D., Novak, G., Schultz, A., and Berkson, M. Maximum voluntary strengths of male adults in some lifting, pushing and pulling activities. *Ergonomics,* Vol. 23 [1], 49–54, 1980.

Waters, T., Putz-Anderson, V., Garg, A., and Fine, L. Revised NIOSH equation for the design and evaluation of manual lifting tasks. *Ergonomics,* Vol. 36 [7], 749–76, 1993.

Waters, T., Putz-Anderson, V., and Garg, A. *Applications Manual for the Revised NIOSH Lifting Equation,* DHHS (NIOSH) Publication No. PB94-176930, 1994.

Winters, J. and Chapanis, A. Thumb push forces exertable by free-standing subjects. *Ergonomics,* Vol. 29 [7], 893–902, 1986.

Wyndham, C. and Heyns, A. Energy expenditures and mechanical efficiencies in pushing a mine-car at various speeds and loads. *Int. Z. agnew Physiol. einschl. Arbeitsphysiol.* Vol. 24, 291–314, 1967.

OVERVIEW

Engineers can select from a wide variety of handtools. The 8 guidelines will help you in your selection. Guideline 1 emphasizes the desirability of specialized tools; 2 using tools with both hands, and 3 powering with motors. Guidelines 4, 5, and 6 focus on the grip. Guideline 7 emphasizes the angles involved and the reduction of repetitive trauma. Guideline 8 focuses on the use of the proper muscle group.

CHAPTER CONTENTS

1 Use Special-Purpose Tools
2 Design Tools to Be Used by Either Hand
3 Power with Motors More Than with Muscles
4 Use the Proper Grip
5 Make the Grip the Proper Thickness, Shape, and Length
6 Make the Grip Surface Compressible, Nonconductive, and Smooth
7 Consider the Angles of the Forearm, Grip, and Tool
8 Use the Appropriate Muscle Group

KEY CONCEPTS

bearing surface	hard joints/soft joints	pulp pinch
benefits/costs	inline grip	semi-power grip
capital/maintenance/utility cost	lateral pinch	special-purpose tool
chuck pinch	leftness stigma	triggering
conductivity	normally open	vibration syndrome
dominant hand	pistol grip	wedge shape
get ready/do/put away	power grip	
grip strength	precision grip	

Handtools extend the capability of the hand. This chapter will help you to select from the many designs available. See Chapter 19 for a related topic, controls—the interface between the body and non-portable machines.

GUIDELINE 1: USE SPECIAL-PURPOSE TOOLS

Return on investment in handtools usually is high due to high output with use of the tool vs. the low cost of the tool.

1.1 Benefits

Tools extend the capability of the hand. The capability can be more grip strength (pliers), impact strength (hammer), torque (wrench), speed (drill rpm), reach (fly swatter), protection (gloves), or even functions that cannot be done with a bare hand (saw, soldering iron). Use of tools is one of the key things that differentiates humans from animals.

One question is how specialized the tool should be. Should a general-purpose tool (knife, hammer, screwdriver) be used or a special-purpose tool (fishing knife, butter knife, razor blade; claw hammer, ball-peen hammer, shoemaker's hammer; hand-powered screwdriver, battery-powered Phillips screwdriver)?

Think of a job as composed of three elements: (1) **get ready,** (2) **do,** and (3) **put away.** A general-purpose tool may save on the get-ready and put-away tasks since one tool does it all and there is less search and select time. The general-purpose tool may be extended into a multifunction tool. For example, a claw hammer combines a hammer and nail claw; a pliers combines a gripper and a wire cutter; a pencil combines writing and erasing. Two tools in one eliminates a reach, grasp, move, and release from labor costs; that is, get-ready and put-away costs are lower. An extreme example of a multifunction tool is my camping shovel, which also is a hammer, saw, ax, bottle opener, and wrench!

Often the multifunction and general-purpose tool doesn't really do any specific job very well. Thus for most industrial tasks (i.e., a task which is repeated tens, hundreds, or thousands of times/week), a **special-purpose tool** is best. This special-purpose tool (e.g., a knife to cut open cardboard boxes, an air-powered nut runner, a highlighter for emphasizing key words) can be selected for the exact characteristics necessary for the specific job. Some users, whose jobs vary more, will have an "arsenal of weapons." For example, a student would have a variety of writing instruments and markers. A mechanic

would consider it foolish to try to do the day's work with only one or two tools.

1.2 Costs

Handtool costs can be divided into capital cost, maintenance cost, and utility cost.

Capital cost of handtools tends to be small. Common nonpowered handtools (scissors, knives, pliers, wrenches) tend to be less than $10. Effectively all cost less than $100. Powered handtools cost from $10 to $1,000, depending on the tool and whether it is sold on the consumer market as well as the industrial market.

Maintenance costs of nonpowered handtools tend to be trivial. Maintenance costs of powered handtools depend upon the tool, of course, but $100/year would be considered high for most powered handtools.

Utility cost is zero for nonpowered tools. For powered tools, it depends upon the power source, energy used per minute, and the duty cycle (% of time the tool is used). A small air motor (such as used on an air-powered screwdriver) uses 12–35 cubic ft/min when operating. A typical duty cycle is 20–25% so 5 cfm or 300 cu ft/h is a reasonable value. Compressed air costs (capital + operating + maintenance) about 25–30 cents/1,000 cu ft of free air ingested by the compressor. Cost would be 8–9 cents/h. A 1/4 hp electric motor uses .746 (1/4) = .187 kWh; at $.07/kWh, this is about 1 cent/h. Clearly, utility costs for both air and electric motors are low.

1.3 Benefits/Costs

The analysis of **benefits/costs** is easiest to understand if put on a benefit/use and cost/use basis. The engineer must estimate how often the handtool will be used. This, in turn, requires two estimates: (1) years of tool use and (2) usage/year. Years of tool use probably will be determined by how long the application lasts rather than the physical life of the tool. Two to five years is a reasonable estimate in most situations. For use/year, estimate usage/day or usage/week and multiply by the number of days or weeks used. A production worker might use a tool once a minute or 450 times/day $\times$ 190 days/yr = 85,000 times/yr. A maintainer might use a tool 5 times/day but only part of the year (say 100 days) or 500 times/yr.

Thus, a common nonpowered handtool might be used by a maintainer 500 times/yr $\times$ a 5-year life = 2,500 uses/life. A powered handtool might be used by a production worker 85,000 times/yr $\times$ 3 years = 250,000 times. (A power handtool should last 250,000 to 500,000 cycles.)

In summary, costs of the nonpowered handtool might be $20 for capital cost, $0 for maintenance, and $0 for utility cost—a total of $20 or $20/2,500 =

$.008/use. Cost of the air-powered handtool might be $500 for capital cost, $100/year × 3 years = $300 for maintenance, and 190 days × 7.5 h/day × 50% duty cycle × $.30/h × 3 years = $640—a total of $1,440/250,000 = $.0058/use.

The benefits depend on how much the proposed tool extends the capability and quality, reduces time, and so forth. See Figure 4.3 for a form showing some things to consider. For the nonpowered handtool, assume it saved $.01/use for quality and 2 s/use for time. If labor cost is $14.20/h, this is $14.20/3,600 = $.004/s × 2 s/use = $.008/use: total savings of $.018/use. For the powered handtool, assume $.005/use for reduced downtime, $.001/use for improved quality, and 2 s/use. Using labor cost of $14.20/h, this is $.004/s × 2 s/use = $.008/use; total savings = $.014/use.

The above is a "quick and dirty" example making two points. First, whether a tool is expensive really depends upon a detailed analysis, not just upon initial capital cost. Second, benefits include savings beyond just labor savings. The above estimate showed how tool benefits far exceed tool costs and why a smart worker has special-purpose tools.

GUIDELINE 2: DESIGN TOOLS TO BE USED BY EITHER HAND

In most circumstances the tool should be in the user's preferred hand. The preferred hand (**dominant hand**) is the right hand for about 90% of the population; the percent seems constant across cultures and for both sexes.

Table 18.1 shows that the nonpreferred hand tends to have 94% of the grip strength of the preferred hand. See also Figure 18.1. Table 18.2 shows, in study 2, that individual fingers on the nondominant hand are weaker than the corresponding finger on the dominant hand.

Dexterity without handtools was measured by Kellor et al. (1971). The nonpreferred hand rate was 96% on a 9-hole pegboard task and 93% on a 50-hole pegboard task. Konz and Warraich (1985) reported that the nonpreferred hand rate was 84% for threading a nut on a bolt. An, Askew, and Chao (1986) reported that the nonpreferred rate for manipulating objects was 87%.

When handtools are used, performance of the nonpreferred hand declines more. Konz and Warraich (1985) had 40 subjects do four tasks with each hand: drilling with an electric drill, hand sawing, hammering a nail, and cutting a pattern with scissors. Kaster et al. (1987) had 15 subjects do five tasks: nut assembly on a bolt, hand sawing, hammering a nail, cutting a pattern with scissors, and drawing a line. Table 18.3 summarizes the time effect for these novices. In addition, errors (accuracy) are much worse with the nonpreferred hand.

Box 18.1 describes gloves.

There is considerable social stigma in being left-handed (**leftness stigma**). The word *left* comes from *lyft*, Anglo-Saxon for weak, broken; in Latin, *sinister*; in French, *gauche*, from which we also get *gawk*; in Spanish, *zurdo* (the word for left-handed) means malicious. In the Bible (e.g., Matthew 25:33) the sheep on the right hand went to heaven, the goats on the left hand went to hell. In medieval plays,

TABLE 18.1 Maximal static **grip strength** (hand squeeze), kg, from various studies.

Hand	Sex	5th Percentile	Standard Deviation	Group
R	M	48.2	9.14	Army personnel
R	M	47.8	8.18	Air force aircrewmen
Pref.	M	44.6	7.64	Air force rated officers
Pref.	M	41.8	6.87	Industrial personnel
R	M	41.4	8.23	Truck and bus drivers
R	M	40.6	9.64	Rubber industry
R	M	33.6	9.55	University men
R	M	19.1	5.46	University men, force for 1 min
L	M	45.0	9.55	Army personnel
L	M	44.6	7.64	Air force rated officers
L	M	43.6	7.27	Air force aircrewmen
L	M	41.8	7.00	Industrial personnel
L	M	39.1	7.46	Truck and bus drivers
L	M	39.1	10.09	Rubber industry
L	M	29.5	8.18	University men
L	M	17.7	4.54	University men, force for 1 min
(R + L)/2	F	26.4	4.00	Navy personnel
Pref.	F	25.9	4.68	Industrial workers

TABLE 18.2   Maximal static finger force for adult males. Study 1 had $N = 100$ (Hertzberg, 1973). Study 2 had $N = 20$ (Dickson et al., 1972). Study 3 had $N = 15$ (Hertzberg, 1973). (See also Table 18.7.)

Finger	Right Hand Mean, kg	%	Standard Deviation, kg	Standard Deviation/ mean, %	Dominant Hand Mean kg	%	Nondominant Hand Mean, kg	%
Thumb vs. object	7.3	124	1.7	23				
2nd vs. object	5.9	100	1.3	22	4.6	100	4.4	96
3rd vs. object	6.4	108	2.0	31	4.3	93	3.9	85
4th vs. object	5.0	85	1.7	34	2.9	63	2.8	61
5th vs. object	3.2	51	1.1	37	2.4	52	2.4	52
Thumb vs. tip of 2nd	9.6•		2.3•	24•	7.5	163	6.9	150
vs. tip of 3rd					8.8	191	8.4	183
vs. tip of 4th					6.4	139	7.8	170
vs. tip of 5th					5.0	109	4.8	104
Thumb vs. tip of 2nd	10.5•		2.2•	21•				

*Study 3

the devil entered from stage left; we give "left-hand" compliments; political radicals are "left-wing." On the other hand, valued assistants are "right-hand" men; honored guests sit at the right of the host; salutes and blessings are given with the right hand; and to be correct is to be right. Right comes from the Anglo-Saxon *rigt*, which means straight or just. In French, *a droit* means to the right; we use *adroit* to mean skillful, nimble in the use of hand. After all, we want to get off on the right foot. Right on!

A tool usable in either hand has two benefits. The first is for the 10% of the population left out. In sports, where emphasis is on maximum performance, both left- and right-hand products usually are available. The same emphasis on performance is desirable for industry. The second benefit is that the nonpreferred hand can be used when the preferred hand is otherwise engaged or is resting. Figure 18.2 shows a foodscoop (Konz, 1975) which can be used only by the right hand. Figure 18.3 shows an alternative version which can be used by both hands. Figure 18.4 shows how a dental syringe was modified so it could be used in either hand. Certain tools are peculiarly right-handed in design (e.g., a scissors) and require a different action when used in the left hand. Rather than force a left-handed person to use a right-handed scissors, buy a left-handed scissors. The goal is to emancipate the left hand.

GUIDELINE 3: POWER WITH MOTORS MORE THAN WITH MUSCLES

First of all, mechanical energy (motors) is 10 to 1,000 times cheaper than human energy for the following reasons:

- People run 24 h/day × 365 days = 8,760 h/yr while machines run only when in use (say 2,000 h).

TABLE 18.3   Time decrement for novices when using the nonpreferred hand. Practice may change the ratios.

Decrement, %	Handtool	Skill and movement
5-10	No	"Simple" hand/arm movements
10-20	No	"Complex" hand/arm movements
25	Yes	"Minimum" skill (saw, hammer)
50	Yes	"Moderate" skill (scissors)

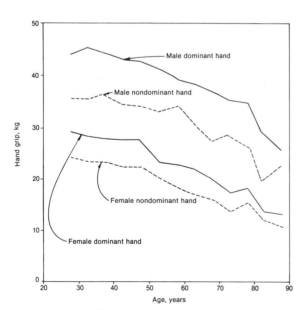

FIGURE 18.1   Handgrip strength is less for the nondominant hand, is less for females than males, and declines with age (Shock, 1962).

- The worker energy cost to the firm is covered by wages. Wages must cover not only food for the worker, but also food for the worker's family. Wages also cover other goods such as housing, entertainment, and transportation.

- People are relatively inefficient as a power source. Bicycle pedaling, the most efficient method of human-generated power, is only 20 to 25% efficient.

BOX 18.1 *Gloves*

Gloves commonly protect the hand against abrasion and impact as well as chemicals (see Chapter 24 on dermatitis), heat (flame, sparks), and cold. Special chainmail gloves are worn by meat cutters to protect against knife cuts. Rubber gloves can protect a surgery patient against infection; rubber gloves (inside protective leather gloves) protect from electricity; and gloves can be used to reduce vibration. Goel and Rim (1987) reported vibration reductions of 24% for leather gloves and 45% for Sorbothane-padded gloves. (An alternative approach would be to pad the handle.)

Of course not all gloves cover the hand the same. Technically, a glove has a separate division for the thumb and each finger. Mittens cover the entire hand with a separate thumb division (giving greater thermal protection with a loss of dexterity). Individual fingers can be guarded with finger guards (cots) and thimbles. Hand pads protect the palm against cuts, abrasions, and hot objects. Partial gloves (covering the hand and the first phalange of the finger) permit dexterity while still protecting against cold (Riley and Cochran, 1984). (See Figure 18.22.) In some extremely cold environments, electrically heated gloves are an option (Scott, 1988). A sock (with the toe cut off) can be placed on the forearm to protect it against abrasion of cardboard boxes when packaging.

Protective work gloves have three wrist styles: (1) knitwrist (a knit cloth fitting snugly about the wrist, keeping loose material from entering; this snugness makes them dangerous near moving machinery as they can pull the hand into pinch points), (2) safety cuff (wide, stiff cuff at the wrist, permitting easy putting on and taking off), and (3) gauntlet (protects some or all of the forearm, depending on its length).

Gloves probably will decrease manual dexterity (vs. bare hands), although receivers in football wear gloves with high coefficients of friction to improve their pass-catching ability. Plumber et al. (1985) reported that assembly time with handtools increased 15 to 37% with gloves. Bradley (1969) reported that times to actuate toggle switches, levers, knobs, and pushbuttons increased 2 to 13% with gloves; the increase tended to be less than .1 s, and so the percent increase may have little practical significance.

Gloves also decrease grip force. Depending on the glove, Cochran et al. (1986) estimated a decrease of 7–17% and McMullin and Hallbeck (1991) estimated 11–21%. The decrease in grip force with gloves (vs. barehanded) is an estimate of the percent reduction in maximum holding time at any force (Cochran et al., 1987). Mital et al. (1994) reported gloves increased the torque exerted on screwdrivers and wrenches even though muscle activity did not differ significantly between wearing and not wearing gloves. Gloves add to hand thickness. The amount depends upon the glove but 15 to 20 mm is a reasonable assumption.

FIGURE 18.2 Right-hand only tools can't be used by left-handers. In addition, the right-handed users can't shift the tool to the left hand and continue working while resting the right hand. Note how all the load from the 1.2 kg spring must be overcome by a single digit, the thumb.

FIGURE 18.3 Either hand can use this foodscoop. In addition, note how the force of the spring is overcome by the muscles of the entire hand, not just the thumb.

FIGURE 18.4   Duplicate controls on an extended control shaft permit this dental syringe to be operated with either hand (Evans et al., 1973). A more common design solution is to put the control on the tool centerline so it is accessible from either side.

As pointed out in Guideline 1, a small air motor, such as used on an air-powered screwdriver or nutrunner, costs about \$.08–.09/h for utilities. A 1/4 hp electric motor costs about \$.01/h. These certainly are a very small percentage of labor costs.

The second reason to replace human muscles with motors is to extend capability. A motor can exert a torque all day—it doesn't get tired! In addition, motors can exert forces and speeds beyond human capabilities. Try to imagine rotating a drill at 3,600 rpm by hand! The development of battery-operated power tools permits more safety (no danger from electrical shorts) as well as improved convenience (no power cord, no need to be near an outlet).

Motors do present a problem; they vibrate. See Box 18.2.

For comments on nonpowered tools for working with dirt, see Box 18.3.

BOX 18.2   *Hand–arm vibration*

Tingling, blanching, and numbness of the fingers of normally healthy people is called Raynaud's disease. Occupationally induced cases were called vibration white finger (VWF) but more recently, as it was realized that nerves and tendons were affected as well as the blood vessels, the recommended term is **vibration syndrome** (VS).

In 1987, threshold limit values were published by a nongovernmental agency (ACGIH, 1987) and in 1989 a governmental agency published a criteria document (NIOSH, 1989), but there is no governmental regulation as of 1994.

It is relatively difficult to estimate the amount of time a person is exposed to vibration. One possibility is occurrence sampling. Teschke et al. (1990) used a noise dosimeter to divide chain saw use into saw off, idling, and cutting.

Engineering procedures to minimize VS are the following:

- Purchase equipment with lower levels of vibration. Some design possibilities are increasing the effective mass, reducing the excitation forces (e.g., balancing), and attenuation (Lindqvist, 1986). Avoiding resonance is especially important.

- Purchase equipment with vibration-isolated handles. See Lindqvist (1986) and Andersson (1990) for design techniques to reduce hand-tool vibration.

- Furnish vibration-absorbing gloves. The vibration damping is primarily a function of material thickness. Even if the damping material does not attenuate much below 500 Hz, the material may insulate from the cold and give a softer grip (i.e., a reduced grip force).

Work practices to reduce the effect of VS include the following:

- Maintain equipment. For example, is the grinding wheel balanced on the shaft? Has the tool become damaged—for example, by dropping it? Are the teeth sharp? Is there a scheduled maintenance program?

- Consider a pad under an impact surface when using an impact tool (e.g., a hammer). A pad will reduce the rate of deceleration.

- Minimize handgrip force on the tool, consistent with safe working (Radwin et al., 1987). For example, can the tool be supported on a balancer so that it is guided with a loose grip instead of held tightly?

- Rest the tool on a support or workplace as much as possible. For example, can a steady rest or balancer be used?

- Avoid continuous vibration (have 10 min of nonvibration/h).

- Keep the body and hands warm and dry (avoiding vasoconstriction). Avoid drafts from exhausts of air-powered tools.

- Avoid smoking (which causes vasoconstriction).

For additional information, see Wasserman (1987) and Griffin (1990). For comments on whole-body vibration (especially for vehicle drivers), see Section 4.31 in Chapter 16.

BOX 18.3   *Shovels, hoes, and rakes*

In agriculture and construction, dirt is moved using human-powered tools.

Freivalds (1986a, 1986b) and Freivalds and Kim (1990) have studied shoveling. Using a criterion of minimization of energy of shoveling sand at 18 scoops/min, the optimum shovel had a ratio of B/W of .068, where B = blade size (m²), and W = shovel weight (kg). However, a shovel is used for digging as well as scooping and throwing. van der Grinten (1987) reported that shovels with a relatively large curvature were good for scooping and throwing but not for digging. Degani et al. (1993) had some success with a two-shaft, two-handle shovel.

Nag and Pradham (1992) recommended Indian agricultural workers use a 2-kg hoe with a blade–handle angle of 65–70°, blade length of 25–30 cm, blade width of 22–24 cm, handle length of 70–75 cm, and handle diameter of 3–4 cm. In a comparison of weeding devices, Tewari, Datta, and Murthy (1991) reported a spade and a three-tine hoe had approximately equal work hours/ha but a khurpi took 2.4 times as long.

Kumar and Cheng (1990) reported a straight rake handle reduced spinal load over a variety of experimental nonstraight handles.

GUIDELINE 4: USE THE PROPER GRIP

Three grips will be discussed: (1) the power grip, (2) semi-power grips, and (3) precision grips.

4.1   Power Grip

Figure 18.5 shows a power grip. Typically the tool handle is perpendicular to the forearm axis. The direction of force, however, may be (1) parallel to the forearm, (2) at an angle to the forearm, or (3) applied as a torque about the forearm.

4.1.1   Parallel to forearm

The muscles can apply force along the forearm axis, as with an electric iron, a saw, a Y-handle shovel. The muscles also can resist force along the forearm axis, as with an electric drill, a suitcase handle, a pistol (the power grip sometimes is called a pistol grip). Typically, the tool handle is approximately perpendicular to the forearm axis—an angle of 80° is common. The two moment arms are caused by (1) tool action force and (2) tool weight. Mital and Kilbom (1992) recommend keeping the weight of the tool supported by the user at about 1.1 kg but no more than 2.3 kg. See Figure 18.6. If tool action force is high and tool weight is

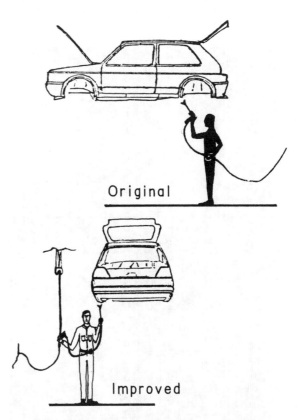

FIGURE 18.6   Tool weight was reduced for Volkswagen workers (Echa, Smolenski, and Zamiska, 1987). In the improved method, the 5-lb gun and hose were replaced by a lightweight wand, the hose and gun were suspended from an overhead trolley, and the hose between the suspension point and the operator was attached to a belt around the operator's waist.

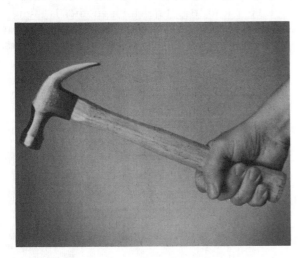

FIGURE 18.5   **Power grips** are for power. The four fingers reach around the handle and are "locked" by the thumb over the first finger.

low (a light tool, a suspended tool, or a tool sliding along a surface), put the handle grip at the rear of the tool. If tool weight is high, place the tool handle under the tool center of gravity (tool balance point) to minimize unnecessary torques as well as to permit sighting along the tool to improve accuracy.

4.1.2 Angle to forearm

The muscles can apply force at an angle to the forearm (hammer, hand ax, ice pick, chisels, reverse grip pliers, and pizza cutter). See Figure 18.7. The tool force angle varies from tool to tool. The tool may be above the hand (hammer) or below (ice pick, pizza cutter); the wrist may be flexed (hammer, fishing rod, tennis racquet) or locked (ice pick, pliers, power screwdrivers). When the tool is above the hand, the wrist is in "high gear"—the top of the hand moves more than the bottom. When the tool is below the hand, the wrist is in "low gear"—movement at the extremity is reduced but power at the wrist is increased.

4.1.3 Torque about forearm axis

Figure 18.8 shows a power grip on a triaxial allen wrench. An extension of the forearm axis projects through the fingers. A corkscrew is another example. (A common

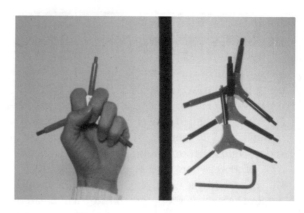

FIGURE 18.8   Power grip with tool axis aligned with forearm. The triaxial allen wrench has three different sizes on each wrench; the wrench is color-coded for size.

problem for torque on controls is insufficient bearing surface and lever arm. In Europe, most doors are opened by handles (i.e., good lever arms) but in the United States architects rely on squeezing the fingers tightly on a polished sphere, a doorknob.)

4.2 Semipower Grips

In the **semipower grips,** the 4 fingers still act as a group but the thumb position changes. The 2 variations are the oblique grip and the hook grip.

In the oblique grip, the thumb is aligned along the tool axis to improve precision but with a loss of power. Strength is about 2/3 of the power grip. Patkin (1969) mentions pointing along a surgeon's needleholder with the thumb. Holding a golf club is a more common example.

In another version, the hook grip, the 4 fingers still wrap around the tool shaft but the thumb is relaxed (passive) and not used. Figure 18.9 shows how Gilbreth improved bricklaying by substituting a hook grip for a pinch grip. One application is carrying a light load (e.g., briefcase); another is lifting a box without handles. Still others are pulling a tray and pulling down on a hook or strap (Rodgers,

FIGURE 18.7   Power grip with tool below the hand and with wrist locked is shown by the pizza cutter. The guard not only protects against injury but permits the four fingers' muscles to be relaxed.

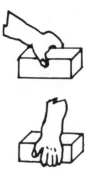

FIGURE 18.9   Gilbreth recommended using a hook grip (lower) to hold a brick in place of a pinch grip (upper).

1986). Of course, resistance to the object being pulled from the hand is not as good as the power grip since the thumb is not acting as a lock.

4.3 Precision Grips

The **precision grip,** used for precise aim, has only about 20% of the strength of a power grip (Swanson et al., 1970). The two divisions of precision grip are internal and external.

4.3.1 Internal precision grip

(This grip is used with a tableknife, toothbrush, or blade razor.) See Figure 18.10. Three characteristics are (1) a pinch grip by the thumb against the first finger (or thumb against first and second fingers), (2) support by the little finger and side of the hand (to reduce tool tremor), and (3) the shaft passes under the thumb (and thus is "internal" to the hand). In many cases the hand itself is supported on a work surface. The arm muscles which control the hand are very sensitive to tremor following overexertion. For hand steadiness, normally cycling women are most steady, men are less steady, and women on oral contraceptives are the least steady (Hudgens et al., 1988). Patkin (1969) recommends that surgeons should not carry suitcases for 24 h before an operation since the seemingly minor exertion reduces hand steadiness.

If tool rotation is required (e.g., small screwdriver), the tool shaft tends to be perpendicular to the work and the end of the tool shaft tends to bore a hole into the palm of the hand. Reduce this problem by (1) using a shaft so long that it extends beyond the palm, (2) reducing palm penetration pressure by using a tool end with a larger **bearing surface** (such as a spherical surface), or (3) reducing penetration pressure by reducing the force required along the tool axis.

4.3.2 External precision grip

(This grip is used with a pencil, a spoon, or chopsticks.) See Figure 18.11. Three characteristics are (1) a pinch grip by the thumb against the first finger (or thumb against first and second fingers), (2) support on the side of the second finger or the skin at the base of

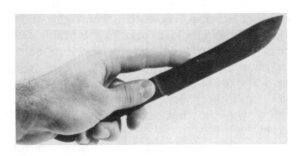

FIGURE 18.10 Internal precision grips have three characteristics: (1) a pinch grip, (2) support by the little finger or side of the hand, and (3) the shaft is "internal" to the hand.

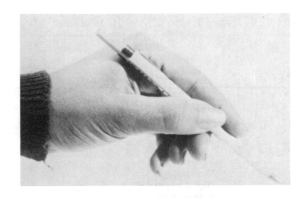

FIGURE 18.11 External precision grips have three characteristics: (1) a pinch grip, (2) support by the side of the second finger or skin at the thumb base, and (3) the shaft is "external" to the hand.

the thumb, and (3) the shaft passes over the thumb and thus is "external" to the hand. The tool shaft usually is at an angle to the work surface.

GUIDELINE 5: MAKE THE GRIP THE PROPER THICKNESS, SHAPE, AND LENGTH

Every tool has two ends: one working on the material, the other on the hand.

Figure 18.12 and Table 18.4 give key dimensions of small and large bare adult hands. Figure 18.13 and

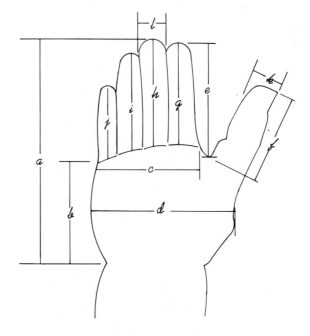

FIGURE 18.12 Small female and large male adult hand dimensions are given in Table 18.4. Gloves increase the dimensions. Remember that hands differ not only in size but also in strength and dexterity. For male adults, the length of the last phalange of the little finger and the width of the thumb approximate one inch. This is known as a "rule of thumb!"

TABLE 18.4   Key hand dimensions, mm, of U.S. adults (Garett, 1971; Rigby, 1973). See also Figure 18.12.

Key	Dimension	Small Women (1st Percentile)		Large Men (99th Percentile)	
		Garrett	Rigby	Garrett	Rigby
a	Hand length*	160	170	218	211
b	Palm length		97		119
c	Metacarpal breadth	69	79	91	99
d	Breadth with thumb		91		117
e	Tip to crotch		97		132
f	Thumb length	43	48	69	69
g	Digit 2 length	56		86	
h	Digit 3 length	66		97	
i	Digit 4 length	61		91	
j	Digit 5 length	46		71	
k	Thumb thickness	15	18	23	23
l	Digit 3 diameter (tip)	13	20	20	25

*Straight and flat. For relaxed hand, subtract 58 mm.

Table 18.5 give estimates of means and standard deviations of various hand dimensions from another source (Eastman Kodak, 1983). Since female hands tend to be smaller than male hands, handtools designed for men tend to be too large for women. Figure 18.14 gives the technical names for various parts of the fingers.

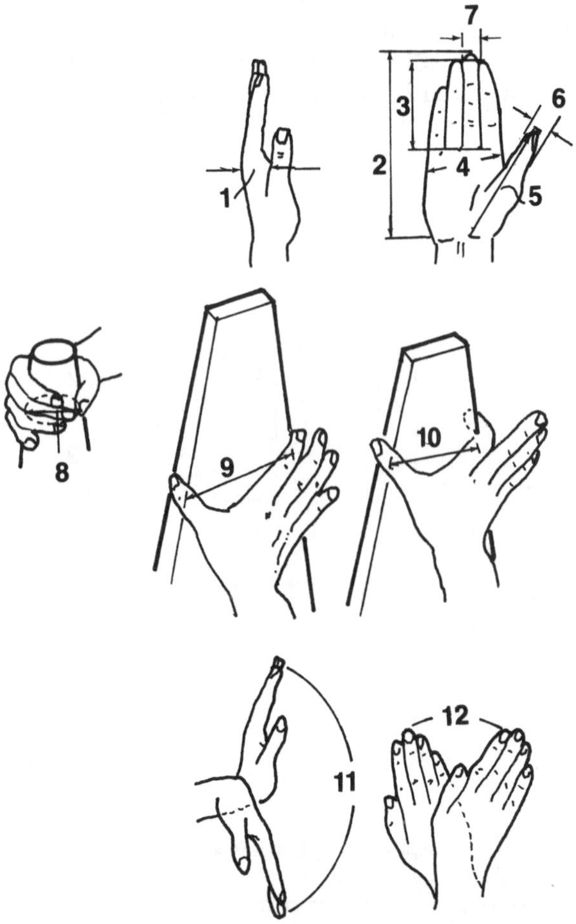

FIGURE 18.13   Hand dimensions for adults given in Table 18.5 (Eastman Kodak, 1983).

5.1   Grip Thickness: Power Grip

Hertzberg (1973) reported grip strength of 43 kg for a 40-mm grip, 65 for a 65-mm grip, 48 for a 100-mm grip, and 36 for a 125-mm grip; strength was about 20% less with gloves. Ayoub and LoPresti (1971) used electromyography and reported that 40 mm was preferred to 50 and 65 mm; if force at 40 is 100%, then force at 50 was 95% and at 65 was 70%.

Greenburg and Chaffin (1977, pp. 51 and 77) recommend that power grip handles be between 50 and 85 mm, with the goal toward the 50. Rogers (1986) recommends 38 mm. The best grip diameter for a power screwdriver is about 50 mm (Johnson, 1988). Saran (1973) reported a T-handle of 25 mm was preferred to handles with either 19 or 32 mm.

If the grip diameter can be customized for an individual, Grant et al. (1992) recommend a diameter 10 mm smaller than the inside grip diameter. Rigby (1973), making recommendations for container handles, gave 6 mm diameter as the minimum for weights of less than 7 kg, 13 mm for weights of 7 to 9 kg, and 19 mm for loads over 9 kg. If handle diameter is too small, there is too much pressure on the hand, which is why you may see workers wrap grips with tape to increase grip diameter. If diameter is too large, the fingers don't overlap, there is no locking with the thumb, and strain is increased sharply.

In summary, power grips between 25 and 50 mm diameter usually will be satisfactory. The most common mistake probably is to use too small diameter handles. Note that if the tool "head" (such as a screw-

TABLE 18.5 Hand dimensions (cm) for American male and female adults (Eastman Kodak, 1983). See Figure 18.13.

Measurement	Males		Females	
	50th percentile	±1 S.D.	50th percentile	±1 S.D.
1. Hand thickness metacarpal 3	3.3	0.2	2.8	0.2
2. Hand length	19.0	1.0	18.4	1.0
3. Digit 2 length	7.5	0.7	6.9	0.8
4. Hand breadth	8.7	0.5	7.7	0.5
5. Thumb length	12.7	1.1	11.0	1.0
6. Breadth of thumb joint	2.3	0.1	1.9	0.1
7. Breadth of digit 3 joint	1.8	0.1	1.5	0.1
8. Grip breadth, inside diameter	4.9	0.6	4.3	0.3
9. Hand spread, 1-2 1st joint	12.4	2.4	9.9	1.7
10. Hand spread, 1-2 2nd joint	10.5	1.7	8.1	1.7
11. Wrist flexion-extension, deg.	132.0	19.0	140.0	15.0
12. Wrist ulnar-radial, deg.	59.9	13.0	66.7	14.0

driver tip) becomes smaller, the hand doesn't shrink; so the handle size should not change.

The span on double-handled tools (see Figure 18.15) should be 50 to 63 mm according to Rodgers (1986). Greenburg and Chaffin (1977) recommend a minimum initial span of about 50 mm (tool force low during closure) and a maximum initial span of about 100 mm (maximum hand size).

5.2 Grip Thickness: Precision Grip
Magill and Konz (1986) and Gainer and Konz (1987) reported that nonpowered screwdrivers which permitted maximum torque required the most time to drive a screw and those which permitted minimum torque had the least time.

Although precision grips are not supposed to require force, diameters less than 6 mm should be avoided as they will cut into the hand if force is required. Kao (1974) reported that boys had better handwriting with 13-mm diameter than with 10- or 6-mm diameter pens. Konz and Oetomo (1988) found pen diameters of 6.3 mm were preferred less than diameters of 9.5 mm or 12.7 mm.

5.3 Shape: Section Perpendicular to Grip Axis
For most situations you want the tool to remain captive in your hands; that is, it does not rotate. Rotation is prevented by a countertorque in the hand. Since a torque is a force times a moment arm, improve the moment arm. Improving the moment arm can be done by having a longer moment arm or by having a good bearing surface to minimize slippage.

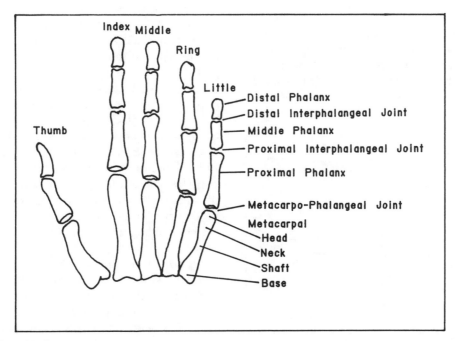

FIGURE 18.14 Parts of the fingers.

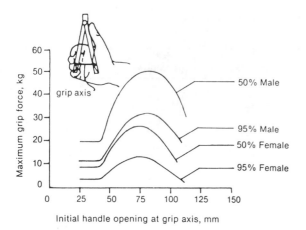

FIGURE 18.15   Two-handle tools have a maximum grip strength when the opening is about 75 mm (Greenburg and Chaffin, 1977). The 50 males and 50 females were electronics manufacturing employees in the United States. Fransson and Winkel (1991) reported that the optimum was 50–60 mm for females and 55–65 for males.

Patkin (1969) showed how a thumb conoid reduced accidental twisting on a surgeon's forceps. Tremor was reduced by using the thumb to rotate the tool while the forearm was supported. Pointing with the thumb aided movement precision and, since the thumb can move up and down, it could open the jaws. A thumb conoid is better than a thumb ring since (1) it fits multiple thumb diameters, (2) it has greater bearing surface, and (3) it has a wider range of up and down movement (since the thumb tip, not knuckle, moves). Figure 18.16 shows how a bayonet forceps was modified to resist turning in the hand.

The second strategy is to improve the bearing surface. A tool with a circular cross section tends to permit slippage. A rectangular cross section gives a good bearing surface (in addition, a rectangular cross-section tool does not roll when placed on a table). A noncircular cross section also allows tactile orientation of the tool. For example, workers in a cellophane plant file a flat area on the bottoms of their knife handles. The flat area permits them to orient the knife without looking at it.

A third strategy is to improve the coefficient of friction of the handle. Gainer and Konz (1987) found that screwdriver handles with soft rubber were preferred over those with hard plastic.

If rotation of the tool is neither good nor bad, then a circular cross section is more forgiving on the hand since there are no sharp edges. (Economics, not ergonomics, is the reason 80% of wooden pencils are hexagonal, as 9 hexagonal pencils can be cut from a slat that yields only 8 round ones.)

Sometimes the tool should rotate within the hand. Make the cross section within the pinch grasp circular to permit rotation by simple finger–thumb movements rather than by the complex regrasps or forearm movements needed for rectangular cross sections. Support the forearm to reduce tool tremor (if the forearm is not required to move during tool rotation).

5.4   Shape: Section Along Tool Grip Axis   The first question is whether the shape's cross section should be constant over the length of the handle or should vary. A change in cross section (1) reduces movement of the tool forward and backward in the hand, (2) permits greater force to be exerted along the tool axis due to the better bearing surface, and (3) can act as a shield if placed at the front.

Many handles have a **wedge shape,** which reduces forward movement of the hand. Since digit 3 is about 25 mm longer than the thumb or digit 5, the diameter at digit 3 can be about 25/3.14 = 8 mm larger. Avoid finger grooves or indentations between the front and rear of the grip since hand width (metacarpal breadth) varies about 18 mm in the population; thus the ridges between the valleys fit no one except the tool designer and become pressure points for other users.

Figure 18.17 shows a guard on the front of a grip. Injury from movement can come from the tool itself (knives, soldering irons) or simple impact of the hand on any unyielding or sharp surface. Greater force can be exerted along the tool axis since the strong muscles of the forearm are not limited by the grip strength of the fingers on the handle. A front shield also can act as a shield against heat or materials (stirring soup, solder splashes). The shield also permits holding the tool farther forward, which may improve accuracy.

A pommel (a shield at the rear of the grip, often used with swords) prevents loss of the tool when the grip is relaxed momentarily, but its most important benefit is to permit greater force to be exerted when the tool is being pulled toward the body. A T-handle permits maximum pulling force.

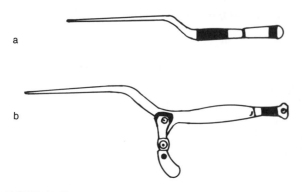

FIGURE 18.16   Accidental twisting was reduced by adding a bearing surface on this surgeon's forceps (Miller et al., 1971).

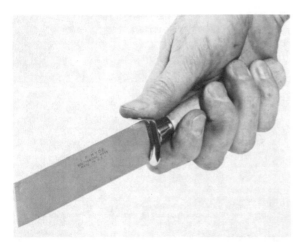

FIGURE 18.17 Guards on grips can improve thrust capability as well as reduce accidents.

5.5 Length

Length depends on the type of grip. For a power grip, all four fingers must make contact. Table 18.5 gave a metacarpal breadth of 79 mm for a small woman and 99 mm for a large man. Thus 100 mm is a reasonable minimum dimension for the population but 125 mm would be more comfortable. If the grip is enclosed (hand saw) or a glove will be worn, use 125 mm as the minimum.

For an oblique grip use 125 mm.

For an external precision grip, the tool shaft must be long enough to be supported at the base of the first finger or thumb. Golf pencils don't comply. Use 100 mm as a minimum for external precision grips.

For an internal precision grip, the tool must extend past the tender palm but not so far as to hit the wrist. The Western Electric pliers shown in Figure 18.18 demonstrate this principle. Screwdrivers

usually are designed to be held in an internal precision grip or oblique grip. However, the user tries to exert force parallel to the tool axis and uses the palm of the hand for the pushing force rather than squeezing the fingers. The result is pressure on the palm, causing pain. Use a large-diameter spherical surface at the handle end and give a good gripping surface on the tool handle to minimize the palm pressure.

GUIDELINE 6: MAKE THE GRIP SURFACE COMPRESSIBLE, NONCONDUCTIVE, AND SMOOTH

6.1 Compressible

Just as a compressible floor (wood or carpet) is easier on the feet and legs than noncompressible concrete, a compressible grip material is easier on the hand. Rubber, compressible plastic, or wood are the best materials. Avoid hard plastic or metal. Compressible materials reduce slipping on the grip as well as reduce tool vibration. A compressible grip with a high coefficient of friction (tennis racquet, golf club, tape on baseball bat) is good. Resistance to absorption of sweat, oil, blood (meat cutting), and so forth is a desirable feature; nonabsorption also tends to improve sanitation. Figure 18.19 shows a screwdriver with a compressible rubber handle.

6.2 Nonconductive

Grips should not conduct heat or electricity. Materials with good electrical **conductivity** tend to be good conductors of heat. Thus wood or rubber is better than plastic, which is better than metal. Note that metal rivets in a wood or plastic handle may conduct even if the rest of the handle is nonconductive. Although the word *heat* is used, in outdoor construction cold surfaces may be the problem.

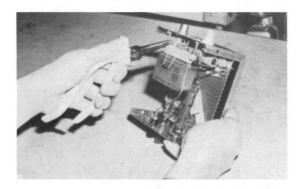

FIGURE 18.18 Improved pliers put the bend in the tool rather than the wrist. In addition, they prevent slippage in the hand and don't dig a hole in the palm since the handle end extends beyond the palm. A spring opens the tool. Because most assembly workers are female, the tool was designed using female hand sizes. Since they were first applied at Western Electric Corp. by Tichauer, they usually are known as Western Electric pliers.

FIGURE 18.19 A compressible rubber grip makes this screwdriver easy on the palm.

Table 25.1 gives the effect of various amps. For a given voltage, the key variable is resistance. Without protective clothing, the key human resistance is of the skin. Dry, clean unbroken skin varies from 100,000 to 600,000 ohms, depending on its thickness; wet or broken, it may be 500 ohms. Contact of 120 volts with a dry finger with 400,000 ohms would give 120/400,000 = .3 milliamps; contact with a wet finger with 15,000 ohms would give 120/15,000 = 8 milliamps. Thus, for safety, keep the skin dry and dirty (dirt can double the skin ohms). Don't increase conductivity with metal objects such as watches, rings, or keys.

For heat, the two essential factors are the material of the contact surface and the duration of contact.

Pain occurs when the temperature at 80 μm under the skin reaches 50°C; maximum pain occurs at 60°C; cell death (through protein denaturalization) occurs at 70°C. Table 18.6 gives the recommended maximum temperatures for short contact periods, that is, for accidental contact. Table 18.7 gives maximum temperatures for long contact periods—for intentional contact. Comfort is maximized at object temperatures of 33–35°C (skin temperature).

Low conductivity materials (wood, rubber) are good for two reasons: (1) They release heat to the hand more slowly and so can be held for a longer time before injury occurs (that is, you can let go of it before you are burned), and (2) they gain heat more slowly and so are less likely to reach a high temperature. With wood, the lower the density and moisture content and the rougher the surface, the higher the

TABLE 18.6   Recommended maximum temperatures (°C) for touchable surfaces to prevent burning with short contact times (Siekman, 1990).

DURATION, S		
1	3–4	Material
65	60	Water
65	60	Metals, uncoated, smooth surface
70	65	Metals, uncoated, rough surface
75	65	Metals, coated with 50 um thick varnish
80	70	Concrete, ceramics, marble
80	75	Ceramics, glazed (tiles)
85	70	Metals, coated with 100 um thick varnish
85	75	Porcelain
85	75	Polyamid 11–12 plastic with glass fiber
—	85	Teflon, plexiglass
95	85	Pertinax (smooth surface); Duroplast with fiber inlay
115	95	Wood (For very dry and very light woods, values may be up to 140 and 120.)

Source: This figure was first published in *Applied Ergonomics,* Vol. 21, No. 1, pp. 69–73, and is reproduced here with the permission of Butterworth-Heinemann Ltd., Oxford, UK.

TABLE 18.7   Recommended maximum temperatures (°C) for touchable surfaces to prevent burning with long contact times (Siekman, 1990).

TEMPERATURE, °C	MATERIAL
Duration: Up to 1 min	
50	Water, metals (coated and uncoated), and other materials with high conductivity
55	Ceramics, concrete, glass, polyamid 11–12
60	Duroplast, pertinax, plexiglass, teflon, wood
Duration: Up to 10 min	
48	All materials
Duration: Up to 8 h	
43	All materials

Source: This figure was first published in *Applied Ergonomics,* Vol. 21, No. 1, pp. 69–73, and is reproduced here with the permission of Butterworth-Heinemann Ltd., Oxford, UK.

burn threshold (Siekman, 1990). Table 18.6 shows the benefits of coating metals with varnish. An air-powered tool can become cold due to compressed air expansion; a plastic sleeve will reduce heat transfer as well as dampen vibrations.

See Section 7.2 of this chapter for comments on clearance.

6.3   Smooth   A knife works by exerting a force over a very small area. Sharp edges on a tool handle act as knives to the hand. Keep tool radii over 3 mm; 6 or 9 is better. See Figure 18.20. Grind away sharp edges (forging lines, casting parting lines, sharp radii) or cover them with tape. Dipping into plastic is another alternative. Smoothness also tends to aid sanitation. Be cautious of very soft grip materials as they may embed chips or splinters.

In general, the hand should not move on the grip. Nonmovement can be achieved by grip shape (see Sections 5.3 and 5.4) or by grip surface (high coefficient of friction).

A poor grip leaves its mark—on the hand.

GUIDELINE 7: CONSIDER THE ANGLES OF THE FOREARM, GRIP, AND TOOL

This topic also is discussed in Section 2 of Chapter 16, Cumulative Trauma. The topic is discussed in two sections: (1) angles and (2) clearance.

7.1   Angle   The goal is to keep the wrist in the neutral (handshake) position. When the wrist is not in the neutral position, not only is grasp strength less (Putz-Anderson, 1988; McMullin and Hallbeck, 1991),

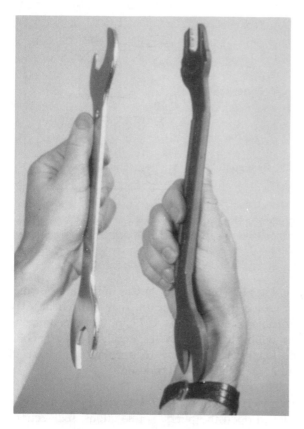

FIGURE 18.20 Sharp radii on tools (left) can cut into the hand.

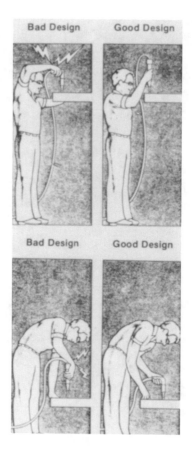

FIGURE 18.21 Optimum tool angle depends on the posture. Ulin et al. (1993) say repetitive screwdriving should not be done on horizontal surfaces above the elbow; if elevated work must be done on horizontal surfaces, avoid pistol grip tools. At midthigh and lower, use pistol grip tools.

but risk of cumulative trauma increases. Achieve wrist neutrality by changing the job or changing the tool. Figure 18.21 shows how the best tool angle varies with the posture. Although power tools tend to have a **pistol grip** (tool axis parallel to forearm) or **inline grip** (tool axis is perpendicular to forearm), there are other possibilities. Figure 18.22 shows a right-angle nutrunner.

7.1.1 Changing the job

Changing the job may simply require a change in worker posture. Standing tends to permit longer reaches than sitting. If the person is sitting, perhaps the chair is too high or too low for that specific person, resulting in an awkward wrist orientation.

The work orientation can be changed also by tilting the entire worksurface, tilting a fixture on the surface, or tilting the work itself. An example of tilting the entire worksurface are some keyboards, which have adjustable angles.

7.1.2 Changing the tool

Figure 18.18 showed a pliers with a bend. Figure 18.23 shows a hammer with a bend. Figure 18.24 shows a soldering iron with a bend. The bend in the tool also may reduce elbow abduction (Konz, 1986).

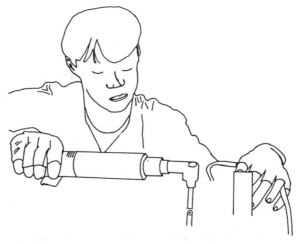

FIGURE 18.22 Right-angle nutrunners are popular for securing fasteners with high (> 20 Nm) torque. **Hard joints** are formed when two solid objects (e.g., pulley to a crankshaft) are brought together. The torque increases (ramps up) quickly, and the joint is completed in perhaps .5 s. **Soft joints** involve two objects having elastic properties. The torque increases more slowly and the joint is completed in perhaps 2 s (Radwin et al., 1989).

FIGURE 18.23 Hammer handles with a small bend (5°–10°) are preferred over a handle with no bend. Schoenmarklin and Marras (1989a, 1989b) reported hammers with a 20–40° bend could be beneficial.

FIGURE 18.24 Bent soldering iron puts the bend in the tool, not the wrist. Note the wrist support to reduce tremor. Also note how the shield permits the operator to hold the iron closer to the tip (since heat and splashes are not a problem), thus giving improved accuracy.

7.2 Clearance
Sufficient clearance minimizes burns and pinch points.

7.2.1 Burns
Increase the distance of the hand from the hot surface to reduce accidental contact and to reduce the effect of radiated heat. See Figure 18.25. A shield permits the actual distance to remain close while increasing protection. However, if the grip is designed so the hand is trapped in the grip, contact time may be increased. Wet clothing from the spill of a hot liquid presents a hazard because it clings to the skin and therefore increases contact time.

Heat gain also can be reduced by handle shape. For example, a narrow neck between a handle and the remainder of the tool reduces conductive heat transfer. Consider this when you buy heat-generating tools such as soldering irons. It also applies to coffee cups!

7.2.2 Pinches
Repetitive-use tools are the real problem. That is, 1 pinch/100 uses is tolerable in

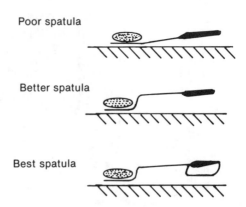

FIGURE 18.25 Increase distance between the hand and the hot surface. A shield increases the protection without requiring a change in the physical distance.

a tool used 10 times per year (say in the home). But if the tool is used 100 times/h, that is 8 pinches/day. Greenburg and Chaffin (1977, p. 124) recommend that double-handle tools (especially locking handles and toggle clamps) have an opening of at least 25 mm in their fully closed position.

Pliers present a specific problem as the operator may insert the first finger into the pinch point in order to open the pliers. For solutions, see Sections 8.1 and 8.2.

GUIDELINE 8: USE THE APPROPRIATE MUSCLE GROUP

8.1 Muscle Direction
Muscles generally are paired. One moves the hand out, one moves the hand in; one rotates it clockwise, another rotates it counterclockwise. Generally the strengths and ranges of motion are not the same in both directions. For example, the muscles which close the hand are much stronger than those which open the hand. Thus when using a pliers or grass clippers or scissors, closing the hand is not a problem, but opening it is relatively tiring. A good idea is to use some type of spring to open such tools so that the tool is **normally open**.

Giving the muscle something to push against helps. For example, loops on the outside of scissors handles allow the muscles to push against something when opening the scissors. However, the tops and sides of fingers are thin (not padded), and so are injured by repetitive pressure. On the conventional pliers without a spring, the fingers have to shift position to open the pliers.

8.2 Muscle Size
Normally bigger is better, but the moment arm also needs to be considered. For example, operators often try to open a pliers by

putting the first finger into the pinch point. Not only does this lead to pinches but there is a very short lever arm. If a spring is not used to open the tool, use the little finger (although it is weak) to open the pliers since it is at the end of a relatively long moment arm of the tool.

A cylindrical power tool (such as shown in Figure 18.26) usually is held in a power grip. Movement of the tool in the hand is prevented by squeezing the fingers. Replace the load on the small muscles of the fingers by using a bearing surface under the little finger. See Figure 18.26. In addition, cover the cold hard metal with rubber or plastic (Johnson, 1988). A push-to-start power screwdriver reduces stress on the trigger finger since the force occurs on the forearm muscles.

A conventional screwdriver is rotated by finger motions. In a human-powered mechanical screwdriver, the bit is rotated by a linear motion of the forearm—a better way.

Designed for reduction of fatigue in drilling or screwdriving applications. Flange is easily attached to the lower diameter of the tool. It makes the tool easier to handle while reducing muscle tension in the operator's arm. Since the heel of the hand exerts downward pressure, the fingers need not be tightly clenched.

Flanges are furnished in three different mounting diameters to fit 0000, 000 and 0-Series straight tool. Order 41364 with proper dash number.

Flange Assembly	Used On
41364.1	0000 Series Straight Tools
41364.2	000 Series Straight Tools
41364.3	0 Series Straight Tools

FIGURE 18.26 Bearing surface under the little finger permits the large muscles of the forearm to resist the tool thrust force instead of the little muscles of the fingers. The specific design shown above also allows the forearm to absorb the tool torque. Simple circular bearing surface accessories are available as add-ons for most tool diameters.

8.3 One vs. Many

As pointed out in Tables 18.2 and 18.8, individual fingers are weaker than the fingers in combination. Very roughly, the strength of an individual finger is 10% of the strength of the hand as a unit. If only one finger is to be used, the thumb is the strongest. The little finger has about 50% of the thumb's strength.

For finger pull strengths, Imrhan and Sundararajan (1992) reported a **pulp pinch** grip (index finger opposes thumb) had a relative strength of 100%, a **chuck pinch** grip (index and middle oppose thumb) had a relative strength of 150%, and a **lateral pinch** grip (side of index finger buttressed by other 3 fingers opposes thumb) had a relative strength of 240%.

For a five-finger static pinch (e.g., squeezing a pliers handle), the mean contributions of the index, middle, ring, and small fingers were 33%, 33%, 17%, and 15%, respectively. For a five-finger static lifting task, the mean contributions of the index, middle, ring, and small fingers were 35%, 26%, 20%, and 19%, respectively. For both situations, the percentages changed as force levels changed (Radwin et al., 1992).

Moving part of the tool independently while the tool is kept steady in the hand is called **triggering.** Triggers often are actuated by a single finger. Trigger strips usually are preferred to trigger buttons. See Figure 18.27. If a single finger must be used, the thumb is preferable (for repetitive motions) over the first finger. If the thumb opposes the other fingers (as with tweezers), move the thumb so that it is aligned with

FIGURE 18.27 Trigger strips are better than trigger buttons because local muscle fatigue can be avoided. In addition, more force can be exerted.

TABLE 18.8 Normal hand strength in newtons (An et al., 1986) (1 *N* = .22 lb or .1 kg). See also Table 18.2.

	Grasp	Tip Pinch	Pulp Pinch	Key Pinch	Radial Deviation		Ulnar Deviation	
					Thumb	Index	Thumb	Index
Male	400	65	61	109	43	43	75	42
Female	228	45	43	76	25	31	43	28

the middle finger; then the ring and little finger can support the middle finger.

When accidental activation is important to avoid (and tool steadiness is not important and the trigger is not used frequently), consider a thumb slide.

8.4 Rotate Inward   With the hand in the neutral (handshake) position (thumb at "11 o'clock"), the right hand and arm can rotate about 70° clockwise about the forearm axis (pronation) and about 150° counterclockwise (supination). Table 18.9 gives the time needed to rotate the hand about the forearm; no distinction is made for direction.

TABLE 18.9   Time for the hand to rotate about the forearm. Work-Factor assumes an "incentive" pace (125%) while MTM is "daywork" (100%).

Angle Rotated, degrees	Work-Factor, min	Methods Time Measurement, min
45	.0017	.0021
90	.0023	.0032
135	.0028	.0044
180	.0031	.0056

REVIEW QUESTIONS

1. How do the 3 elements of jobs affect the decision to use a general-purpose or special-purpose tool?

2. Calculate the cost/use for a specific handtool. Show calculations and assumptions.

3. The preferred hand is the right hand for approximately what percent of the population?

4. What percent performance would you expect for the nonpreferred hand (using the preferred as 100%) for bench assembly work without tools? What percent for use of a manual screwdriver?

5. What are the 2 reasons a tool should be usable with either hand?

6. Give an example of a tool using: (1) power grip with force parallel to the forearm, (2) power grip with perpendicular angle to the forearm, (3) torque about the forearm, (4) an oblique grip, (5) a hook grip, (6) an internal precision grip, and (7) an external precision grip.

7. Approximately what diameter should power grip handles be?

8. How long should a power grip be?

9. Give 3 reasons why a thumb conoid is better than a thumb ring.

10. What is a pommel?

11. Give 2 reasons why wood is a good handle material.

12. What is the maximum temperature for a tool handle to be held continuously?

13. Why does a bent wrist cause problems?

14. Are the hand-opening or hand-closing muscles stronger?

15. From the neutral position, can the hand rotate more in pronation or supination?

16. Briefly describe the 3 wrist styles on gloves.

17. What is the difference between a glove and a mitten? When should a mitten be used in place of a glove?

REFERENCES

ACGIH. *Threshold Limit Values and Biological Exposure Indices for 1987-1988.* Cincinnati, Ohio: American Congress of Governmental Industrial Hygienists, 1987.

An, K., Askew, L., and Chao, E. Biomechanics and functional assessment of upper extremities. In *Trends in Ergonomics/Human Factors III.* Korwowski, W. (ed.). Amsterdam: Elsevier, 1986.

Andersson, E. R. Design and testing of a vibration attenuating handle. *Int. J. of Ind. Ergonomics,* Vol. 6, 119-25, 1990.

Ayoub, M. and LoPresti, P. The determination of an optimum size cylindrical handle by use of electromyography. *Ergonomics,* Vol. 14 [4], 509-18, 1971.

Bradley, J. Effect of gloves on control operation time.

Human Factors, Vol. 11 [1], 13-20, 1969.

Cochran, D., Albin, T., Bishu, R., and Riley, M. An analysis of grasp force degradation with commercially available gloves. *Proceedings of the Human Factors Society,* 852-55, 1986.

Cochran, D., Bishu, R., and Riley, M. The effects of gloves on holding time. *Proceedings of the Human Factors Society,* 894-97, 1987.

Degani, A., Asfour, S., Waly, S., and Koshy, J. A study of two shovel designs. *Applied Ergonomics,* Vol. 24 [5], 306-12, 1993.

Dickson, A., Petrie, A., Nicolle, F., and Calnan, J. A device for measuring the force of the digits of the hand. *Biomedical Engineering,* 270-73, July 1972.

Eastman Kodak. *Ergonomic Design for People at Work,*

Vol. 1. New York: Van Nostrand Reinhold, 1983.

Echard, M., Smolenski, S., and Zamiska, M. Ergonomic considerations: Engineering controls at Volkswagen of America. In *Ergonomic Interventions to Prevent Musculoskeletal Injuries in Industry*, 117-31. Chelsea, Mich.: Lewis Publishers, 1987.

Evans, T., Lucaccini, L., Hazell, J., and Lucas, R. Evaluation of dental hand instruments. *Human Factors*, Vol. 15 [4], 401-406, 1973.

Fransson, C. and Winkel, J. Hand strength: The influence of grip span and grip type. *Ergonomics*, Vol. 34 [7], 881-92, 1991.

Freivalds, A. The ergonomics of shovelling and shovel design—A review of the literature. *Ergonomics*, Vol. 29, 3-18, 1986a.

Freivalds, A. The ergonomics of shovelling and shovel design in experimental study. *Ergonomics*, Vol. 29, 19-30, 1986b.

Freivalds, A. and Kim, Y. Blade size and weight effects in shovel design. *Applied Ergonomics*, Vol. 21 [1], 39-42, 1990.

Gainer, G. and Konz, S. An evaluation of six screwdriver handles. *Proceedings of IX Int. Conf. on Production Research*, Amsterdam: Elsevier, 1987.

Garett, J. The adult human hand: Some anthropometric and biomechanical considerations. *Human Factors*, Vol. 13 [2], 117-31, 1971.

Goel, V. and Rim, K. Role of gloves in reducing vibrations: An analysis for pneumatic chipping hammer. *American Industrial Hygiene Association J.*, Vol. 48 [1], 9-14, 1987.

Grant, K., Habes, D., and Steward, L. An analysis of handle designs for reducing manual effort: The influence of grip diameter. *Int. J. of Industrial Ergonomics*, Vol. 10, 199-206, 1992.

Greenburg, L. and Chaffin, D. *Workers and Their Tools*. Midland, Mich.: Pendall Publishing, 1977.

Griffin, M. *Handbook of Human Vibration*. San Diego, Calif.: Academic Press, 1990.

Hertzberg, H. Engineering anthropometry. In *Human Engineering Guide to Equipment Design*, Van Cott, H. and Kincaid, R. (eds.), Chapter 11. Washington, DC: Supt. of Documents, 1973.

Hudgens, G., Fatkin, L., Billingsley, P., and Muzurczak. Hand steadiness: Effects of sex, menstrual phase, oral contraceptives, practice, and handgun weight. *Human Factors*, Vol. 30 [1], 51-60, 1988.

Imrhan, S. and Sundararajan, K. An investigation of finger pull strengths. *Ergonomics*, Vol. 35 [3], 289-99, 1992.

Johnson, S. Evaluation of powered screwdriver design characteristics. *Human Factors*, Vol. 30 [1], 61-69, 1988.

Kao, H. Human factors design of writing instruments for children: The effect of pen size variations. *Proceedings of 18th Annual Meeting of the Human Factors Society*, 1974.

Kaster, E., Katagihara, A., and Konz, S. Performance of preferred vs. non-preferred hand and arm. *Proceedings of IX Int. Conf. on Production Research*, Cincinnati, 431-36, 1987.

Kellor, M., Kondrasuk, R., Iverson, I., Frost, J., Silberberg, N., and Hoglund, M. *Hand Strengths and Dexterity Tests*, Manual 721. Minneapolis, Minn: Sister Kenny Institute, 1971.

Konz, S. Design of foodscoops. *Applied Ergonomics*, Vol. 6 [1], 32, 1975.

Konz, S. Bent hammer handles. *Human Factors*, Vol. 28 [3], 317-23, 1986.

Konz, S. and Oetomo, I. Comfort of various size and shape pens. *Proceedings of South East Asian Ergonomics Society Conference*, Denpasar, Bali, 1988.

Konz, S. and Warraich, M. Performance differences between the preferred and non-preferred hand when using various tools. *Ergonomics International* 451-53. (Proc. of 8th Int. Congress of IEA, Southampton). London: Taylor and Francis, 1985.

Kumar, S. and Cheng, C. Spinal stresses in simulated raking with various rake handles. *Ergonomics*, Vol. 33 [1], 1-11, 1990.

Lindqvist, B. *Ergonomic Tools in Our Time*. Stockholm: Atlas Copco, 1986.

Magill, R. and Konz, S. An evaluation of seven industrial screwdrivers. In *Trends in Ergonomics/Human Factors III*, Karwowski, W. (ed.). Amsterdam: Elsevier, 1986.

McMullin, D. and Hallbeck, M. Maximal power grasp force as a function of wrist position, age and glove type. *Proceedings of the Human Factors Society*, 733-37, 1991.

Miller, M., Ransohoff, J., and Tichauer, E. Ergonomic evaluation of a redesigned surgical instrument. *Applied Ergonomics*, Vol. 2 [4], 194-97, 1971.

Mital, A. and Kilbom, A. Design, selection and use of hand tools to alleviate trauma of the upper extremities: Part I—Guidelines for the practitioner and Part II—The scientific basis (knowledge base) for the guide. *Int. J. of Industrial Ergonomics*, Vol. 10, 1-5 and 7-21, 1992.

Mital, A., Kuo, T., and Faard, H. A quantitative evaluation of gloves used with non-powered hand tools in routine maintenance tasks. *Ergonomics*, Vol. 37 [2], 333-43, 1994.

Nag, P. and Pradham, C. Ergonomics in the hoeing operation. *Int. J. of Ind. Ergonomics*, Vol. 10, 341-50, 1992.

NIOSH. *Criteria for a Recommended Standard: Occupational Exposure to Hand-Arm Vibration*, NIOSH Pub. 89-106. Cincinnati: NIOSH, 1989.

Patkin, M. Ergonomic design of a needleholder. *Medical J. of Australia*, Vol. 2, 490-93, September 6, 1969.

Plumber, R., Stobbe, T., Ronk, R., Myers, W., Kim, H., and Jaraiedi, M. Manual dexterity evaluation of gloves used in handling hazardous materials. *Proceedings of the Human Factors Society*, 819-23, 1985.

Putz-Anderson, V. *Cumulative Trauma Disorder—A Manual for Musculo-Skeletal Disease of the Upper Limbs*. London: Taylor and Francis, 1988.

Radwin, R., Armstrong, T., and Chaffin, D. Power hand tool vibration effects on grip exertions. *Ergonomics*, Vol. 30 [5], 833-55, 1987.

Radwin, R., Oh, S., Jensen, T., and Webster, J. External finger forces in submaximal five-finger static pinch prehension. *Ergonomics*, Vol. 35 [3], 275-88, 1992.

Radwin, R., VanBergeijk, E., and Armstrong, T. Muscle response to pneumatic hand tool torque reaction forces. *Ergonomics,* Vol. 32 [6], 655-73, 1989.

Rigby, L. Why do people drop things? *Quality Progress,* 16-19, September 1973.

Riley, M. and Cochran, D. Partial gloves and reduced temperatures. *Proceedings of the Human Factors Society,* 179-82, 1984.

Rodgers, S. (ed.). *Ergonomic Design for People at Work: Vol. 2.* New York: Van Nostrand-Reinhold, 1986.

Saran, C. Biomechanical evaluation of T-handles for a pronation supination task. *J. of Occupational Medicine,* Vol. 15 [9], 712-16, September 1973.

Schoenmarklin, R. and Marras, W. Effects of handle angle and work orientation: I. Wrist motion and hammering performance. *Human Factors,* Vol. 31 [4], 397-411, 1989a.

Schoenmarklin, R. and Marras, W. Effects of handle angle and work orientation: II. Muscle fatigue and subjective ratings of body discomfort. *Human Factors,* Vol. 31 [4], 413-20, 1989b.

Scott, R. The technology of electrically heated clothing. *Ergonomics,* Vol. 31 [7], 1065-81, 1988.

Shock, N. The physiology of aging. *Scientific American,* Vol. 206, 100-10, January 1962.

Siekman, H. Recommended maximum temperatures for touchable surfaces. *Applied Ergonomics,* Vol. 21 [1], 69-73, 1990.

Swanson, A., Matev, I., and Groot, G. The strength of the hand. *Bulletin of Prosthetics Research,* 145-63, Fall 1970.

Teschke, K., Brubaker, R., and Morrison, B. Using noise exposure histories to quantify duration of vibration exposure in tree fallers. *Am. Ind. Hygiene Association J.,* Vol. 51 [9], 485-93, 1990.

Tewari, V., Datta, R., and Murthy, A. Evaluation of three manually operated weeding devices. *Applied Ergonomics,* Vol. 22 [2], 111-16, 1991.

Ulin, S., Armstrong, T., Snook, S., and Franzblau, A. Effect of tool shape and work location on perceived exertion for work on horizontal surfaces. *Am. Ind. Hygiene Association J.,* Vol. 7, 383-91, 1993.

van der Grinten, M. Shovel design and back load in digging trenches. In *Musculoskeletal Disorders at Work,* Buckle, P. (ed.). London: Taylor and Francis, 1987.

Wasserman, D. *Human Aspects of Occupational Vibration.* Amsterdam: Elsevier, 1987.

19 | Controls

OVERVIEW

This chapter presents six control guidelines. Detailed information and examples accompany each guideline.

CHAPTER CONTENTS

1 Select the Proper Type of Control
2 Select the Proper Control Characteristics
3 Prevent Unintended Activation
4 Prevent Incorrect Identification
5 Make Accomplishments Equal Intentions
6 Properly Locate and Arrange the Controls

KEY CONCEPTS

color stereotypes	interlocks	reaction time
control/response ratio	lockout/tagout	shape coding
discrete/continuous	manipulative controls	unintended activation
exclusion percent	open/closed loop	zero mechanical state
human–machine system	population stereotypes	

Six design guidelines will be discussed. See also section H of Table 5.2.

1 SELECT THE PROPER TYPE OF CONTROL

1.1 Control Systems   Figure 19.1 shows a **human–machine system.** People impose their will on the machine through controls; the machine communicates to the people through displays. The entire system exists in an environment with temperature, noise, illumination, chemicals, and so forth. Example controls are levers, wheels, pedals, cranks, switches, buttons, and knobs. A control is a tool attached to a machine, and so Chapter 18 can provide additional comments.

Figure 19.1 shows the human as an operator of the machine. Using an automobile example, the human uses a manual transmission to shift the gears of the machine. But it is also possible for the human to function as a supervisor of the machine, giving only general orders and having the machine use a servomechanism to carry out the details. (After all, the sole purpose of machines is to serve people.) Camera examples are automatic light adjustments, focusing, and film advancing. Automobile examples are cruise control and automatic transmission. With microprocessors, considerable ability to carry out orders can be built into the machine.

Systems can be divided into two general types: open loop and closed loop. In an **open-loop** system, there is a desired input or setting of the device but there is no feedback of what the device is doing to affect the desired input. In a **closed-loop** system, feedback affects the desired input. Consider turning a valve to let water flow. In an open-loop system, the

valve is turned and it is assumed that the water flows. In a closed-loop system, the flow of the water is "fed back" to influence the valve opening. For example, if no water flows, an alarm may sound and an operator fixes the system. If the flow is too small or too large, the feedback element adjusts the input (valve position). In many systems, a human is the feedback element. See Box 23.2 for comments on human thermoregulatory control.

Before designing a control, get information on what specifically is to be controlled, the control task requirements (precision, force, etc.), the operator's information needs, the workplace restrictions, and consequences of accidental operation.

1.2 Types of Controls   First, decide whether the command to the machine will be **discrete** or **continuous.** Example discrete automobile controls are the ignition (on-off), headlight (low-high), gear shift (1st, 2nd, park, reverse). Continuous automobile controls are the radio volume, steering wheel, window crank, and accelerator.

Second, decide what part of the body will implement the mental command. In most cases, it is the hand–arm; occasionally it is the foot. Rare applications include voice (box shifting on conveyor lanes), eye focus point (helicopter guns), and body temperature (elevator call buttons). (Some elevator call buttons are actuated by temperature rather than pressure. Check this by pushing a button with a pencil to see if it responds.) This text focuses on hand–arm controls.

Third, decide on the mechanical interface between the human body and the machine interior. This depends both upon body biomechanics and upon the machine mechanism. In some cases, it is better to use a linear control motion and in some cases a rotary control motion. For example, if the human motion is to cause a gear or screw to turn, it is easier mechanically to rotate a crank or knob than to make a linear control motion. Both linear and rotary motions can be in the X, Y, or Z axis.

Table 19.1 shows example applications for discrete controls; Table 19.2 shows example applications for continuous controls.

When selecting the control type, consider the amount of human power required. Some controls permit transmission of human force to the machine while others just are switches. Using automotive examples, are the brakes conventional or power? Is the window moved by a crank or a motor? If the power is furnished by muscles, then consider controls with mechanical advantage—that is, a lever arm. The longer the lever arm, the less muscle force needed. When considerable force is needed, consider two-handed controls such as wheels. For

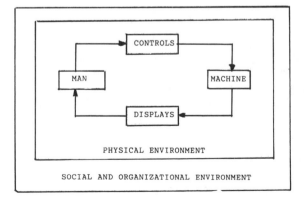

FIGURE 19.1 In the human–machine system, information flows from human to machine through controls. Information flows from machine to human through displays. Displays include not only instruments but also labels and accompanying printed instructions and warnings.

TABLE 19.1 Guide to discrete control selection (Hutchinson, 1981).

Control Requirement	Linear or Rotary	Type	Examples
• Select two discrete settings or states: stop–start on–off	L	Manual pushbutton	Keyboards, vending machines Elevator floor buttons Doorbell, horn (auto) Cruise control (auto)
insert momentary signal set state	L	Rocker switch	Hard-copy word processor
	L	Toggle switch	Wall lights
	L	Foot-actuated button	Headlight dimmer (older auto)
	L	Slide switch	Flashlight
	L	Pull cords	Houselamps
	L	Push/pull switch	Headlights-on/off (auto), TV-on/off
	R	Key-operated switch	Ignition (auto)
	R	Knob (volume/on combo)	Radio
• Two settings–larger force	L	Detent lever	Landing gear (aircraft)
	L	T-handle or stirrup	Hand brake (auto)
• Select three or more settings	L	Groups of legend switches or back-lighted pushbuttons	Car radio Keyboards, TV channels, dialing
	L	Lever	Wiper/washer (auto), temperature control, mode selection, seat position, steering wheel position
	L	Lever with detent	Throttle (aircraft), gear shift (auto)
	R	Circular dial (poor)	Conventional telephone
	R	Rotary selectors	Microwave timer

TABLE 19.2 Guide to continuous control selection (Hutchinson, 1981).

Control Requirement	Linear or Rotary	Type	Examples
• Precise adjustment small range, little force	R	Knob	Volume controls
	L	Continuous lever	Joy stick (aircraft)
	R	Continuous thumbwheel	Air vent vanes (auto)
• Rapid adjustment large range, small force	R	Spinner on multirotational knob	Music box windup
	R	Small crank	Manual windows (auto)
	R	Toggle or bat-handle	Slew command marker (USAF aircraft)
• Gross adjustment large force, small range	R	Hand wheels	Steering wheels (auto) Valves
	L	Translatory pedal	Brake (auto) Accelerator (auto)
	L	Reciprocating pedal	Brake (auto)
	R	Rotary pedal	Bicycle
	L	Continuous lever	Lawnmower throttle
large force, large range	R	Large crank	Artillery
• Multiple continuous positioning	L	Joy stick	Attitude (aircraft)
	L	Wheel/yoke and joy stick combo	Attitude (aircraft)
	L	Pantograph	Remote manipulators

maximum force, replace the arm with the leg and use pedals.

When selecting the control type, consider errors. For example, the error rate/1,000,000 characters is abut 30,000 for a keyboard, 900 for an optical character recognition wand, and .3 for a bar code and hand-held wand (Allais, 1982).

Table 19.3 shows control recommendations using criteria of speed of operation, accuracy, mounting space, operation in an array, and ease of check reading.

The control also may have special features. Push buttons, for example, may be alternate action (push on, push off), momentary contact (doorbell), touch-sensitive (calculator keys), interlocking (when button A is depressed, button C is released), connected to a display (light when depressed), and so forth.

2 SELECT THE PROPER CONTROL CHARACTERISTICS

Once the proper type is selected, specify the control size and shape, amount of control movement, and actuation force.

2.1 Force
The amount of force (torque) which can be exerted on a control depends on people variables and control variables.

2.1.1 People variables
People variables include the muscles used, the user population, and the percentile of the population designed for.

Larger muscle groups can exert more force. The general sequence from weaker to stronger is finger, combination of fingers, arm, both arms, foot, both feet. See Table 18.1 for grip strength, Table 18.2 and Table 18.8 for finger strengths and Table 11.6 for arm strengths. Summarizing: (1) The leg is approximately 3 times stronger than the arm. (2) Direction is very important, with arm forces at the nonoptimum angles being 50 to 80% of the force at the optimum angle. (3) The nonpreferred arm averages 60 to 150% of the strength of the preferred arm (depending upon the angle and the direction). (4) There seems to be no appreciable difference between the strength of the left and right legs. (5) All force recommendations assume that the fingers/hand/arm/leg have sufficient clearance to operate.

Figure 19.2 shows how pedal force varies between populations and within populations. Figure 11.11 showed that leg strength of women not only is lower than that of men but also showed that women begin to lose strength at an earlier stage.

In general, do not design for the mean of a population, as that results in 50% of the population not being able to do the task. The percent of the population to exclude (the **exclusion percent**) is a design decision which depends upon the cost of excluding weak people and the benefits of including weak people. In addition, just because most people can exert 7 N-m of torque on a particular control is not a reason to use 7 N-m in your design—less is better. Thus the percent to exclude should be quite small, perhaps 1 in 1,000 or less.

Assuming the normal distribution, to exclude 1% use 2.33σ, to exclude .1% use 3.09σ, to exclude .01% use 3.84σ.

Thus, if the population mean could exert 10 N-m torque on a knob and the standard deviation is

TABLE 19.3 Characteristics of common controls (Eastman Kodak, 1983).

Control	Suitability Where **SPEED** of Operation is Required	Suitability Where **ACCURACY** of Operation is Required	**SPACE** Required to Mount Control	Ease of **OPERATION IN ARRAY** of Like Controls	Ease of **CHECK READING IN ARRAY** of Like Controls
Toggle switch (On-Off)	Good	Good	Small	Good	Good
Rocker switch	Good	Good	Small	Good	Fair[1]
Pushbutton	Good	Unsuitable	Small	Good	Poor
Legend switch	Good	Good	Small	Good	Good
Rotary selector switch (discrete steps)	Good	Good	Medium	Poor	Good
Knob	Unsuitable	Fair	Small-Medium	Poor	Good
Crank	Fair	Poor	Medium-Large	Poor	Poor[2]
Hand wheel	Poor	Good	Large	Poor	Poor
Lever	Good	Poor (Horizontal) Fair (Vertical)	Medium-Large	Good	Good
Foot pedal	Good	Poor	Large	Poor	Poor

[1]Except where control lights up for "on."
[2]Assumes control makes more than one revolution.

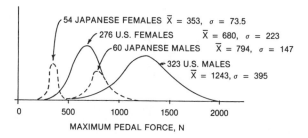

54 JAPANESE FEMALES $\bar{X} = 353$, $\sigma = 73.5$
276 U.S. FEMALES $\bar{X} = 680$, $\sigma = 223$
60 JAPANESE MALES $\bar{X} = 794$, $\sigma = 147$
323 U.S. MALES $\bar{X} = 1243$, $\sigma = 395$

MAXIMUM PEDAL FORCE, N

FIGURE 19.2 Maximum pedal force varies depending upon both the population and the individual in the population (Aoki, 1960; Mortimer et al., 1970). That is, although Americans could, on the average, exert more force than Japanese, there was considerable variability in both populations. The 20% coefficient of variation in the Japanese study (vs. a 32% in the American study) probably reflects better experimental technique.

1 N-m, then a $10 - 3.09 (1) = 6.9$ N-m knob would exclude .1% of this population, assuming the normal distribution is valid. However, human finger–arm strength measurements tend to have a positive skew, and so the normal distribution estimates lower values than the actual values. The above exclusion estimates therefore tend to predict more exclusions than would really occur.

Consider that someone from a weaker population may use the control and that many controls are used repetitively. In addition, the control may not be in a good position but may be at knee height, around a corner of a panel, or behind some apparatus; it may be covered with grease; the operator may wear slippery or bulky gloves, and so on. Therefore, strive to reduce the force required to actuate the control.

2.1.2 Control variables

The force required to actuate the control depends upon what the control is to accomplish and the control design. Assume, for example, that the designer has a leadscrew to be rotated and decides to use a knob. First, information is needed on how much torque is needed to rotate the leadscrew. If it is large, consider redesigning the mechanical system or using power assist. Only if it is well within human capability should a muscle-powered system be considered.

2.2 Keys and Pushbuttons

A special type of pushbutton (a key) is used for data entry. Keys generally should not be round. A concave top helps center the finger; a "pimple" on a frequently used key allows a user to locate a key by touch. Because of the repetitive action and multiple fingers, displacement should be consistent between keys. Key interlocks prevent simultaneous activation of two keys by preventing triggering until 75% of displacement has occurred. They improve speed and decrease errors. (Reportedly when Samuel Soule in 1867 devised the QWERTY key layout that is used today his intent was

to make it awkward, so a fast typist couldn't jam the levers and gears!) Feed back (tactile, auditory, or visual) the key activation to the operator. If space is very limited, it is possible to reduce the number of keys by having a key do multiple functions, with the function selected with a shift key: Examples are typewriter and calculator keyboards. Membrane keypads need tactile and/or kinesthetic feedback. Rectangular or square buttons are easier to label than round ones. Different-shaped buttons can indicate different functions. On handtools, triggers (i.e., multifinger operation) are better than buttons. Avoid large pushbuttons (palm buttons) in repetitive operations because the constant blow on the palm of the hand from the hard surface may cause carpal trauma. See Chapter 16.

2.3 Cranks

For one-hand controls, Table 19.4 and Figure 19.3 give design recommendations for cranks. (Also see Table 29.4.) Since the grip normally does not move within the hand, a high coefficient of friction is good; however, the grip should rotate on the crank to allow good hand–wrist orientation throughout the rotation. Another grip option is a 50-mm diameter sphere, because a sphere does not limit hand position as much as a cylinder. When power output is measured, there is less than a 5% advantage for a vertical axis of turning (vs. horizontal) and clockwise rotation (vs. counterclockwise); there is

TABLE 19.4 Crank design recommendations for Figure 19.3 (*Human Engineering Design Data Digest*, 1984).

Variable	Minimum	Preferred	Maximum
Light Loads (Wrist and finger movements; less than 22 N)			
Handle length (L), mm	25	38	75
Handle diam. (D), mm	10	13	16
Turning radius (R), mm			
Rate below 100 rpm	38	75	125
Rate over 100 rpm	13	65	115
Heavy Loads (Arm movement; over 22 N)			
Handle length (L), mm	75	95	
Handle diam. (D), mm	25	25	38
Turning radius (R), mm			
Rate below 100 rpm	190		510
Rate over 100 rpm	125		230

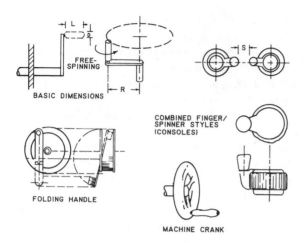

FIGURE 19.3   Crank dimensions for Table 19.4.

approximately a 20% advantage for a 100-mm radius over a 300-mm radius (Raouf, Imanishi, and Morooka, 1986).

2.4 Knobs   With circular knobs, you depend upon hand friction along the circumference; with the other shapes, you have a bearing surface. Figure 19.4 shows how knob shape can compensate for a slippery grip. Figure 19.5 shows that even circular knobs

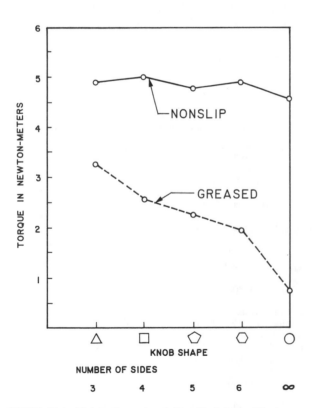

FIGURE 19.4   Minimize the number of sides of knobs—especially when gripping friction is poor (Kohl, 1983). For torque applications, clamping screws and knobs should have a turning circle diameter of about 90 mm; a rubber- or plastic-coated surface reduces abrasion on the fingers.

FIGURE 19.5   Better knobs (lower figure) have a bearing surface.

can have a bearing surface. Eastman Kodak (1983) recommends a maximum torque of 7 N-m for a 35-mm handle diameter or length. With both types, eliminate sharp edges and small radii, which can exert high pressure on the hand.

For the circular design, if the rotation axis is parallel to the base of the fingers, increasing diameter increases the torque up to a diameter equal to the hand grip span (fingers and palm just touch). This is about 50 mm (Replogle, 1983). If the axis is perpendicular to the palm, torque increases with diameter up to about 50% of hand length, which is about 90 mm (Imrahan and Loo, 1986); knurling does not improve torque capability up to 90 mm but it may above 90 mm (Imrahan and Loo, 1986; Nagashima and Konz, 1986). With a lever, any length is possible. Be sure to allow sufficient hand–arm clearance.

A task somewhat similar to turning of knobs is turning jar and bottle lids. The finger position will vary depending on the lid diameter and whether the lid also requires pushing. If pushing is not required, knurls permit as much as 70% more torque (Konz and Ravishankar, 1989). Sharp edges on the lid or knurl injure the hand. If lids are opened often, use a V-shaped gripper (to fit all sizes). The gripper can be hand held or permanently mounted.

2.5 Hand Wheels   For greater torque, use two hands on hand wheels. The wheel diameter should be 180 to 530 mm. The grip diameter should be 20 to 50 mm. McMulkin et al. (1993) reported the grip has a major influence on the amount of force a person can exert on a hand wheel (such as for railroad car brakes). On the standard circular wheel with 28-mm diameter grip, torque was 95 N-m (100%). Changing the grip diameter to 43 mm increased torque to 132 N-m (139%). Mounting 65-mm spherical knobs on the rim gave 147 N-m (155%). Making the rim a zigzag gave 191 N-m (201%).

If the wheel is oriented vertically, place it between 950 and 1,200 mm above the floor; if oriented horizontally, place it between 1,250 and 1,400 mm above the floor (Eastman Kodak, 1983).

Corlett (1982) says when maximum force is to be overcome, use a hand wheel set in the horizontal plane at chest height for the shortest operator (waist height for the tallest operator) at the largest diameter feasible but not greater than 500-mm diameter. For maximum precision, the key variable is the dial or display being set, and the larger the display the higher the precision.

Hand wheels also can be used to control vehicles. Vehicle steering wheels should have a wheel diameter of 350 to 400 mm for power steering and 400 to 510 mm for nonpower steering. Grip diameter should be 20 to 32 mm. For power steering, preferred orientation vs. vertical is 30°; for nonpower steering, preferred orientation is 45° (*Human Engineering Design Data Digest*, 1984).

2.6 Foot/Leg Controls

For still greater force, use the leg. Use of the leg also frees the hand. Table 19.5 and Figure 19.6 give design recommendations for foot switches; Table 19.6 and Figure 19.7 for foot pedals. Figure 19.8 shows the relative efficiency of various types of pedals.

Bullinger et al. (1991) discuss design variables of foot controls. Also see Guideline 6 of Chapter 15. Note that the contact surface of a pedal usually will be a shoe sole, so slipping may be a problem. In addition, pedals, being on the floor, may have dirt or other debris on them, further reducing friction.

For continuous control (such as with an auto accelerator), it is better to bend the ankle by depressing the toe rather than depressing the heel or moving the entire foot and leg. The foot of a 70-kg person weighs 1 kg vs. 4.1 kg of the leg + calf or the 11.3 kg of the entire leg. Moving just the toe permits the heel to rest on a support and reduces the amount of weight supported by muscles. The range of movement at the ankle should be between 80° and 115° (Nowak, 1972).

On–off controls (such as faucets, clamping fixtures) can be actuated by lateral motion of the knee as well as by the vertical motion of the foot. The knee should not have to move more than 75 to 100 mm; force requirements should be light. The advantage compared to the foot is that the weight of the foot need not be lifted. Hospitals use knee switches to actuate faucets to improve germ control on the hands.

Use large-force controls only for occasional or emergency use.

3 PREVENT UNINTENDED ACTIVATION

The more severe the consequences of **unintended activation** (activating a control unintentionally), the greater should be the precautions taken. However, if a protective measure makes it difficult to use the

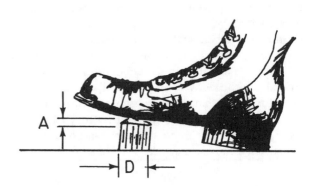

FIGURE 19.6 Foot switch dimensions for Table 19.5.

TABLE 19.5 Foot-operated switch design recommendations for Figure 19.6 (*Human Engineering Design Data Digest,* 1984; Eastman Kodak, 1983).

| Variable | Human Eng. Digest | | Eastman Kodak | | |
	Min	Max	Min	Preferred	Max
Diameter, mm	13		12	50-80	
Displacement, mm					
Normal operation	13	65			
Heavy boot	25	65			
Ankle flexion only	25	65	12		65
Total leg movement	25	100	25		180
Resistance, N					
Foot doesn't rest on control	18	90	15		75
Foot rests on control	45	90			

TABLE 19.6   Pedal design recommendations for Figure 19.7 (*Human Engineering Design Data Digest,* 1984; Eastman Kodak, 1983). The ankle should stay in the range from the neutral position to 20° up and 30° down.

Variable	Human Eng. Digest			Eastman Kodak	
	Min	Preferred	Max	Min	Max
Height (H), mm	25			80*	
Width (W), mm	75			90	
Separation (S), mm					
One foot random	100	150			
One foot sequential	50	100			
Displacement (D), mm					
Normal operation	13		65		
Heavy boots	25		65		
Ankle flexion only	25		65	12	65
Total leg movement	25		180	25	180
Resistance, N					
Foot doesn't rest on pedal	18		90	15	90
Foot does rest on pedal	45		90	15	90
Ankle flexion only			45		
Total leg movement	45		800		

*250 mm if used constantly.

control in normal circumstances, operators will be tempted to bypass the guard. Therefore, judgment must be used. A management policy should be established on who can adjust various process controls. That is, can operators change the settings on their own or is management approval required? In general, all equipment should have an "emergency stop" control; in some cases (e.g., gasoline pumps) they should turn off or stop when a control is released—a "deadman" switch. With greater danger, the control should be more prominent and even may be at multiple locations (e.g., cord along machine so it can be stopped by people near the cord). Of course, the control should not be activated by natural events such as vibration. Note that there are two **reaction times**—the human and the machine. Thus, it may take 1.0 s for a person to decide to stop a car and depress the pedal. Then the car may take 4.5 s to stop.

The human reaction time can be further divided into sensing, making a decision, and carrying out the action. For example, Sivak and Flannagan (1993) recommend using a brake light filament that is continuously preheated at a 2-volt level (i.e., below the visible). Then when such a lamp is activated, it reaches full brightness in 50 ms instead of the 250 ms required for the nonpreheated lamp; sensing reaction time (and total reaction time) is reduced by about 115 ms.

Have the control indicate (by position, light, sound, touch, etc.) when it has been activated. Then the person may be able to reverse the control if it is activated accidentally. Rocker switch activation is difficult to detect unless connected to a light.

Seven methods of reducing accidental activation follow. Often these methods are combined.

3.1 Key or Special Tool Activation (Locks)
Locks prevent activation by unauthorized people as opposed to accidental activation. Generally locks should not be used for emergency controls since the key or tool may not be available. Designing a device

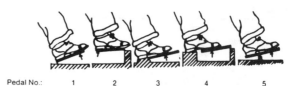

FIGURE 19.7   Pedal dimensions for Table 19.6.

MAXIMUM 60 DEG

FIGURE 19.8   Pedal strokes/minute vary with pedal design. Barnes et al. (1942) reported 187 strokes/min for design 1, 178 for 2, 176 for 3, 140 for 4, and 171 for 5. Trumbo and Schneider (1963) reported the time to depress the pedal through 15° of arc was 346 ms for 1, 395 for 2, 350 for 3, 370 for 4, and 355 for 5. Using either criterion, design 1 is best.

Pedal No.:   1   2   3   4   5

(such as a thermostat) to be actuated with an Allen wrench or Bristol wrench makes it difficult for a casual user to use the control. Maintenance also can be controlled if covers, modules, and so forth have Bristol heads and other nonstandard fasteners. Locks function by (1) what you have (e.g., a key), (2) what you know (e.g., a number), and (3) who you are (e.g., a photograph, fingerprints). More secure locks use methods 1 plus 2 (e.g., key plus number) or method 3.

A related control problem is an authorized person turning on a machine when it should remain off (e.g., during equipment maintenance). A **lockout/tagout** system in which the maintainer has the only key is the best solution. If the system is maintained by a crew, each individual should attach a personal lock. Each lock should be accompanied by a tag which contains the employee's name, date, and purpose of the lockout (Bimonte, 1990). Remember that electrical lockout is not sufficient because there may be energy stored in hydraulic lines, compressed air, or springs, or there may be potential energy of the suspended parts. The machine should be in a **zero mechanical state** (have zero potential energy). Emergency stop controls should be properly located, perhaps at multiple locations. A visual lock is a tag attached to controls during maintenance, indicating that the control should not be activated.

3.2 Interlocks

Interlocks can be mechanical or electrical. Generally a specified sequence must be followed. For example, a car transmission must be in neutral or park before the ignition will work. Some controls are sequential in two directions. The first direction move releases the control; the second permits operation. A detent is a slot with short perpendicular slots for a lever. Toggle switches and rotary handles can be designed so they require a pull motion before the linear or rotary motion; container caps for dangerous products commonly require turn and pull or depress and turn.

A parameter (temperature, light, time, flow, etc.) may have to be within specifications before the control will operate. Controls may be dual (require both hands or multiple people) to be operated.

Enabling controls are another option. Control A must be activated before controls B or C can be activated. An example is a button release on an auto transmission lever. If the enabling switch is remote or hidden, it can serve as a lock.

3.3 Barriers or Covers

A guard rail on the front edge of a panel may prevent falling onto a panel but tends to get in the way for normal operation. A nonsharp strip barrier can be placed between adjacent controls (e.g., pushbutton, toggle switches), a full barrier can be placed on all sides of the button, or a hinged cover can be placed over the control. Covers give considerable protection, with some operator inconvenience. A barrier around a button (i.e., the button is in a "collar") not only reduces accidental activation; it can prevent deliberate activation by a part of the body other than a finger (e.g., using an elbow to hold a button down while reaching into a press). Barriers, which can be retrofitted to existing protruding buttons as well as installed originally, are an alternative to recessing.

3.4 Recessing

The control (lever, pushbutton, toggle) is below the panel surface. Emergency buttons, in contrast should protrude, and often they are "mushroomed" so that they operate if hit from any angle.

3.5 Spacing

Clearances between controls must be sufficient so the worker does not operate one control when reaching for another. See Table 19.7 and Figure 19.9. If controls may be added later, reserve space for them. Spacing and painted lines can be used to group and differentiate controls. Try to keep spacing consistent within a panel; that is, don't put three related pushbuttons 50 mm apart in one part of the panel and 100 mm apart in another. Generally the problem is insufficient distance between controls—especially controls used without vision (e.g., pedals). Occasionally the control is too far from its label or associated display. Miniature switches and buttons are available, but don't forget that sizes of human fingers haven't shrunk also. In addition, leave panel space for labels.

3.6 Resistance

The control should require sufficient resistance in its designed line of travel so that light touches do not activate the control. However, keep the force reasonable, especially at the limits of reach.

3.7 Direction

The line of travel can be selected (in some situations) to minimize accidental activation. For example, a lever might move up–down so that contact force if a person walking by the control brushed it might be normal to the activation direction. Snagging on clothing also may be a problem.

4 PREVENT INCORRECT IDENTIFICATION

Six methods of reducing control identification errors are labeling, color, shape, size, mode of operation, and location. Often these methods are combined.

TABLE 19.7   Hand-control envelope minimum edge-to-edge spacing (mm) between various types of hand controls shown in Figure 19.9 (Kinkade and Anderson, 1984). Greater spacing is better—especially if gloves are worn or labeling is used.

Type of Control	Numbered Type										
	1	2	3	4	5	6	7	8	9	10	11
1. Key-operated controls	25	13	38	25	19	19	13	19	19	125	50
2. Pushbuttons (not in an array)	13	13	50	50	13	13	13	13	13	150	75
3. Pushbutton arrays*	38	50	50	50	38	38	38	50	50	150	75
4. Legend switches or legend switch arrays†	25	50	50	50	38	38	38	50	50	150	50
5. Slide switches or rocker switches	19	13	38	38	13	19	13	13	13	125	50
6. Toggle switches‡	19	13	38	38	19	19	13	13	13	150	75
7. Thumbwheels or thumbwheel arrays	13	13	38	38	13	13	13	19	19	125	50
8. Rotary selector switches	19	13	50	50	13	19	19	25	25	125	50
9. Continuous rotary controls	19	13	50	50	13	19	19	25	25	125	50
10. J-handles (large)	125	150	150	150	125	150	125	125	125	75	125
11. J-handles (small)	50	75	75	75	50	75	50	50	50	125	25

*Pushbuttons within an array, 19 mm center-to-center.

†Legend switches within an array, no minimum distances, but should be separated by a barrier that is at least 3 mm wide, 5 mm high, with rounded edges. Legend switches manufactured as elements of a module or modular array may be mounted as closely as engineering considerations permit.

‡Toggle switches arrayed in a horizontal line, 19 mm center-to-center.

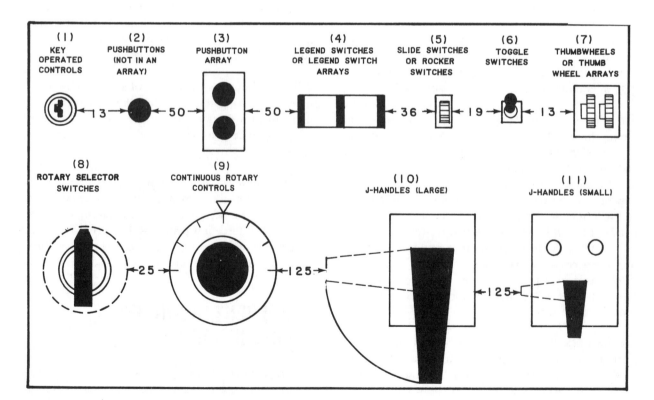

FIGURE 19.9   Minimum separable distances (mm) for Table 19.7.

4.1 Labeling

In general, labels should be

- Understandable by the operator. Use consistent wording and avoid abbreviations if possible. For example, a *C* on a faucet stands for cold in English but hot in Spanish and French (caliente, chaud).
- Legible, with good contrast and sufficiently large characters.
- Brief and precise (e.g., "pump on–off," not "pump control"; "hold for 2 s," not "spring-loaded").
- Located horizontally above the control so they are readable while operating the control. Labels below the control are obscured by the hand.

4.2 Color

About 4% of healthy adult males have color perception problems. Color identification requires standard white illumination. Only a relatively few colors (8 or fewer) should be used. Red, orange, green, yellow, white, and blue are good. Some colors have special connotations **(color stereotypes)**—for example, red for stop, halt, danger, off; yellow for caution, marginal situation; green for OK, safe, go, on. Color stereotypes may not be constant across cultures.

4.3 Shape

Knobs have been studied extensively for differentiation by shape **(shape coding).** Up to 9 different shapes can be used. Excluding shapes with special meanings (swastika, cross, crescent, and airplane), a tactually discriminable set of shapes is circle, star, ellipse, and square (Eastman Kodak, 1983). Tactual shape coding, which depends on a bare or lightly gloved hand, is not as fast as color coding, but controls can be used without vision. Shapes also can be a visual signal. Thus, keep shapes consistent for the same type of control. For example, all stop buttons could be round.

A tactile signal may help. For example, on some keyboards, there is a "pimple" on the F and J keys, which allows the operator to identify the home keys by feel. On the Kodak slide projector remote control, the forward button is smooth but the reverse button has a ridge on it so you can identify each button by feel.

4.4 Size

Only a few sizes (2 to 3) can be differentiated. A large size gives a mechanical advantage. Larger sizes are needed for cold environments (non-heated areas and outside) because gloves may be used.

4.5 Mode of Operation

Controls can have distinctive ways of operation—push/pull vs. rotate vs. slide. However, only when the operator tries to actuate the control by turning and it won't turn does the operator know the wrong control was used. Thus, mode of operation is a method to be used as a backup for other methods.

4.6 Location

Both individual controls and groups of controls can be identified by location.

5 MAKE ACCOMPLISHMENTS EQUAL INTENTIONS

Be sure that the operation accomplishes what the operator intended. For example, for critical computer commands, such as delete file, the computer should be programmed to require verification of the command. In effect, this is a version of an interlock or enabling switch, requiring critical actions to be two-stage instead of one-stage. Figure 19.10 shows a badly designed control and a suggested alternative.

Complex sequences of actions (such as computer or microprocessor inputs) need to be user-friendly. There is a possibility that the designer has designed so many options into the controls (or one control with multiple functions) that a novice user is not able to use the device. VCR controls drive many to tears. I had to remove an automated thermostat from my house because I was unable to turn the heat up on vacation days. The clocks on some Dodge autos were so complicated to adjust that people brought them into the dealer to adjust for daylight savings time! You may be familiar with other examples of poorly designed controls. The key is to field test a control design before going into production.

More care and testing is needed if the action has multiple steps (as with computer commands). Novices will need more help (well-thought-out and field-tested documents as well as displays giving feedback of the results of actions) than will experienced operators. The detail of the "prompt" should be adjustable by the user so the novice can have lots of detail and the experienced user can just hit the highlights. A good computer program also will check input data for validity. For example, all social security numbers entered must have 9 digits, all ZIP codes have 5 (or 9) digits, all part numbers will have the proper number of letters and numbers in the specified locations, part numbers entered must be numbers in use, and so forth.

Another challenge is **population stereotypes** (habit patterns). The habit patterns of the engineer should not be considered to be the habit patterns of the users. Table 19.8 gives common habit patterns in the United States. The designer cannot depend upon

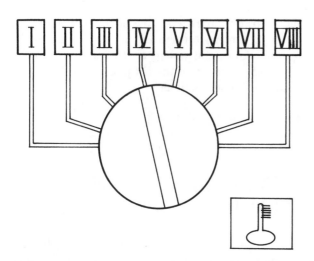
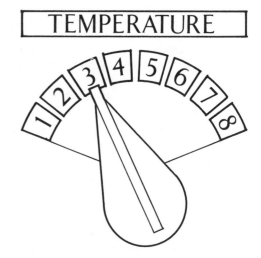

FIGURE 19.10   Five improvements to the poor design control/display (on the left) are: (1) improving the knob shape so its direction is obvious, (2) replacing the abstract thermometer symbol with the word TEMPERATURE (although automobiles must follow international symbol coding, other manufacturers should give customers in each country the courtesy of their own language), (3) moving the TEMPERATURE legend to above the control so it won't be obscured by the hand, (4) putting the heat levels on a semicircle instead of the ambiguous lines (do you point the knob at the line or the number?), and (5) replacing the Roman numerals with Arabic numerals.

the control user following the habit pattern but if the control violates the operator's expectation, there is more chance for error. Lewis (1986) points out that the U.S. expectation for power switches to be up = on and down = off is very strong, but stereotypes for left–right and front–back are mixed. Thus, the power switch should be on the front of the machine and up = on. If it must be on the side, put it on the right side. Unfortunately, such stereotypes are often country-specific. For example, in England flipping a light switch up turns the light off.

Displays also have habit patterns, which also may differ by population. For example, in Mexico, using the left blinker light on your car does not indicate that you want to turn left; it indicates to cars behind you that they can pass *you* on the left! In Mexico, the hot water faucet is labeled *C*, which is logical since *C* stands for caliente (hot). In China, the color for funerals is white.

TABLE 19.8   Conventional control movements in the United States (*Human Engineering Design Data Digest*, 1984).

Function	Control Action
On	Up, right, forward, clockwise, pull
Off	Down, left, rearward, counterclockwise, push
Increase	Up, right, forward, clockwise
Decrease	Down, left, rearward, counterclockwise
Raise	Up
Lower	Down
Right	Right, clockwise
Left	Left, counterclockwise
Retract	Up, rearward, pull
Extend	Down, forward, push

Labeling may help. For example, have two numbers on a thumb-wheel to show the operator which direction is increase or decrease.

For continuous controls, consider the **control/response ratio** (the distance the control moves/the distance the display indicator moves). Two basic movements are gross adjustment to the vicinity of the target and precision adjustment to the exact target. See Sanders and McCormick (1987) for a more extensive discussion.

A control does not exist in isolation. Controls need to be compatible with other controls and with displays. Control positions often act as displays; therefore, have good color contrast between the control and the panel.

6 PROPERLY LOCATE AND ARRANGE THE CONTROLS

6.1 Location   Location depends primarily on whether the control will be used by the hands or the feet.

6.1.1 Foot controls   Avoid foot controls for standing operators because they are tiring. In addition, if used with hand motions, there are potential hand safety problems (e.g., in presses). Foot controls can be for continuous power (e.g., bicycle) or discrete power (brake pedal).

For continuous power, the seat-to-pedal distance should be adjusted so that the leg is fully extended at the bottom of the stroke. The crank length should be approximately 20% of leg length, that is, 10% of stature height (Gross and Bennett, 1974). The pedal should be in line with the axis of the lower leg so the force is exerted by the leg muscles rather than the ankle muscles.

Discrete power usually is furnished by one leg since application time is usually less than 10 s and thus fatigue is not a problem. There does not seem to be any power advantage to using the right or left foot (Mortimer et al., 1970; Von Buseck, 1965). Adjusting Von Buseck's data for learning, force using both feet is 106 to 118% greater than using a single foot, but people will not always use both feet and the designer should not depend upon use of both feet.

In general, maximal force can be exerted if there is a straight line between the pedal and back support (Rees and Graham, 1952). That is, if the pedal is 250 mm below the seat, then the back support should be 250 mm above the seat. Aoki (1960) reported maximal pedal force when the calf–thigh angle was 110° and the thigh–back angle was 73°. Hugh-Jones (1947) reported maximum at a knee angle of 160°. Figure 19.11 shows the mean results of 155 males each

exerting force at 26 different positions. The mean efficiency index is defined as "force at a specific position for an individual/force at the individual's best position."

Note that maximum force capability may not be as important as comfort. Figure 19.12 shows the preferred seat reference point (intersection of the planes of the seat and back) relative to the accelerator (Martin and Johnson, 1952).

If the pedal will be used repeatedly, muscle fatigue will become a problem. Four design solutions are:

1. a wide pedal so either foot can be used at the operator's option
2. a pedal for each foot
3. lateral movement of the chair
4. a wide chair (a bench) upon which the operator can change position from time to time

In most industrial applications, the time required to move the foot from one location to another is not critical as the movement can be done at the operator's leisure or can be done simultaneously with some other motion. However, in some situations (such as auto braking) minimum reaction time is important. If your foot is already on the control, you can save about .25 s over moving your foot to the control. But if one foot is poised over the brake and the other is on the accelerator, you are in a "straight jacket" and this position restriction is tiring. Within the distance of 5 to 13 cm of movement, foot reaction time is approximately the same. Casey and

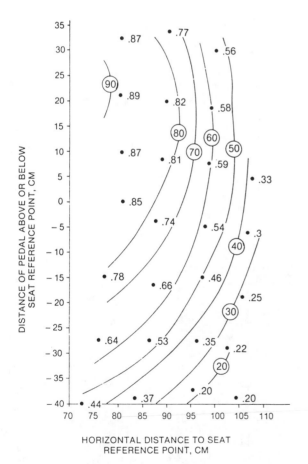

FIGURE 19.11 Pedal force decreases as distance from the pedal increases (Martin and Johnson, 1952). Note the maximum force is when the foot is above the seat reference point (SRP is the intersection of the planes of the seat and the seat back).

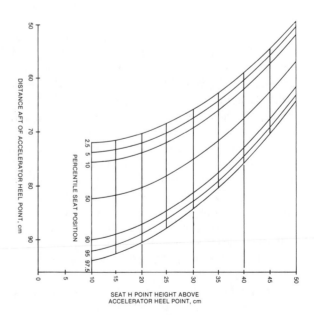

FIGURE 19.12 Preferred seat height above the accelerator pedal rises as you move closer to the pedal. The values are for the U.S. driving population in 1973 (Roe, 1975).

Rogers (1987) point out that some separation of controls (e.g., brake and accelerator) is necessary or both controls might be actuated. On the other hand, if the separation is too great, the brake might be missed! Thus, multiple criteria need to be considered.

6.1.2 Hand controls

Hand controls will be divided into keyboards, manipulative controls, and force controls.

Keyboards   Keyboards have two special characteristics. They are used repetitively and they are used (generally) with both hands. It is assumed that the keyboard location is independent of the location of a display for the eyes; however, some devices do have both the keyboard and a display in the same housing and the choice becomes whether to locate the display for the eyes or the keyboard for the hands.

The approximate location of the keyboard should be at elbow height. Since elbow height varies with sitting and standing and with individuals, a keyboard location should not be specified from the floor. Desks often are designed so a large proportion of the population can sit with their feet on the floor and not have the upper portion of the thigh touch the underside of the desk. This tends to result in a relatively high keyboard. The solution is a chair that can be raised to position the elbow properly so that it is even with the bottom row of the keyboard. This in turn, however, often requires a footrest. Having no footrest will result in uncomfortable legs (since the feet dangle without support). The alternative is to keep the feet on the floor, but now the keyboard will be too high and the neck and shoulders will hurt.

If you study the keyboard hand position in detail, as Kroemer (1990) and Nakeseko et al. (1985) did, you will find that hand position is important. The minimum stress position for the hand is the handshake position. However, when operating most keyboards, the hands must rotate to be approximately horizontal, with the knuckles above the wrist. In addition, the elbow tends to be lifted outward and forward (which is bad). Both Kroemer and Nakeseko et al. recommended splitting the keyboards and orienting them to minimize deviations. When using a conventional nonsplit keyboard, reduce this cumulative trauma (repetitive strain) through job rotation and task variety. See Table 5.4. A wrist support below the keyboard for use in the pauses between entries is recommended.

Ideally the keyboard orientation on the work surface is adjustable for all three axes by the user.

Manipulative controls   Manipulative controls (low-force controls such as knobs and switches—see

Tables 19.1 and 19.2) should be within reach. Define reach distance as the first percentile of the population, not the design engineer's reach. Define reach to the center of the palm, not the fingertip. Ideally, the operator should not have to change posture when reaching (e.g., kneel or bend over when standing, rise when sitting). Avoid blind reaches (where the operator cannot look at the control) if possible. Distinctive control shape or location may reduce some of the difficulty of blind reaches.

A design question is how much attention to pay to left-handers. Discrete controls (e.g., toggle switches) reflect elementary control tasks and handedness is not relevant. But continuous controls (e.g., joysticks) are relevant. One possibility is to place the control in the center; however, this uses valuable space and compromises the majority of the population. Garonzik (1989) has a design guide for control location which considers handedness.

Force controls   Force controls, as other manipulative controls, should be within reach to reduce user stress. See Figure 19.13. However, some pos-

FIGURE 19.13   Place controls in easy reach. Repeated overhead motions (a) can cause "frozen shoulder," causing severe pain and functional impairment (Putz-Anderson, 1988).

tures permit greater muscle force and torque on the control due to human biomechanics.

In general, force capability while standing is better at waist height than knee or shoulder height and is better when the person is braced. At knee height, pull is better than push and up is better than down. At waist height, push is better than pull and up and down are equal. At shoulder height, push is better than pull and down is better than up (Collier et al., 1986).

For sitting cross-legged, place controls 10 to 25 cm above the floor; for kneeling, place controls 30 to 50 cm above the floor (Conway et al., 1981). For kneeling, exertable force is about 20% less when the arm is fully extended; lateral force exertions toward the sides are quite awkward and are relatively weak (Haslegrave, Tracy, and Corlett, 1987).

Grieve (1984), in studying the effect of height on pull for outboard motor starting, reported great differences for different postures.

6.2 Arrangement

Spacing of controls was discussed in Section 3.5, Spacing. See Table 19.7. See also Section 4.2, Arrangement of Displays, in Chapter 20.

If the control action is reported on a display (as it often is), then the arrangement of the control vs. the display must be obvious so the operator knows which control is related to which display.

From the concepts of Tullis (1983):

- Group related controls and displays together.
- Minimize layout complexity by aligning controls vertically and horizontally instead of randomly.
- Have blank space on the panel. Blank space not only reduces inadvertent operation (Table 19.7) but also provides visual structure. Make spacing between *groups* of related controls greater than between related controls.

REVIEW QUESTIONS

1. Sketch a schematic (see Figure 19.1) of human, machine, controls, displays, and environment.
2. For an automobile, give 5 continuous controls and 5 discrete controls.
3. Assume you are designing a control for a milling machine. What percentage of the population should be excluded from using the lathe because they are too weak to operate the control? Justify your answer.
4. List the 7 methods of reducing accidental activation of controls.
5. What is an enabling control? What is an interlock?
6. Should the control label be above or below the control? Why?
7. List the 6 methods of preventing incorrect identification of controls.
8. Briefly discuss population stereotypes.
9. Give 4 design solutions which permit a person to use either foot to operate a pedal.
10. Discuss how use of a footrest affects desired keyboard height.
11. Briefly summarize force capability at knee, waist, and shoulder heights.

REFERENCES

Allais, D. *Bar Code Symbology.* Lynwood, Wash.: Intermec, 1982.

Aoki, K. Human factors in braking and fade phenomena for heavy application. *Bulletin of Japan Society Mechanical Engineers,* Vol. 3 [12], 587-94, 1960.

Barnes, R., Hardaway, H., and Podolsky, O. Which pedal is best? *Factory Management and Maintenance,* Vol. 100 [98], January 1942.

Bimonte, A. Lockout procedures protect employees from harm. *Plant Engineering,* 80-82, January 11, 1990.

Bullinger, H., Bandera, J., and Muntzinger, W. Design, selection and location of foot controls. *Int. J. of Ind. Ergonomics,* Vol. 8, 303-11, 1991.

Casey, S. and Rogers, S. The case against coplanar pedals in automobiles. *Human Factors,* Vol. 29 [1], 83-86, 1987.

Collier, S., Chan, W., Mason, S., and Pethick, A. *Ergonomic Design Handbook for Continuous-Miners,* Report TM/86/11. Edinburgh, Scotland: Institute of Occupational Medicine, 1986.

Conway, E., Helander, M., and Curtin, R. Optimum control heights for sitting cross-legged and kneeling. *Proceedings of the Human Factors Society,* 767-71, 1981.

Corlett, N. Design of handtools, machines, and workplaces. In *Handbook of Industrial Engineering,* Salvendy, G. (ed.). New York: Wiley & Sons, 1982.

Eastman Kodak. *Ergonomic Design for People at Work,* Vol. 1 Belmont, Calif.: Lifetime Learning Publications, 1983.

Garonzik, R. Hand dominance and implications for left-handed operation of controls. *Ergonomics,* Vol. 32 [10], 1185-92, 1989.

Grieve, D. The influence of posture on power output generated in single pulling movements. *Applied Ergonomics,* Vol. 15 [2], 115-17, June 1984.

Gross, V. and Bennett, C. Bicycle crank length. *Proceedings of the 6th International Ergonomics Meeting,* College Park, Md., 1976; see also Gross, V., Bicycle crank length and load, M.S. thesis, Manhattan: Kansas State University, 1974.

Haslegrave, C., Tracy, M., and Corlett, N. Biomechanical effects of force exertions while kneeling. *Proceedings of the Human Factors Society,* 318-22, 1987.

Hugh-Jones, P. The effect of limb position in seated subjects on their ability to utilize maximum contractile force of the limb muscles. *J. of Physiology,* Vol. 105, 332-44, 1947.

Hutchinson, R. *New Horizons for Human Factors in Design.* New York: McGraw-Hill, 1981.

Human Engineering Design Data Digest. Redstone, AL: Human Engineering Laboratory, U.S. Army Missile Command, 1984.

Imrahan, S. and Loo, C. Torque capabilities of the elderly in opening screw top containers. *Proceedings of the Human Factors Society,* Santa Monica, 1167-71, 1986.

Kinkade, R. and Anderson, J. (eds.). *Human Factors Guide for Nuclear Power Plant Control Room Development.* Palo Alto, Calif.: EPRI, 1984.

Kohl, G. Effects of shape and size of knobs on maximal hand-turning forces applied by females. *Bell System Technical T.,* Vol. 62 [6], 1705-12, July–August 1983.

Konz, S. and Ravishankar, H. Knurls on pop bottle lids. *Proceedings of the Human Factors Society,* 483-85, 1989.

Kroemer, K. Cumulative trauma disorders. *Applied Ergonomics,* Vol. 20, 274-80, 1990.

Lewis, J. Power switches: Some user expectations and preferences. *Proceedings of the Human Factors Society,* 895-99, 1986.

Martin, W. and Johnson, E. *An optimum range of seat positions as determined by exertion of pressure upon a foot pedal,* AMRL Report 86. Fort Knox, Kentucky: Army Medical Research Laboratory, June 1952 (AD 21654).

McMulkin, M., Woldstad, J., McMahan, P., and Jones, T. Wheel turning strength for four wheel designs. *Proceedings of the Human Factors and Ergonomic Society,* 730-34, 1993.

Mortimer, R., Segel, L., Dugoff, H., Campbell, J., Jorgeson, C., and Murphy, R. *Brake Force Requirement Study,* National Highway Safety Bureau Final Report FH-11-6952. Washington, DC, April 1970.

Nagashima, K. and Konz, S. Jar lids: Effect of diameter, gripping material and knurling. *Proceedings of the Human Factors Society,* 672-74, 1986.

Nakeseko, M., Grandjean, E., Hunting, E., and Gierer, R. Studies on ergonomically designed alphanumeric keyboards. *Human Factors,* Vol. 27 [2], 175-87, 1985.

Nowak, E. Angular measurements of foot motion for application to the design of foot pedals. *Ergonomics,* Vol. 15 [4], 407-15, 1972.

Putz-Anderson, V. (ed.). *Cumulative Trauma Disorders.* London: Taylor and Francis, 1988.

Raouf, A., Imanishi, H., and Morooka, K. Investigations pertaining to continuous and intermittent cranking motion. *Int. J. of Industrial Ergonomics,* Vol. 1, 29-36, 1986.

Rees, J. and Graham, N. The effect of backrest position on the push which can be exerted on an isometric foot-pedal. *J. of Anatomy,* Vol. 86, 310-19, 1952.

Replogle, J. Hand torque strength with cylindrical handles. *Proceedings of the Human Factors Society,* 412-16, 1983.

Roe, R. Describing the driver's workspaces eye, head, knee and seat positions. Society of Automotive Engineers paper 730 356, February 1975.

Sanders, M. and McCormick, E. *Human Factors in Engineering and Design,* 6th ed. New York: McGraw-Hill, 1987.

Sivak, M. and Flannagan, M. Fast-rise brake lamp as a collision-prevention device. *Ergonomics,* Vol. 36 [4], 391-95, 1993.

Trumbo, D. and Schneider, M. Operation time as a function of foot pedal design. *J. of Engineering Psychology,* Vol. 2 [4], 139-43, 1963.

Tullis, T. The formatting of alphanumeric displays: A review and analysis. *Human Factors,* Vol. 25 [6], 657-82, 1983.

Von Buseck, C. Excerpts from maximal brake pedal forces produced by male and female drivers, Research Report EM-18. Warren, Mich.: GM, January 1965.

OVERVIEW

Both passive displays (alphanumeric characters and their arrangements) and active displays (instruments) will be discussed. Many specific design recommendations are given to improve communication through improved legibility and reduced errors.

CHAPTER CONTENTS

1 Alphanumeric Characters
2 Arrangement of Characters and Symbols
3 Instruments
4 Location and Arrangement of Displays

KEY CONCEPTS

analog/digital	explicit/implicit	pictographs
codes	fonts	pie chart
conversion line	formulas	points
discrete/continuous/	graphs	reverse image
representational	icons	serifs
divided bar graph	justified	significant digits
dot charts	line of sight	slide/transparency
dot matrix	menus	stroke width
dot pitch	outside-in/inside-out	time series
doughnut chart	picas	visual angle

Failure of a display to communicate can be at two levels. First, and perhaps most obvious, is legibility or detectability. The letters on the display may be too small or the contrast too poor. Second is the problem of understanding. For example, what is the meaning of the word *subsequent* or *prior?* What is the meaning of a pointer pointing to a value of 220? Although this chapter focuses on solving the first-level (legibility) problems, the designer should not forget level-two (understanding) problems.

First we will discuss passive displays (alphanumeric characters and their arrangements), then active displays (instruments and their arrangements).

1 ALPHANUMERIC CHARACTERS

1.1 Font   Handwritten characters tend to be less legible than printed characters due to the wide variation in individual concepts of what various letter shapes should be and poor execution of the ideal concept. Tables 20.1 and 20.2 give recommended handwriting styles for capital letters and numbers. Commas and periods can be confused easily as can the capital letters D and O and the numbers 6 and 0 and 4 and 9.

For print, there is relatively little difference in legibility for most reasonable type styles **(fonts).** Readers not only find extreme styles less legible, they consider them less attractive. See Figure 20.1. Roman font has **serifs** (little flourishes and embellishments); sans serif fonts are without serifs. The **stroke width** (thickness of lines within a letter) is almost constant in sans serif, varies some in Roman, and varies considerably in modern. Cooper et al. (1979) found that Press Roman, Theme, and Univers fonts were preferred over Baskerville, Bodoni Book, Aldine Roman, and Century. Don't use Roman numerals. This applies everywhere—chapter numbers in books, volume and table numbers, dates. Perry (1952) reported that Roman numerals took 50 to 100% more time to read and produced from 3 to 30 times as many errors.

Characters printed on signs often must be read under difficult conditions. The optimum stroke width is 12 to 18% of the height when it is black on white and 8 to 10% for white on black and transilluminated characters (Konz and Mohan, 1972). (Note that dot matrix displays automatically define stroke width: a 7×9 matrix has a stroke width of $1/9 = 11\%$ and a 5×7 has a stroke width of $1/7 = 14\%$.) Character width (for most letters) has an aspect ratio of $L/W = 1.67$. That is, width (W) is 60% of height

TABLE 20.1   Recommended handwriting characters and guide for understanding the nuances of each character and the rationale for selection (Association for Computing Machinery, 1969).

Letter	Guide	Letter	Guide
A	Use of squared top not supported by sufficient evidence of confusion.	N	Parallel legs.
B	Overhang top and bottom is used to reduce possibility of confusion with numeral 8 or 13. Distinct center division required to avoid similarity to letter D.	O	Loop added at top by arbitration to avoid virgule, now too confusing.
C	No evidence of confusion; there is some similarity to left paranthesis if curve is not deep enough.	P	Overhang at top added for consistency with letters B, D, and P.
D	Overhang top and bottom is used to reduce possibility of confusion with numeral zero. This convention is similar to that for letter B.	Q	No special convention.
		R	Overhang at top added for consistency with letters B, D, and P.
E	Rounded left side is to be avoided to reduce confusion with ampersand.	S	Serif added at top only for ease of preparation and to distinguish from numeral 5 and special character dollar sign.
F	Similar to letter E above.	T	No special convention.
G	Strong, emphasized serif reduces possibility of confusion with letter C or numerals 6 and 10.	U	This convention adopted to distinguish from letter V and lowercase letter u.
H	Parallel sides.	V	No special convention required if the letter U has an identifying characteristic.
I	Serifs top and bottom are de facto standards.	W	Center division extends to top of letter. Rounded bottom should be avoided.
J	Top serif reduces confusion with letter U.	X	No special convention.
K	Slanting legs are joined at center.	Y	Vertical leg bisects angle formed by top legs to avoid confusion with numeral 4.
L	No special convention.	Z	Horizontal bar is de facto standard.
M	Legs spread at bottom; center division extends to bottom of letter. Rounded tops should be avoided.		

TABLE 20.2 Recommended handwriting numbers and guide for understanding of the nuances of each number and of the rationale for selection (Association for Computing Machinery, 1969).

Numbers	Guide
0	Closed circle with no added identifying characteristic
1	Single vertical bar, no added identifying characteristic
2	No loop at bottom
3	Curved lines, no straight top line
4	Open top to reduce confusion with 9
5	Vertical and top lines joined at right angles
6	Loop closed at bottom to avoid confusion with zero or lower case b
7	Crossbar used in Europe considered confusing with letter A, and does not have support in United States
8	Made with two circles adjoining vertically to avoid confusion with special characters ampersand and dollar sign
9	Straight leg from common usage

(L); on curved surfaces width = 100% of height. Minimum character spacing is 1 stroke width within words; 1 character width between words. Line spacing = 50% of character height (Hutchinson, 1981).

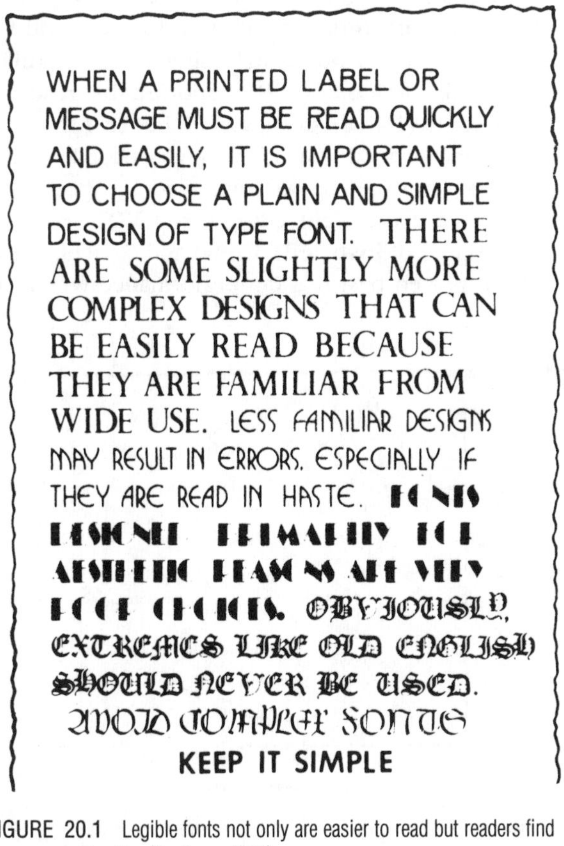

FIGURE 20.1 Legible fonts not only are easier to read but readers find them more attractive (Faulkner, 1972).

For computer screens, type style is more critical (Shurtleff, 1980). Reading speed is about 25% slower on VDTs than print, primarily due to the poor contrast and sharpness of characters on screens (Gould et al., 1987a, 1987b; Dillon, 1992). This reading differential can be reduced with better quality screens. A VGA (picture quality) has 16 colors; a super VGA (photo quality) with 800×600 resolution has 16 to 256 colors; a high-resolution VGA has $1,024 \times 768$ resolution with 16 to 256 colors. A **dot pitch** (how close the pixels are to each other) of .41 mm is typical, but .28 or even .26 are available. To avoid flicker, use a noninterlaced monitor in which the lines on the screen are painted every cycle; on interlaced monitors they are painted every other pass.

Screen characters usually are formed by dot matrices. Dot matrices typically are 5×7, 7×9, or 9×11 dots. Two extra dots are required for "true" descenders (e.g., g or y). Adding 1 dot for spacing between lines thus requires 10 vertical dots for a 5×7 matrix (7 for character, 2 for descender, 1 for spacing). The closer a dot character resembles a regular stroke character the easier it is to read. In addition, because they fill more space, square dots are better than oblique or round dots (Helander, 1987).

1.2 Size

The optimum size of a character depends upon a number of environmental factors. See Chapter 21 for more on illumination.

The minimum visible size (character height) depends upon the distance of the character from the eye. However, just because someone with perfect eyes can read a character does not mean the character should be that small. Bigger is better. As size increases, economics becomes important because more pages are required for a book with large print, a highway sign has more area, and so forth. A rule of thumb is that an object should be 2.5 times larger than threshold size. Equation (20.1) gives a rule of thumb for character height vs. distance (See also Figure 20.2.) K is in radians. K_1 is from Kinkade and Anderson (1984); K_2 is from Eastman Kodak (1983).

$$CH = K(D) \qquad (20.1)$$

where

CH = Character height

D = Distance from eye

K_1 = .004 radians for minimum character height at $\geq$300 lux

= .006 for preferred character height at $\geq$300 lux

K_2 = .011 700 to .005 500 for routine viewing at $\geq$100 lux

= .003 500 to .007 300 for critical viewing at $\geq$100 lux

Eastman Kodak (1983) recommends, for low-light situations (such as darkrooms), multiplying K_2 by using 1.5 and using white characters.

Size also can be expressed in terms of **visual angle** to eliminate the effect of distance. (See Figure 20.2 and Section 1.1.2 of Chapter 21.) For characters on VDT screens, Grandjean (1987) recommends 16–25 min of arc and the American National Standard (1988) recommends 20–22 min of arc. Chang and Konz (1993) confirmed these values but showed severe penalties for characters less than 15 min of arc. For characters viewed from less than 1 m, use values at the upper end of the range (Smith, 1979; Sherr, 1979; Winkler and Konz, 1980).

For short messages such as signs, figure legends, words on projected transparencies, and the like, use all capitals since, for the same type size, capital letters are about 30% larger than lower case letters. Capital letters also can be used to HIGHLIGHT a word on a monochromatic display. For longer messages (i.e., text), use a mixture of capitals and small letters as the characters should be far above threshold size and the mixture gives structure to the text to improve readability.

2 ARRANGEMENT OF CHARACTERS AND SYMBOLS

2.1 Text   Printers measure type height in **points** (where 1 point is 1/72 inch = .353 mm) and type width in **picas** (where 1 pica = 1/6 inch = 4.233 mm). (However, this is the height of the slug on which the type is set—use 1 point = .01 inch for the character height.) For typewriters, elite type is 10 points high with pitch of 12 characters/linear inch; pica type is 12 points high with pitch of 10 characters. (Computerized printers often allow changing the pitch.) This book is set in 10 point Garamond Book type with 2 points leading (space between lines), 19 picas (228 points) wide. Figure 20.3 gives changes in reading speed for 10-point type. The easiest reading is with text in columns (magazines, newspapers) rather than across the entire page. Table 20.3 compares the optimum for each type size. Tinker (1963) recommends 11-point type as readers prefer it and it gives greater flexibility in line width and leading. Tinker also says that, although readers generally prefer some leading, they prefer a 10-point type set solid (minimum spacing) over an 8-point type with 2 points leading.

Put a space between a number and the units, for example, 1.5 mm, not 1.5mm.

For characters on VDTs, typical center-to-center distance of characters is 2.8 mm horizontally and 5.6 mm vertically (15.9 points including spacing) (Tullis, 1983). Kruk and Muter (1984) reported that readability of text on VDT screens was 11% better if the lines were double-spaced instead of single-spaced. Trollip and Sales (1986) reported that reading speed was 11% greater if the right edge of the VDT text was uneven (i.e., not **justified**). Gould and Grischkowsky (1986) reported best reading performance when the

FIGURE 20.2   Trade-off between letter size and legibility indicates that at .002 radians 70% of the letters were legible, at .004 radians 95% were legible, and at .007 radians about 100% were legible (Smith, 1979).

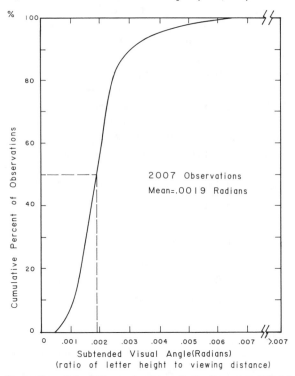

Source: Figure 3 (p. 667) reproduced from "Letter Size and Legibility" by Sidney L. Smith. Reprinted with permission from *Human Factors,* Vol. 21, No. 6, 1979. Copyright 1979 by the Human Factors and Ergonomics Society. All rights reserved.

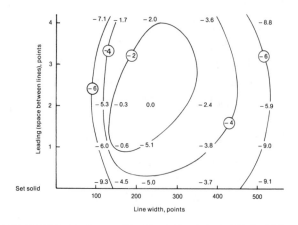

FIGURE 20.3   Reading speed for 10-point type is best at 19 pica (228 points) and 2-point leading (Tinker, 1963).

TABLE 20.3   For each type size a "near optimum" line length was used; the type was Granjon with 2-point leading (504 subjects) (Tinker, 1963).

Type Size, Points	Line Length, Pica	Points	Reading Speed, Percent
12	24	288	100.0
11	22	264	99.0
9	18	216	98.7
10	20	240	97.3
8	16	192	95.6
6	14	168	94.0

line width of text subtended about 20°; there was little difference within the range of 16° and 36°.

Long lines of text form a pattern of stripes which causes reading problems (Wilkins and Nimmo-Smith, 1987). This pattern can be broken up by relatively short paragraphs and by using short line lengths (as in newspapers or magazines); short lines also help the eye return to the beginning of the following line. Typed material typically should be double-spaced, although 1½ spacing can be used. Reserve single spacing for short items (summaries, footnotes, references, table headings), provided no difficulty arises due to sub- or superscripts.

For manuals, people like a good table of contents, labeled tabs, and distinctive headings; they also appreciate color to highlight types of information, and they like lots of figures and examples (Angiolillo and Roberts, 1991).

2.2   Codes

Part numbers, telephone numbers, voucher numbers, charge numbers, and so forth, are **codes.** The goal is to make it difficult to generate an error and easy to detect an error. To minimize errors in coded transmission:

- *Automate the code transfer.* Examples of automated transfer are bar codes, magnetic ink bank identifiers on checks, and other machine-readable devices. Automated transfer tends to be very error-free as well as fast.

- *Make codes checkable.* Codes transferred by computer can be made self-checking by adding check digits to the code and by having the computer check its memory to see if the transferred number is in its memory. Table 20.4 shows some additional checks which can be done either with computer software or manually.

- *Use short codes.* With manual transfer, human error rate increases exponentially if there are over 5 or 6 symbols in the code (Konz et al., 1968; Stanhagen and Carlson, 1970); 5 numbers or 4 letters will have an error rate of about 5% and 4 numbers about 2% (Caplan, 1975). Note that 1 letter gives 26 possible characters while 1 number gives 10 possible characters. Break long codes into shorter segments of 2 or 3. For example use 399-29-7456, not 399297456; use 660 50 009, not 66050009. If the product code is 2562 and style is 0148, present the information as 2 separated numbers instead of 25620148. When using voice transmission, use pauses every 2 or 3 symbols; in addition, 8 and H often are confused, so eliminate 8 or H as possible characters.

- *Make the code more meaningful.* Keep a code all upper case (ABCD) or lower case (abcd) but not a mixture (Abcd). Placing key information in a prominent position in the code (i.e., first or last) is beneficial. If the user can chunk the information, errors are fewer. For example, the telephone number has a 3-digit exchange segment

TABLE 20.4   Error opportunities in coding (Eastman Kodak, 1983; Caplan, 1975). The key is to add structure to the code.

Error Type	Preventive Feature in Code Design	Reason
Omission or addition of characters to code	Use uniform length and composition	Omission or addition will result in code immediately recognized as nonexistent
Substitution between numbers and letters	Use consistent location for numbers and letters	Substitution will result in code immediately recognized as nonexistent
Transposition of letters	Use a familiar acronym or pronounceable word (instead of random letters) that is visually and audibly distinct	It will be remembered as one element rather than individual elements, as random letters are remembered
Transposition of numbers	Introduce a rule for the relationship between adjacent numbers in the string	Transposition will yield a code with a pair of digits out of order
Illegibility	Use consistent number and letter locations. Control handwriting by providing individual box for each character	Poor handwriting more easily deciphered

followed by a 4-digit random number. Be careful about including the letters O and L as they can be confused with the numbers 0 and 1. However, letters can make codes meaningful. When using letters, keep them together rather than spreading them over all positions. RL J2042 on a Kansas license plate means the car is from Riley county and the owner's last name starts with J, K, or L, so the license expires in July. Pronounceable or meaningful letter groups (PAR) are better than random (FXT) groups. Good examples are the mnemonics for hotel phone numbers (1-800-HOLIDAY and 1-800-2-RAMADA), the Texas Instruments service number (1-800-TI-CARES), and the Army recruiting number (1-800-USA-ARMY). Some single-letter mnemonic codes are: M = male, F = female, U = unknown; U = up, D = down; L = left, R = right; M = married, S = single, D = divorced, W = widowed; Y = yes, N = no; H = husband, W = wife, J = joint; H = hourly, S = salaried.

As an example, consider showing the three dimensions of a storage location. First identify the rack with a letter (2 letters if there are over 26 racks). Then identify the position in the rack with a number (i.e., a contrast to the letter). Start at one wall and keep numbers aligned for all racks—possibly skipping numbers in some racks if necessary. Finally, identify the vertical dimension with a letter (a contrast to the number)—A is on the floor. Thus, a code M62C means rack M, 62 spaces from one wall, and the third level up.

Designing codes for word processor files is another example of the desire to code several dimensions in a single code. 625OUT might be an outline for course 625. L12SEP could indicate a letter sent on September 12th; note how the letter–number contrast is better in L12SEP than LSEP12. On the other hand, the computer probably sorts in sequence so the LSEP12 code would have all September letters together and then sorted by date within September.

2.3 Abbreviations

Words can be abbreviated to speed data or command entry (i.e., reduce number of keystrokes); this is encoding. Conversely, if abbreviations are used to reduce display space, the operator must translate the abbreviation back into a word; this is decoding. Abbreviations should not include periods. (To avoid confusion, don't abbreviate inch.) Abbreviations should be used with caution; when in doubt, spell it out.

Some research results (Ehrenreich, 1985) follow:

- For encoding, results are better when rules are followed than when people can "freewheel." Truncation (e.g., changing DELETE to DEL)

seems to give abbreviations as well or better than other rules. Encoding performance can be improved by using a simple rule and teaching that rule to the user. For example, when looking up your flight reservation, the gate agent just enters the first 3 letters of your last name.

- For decoding, there are no consistent differences among techniques. The most popular are truncation (TRANSFER becomes TRA) and vowel deletion (TRANSFER becomes TRN).

- Rules best for encoding are not necessarily best for decoding.

My personal decoding preference is vowel deletion for, as any "Wheel of Fortune" fan knows, it is easier to guess vowels than consonants. Note that a computer can be programmed to accept multiple abbreviations for the same command. For example, for DELETE the computer could accept DELETE, DEL, DTE, or even DE. However, the computer should be consistent in its display of the abbreviation, for example, always display DEL. Note also that just because you key an abbreviation into the computer doesn't mean it can't be programmed to print out the full word.

2.4 Formulas

Formulas permit exact calculations while **graphs** give approximate values and show relationships. Formulas can also be used for multiple input variables. In addition, formulas (and tables) can show many independent variables in relation to the criterion (Y is a function of A, B, C, D, and E), whereas it is difficult to show more than two variables in relation to the criterion on a graph. Formerly, hand calculations presented a problem in using formulas, but the widespread use of calculators and computers has eliminated this problem.

To reduce computational errors, present the formula in units that the user will enter; the answer also should be in the desired units. For example, if the user will enter cm and wants the answer in TMU, use equation (20.3), not equation (20.2). Reduce decimal point errors by presenting numbers in groups of 3 with an intervening space and following zeros. Decide how many significant digits are needed. Equation (20.4) probably has sufficient accuracy for most applications.

$$R - A = 3.488 + .482\ 574\ (DI) \tag{20.2}$$
$$R - A = 3.488 + 1.225\ 738\ (DCM) \tag{20.3}$$
$$R - A = 3.5 + .5\ (DI) \tag{20.4}$$

where

$R - A$ = Time for R-A reach, TMU

DI = Distance, inches

DCM = Distance, cm

2.5 Menus   Lists of options are called **menus.** The information may be computerized and the menu presented in a hierarchical structure (after selecting baseball, you have a list of teams; after selecting the Royals, you have a list of players). Provide shortcuts for experienced users. Options may be highlighted to reduce search time in a number of ways: underlining, written in boldface, reverse video, color, and blinking. Tullis (1983), after a survey of the literature, grouped his recommendations into four rules:

1. Minimize frame density. Have less than 25% of all the possible character spaces in a frame filled. To keep the display uncluttered, this may require strict attention that only relevant information is displayed.

2. Have ample blank space between items. Blank space provides structure to the user. Text should be double-spaced. Groups of items should be separated by 3 to 5 rows or columns of blank space.

3. Group related items together.

4. Minimize layout complexity. If working with a paperwork form, the form and the screen should have the same format. Words and alphanumeric data should be left-justified; numeric data should be right-justified on the decimal point. Alphabetical lists are better than random lists, but alphabetical within categories is better yet. Another possibility (for a short list) is the common options listed under rule 1.

2.6 Tables   Although the distinction is not always clear, tables can be divided into travel information tables and data tables.

Travel information tables present travel information. These tables are sometimes near elevators, in plane/train/bus terminals, or other public areas. The best format (Verhoef, 1993) is to structure the table so the user enters the table with known information (such as destination, sequence of alphabet) to determine the unknown (flight gate, departure time, floor of building), rather than the converse. Butler et al. (1993) reported that a series of signs is better than "you are here" maps (which tend to require too much memory and interpretation of directions).

The goal of a data table is to make patterns; exceptions should be obvious at a glance, at least if you know what they are (Ehrenberg, 1977). Eight guidelines for good table design follow.

2.6.1 Round data to two significant digits
Final zeros don't count. When rounding to two **significant digits,** 5311 becomes 5300 and .0511 becomes .051. The reason is that a difference of less than 1% (the third digit) rarely is important and the surplus digits make it difficult to compare the data. This rounding rule may make totals of columns and rows not exactly equal to their components, but people who can't cope with rounding probably won't be using statistical tables anyway. Note also that rounding is well accepted on graphs and that the basic data (in the computer, an equation, or your detailed records) need not be rounded—just the data that you present to others.

2.6.2 Use explicit tables
Explicit tables give all the information directly; **implicit** tables, however, require the user to make calculations. Table 20.5 is an explicit table of the normal distribution. In the implicit form, only the upper half of the table is printed; the reader is assumed to know that the distribution is symmetrical and to be able to "mirror" calculations. Table 20.6 shows explicit and implicit tables for converting C to F. In this implicit table, you must make additions; in other implicit forms, you interpolate.

Although implicit tables save space, they cause more errors and take more time, especially with novice users (Wright and Fox, 1970, 1972).

2.6.3 Avoid matrix tables
Matrix tables on a diagonal, such as Table 20.7, are bad. Use an explicit linear table, such as the telephone book or

TABLE 20.5 An explicit table of the normal distribution. In an implicit table only the negative or positive values of *z* would be given.

NUMBER OF STANDARD DEVIATIONS *z*	CUMULATIVE AREA FROM NEGATIVE INFINITY
−3.0	.001 300
−2.33	.01
−2.0	.022 800
−1.96	.025
−1.64	.05
−1.28	.10
−1.00	.159
−0.80	.212
−0.60	.274
−0.40	.345
−0.20	.421
0	.5
+0.20	.579
+0.40	.655
+0.60	.726
+0.80	.788
+1.00	.841
+1.28	.90
+1.64	.95
+1.96	.975
+2.0	.977 200
+2.33	.99
+3.0	.998 700

TABLE 20.6   Examples of implicit and explicit tables. The less desirable implicit table saves space but requires the user to make calculations.

Implicit Format				Explicit Format			
Degrees, C	Degrees, F	Degrees, C	Degrees, F	Degrees, C	Degrees, F	Degrees, C	Degrees, F
0	32.0	1	1.8	0	32.0		
10	50.0	2	3.6	1	33.8	11	51.8
20	68.0	3	5.4	2	35.6	12	53.6
30	86.0	4	7.2	3	37.4	13	55.4
40	104.0	5	9.0	4	39.2	14	57.2
50	122.0	6	10.8	5	41.0	15	59.0
		7	12.6	6	42.8	16	60.8
		8	14.4	7	44.6	17	62.6
		9	16.2	8	46.4	18	64.4
		10	18.0	9	48.2	19	66.2
				10	50.0	20	68.0

Table 20.8. Note that the table can be "folded" (have multiple columns) to save space. If a matrix table is used, present the information vertically and horizontally, as in Table 20.9.

2.6.4 Make the primary comparison down the columns

The eye can make comparisons more easily when reading down a column than across a row (Wright and Fox, 1970; Ehrenberg, 1977). For values, use decimals rather than fractions. Align values along a common decimal point. Within a column put the large number on top because it is easier to do mental subtraction between rows if the large number is on top. Put the units in the column heading, not in the table itself.

2.6.5 Reduce row alignment errors

One technique is to divide rows into groups. (Generally, rows should be single-spaced.) Groups of 5 are better than groups of 10 which are better than no groups at all (Tinker, 1963). Grouping can be with spaces or light lines. Logical grouping, as in Table 20.5, if possible, is good. Another technique is to reduce the horizontal distance between the two columns; still another is to use dots to lead the eye. See the lower part of Table 34.1.

TABLE 20.7   Don't use matrix tables with diagonals. If a matrix table must be used, present the information horizontally and vertically as Table 20.9.

Office			
	50		
Punch press			
	50		
Drill press	100	75	
	80	75	
Sheet metal	75	30	25
	80	80	
Die storage	30	60	
	50		
Shipping dock	25		

The conventional table arrangement is text on the left and data on the right (see Table 20.8). There are eye travel advantages for the numbers on the left and text on the right because the variation in the length of the numbers usually is small (see Table 17.5, Table 16.3, and Table 22.2). Allen et al. (1991) demonstrated it took about 10% less time to use the conventional table arrangement (with dots connecting the row items in the conventional arrangement); perhaps this is due to the need to reverse eye travel direction when the text is on the right.

2.6.6 Facilitate comparisons

Arrange the sequence of rows and columns. A logical grouping (categories) is better than an alphabetic sequence (McDonald, Stone, and Liebelt, 1983). The mean value in the columns can increase or decrease as you go to the right in the table without any substantial effect on performance. People generally start on the left column. In general, the lower the usage of a column, the farther to the right it should be.

2.6.7 Reduce column selection errors

Tinker (1963) reported better legibility for columns separated by a pica of space than by a vertical line (a rule); legibility was equal for columns separated by a pica and by a pica + rule. Columns also can be distinguished by printing them in different type fonts or darkness. Spacing can be used to set off *related pairs* of columns (e.g., observed vs. theoretical values, last year vs. this year).

2.6.8 Include averages for rows and columns

The average gives a frame of reference and allows the reader to make easy comparisons between rows and between columns (i.e., the "main effects"). The average also facilitates comparisons of individual values vs. the row or column mean.

TABLE 20.8   Fewest errors as well as least time are the advantages of linear explicit tables, but they require more space than matrix tables. "Fold" (have multiple columns) to save space.

Office to		Sheet Metal to	
Punch press	50	Office	75
Drill press	50	Punch press	80
Sheet metal	75	Drill Press	75
Die storage	70	Die storage	30
Shipping dock	25	Shipping dock	80
Punch Press to		**Die storage to**	
Office	50	Office	70
Drill press	50	Punch press	30
Sheet metal	75	Drill press	80
Die storage	70	Sheet metal	30
Shipping dock	25	Shipping dock	25
Drill Press to		**Shipping Dock to**	
Office	50	Office	25
Punch press	100	Punch press	50
Sheet metal	75	Drill press	60
Die storage	70	Sheet metal	80
Shipping dock	25	Die storage	25

TABLE 20.9   Matrix tables such as the one shown below are difficult for the general public to use without error, even when they do not contain diagonals. Use linear explicit tables such as Table 20.8 if possible even though they do use more space.

	1	2	3	4	5	6
1. Office	—	50	50	75	70	25
2. Punch press	50	—	100	80	30	50
3. Drill press	50	100	—	75	80	60
4. Sheet metal	75	80	75	—	30	30
5. Die storage	70	30	80	30	—	25
6. Shipping dock	25	50	60	80	25	—

2.7 Graphs   Graphs should be used to present comparisons between complex relationships. They are not good for determining the exact values. Tables almost always outperform graphs for reporting on small data sets of less than 20 numbers. Simple things belong in tables or the text. Statistical graphics should help people reason about numbers; they should encourage comparisons (Tufte, 1983).

Figure 20.4 shows the relationships of time vs. distance for MTM reaches; they also are given in Table 29.2. In this case, the table is superior to the graph as the numerical values can be obtained more easily and more accurately from the table and the graphical relationship is quite simple.

Figure 20.5 shows a conversion line. This is a very poor device as it does not have the advantage of a graph (showing a relationship) or of a table (easily obtaining exact numbers).

Certain types of graphs are better. The most common graph is a **time series** (see Figure 3.2), but the relational form (Figures 4.1, 11.2, 11.3) also is excellent. Cleveland and McGill (1985) recommend dot charts (see Figure 20.6) over divided bar graphs. Figure 20.7 shows some techniques of indicating data variability. Line charts have an advantage over vertical bar graphs (where the data point is the end of a column from the axis) in that the line chart emphasizes a key feature, the slope, without the distractions (clutter) of the vertical lines of the bars. However, line charts require a continuous axis. Scales on the axes either are continuous (time, temperature, weight) or discrete (cities, people,

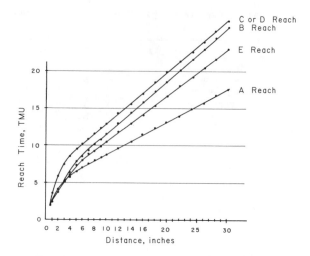

FIGURE 20.4   General relationships are shown well by a graph; however, it is difficult to determine accurate values.

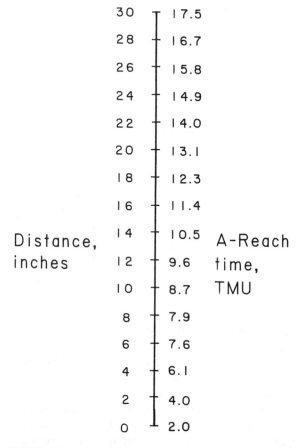

FIGURE 20.5   A **conversion line** reduces the distance from the scale to the line that occurred in Figure 20.4. Thus accuracy is improved but you don't know what the relationship is (i.e., curve or straight line). Conversion lines are really two-column tables.

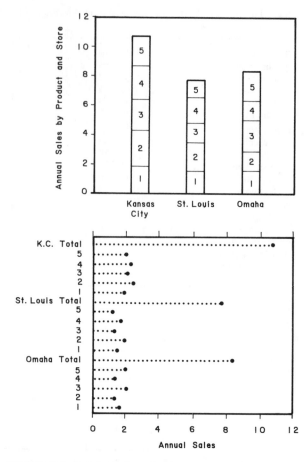

FIGURE 20.6   **Dot charts** (lower figure) communicate better than **divided bar graphs** (upper figure) (Cleveland and McGill, 1985) because they minimize judgment of length and position.

experimental conditions). You can connect the data points (i.e., have a line) for continuous axes (see Figure 20.4) but not for discrete axes (see Figure 20.6).

Lines also are better for comparison of multiple data sets. If a bar chart must be used, make the bars horizontal as they are much easier to label than vertical bars. A table is almost always better than a **pie chart,** a circular chart divided into segments illustrating relative magnitudes or frequencies (which has low data density and a failure to order numbers along a visual dimension). If a pie chart is used, give the percentages in numbers on the figure. A variation on the pie chart is the **doughnut chart,** in which the sample size is displayed in the circular center section (doughnut hole).

Do not make the graph lie by using area or volume of an object to represent a change in a one-dimensional value (a pictograph). For example, if you want to represent the height of people at different ages, don't use figures of people as the figure will increase both the height and width (i.e., area); showing a cube (such as an isometric sketch of barrels of oil of varying height to show change in oil price vs. time) is even more deceiving. Another common technique of distorting data is to start the vertical axis at a nonzero value. For example, if sales increased from

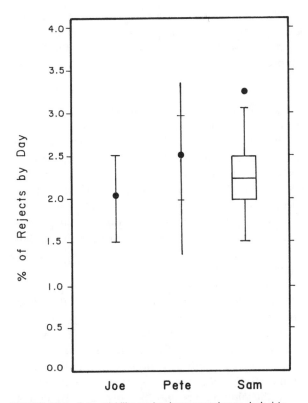

FIGURE 20.7 Data variability can be shown several ways. Joe's data show mean and − and + 1 standard deviation (i.e., 68% confidence if data is normally distributed). However a high confidence (such as 95%) usually is more useful. Pete's data show the mean, the inner bars at 1σ and the ends of the line at 2σ; this is 68% and 95% confidence if the data are normally distributed. Sam's data are shown as a "Tukey plot." The horizontal line is the 50th percentile of the data; the lower and upper box edges are the 25th and 75th percentile. The ends of the lines are "adjacent values." The box length is t and adjacent values are 1.5 t from the mean. Individual outliers beyond the adjacent values are shown as small circles. If the data are distributed normally, the 50% is the mean, the 25% and 75% are − and + .675σ, and the adjacent values (.4% and 99.6%) are − and + 2.67σ.

$5,400,000 to $6,000,000, then starting the axis at $5,000,000 is an attempt to deceive. Be careful with the semi-log graph (log of quantity on the Y axis and time on the X axis) as it is deceiving to most viewers (Taylor and Taylor, 1986). However, it does have the advantage that a constant percent change is a straight line.

For good graphs, do the following:

- Make graphs about 50% wider than tall (Tufte, 1983) as the X axis normally is the axis to emphasize and labeling is easier than if the Y axis is longer than the X axis. (Transparency graphs, however, will fit the projection area better if they are taller than they are wide.)

- Make scales on axes in units of 5 or even units of 2. That is, use 5, 10, 15 or 50, 100, 150 or 20, 40, 60. Avoid 1, 3, 5, 7 or 10, 30, 50. Worst of all are computerized scales where the decision to have

the graph use as much of the page as possible results in scales such as 1.31, 2.62, 3.93, or the like. Under no circumstances should you change the scale axis during the scale. If you see years 1960, 1970, 1980, 1985 plotted at equal intervals or defects 1, 2, 3, 4, 5, 10 plotted at equal intervals, you know they are lying to you. Start each axis at zero or show a "break" symbol. Don't forget to give the units of each axis. If the scales are chosen so the data have a slope over 30°, it tends to be interpreted as being significant; a slope of 5–10° tends to be interpreted as nonsignificant.

- Show scale subdivisions with tick marks—especially for nonequal interval scales (log scales, probability scales). A "light" grid (such as in Figure 28.2) is helpful.

- Don't use hatching as it gives a "moire vibration" (Tufte, 1983). Instead replace it with screens of varying density and shades of gray. In general, make the lower areas darkest and the top areas lightest. See Figure 20.8.

- Use only a few curves (four is maximum). Make curves distinctive by using color, thin vs. thick lines, solid vs. dashed, or another method. Using symbols, such as circles vs. triangles, is a last resort, but if you do use them, make the symbols big and bold.

- Indicate data points with open circles, squares, and triangles. Worst are dots, crosses, and Xs.

- Use brief labels rather than keys (Milroy and Poulton, 1978).

- Make labels and legends sufficiently large. Graphs often are reduced; therefore use all capital letters rather than caps and lower case, because capital letters are about 30% larger.

2.8 Projected Images Presentations to groups often include projected images. Examples are movies, TV, slides, and overhead transparencies. The projected image will not have the sharpness of printing. In addition, because the presentation is often needed in a short time, the designer of the slide or transparency often does not take the necessary steps to present the best quality image.

For legible displays and accurate viewing, the screen's vertical height should be ⅙ of the maximum viewing distance as measured from the far side of the display to the furthest off-axis viewer. For relatively unobstructed and comfortable viewing, the bottom of the projection should be about seated eye height (assumed to be 48 inches above the finished floor). The screen should extend upward a distance equal to ⅙ of the maximum viewing distance. Add at least 6 inches for trim above the screen (Konz, 1994).

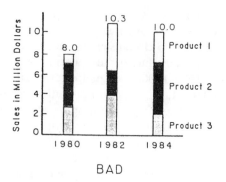

FIGURE 20.8   Time series should not be given as bar charts. Use a dot chart (Figure 20.6) or lines (lower part of this figure). In addition, dividing the bar makes it difficult to compare components. If you insist on a bar chart, make the bars wider than the spaces between the bars, have shading go from dark on bottom (or left) to light on top (right), and make the lettering horizontal. Give the value at the end of the bar. Numbers inside columns or bars (middle figure) permit a more compact figure. Columns or bars with numbers permit omitting the scale. Best is a time series with the individual components and total displayed. Be sure that the line is much darker than the grid. Each line is labeled rather than using keys; the coding of products is made redundant through use of symbols and line characteristics. Note also that the legends are within instead of outside the figure. The reproduced size must consider the figure and legends; so inside labeling allows a larger reproduced size. Be sure to make the numbers large enough.

$$MCH = 54 + MVD/6$$

where

    MCH = Maximum ceiling height, inches

    MVD = Maximum viewing distance, inches

Note that maximum viewing distance may be less than room length. Table 33.3 summarizes information of effective overhead and slide presentations.

The first decision is whether to use a **slide** or a **transparency**. The slide is the high visual quality alternative; it also costs the most and has the longest preparation time.

2.8.1 Slides

Slide film (typically 35 mm) is normally color but what if the object (such as text) is black and white? Then the basic decision is whether to emphasize legibility (black on white) or esthetics (colored text and/or background). Color also permits you to emphasize and organize portions of the image. For example, the heading could be one color and the subpoints another color; the advantages could be in green and the disadvantages in red. Multicolor text can be done relatively easily with various computer graphics packages.

A slide of black type or print on white paper can be made more esthetic by covering the developed film with an overlay; the result is black print with a yellow, blue, red, or green background. The overlays reduce the brightness of both the text and the background and thus reduce contrast and legibility. In my opinion, only yellow overlays are worth considering, since the pink, green, and blue overlays reduce legibility too much.

Some film will color the text red, green, blue, or other color while leaving the background white. This reduces contrast (compared to black text) but not as much as an overlay.

You may use a **reverse image,** that is, light letters on a dark background. Although reversed images may give better legibility in a darkened room, the white letters on the film also can be "painted"; the result is yellow, green, orange (or whatever color) letters on black. Painting allows different colors on different parts of the slide. An overlay can be used which makes all the text one color with a black background.

You may use a solid color overlay (e.g., yellow) while making the background another color. An example result is yellow letters on a red background.

Reduced contrast (and thus reduced legibility) is more critical if the room is more brightly lighted, the audience is farther away, or the screen quality is poorer (e.g., if a back projection screen is used).

The next problem is size of individual projected characters, which will depend upon the distance of the audience from the screen. Text should be double-spaced. Graphs and tables that are satisfactory in print should be reformatted and simplified and lines emphasized for slides.

2.8.2 Transparencies (overheads)

The most common transparency is black text on clear film, giving a black image on a white screen. The background can be colored, leaving the text black, or the text can be colored (leaving the background white) either by the film or by special colored pens. Dark blue text on a white background gives a good appearance with little loss in legibility (Konz et al., 1988). Multicolor graphs can be made with computer graphics; color transparencies can be made from color photographs.

Because the projected image from a transparency is lower quality than from a slide, even more attention must be paid to character size. As a general rule, use characters larger than 18 points (typewriter type is 9 points). Use all upper case letters with a simple font such as sans serif (Verschelden and Konz, 1988). Be sure the transparency is readable from the farthest viewer position in the room you will use. Simplify graphs and tables. Note that the projected shape is taller than it is wide. Computer printouts and detailed tables can be used by giving each of the audience a printed copy and using a transparency as a map to show the audience the information of interest.

An alternative to a transparency is the flip chart—a large paper pad on which to write with a thick marker. Information is presented one frame at a time and can be prepared in advance in multiple colors. Flip charts do not require a projector, screen, or power but do require a pad and easel. Two disadvantages are lack of magnification and difficulty of duplication.

2.9 Symbolic Messages

"A picture is equal to 1,000 words." Although this is not exactly true, there is a germ of truth in the concept. Symbolic messages can be language free, can be read more quickly than the group of words they replace, can conserve space, and can be read at greater distances than text signs. Symbolic messages do not always meet all four of these ideals, however. The primary problem is errors in understanding what the symbol means, especially by novices.

There has been considerable use of symbolic interfaces (**icons**) on computers to replace text. For example, a traffic light with the green light on replaces "GO." Benbasat and Todd (1993) review the icon literature and demonstrate experimentally that there is no advantage to icon representation over text-based representation of actions and objects.

There has been considerable research on symbolic messages in the field of transportation; examples are on vehicle controls and displays, directions at transportation terminals, and highway signs.

The symbolic messages can be geometric shapes and colors (octagonal red stop sign, shield-shaped blue interstate highway sign), diagrams (left exit from interstate highway), or pictorial silhouettes, also called **pictographs** (school crossing showing people walking, deer warning, male and female symbol on toilet doors, baggage doors at airport). As Figure 20.9 shows, in practice, words often accompany the symbols in the messages.

For highway use, the fact that symbolic signs can be recognized at 1 to 4 times the distance of text signs is important (Paniati, 1988); the average was 2.8 times as far. The wide range indicates that the legibility of some symbolic signs can be improved. Kline and Fuchs (1993) modified symbolic signs after studying them in a "low pass filter" and found recognition distance could be increased about 50% over the standard symbolic signs.

Arrows can indicate direction. An arrow has both a head and a shaft; both need to be present and distinguishable, and the head needs to be distinctive from the shaft.

Before using a symbolic message, decide if the benefits are worth the cost. Use words instead of symbols if (1) message readers are novices to the message, (2) readers all use the same language, (3) space is plentiful, and (4) reading time (distance) is not critical. Field test the proposed symbols with actual users before deciding to use the symbols. If the symbols will be used internationally, test in multiple countries.

3 INSTRUMENTS

Instruments can be divided into three categories. (1) **Discrete** (also called qualitative) displays indicate a status among a finite choice of options, such as go/slow/stop or on/off. (2) **Continuous** (also called

FIGURE 20.9 Pictographs in theory do not need the accompanying words, but real signs often retain the words.

quantitative) displays indicate a point on a scale, as with a clock. (3) **Representational** (also called pictorial) displays show a diagram or picture of the system being measured. Carefully analyze the operator's specific task before selecting a specific instrument among or within the three categories. For a short checklist, see sections G and I of Table 5.2.

With increasing computerization, the operator often becomes more of a supervisor than an active operator, and so warning messages have become more common. Berson et al. (1984) recommend that emergency messages (1) be both visual and aural, (2) be within the primary field of view, and (3) give guidance information rather than status information (i.e., "Turn off pump 3 due to overheating" rather than "Pump 3 is overheating.")

Lighting problems with instruments are often due to too much light (i.e, from the sun) rather than too little. This veiling luminance (glare) tends to desaturate colors, decreasing any color-coding effect, as well as decreasing legibility. Careful shielding and nonreflective glass covers help.

3.1 Discrete

Discrete status (motor is running or not running, power is on or off, system status is go or no go) often is indicated by indicator or warning lights. Traffic lights (red = stop, yellow = caution, green = go) are an example of multiple discrete levels with color coding. Table 20.10 summarizes some indicator and warning light recommendations.

A control position also can act as a display. It is better if the meaning of the position is confirmed by a legend. On automobiles, a legend confirms the position of the automatic transmission lever but no legend confirms the position of manual door locks (you must not only know that up = unlocked and down = locked but also must be able to visually recognize which state the control is in).

3.2 Continuous

Continuous scales are divided into two versions: analog and digital.

The **analog** instrument typically has a pointer and a scale. In some designs, the scale is omitted (e.g., a clock with hands on a blank background). In other designs, the scale is omitted but the background is color coded (e.g., background for automobile oil pressure is green, yellow, or red). This is an attempt to communicate meaning instead of a mere number. The pointer does give more information than discrete "idiot lights" but the skilled specialist likes to have numbers as well as just colored backgrounds.

The goal is to maximize legibility of the pointer and the scale.

Generally, there should be only one pointer per instrument (in spite of generations of experience with clocks, two pointers often lead to confusion).

Good pointers (Sanders and McCormick, 1987) have:

- a point (about 20° tip angle), not a "fancy" tip
- a tip which doesn't overlap the scale but doesn't meet it either (maximum gap of 2 mm)
- the pointer close to scale surface to avoid parallax
- good color contrast vs. scale, preferably same color as lettering
- the pointer extending one direction from center (for circular scale)
- the pointer moving off scale if the meter fails

The scale should be legible:

- Scale numbers should progress by 1s, 2s, or 5s. If decimals are needed, omit the zero before the decimal point.

TABLE 20.10 Indicator and warning light recommendations (Eastman Kodak, 1983; Hutchinson, 1981; Sanders and McCormick, 1987).

- Label the light so people know what it is indicating.

- Indicator and warning lights need to be detectable. First of all, they must be close to the operator's line of sight (say 30°)—certainly not behind the operator or in another room. For remote warnings, use an auditory warning plus indicator lights. Next, they need to be reasonably bright—especially in sunlight or well-lighted areas. Third, consider reducing background clutter to improve detectability; one example is a solid black panel around traffic lights.

- Use steady state for continuous, ongoing conditions; use flashing for warnings requiring immediate attention. For nonemergency warnings, a trade-off needs to be made between alerting the user and the annoyance of a flashing light.

- Flash rate (if used) should be approximately 4/s (range 2-10/s) with approximately 50% on and 50% off time. If the flashing light is connected to a time (e.g., self-timer on a camera), an increase in the flash rate can indicate when the period is ending.

- Warning lights generally should be red (due to population stereotype) or possibly yellow. A different color can be used to differentiate types of warning (blue lights for police cars, red lights for ambulances).

- Warnings should be indicated by the light coming on, not the light going off. Either a bulb-test capability or 2 bulb assemblies should be used.

- They should be large enough (see equation 20.1).

- They should increase clockwise, left to right, bottom to top.

- The 0 should be at a logical position. For round dials with a continuous scale, put 0 at 9 or 12 o'clock. If the scale has a gap, put the 0 at 6, 7, or 12 o'clock.

- Numbers should be oriented vertically. (For a moving circular scale and fixed pointer, align the numbers radially so they are vertical when opposite the pointer.)

- Two numbers should appear in a window simultaneously to indicate movement direction (if a window is used).

The pointer and scale arrangement can be the fixed scale and moving pointer (in both circular and linear versions) or the moving scale and fixed pointer (also in both circular and linear versions). See Table 20.11.

The **digital** instrument gives the numerical value directly. Examples are digital clocks, digital speedometers, digital radio frequencies, digital television channels, and digital odometer mileage. For digital counters, numbers should:

- be large enough (see equation 20.1)

- have a character height-to-width ratio of 1:1 (due to distortion of curved drum surface)

- have no more than one digit appear in a window at a time (In addition, numbers should change by snap action, not continuous movement.)

- not be shielded from view or shadowed by the sides of the instrument

- have spacing (commas, decimal point, or space every 3 numbers) if more than 4 numbers are to be read

- advance on manual counters about 50 counts for one revolution of the control knob

3.3 Representational These pictorial displays (graphic panels) mimic the status of a system. One example is the light display on auto control panels showing which doors are not closed. Chemical process plants, railroad switch yards, factory conveyor networks, and HVAC systems also often have a series of lights indicating status of various components of a system. Pilots can be given information concerning the angle of the wings vs. the ground. Pictorial views usually are **outside-in** (i.e., view from outside the system) although occasionally **inside-out** (pilot's eye view) is used.

4 LOCATION AND ARRANGEMENT OF DISPLAYS

4.1 Location Since visual displays are visual, locate them where they can be seen easily. Although

TABLE 20.11 Comparison of mechanical displays (*Human Engineering Design Data Digest*, 1984).

Use	Fixed Scale Moving Pointer	Moving Scale Fixed Pointer	Counter
Continuous information	*Fair:* May be difficult to read when pointer is in motion.	*Fair:* May be difficult to read when scale is in motion.	*Good:* Minimum time and error for exact numerical value.
Discrete information	*Good:* Easy to locate pointer; numbers and scale need not be read; position change easily detected.	*Poor:* Hard to judge direction and amount of deviation without reading numbers and scale.	*Poor:* Numbers must be read; position changes hard to detect.
Setting	*Good:* Simple, direct relation of pointer motion to setting knob; position change aids monitoring.	*Fair:* Possible ambiguous relation of motion to setting knob; no pointer position change to aid monitoring; not readable during rapid setting.	*Good:* Most accurate monitoring of numerical setting; relation of display to setting knob not as direct as moving pointer; not readable during rapid setting.
Tracking	*Good:* Pointer position easily controlled and monitored; simplest relation to manual control motion.	*Fair:* No position change to aid monitoring; ambiguous relation to control motion.	*Poor:* No gross changes to aid monitoring.
General	Largest exposed and lighted panel area; scale length limited unless multiple pointers used.	Only small section of scale need be exposed and lighted; use of tape allows long scale.	Least space and light. Scale length limited only by number of counter drums.

that sounds obvious, there are many examples of displays which are "around the corner," too high, too low, and so forth. As a small example, the architect for my building put the room numbers above the office doors; since people rarely look at the ceiling, visitors have difficulty in finding specific offices.

Eye height varies—primarily due to posture (sit vs. stand) but also due to stature (adult vs. child, tall vs. short adult). If the height of the eye can be fairly well specified (e.g., eye height of a seated adult), then locate the display in a 20° circle about the line of sight. Many previous recommendations have given the preferred **line of sight** as 15° below the horizontal. Hill and Kroemer (1986) suggest these recommendations derived from a single study in the 1940s "akin to holding a pair of binoculars for somebody else." Hill and Kroemer's own study reported that the preferred angle was about 30° below the Frankfurt Plane (line between earhole and bottom of eye socket). The Frankfurt Plane coincides with the horizontal when the head is held straight up. See Figure 20.10. Also see Heuer et al. (1991).

Thus, if you believe the display viewer will sit (or stand) straight up, put the display about 30° below the horizontal. Kroemer and Hill (1986) suggest, however, that a typical seated operator probably has a forward head tilt of 10° to 15°. Using 10°, this would be a line of sight 40° below the horizontal.

Kroemer and Hill (1986) report that 90% of all their data were within a cone 20° above and below the mean preferred line. The display for most personal computers is in one unit with the disc drive in another. This furnishes an opportunity for the operator to vary screen height by either placing the screen on the drive or placing the screen on the table and turning the drive on its side.

Naturally, as the number of displays increases, not every display can be at the optimum location.

4.2 Arrangement of Displays (Instruments)

For an excellent detailed discussion of panel design, see Kinkade and Anderson (1984). The panel of instruments may include some controls. The goal is to organize the items on the panel to aid operator performance. Therefore, the first requirement is to determine what the operator is required to do. Consistency is important in tying items together or separating them and in establishing visual patterns. Hence, there needs to be an overall plan for the panel as well as all panels in the facility.

An early decision is the grouping logic. For example, assume the facility had 3 packaging lines, each with 6 different machines, and each machine had a number of displays of motor status, process temperature, and so forth. One possibility is to use 3 panels—1 for each line, with subpanels for each machine. A second possibility would be 6 panels—1 for each machine with subpanels for each line. Another possibility would be grouping panels by function, that is a panel with motor status of all 18 motors on the 6 lines, another panel with process temperature at all locations, and so forth. There is no general rule for which logic is best but some logical format must be decided upon and implemented consistently.

Computer programs are now being developed which allow the designer to simulate the grouping of displays on a panel (Metz et al., 1987; Palmiter and Elkerton, 1987). Consider "link analysis" of the displays (see Section 2 of Chapter 7). The displays are the areas to arrange.

Palmiter and Elkerton's program uses four criteria for panel layouts:

1. Overall density—the ratio of the free space (space in the panel not occupied by controls or displays) to occupied space.

2. Local density—how close the controls or displays are to each other. The general goal is to group related items closely and then separate them by blank space or lines.

3. Layout complexity—the irregularity of control and display arrangement. If the upper left edges of controls or displays are in horizontal or vertical lines, complexity is low.

4. Display grouping—the number of displays and display groups. The goal is to group like-functioned entities.

FIGURE 20.10 Head landmarks are the ear canal (meatus), the corner at which the eyelids meet (canthus), and the lower rim of the eye socket (orbis) (Kroemer, 1993). The ear–eye line passes through the ear canal and the eyelid corner. The Frankfurt line passes through the ear canal and the lower rim of the socket; it is about 11° below the ear–eye line. Using the ear–eye line as a reference, the line of sight (LOS), projected into the sagittal plane, gives the angle LOSEE.

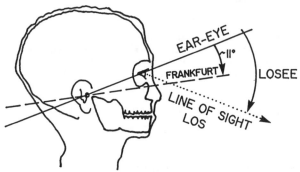

REVIEW QUESTIONS

1. Give the 2 levels at which a display can fail to communicate.

2. What is the recommended handwriting form for the capital letters U and V? For the numbers 4 and 9?

3. What is the optimum stroke width for black on white? For white on black?

4. What is the preferred character height when viewing at a distance of .5 m and 400 lux? What is the character height for critical viewing at .6 m and 50 lux? What character height would you recommend for .7 m and 10 lux?

5. At 6 m (normal viewing distance for a Snellen chart), how many min of arc does the E subtend?

6. What are two ways of improving readability of text on a VDT screen by changing the format?

7. Approximately how much higher are capital letters than lower case letters?

8. Give some examples of how a telephone number can be presented so as to reduce errors.

9. Assume you have 5,000 bin locations in your stock room. Describe your recommended coding system, giving at least 2 example bin codes.

10. For abbreviations, what is the difference between encoding and decoding?

11. For the command COMPRESS, give a 3-digit abbreviation using the truncation rule and using the vowel deletion rule.

12. Give Tullis's 4 rules for menus.

13. Should data in tables be rounded to 2 significant digits? Why?

14. Give an example of an implicit and an explicit table. Which is best to minimize errors?

15. When should a graph be used and when a table?

16. Should lettering on graphs be all capitals or capitals and lower case?

17. Give 3 different ways to make a figure deceiving.

18. Should graphs be wider than tall or vice-versa?

19. Should lines on graphs be labeled or keyed?

20. When should a word message be used in place of symbols?

21. Give the 3 categories of instruments and give an example of each.

22. What should the flash rate be for indicator lights?

23. Give 6 characteristics of good instrument pointers.

24. When should numbers on a scale be aligned radially?

25. Briefly discuss the angle for the preferred line of sight.

REFERENCES

Allen, R., Bailey, R., McIntyre, G., and Bozza, M. Placement of menu choices. *Proceedings of the Human Factors Society,* 379-82, 1991.

American National Standard for Human Factors Engineering of Visual Display Terminal Workstations. Santa Monica, Calif.: The Human Factors Society, ANSI/HFS 100, 1988.

Angiolillo, J. and Roberts, L. What makes a manual look easy to use? *Proceedings of the Human Factors Society,* 222-24, 1991.

Association for Computing Machinery. *Communications of ACM,* Vol. 12 [12], 697-98, December 1969.

Benbasat, I. and Todd, P. An experimental investigation of interface design alternatives: Icon vs. text and direct manipulation vs. menus. *Int. J. Man-Machine Studies,* Vol. 38, 369-402, 1993.

Berson, B., Boucek, G., and Hanson, D. Evaluating new concepts for aircrew alerting. *Proceedings of the Human Factors Society,* 921-25, 1984.

Butler, D., Acquino, A., Hissong, A., and Scott, P. Wayfinding by newcomers in a complex building. *Human Factors,* Vol. 35 [1], 159-73, 1993.

Caplan, S. Guidelines for reducing human errors in the use of coded information, *Proceedings of the Human Factors Society,* 154-58, 1975.

Chang, P.T. and Konz, S. Character size vs. viewing distance on VDTs. In *Work with Display Units,* Luczak, H., Cakir, A., and Cakir, G. (eds.). Amsterdam: Elsevier, 1993.

Cleveland, W. and McGill, R. Graphical perception and graphical methods for analyzing scientific data. *Science,* Vol. 229, 828-33, August 30, 1985.

Cooper, M., Daglish, H., and Adams, J. Reader preferences for report typefaces. *Applied Ergonomics,* Vol. 10 [2], 66-70, 1979.

Dillon, A. Reading from paper versus screens: A critical review of the empirical literature. *Ergonomics,* Vol. 35 [10], 1297-1326, 1992.

Eastman Kodak. *Ergonomic Design for People at Work,* Vol. 1. Belmont, Calif.: Lifetime Learning Publications, 1983.

Ehrenberg, A. Rudiments of numeracy. *J. Royal Statistical Society,* Vol. 140 [3], 277-97, 1977.

Ehrenreich, S. Computer abbreviations: Evidence and synthesis. *Human Factors,* Vol. 27 [2], 143-55, 1985.

Faulkner, T. Keep it simple. *AIIE Ergonomics News,*

April 1972.

Gould, J., Alfaro, L., Barnes, V., Finn, R., Grischkowsky, N., and Minuto, A. Reading is slower from CRT displays than from paper: Attempts to isolate a single-variable explanation. *Human Factors,* Vol. 29 [3], 269-99, 1987a.

Gould, J., Alfaro, L., Finn, R., Haupt, B., and Minuto, A. Reading from CRT displays can be as fast as reading from paper. *Human Factors,* Vol. 29 [5], 497-517, 1987b.

Gould, J. and Grischkowsky, N. Does visual angle of a line of characters affect reading speed? *Human Factors,* Vol. 28 [2], 165-73, 1986.

Grandjean, E. *Ergonomics in Computerized Offices.* London: Taylor and Francis, 1987.

Helander, M. Design of visual displays. In *Handbook of Human Factors,* Salvendy, G. (ed.), Chapter 7. New York: Wiley & Sons, 1987.

Heuer, H., Bruwer, M., Romer, T., Kroger, H., and Knapp, H. Preferred vertical gaze direction and observation distance. *Ergonomics,* Vol. 34 [3], 379-92, 1991.

Hill, S. and Kroemer, K. Preferred declination of the line of sight. *Human Factors,* Vol. 28 [2], 127-34, 1986.

Human Engineering Design Data Digest. Redstone, AL: Human Engineering Laboratory, U.S. Army Missile Command, 1984.

Hutchinson, R. *New Horizons for Human Factors in Design.* New York: McGraw-Hill, 1981.

Kinkade, R. and Anderson, J. *Human Factors Guide for Nuclear Power Plant Control Room Development* (NP-3659). Palo Alto, Calif.: EPRI, 1984.

Kline, D. and Fuchs, P. The visibility of symbolic highway signs can be increased among drivers of all ages. *Human Factors,* Vol. 35 [1], 25-34, 1993.

Konz, S. *Facility Design: Manufacturing Engineering,* 2nd ed. Scottsdale, Ariz.: Publishing Horizons, 1994.

Konz, S., Braun, E., Jachindra, K., and Wichlan, D. Human transmission of numbers and letters. *J. of Industrial Engineering,* Vol. 19 [5], 219-24, 1968.

Konz, S., Jackson, R., Knowles, J., and Verschelden, M. Legible and attractive transparencies. *Proceedings of Int. Ergonomics Association 1988.* London: Taylor and Francis, 1988.

Konz, S. and Mohan, R. The effect of illumination level, stroke width and figure ground on legibility of NAMEL numbers. *Proceedings of the Human Factors Society,* 431-35, 1972.

Kroemer, K. Locating the computer screen: How high, how far. *Ergonomics in Design,* 7-8, October 1993.

Kroemer, K. and Hill, S. Preferred line of sight angle. *Ergonomics,* Vol. 29 [9], 1129-34, 1986.

Kruk, R. and Mutter, P. Reading of continuous text on video screens. *Human Factors,* Vol. 26 [3], 339-45, 1984.

McDonald, J., Stone, J., and Liebelt, L. Searching for items in menus: The effects of organization and type of target. *Proceedings of the Human Factors Society,* 834-37, 1983.

Metz, S., Richardson, R., and Nasirudden, M. Rapid software for prototyping user interfaces. *Proceedings of the Human Factors Society,* 1000-1004, 1987.

Milroy, R. and Poulton, E. Labelling graphs for improved reading speed. *Ergonomics,* Vol. 21 [1], 55-61, 1978.

Palmiter, S. and Elkerton, J. Evaluation metrics and a tool for control panel design. *Proceedings of the Human Factors Society,* 1123-27, 1987.

Paniati, J. Legibility and comprehension of traffic sign symbols. *Proceedings of the Human Factors Society,* 568-72, 1988.

Perry, D. Speed and accuracy of reading: Arabic and Roman numerals. *J. of Applied Psychology,* Vol. 36, 346-47, October 1952.

Sanders, M. and McCormick, E. *Human Factors in Engineering and Design,* 6th ed. New York: McGraw-Hill, 1987.

Sherr, S. *Electronic Displays.* New York: Wiley-Interscience, 1979.

Shurtleff, D. *How to Make Displays Legible.* La Mirada, Calif.: Human Factors Design, 1980.

Smith, S. Letter size and legibility. *Human Factors,* Vol. 21 [6], 661-70, 1979.

Stanhagen, J. and Carlson, J. Identifying and controlling coding errors in information systems. *Ergonomics,* Vol. 22 [4], 441-52, 1970.

Taylor, B. and Taylor, W. Graphs that result in erroneous conclusions. *Proceedings of Int. Industrial Eng. Conference,* 470-76, 1986.

Tinker, M. *Legibility of Print.* Ames, Iowa: Iowa State Press, 1963.

Trollip, S. and Sales, G. Readability of computer-generated fill-justified text. *Human Factors,* Vol. 28 [2], 159-63, 1986.

Tufte, E. *The Visual Display of Quantitative Information.* Cheshire, Conn.: Graphics Press, 1983.

Tullis, T. The formatting of alphanumeric displays: A review and analysis. *Human Factors,* Vol. 25 [6], 657-82, 1983.

Verhoef, L. A new conceptual structure for travel information. *Applied Ergonomics,* Vol. 24 [4], 263-69, 1993.

Verschelden, M. and Konz, S. Absolute and relative ratings of type fonts and styles. *Proceedings of Ergonomics Society 1988.* London: Taylor and Francis, 1988.

Wilkins, A. and Nimmo-Smith, M. The clarity and comfort of printed text. *Ergonomics,* Vol. 30 [12], 1705-20, 1987.

Winkler, R. and Konz, S. Readability of electronics displays. *Proceedings of SID,* Vol. 21 [4], 309-13, 1980.

Wright, P. and Fox, K. Presenting information in tables. *Applied Ergonomics,* Vol. 1, 234-42, 1970.

Wright, P. and Fox, K. Explicit and implicit tabulation formats. *Ergonomics,* Vol. 15 [2], 175-87, 1972.

21 EYE AND ILLUMINATION

OVERVIEW

This chapter explains the eye and light, provides specific recommendations for reducing visual problems, and then makes design recommendations for general and special lighting.

CHAPTER CONTENTS

1 The Eye and Light
2 Reduction of Visual Problems
3 General Lighting
4 Special Lighting

KEY CONCEPTS

accommodation	esthetics	nearsightedness
astigmatism	farsightedness	negative contrast
bifocals/trifocals	general/task lighting	night vision
coefficient of utilization	hue/brightness/saturation	optical aids
color of light	illumination cost	orientation of lights
color weak	indirect/direct glare	polarization
cones/rods	lamp lumen depreciation	presbyopia
contrast	lumens	reflectance
convergence	luminaire dirt depreciation	restrike time
dark adaptation	luminance/brightness	Roy G. Biv
diffuse/specular reflections	luminance ratio	sitting ducks
diopters	lux/footcandles	visual acuity
direct/indirect	mirror test	wavelength (color)
dynamic visual acuity	near point	zonal cavity method

1   THE EYE AND LIGHT

1.1   The Eye

First the eye anatomy will be discussed, then vision.

1.1.1   Anatomy

Figure 21.1 shows the details of the eye, a 25-mm sphere. Light from the air enters through the transparent cornea. The cornea protects the eye and performs about 70% of the focusing required to produce an image on the retina. In addition, the change in refraction from air to the cornea permits the lens–fovea distance to be only 15 mm.

Next is the aqueous humor (watery fluid, in Latin) which nourishes both the cornea and lens. Glaucoma is a high pressure in the aqueous humor.

Next the light passes through the pupil of the biconvex lens. (*Pupil* is Latin for doll since you can see a small image of yourself reflected in the pupil.) The iris (rainbow, in Greek) expands and contracts to control the amount of light admitted. The iris tends to be blue in northern climates where sunlight is weak; in the tropics more melanin results in brown eyes, which are less sensitive to glare. Pupil diameter varies from about 1.5 mm to 9 mm—a factor of 6. Since light admitted is proportional to area, that is, diameter squared, the maximum amount of light admitted is about 40 times greater than minimum. The pupils expand with emotion and interest as well as light. (Belladonna, beautiful lady in Italian, causes the pupils to dilate; the drug got its name because Italian men consider any lady interested in them to be beautiful.) Changes in pupil size reflect changes in attitude and can affect the attitude and responses of the person observing the pupil, even if the observer doesn't realize it is pupil size being reacted to (Hess, 1975). The lens of the eye differs from a camera lens in that the ciliary muscle changes the lens shape to vary focal distance; this is called **accommodation.**

This distance is measured in meters or, more commonly, the inverse—**diopters.** The resting state of the eye, dark focus distance, is .6 m (1.67 diopters). Accommodative insufficiency can be due to weakness of the ciliary muscle or hardening of the lens with age. When viewing an object at infinity, "parallel" rays of light are focused on the retina. However, if the object is close to the eye, the light rays are not parallel; the lens thickens to maintain the focus at the same point (see Figure 21.2). The **near point,** the closest point at which you can focus, moves out as the lens hardens with age. Table 21.1 shows it is about .1 m at age 20 and 1 m at age 70. (Determine your own near point by bringing a printed page toward the eyes until the letters blur.) There is relatively little strain as long as work is done at over twice the near point.

Convergence, the act of aiming both eyes at a single point, is done by the 12 extraocular muscles. It is not needed for distances over 6 m (20 ft); that is why eye tests are set for that distance. Predetermined time systems (Chapter 29) include eye focus, convergence, and eye travel times.

After the lens, the light rays pass through the vitreous humor, a jellylike substance whose primary function is to maintain the eye shape. (If you see moving things when you close your eyes, it may be red blood cells, which escaped from the retina, in the vitreous humor.)

Finally, the light strikes the retina, about the size and thickness of a postage stamp, which is attached to the sclera, the "white" of the eye. The retina, an extension of the brain, preprocesses the information before sending it through the optic nerve to the brain. The retina not only has two different transducer systems (cones and rods) to convert light to an electrical signal, it also changes the transducer's sensitivity with incoming light. This is analogous to having a camera with both black and white and color

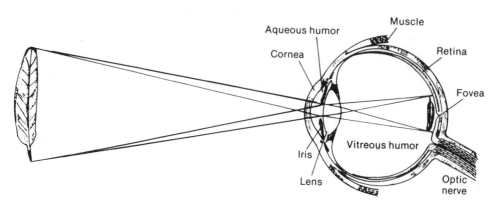

FIGURE 21.1   Human eyes, 25-mm spheres, have an iris to adjust the light admitted, a lens which adjusts focal length, and two "photographic films"—black and white and color. In addition, the films change their "speed" as required.

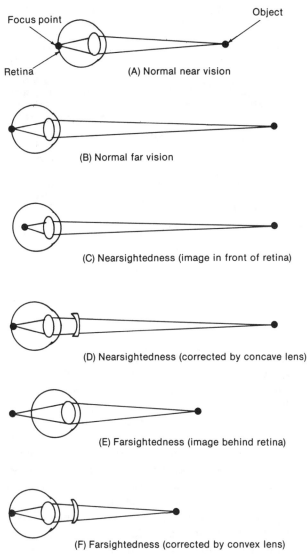

FIGURE 21.2 Eye problems and corrections are shown. A and B show the normal eye at near and far vision. For normal far vision (B) the light rays are close to parallel and the eye is relaxed. As the object comes closer, the rays become less parallel and the lens is accommodated by the ciliary muscle to maintain the focus on the retina. C and D show **nearsightedness** (myopia) and its optical correction. E and F show **farsightedness** (hyperopia) and its optical correction.

film with the film also changing speed as required. Figure 21.3 shows the effect of decreasing the light— **dark adaptation.** If, on the other hand, the light suddenly increases (the eye sees a bright object such as an electric light, window, or reflecting table top), the sensitivity of the entire retina decreases to about 20% of its previous value in about .05 s (alpha adaptation); it continues its decline in sensitivity (beta adaptation) at a slower rate for about 30 min (Grandjean, 1969). Thus, when changing from light to dark or dark to light, it takes a while until you see your best.

When looking at objects, the retina emphasizes changes in brightness more than absolute levels; the minimum detectable brightness difference between

objects is about 1 to 2%. If an object oscillates in brightness, it will appear to flicker. This flicker will fuse or disappear at the critical fusion frequency (CFF). For optimum flicker detection situations (large, uniform fields, 100% on–off, high adaptation luminance), CFF is about 75 Hz.

Table 21.2 shows brightness for the two different transducers. Above .001 lux, the **cones** (color, daylight, photopic system) are used; from .001 to .0001 both systems are used, and below .0001 lux the **rods** (black and white, night, scotopic system) are used. Figure 21.4 shows that the rods are more sensitive to light than the cones.

The 7,000,000 cones are concentrated about a small pit (*fovea,* in Latin) in line with the lens. Only a small portion (about 1°) of the visual field is covered by the fovea at any one instant; beyond this area of sharp vision you detect only movement and strong contrast. At a distance of .5 m from the eye, this sharp vision circle has an 8-mm diameter. The larger area we actually observe is due to the saccadic eye movement as well as voluntary shifting of the focus point. Figure 21.4 shows that the cones are more sensitive to some **wavelengths** (i.e., colors) than others. (The rods also are more sensitive to some wavelengths than others but they don't respond to different wavelengths with different colors.) What color would you make an object if you wanted it to be very noticeable? If you had a lamp emitting light at only one wavelength, what would be the wavelength to use? Note that light above 650 nm (red light) does not affect the rods so dark adaptation is maintained while the cones can see. One "red light" approach is to have the operator wear goggles with a filter for light below 650 nm; the other is to have lights which only give light above 650 nm. Note that color discrimination is very poor with red light. Figure 21.5 shows that green, red, and yellow are not seen well if they are peripheral.

The 125,000,000 rods are scattered over the entire retina (except the fovea). Thus, for night vision, look out the corner of your eye. Rhodopsin, a chemical in the rods manufactured by the body from vitamin A, is bleached by light to give the photochemical effect for nerve excitation. See Figure 21.3 for dark adaptation time.

1.1.2 Normal vision

Visual acuity, the ability of the eye to distinguish detail, is:

$$VA = \frac{1}{\text{Visual angle of minimum object detectable, min of arc}}$$

Table 21.3 shows the formula geometry. By expressing visual acuity in terms of the visual angle, any size target (dimension *h*) can be used at any distance

TABLE 21.1   Age affects speed of perception, dark adaptation time, ability to detect peripheral movement, resistance to disability glare, and luminance and contrast thresholds. The closest point at which the eye can focus, the near point, increases with age; visual acuity decreases. The increase in near point is called **presbyopia** (old man's vision, in Greek).

| Age, Years | Percent Visual Acuity (6/6 = 100) | | | Mean Near Point | | Glare Borderline Between Comfort Discomfort,† nits |
	Best Refractive Correction	Actual Correction	Without Correction	Meters	Diopters*, 1/m	
10	—	—	—	.077	13.0	6900
20	100	83	75	.091	11.0	3100
30	99	83	60	.111	9.0	1900
40	96	82	50	.167	6.0	1400
50	90	68	25	.500	2.0	1050
60	85	60	20	.833	1.2	850
70	70	40	15	1.000	1.0	700

*The power of a lens is given in diopters:

$$P\,(\text{diopters}) = \frac{1}{f\,(\text{meters})} = \frac{1}{S_1} + \frac{1}{S_2}$$

where S_1 = distance from light source to lens node, m
S_2 = distance from lens node to focal point, m
For normal viewing, S_1 = infinity and S_2 = .015 m, so the power of the eye is 67 diopters.
†BCD = 103,000 (age, yrs)$^{-1.17}$ from Bennett (1977).

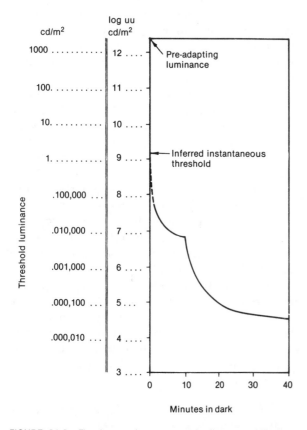

FIGURE 21.3   Time in an environment strongly affects your ability to see. Adaptation takes place with a pupillary/neural system (responds within a fraction of a second but with limited range) and a photochemical system (responds over several minutes but with a large range). The curve shows ability to detect a target vs. time. Initially exposed to a luminance of 3,400 cd/m2, the eye requires a target to be 3 cd/m2 to be detected. After 2 min of dark, the target need only be .03 cd/m2 to be detected; after 40 min of dark, the target need only be .000 030 cd/m2 to be detected. (Modified from Figure 13.33, Parker and West, 1973.)

from the eye (dimension *d*). The tangent of the visual angle, θ, divided by 2 is equal to *d*/2 divided by *h;* since the angle is small, the approximation tan θ = *h/d* may be used.

By convention, "normal" vision is ability to detect an object 1 min of arc at 6 m (20 ft). This is expressed as 6-6 vision (20-20 in the United States). That is, you see at 20 ft what the normal person sees at 20 ft; 20-100 vision means you see at 20 ft what the normal person sees at 100 ft.

There are several kinds of visual acuity measurements: minimum separable (gap detection), minimum perceptible (spot detection), vernier (lateral displacement of two lines), and dynamic. The minimum gap detectable is not 1 min of arc but is about 25 s for very good contrast, long viewing times, and high luminance. Squares and spots (dark and light) can be detected even if they are very small. A dark square of 14 s of arc against a bright sky can be detected 75% of the time; a star subtending .06 s of arc can also be detected. Vernier acuity, the ability to detect changes in the alignment of lines, is about 2 to 3 s of arc. Since the diameter of a single foveal cone is 10 to 40 s of arc, these acuities demonstrate the processing ability of the retina.

Dynamic visual acuity, the ability to discriminate detail in a moving target, is important for some inspection tasks. Unfortunately an individual's static visual acuity (gap detection as measured on an eye chart) is not a good predictor of dynamic acuity; however, dynamic visual acuity can be improved with practice (Long and Rourke, 1989). Under most favorable conditions an object is detectable when it moves over 2 min of arc/s. Discrimination of detail in

TABLE 21.2 Luminance (cd/m²) of various surfaces.

Approximate Brightness, cd/m²			
to	.000 001 .000 010	Absolute threshold of seeing	
to	.000 010 .000 100	Grass in starlight	
to	.000 100 .001	Snow in starlight	rod vision
to	.001 .010	Earth in full moon	rod + cone vision
to	.010 .100	Snow in full moon	
to	.100 1.0	White paper 1 foot from a 1 candela source	rod + cone vision
to	1. 10.		cone vision
to	10. 100.	White paper in good light, indoor movie screen, TV screen	
to	100. 1000.	Average sky on cloudy day, full moon	
	1000.	Average sky on clear day, instant-start cool white fluorescent tube (900-1150) for medium-loaded lamps, 1350 to 1800 for high-load lamps; preheat starting 500 to 1200	

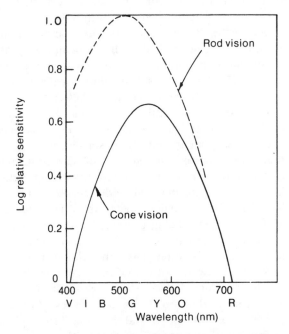

FIGURE 21.4 Rods, the **night vision** system, require only 1% of the light required for the cones, the day system. Both rod and cone sensitivity vary with wavelength; the maximum sensitivity for cones is 555 nm (yellow-green). From 440–450, light is violet and indigo; from 450 to 500, blue; from 500 to 545, green; from 545 to 590, yellow; from 590 to 610, orange; and from 610 to 760, red. Remember this rainbow sequence by the name **Roy G. Biv**. Also remember to reverse the sequence since red is at 700 nm, not 460.

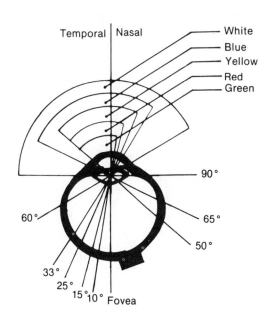

FIGURE 21.5 Peripheral color detection is difficult for red, yellow, and green. The figure shows that the total field of view is about 150° for white light and is less for colors.

TABLE 21.3 Visual acuity, the ability to discriminate detail, is calculated from the ratio of the minimum detectable gap or object (h) over the distance from the eye to the target (d). $\theta = 3400\ h/d$; visual acuity $= 1/\theta_0$.

Tan θ, h/d	θ, minutes of arc	Visual Acuity
.017 455	60.00	0.02
.001 000	3.43	0.29
.000 500	1.72	0.58
.000 241	1.00	1.00
.000 200	0.68	1.45
.000 100	0.34	2.90
.000 050	0.17	5.81

a moving target is satisfactory if the eye can lock onto the target. Boyce (1981, p. 58) says ability to lock on worsens rapidly beyond velocities of 50°/s. As any target shooter can tell you, it is easier to track a predictable target than a varying target.

Just because you can detect a small object doesn't mean that it is desirable to have objects you look at as small as possible—bigger is better. Figure 21.6 shows the ratio of object size (θ) to the minimum object size detectable (θ_0) vs. contrast. Fortuin (1970), summarizing a lifetime of work, recommends $R = 2.5$ for "easy seeing." That is, for easy seeing, an object should be at least 2.5 times threshold size. If R

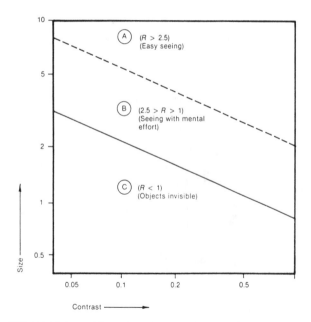

FIGURE 21.6 Visibility varies with both size (min of arc) and contrast of the target (Fortuin, 1970). The figure shows detectability of targets with a background of 10 cd/m². Below the threshold (area C), objects are invisible. Above the threshold in area B they are visible but with considerable effort; in area A, they are visible with little effort. Duncan and Konz (1976) found that people could read light-emitting diodes and liquid crystal displays with no errors if they were 4–6 min of arc; however, the subjects preferred the displays to be 20–30 min of arc and thus preferred them to be about 5 times the minimum size.

> 2.5, easy seeing; $1 < R < 2.5$, strenuous seeing, resulting in eye fatigue; $R < 1$, invisible. Note that the eye is not hurt by insufficient light just as a camera isn't hurt. Insufficient light causes eye muscle fatigue, bloodshot eyes, and headaches, but not permanent damage.

Unfortunately, most people do not have perfect eyes.

1.1.3 Eye problems
Figure 21.2 shows two common eye problems and their optical solutions. A nearsighted person (myopia) has a long eyeball; the light rays from distant objects begin to diverge before hitting the retina. The solution is a concave external lens to bring the rays farther apart on the eye lens. A farsighted person (hyperopia) has the opposite problem—a short eyeball. The light rays from close objects meet behind the retina. The solution is a convex lens. As you become older, your accommodation range decreases so you are forced to use bifocals, trifocals, or progressive focal length lenses. (With **bifocals,** there are two focal length lenses; with **trifocals** there are three focal length lenses; with progressive, there is one lens with the focal length varying continuously from top to bottom.) Some people have football-shaped eyeballs with unequal radii of curvature in two axes; the result, a line focus instead of a point focus, is called **astigmatism** ("no point," in Greek). Astigmatism can be corrected by an external lens with unequal curvature.

Spectacles can be made impact resistant but note that any glass will break if hit hard enough. Polycarbonate lenses are lighter than glass but scratch easily. Many goggles have side shields; dust goggles have a fuzzy cloth next to the skin to give a better seal. Face shields protect the entire face from liquids and impact. Spectacles with chemicals in them to reduce light (either permanently as in sunglasses or temporarily as in lenses that automatically darken in sunlight) may present a safety hazard in welding areas. The chemicals probably will be designed only to cut off radiation in the visible spectrum. Thus, workers will think they are being protected, but the dangerous ultraviolet radiation is still hitting the eye. For that reason, use welding glasses in welding areas.

Spectacles were not considered manly for Prussian officers. Monocles were not very satisfactory, so the Germans developed contact lenses. Contact lenses support the cornea and thus retard the typical lengthening of the eyeball with age. The cornea is furnished with oxygen from the air instead of from inside the eye. Hard lenses prevent this breathing and so must be removed in a short time. Newer materials breathe better, so wearing time has been considerably extended. However, contact lenses do not provide eye protection against impact as a spectacle can. They also tend to seal the surface of the cornea; thus, they should not be worn where people are exposed to chemical fumes, vapors, splashes, or dusty atmospheres (even inside a respirator) as they tend to hold the irritant against the eye and not let the normal tears wash it out. Intense heat (e.g., furnaces) also can be a problem.

Some people have color perception deficiency. Cones specialize in detecting red, green, and blue. If your red cone pigment is defective, you cannot distinguish red or blue-green from gray; if your green cone pigment is defective, you cannot distinguish green or reddish-purple from gray. The X chromosome influences cone pigments. Since females have two X chromosomes, they have a spare. About .4% of females and 8% of males have some color perception

deficiency. Usually they are not completely deficient; consider them **color weak,** not color blind (Rushton, 1975). Only about .003% of males are truly color blind.

1.2 Light Definitions Table 21.4 gives the definitions of wavelength, polarization, luminous intensity, illuminance, luminance, reflectance, and brightness contrast. Wavelength and **polarization** describe the light. The key equation is:

$$\text{Incident flux (Reflectance)} = \text{Reflected flux}$$

That is,

$$\text{Illuminance (Reflectance)} = \text{Luminance}$$

Radiant power (radiant flux) is the time rate of flow of radiant energy; it is measured in watts.

TABLE 21.4 Units and definitions of illumination.

Quality	Unit	Definition and Comments
Luminous flux, lm	lumen	Light flux, irrespective of direction from a source
Luminous intensity, I	candela	Light intensity within a very small angle, in a specified direction (lumen/steradian) Candela = 4π lumens
Illuminance, $d\text{lm}/dA$	lux	1 lumen/m² = 1 lux = .093 footcandle 1 lumen/ft² = 1 footcandle = 10.8 lux
Luminance	nit	Luminance is independent of the distance of observation as candelas from the object and area of the object perceived by the eye decrease at the same rate with distance 1 candela/m² = 1 nit = .29 footlambert 1 candela/ft²π = 1 footlambert = 3.43 nits
Reflectance	unitless	Percentage of light reflected from a surface

Typical Reflectance		Recommended Reflectances	
Object	%	Object	%
Mirrored glass	80-90	Ceilings	80-90
White matte paint	75-90	Walls	40-60
Porcelain enamel	60-90	Furniture and equipment	25-45
Aluminum paint	60-70	Floor	20-40
Newsprint, concrete	55		
Dull brass, dull copper	35	Munsell Value	Reflectance (%)
Cardboard	30		
Cast and galvanized iron	25	10 =	100
Good-quality printer's ink	15	9 =	100
Black paint	3-5	8 =	78
		7 =	58
		6 =	40
		5 =	24
		4 =	19
		3 =	6

Quality	Unit	Definition and Comments
Brightness contrast	unitless	$C = \dfrac{\text{Luminance of brighter - Luminance of darker}}{\text{Luminance of brighter}}$
Wavelength	nanometers	The distance between successive waves (a "side view" of light). Wavelength determines the color hue. Saturation is the concentration of the dominant wavelength (the degree to which the dominant wavelength predominates in a stimulus). Of the 60 octaves of the electromagnetic radiation, the human eye detects radiation in the octave from 380 to 760 nanometers.
Polarization	degrees	Transverse vibrations of the wave (an "end view" of the light). Most light is a mixture; horizontally polarized light reflected from a surface causes glare.

Luminous power, the visible part of radiant power from a source, is measured in **lumens** (lm). For example, a lamp may give 1,200 lumens. Luminous intensity, *I*, the luminous flux in a specified direction, is given in candelas (cd). A sphere subtends an angle in space of 4π steradians; thus the total flux emitted by a 1-candela uniform light source is 12.57 lm. The illuminance on a 1-m sphere by a 1-candela source is $1 \text{ lm/m}^2 = 1$ lux. In the SI system, lumens/m² is called **lux** (lx); in the U.S. system, lumens/ft² is called **footcandles** (fc). Illuminance transmitted through a surface is called transmittance, *t*, unitless, with no transmittal = 0 and perfect transmittal = 1.

Luminance, or reflected luminous flux, in the U.S. system is lumens/ft², called foot lamberts (fl). However, in the SI system, luminance is measured in candelas/m², called nits, not lumens/m². Since candelas involve a spherical surface, a π term must be included for conversion to flat surfaces. Thus, for 1,000 lux incident on a 10% reflectance surface, the illuminance is $1,000\ (.10)/\pi = 31.8$ nits. Although often used interchangeably, technically luminance is a measure of the physical intensity of light while **brightness** is a measure of the intensity of the sensation perceived by the observer.

In both the U.S. and SI systems, **reflectance** applies only to perfectly diffuse reflection and is the ratio of reflected luminous flux to incident luminous flux.

Reflectance (in SI system)

$$= \frac{\text{Reflected flux}}{\text{Incident flux}} = \frac{\text{Luminance (nits)} \times \pi}{\text{Illuminance (lx)}}$$

Reflectance (in U.S. system)

$$= \frac{\text{Reflected flux}}{\text{Incident flux}} = \frac{\text{Luminance (fl)}}{\text{Illuminance (fc)}}$$

Brightness contrast, the difference in the amount of light reflected, is the ratio of two luminances. Color contrast differs with the composition of the light reflected.

The three attributes of color are **hue** (red, blue), related to wavelength of light; **brightness** (light red, dark red), related to intensity of light; and **saturation** (rich full red, pale red), related to purity of light.

Light should be measured with a color- and cosine-corrected meter. Color-corrected means the reading is corrected for the sensitivity of the human eye. Cosine-corrected means the reading will not be affected by holding the sensor area at an angle to the light flux. The better meters give a digital output and have a "freeze" button so the last reading remains on the display even though the meter is moved.

2   REDUCTION OF VISUAL PROBLEMS

2.1   Criteria
The basic criteria for lighting a task are the following:

1. Have satisfactory visual performance.
2. Minimize cost of the lighting.
3. Have satisfactory esthetics.

Cost of illumination tends to be low in relation to labor costs, although it still has some absolute cost. **Illumination cost** can be divided into energy cost, fixture cost, and lamp cost. As a rule of thumb, energy cost is 80 to 90% of the total cost. Annual lighting cost in 1994 for an office or manufacturing area was about \$.50/ft² (for an 8-h day).

Labor cost/ft² depends upon the labor cost and the worker density. Assume a wage rate of \$8/h and fringe costs of 35% for a labor cost of $8 \times 1.35 = \$10.80/h$. Assuming 1,900 h/year, annual cost is \$20,520/yr. Assume 1 worker/300 ft². Then annual labor cost is \$68.40/ft²; labor is \$68.40/.50 = 137 times as expensive as lighting costs. The ratio will vary with labor cost, worker density, and type of lighting, but ratios usually are over 100 and often are over 300. Don't save a penny by reducing lighting costs if it costs a dollar in reduced labor productivity.

In general, performance in detecting small targets increases with increased illumination up to an asymptote; the exact asymptote depends upon many factors but is about 100 lux for easy tasks and about 1,000 lux for difficult tasks. The problem, of course, is to define easy and difficult! Second, performance generally can be improved more by changing target size and contrast than by changing illumination. Third, regardless of the amount of illumination, performance on easy visual tasks far exceeds performance on difficult visual tasks; that is, more light will not change a difficult task into an easy one (Cushman and Crist, 1987).

Visual performance is affected by (1) individual differences, (2) quantity of light, (3) quality of the light, and (4) task requirements.

2.2   Individual Differences
People vary. Table 21.1 shows that visual acuity declines with age even with the best refractive correction. Resistance to glare and color discrimination also decline with age. Table 21.1 also shows how many people do not have the best possible correction. For example, Ferguson, Major, and Keldoulis (1974) reported that 69% of workers without glasses needed glasses and 37% with glasses needed a new prescription. In addition, there are great differences in individual accommodation and convergence capabilities, so people may be

able to do a task for a short time but prolonged performance results in fatigue and headaches.

Thus, if employees have visual problems on a task, have their visual abilities checked. Jobs may require good vision at 20 to 40 inches (.5-1 m); the plant nurse can test an employee's capability with a vision screener—a standard wall chart tests only far vision.

2.3 Quantity of Illumination
The recommended amount, sources, and fixtures will be discussed.

2.3.1 Recommended amount
The Illuminating Engineering Society (IES) recommendations are given in Table 21.5; Table 21.6 gives example situations for the 9 levels in Table 21.5. However, Table 21.7 shows how, for **general lighting** (lighting from the ceiling), the amount of illumination depends upon the task and occupant age and room reflectance. For **task lighting** (lighting from the floor or table), the amount of illumination depends on occupant age, speed and accuracy, and room reflectance. More light is needed for people over 40, when speed or accuracy is important, or when there is insufficient room reflectance.

2.3.2 Sources
The ideal illumination source (which does not exist) would be free, give the desired amount of light on demand, and have high quality (color, glare, highlighting, contrast).

Under normal circumstances, there is no physiological need for daylight in workplaces. However, close to the north and south poles, lack of sunlight in the winter exacerbates depression, resulting in a condition called seasonal affective disorder (SAD). Although sunlight can act as a supplement for artificial light in some situations, there are many problems with sunlight as an illumination source. See Box 21.1. Thus the choice generally is among artificial sources.

Table 21.8 gives some lamp characteristics. Which lamp to use depends upon cost, convenience, and color.

Cost is primarily (about 90%) the cost of the energy with about 10% for the fixture, lamp, and cost of replacing the lamp. Lumen output decreases with age, so for cost calculations use mean lumens/watt rather than initial. Don't forget ballast losses. Lumens/watt are higher for bigger bulbs; often the advantage is over 25%.

Convenience primarily is the replacement frequency and restrike time. All bulbs except incandescent tend to have lives of over 10,000 h; mercury tends to last especially long. If there is a power interruption, when the power resumes there is a delay (**restrike time**) before the lamp lights for sodium, mercury, and metal halide lamps; this may require supplementary fluorescent or incandescent lighting for emergencies.

Color of light of the low-pressure sodium lamp (LPS) is very poor; color of high-pressure sodium (HPS) is poor. In addition, Lin and Bennett (1983)

TABLE 21.5 Recommended target maintained illuminance (lux) for interior industrial lighting (IES, 1983). See Table 21.6 for task examples.

Reference Work Plane	Type of Activity	Illuminance Category	Illuminance (lux)		
			Total Factor[a] -3 or -2	Total Factor[a] -1, 0, +1	Total Factor[a] +2 or +3
General lighting throughout spaces	Public spaces with dark surroundings	A	20	30	50
	Simple orientation for short temporary visits	B	50	75	100
	Working spaces where visual tasks are only occasionally performed	C	100	150	200
Illuminance on task	Performance of visual tasks of high contrast or large size	D	200	300	500
	Performance of visual tasks of medium contrast or small size	E	500	750	1000
	Performance of visual tasks of low contrast or very small size	F	1000	1500	2000
Illuminance on task obtained by a combination of general and local (supplementary) lighting	Performance of visual tasks of low contrast and very small size over a prolonged period	G	2000	3000	5000
	Performance of very prolonged and exacting visual tasks	H	5000	7500	10,000
	Performance of very special visual tasks of extremely low contrast and small size	I	10,000	15,000	20,000

[a]See Table 21.7 for factor calculations.

TABLE 21.6   Selected task examples of illuminance categories for use with Table 21.5. The IES also has examples for sports and recreation, outdoor facilities, and transportation vehicles.

ILLUMINANCE	TASK/ACTIVITY/AREA	ILLUMINANCE	TASK/ACTIVITY/AREA

A	Nursing station corridors: night
	Pharmacy: night light
	Sawmills: basement area
		E	Barber shops and beauty parlors
			Laboratories: science
			Restaurant: kitchen
			Nursing desk
			Hotel: front desk
			Library: card files
			Offices: mail sorting, reading 6-point type or
			  phonebooks
			Assembly—moderately difficult
			Inspection—moderately difficult
			Machine shop: medium bench or machine
			  work
			Painting: fine hand painting and finishing
			Sheet metal work: most areas
			Woodworking: fine bench and machine work,
			  fine finishing

B	Dance halls and discotheques Restaurant: dining Corridors in nursing areas at night Library: inactive stacks Offices: microfiche reader, video display terminal Farm shed: machine storage		

C	Lobbies: bank
	Drafting: light table
	Elevators
	Corridors in nursing areas in daytime, hotel
	  corridors and stairs
	Locker rooms, toilets
	Offices: lounges, lobbies, reception areas
		F	Operating room—general
			Classroom: demonstration
			Graphic design: charting, mapping, layout,
			  artwork
			Assembly—difficult
			Inspection—difficult
			Sheet metal: scribing

D	Conference rooms
	Restaurant: cashier
	Nursing station
	Hotel: lobby, reading area
	Library: active stacks, circulation desk, book
	  repair
	Office: AV area, duplicating area, reading ink
	  writing, newsprint, typed originals, 8- to 10-
	  point type
	Assembly—simple
	Inspection—simple
	Machine shop: rough bench or machine work
	Material handling: wrapping, labeling, stock
	  picking
	Painting: dipping, simple spraying
	Welding: orientation
	Woodworking: most areas
		G	Autopsy table
			Cloth products: sewing, cutting
			Inspection—very difficult
			Machine shop: fine bench or machine work,
			  fine polishing
			Painting: extra fine hand painting and
			  finishing

		H	Dental suite: oral cavity Surgical task lighting Inspection—exacting Machine shops: extra fine work Welding: precision manual arc-welding

concluded "Don't use sodium sources where there is substantial viewing of faces." Fluorescent lamps can be mounted among LPS lamps; the result is equivalent to HPS lighting—colors but not shades of colors are distinguishable. Metal halide gives good color (discriminate shades of colors); thus metal halide lamps often are mingled with HPS in high bays to get acceptable color and high lumens/watt. Fluorescent lamps not only give good color rendition but also permit you to select the color. In the United States cool white is the favorite. If you prefer a "warmer" (more red) light, use cool white deluxe. Warm white gives the same color as incandescent. There also are specialized fluorescents to approximate spring

TABLE 21.7   Factors for illuminance table (Table 21.5) (IES, 1983). If the task was in illuminance category D, the worker age was "under 40," speed/accuracy were "important," and task background reflectance was "30 to 70," then the components would be −1, 0, and 0; the total factor would be their sum, −1, so the middle illuminance column in Table 21.5 would be used and the recommended illuminance would be 300 lux.

Variable	Factor		
	-1	0	+1
For Illuminance Categories A, B, C (General Lighting Throughout Spaces)			
Occupants' ages, year	Under 40	40-45	Over 55
Average weighted room surface reflectance (%)	Over 70	30-70	Under 30
For Illuminance Categories D, E, F, G, H, I (Illuminance on Task)			
Workers' ages, year	Under 40	40-55	Over 55
Speed and/or accuracy	Not important	Important	Critical
Task background reflectance (%)	Over 70	30-70	Under 30

BOX 21.1 *Windows (Konz, 1994)*

Windows are primarily a conflict of esthetics and a view vs. energy conservation. If windows are used, they should allow a seated worker to see the view. Ideally, the horizon should bisect the window and the view should be of nature (such as water, foliage) rather than manufactured objects (such as a brick wall). To maximize the number of viewers, place windows in break rooms and at ends of corridors and use interior windows between windowed perimeter rooms and the interior core so people from the core can see through both sets of windows.

Window disadvantages (compared to solid walls):

- They cost more in both capital cost and operating cost (heat entry in summer, cold entry in winter, washing and repair costs). A single-pane glass window has a U value of about 1 BTU/h − ft^2 − °F compared to .5 for a double pane, .4 for a triple pane, and .2 for a normal wall. Thus, a single-pane window transmits 5 times as much as a wall.

- They are a source of glare. Many windows are almost permanently covered with shades, drapes, or curtains. Glare is a special problem in computer areas for viewing monitors and in conference rooms for visual aids. Windows also restrict the arrangement of nearby workstations.

- Windows are not a practical source of illumination in factories because the light is too variable (both in amount and color), depending upon time of day, season, and weather, and because the light decreases by distance squared from the window. Work close to the window gets too much light and work far away gets too little. If you depend on sunlight through windows for illumination, equipment and workstations must be within about 20 ft (6 m) of a window, making maximum building width about 40 ft (12 m). Artificial illumination permits varied building size, shape, and layout.

- Windows may admit air. Exterior windows are not desirable sources of ventilation in industrial buildings: (1) The air volume passing through a window is too variable because of wind velocity and direction, and is difficult to use in hot, cold, or wet weather. (2) The air admitted and released is not controlled. Thus you get low velocities at locations far from windows. If pollutants are being discharged, they are not controlled (e.g., passing through a filter). (3) Mechanical ventilation is relatively cheap and is easily controlled.

- Windows pass noise and distractions from outside to in, and inside to out.

Window characteristics that can be good or bad:

- They decrease privacy for people on both sides of the window.

- If openable, they allow air passage. If the heating, ventilating, and air conditioning (HVAC) system fails, either locally or totally, air passage may be useful. However, for normal operation of a HVAC system, open windows decrease system control and probably increase energy costs.

Window advantages:

- They permit a view.

Window disadvantages can be decreased by doing the following:

- Maximize the view. Use windows with a long horizontal axis and a short vertical axis. Keep the sill low (.5 m above the floor) so that a seated person can see the view. Make windows public rather than private by placing them in common areas.

- Have minimum surface area. Reduce area of existing windows with opaque insulated panels.

- Use double panes to reduce heat (and noise) transfer. Consider different glazing treatments for the south and west side than for the north and east side.

- Shade windows from solar heat and glare. Permanent treatments include fenestration, recessed windows, and windows on enclosed courtyards (atriums). A semipermanent treatment is shading with deciduous trees. Manual treatments include blinds and curtains. Ease of adjustment is important; if the blinds and curtains are never opened, the window should be boarded up.

- Decide whether the window will be permanently sealed, openable with a key, or openable anytime. Since openable windows leak air (even when closed), consider making only a few (e.g., 1/3) of the windows openable. Another option is a large sealed window and a small openable window.

- If using skylights, make them vertical rather than horizontal (reduces dirt and breakage), plastic rather than wired glass (better light transmission and less breakage), translucent rather than transparent (less glare), and openable (to supplement mechanical ventilation).

TABLE 21.8   Characteristics of typical industrial lamps (Courtesy of General Electric Lighting Business Group)

TYPE OF LAMP	WATTS	LUMENS/WATT		LUMEN MAINTENANCE (%)	RATED LIFE (H)	RESTRIKE (Min)	RELATIVE COST
		Initial	Mean				
High-pressure sodium	35–1,000	64–140	58–126	90–92	24,000	1–2	Low
Metal halide	175–1,000	80–115	57–92	71–83	10,000–20,000	10–15	Medium
Fluorescent	28–215	74–100	49–92	66–92	12,000–20,000+	Immediate	Medium
Mercury	50–1,000	32–63	24–43	57–84	16,000–24,000+	3–6	High
Incandescent	100–1,500	17–24	15–23	90–95	750–2,000	Immediate	High

outdoor daylight in northern Europe (the "standard" color).

Light may be important for inspection as well as esthetics. The perceived color of an object depends not only upon the spectral characteristics of your eyes and the actual color of the object; it also depends upon the spectrum of the ambient light. When inspecting the color of an object, specify whether the ambient light should be cool white, incandescent, high-pressure sodium, or other, because perceived colors are quite different (Misra and Bennett, 1981). "Biological" inspection (e.g., grain, meat, cotton) should have a lighting source specified with a high color rendering index (CRI). Metameric colors are colors which look the same under one light spectrum but different under another. The "spectral power" of metal halide and sodium lamps is quite peaked and so they are questionable for precise color discrimination.

2.3.3 Luminaires (fixtures)

Lamps can be put into a wide variety of fixtures. A fixture will release 50 to 80% of the light from the lamp to the room. If a fixture's **coefficient of utilization** is .70, it means 70% of the lamp light inside the fixture is distributed to the work plane. The light distribution from a fixture (up vs. down) is in five categories: **direct** (90% down), semidirect (90–60% down), general diffuse (60–40% down), semi-indirect (40–10% down), and **indirect** (10–0% down). Direct and semidirect are most used. Some uplight tends to be best since (1) the light on the ceiling reduces brightness contrast and (2) nondirect luminaires tend to stay cleaner (i.e., lose less light) as air can move upward through the fixture. The downward component is further described by beam spread (highly concentrating, concentrating, medium spread, spread, widespread). Wider beam spreads give more overlapping (better illumination on vertical surfaces and less dependence on a single lamp). At high mounting heights, use narrower beams. The shielding angle (the angle between a horizontal line and the line of sight at which the source becomes visible) should be greater than 25°, preferably approaching 45°. The lighting distribution from linear sources (fluorescent and low-pressure sodium) tends to "batwing" (be emitted at a 45° angle downward, as viewed from the tube end).

Since it is difficult to foresee the future, fixtures should be selected that are relatively easy to relocate within the area. A variety of plug-in designs are available. Some fixtures have a disconnect feature which permits maintenance without shutting off the main circuit.

2.4 Quality of Illumination

In addition to the color of the light (mentioned above), consider the glare, orientation, and variability of the light.

2.4.1 Glare

Glare is any brightness within the field of vision which causes discomfort, annoyance, interference with vision, or eye fatigue. Glare is divided into direct and indirect glare.

Direct glare   Because **direct glare** is caused by a light source within the field of view, windows are the most common problem. One approach is to change the window brightness with louvers, curtains, and so forth. Another approach is to reduce window transmittance by using films or translucent glass. See Figure 21.7 for another technique of reducing the contrast between a bright window and its surrounding wall. Alternatively, you could reorient the worker at the workstation to face away from the window. See Box 21.1. Another source of direct glare occurs when inspecting backlighted objects such as microfilm readers. Use a mask to cover the portion of the screen not covered by the object. A third source of direct glare is lights (especially point sources and particularly with dark backgrounds). Shield to at least 25° from the horizontal; 45° is better. For example, 24% of people were comfortable with a 21° shielded luminaire while 90% were comfortable with it shielded at 33%; for equal performance, calculations

FIGURE 21.7 Splayed recesses have an intermediate brightness between the window and the wall, which reduces luminance contrast (Konz, 1992).

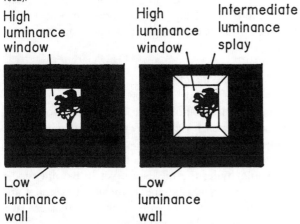

Source: S. Konz, "Vision at the Workplace: Part I—Guidelines for the Practitioner"; and "Vision at the Workplace: Part II—Knowledge Base for the Guide," *Int. J. of Industrial Ergonomics,* Vol. 10, pp. 139–160. Copyright © 1992 by Elsevier Science, Amsterdam, Netherlands. Reprinted with permission.

showed the 21° unit would require 28% more light on the task than the 33° unit (Illuminating Engineering Society, 1970). Have some light on the ceiling to reduce contrast. A fourth source is incandescent objects. Sliney and Wolbarsht (1980) describe in 1,000 pages the safety problems of lasers, bright lights, welding arcs, and hot metal. Use specialized glasses for this type of source. Ordinary sunglasses are dangerous because they transmit too much infrared and ultraviolet light; the sunglass reduces visible radiation and so the person has a false sense of security.

Indirect (reflected) glare Caused by high luminance from a surface, **indirect glare** is horizontally polarized light.

One approach is to decrease the incoming light so less is reflected. Lion et al. (1968) reported better inspection performance when line sources (fluorescent) were used instead of point sources (incandescent). In other words, increase the area of the incoming light so the lamp brightness is lower. Another possibility is to use multiple low-powered sources instead of a single high-powered source. Matsushita puts curtains on the inside of inspection booths to reduce glare off objects. Blackwell (1963) recommended filtering the light source with a multilayer polarizer to minimize the horizontal reflection.

If the reflected glare is specular or directional (as in a mirror), try to reorient the reflecting surface or to reposition the glare source so the glare misses the eye.

A third approach is to decrease reflectance through matte finishes. Car dashboards and satin

finish chrome and stainless steel are examples. Psychologically, people prefer matte blue, green, or brown over gray.

A fourth approach is to put the filter at the eye—sunglasses. Sunglasses reduce luminance but not contrast, so they reduce visual acuity. However, in most situations visual acuity is not critical as capability far exceeds task requirements; thus people trade off a reduction of surplus capacity for an increase in comfort. Polarized sunglasses, however, filter horizontally polarized light and improve both visual acuity and comfort (Mehan and Bennett, 1973).

A fifth possibility is to have people move their heads. However, to avoid the glare, operators may adopt a bad posture and thereby get neck, shoulder, or back pain.

2.4.2 Orientation
Orientation of lights to sharpen or blur the surface texture or form of an object is called modeling. For example, inspect bottles of liquid with light through the bottom. Back lighting may be useful for transparent materials. Low-angle lighting helps detect surface flaws. Figure 21.8 gives five different luminaire locations.

Orientation can be used to modify people's facial appearance. On the stage, light from below is used to make a person look evil. Lighting of a speaker is best with a light 20° horizontally on each side of the face; vertically they should be above 45° to reduce glare for the speaker but at 30° to improve audience impression (Golden, 1985). To emphasize the brightly lighted speaker, consider a dark background; fill light

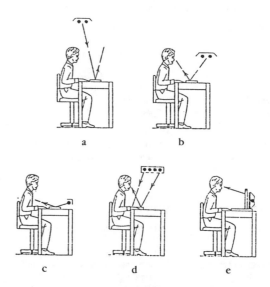

FIGURE 21.8 Placement of supplementary luminaires can use 5 techniques: (a) luminaire located to prevent reflected glare (reflected light does not coincide with angle of view); (b) reflected light coincides with angle of view (it is in the "offending zone"); (c) low-angle lighting to emphasize surface irregularities; (d) large-area surface source and pattern reflected toward the eye; and (e) transillumination from diffuse source.

(even from a light-colored podium or podium light) also can help.

2.4.3 Esthetics

Light not only facilitates visibility of a task; it also contributes in other ways to the visual quality **(esthetics)** of the space. Flynn (1977) and Flynn et al. (1979) did key studies showing that light also must be seen as a form of communication, evoking a perceptual response. Light communicates subjective impressions of the environment and also provides suggestions for behaviors.

Some research indicates which lighting results in various perceptions. See Table 21.9. Uniform supplementary wall lighting encourages perceptions of clarity. Nonuniform wall lighting, with its play of light and shadow, implies relaxation, privacy, and pleasantness.

Light also suggests behaviors. Orient user attention with a higher illuminance level than the surroundings (an extreme example is a spotlight on a stage; a subtle example is a display case in a store). Lighting also can suggest a circulation pattern because when following a path, people tend to follow the brighter path.

In general, esthetic lighting design should consider the design of shadows as well as the design of light (Gordon, 1987)—emphasizing asymmetry and variability. Without shade or darkness, light loses much of its meaning. A cloudy overcast day is bland; a bright sunny day has bright highlights and sharp shadows.

2.5 Task Requirements

The task can be improved by increasing size of objects, by increasing the contrast, and by increasing the time available for viewing. If the object is the "signal" and the environment is the "noise," design a task with a high signal/noise ratio. The goal is to enhance the object so it emerges from its environment like an island from the sea.

2.5.1 Increasing size

One possibility is to increase the size of the object itself. An example might be to use 20-point type instead of 9-point type on transparencies viewed by people at a meeting; another might be to use 10-point print in a book instead of 8-point print. (The text of this book is set in 10-point type; the tables are set in 8-point type.) Size can be increased mechanically. Examples are pantographs used in engraving or master/slave manipulators used in microscope work.

Another possibility is to bring the object closer to the eyes. This does present a conflict, however, because the hands usually are involved and holding objects close to the eyes becomes tiring. It is possible to support the arms with arm supports (see Figure 15.14). Another problem, however, is that the eye muscles (especially of older workers) become tired due to the efforts of accommodation and convergence.

An alternative is the use of **optical aids** (magnifiers). A 2X lens will double the size of the object while giving good depth of field. As magnification increases to 4X and more, depth of field becomes quite critical. A lens with a local lighting unit mounted on a swinging arm can be used in many tasks. When optics are used, visual clarity will decline due to light scattering in the lens and reduction of the visual field; however, the apparent size of the

TABLE 21.9 Lighting influences for perceptions of various spaces.

Perceived Illumination Has	Candidate Spaces	Reinforcing Lighting Modes
Perceptual clarity (clear, distinct vs. hazy, vague)	Offices, classrooms, laboratories industrial space	• High luminance on the horizontal plane in the central part of the space • Additional emphasis on the peripheral surfaces • Cool, continuous spectrum light sources
Spaciousness	Corridors	• Peripheral wall emphasis of a uniform nature
Relaxation	Lounges, some restaurants	• Nonuniform lighting with peripheral emphasis • Warm color tones of white light
Privacy/public	Private are intimate restaurants, reading areas Public are circulation areas, lobbies	For private: • Low intensity levels in the immediate user area with higher intensities further away • Nonuniform distribution patterns
Pleasantness	Lobbies, retail space	• Nonuniform lighting with a peripheral emphasis

object will increase. For good targets, young eyes, less than 2-h inspection, and over 1,000 lux on the target, magnification probably will help if the targets are less than 2.5 min of arc (Wei and Konz, 1978). If the target is to be magnified, the question is how much? Wei and Konz (1978) found 4X better than 2X magnification; increasing the target size (size after magnification) to the eye beyond 7 min of arc gave little benefit. Smith and Adams (1971), using microscopes, reported that time/correct inspection was minimum at 8 to 12 min of arc. A higher magnification, such as a microscope, has an additional problem in that it tends to put the operator in a very fixed position—a "straight jacket." Reduce static muscle fatigue of such workers by building working breaks into their job every 20 to 30 min. For example, when they finish with an assembly, have them carry it to a storage rack 10 m away or walk to get a supply of components. Magnification without the fixed posture is possible by using projection, such as with TV or optical comparators.

Another possibility is to put the magnification on the operator's head. The watchmaker's glass is an example which has the disadvantage of no biocular vision but the advantage of no accommodation strain. Another example is a surgeon's loupe which fits on glasses and gives 2.5X magnification at 18 inches. Alternatively, use magnifying lenses in spectacle frames. Usually they should be half-lenses (have the top omitted) so the operator can have normal distance vision. It also is possible to use glasses which cause the depth of the "stare point" to recede for close work (i.e., the object actually is .2 m from the eye but the eye through the lens thinks of it as .3 m). The advantage is minimized accommodation and convergence effort. Another example is VDT operators using prescriptions ground for the focal length of their task (typically about .5–.6 m) rather than a general-purpose prescription. The point is that glasses can be considered as an effort-saving device rather than just as a defective vision correction device.

Finally, the task also might be improved by reorienting the object since reorientation might improve visual size of a defect or improve contrast.

2.5.2 Increasing contrast

The art of camouflage consists of obliterating contrasts; we are concerned with anticamouflage. Improved **contrast**, which can be either in color or brightness, is especially helpful when the contrast is less than 30%. Attention to color contrast is especially useful. Seize any opportunity to modify target color. For example, color code file folders (Konz and Koe, 1969), tool handles, areas within a building, equipment, and worker clothing (to aid in identification of type of

worker). Use color in text and sign printing, photographs, and video displays. Some examples are "bluing" a piece of metal before scribing lines on it, bluing bearings before fitting, staining a tissue in a microscope slide, and painting a white line to differentiate the highway shoulder and pavement. A simple reflector under the task often makes viewing much easier. Therefore, use a light color for table tops. For reading materials, use new printer ribbons, clear photo-duplications, pens not pencils, black pens not blue, felt tip pens not ball points, white paper for computer printouts instead of green, and white paper instead of brown or manila. Spacing and grouping also is helpful. See the discussion of printed displays in Chapter 20. Fox (1977) reported that coin inspection was more accurate when presented in a standard array than in a random array—the good coins provided a background against which the defective coins stood out.

For more on contrast, see Section 4.2.3 of this chapter.

2.5.3 Increasing time

Shoot **sitting ducks.** That is, work on or inspect stationary items rather than moving items.

A less desirable alternative is working on an object moving past the operator at a constant velocity. Cochran, Purswell, and Hoag (1973) found that inspection performance was degraded at viewing times of .25 s but not at .5 s. In addition to time, consider angular velocity. Values between 10 and 30 degrees/s will not affect performance too much if the worker has satisfactory dynamic visual acuity (Ludvigh and Miller, 1959).

One solution technique is to have the operator face "upstream" to maximize viewing time. Removing visual obstructions is another. If the operator is viewing randomly spaced objects on a fixed-velocity conveyor, there is a large variability in available viewing time. Some objects might be available for viewing for .2 s while others are available for 2.0 s. If the operator must occasionally make a motor movement (e.g., to remove a defective item), the next 10 items may not be inspected at all. With this fixed-pace system, the conveyor speed must be set for the worst performance level of the slowest operator.

A serious problem with some machine-paced conveyors or index tables is that the operator must make a positive action to prevent the departure of a defective unit to the next station. If it is necessary to use machine pacing, have the operator make a positive action to send the units to the next station. However, it is better to use operator-paced stations so the operator may vary viewing time in relation to viewing requirements.

3   GENERAL LIGHTING

3.1   Uniform Ceiling Lighting

One possibility is to light an entire area uniformly. Advantages include maximum flexibility in arranging the machines and workstations in the area, which eliminates the need to move fixtures if the area is rearranged and allows use of large lamps (which have higher lumens/watt than small lamps). Nonuniform lighting (i.e., low general lighting supplemented with task lights) has advantages of lower cost, more precise control of the light, and more esthetic appeal.

For uniform lighting, the basic **zonal cavity method** equation is

$$I (A) = (N_1) (N_2) (L)$$

where

I = Illuminance in area, lux
A = Area illuminated, m²
N_1 = Number of fixtures
N_2 = Number of lamps/fixture
L = Lumens/lamp

For example, if you wished 750 lux evenly in a 2,500 m² area and were going to use fixtures with 1 lamp/fixture and each lamp was rated as 22,500 lumens, then N_1 = 83 fixtures needed.

However, the equation needs to be modified for three types of losses. The first loss, **coefficient of utilization** (CU), considers absorption of the light by the room surfaces. The second loss, **lamp lumen depreciation** (LLD), considers the loss of light output of a lamp with age. The third loss, **luminaire dirt depreciation** (LDD), considers the loss of light output due to dirt on the fixture.

The resulting equation is

$$I (A) = CU (LLD) (LDD) (N_1) (N_2) (L)$$

Detailed values for *CU, LLD,* and *LDD* along with N_2 and L for various fixtures are in the IES Handbook (1987) under zonal cavity method. Lamp and fixture manufacturers also can furnish values, generally with a computer program to solve the equation for any design you suggest.

As a rule of thumb, assume *CU (LLD) (LDD)* = .5. That is, the space would need 166 fixtures instead of 83.

The next step is to locate the fixtures in the space. For uniform lighting on the worksurface, the checkerboard pattern with the fixtures on the same color squares is reasonably efficient. However this does tend to overlight the center and underlight the room perimeter, so widen the spacing in the center and reduce it on the perimeter. End-to-end fixtures are an inefficient pattern.

For efficient uniform ceiling lighting, there are three guidelines:

1. *A distant light is dim.* The farther the light has to travel to the worksurface the less desirable it is. Thus, uniform ceiling lighting tends to be used with low ceilings.

2. *Reuse the light.* The light that travels directly from the fixture to the worksurface is called direct light. However, light which first bounces off a surface (indirect light) helps. Thus, light-colored ceilings, walls, floors, and furnishings give more indirect light.

3. *Use efficient fixtures.* Some fixtures trap as much as 30% of the light inside the fixture while others trap only 5–10%. In addition, some fixtures can direct the light better than others. Fixtures with some light going up (uplight) tend to be better than those with zero uplight because they permit air flow through the fixture (which keeps it cleaner); light on the ceiling gives better brightness contrast for the people in the room.

3.2   Energy Conservation

As was pointed out in Section 2.1, labor costs far exceed lighting costs. It is important not to reduce labor productivity dollars while saving a few pennies on lighting. However, excess energy costs can be reduced; payoffs often take less than two years. The energy savings is not only in reduced lighting costs but in lower air conditioning expense.

The first basic decision is on (1) high uniform ceiling lighting vs. (2) low uniform ceiling lighting supplemented with task lighting vs. (3) nonuniform ceiling lighting supplemented with task lighting.

The next basic decision is on lamps and fixtures. Replacing incandescent and mercury lighting with fluorescent or high-pressure sodium lamps generally pays off in less than two years. (The color disadvantage of HPS lamps can be overcome by mixing some metal halide lamps in with them.) Newer lamp designs tend to give more lumens/watt than the older designs. For example, new fluorescent lamp designs (vs. older designs) give more lumens/watt with (sometimes) a higher initial cost. It also is possible to use HPS or metal halide lamps in existing mercury fixtures. It isn't quite as efficient as in new fixtures but you don't have to replace the fixtures. Present-day fixtures tend to emit more of the light than fixtures of 15 to 20 years ago, so it may be worthwhile to replace old fluorescent or mercury fixtures.

The next possibility is to turn the lights off when there is no one in the area. Although this sounds obvious, it is difficult to implement. Ask any parent! Many years ago, it was cheaper to leave fluorescent lamps on when someone left the area since the energy cost was less than the cost of the reduced lamp life. With improved lamp designs and increased energy costs, this policy is no longer appropriate. Turn the lamps off when no one is there.

How to get the lights turned off is the problem. More switches help. That is, control 10 fixtures/switch instead of 100, which will permit people to turn off the lights in their areas without affecting people in other areas.

Automation is another alternative; that is, a machine turns off the lights. Sensors can detect body heat or motion in an area; if there is no heat or motion within a specified time period, the system turns off the lights. (A system may "flick" the lights 30 s before shutting them off to give any people it missed a chance to run to the wall switch.) You can replace the sensor with a timer for occasional-use areas such as toilets and storage areas. In this case, the lights go off 10 min (or some other time) after they are turned on.

Another possibility is to dim the lights instead of turning them off. This system is used primarily when lighting from the sun (daylighting) is supplementing artificial lighting. Note that the fixtures near the windows need to be on a separate circuit controlled by a light sensor.

4 SPECIAL LIGHTING

4.1 Lighting for VDT Areas
The following will discuss lighting for video display terminal (VDT) areas. For a discussion of workstation furniture for VDT areas, see Box 15.1.

4.1.1 Paper vs. screens
Until approximately 1980, the vast majority of tasks in offices involved working with paper. The paper tended to be relatively horizontal, and more light improved the ability to do the task. However, by 1990, there were over 40,000,000 VDTs in the United States. The screen is relatively vertical but more importantly, ambient light washes out the contrast on the screen. At present, we have not gone to the paperless office, so lighting must be designed for vertical screens, vertical paper (in document holders), and horizontal paper.

If the light is uniform throughout the area, the amount of light is often too much for the screen and too little for the paper. A nonuniform alternative is task lighting. A further consideration is that, because most VDTs are in offices, there is considerable interest in esthetically pleasing solutions rather than just functional solutions.

Of the two tasks (paper and screen), the screen task tends to be the most difficult as the electronic characters are not as sharp as print, the letter/background contrast is worse, and there is some flicker. Reading rates are about 25% slower on screens than print (Gould and Grischkowsky, 1986). Thus, if the lighting will be uniform, design for the screen since it is the more difficult task.

Light below 100 lux enhances screen legibility; light above 500 lux enhances paper legibility. For uniform lighting, the Human Factors Society (1988) recommends 200–500 lux. The IES recommends that the light level for paper tasks should not exceed 750 lux if there is uniform lighting.

If task lighting is used, then there can be relatively high illumination on the document (see Table 21.6) and relatively low illumination on the screen—assuming the task light is properly directional. The general lighting should not be too low. The IES recommends 200–300 lux—assuming proper shielding of glare sources. Yearout and Konz (1987b) reported that when VDT operators had general lighting of 350 lux supplemented by task lighting, the operators wanted their general view of the rest of the office to be much brighter (770 lux) and to be visually interesting. To solve this challenge, use light to "wash" a wall, use illuminated artwork, and so forth. For esthetic reasons, therefore, avoid "cavelike" offices for VDTs.

Operators prefer the general lighting to be a mixture of indirect (uplight) and direct (downlight) rather than all downlight. Hedge (1991) reported, from a long-term field study of office workers, that both direct-parabolic and lensed-indirect lighting were preferred over recessed fluorescent luminaries with prismatic diffusers; the lensed indirect was preferred over the direct parabolic. The direct parabolic had recessed luminaries shielded with a grid of 18 × 10 cm parabolic louvers and provided 750 lux to the workplace. The lensed indirect had ceiling-suspended (pendant) luminaries that provided upward light which then was reflected down to the workplane. It provided 500 lux to the workplace.

VDTs occasionally are used in a paperless office (i.e., computer display only) environment. In such an environment with low ambient light levels (e.g., 10–30 lux), use indirect lighting (Krois et al., 1991).

4.1.2 Reflections
Reflections on the screen can be diffuse or specular.

Diffuse (veiling) **reflections** are caused by ambient light as well as the phosphor. Diffuse reflections increase the luminance of both the screen background and characters, thereby reducing the contrast ratio. It is possible to improve the contrast with some types of screen filters. Their general characteristic is that the ambient light goes through the filter twice (in and back out), whereas the light from the characters goes though the filters once (out). Thus, although the luminance is reduced for everything, the luminance is reduced less for the characters, so contrast is improved. If the filter is neutral density (gray), it reduces all energy of all wavelengths equally. A color filter (the color of the phosphor) will pass most of the phosphor color but will not pass much of the ambient light (which typically is white); thus, contact is enhanced. **Specular reflection,** a mirrorlike image, also occurs. Coatings such as quarter-wavelength filters (like the coating on camera lenses) and matte treatment (frosting) can reduce specular reflection.

In addition, glare can be reduced by reducing bright objects in the environment. Detect these bright objects with the **mirror test**—holding a mirror in front of the screen and looking at it from the operator's position.

One potential glare source is ceiling fixtures used for direct lighting, especially in larger offices. (In smaller offices, the angle of the eye to the screen and the light from the fixture to the screen do not match.) The reflected luminance of the fixture seen on the screen can be reduced by applying diffusers, prismatic lenses, polarizers, egg-crate louvers, and parabolic louvers to the fixtures.

Another common brightness source is windows. Solutions include reorienting the screen by tilting it vertically or reorienting the screen so there is no window behind the operator. If windows are present in a room with VDTs, put them at the operator's side. Windows placed in front of the screen cause problems due to the eye adjustment from the bright window to dim screen. Windows also can be covered with films, blinds, shades, or dark curtains.

Other brightness sources must be considered. For example, replace white shirts with dark shirts, light-colored table tops with dark table tops, light walls and vertical surfaces with darker walls and surfaces. Also, move bright objects from shelves behind the operator.

Another alternative (although not commonly used) is a **negative contrast** screen—dark characters on a light background. Specular reflections are less noticeable on the light background. Negative contrast screens may be more expensive, software may have to be modified, flicker may be more apparent, and negative contrast depletes the phosphor (IES, 1989). In some word-processing programs (e.g., WordPerfect), the user has an option of positive or negative characters and a choice of colors. For example, the user can have blue characters on a tint blue background with red highlighting or white characters on a blue background with yellow high-

lighting. Some graphics cards permit additional enhancements such as bold face, italics, and different font sizes.

See Table 21.10 for a summary of measures to reduce screen reflections.

4.1.3 Luminance ratios

For a **luminance ratio** (characters to screen background), the Human Factors Society (1988) recommends characters be at least 7 times brighter than the screen background. The IES (1989) says the ratio (for both positive and negative displays) should be between 5:1 and 10:1.

Does the eye have a problem looking at a bright surface (such as a document) and then at a dim surface (such as a screen)? The HFS (1988) says this is not a problem since the brightness within a workstation can vary as much as 20 to 1 with no problem.

For additional comments on characters, see Chapter 20 on displays and Section 2.5 of this chapter on how to improve the task. See Section 2.2 of this chapter on vision problems of individuals.

4.2 Inspection

This section will consider inspection lighting in three areas: amount of light, color of light, and contrast. See Chapter 12 for error reduction in general and Box 12.2 for comments on inspection.

4.2.1 Amount of light

The amount of light on the inspection task may be insufficient. Inspection areas often are placed in an area with general lighting, which is sufficient for walking around and general seeing but not for inspection. (In one of my consultation jobs, an inspector took items off a conveyor over which a fluorescent lamp was suspended. The lamp was moved 2 ft toward the inspector so that it lighted where the object was *inspected,* not where the object was *grasped.*) See Table 21.5 for illuminance recommendations.

4.2.2 Color of light

Perceived object color is affected by color of the light. Color differences on red surfaces are emphasized by sources strong in blue light, and on blue surfaces by sources strong in red (Misra and Bennett, 1981). When designing areas for inspecting color, specify whether the lamp to be used is incandescent, cool-white fluorescent, HPS, or whatever, since colors appear quite different under different lamps.

4.2.3 Contrast

The object may be obscured by reflected glare. Figure 21.9 shows how lamps are tilted so the reflection is not into the inspector's eyes.

For color contrast, the inspection can be for the object shape or the object surface characteristics.

TABLE 21.10   Measurements for reducing screen reflections. The optimal combination of measures depends on the specific office, office layout, office equipment, and VDT (Helander, 1987).

MEASURE	ADVANTAGE	DISADVANTAGE
At the source		
Cover windows		
Dark film	Reduces veiling and specular reflections	Difficult to see out
Louvers or miniblinds	Excludes direct sunlight, reduces veiling and specular reflections	Must be readjusted in order to see out
Curtains	Reduces veiling and specular reflections	Difficult to see out
Lighting control		
Control of location and direction of illumination	Reduces veiling reflections, may eliminate specular reflections	None
Indirect lighting	Reduces specular reflections, economy of office space by moving workstations closer	None
Task illumination	Reduces veiling reflection, increases visibility of source document	None
At the work station		
Move workstation	Reduces veiling and specular reflection	None
Tiltable screen	Reduces specular reflection	Readjustment necessary
Tilted screen filter	Eliminates specular reflection	Bulky arrangement for large screens
Screen filters and treatments		
Neutral density (gray) filter	Reduces veiling reflection, increases character contrast and visibility	Less character luminance
Color filter (same color as phosphor)	Reduces veiling reflection, increases character contrast and visibility	Less character luminance
Micromesh, microlouver	Reduces veiling reflection, increases contrast	Limited angle of visibility, nonembedded filters get dirty
Polaroid filter	Reduces veiling reflection, increases contrast and visibility	Less character luminance
Quarter wavelength anti-reflection coating	Eliminates specular reflection	Expensive, difficult to maintain
Matte (frosted) finish of screen surface	Decreases specular reflection	Increases character edge spread (fuzziness, increases veiling reflection)
CRT screen hold	Reduces veiling and specular reflection	Difficult to avoid shadows on screen
Sunglasses (gray, brown)	None—contrast unchanged	Less character luminance and visibility
Reversed video	Reduces specular reflection	Increased flicker sensitivity
Screening of luminaries and windows	Reduces specular reflection	Might create isolated workplaces

Source: M. Helander, "Design of Visual Displays," in *Handbook of Human Factors*, G. Salvendy (Ed.). Copyright © 1987 by John Wiley & Sons. Reprinted by permission of John Wiley & Sons, Inc.

To detect shape, maximize the contrast of the task and the background. For example, if buttons are inspected for holes, the table color should contrast with the button color. Another technique is a mask with the specified item shape; transillumination allows a thin border of light to show for a part in tolerance. When you are looking for shape, object orientation is critical. To demonstrate this, try to recognize a print of a face turned upside down. For printed material, letters and background should have maximum contrast. Avoid "arty" low-contrast letters (black on red, red on brown, etc.). They are not only less legible but less attractive (Konz et al., 1972).

To detect surface characteristics such as color and texture, minimize the contrast of the task against the background. Pearls sold at retail are displayed on black velvet; the maximum contrast makes it difficult to detect color differences between pearls. When pearl merchants buy from each other, they display the pearls on white cloth to maximize color differences. Thus if you are sorting green beans for color, sort on a table painted the green of a good bean.

Orientation of the lights may be important, both to reduce glare and to emphasize features. See Sections 2.4.1 and 2.4.2. It is good practice to have both the brightness and angle of incidence of inspection lighting adjustable by the inspector so performances can be maximized. See Cushman and Crist (1987) for a good discussion of inspection lighting.

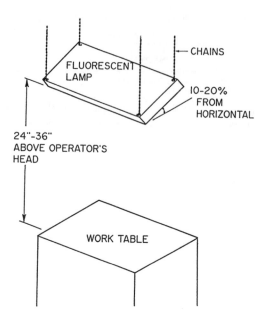

FIGURE 21.9   Tilt lamps 10–20° so the reflection is not into the inspector's eyes. Tilting also reduces directional noise coming from the workstation. An alternative is to tilt the worksurface.

4.3 Warehouse Aisle Lighting

In a warehouse, items are stored on racks with long, narrow access aisles. The visual requirement is on the vertical plane (box sides) rather than the horizontal plane. If the stacking height is less than aisle width, treat the area as an open bay rather than a warehouse aisle.

The use of high racks makes high-intensity discharge (HID) lamps preferred over fluorescent lighting, since HID luminaries have better directional characteristics. Different directional patterns are available, such as medium symmetrical, wide symmetrical, and directional asymmetrical (4 lobes, 2 up and down the aisle and 2 into racks). For luminaire mounting heights below 16 ft, type of HID luminaire makes relatively little difference. However, at low mounting heights, a high-watt point source may be too bright. A high-reflectance aisle floor helps. Frier and Frier (1980a) recommend mounting the luminaires above the aisle instead of above the racks. Use a maximum distance above the rack top of .5 (aisle width). Otherwise, too much light is wasted on the rack tops. A large number of low-watt sources will give a more even distribution than a small number of high-watt sources. If the aisle or rack locations are not permanent, mount the fixtures on tracks perpendicular to the aisles and make the wires plug-in instead of permanent. To save energy, have a small number of lamps per switch so only the area needs to be lighted.

4.4 Emergency Lighting

Emergency lighting is designed to help people leave the building if the normal power supply fails. Although local codes may also apply, the Life Safety Code specifies 10 lux on the floor along the route people might take leaving the building (Life Safety Code, 1988). Frier and Frier (1980b) recommend 30 lux for congested or critical areas such as corridor intersections, top of stairs, dangerous machinery, and so forth. The vertical surface around the exit door should have 20 to 50 lux. See Table 21.11.

The emergency lighting can be either completely separate lights with batteries or the normal lights with a backup power source. Separate lights tend to be projector and reflector (PAR) lamps, a battery, and a device to turn the light on when the power fails. Use PAR with wide horizontal patterns for open areas and with long narrow beams for corridors and stairs. The normal light can have a local battery (e.g., units in AC fluorescent fixtures) or a central battery. There also can be a motor generator. Note that HID lamps need 1 to 15 min restrike time (see Table 21.8) and another 3 to 5 min before coming to full intensity; thus, you may need some fluorescent lamps in HID areas. Another option is a quartz lamp (about the size of a lipstick case) in some HID fixtures. After a momentary power interruption, the quartz lamp automatically lights until the HID lamp cools down, restrikes, and regains 60% of its full light output.

Exit signs can be self-powered but, in much of the country, codes permit the exit sign to be illuminated externally by the emergency lighting, which should be less expensive than having both emergency lights and internally powered exit signs.

Note that the restrike problem of some lamps may make them unsatisfactory for emergencies, even with a backup power system. Some solutions include emergency fluorescents, a quartz lamp in the HID fixture, and dual-filament HID lamps (one filament is on "standby").

4.5 Security Lighting

The purpose of security lighting is crime prevention. The light should discourage intruders (reduce "offense") and improve detectability when entry is effected (improve "defense").

TABLE 21.11   Minimum illuminance (lux) levels for safety (IES, 1983). These are the absolute minimums at any time and at any location on any plane where safety is related to seeing conditions.

Hazards Requiring Visual Detection	Normal Activity Level	
	Low	High
Slight	5	11
High	22	54

If there is a guard and/or TV camera on-site, consider them as the "audience" and the intruder as the "actor." That is, the audience should have a good view of the actor; conversely, the actor should not be able to see the audience. The defender (the guard) should not be visible to the intruder. A guard station should be lit so it is not obvious whether it is occupied. It should be relatively dim within the station, relatively bright outside. Design the stage lighting to maximize glare for the intruder and minimize it for guards, cameras, and neighbors in the plant vicinity. If there is no on-site defender and you depend upon outside police observers, turn the lighting system around and light the building exterior.

Table 21.12 gives recommended security lighting illuminance (Baker and Lyons, 1978). High-pressure sodium (HPS) is the recommended source due to its low cost and its compatibility with most TV cameras. (Generally, low-light cameras are used except in well-lighted areas.)

Garages and parking areas need about 10 lux. For garages, place the fixtures to illuminate the between-car walk spaces rather than the area over the car stalls.

TABLE 21.12 Recommended area illuminances (lux) for security lighting, measured on the horizontal plane (Baker and Lyons, 1978).

	District Brightness		
Risk	High (Adjacent Main Road Lighting; Lighted Adjacent Land; Floodlighting)	Medium (Adjacent Secondary Road Lighting)	Low (No Adjacent Lighting on Adjoining Property or Nearby Roads)
Extreme	20-30	10-20	5-15
High	10-20	5-15	2-10
Moderate	5-15	2-10	1-5

REVIEW QUESTIONS

1. Sketch an eye, labeling the cornea, aqueous humor, lens, pupil, iris, vitreous humor, and retina.

2. At what wavelength (and color) are the cones most sensitive?

3. Define visual acuity with a formula.

4. Does insufficient light permanently damage the eye?

5. What is astigmatism? What is accommodation?

6. Why is it dangerous to wear sunglasses in welding areas?

7. Do males or females tend to have more color weakness?

8. What is a lumen/square m? A lumen/square ft?

9. What are the three attributes of color?

10. What is a typical ratio, in an office, of labor cost to illumination cost? Give assumptions.

11. What illumination level would you recommend for the classroom in which the course is taught? Indicate how you got the number.

12. Rank incandescent, mercury, fluorescent, metal halide, and high-pressure sodium lamps for lumens/watt, life, and restrike time.

13. What is the fluorescent color which is the same as incandescent?

14. Is reflected glare horizontally or vertically polarized light?

15. Give three different ways (including examples) of how to increase visual size.

16. Give three examples of how glasses can be used as an effort-saving device.

17. Give three examples of anticamouflage.

18. Using the example of pearls, when inspecting for surface characteristics such as color, should the contrast of the object and background be high or low?

19. Discuss the advantages/disadvantages of uniform lighting vs. nonuniform lighting.

20. Give the three guidelines for uniform ceiling lighting.

21. Give three different ways of increasing the chances of lights being turned off.

22. Briefly discuss why filters on VDTs improve legibility.

23. Briefly describe the mirror test.

24. Briefly discuss ambient lighting levels in a VDT area.

25. Discuss search for items on a conveyor belt.

26. Discuss warehouse aisle lighting.

27. What lux level do you recommend for emergency lighting?

28. Discuss security lighting in terms of offense and defense.

REFERENCES

Baker, J. and Lyons, S. Lighting for the security of premises. *Lighting Research and Technology,* Vol. 10 [1], 10-18, 1978.

Bennett, C. The demographic variables of discomfort glare. *Lighting Design and Application,* Vol. 7 [1], 22-25, 1977.

Blackwell, H. Visual benefits of polarized light. *American Institute of Architects Journal,* 87-92, November 1963.

Boyce, P. *Human Factors in Lighting.* New York: MacMillan, 1981.

Cochran, D., Purswell, J., and Hoag, L. Development of a prediction model for dynamic visual inspection tasks. *Proceedings of the 17th Annual Meeting of the Human Factors Society,* 31-43, 1973.

Cushman, W. and Crist, B. Illumination. In *Handbook of Human Factors,* Salvendy, G. (ed.), Chapter 6.3. New York: Wiley & Sons, 1987.

Duncan, J. and Konz, S. Legibility of LED and liquid-crystal displays. *Proceedings of the Society for Information Display,* Vol. 17 [4], 180-96, 1976.

Ferguson, D., Major, G., and Keldoulis, T. Vision at work. *Applied Ergonomics,* Vol. 5 [2], 84-93, 1974.

Flynn, J. A study of subjective responses to low energy and nonuniform lighting systems. *Lighting Design & Application,* Vol. 7 [2], 6-15, February 1977.

Flynn, J., Hendrick, C., Spencer, T., and Martyniuk, O. A guide to methodology procedures for measuring subjective impressions in lighting. *Journal of the Illuminating Engineering Society,* Vol. 8 [2], 95-120, 1979.

Fortuin, G. Lighting: Physiological and psychological aspects—Optimum use—Specific industrial problems. In *Ergonomics and Physical Factors.* Geneva: International Labour Office, 1970; pp. 237-59.

Fox, J. Quality control of coins. In *Human Factors in Work, Design and Production,* Weiner, J. and Maule, H. (eds.). London: Taylor and Francis, 1977.

Frier, J. and Frier, M. Design techniques for long, narrow areas. In *Industrial Lighting Systems,* Chapter 7. New York: McGraw-Hill, 1980a.

Frier, J. and Frier, M. Emergency lighting systems. In *Industrial Lighting Systems,* Chapter 14. New York: McGraw-Hill, 1980b.

Golden, P. The effects of lighting a public speaker upon observer impression. *Lighting Design + Application,* Vol. 15 [12], 37-43, December 1985.

Gordon, G. The design department. *Architectural Lighting,* Vol. 1 [1], 53-54, January 1987.

Gould, J. and Grischkowsky, N. Does visual angle of a line of characters affect reading speed? *Human Factors,* Vol. 28 [2], 165-73, 1986.

Grandjean, E. *Fitting the Task to the Man.* London: Taylor and Francis, 1969; p. 94.

Hedge, A. The effects of direct and indirect office lighting on VDT workers. *Proceedings of the Human Factors and Ergonomics Society,* 536-40, 1991.

Helander, M. Design of visual displays. In *Handbook of Human Factors,* Salvendy, G. (ed.), Chapter 5.1. New York: Wiley & Sons, 1987.

Hess, E. The role of pupil size in communication. *Scientific American,* Vol. 233, 110-15, November 1975.

Human Factors Society. *American National Standard for Human Factor Engineering of Visual Display Terminal Workstations.* Santa Monica, Calif., 1988.

IES. *IES Lighting Handbook: Application Volume.* New York, 1987.

IES. *The IES Recommended Practice for Lighting Offices Containing Computer Visual Display Terminals.* New York: Illuminating Engineering Society of America, 1989.

IES Industrial Lighting Committee. Proposed American national standard practice for industrial lighting. *Lighting Design and Application,* Vol. 13 [7], 29-68, July 1983.

Illuminating Engineering Society. *Industrial Lighting,* American National Standard Practice for Industrial Lighting, RP-7. New York, 1970.

Konz, S. *Facility Design: Manufacturing Engineering.* Scottsdale, Ariz.: Publishing Horizons, 1994.

Konz, S. Vision at the workplace: Part I—Guidelines for the practitioner, and Part II—Knowledge base for the guide. *Int. J. of Industrial Ergonomics,* Vol. 10, 139-60, 1992.

Konz, S. and Koe, B. The effect of color coding on performance of an alphabetic filing task. *Human Factors,* Vol. 11 [3], 207-12, 1969.

Konz, S., Chawla, S., Sathaye, S., and Shah, P. Attractiveness and legibility of various colors when printed on cardboard. *Ergonomics,* Vol. 15 [2], 189-94, 1972.

Krois, P., Lenorovitz, D., McKeon, P., Snyder, C., and Tobey, W. Air traffic control facility lighting. *Proceedings of the Human Factors Society,* 551-55, 1991.

Life Safety Code, An American National Standard. National Fire Protection Association, 1988.

Lin, A. and Bennett, C. Lamps for lighting people. *Lighting Design & Application,* Vol. 13 [2], 42-44, February 1983.

Lion, J., Richardson, E., and Browns, R. A study of industrial inspectors under two kinds of lighting. *Ergonomics,* Vol. 11 [1], 23-24, 1968.

Long, G. and Rourke, D. Training effects on the resolution of moving targets. *Human Factors,* Vol. 31 [4], 443-51, 1989.

Ludvigh, E. and Miller, J. Study of visual acuity during the ocular pursuit of moving test objects. *J. of the*

Optical Society of America, Vol. 48 [11], 799-802, 1959.

Mehan, R. and Bennett, C. Sunglasses—Performance and comfort. *Proceedings of the 17th Annual Meeting of the Human Factors Society,* 174-77, 1973.

Misra, S. and Bennett, C. Lighting for a visual inspection task. *Proceedings of the Human Factors Society,* 631-33, 1981.

Parker, J. and West, V. *Bioastronautics Book.* Washington, D.C.: Supt. of Documents, 1973.

Rushton, W. Visual pigments and color blindness. *Scientific American,* 64-74, March 1975.

Sliney, D. and Wolbarsht, M. *Safety with Lasers and Other Optical Sources.* New York: Plenum, 1980.

Smith, G. and Adams, S. Magnification and microminiature inspection. *Human Factors,* Vol. 13 [3], 247-54, 1971.

Wei, W. and Konz, S. The effect of lighting and low power magnification on inspection performance. *Proceedings of the Human Factors Society* 196-99, 1978.

Yearout, R. and Konz, S. Illumination levels in offices with visual display units. *Proceedings of the Human Factors Society,* 1113-15, 1987a.

Yearout, R. and Konz, S. Task lighting for visual display unit workstations. *Proceedings of IX Int. Production Engineering Conf.,* 1862-66, 1987b.

CHAPTER

22 | EAR AND NOISE

OVERVIEW

Although the ear is well designed, noise causes problems. The noise scale, decibels, confuses the general public because a doubling of the noise level gives an increase of only 3 dB.

Noise control at lower levels (55 to 80 dBA) primarily is to eliminate annoyance; noise control at levels of 90 dBA and up is to protect hearing. Under most circumstances, noise does not affect productivity.

Noise reduction is relatively inexpensive if you plan ahead. It calls for some ingenuity and expense to modify the noise source; it requires considerable ingenuity and greater expense to modify the sound wave. Use of personal protective equipment requires day-after-day selling, motivation, and supervision.

CHAPTER CONTENTS

1 The Ear
2 Noise
3 Effects of Noise
4 Noise Reduction

KEY CONCEPTS

acoustic glare
audiograms
confine/absorb
dBA
decibel
dosimeters
earmuffs/earplugs
equal energy/equal pressure
free field
frequency

hearing impairment
impulse sound
inverse square law
masking noise
octave band
outer/middle/inner ear
permanent threshold shift (PTS)
power watt level (PWL)
pure tone
resonance

sones
sound pressure level (SPL)
speech interference
speech interference level (SIL)
temporary threshold shift (TTS)
turbulence
vestibular system
white noise

1   THE EAR

1.1   Anatomy of the Ear

Figure 22.1 gives an overview of the ear, Figure 22.2 a detail of the cochlea, and Figure 22.3 a detail of the organ of Corti.

The outer part of the ear serves as a collector of sound vibrations in the air and funnels them to the eardrum. The eardrum is extraordinarily sensitive and will move in response to changes of as little as .000 02 N-m² and then will move only .000 000 001 cm! (And .000 000 001 cm = 1/2 the diameter of a hydrogen molecule.)

In the middle part of the ear, the vibration of the eardrum (tympanic membrane) is transmitted to the oval window through three small bones (ossicles) known as the hammer (malleus), anvil (incus), and stirrup (stapes).

The hammer "handle" is connected to the eardrum and the head to the top of the anvil. The base of the anvil is connected to the top of the stirrup. The baseplate of the stirrup moves the oval window leading to the inner ear. The ligament of the stapes also dampens loud impulse noises and thus protects the inner ear. This acoustic reflex occurs after a short delay (.1 s), so it is not much protection

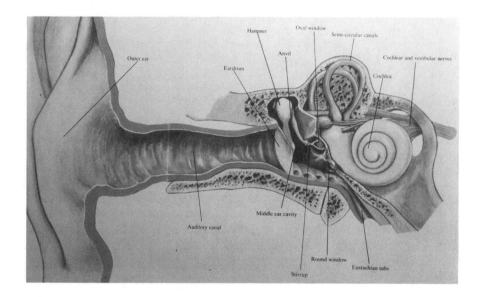

FIGURE 22.1   Three parts of the ear are the outer, middle, and inner. The **outer ear** transfers vibrations in air to vibrations of the eardrum, the **middle ear** amplifies and transmits the vibrations to the oval window, and the **inner ear** transmits vibrations in the basilar membrane into electrical pulses in the auditory nerve. Figure courtesy of Bilsom.

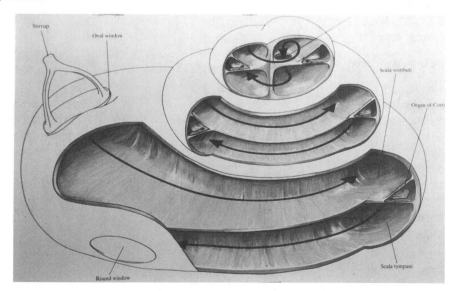

FIGURE 22.2   Vibrations of the oval window are transmitted by fluid to become vibrations in the basilar membrane which, in turn, affects the hair cells in the organ of Corti. Figure courtesy of Bilsom.

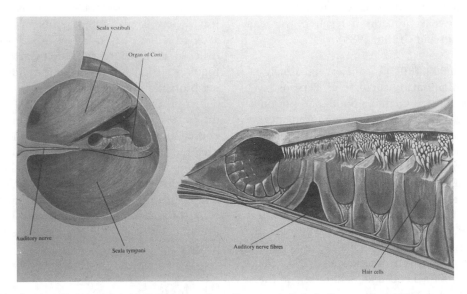

FIGURE 22.3   Hair cells, when vibrated, send electrical signals along the auditory nerve to the brain. Figure courtesy of Bilsom.

against one sudden noise impulse but helps against steady loud impulsive noise (e.g., punch press area). The original vibration in air has now been transferred to a vibration of a second membrane. The signal is magnified, as the eardrum area is 14 times larger than the oval window area (gain 23 dB) and by a 1.3 lever arm ratio for the 3 bones (gain 2.5 dB) for a total gain of 25.5 dB. The eardrum will be most efficient if the air pressure is the same on both sides, so the Eustachian tube comes in handy. One end is in the middle ear and the other end in the top of the mouth. If this becomes plugged and pressure is changing rapidly (jet descent or elevator descent), hold your nose, close your mouth, and "blow" gently.

The inner ear is the most interesting of all. The vibrations of the oval window set up vibrations in the fluids of the cochlear duct. The duct upper passage (scala vestibuli) starts at the oval window; the lower passage (scala tympani) ends at the round window. Connecting the two is a small gap (helicotrema). The round window bulges out when the oval window bulges in, and so the vibrations are not dampened. The basilar membrane divides the two passages. Now the noise has become a wave traveling in the fluid of the inner ear.

The basilar membrane vibrates with the frequency of the fluid vibration. The basilar membrane is narrow and stiff near the oval window and wide near the gap. High-pitched sounds travel only a short distance along the membrane before they die out, so the end near the oval window detects high-frequency sounds; low-pitched sounds are detected near the gap.

On top of the basilar membrane is the organ of Corti, which has about 30,000 hair cells. As the wave goes through the tympanic chamber it deflects the vestibular membrane down into the cochlear duct. This in turn deflects the basilar membrane down into the tympanic chamber. This pulls down the hair cells in the organ of Corti. The hair cells are supported at both ends, so the pulling makes the hair cells send out an electrical pulse which goes to the brain by way of the cochlear nerve. The brain then decides that the muffler in your car either has holes in it or has a pleasant throb. The brain also is able to detect the very slight difference in time that it takes for sound to travel to each ear and so identify whether the sound came from the left or right—auditory localization. This can be demonstrated by placing yourself between two speakers; by adjusting the volume of one speaker you can make the apparent sound location move.

Conductive hearing loss occurs in the outer or middle ear from wax, punctured eardrums, and corrosion of the bones, among other causes. Often it is possible to cure with medical or surgical treatment. Nerve loss in the inner ear is rarely curable. Nerve loss can be caused by old age, viruses, drugs, and noise.

In addition to hearing, the ear provides a sense of balance through the **vestibular system,** which is subdivided into the utricle and the semicircular canals. The utricle has "pebbles" (otoliths) resting on a bed of hairs (cilia); the pressure tells you which way gravity is operating. The three semicircular canals (one for each axis) are mutually perpendicular; they sense acceleration through movement of liquid in the canals. Seasickness is primarily a result of vertical vibration (Lawther and Griffin, 1986).

1.2 Hearing Measurement

Hearing usually is measured on an automatic audiometer. Persons having their hearing tested go into a booth and put on a set of earphones. Then the audiometer automatically goes through a test cycle for the left ear and the right ear and plots the results as in Figure 22.4. During the test cycle the machine will present a tone at 250 Hz that increases in loudness until the person indicates sound perception by pushing a button. It then decreases until the person releases the button. After 3 or 4 tests for a specific frequency, it will index to the next higher frequency and repeat the tests until it has completed all seven frequencies (250, 500, 1,000, 2,000, 4,000, 6,000 and 8,000 Hz); it then repeats the test for the other ear.

Hearing loss from noise can be temporary (recover overnight) or permanent. **Temporary threshold shift (TTS)** is measured two minutes after the end of exposure. With repeated exposure, TTS becomes **permanent threshold shift (PTS)** or noise-induced permanent threshold shift (NIPTS). Think of hair cells as grass that is being walked on. TTS is walking on the grass occasionally; the grass springs back up. PTS is walking on the grass so much that a path is worn and the grass killed.

Audiograms should be performed annually; if the audiogram shows a significant shift, it should be repeated within 30 days. If it still shows a shift, medical review is warranted. Problems of TTS can be avoided if the person has at least 14 h of no noise over 80 dbA before the audiogram. Typically, this requires testing at the start of the shift or use of hearing protection devices (HPD).

From the output in Figure 22.4, it appears that the tested person had a neurosensory loss, not entirely due to noise. Occupational hearing loss damages the inner hair cells first (3,000–6,000 Hz) and later the farther hair cells. It almost always occurs on both sides equally and develops gradually. That is, not all hearing loss is occupationally related, and a series of audiograms over the years will help establish whether or not the occupation is responsible for the loss.

2   NOISE

2.1 Noise Definitions

Absolute levels of sound pressure detectable by the human ear vary by 1,000,000,000,000 to 1. This scale is too large to use conveniently. The scale was compressed by using a log ratio (originally called the Bell after Alexander Graham Bell, but since a smaller measure was needed, it was divided by 10 and called the deciBel or **decibel**). There are two relationships: sound pressure and sound power. The relation between sound pressure and sound power is analogous to temperature and heat.

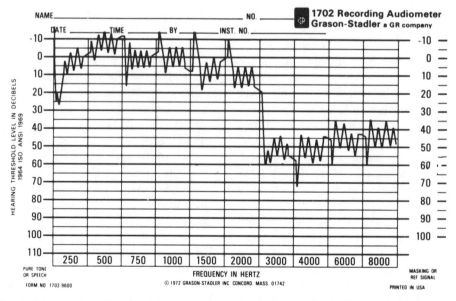

FIGURE 22.4 Audiograms show hearing ability. Plotted at each of 10 bands is the right ear of a 47-year-old-male with 7,700 hours of flying time. Zero dB indicates normal hearing. Hearing loss due to noise usually starts with a small dip of 20 to 30 dB at 4,000 Hz before progressing to the severe loss of this pilot. Recreational noise causes problems also. A target shooter often has a 40 to 80 dB loss at 4,000 Hz. Since nonvoiced consonants (p, t, k, s) have frequencies of 2,000–8,000 Hz, they are the first to be lost. Next lost are the voiced consonants (b, d, g, z) and finally the vowels.

$$SPL, \text{dB} = 10 \log_{10} \left(\frac{P}{P_0}\right)^2 = 20 \log_{10} \left(\frac{P}{P_0}\right) \quad (22.1)$$

where SPL = **Sound pressure level,** decibel

P = Sound pressure level of the noise, N-m² (the 2 reflects the effect of area of the eardrum)

When first measured, 0 dB was the minimum level of hearing. With more refined measurements, the mean minimum of the population with unimpaired hearing is now given as 4 dB.

P_0 = Reference sound pressure level

= .000 020 N-m² = 0 dB = 20 μ N-m²

= .000 200 dynes/cm² = .000 200 μ bar (approximately the minimum level of hearing of a young person)

The power level of a noise is:

$$PWL, \text{dB} = 10 \log_{10}\left(\frac{W}{W_0}\right) \quad (22.2)$$

where PWL = **Power watt level** of the noise, decibel

W = Acoustic power of the noise, watts

W_0 = Reference power level

= 1×10^{-12} watt

Since $1 \times 10^{-12} = -120$ dB, equation (22.2) can be expressed as:

$$PWL = 10 \log W + 120$$

That is, 120 dB of power corresponds to 1 W, 110 to .1 W, 100 to .01 W, and so on. The decibel scale, although giving a small range, confuses the public since it is not linear; it also requires "new math." Most confusing is that 100 dB + 100 dB is not equal to 200 dB! When combining or subtracting noises with the formulas, we use the power formulas, not the pressure formula. Consider a machine generating noise with a PWL = .01 W (i.e., 100 dB) and with an SPL of 1 N-m² (94 dB) at a location 5 m from the machine. Now add another identical machine.

$$\begin{aligned} PWL &= 10 \log .020 + 120 \\ &= 10 \log 2 \times .010 + 120 \\ &= 10 \log 2 \times 10^{-2} + 120 \\ &= 10 (-2 + .3) + 120 \\ &= -17 + 120 = 103 \text{ dB} \end{aligned}$$

Thus adding 100 dB of power to 100 dB of power gives 103 dB. The corresponding SPLs are 94 + 94 = 97.

In addition to 94 + 94 = 97, we also have 74 + 80 = 81, and 80 + 95 = 95! Ten identical sound sources have an SPL 10 dB louder than just one source! In other words, if we have two noisy machines, each generating at 80 dB, and we completely silence one machine, the noise level drops from 83 to "only" 80. Rather than use the formulas, use Figure 22.5, a quick graphical technique.

It will be emphasized that the use of Figure 22.5 is the theoretical addition in a **free field.** In practice most fields are not free. Outdoors, the wind can have a substantial effect, as well as reflections from reflectors such as brick walls. The noise might even be attenuated slightly by passing through intervening shrubbery. Indoors, there are many reflecting surfaces as well as solid paths to conduct the noise; occasionally some of the indirect noise is reduced by absorbent materials on the floor, wall, or ceiling.

$$\text{Wavelength} = \frac{\text{Speed of sound}}{\text{Frequency}} \quad (22.3)$$

The speed of sound is 770 miles/h, 1,238 km/h and 344 m/s. **Frequency** (the rate of oscillation of the sound) is in Hertz (cycles/second). Thus, a low-frequency sound (20 Hz) has a wavelength of 344/20 = 17 m. A high-frequency sound (20,000 Hz) has a wavelength of 344/20,000 = .017 m = 1.7 cm. A low-frequency sound (long wavelength) easily travels around corners and through openings. A high-frequency sound (short wavelength) behaves like light. It does not turn corners well and it can be reflected. High-frequency noise is more attenuated by distance traveled in air than low-frequency noise. High-frequency noise is more annoying.

A **pure tone** is a one-frequency sound. Most industrial noise is a mixture of frequencies known as

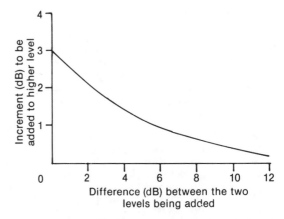

FIGURE 22.5 Add decibels in a free field from the figure. Take the difference between levels, read increment from graph, and add increment to higher level. For example: 80 − 80 = 0; 80 + 3 = 83. For example: 86 − 80 = 6; 86 + 1 = 87.

broadband noise. If the frequencies are equally distributed throughout the audible range, it is known as **white noise,** which sounds like rain.

If a sound has a duration of <1 s, it is known as **impulse sound.**

Tables 22.1 and 22.2 give example noise levels. Clark (1991) points out that rock concerts and personal stereo systems may present some hazard but that the primary nonoccupational hazard is noise from gunfire. For a right-handed person, the stock protects the right ear, so the difference in hearing level of the two ears can be used as an index of gunfire loss. Females have fewer problems because they tend to use smaller caliber guns. For males, the ear differential may be greater than 15 dB.

2.2 Noise Measurement

Measuring the sound pressure level is not sufficient since the ear is more sensitive to sound at some frequencies than at others. That is, from a physiological judgment viewpoint, 80 dB at one frequency does not sound as loud as 80 dB at another frequency. The unit of loudness,

the phon, equates loudness at other frequencies with the sound pressure level of a 1,000 Hz tone; thus, 60 dB at 1,000 Hz = 60 phons, and 68 dB at 100 Hz = 60 phons. See Figure 22.6. The curves show that the number of decibels to cause sounds that are equally loud varies considerably with the frequency of the sound. What makes the situation even more complicated is that the curves at various loudness levels (60, 80, 100, etc.) are not parallel.

The range of human hearing is approximately 20 to 20,000 Hz. See Figure 22.7. The ear is most sensitive from approximately 600 to 4,800 Hz. The frequency range of telephones is about 200 to 3,600 Hz. The range of frequencies on a piano is from 27 to 4,186 Hz; "middle C," which is 2^8, is 256 Hz. The reader can simulate this frequency by singing the musical note "do." As you progress through the 8 notes (an **octave band**) of the song from a female deer to a drop of golden sun to a name I call myself and back to "do," you have completed an octave (doubling of frequency) to 512 Hz. Going through the notes again through a second octave will demonstrate 1,024 Hz (as well as a remarkable voice). When Figure 22.6 is used, the reader can see that the high-pitched whine of a jet engine or saw is worse than the rumble of an engine due to the ear's greater sensitivity to high-pitched noise.

As a rule of thumb, a sound will be "twice as loud" when noise increases 6 to 10 dB. Thus, eliminating one of two identical noise sources (a drop of 3

TABLE 22.1 Example noise levels.

DECIBEL LEVEL, dbA	EXAMPLE
30	Quiet library, soft whisper
40	Living room, refrigerator, bedroom away from traffic
50	Light traffic, normal conversation, quiet office
60	Air conditioner at 20 ft, sewing machine
70	Vacuum cleaner, hair dryer, noisy restaurant
80	Average city traffic, garbage disposal, alarm clock at 2 ft
90	Subway, motorcycle, truck traffic, lawn mower
100	Garbage truck, chain saw, pneumatic drill
120	Rock concert in front of speakers, thunderclap
140	Gunshot blast, jet plane
180	Rocket launching pad

TABLE 22.2 Noise levels of leisure activities (Brown and Yearout, 1991).

MEAN, dbA	ACTIVITY
90	Woodcutting, rough terrain driving
92	Motorcycling
94	Farming
95	Powerboating
96	Powered lawn equipment
98	Woodworking
99	Discotheques
101	Stock car races
110	Concerts (rock), hunting/target shooting
121	Drag races

Source: P. Brown and R. Yearout, "Impacts of Leisure Activity Noise Levels on Safety Procedures and Policy in the Industrial Environment," *Int. J. of Industrial Ergonomics,* Vol. 7, pp. 341–346. Copyright © 1991 by Elsevier Science, Amsterdam, Netherlands. Reprinted with permission.

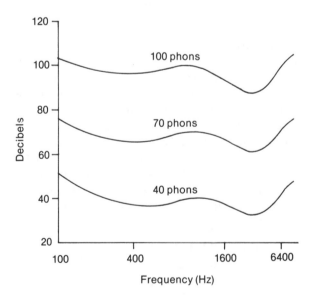

FIGURE 22.6 Equal loudness curves (free field) for pure tones are shown. The number of phons (loudness level of a sound) equates numerically with the sound-pressure level of a 1,000 Hz tone that sounds as loud as the sound being described. Note that not only does subjective loudness vary with frequency (curve is not horizontal) but also that the curve shapes vary with intensity (curves are not parallel). The 40-phon curve corresponds to dBA, the 70-phon to the B scale, and the 100-phon to the C scale.

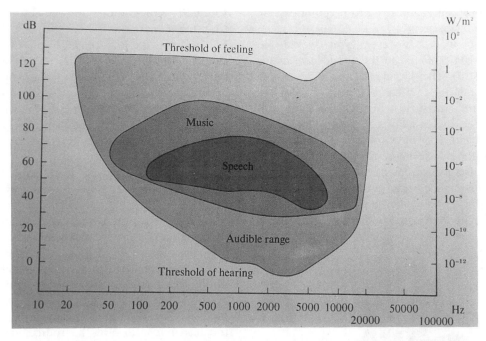

FIGURE 22.7 Ear sensitivity varies with frequency and volume. The threshold of feeling (pain) is relatively independent of frequency at about 120 dB. Very low frequencies (infrasound) may affect organs of the body other than the cochlea. The brain, for example, is especially sensitive to a frequency of 7 Hz, the frequency of the brain's alpha waves. Figure courtesy of Bilsom.

dB) will not cut "loudness" in half; subjectively, a change of 3 dB is "just noticeable"!

Loudness of a sound (in pure tones) is measured by the **sones.**

$$S = 2^{(P - 40)/10} \qquad (22.4)$$

where S = Loudness, sones

P = Phons

What the sones equation (which seems to be valid for nonpure tone sounds also) shows is that apparent loudness doubles only when noise increases by 10 dB (i.e., P changes by 10). A change of 3 dB is just noticeable.

How then to record noise? First the instrument makers standardized measuring noise by octave band. See Table 22.3.

With this type of instrument (an octave band analyzer), you could report to your boss that the noise level of a machine was 76 dB at 31.5 Hz, 77 dB at 63 Hz, 79 dB at 125 Hz, 83 dB at 250 Hz, 82 dB at 500 Hz, 86 dB at 1,000 Hz, 85 dB at 2,000 Hz, 82 dB at 4,000 Hz, and 77 dB at 8,000 Hz. This type of detailed analysis is quite useful when doing noise

TABLE 22.3 Octave band center frequency, lower limit, upper limit, and adjustment for dBA. The octave band center frequency is the geometric mean of the lower and upper limits; the upper limit is twice the lower limit. Octave bands double in width for each successive band. White noise has a continuous frequency spectrum with equal energy/Hz over the band; thus, the energy in a band doubles in each successive band. Pink noise has a constant energy per band. The energy profile is sloped 3 dB per octave.

Center Frequency, Hz	Lower Limit, Hz	Upper Limit, Hz	Phon dBA Adjustment, dB
31.5	22	44	-39
63	44	88	-26
125	88	177	-16
250	177	355	-9
500	355	710	-3
1000	710	1420	0
2000	1420	2480	1
4000	2840	5860	1
8000	5680	11,360	-1

reduction work as it enables you to pinpoint exactly where the problems are.

For most work, however, you would be swamped in data. This led to the sound-level meter (see Figure 22.8). The sound-level meter gives just one number for noise. What it does is combine—inside the meter—the various frequencies. Table 22.3 shows how the meter adjusts the actual noise for each octave band. For the A setting, 39 dB are subtracted from the octave band measurement for 31.5 Hz, 26 from the measurement for 63 Hz, etc. The adjusted band readings then are averaged and the result displayed on the dial. The A adjustment corresponds to the 40-phon equal-loudness contour, the B

to the 70-phon contour, and the C to 100-phon contour. On some meters there is a D scale which not only reduces the importance of the frequencies outside the speech range but also increases the importance within the critical 1,000 to 10,000 Hz range. It approximates the "perceived noise level" used to appraise aircraft noise. The A scale, however, has become the worldwide standard for reporting noise regardless of the intensity level. Report values using the A scale as **dBA**.

Noise is not always a steady state. The speed of response of the meter can be set slow or fast. The dBA slow setting that is usually used knocks the tops off spikes. It is possible to purchase special meters (impulse meters) to accurately measure peaks.

The American National Standards Institute (ANSI) requires use of a random-incidence microphone (responds equally to sounds arriving simultaneously from all directions, as in a diffuse field). If used in a free-field environment, instead of pointing the instrument toward the source, orient it at an included angle of 70–80°. A porous ball of sponge on the microphone will reduce wind noise. Hold the meter at arm's length to avoid sound reflections from your body.

Measured noise will be the machine noise (signal) plus the background noise. To determine the machine noise:

1. Measure noise level ($L_{S + N}$) with machine running.
2. Measure noise level (L_N) with the machine off.
3. Calculate the difference in the two levels. If the difference is < 3 dB, the background noise is too high for an accurate measurement.

With the passage of noise legislation permitting 85 dBA for 16-h exposure, 90 dBA for 8-h exposure, 95 for 4-h, 100 for 2-h, and so on, many people wonder how long an employee is exposed to each level of noise. One way to find out would be to make an occurrence sampling study, but this tends to cost too much. Manufacturers have developed devices which record in proportion to the noise standard; that is, when noise is at 90 dB, it records at 100%, when at 95 at 200%, when at 100 at 400%, when below 85 at 0%, and so on. Then, when the dosimeter (see Figure 22.9) is read at the end of any period of exposure, it is known whether the person was exposed to too much noise. Some devices have a light that is activated if there is any exposure over the 115 dBA limit.

Note that noise readings, whether on a meter or a dosimeter, are a sample from a population. A single estimate is not as good as multiple observations. In addition, consider whether you wish to use the mean of the observations (i.e., protect 50%) or a value

FIGURE 22.8   Sound level meters usually use the A scale and the slow setting. When recording sound, remember: (1) Don't shield the sound with your body; put the meter on an extension cable or at arm's length. (2) Microphones are directional. For a perpendicular-incidence microphone, sound waves not perpendicular to the microphone are attenuated at high frequencies; therefore, orient the microphone to get the maximum reading. For grazing-incidence microphones, sound waves not grazing the microphone can amplify or attenuate the sound, depending on the angle. Orient the microphone so the dominant sound grazes the microphone. Photograph courtesy of GenRad, Inc., Concord, Massachusetts.

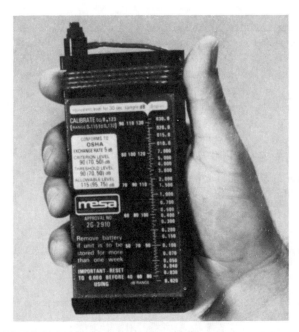

FIGURE 22.9 Noise **dosimeters** give exposure over a time period. Photograph courtesy of GenRad, Inc., Concord, Massachusetts.

which would protect a higher percent of the population (e.g., 90%).

3 EFFECTS OF NOISE

3.1 Comfort and Annoyance
Noise reduces comfort because the workers must increase their concentration; this tends to increase fatigue. Annoyance with noise has increased over the years because people now expect that technology can do anything and someone else will pay the cost. Noises which our parents experienced no longer seem tolerable. There also is some evidence that our environment is noisier than our parents' environment. The point to remember is that noise reduction may be required even if the costs are high and the economic and health benefits are small or negligible.

Epp and Konz (1975) made a study of annoyance and speech interference as a function of noise level for appliances; Figure 22.10 shows the results. If you measure a noise level of 80 dBA for an appliance, then the vote could be predicted graphically as 3.7 where 5 = extremely annoying, 4 = quite annoying, 3 = moderately annoying, 2 = slightly annoying, and 1 = not annoying. The vote can be predicted also from the equation

$$\text{Vote} = -4.798 + .1058 \text{ dBA}$$

in which case the predicted vote would be $-4.798 + .1058 (80) = 3.7$. Speech interference in percent of words missed could be predicted from the graph as 4.7% or could be predicted from the equation

$$\% \text{ missed} = 11.17 + .1989 \text{ dBA}$$

as $-11.17 + .1989 (80) = 4.7\%$. It should be noted that the subjects were college females and that people with higher education levels make more complaints than those with less education. Expectations also might vary for offices, factories, stores, and so forth, rather than homes. See Table 22.4. (For appliance noise, it should be noted that some people associate noise with power, so noisy appliances are considered good, not bad.)

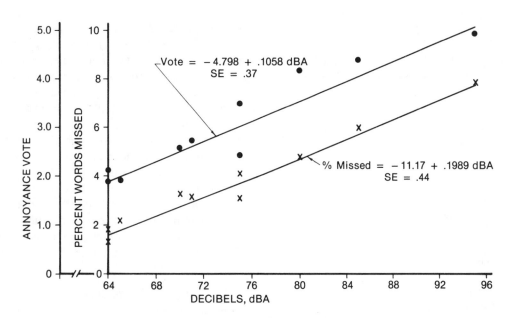

FIGURE 22.10 Annoyance and speech interference can be predicted from dBA (Epp and Konz, 1975).

TABLE 22.4   Tolerable limits (dBA) in various rooms for noise continuously present from 7 A.M. to 10 P.M.

dBA	Type of Space
28	Broadcast studio, concert hall
33	Theaters for drama (500 seats, no amplification)
35	Music rooms, schoolrooms (no amplification). Very quiet office (telephone use satisfactory), executive offices, and conference rooms for 50 people
38	Apartments, hotels
40	Homes, motion picture theaters, hospitals, churches, courtrooms, libraries
43	"Quiet" office; satisfactory for conferences at a 5 m table; normal voice 3 to 10 m; telephone use satisfactory; private or semiprivate offices, reception rooms, and small conference rooms for 20 people
45	Drafting, meeting rooms (sound amplification)
47	Retail stores
48	Satisfactory for conferences at a 2 to 2.5 m table; telephone use satisfactory; normal voice 2 to 6 m. Medium-sized offices and industrial business offices
50	Secretarial offices (mostly typing)
55	Satisfactory for conferences at a 1 to 1.5 m table; telephone use occasionally slightly difficult; normal voice 1 to 2 m; raised voice 2 to 4 m; large engineering and drafting rooms, restaurants
63	Unsatisfactory for conferences of more than 2 or 3 people, telephone use slightly difficult; normal voice .3 to .6 m; raised voice 1 to 2 m. Secretarial areas (typing); accounting areas (business machines), blueprint rooms
65	"Very noisy," office environment unsatisfactory, telephone use difficult

Source: K. Kryter, *The Effects of Noise on Man,* 2nd ed., (New York: Academic Press), 1985.

Community reaction to industrial noise is highly variable. The biggest factor well may be whether there is a local lawyer trying to get some free publicity! It also depends strongly on the history and background of the community. Expect more complaints from those with clout. Table 22.5 is based on the adjustments for composite noise rating curves and

TABLE 22.5   Adjustments to noise levels when attempting to predict community annoyance to noise (Goodfriend, 1973).

Situation	Adjustment in dBA level, dBA
Very quiet suburban	+ 5
Suburban	0
Residential urban	- 5
Urban near some industry	- 10
Heavy industrial area	- 15
Daytime only	- 5
Nighttime	0
Continuous spectrum	0
Pure tone(s) present	+ 5
Smooth temporal character	0
Impulsive	- 5
Prior similar exposure	0
Some prior exposure	- 5
Signal present 20% of the time	- 5
5% of the time	- 10
2% of the time	- 15

may give the reader an idea of some of the variables. Variability of noise generally increases annoyance.

The adjustments are independent, so if the noise is in a heavy industrial area and only during the daytime you could (on the average!) expect the same community reaction at another factory with noise level 20 dBA lower but located in a very quiet suburb and making noise at night.

3.2   Performance

There is no firm evidence that productivity is lower when work is done in high-level noise (say 100 dBA) unless the person is working at maximum mental capacity (Kjellberg, 1990). For reference by the reader, conversation at 1 m with a normal voice is 55 to 68 dBA; in a 90-dBA environment, you must shout to be understood at 1 m. To simulate a 40-dB hearing loss, firmly block both ear canals with your fingers and try to carry on a conversation. The "no loss in productivity" has an important assumption—speech communication is not an important part of the job. Hartley, Boultwood, and Dunne (1987), in an interesting experiment, suggest that 95-dBA noise *helps* when following written instruction and *hurts* when following pictorial instructions.

3.2.1   Speech interference

See Kryter (1985) or Sanders and McCormick (1987) if you wish to dig deeper.

The usual criterion for **speech interference** is the percent of words missed. "Words" are quite specifically defined. Single-syllable phonetically balanced words (are, bad, bar, bask, box) or two-syllable words, called spondees (airplane, armchair, backbone, bagpipe), are used.

The most refined index is noise criteria (NC) curves; articulation index (AI) is not quite as good although it uses 20 frequency bands. **Speech interference level (SIL)** is the arithmetic mean of the dB readings in 3 octave bands centered at 500, 1,000, and 2,000 Hz. Figure 22.11 gives the SIL as a function of distance and voice level. In general, SIL is about 7 dB lower than dBA for most common noises.

Figure 22.12 gives speech interference at various levels of noise for earplugs and no earplugs (Kryter, 1946). Note that in high-level noise (greater than 85 dB), intelligibility is improved with earplugs or earmuffs since the ear is not overloaded and can better discriminate the signal from the noise. In low-level noise, however, speech as well as noise is reduced to a level below the listener's threshold of hearing, so intelligibility is reduced. An interesting point is that speakers wearing hearing protective devices (HPD) lower their voices by 3–4 dB, which, of course, makes it more difficult for *listeners* to hear.

3.2.2   Reduction of speech interference

In addition to reducing the noise, there are four other stages at which speech transmission can be improved: the message, the speaker, the transmission system, and the listener.

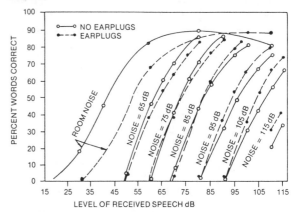

FIGURE 22.12 Earplugs improve speech transmission in loud environments. However, in environments with noise below 85 dB earplugs degrade intelligibility.

Source: K. Kryter, *The Effects of Noise on Man,* 2nd ed., (New York: Academic Press), 1985.

If the possible vocabulary is limited, with only certain words and sequences of words being permitted, intelligibility improves. Table 22.6 gives the international aviation alphabet as interpreted in a French pilot-training manual (David, 1974).

Intelligible talkers have longer average syllable duration, speak louder, have fewer pauses, and vary pitch more often (Sanders and McCormick, 1987). Nontransmission of various frequencies reduces intelligibility. Filtering below 600 Hz or above 4,000 Hz has relatively little effect; filtering between 1,000 and 3,000 Hz degrades speech severely.

FIGURE 22.11 Speech interference levels (SILs) vary with the level of the speaker's voice as well as the distance from the speaker to the ear. The figure is based on males, with average voice strengths, facing the listener, no reflecting surfaces nearby, and the spoken material not being familiar to the listener. Maximum permissible speech interference level (PSIL) = SIL + 3. The masking effect of a sound is greatest upon sounds which are close to it in frequency. At levels above 60 dB, the masking spreads to cover a wider range, mainly for frequencies above the dominating components. For U. S. telephones, telephone use is satisfactory when SIL is less than 65 dB, difficult from 65 to 80, and impossible above 80. Subtract 5 dB for calls outside a single exchange.

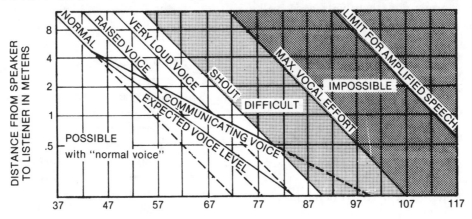

SPEECH INTERFERENCE LEVEL, dB

Source: A. P. G. Peterson and E. Gross, *Handbook of Noise Measurement* (Concord, Mass.: GenRad, Inc.), 1972.

TABLE 22.6 International aviation alphabet as interpreted in a French pilot-training manual (David, 1974).

Lettre a Identifier	Mot de Code	Prononciation du Mot de Code*
A	Alfa	*AL* FAH
B	Bravo	*BRA* VO
C	Charlie	*TCHAH* LI (*CHAR* LI)
D	Delta	*DEL* TAH
E	Echo	*EK* O
F	Foxtrot	*FOX* TROTT
G	Golf	GOLF
H	Hotel	HO *TELL*
I	India	IN DI AH
J	Juliett	*DJOU* LI *ETT*
K	Kilo	KI LO
L	Lima	*LI* MAH
M	Mike	MAIK
N	November	NO *VEMM* BER
O	Oscar	*OSS* KAR
P	Papa	PAH *PAH*
Q	Quebec	KE *BEK*
R	Romeo	RO MI O
S	Sierra	SI *ER* RAH
T	Tango	*TANG* GO
U	Uniform	*YOU* NI FORM (*OU* NI FORM)
V	Victor	*VIK* TAR
W	Whiskey	*OUISS* KI
X	X-ray	*EKSS* RE
Y	Yankee	*YANG* KI
Z	Zulu	*ZOU* LOU

The listeners should have normal hearing and know the various messages they may receive. Repeating the message back to the speaker in different words is a desirable check.

3.3 Hearing

Hearing loss is an example of repetitive trauma, of cumulative strain. If your employees are not to lose hearing just because they work for your organization, then noise needs to be kept as low as possible. How low is sufficient? Kryter (1985) says that hearing loss begins to increase faster than it would from age alone when people are exposed to noises over 67 dBA.

However, hearing loss depends upon many factors. The following is a simplified version of the procedure given by Burns and Robinson (1970); it applies only to steady noise without any peaks at any frequency.

Given   Noise level = 85 dBA

Exposure   = 250 days/yr of 8 h for 25 years

To Find   Hearing loss at 1,000 Hz for 25th percentile of population

Solution

Step	Value

1. Determine noise emission level

   A. Determine dBA for environment   85 dBA

   B. Correct for *duration* of exposure, D

   D, dB $= 10 \log_{10} T$

   T = Years of exposure $D = 14$

   $85 + 14 = 99$

   C. Adjust for *gender, G*

Group	G Noise Adjustment, dB
Female	−1.5
Mixed	0
Male	+1.5

   $G = 0$

   $99 + 0 = 99$

   D. Adjust for *frequency* at which ear will lose hearing, F

Frequency Hz	F Noise Adjustment, dB
500	− 5
1,000	0
2,000	+ 7
4,000	+15

   $F = 0$

   $99 + 0 = 99$

2. Predict noise-induced permanent threshold shift *(NIPTS)* measured on audiometer, dB

   A. Read value H for population median from Figure 22.13 $H = 2$ dB

   B. Adjust for percentile of population, P

Percentile Losing Hearing	P Audiometric Adjustment, dB Emission at 1,000 Hz		
	85	95	105
10	+8	+9	+11
25	+4	+5	+ 6
50	0	0	0
75	−4	−5	−6
90	−8	−9	−11

   $P = 5$

   $2 + 5 = 7$

   C. Adjust for *age (A)*

   $A = cT^2$

   c = Frequency constant = .0043

   T = Years of exposure

Audiometric Frequency, Hz	c
500	.0040
1,000	.0043
2,000	.0060
4,000	.0120

   $A = .0043 (25)^2 = 3$

   $7 + 3 = 10$

   Predicted *NIPTS* = 10 dB

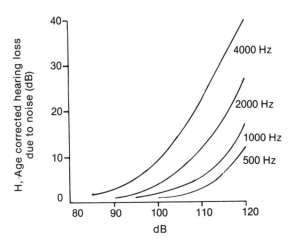

FIGURE 22.13 Predicted hearing loss caused by noise can be calculated from this simplified adaptation of Burns and Robinson's (1970) Figure 10.17. Noise emission level and audiometric frequency determine the predicted noise loss for the population median.

As the reader can see from the example solution of the model, the really critical adjustment is *D*, duration of exposure. Burns and Robinson (1970) say the coefficient in the equation should be 10 (the **equal energy** principle); Kryter (1985), summarizing many studies by many authors, says it should be 20 for noise (the **equal pressure** principle) but 10 for recovery. A 10 means a doubling (or halving) of time changes *D* by 3 dB; a 20 changes *D* by 6 dB. Thus, Kryter would add 28 dB instead of 14 for 25 years of exposure. On the other hand, if the noise was for 4 h/day instead of 8, Kryter would subtract 6 from the 85 while Burns and Robinson would subtract only 3. The U. S. adjustment from the Occupational Safety and Health Administration (OSHA) is 5; the U. S. Environmental Protection Agency (EPA) and some other countries use 3. (There must be more pressure than energy in Washington.)

Determining *P*, the percentile effect, is another problem. Kryter, to protect 75% of the population, would give *P* a value of −10 instead of −4 to −6. Burns and Robinson, in their detailed calculations of their Figure 10.17, give *P* = −6 for a noise emission level of 105 dBA and *P* = −5 at 95 dBA. This sensitivity of the ears of some of the population means that some people, if exposed for many years, start to lose hearing at some frequencies at noise levels below 70 dBA; Kryter says 67 dBA. There still is no technique which can identify the sensitive ears before the hearing loss. A pre-employment audiogram can demonstrate whether hearing loss occurred before start of employment. Periodic audiograms also can identify sensitive ears by noting employees who have lost some hearing so they can be transferred before the hearing loss becomes too large. The

decision of whether to ignore sensitive ears depends on whose ears are going to be ignored.

The effect of gender is controversial. Some experts agree, others don't, that noise affects females less and that women have better hearing. Some feel it is reduced exposure to noise or reduced ear diseases rather than reduced susceptibility that affects the value of *G*.

Even the value of *A* is not agreed upon. For 1,000 Hz, Burns and Robinson give *A* = 3, 5, and 9 at 45, 55, and 65 years, while the American Industrial Hygiene Association (1966) gives 3, 6, and 11 at the same ages. Part of the effect of aging (presbycusis) is poor blood circulation in the ears. Vasodilators (such as niacin and other B vitamins) are good for circulation; vasoconstrictors (such as nicotine and caffeine) hinder it.

Another problem is to define **hearing impairment.** Everyone starts with normal hearing values from the International Standards Organization (ISO) as a reference, but from there they differ. OSHA defines material impairment of hearing as a "25-dB hearing level at 1,000, 2,000, and 3,000 Hz." The meaning of this obscure but vital statement is that you have a "25 dB deductible." The hearing loss also is not in the speech frequencies. Kryter (1973) proposes "an increase of 10 percentage points on the number of people to suffer average hearing losses of greater than 25 dB at 500, 1,000, and 2,000 Hz because of noise rather than aging." The American Academy of Otolaryngology (1979) and the American Medical Association (1973) specify a mean loss of 25 dB at 500, 1,000, 2,000 and 3,000 Hz as "the beginning of slight impairment for the understanding of spoken English." Kryter (1985) says that if the three bands were 1,000, 2,000, and 4,000 instead of 500, 1,000, and 2,000, the NIPTS would be 10 dB higher. Kryter (1973) says the "slight impairment" of the AAO and AMA can be more precisely defined as "a person could understand but 90% of the sentences and 50% of the monosyllabic phonetically balanced (PB) words in the quiet, uttered at a normal conversational level of effort by a person one meter from him." However, there is considerable redundancy in everyday speech, so should we be concerned with the unexpected message?

Note that all the foregoing assumes that the noise was continuous and broadband, and important factors such as pre-exposure hearing, general health, and drug effects were not considered. However, one of the virtues of a math model is the explicit statement of the coefficients.

As you might surmise from the above, setting a legal standard for noise is not simple. The occupational standard in the United States is 90 dBA for 8 h of exposure/day with a 5-dBA trade-off vs. time during the day (the EPA uses a 3-dBA trade-off vs. time). Table 22.7 gives the resulting standard. There is no adjustment for years of exposure, G, or F. All noise below 85 dBA is assumed to have 0 effect. Beyond 115 dBA, 0 exposure is permitted. Table 22.8 gives recommendations for ultrasonic noise. For a general summary of the effect of noise on health, see Fay (1991).

4 NOISE REDUCTION

To make conversation more private, you may wish to increase the environmental noise—that is, use **masking noise.** Examples are fountains, background music, fans, air conditioners, and fluorescent ballast hum. Firms even sell white noise generators, which make a sound like rain or static.

However, in most cases, we are interested in reducing the cumulative trauma on hearing and so will reduce noise. Use the following sequence: (1) plan ahead, (2) modify the existing noise source, (3) modify the sound wave, and (4) use personal protection.

TABLE 22.7   Maximum daily noise exposure in the United States. OSHA values are the legal values; the American Congress of Governmental Industrial Hygienists (ACGIH, 1993) values are recommendations. The trade-off of exposure time vs. noise is 5 dBA for each doubling (halving) in each column. OSHA ignores noise below 85 dBA while ACGIH ignores noise below 80 dBA. Neither permits noise above 115 dBA.

DURATION/DAY, H	NOISE, dbA	
	OSHA	ACGIH
16	85	80
8	90	85
4	95	90
2	100	95
1	105	100
.5	110	105
.25	115	110
.125	—	115

TABLE 22.8   For ultrasonic noise, the American Congress of Governmental Industrial Hygienists in 1987 recommended the following threshold limit values.

1/3 Octave Band, kHz	Noise, dB
10	80
12.5	80
16	80
20	105
25	110
31.5	115
40	115
50	115

4.1 Plan Ahead

An ounce of prevention is worth a pound of cure. The four subcategories of the *ounce* are to substitute less noisy processes, purchase less noisy equipment, use quieter materials and construction, and separate people from equipment.

4.1.1 Substitute less noisy processes

Three examples of substitution are (1) reducing the use of impact tools (e.g., welding, not riveting) and chipping (by using grinding), (2) replacing internal-combustion engines with electric motors (lift trucks, lawn mowers), and (3) replacing gear transmissions with belt transmissions.

4.1.2 Purchase less noisy equipment

Include noise levels in equipment purchase specifications. A specification must include three things: units, levels, and conditions. For example, "SPL for machines with auxiliary equipment shall not exceed 85 dBA (slow response) at the operator location when installed as specified." Ask vendors to include silencers and sound-damping devices in their quotes. Buy noise-suppression equipment (such as mufflers) from the manufacturer on the original requisition. It is much easier to get these low-cost items approved then than as a separate requisition later.

Certain types of equipment are quieter than their alternatives. Bevel gears are quieter than spur; nylon gears are quieter than metal; belt drives are quieter than gear drives; V-belt drives are quieter than toothed-type belts; reinforced rubber belts are quieter than leather or canvas belts; electric handtools are quieter than pneumatic tools; electrically operated valves and solenoids are quieter than air-powered ones. In drilling concrete, use carbide or diamond-tipped drills since star drills and air hammers create both noise and dust. Cast aluminum vibratory feeder bowls may be 15 dBA quieter than fabricated sheet metal bowls. Bearings with less clearance and higher finish make less noise.

Squirrel-cage fans are quieter than propeller fans. Properly sized fans (i.e., running at their peak efficiency) are quieter since undersize fans have high rpm whereas oversize fans have low rpm but separation of air flow over the blades. For the same capacity, large slow fans are quieter than small fast fans. A low-speed, multibladed fan is quieter than a high-speed, two-bladed fan. Thermoplastic fan blades have high inherent damping and are poor resonators and thus are quieter than metal blades. ASHRAE (1991, Chapter 42) has detailed design suggestions on how to reduce fan and duct noise for HVAC systems.

For manually operated punch presses, a pin-type clutch contributes about 70% of the total sound energy at the operator's ear; therefore, buy presses with hydraulic drives or air-actuated clutches.

4.1.3   Use quieter construction and materials

Some reasons for using quieter construction and materials are to reduce impact, vibration, and turbulence. Another is to reduce noise transmission.

Impact   Wood block floors are easier on the feet than concrete; carpet is easier than tile; both are quieter also. (Carpet has an additional advantage over tile of lower maintenance.) Avoid metal-to-metal contact. Chutes and hoppers should be wood or plastic instead of metal; if metal is required for wear, back it up with wood. Cover the inside of chute tops with sound-absorbent material. Have products slide rather than drop onto chutes (e.g., change chute angle). When two dies register on each other, can one of the impact surfaces be nonmetal (such as a plastic)? Line walls of tumbling barrels. Buy plastic or fiberglass tote pans. They weigh less and are quieter. Cover metal conveyor rollers with rubber, plastic, or carpet. Replace metal wheels on vehicles with tires. In kitchens, rubber mats on sinks and drain boards reduce noise and breakage.

Vibration   Figure 22.14 shows the isolation of pipes. Pipe noise also can be reduced by using bends in pipes, flexible couplings, and vibration isolators. Machine guards should be designed for noise control as well as safety. Reduce guard vibration by selection of material or shape. Encourage use of rubber or flexible plastic; discourage sheet metal and hard plastic. Encourage sheets with corrugations or holes (see Figure 22.15); discourage thin, flat sheets supported on one side. Drums make noise by vibration of the drumheads. Reduce vibration by modifying the drumhead. Figure 22.15 shows that one approach is to poke holes in the drum head.

Another approach is illustrated by hitting a hammer against a concrete wall and against a brass gong. This demonstrates the effect of **resonance.** Reduce resonance by damping—either by using material such as lead or sand or by stiffening the physical structure of the device (adding ribs). The ability of materials to amplify vibration (given as Q, the magnification ratio) is 1,000 for steel and aluminum, 500 for reinforced plastics, 200 for concrete, 100 for plywood, 50 for cast iron, and 10 for steel with polymer damping compound. For example, the screech of subway wheels was reduced 34 dB by adding polymer coating to the steel rim.

Adding ribs to the surface can reduce vibrations; Figure 22.16 illustrates this for saws. Flat sections of sheet metal can be backed with wood, felt, lead, or elastomers—making a "sandwich." Clamp items being machined or riveted.

Increasing the mass is another option. For example, fill the base of a machine with sand.

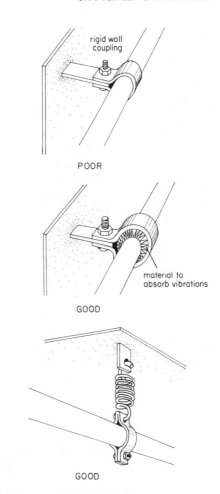

FIGURE 22.14   Isolate vibrations in pipes.

Turbulence   Smooth flow makes less noise than **turbulent** (non-smooth) flow, for both liquids and air. See Figure 22.17 and Section 4.2.3.

Transmission   Double doors and double walls with mineral wool or other insulation between walls

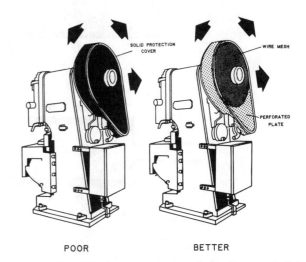

FIGURE 22.15   Holes in a drumhead reduce vibration. Thus, a perforated guard gives less noise than a solid one.

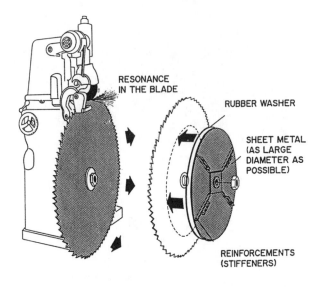

FIGURE 22.16   Dampen vibration by adding stiffeners. Damping also can be done by material selection—plastic vs. metal bins. Damping also can be done by both shape and material. See the "sandwich" of the blade, rubber washer, and sheet metal.

reduce sound transmission and conserve heat; double windows with 100 mm air space do the same. Put an airlock between the office and the shop so shop noise doesn't come into the office when someone opens the door. Figure 22.18 shows how vibration transmission through the floor is reduced.

Quiet air ducts have large cross sections to give low velocity, have several elbows, have internal insulation after elbows (before and after elbows is better), use parallel or staggered baffles, have vibration breaks, and are not placed to produce the megaphone effect. If it is not possible to modify a duct, put a plywood baffle in front of the room air supply grille. The baffle (50% larger than the grille) should be surfaced with fiberglass on the grille side and should be placed far enough from the grille to not restrict air flow. When a machine cannot be totally enclosed because it is necessary to supply air (e.g., for cooling), have the passage make several right-angle bends and line the passage with acoustic material.

Arched ceilings convey sound from remote locations. One Italian cathedral has a spot worn in the

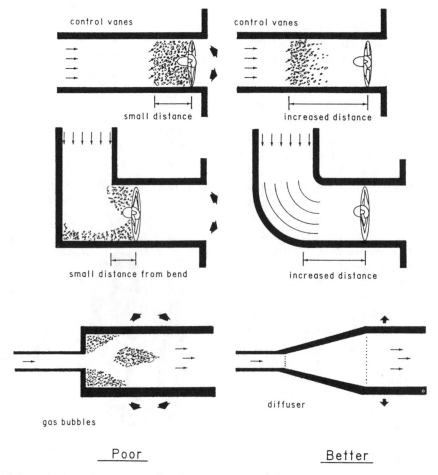

FIGURE 22.17   Smooth flow makes less noise than turbulent flow. Smooth flow requires (1) no abrupt directional changes, (2) no abrupt volume changes, and (3) distance for turbulence to die down.

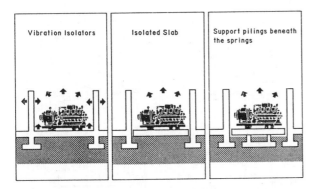

FIGURE 22.18 Machine foundations transmit vibrations and thus noise. Lighter machines can rest on various vibration isolators. Heavier machines need the slab itself to be isolated. If the ground is clay, it may be necessary to have the slab rest on pilings.

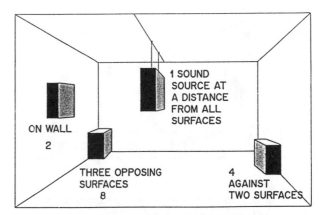

FIGURE 22.19 Noise radiates through a whole sphere to 1/8 of a sphere depending upon where the source is located. A noise source located in the room center radiates through the entire space. A source located in the corner radiates through 1/8 of a sphere; therefore, this is a good loudspeaker location.

marble floor. People could stand there and eavesdrop on conversations from a remote confessional.

If a workstation is lit by suspended fluorescent fixtures (see Figure 21.9), tilt the fixture so sound is directed away from the person.

4.1.4 Separate people and noisy equipment
Mechanical equipment (pumps, boilers, fans, etc.) for heating, ventilating, and air conditioning is inherently noisy; locate such equipment in an isolated room. Penetrations of the mechanical room by ducts, pipes, and so forth, should be sealed airtight with a nonhardening sealant. The key number is 6; the **inverse square law** says 6 is the reduction in dB from a point source in a free field with each doubling of distance (provided the initial measurement is at a distance several times the longest dimension of the source). With absorbent material and good design, this can be increased to 8 dB or even more, even in an open-plan office. In an open space, sound flows in all directions, so the direction factor (magnification ratio), Q, equals one. See Figure 22.19. On a wall, sound can flow in only $1/2$ of a sphere, so $Q = 2$. At a junction of two walls $Q = 4$ and in a corner $Q = 8$. Higher frequency sound waves have a high Q factor even if the source is in an open space. Therefore, don't put noisy equipment in a "megaphone" by locating it in a corner or at the end of a corridor. Supervise noisy equipment by TV rather than in person.

4.2 Modify the Noise Source
The pound of cure is more expensive than the ounce of prevention. Because of the way decibels add, always start with the loudest noise first. Three noise sources of 90, 85, and 101 dB combine to 102 dB. If the 90 dB noise is completely eliminated, the 95 and 101 still combine to 102 dB. The 3 ways to modify the source are to reduce the driving force, change the direction

of the noise, and minimize velocity and turbulence of air.

4.2.1 Reduce driving force
Step one is maintenance. Sharpen tools. Tighten screws and bolts. Lubricate bearings. Greasing and oiling equipment and replacing worn parts reduce noise as well as wear. Rebalancing rotating equipment reduces wear and noise and improves quality. Replace leaky compressed air valves to save air as well as reduce noise.

Use reducing valves when full shop pressure is not needed (e.g., when blowing off dirt) to improve safety as well as reduce noise. When drills, mills, and taps are under heavy load, cutting compounds reduce noise while they lubricate. A rake angle on punches lets the punch hit the work gradually, so the same energy is expended over a longer time period. See Figure 22.20. Stagger punch lengths when punching several holes at one stroke. Because of the reduced impact peak, noise is reduced and smaller presses can be used. The same principle works for shears. Spreading energy over a longer period also can be used to quiet diesel engines. Increase the turbulence of the fuel-air mixture (by bouncing it off the walls) so ignition occurs over a longer time; set the ignition timing to minimize peak pressures. Turn down the paging system volume at lunch.

4.2.2 Change the direction of the noise
Since sound waves have a longer wavelength than light waves, they are not quite as directional as light. Yet (especially for high-pitched noise) turning a machine or exhaust 90 degrees often reduces noise as much as 5 dB. Gaseous jet noise is very directional. The best direction to turn it is up if no one is above the machine. The same 5-dB benefit can be

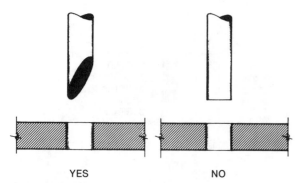

FIGURE 22.20   Exert force over a longer time. Examples are angles on punches, staggering multiple punches, helical and bevel gears vs. spur gears, and standard vs. progressive shears.

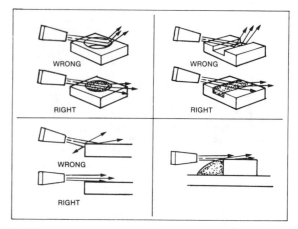

FIGURE 22.21   Air-jet noise is increased greatly whenever the jet blows on a sharp edge. Filling in cavities or redirecting the jet can reduce noise levels as much as 7 dBA (Olivatt, 1981).

obtained if the operators face the sound so the sound grazes their ears rather than hitting them directly. Changing the hinge location on windows and doors so they open in a different direction may help.

4.2.3   Minimize the velocity and turbulence of air
Can a vacuum hold-down (clamp) be replaced with a mechanical hold-down? It would not only reduce noise but save compressed air (which commonly costs $.20–.30/1,000 cu ft). Avoid sonic velocities. Cutting air velocity in half can reduce dB by 15. (Dispersive mufflers work by reducing air velocity.) To air-dry objects, use a large number of low-velocity nozzles since it is volume, not velocity, which is desired. Furnace noise has been cut by using additional burners to reduce gas flow per burner. For air ejection, use a multiple-opening nozzle at low line pressure and place close to the part. Since the high-velocity portion of the jet is only 2 jet diameters wide, accurate aiming of the jet permits lower velocities. Halving the distance between the jet and the part permits a 30% reduction in velocity which will reduce noise by 8 to 10 dB. Jet exits should be designed to give laminar (not turbulent) flow.

Since many operations do not require full line pressure, use pressure regulators. (The OSHA requirement limiting air pressure for cleaning equipment to 30 psi is frequently cited for violation.) Automatic shutoffs save compressed air as well as reduce unnecessary noise; since the noise is intermittent, hearing loss is less. To make a whistle, blow across a sharp edge. On vacuum clamps, jigs, fixtures, and dies used with air ejection, streamline the sharp edges upstream of the part with fillets and chamfers. See Figure 22.21 for ways to reduce sharp edges. If you didn't plan ahead and can't reduce the noise at the source, then you can try to modify the sound wave on its path to the ear.

4.3   Modify the Sound Wave
Confine and **absorb** the sound wave. Unfortunately these are expensive procedures which often reduce dB very little.

4.3.1   Confine
For confinement the problem is the long wavelength. Any sound will escape even through a small opening, and so, for confinement to work, enclosures must be total. Holes amounting to as little as .01% of the total area of the enclosing structure will transmit more than half the sound energy, lowering the total reduction by 3 dB. For a barrier reducing noise 40 dB, openings of .01%, .10%, 1% and 10% reduce attenuation to 37 dB, 30 dB, 20 dB, and 10 dB. Access openings are the weak points in an enclosing structure.

The first goal is to confine and absorb before the sound gets out. Light, porous materials (fiberboards, cork, foam rubber, cloth, mineral wool) absorb sound but also transmit with little attenuation of sound. Hard, massive materials (brick, concrete) reflect and prevent transmission but absorb little (see Table 22.9 and Figure 22.22). Double-wall construction (see Figure 22.23) can help. Transmission is less for frequencies over 1,000 Hz. The ideal is a heavy wall of brick to prevent transmission, with an inner lining of foam rubber to absorb. A compromise is a thin sheet of lead with a layer of foam. Note that not all foams are acoustical foams, so just because a foam is cheap does not mean it is a good buy. Unglazed brick absorbs 3 times as much as glazed brick; unpainted concrete block absorbs 3 to 5 times as well as painted block; a wood floor absorbs 3 to 15 times as well as concrete, while a heavy carpet can absorb 30 times as much as concrete at some frequencies. Carpeting office walls is effective.

TABLE 22.9 Sound absorption coefficients (α) of typical surfaces (α is the ratio of sound energy absorbed by the surface to sound energy incident). Note the effect of frequency. (Simplified from Table 37–3 of Hill, 1973).

Material	Frequency (Hz)			
	125	500	1000	4000
Brick: glazed	.01	.01	.01	.02
Brick: unglazed	.03	.03	.01	.07
Concrete block: coarse	.36	.31	.29	.25
Concrete block: painted	.10	.06	.07	.08
Floor: carpet, heavy with 40-oz pad	.02	.14	.37	.65
Floor: linoleum, rubber, or cork tile on concrete	.02	.03	.03	.02
Floor: wood	.15	.10	.07	.07
Glass fiber: mounted with impervious backing, 3 lbs/cu ft. 1 inch thick	.14	.67	.97	.85
Glass fiber: mounted with impervious backing, 3 lbs/cu ft. 3 inch thick	.43	.99	.98	.93
Glass: window	.35	.18	.12	.04
Plaster on brick or tile	.01	.02	.03	.05
Plaster on lath	.14	.06	.04	.03
Plywood paneling, ⅜ inch	.28	.17	.09	.11
Steel	.02	.02	.02	.02

Complete enclosure of single-source noise such as motors and gearboxes is sometimes worthwhile but, as the noise source gets larger, the enclosure volume (and therefore material required and cost) increases rapidly. Temperature rises may be a problem. Line the enclosure inside (not outside) with material with an absorption coefficient of at least .7 (see Table 22.9 and Figure 22.23).

In offices, noise often is transmitted through air conditioning ducts. See Chapter 42 of ASHRAE (1991) for some duct treatments. Sound flows through small cracks, so use gaskets on doors and shafts. In some cases enclosure of the operator is feasible but it tends to be expensive due to working space requirements, door seals, provision of air for breathing, access for maintenance, and other costs. Partial enclosures, such as for shop telephones, printers, and the like are better than nothing, but not by much. In some cases, complete sealing isn't necessary because a series of noise-trapping baffles can be used. A shield of auto safety glass between the noise and the employee can be used to reflect high-frequency sound such as air jets while permitting vision if both the noise and employee are close to the shield. Have the "acoustical shadow" extend a meter beyond the head. Higher frequency noise with its shorter, more directional waves can be absorbed if the absorbent is close to the noise source.

If the sound can't be confined and absorbed directly at the source, it can be dissipated by a muffler. Muffle intakes and outlets. Use the proper design and don't forget that mufflers wear out. Resonance mufflers are effective for specific frequencies. Be sure they are designed by an expert. Check that the operator doesn't remove the muffler. (Most motorcycle violations are due to failure to use the manufacturer's muffler.) A relatively ineffective (but cheap) shield can be created by storing work in process between the noise and the operator.

The discussion has assumed enclosing the noise and allowing the operator to be mobile. Another possibility is to create a noise refuge for the operator so

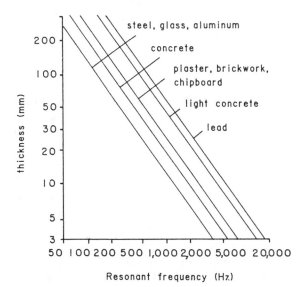

FIGURE 22.22 Resonant frequency depends upon both wall thickness and material. A 50-mm (2-inch) concrete wall would have a resonant frequency of about 500 Hz.

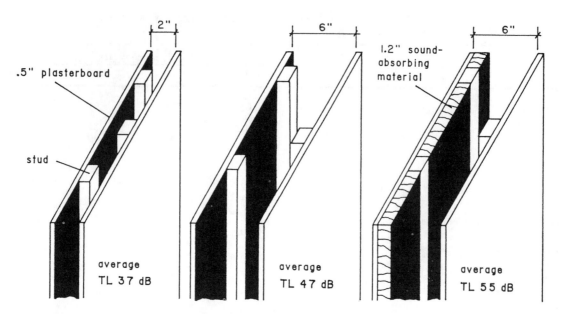

FIGURE 22.23 Double-wall construction is more effective than single. Air gaps give good transmission losses (TL). Note the penalty when the stud contacts both walls.

that the operator supervises the equipment though a window or by TV. The enclosure need not be total. Three walls and a glass roof give some noise attenuation with minimum ventilation, lighting, and access problems.

4.3.2 Absorb

Absorbing the sound once it is "out" (that is, the ear is between the noise source and the absorbent) by using absorbent panels on the walls or ceiling is not cost-effective. To reduce noise by 20 dB we need to drop intensity to 1 part in 100; a 30 dB drop requires reduction to 1 part in 1,000. Achieving these magnitudes is not economic. Reductions of 7 to 10 dB in the higher frequencies are practical if the room is reverberant, but in most rooms 5 dB is the best that can be accomplished. Put the absorbent on the walls if the minimum floor dimension is less than four times the room height. Carpeting a cinder block wall may be useful—especially if the wall is an outside wall and the carpet can therefore act as thermal insulation also. In classrooms and offices with multiple noise sources, soft surfaces will help until about 50% of the surfaces are soft. Flush-mounted fluorescent lamp fixtures act as sound reflectors and thus negate the effect of acoustical ceilings.

If you didn't plan ahead, couldn't modify the source, and found modifying the sound wave was too expensive, then there remains, as a last resort, protection of the receiver, the ear.

4.4 Personal Protection

The ear can be protected with time or with equipment.

4.4.1 Time

Higher noise levels can be tolerated if exposure time for a specific ear is short. In the United States the trade-off is 5 dBA for each doubling of time instead of the 3 dBA used in most of Europe. Since our present maximum is 90 dBA for 8 h, a person can be exposed to 95 for 4 h, 100 for 2 h, 105 for 1 h, 110 for ½ h, and 115 for ¼ h. Beyond 115 dBA is not permitted.

Reduce exposure time by making the source noise intermittent. Then minimize the exposure of the worker. Redesign jobs so that noisy areas are inspected by TV. Provide a quiet area for other work which can be done before and after work in the noisy area (blueprint reading, paperwork). Modules and quick disconnects minimize downtime as well as exposure of maintenance personnel. Consider job rotation within the day.

4.4.2 Equipment

Earmuffs and **ear plugs** attempt to totally enclose the hearing system. They are the only realistic protection against gunfire and explosive tools. See Figure 22.24 for typical attenuations of earmuffs and ear plugs. See Box 22.1 for some recent developments.

In general, hearing protectors give a greater dB attenuation of higher frequency noise. But, because many purchasers of hearing protectors don't know the frequency spectrum of the noise and don't want to have 9 different numbers (1 for each of the 9 octave bands from 125 to 8,000 Hz) for the attenuation of the protector, hearing protectors have a noise reduction rating (NRR) which summarizes the attenuation regardless of frequency. In general, the NRR

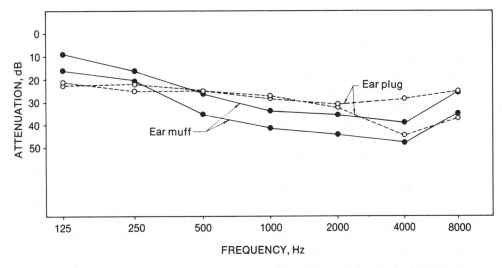

FIGURE 22.24 Earmuffs tend to give more protection than ear plugs. However, as with any other product, some models are better than others. The reported attenuations were from ANSI Z24.22–1957 standard. The ANSI S.19–1974 test standard uses a different test methodology and averages 5 dB less attenuation than the Z24.22–1957 test. Earmuffs can be designed to be worn with the band over the head, behind the head, or under the chin. One manufacturer reported, for its model, 2–8 dB less attenuation in the speech frequencies when behind the head and 5–8 dB less attenuation when under the chin. Wearer-molded (plastic foam) plugs probably will give attenuations in practice equal to the manufacturer's test data but premolded plugs may give only 50% of their attenuation potential due to poor fitting by the users. Derate earplugs to about 50% of the manufacturer's NRR rating (Casali, 1992).

BOX 22.1 *Technology advances in noise control (Casali, 1992)*

Conventional hearing protective devices (HPDs) do not improve the speech/noise ratio, the most important factor for improving speech intelligibility, although they do improve speech intelligibility. They lower the total incident energy of both speech and noise. **Acoustic glare** is therefore reduced. (Sunglasses improve visual performance by reducing visual glare.)

There are some modifications to the conventional passive HPDs.

Passive, uniform attenuation. A conventional HPD increases attenuation with increased frequency. This "colors" the sound in a spectral sense. Uniform-attenuation HPDs have been developed which give the same attenuation, regardless of frequency. However, they give less total attenuation than conventional HPDs.

Passive, frequency-sensitive attenuation. The attenuation is reduced at low frequencies.

Passive, adjustable attenuation. The user adjusts the amount of attenuation by adjusting a setting on the HPD.

Passive, amplitude-sensitive attenuation. These HPDs provide minimum attenuation at low noise amplitudes but switch to increased attenuation at high noise amplitudes. At present, they work best for impulse noise such as gunfire.

Active. Rather than just passively accepting the noise vibrations, active noise control senses the noise vibration and cancels it by adding noise that is 180° out of phase—giving a constant pressure, which is perceived as silence. The concept depends on the much higher speed of electronics compared to the relatively slow speed of sound in air.

A microphone detects the incoming noise and adds noise to cancel it. The speech signal typically is pre-emphasized so the cancellation does not destroy the speech signal. At the present state of technology, the active systems do not perfectly cancel the noise, but 5–10% improvements in speech intelligibility have been achieved.

exceeds the attenuation at the worst of the octave bands.

A laboratory comparison (Berger, 1983) showed an NRR of 8 for fiber plugs, 14 for premolded plugs, 17 for partially inserted (20% of plug in ear canal) foam plugs, and 31 for deeply inserted (100% of plug in ear canal) foam plugs; NRR for two different earmuffs were 21 and 25. When the muffs were combined with the plugs, NRR ranged from 25 to 35.

A number of studies have shown that actual noise reduction to the ear is not as good as implied by the laboratory-developed NRR values (Park and Casali, 1991). Thus, an NRR derating (of say 50%) should be considered, especially for ear plugs.

Very rarely is the noise environment so loud that noise cannot be reduced to a safe level with personal protective equipment. Note that 15-dB attenuation is usually sufficient since most noise is below 105 dBA. Note also that since ear protectors work best in the high-frequency range, it may be more desirable to reduce low-frequency components of the source noise if ear protectors will be worn anyway. For rockdrills, the low-frequency exhaust noise is dominated by the high-frequency piston and drill noise. It is worthwhile to reduce exhaust only if the high-frequency noise is also eliminated (e.g., by earplugs).

The problem of personal protective equipment is not bad design but bad application. Workers often are fitted with the wrong size, seals no longer seal with age, and, worst of all, workers refuse to wear the device at all due to the gradual onset of the hearing disability and thus their lack of a feeling of danger. It is probably desirable to make wearing of hearing protection a part of the job description and a condition of employment. Emphasize to the employees that the employer is helping them; there is no need to lose the enjoyment of speech, music, and television. The most effective hearing protector is the one that is worn.

Earmuffs are advantageous in that one size fits almost everyone, and they are visible from a distance; a supervisor can tell who is wearing personal protective equipment. But they are hot, heavy, and muss curly locks; glasses reduce their seal, and the head is within a spring.

Earplugs should be individually fitted because ear canal sizes and shapes differ. Often the right ear is not the same size or shape as the left. Acceptance is increased considerably if employees can get earplugs molded to their ears or can try four or five sizes and models until they find plugs that are comfortable to their individual ears. You can hear better in noise with earplugs than without earplugs! (See Figure 22.12.) Forbid use of useless dry cotton wads as earplugs. Good quality earplugs are light, compact, and do not affect appearance. However, the supervisor can't tell if they are being worn, and they get lost. Reduce lost plugs by having the plugs connected with a cord. Be sure to have spares.

In very high noise levels, have workers wear earmuffs over their earplugs. Getting workers to wear earplugs is a serious problem, but some strategies to improve usage have been reported. Casali and Epps (1986) reported that noise attenuation of some earplugs tripled when revised donning instructions were given. Zohar et al. (1980) gave employees audiometric tests on two successive days. On one day employees wore hearing protection and one day they didn't. Thus, the nurse could show specific evidence of the benefit of the protectors. Harns (1980) gives another technique. Workers are asked to set their car radio volume at a just-audible level upon arriving for work and then turn off the ignition. If, upon leaving work, they cannot hear the radio, they need more hearing protection.

Reduction of noise is not a very difficult problem if you plan ahead. It calls for some ingenuity and expense if you modify the noise source. It requires considerable ingenuity and expense to modify the sound wave. It requires day-after-day selling, motivation, and supervision to get personal protective equipment used.

The goal is "Ears alive at 65!"

REVIEW QUESTIONS

1. What is the purpose of the Eustachian tube?
2. What is the total dB from 70 dB to 70 dB? From 70 + 75? From 70 + 90?
3. What is the noise level standard in the United States for 8 h of exposure? What is the trade-off for each doubling or halving of time? What is the maximum permitted level?
4. List in sequence the four major approaches to noise reduction.
5. Why would you try to eliminate the loudest noise first? Discuss using an example of a 100-dBA and a 90-dBA noise.

REFERENCES

ACGIH. *1986-1987 Threshold Limit Values.* Cincinnati: American Congress of Governmental Industrial Hygienists, 1987.

ACGIH. *1993-1994 Threshold Limit Values and Biological Exposure Limits.* Cincinnati: American Congress of Governmental Industrial Hygienists, 1993.

American Academy of Otolaryngology. Guide to the evaluation of hearing handicap. *J. of American Medical Association,* Vol. 241, 2055-59, 1979.

American Industrial Hygiene Association. *Industrial Noise Manual.* Detroit, Mich., 1966.

ASHRAE. *HVAC Applications.* Atlanta: American Society of Heating, Refrigeration and Air Conditioning, 1991.

Berger, E. Laboratory attenuation of earmuffs and earplugs both singly and in combination. *American Industrial Hygiene Association J.,* Vol. 44 [5], 321-29, 1983.

Brown, P. and Yearout, R. Impacts of leisure activity noise levels on safety procedures and policy in the industrial environment. *Int. J. of Ind. Ergonomics,* Vol. 7, 341-46, 1991.

Burns, W. and Robinson, D. *Hearing and Noise in Industry.* London: Majesty's Stationery Office, 1970.

Casali, H. Technology advances in hearing protection. *Proceedings of the Human Factors Society,* 258-62, 1992.

Casali, J. and Epps, B. Effects of user insertion/donning instructions on noise attenuation of aural insert hearing protectors. *Human Factors,* Vol. 28 [2], 195-210, 1986.

Clark, W. Noise exposure from leisure activities: A review. *J. Acoustical Society of America,* Vol. 90 [1], 175-80, July 1991.

David, H. French version of alphabet. *Ergonomics Research Society Newsletter,* March 1974.

Epp, S. and Konz, S. Appliance noise: Annoyance and speech interference. *Home Economics Research J.,* Vol. 3 [3], 205-208, 1975.

Fay, T. (ed.). *Noise and Health.* New York: New York Academy of Medicine, 1991.

Goodfriend, L. Control of community noises from industrial sources. In *The Industrial Environment—Its Evaluation and Control,* Chapter 46. Washington, D. C.: Supt. of Documents, 1973.

Harns, D. Combatting hearing loss through worker motivation. *Occupational Safety and Health,* 38-40, March 1980.

Hartley, L., Boultwood, B., and Dunne, M. Noise and verbal or spatial solutions of Rubik's cube. *Ergonomics,* Vol. 30 [3], 503-509, 1987.

Hill, V. Control of noise exposure. In *The Industrial Environment—Its Evaluation and Control,* Chapter 37. Washington, D. C.: Supt. of Documents, 1973.

Kjellberg, A. Subjective, behavioral and psychophysiological effects of noise. *Scand. J. Work Environ. Health,* Vol. 16 (suppl. 1), 29-38, 1990.

Kryter, K. Effects of ear protective devices on the intelligibility of speech in noise. *J. of Acoustic Society of America,* Vol. 18, 413-17, 1946.

Kryter, K. *The Effects of Noise on Man,* 2nd ed. New York: Academic Press, 1985.

Kryter, K. Impairment to hearing from exposure to noise. *J. of Acoustic Society of America,* Vol. 53 [5], 1211-34, 1973.

Lawther, A. and Griffin, M. The motion of a ship at sea and the consequent motion sickness amongst passengers. *Ergonomics,* Vol. 29 [4], 535-52, 1986.

Olivatt, M. Energy conservation and noise control in pneumatic devices and systems. *Plant Engineering,* 116-18, August 6, 1981.

Park, M. and Casali, J. A controlled investigation of infield attenuation performance of selected insert, earmuff and canal cap hearing protectors. *Human Factors,* Vol. 33 [6], 693-714, 1991.

Peterson, A. and Gross, E. *Handbook of Noise Measurement.* Concord, Mass.: GenRad, Inc., 1972.

Sanders, M. and McCormick, E. *Human Factors Engineering and Design,* 6th ed. New York: McGraw-Hill, 1987.

Zohar, D., Cohen, A., and Azar, N. Promoting increased use of ear protectors in noise through information feedback. *Human Factors,* Vol. 22 [1], 69-79, 1980.

OVERVIEW

The volume of air required depends primarily on the purity of the air supplied because more pollution requires more air for dilution. A comfortable climate is primarily a function, on the hot side, of temperature, humidity, and air velocity and, on the cold side, of temperature and air velocity.

Performance effects of hot and cold environments are difficult to obtain due to the effect of motivation. Prediction of the physical effects on the body can be done with various equations. The primary defense against heat stress is adequate air velocity to improve evaporation of sweat. The primary defense against cold stress is properly designed clothing.

CHAPTER CONTENTS

1 Air Volume and Quality
2 Comfort
3 Heat Stress
4 Cold Stress

KEY CONCEPTS

area/local ventilation
Botsball temperature
circadian rhythm
clean rooms
clo
comfort zone
convective heat transfer
core temperature
dehydration
dry bulb temperature
effective temperature (new)

evaporative heat transfer
heat stroke
homeostasis
lower explosive limit (LEL)
mean radiant temperature
mechanical efficiency
natural wet bulb
open loop/closed loop
percent people dissatisfied (PPD)
prescriptive zone
psychrometric chart

radiant heat transfer
relative humidity
setpoint
sick building syndrome
source downwind
threshold limit values (TLVs)
vapor pressure
wet bulb globe temperature
    (WBGT)
wet bulb temperature
wind chill index

1   AIR VOLUME AND QUALITY

1.1   Clean (Office) Environments

The volume of air required generally is proportional to the local contaminants; that is, outside air is used to dilute contaminated air. Contaminants can be defined a number of ways.

Generally, lack of oxygen is not a problem since a sedentary person (met = 1) has oxygen requirements of about .006 L/s. Assuming oxygen is 21% of the air and 25% of the air breathed is consumed, air requirements are only .12 L/s.

Carbon dioxide may be limiting. Outside air is about .03% carbon dioxide. In inspired air, carbon dioxide equals $.03 + .52/V$, where V = ventilation rate, L/s-person. ASHRAE (1988) uses .25% as the maximum allowable carbon dioxide. Using CO_2 as a criterion, for a 1.0 met task, recommended $V = 2.5$ L/s-person since $.03 + .52/2.5 = .24\%$ carbon dioxide. For 1.5 met, $V = 3.5$ L/s; for 2.0 met, $V = 5$ L/s; for 2.5 met, $V = 10$ L/s.

Body odors and cigarette smoke also can be contaminants. In addition, there may be odors from the materials in the workspace and from the ventilation system itself. See Box 23.1.

ASHRAE, recognizing the health dangers and cognitive effects (Oborne, 1983) of passive smoking (breathing tobacco smoke fumes), sets higher limits for rooms where smoking is permitted; for 1.0 met, $V = 10$ L/s and for 1.5–2.5 met, $V = 17.5$ L/s.

Fanger (1988a) gives the following formula to estimate the percentage of people dissatisfied with indoor air quality:

$$PD = 395 \exp(-3.25/P^{.25})$$

where PD = Percentage of people dissatisfied

P = Perceived air pollution, decipols

For 10% dissatisfied, $P = .7$; for 25% dissatisfied, $P = 2$; for 50% dissatisfied, $P = 6$; and for 100% dissatisfied, $P = 31$. If Fanger's formula is used for ventilation requirements, ventilation volumes might be quite high as he defines 10 L/s-person of unpolluted air (and assuming no ambient odors) as $P = 1$.

ASHRAE (1989), based on odor detection by 80% or more of visitors, recommends 8 L/s (15 cfm) of outside air/person. For office space and conference rooms, the recommendation is 10 L/s (20 cfm) of outside air/person.

BOX 23.1   *Odor perception*

Professor P. O. Fanger of the Technical University of Denmark developed two units (the olf and the decipol) to quantify how air quality is perceived by humans. The olf is the pollution emission rate; it is analogous to lumen (light) and watt (noise). The decipol is the perceived level; it is analogous to lux (light) and decibel (noise).

One olf (from the Latin "olfactus" = olfactory sense) is the emission rate of air pollutants (bioeffluents) from a standard person (an average adult working in an office or similar nonindustrial workplace, sedentary, and in thermal comfort with a hygienic standard equivalent to 0.7 bath/day). Measurement of olf values requires a panel of judges and a measurement of the supply of outside air to the space. Fanger (1988a) estimates olf values as:

Olf Value	Source
0–5/m² of floor	Materials in office
1	Sedentary person, 1 met
5	Active person, 4 met
6	Smoker, average
11	Active person, 6 met
25	Smoker, when smoking

One decipol (from the Latin "pollutio" = pollution) is the pollution caused by 1 olf ventilated by 10 L/s of unpolluted air. That is:

1 decipol = 1 olf/10 L/s = .1 olf/L/s

In a study of 15 Copenhagen offices (Fanger, 1988b), there were an average of 138 olfs, although there were an average of only 17 people. That is, the people themselves contributed only 17/138 = 12% of the olfs. The remaining 121 olfs were from smokers (35), materials in the space (28), and the ventilation system (58). Fanger felt that the 28 olfs from materials and 58 olfs from the ventilation system were important findings, as the typical assumption is that the building is perfect and all indoor air pollution comes from people. See ASHRAE (1993) for more on odors (Chapter 12) and air contaminants (Chapter 11).

Higher humidity reduces odor intensity. For minimum odor perception and irritation, keep air water vapor pressure between 10 and 15 torr. (1 torr = 1 mm Hg.)

For forced-air heating and cooling, the ventilation volume also may be determined by the room temperature. That is, to keep the temperature up in winter, warm air needs to be added and to keep the temperature down in summer, cold air needs to be added.

Note that if the space is not occupied, ventilation requirements change drastically as odors no longer are a consideration and temperatures can drift (higher in summer, lower in winter) from the thermal comfort values. Thus, since the typical office is occupied fewer than 50 of the week's 168 h, there is great potential for reduction of ventilation (and thus saving of energy) during the time the space is not occupied.

Exhaust air also can be recycled. For example, electronic precipitators are used to remove cigarette smoke and odors from the exhaust air; the purified air then is mixed with outside air to furnish supply air. The primary advantage is that the purified air does not have to be conditioned (brought to desired temperature and humidity levels).

The emphasis on energy conservation has led to relatively low ventilation rates ("tight" buildings), but sometimes this leads to **sick building syndrome** since interior pollutants are not dispersed. Example pollutants include cigarette smoke, cleaning compounds, hydraulic elevator fluid, ozone (from copiers), and even emissions from new furniture. Carbon monoxide (from cars) can enter through supply air ducts located near busy streets or through infiltration (e.g., elevators from underground parking garages leading to the building).

1.2 Dirty (Shop) Environments

An environment may become polluted through a local contaminant such as welding and solder fumes, solvent evaporation, smoke from ovens, and a variety of others. Here the problem is worker health rather than odors.

As pointed out in Chapter 24, the first strategy should be decreasing the concentration of the airborne contaminant (containment, isolation, substitution, and change of operating procedures). Then, through administrative control, reduce worker exposure duration. Finally, consider ventilation.

A key goal is to remove pollutants from the ventilation air. The capital cost of most ventilation systems over their life is relatively small compared with the operating cost. The operating cost is twofold: (1) the direct cost of electrical power for the fan and (2) the hidden cost of replacing the conditioned air (heated or cooled and humidified and purified to desired values) with new conditioned air. Thus, consider energy-recovery devices such as rotary wheels, fixed plates, heat coils, and runaround coils.

There are two general approaches: area ventilation and local ventilation. If the contaminant source is discrete, local ventilation usually is best.

1.2.1 Area ventilation

Area ventilation usually is used when contaminant sources are diffuse. The design rule is to keep the contaminant source between the person and the exhaust—that is, keep the **source downwind** from the person. In a multiple-source room, the air inlet and exhaust locations usually are fixed. The things that can be varied easily are (1) the orientation of the equipment relative to the exhaust and (2) the location of the equipment in the room. Therefore, try to locate the workstation so the source is downwind from the operator. Try to locate the source as close to the exhaust as possible.

Some processes (usually electronic or pharmaceutical) require very clean environments; these processes can be done in **clean rooms** where the input air is specially filtered, the air-flow direction is specially controlled, and local contamination (such as from worker hair and hands) is carefully controlled.

1.2.2 Local ventilation

With **local ventilation,** the concept is direct capture of the contaminant with a duct and local exhaust. The contaminated exhaust air can be dumped into the outside air or can be filtered and recycled into the work area. Be sure that any filter system is fail-safe. That is, if the filter fails, the exhaust must not be recycled into inhabited space.

Air volume sufficient to remove explosive substances such as acetone, ethanol, or xylene usually is not sufficient to prevent health problems as the **threshold limit values (TLVs)** are 1% to 3% of the **lower explosive limit (LEL).** For example, the LEL for acetone is 25,500 ppm while the TLV is 750 ppm. Thus the TLV usually is limiting. See Chapter 24 for more on TLVs.

Office air often is recycled. For example, the exhaust air from the office can be part of the supply air for the paint booth or ovens.

Contaminant removal techniques include particle removal control (mechanical filters and electronic air filters) and gas and vapor removal (activated charcoal). Since dumping contaminated air "out the window" is not considerate of those downwind, you may as well reuse the air yourself. For more on ventilation, see ASHRAE's *HVAC Applications* (1991, Chapter 25) and ASHRAE's *Handbook of Fundamentals* (1993, Chapter 23).

2 COMFORT

2.1 Psychrometric Chart
ASHRAE defines comfort as "that state of mind which expresses satisfaction with the thermal environment." However, specifying a comfortable environment is difficult since comfort is influenced by 7 major factors. Four are environmental (dry bulb temperature, water vapor pressure, air velocity, and radiant temperature) and two are individual (metabolic rate and clothing); the seventh is time of exposure.

The **psychrometric chart,** Figure 23.1, gives the relation between adjusted **dry bulb temperature** (mean of air temperature and radiant temperature) on the horizontal axis and water vapor pressure on the vertical axis.

2.2 Comfort for Standard Conditions
This simplified psychrometric chart (mechanical engineers use a more complicated one) has a cross-hatched area giving the American Society for Heating, Refrigeration and Air Conditioning Engineers' (ASHRAE) standard **comfort zone.**

The standards are based on studies on several thousand subjects at Kansas State University. Subjects voted their comfort: $TS = 1 =$ cold; $2 =$ cool; $3 =$ slightly cool; $4 =$ comfortable; $5 =$ slightly warm; $6 =$ warm; $7 =$ hot. This vote can be predicted from the following (Rohles et al., 1975):

$$TS = -1.047 + .158\ ET^* \qquad ET^* < 20.7$$
$$TS = -4.444 + .326\ ET^* \qquad 20.7 < ET^* < 31.7$$
$$TS = 2.547 + .106\ ET^* \qquad ET^* > 31.7$$

where TS = Thermal sensation vote for sedentary activity and .60 clo

ET^* = New effective temperature, C

It must be emphasized that you can't satisfy all the people any of the time. For the conditions within the crosshatch, none of the 1,600 subjects voted hot (7) or cold (1) but 3% were warm (6) or cool (2); 94% were slightly warm (5), slightly cool (3), or comfortable (4); the mean vote was 4.0 and standard deviation was .7 (Rohles and Nevins, 1973). This variability occurred in spite of three hours of exposure, standard clothing (0.6 **clo**), and standard metabolic rate (sedentary sitting).

$$1\ \text{clo} = \frac{.155\ \text{m}^2 \times \text{C}}{\text{W}} = \frac{.18\ \text{m}^2 - \text{h} \times \text{C}}{\text{kcal}}$$

The **percent people dissatisfied (PPD)** can be calculated (Rohles et al., 1980).

PPD = Percentage of people dissatisfied (voting other than 3, 4, 5), corresponding to cumulative area from negative infinity for $CSIG$ or $HSIG$

$CSIG$ = Number of standard deviations from 50% for cold conditions ($< 25.3\ ET^*$) for sedentary activity and .5-.6 clo

$\quad = 10.26 - .477\ (ET^*)$

$HSIG$ = Number of standard deviations from 50% for hot conditions ($> 25.3\ ET^*$) for sedentary activity and .5-.6 clo

$\quad = -10.53 + .344\ (ET^*)$

ET^* = New effective temperature, C

For example, for ET^* of 18 C, $CSIG = +1.67$; from a normal table, 95% are dissatisfied. For ET^* of 30 C, $HSIG = -.21$, and 42% are dissatisfied. At 25.3 ET^*, the minimum (6%) are dissatisfied.

The comfort temperature with a PPD of 6% and amount of clothing are related as follows:

$$ET^*6 = 29.75 - 7.27\ (ICL)$$

where ET^*6 = ET^* temperature (C) at which 6% will be dissatisfied

$\quad ICL$ = Insulation value of clothing ensemble, clo $\qquad ICL < 1.1$

$\quad = .82\ (\Sigma\ ICLI)$

$\quad ICLI$ = Insulation values of individual clothing items, clo (see Table 23.1)

For example, 6% will be dissatisfied with .8 clo at 23.9 and with .95 clo at 22.8 C.

The former standard was valid for persons dressed with 0.7 to 1.0 clo with a metabolic rate of "office work." The new standard (parallelogram) is valid for persons dressed with 0.5 to 0.7 clo and a metabolic rate of "sedentary sitting." The assumption is that people will wear more clothing in the winter than the summer. That is, clothing worn inside is affected by outside weather. Table 23.1 gives the clo values for various garments. Figure 23.2 shows clo values for three typical ensembles for men and women.

Engineers should remember that sweating to remove heat is not considered "comfortable." That is, a wet skin is not acceptable for comfort. (Ingredients in old ink recipes include salt as mold-retardant and a few drops of brandy as antifreeze—reminder of the days when climate control was crude.)

Fanger (1973a) reported that during comfort:

$$TSKIN = 35.7 - .027\ 6\ M$$
and $$ESWEAT = .42\ (M - 58)$$

where $TSKIN$ = Mean skin temperature, C

$\quad M$ = Metabolic rate, W/m^2

$\quad ESWEAT$ = Evaporative sweating rate, W/m^2

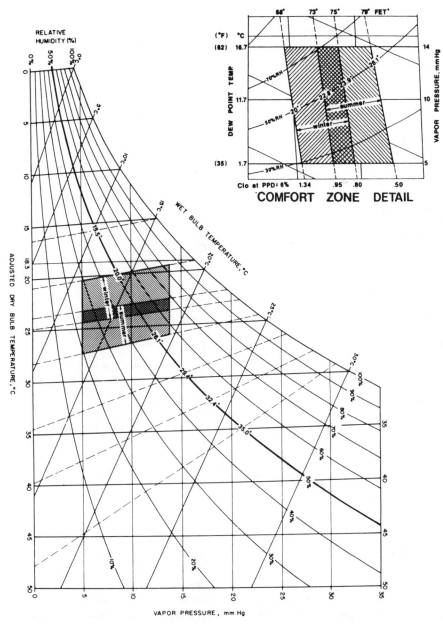

FIGURE 23.1 Psychrometric charts show that a dry bulb temperature of 25 C and **vapor pressure** of 15 mm of Hg (i.e., 15 torr) identifies a specific point. Use the curve coming from the left vertical axis to read 64% **relative humidity** (23.5 torr is the maximum water vapor pressure in 25 C air; 15/23.5 = 64%). Move horizontally to the left to read dew point temperature of 17.5 C. Move to the left using the solid slanting lines to read psychrometric **wet bulb** of 20.2 C. Move up on the dashed line at a steep angle to read effective temperature *(ET)* of 24 C when the dashed line reaches 100% humidity. Any point along a dashed line has approximately the same skin wetness and gives approximately the same comfort. In 1971 the number attached to each dashed line was changed to be the intersection with the 50% rh line (rather than the 100% line); the line was named **"new effective temperature"** or *ET** (rather than effective temperature, which is *ET)*; the weather reporters call *ET** the temperature index. Move down on the dashed line to read *ET** = 25.5 C.

2.3 Adjustments for Nonstandard Conditions
Nevins and Gorton (1974) and Fanger (1973b) give the following recommendations for adjustments to achieve comfort.

2.3.1 *Clothing*
Nevins (1975) says dry bulb temperature (DBT) decreases .6 C for every .1 clo increase. Nevins and Gorton (1974) say DBT decreases .6 C for every .1 clo increase from .6 clo

TABLE 23.1   Garment insulation values *(ICLI)*. Ensemble clothing, I$_{cl}$ = Σ *(ICLI)*. Thin garments are made of light fabrics worn in the summer; thick garments are made of heavy fabrics worn in the winter.

ICLI, clo	GARMENT DESCRIPTION	*ICLI,* clo	GARMENT DESCRIPTION
Underwear		**Dresses and skirts (knee length)**	
.01	Bra	.14	Skirt, thin
.03	Panties	.23	Skirt, thick
.04	Men's briefs	.23	Thin, sleeveless, scoop neck
.08	T-shirt	.27	Thick, sleeveless, scoop neck (jumper)
.14	Half slip	.29	Shirtdress, thin, short-sleeve
.15	Long underwear—bottoms	.33	Shirtdress, thin, long-sleeve
.16	Full slip	.47	Shirtdress, thick, long-sleeve
.20	Long underwear—top		
		Sweaters	
Footwear		.13	Vest, thin, sleeveless
.02	Socks, ankle-length athletic	.22	Vest, thick, sleeveless
.02	Panty hose	.25	Thin, long-sleeve
.02	Sandals, thongs	.36	Thick, long-sleeve
.03	Slippers, quilted, pile-lined		
.03	Socks, calf-length	**Suit jackets and vests (lined)**	
.06	Socks, knee-length (thick)	.10	Vest, thin, sleeveless
.10	Boots	.17	Vest, thick, sleeveless
		.36	Single-breasted, thin
Shirts and blouses		.42	Double-breasted, thin
.12	Blouse, sleeveless, scoop-neck	.44	Single-breasted, thick
.17	Shirt, sport, knit, short-sleeve	.48	Double-breasted, thick
.19	Shirt, dress, short-sleeve		
.25	Shirt, dress, long-sleeve	**Sleepwear and robes**	
.34	Shirt, sweat, long-sleeve	.18	Gown, thin, short, sleeveless
.34	Shirt, flannel, long-sleeve	.20	Gown, thick, long, sleeveless
		.31	Gown, hospital, short-sleeve
Trousers and coveralls		.34	Robe, thin, short, short-sleeve
.06	Shorts, short	.42	Pajamas, thin, short-sleeve
.08	Shorts, walking	.46	Gown, thick, long, long-sleeve
.15	Trousers, thin, straight	.48	Robe, thick, short wrap, long-sleeve
.24	Trousers, thick, straight	.57	Pajamas, thick, long-sleeve
.28	Sweatpants	.69	Robe, thick, long wrap, long-sleeve
.30	Overalls		
.49	Coveralls		

Source: ASHRAE, "Physiological Principles and Thermal Comfort" in *Handbook of Fundamentals*, R. A Parsons (Ed.). Copyright © 1993 by American Society of Heating, Refrigeration and Air Conditioning Engineers. Used with permission.

when total metabolism is less than 225 *W* but DBT decreases 1.2 C/.1 for over 225 *W.*

2.3.2 Activity

For each 30 *W* increase in total metabolism above 115 *W,* decrease *DBT* by 1.7 C. To permit sweat evaporation, keep relative humidity below 60% (15 torr).

2.3.3 Air velocity

For each .1 m/s increase in velocity up to .6 m/s, increase *DBT* by .3 C; for each .1 m/s between .6 and 1.0, increase *DBT* by .15 (Nevins, 1975). Rosen (1982), using a box fan (i.e., turbulent flow) at .8 and 1.3 m/s, found .1 m/s offsets a 0.3 C increase in temperature. Keep maximum velocity less than .7 m/s for sedentary occupations. The air can come from the front, rear, side, or above or below. Air flow from above interferes with the heat rising from the body, causing turbulent flow; air from below gives a more laminar flow. Bring warm air for heating in at the bottom of the people as they will be comfortable at a lower air temperature; bring air for cooling from above as they will be comfortable at a higher air temperature. Discomfort is affected by velocity fluctuations, so for comfort keep fluctuations low, that is, use nonturbulent air (Fanger and Christensen, 1986). For fixed work positions with light activity, velocity should be about .2 to .3 m/s. Maximum velocity for continuous exposure is 1.0 m/s. For intermittent exposure and high work levels, use 5 to 20 m/s; 5 to 10 are common; 0.5 to 1.5 m/s are most common for spot cooling of workplaces.

2.3.4 Mean radiant temperature (MRT)

MRT is the average radiant temperature coming from all directions; it depends on the shape presented to each direction. For each 1 C deviation of *MRT* from dry bulb temperature, change the *DBT* 1 C in the

(continued)

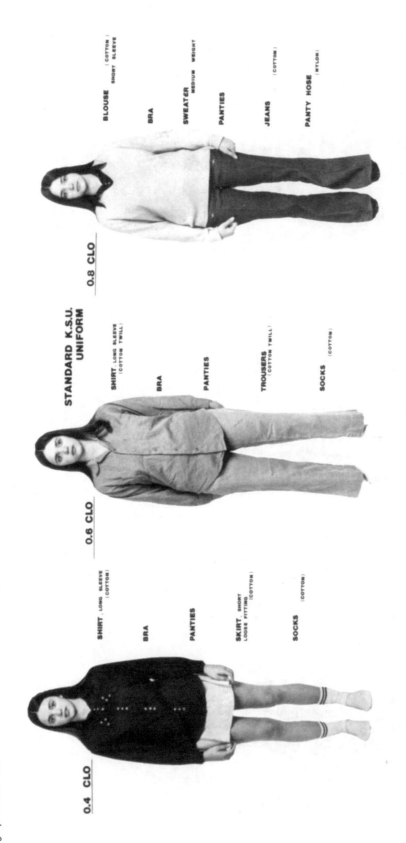

FIGURE 23.2 Clothing insulation is given in **clo** units (1 clo = (.155 m² − C)/W). As heating and air conditioning technology have improved, people have tended to wear less clothing. Zero clothing equals 0 clo; arctic clothing equals about 4 clo.

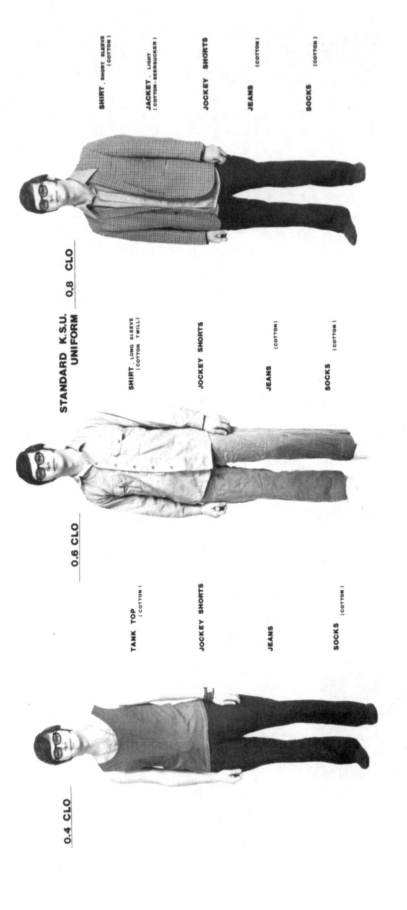

FIGURE 23.2   (continued)

opposite direction. For example, 25 C is a comfortable *DBT* when *MRT* also equals 25. But if *MRT* equals 27, then *DBT* should be 23.

2.3.5 Time of exposure
The standard values are based on the vote after 3 h of exposure. Males voted .5 vote warmer and females .2 warmer at the end of the first hour than they did after 3 h. At present, it is recommended that you use the 3-h values and not compensate for actual occupancy time.

2.3.6 Time of day
Even though core temperature is on a 24-h cycle, thermal comfort conditions do not differ with time of day (Fanger et al., 1974).

2.3.7 Season of year
To most people's surprise, comfort temperature does not vary with season of the year even for different ethnic groups—if clothing and metabolic rate are standardized. Fanger (1973b) found that Nigerians who had just arrived by jet in Copenhagen had the same comfort temperatures as Danes, Danes who swam in the sea in winter, Danes in cold meat packing jobs, and Americans.

Humidity control becomes a problem in winter. Vapor pressures above 12 torr may cause condensation problems and affect the glue from boxes in the warehouse. A vapor pressure of 7 torr can still fit within the comfort zone. Cold outside air could have an 80% relative humidity but still only have 7 torr. When this warmed air with the low vapor pressure passes a nose or respiratory tract (with moisture at about 45 torr), the 38 torr driving force transfers water from the person to the air. The person gets a sore throat although the room temperature is comfortable. Green (1974) found, for indoor climates in the winter, that colds and upper respiratory infections increase as humidity drops below 50% (about 12 torr). In general, for comfort, .1 torr is equivalent to .10 *CDB*.

2.3.8 Gender of occupant
If men and women have the same clothing insulation values, they have the same comfort temperature. Women's skin temperature at comfort is about 0.2 C lower and their evaporative loss during comfort is about 4 g/m²-h less; these characteristics balance women's lower basal metabolic rate (Fanger, 1972). Table 23.1 shows that women tend to wear clothing with lower clo values.

2.3.9 Age of occupant
If the occupants have the same clothing insulation values and the same metabolic rates, they have the same comfort temperature. However, since activity level tends to decline with age, older people usually prefer warmer rooms than younger people do.

The decline in basal metabolic rate with age seems to be compensated for by a lower evaporative loss during comfort (19 g/m²-h at age 23; 15 at 68; and 12 at 84) (Fanger, 1973b).

3 HEAT STRESS

3.1 Criteria of Stress
In the previous section, comfort was the criterion. For more extreme environments the problem becomes the effect on performance (both physical and mental) and on health.

The most serious of heat-induced illnesses is **heat stroke,** which has the potential to be life threatening or result in irreversible damage. Other heat-induced illnesses include heat exhaustion, heat cramps, and heat disorders (dehydration, rashes, etc.). If, during the first trimester of pregnancy, a female worker's core temperature exceeds 39 C for extended periods, there is a risk of fetus malformation. Additionally, core temperatures above 38 C may cause temporary infertility in both males and females (ACGIH, 1993).

Wyndham and Strydom (1965) gave Figure 23.3 as the temperature–time trade-off for physical work (8.7 W/kg). Wing and Touchstone (1963) compiled a 162-reference bibliography on the effects of temperature on human performance. Wing (1965), summarizing 15 different studies of sedentary work in heat, gave Figure 23.4 as the temperature–time trade-off for mental performance; he noted that human performance deteriorates well before physiological limits have been reached. Hancock and Vercruyssen (1988) show the effect of heat on different tasks in Figure 23.5. Ramsey and Kwon (1988) report that perceptual motor tasks show an onset of performance decrements when the environment is in the 30 to 33 C *WBGT* range; mental tasks and simple tasks show little effect of heat.

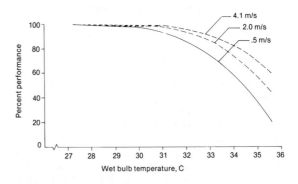

FIGURE 23.3 Physical work (shoveling) was reduced in high temperatures; the decline was reduced with higher air velocity; 27.2 C *WB* was 100% (Wyndham and Strydom, 1965).

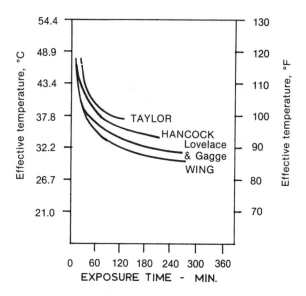

FIGURE 23.4   Sedentary mental performance in 15 studies was summarized by Wing (1965). The curves give the upper limit. Above the curve some performance decrement should be expected. Grether (1973), ignoring exposure time, gives 35 C *ET** (29.5 *ET)* as the point at which performance begins to deteriorate.

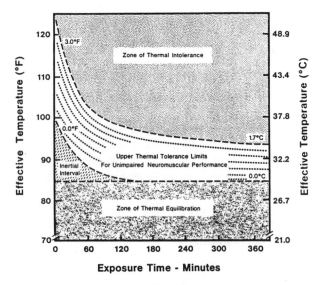

FIGURE 23.5   Performance in heat of sedentary workers is divided into three zones (Hancock and Vercruyssen, 1988). When people exercise, equilibrium body temperature rises (see Figure 23.6) and the exercise equilibrium body temperature may become the floor level from which the changes described in Figure 23.5 occur. In the zone of thermal equilibration, body temperature of sedentary workers (seated with metabolic rate up to 180 kcal/h) will not rise and performance will not be affected by heat. In the inertial interval, the body temperature will rise (and performance will drop) as the body heats up. The dashed lines indicate combinations of time and environmental temperature which will give a constant rise in body temperature. The physiological tolerance limit implies a 1.7 C rise in core temperature above a resting threshold for a sedentary worker. Hancock (1981) reported breakdowns in dual task performance with a deep body temperature rise of .2 C, psychomotor performance with a rise of .9 C, and mental performance with a rise of 1.3 C.

A difficulty of using performance criteria from short-run situations (such as experiments) is that motivation seems to increase during experimental stress situations. Performance is *higher* in a heat-stress environment than in comfortable conditions. Meese et al. (1984) studied approximately 1,000 people working for a day and found that performance improved up to 32 C, even though comfort decreased. It can't be entirely coincidence that all the developed countries are in the temperate zone, not the tropics. Ramsey et al. (1983) report that experimental evidence of 17,000 observations shows safe working behavior is best when at comfort conditions. Therefore, design using comfort as the criterion, realizing that motivated people can work in hot environments but the ordinary worker in a hot environment will work at a low efficiency.

For health, increase in heart rate or systolic blood pressure could be the criterion. During physical exercise in the heat, the effect of heat is reflected in increased skin blood flow up to about seven times basal skin blood flow. Pirnay, Petit, and Deroanne (1969) report that the heart rate increases 32 beats/min for every C increase in body temperature. For a single criterion, most experts recommend maximum body **core temperature** as the best. The National Institute for Occupational Safety and Health (NIOSH) set 38 C as the maximum core temperature. Core temperature usually is measured as rectal temperature, although ear canal and esophageal temperature also are used. Rectal temperature = brain temperature = liver temperature = right heart atrium + .6 C = esophageal + .6 C = mouth + .4 C. Rectal temperature represents the temperature of 20% of body mass while esophageal represents the temperature of 80% of the body (Minard and Copman, 1963).

My recommendation, based on extensive experimental evidence (Konz et al., 1980; Konz et al., 1983), is to use multiple criteria rather than just body temperature. My recommendation is to use body temperature, heart rate, and also subjective feelings of illness or faintness.

3.2   Environmental Limits
To set limits, there are two possible strategies: (1) measuring physiological responses during work and (2) predicting the stress on average people beforehand based on predicted environments and tasks.

Until recently, measuring during work was not practical. However, the Electric Power Research Institute has sponsored work which has led to a personal monitor that monitors heart rate and body temperature and beeps when the criteria levels are exceeded. Stay time at work was 50 to 100% longer when wearing the monitor over standardized prediction times (O'Brian, Bernard, and Kenney, 1988; Cohen, 1988).

For standardized predictions, Lind (1963) presented the critical concept of the prescriptive zone (see Figure 23.6). At lower environmental temperatures the body remains in thermal equilibrium, although the set point temperature is higher for higher metabolic rates. The point at which rectal temperature begins to rise is a function of metabolic rate as well as the environment. In addition, at higher metabolic rates there is little margin before rectal temperature reaches 38 C. Nielsen (1994) says people are exhausted (unable to continue) when their core temperature is 40 C.

Wyndham and Strydom (1965) remark that under compulsion (e.g., military training), heat stroke may occur at 28.5 C *ET,* whereas industrial workmen in Australian mines have worked without heat stroke at 32 C *ET* since they slow down in the heat.

NIOSH (Ramsey, 1975) took Lind's concept and the idea that a small amount of heat storage is not harmful (i.e., as long as core temperature is below 38 C) to get the values given in Figure 23.7, which shows threshold limit values for heat. ACGIH (1993) recommends the following adjustments: 0 C for summer work uniform (.6 clo), −2 C for cotton coveralls (1.0 clo), −4 C for winter work uniform (1.4 clo), and −6 C for water barrier, permeable (1.2 clo); also −2.5 C for unacclimatized workers. The values protect 95% of the workforce; it was assumed that the 5% of heat-intolerant individuals would not be

working on hot jobs. It is essential to note that these are not heat *limits;* these are values at which precautions (provision of adequate drinking water, annual physical examinations, training in emergency aid for heat stroke) should begin to be taken.

The recommendations are expressed in a specific index called **wet bulb globe temperature (WBGT),** which attempts to combine into one number the effect of dry bulb temperature (DBT), water vapor pressure, air velocity, and radiant temperature. *WBGT* "probably exaggerates the danger of heat stress in medium and low humidity environments" (Gagge and Nishi, 1976).

For an environment in which radiant temperature is close to air temperature, indoors or outdoors, with no solar load:

$$WBGT = .7\ NWB + .3\ GT$$

For an environment in which radiant temperature is not close to air temperature, outdoors with solar load:

$$WBGT = .7\ NWB + .2\ GT + .1\ DBT$$

where *WBGT* = Wet bulb globe temperature

NWB = **Natural wet bulb** temperature (temperature of a sensor with a wet wick exposed to natural air currents). Don't confuse it with *WB,* **wet bulb temperature,** also called psychrometric wet bulb, which is the temperature of a sensor with a wet wick exposed to

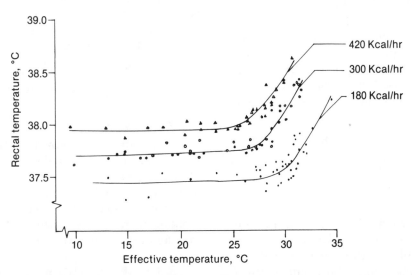

FIGURE 23.6   Rectal temperature remains constant at lower environmental temperatures. In this nonstress zone, rectal temperature = 37.0 + .0038 (metabolic rate, *W*) (Berenson and Robertson, 1973). Nielsen (1938) made the original discovery that rectal temperature varied with metabolic rate in the nonstressed zone. However, as environmental temperature rises, eventually body temperature also begins to rise due to the environment. This environmentally driven zone, beginning about 26–29 C *ET,* is called the **prescriptive zone.**

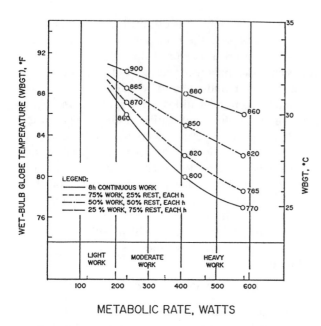

FIGURE 23.7   Threshold limit values for heat (ASHRAE, 1991, Chapter 25) for heat-acclimatized people. The rest area is assumed to have the same *WBGT* as the work area. If the rest area *WBGT* is below 24 C (75 F), reduce resting time by 25%.

high (3 m/s) air velocities. *WB* = *NWB* for air velocity > 2.5 m/s; for .15 < *V* < 2.5, *NWB* = .1 *DBT* + .9 *WB* (Gagge and Nishi, 1971).

GT = Globe temperature (temperature at the center of a 15-cm diameter black sphere)

DBT = Dry bulb temperature (sensor shielded from radiation)

Rather than measuring *WBGT* directly with the expensive and cumbersome WBGT apparatus, many people use the inexpensive and simple Botsball device. Then either the **Botsball temperature** can be used directly or the WBGT temperature calculated. A good conversion equation (r^2 = .90) (Beshir, 1981) is:

$$WBGT, C = 0.80 + 1.07 \text{ (Botsball Temp., C)}$$

See Guideline 6 of Chapter 18 for a discussion of burn temperatures.

3.3   Reduction of Heat Stress

Box 23.2 discusses homeostasis and control systems.

Solutions for various environments are complex due to the many feedback loops, but solutions can be found by computer simulation (Haslam and Parsons, 1994). For the person concerned with reducing heat stress, examination of the heat storage equation

(with the realization that the equations treat the body as an open-loop system in a static environment) will solve most problems. For the student interested in digging deeper, consult ASHRAE's *Handbook of Fundamentals* (1993).

The key equation is the heat balance equation:

$$S = M - (\pm W) + (\pm R) + (\pm C) + (\pm E) + (\pm K)$$

where S = Heat storage rate, watts

M = Metabolic rate, watts

W = Mechanic work accomplished rate, watts (walk up steps = +; down = −)

R = Radiation rate, watts (gain = +; loss = −)

C = Convection rate, watts (gain = +; loss = −)

E = Evaporation rate, watts (condensation = +; loss = −)

K = Conduction rate, watts (gain = +; loss = −)

3.3.1   Storage

Human body heat storage is:

$$S = 1.15 \, m \, C_p \, (MBT_f - MBT_i)/t$$

where S = Storage gain (+) or loss (−), watts

m = Weight of body, kg

C_p = Specific heat of body = .83 kcal-kg/C

MBT_i = Initial mean body temperature, C

MBT_f = Final mean body temperature, C

t = Time, h

Mean body temperature is calculated by weighing skin temperature and core temperature—usually with coefficients of .33 and .67, although in the heat some experts recommend .3 and .7 and others .2 and .8 as the skin "shell" becomes thinner.

$$MBT = .33 \text{ (skin temperature)}$$
$$+ .67 \text{ (core temperature)}$$

3.3.2   Metabolism

Table 11.2 gave metabolic rates for various activities. (Although you have more heat than a light bulb, you're not so bright!) (100 kg-m/min = 723 ft-lb/min = 16.35 watts = 14.04 kcal/h). Reducing metabolic rate can be done by mechanization or by working more slowly. Since working more slowly usually reduces productivity, look for other alternatives first.

3.3.3   Work

When the body accomplishes mechanical work such as walking stairs or pedaling a bicycle, this energy must be subtracted from *M* to determine the net heat within the body core. The

BOX 23.2 *Homeostasis*

Interior human body temperature is maintained at a constant value even though the environment varies; this is called **homeostasis.** (In Greek, *homoios* = similar and *stasis* = position, standing.) Other variables (blood sugar, water balance) also are maintained in homeostasis.

However, the body really does not maintain these values at a fixed point **(setpoint).** There is a biological rhythm. Figure 13.2 shows the biological rhythm for body temperature. Since it is a 24-h rhythm, it is called a **circadian** (*circa die* in Latin) **rhythm.** Menstrual cycles are an example of a lunar rhythm.

In addition, the setpoint for body temperature can change due to exercise.

Control systems can be **open loop** or **closed loop.** For a mechanical example, a traffic light controlled by a timer is an open-loop system. A traffic light controlled by a magnetic presence sensor in the pavement (feeding back the presence of a vehicle) is a closed-loop system.

Figure 23.8 depicts the human thermoregulatory system—a closed-loop system. The schematic has two portions: the controlling system and the controlled system. Body temperature (both core and skin) is fed back to the comparator (hypothalamus) where it is compared with the setpoint (reference). The difference (error) triggers release of hormones, which actuate one or more of three controllers (muscles, sweat glands, blood vessels). Muscle activity adds heat; sweat gland activity increases heat loss; blood vessel vasodilation or vasoconstriction fine tunes convective heat loss from the skin. The resulting body temperature is fed back to the comparator where the cycle repeats. External events (metabolic work or change in environmental conditions) act as disturbances.

Since the system attempts to reduce the error it is called a negative feedback system. The response (e.g., sweat) is proportional to the error (more sweat when you are hotter). Most mechanical systems use a discrete response. For example, a furnace turns on when you need heat but does not run at different levels of heat. It just runs for a longer or shorter time.

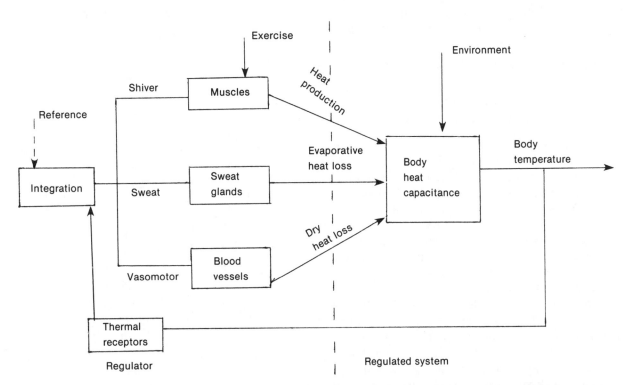

FIGURE 23.8 Human thermoregulation, a closed-loop system, can be divided into the controlling system and the controlled system (body temperature). Many of the problems associated with setting work limits in hot environments are due to using open-loop equations rather than closed-loop equations with individual variation in equation coefficients.

ratio of W/M is the **mechanical efficiency** of the body when performing work. It is about .2 for pedaling and .06 for repetitive lifting, and is considered 0 for most activities.

3.3.4   Radiation   Radiant heat transfer is:

$$R = \sigma A f_{eff} f_{cir} F_{cir} e(T^4_{mrt} - T^4_{skin})$$

where R = Radiant gain (+) or loss (−), watts

σ = Stefan-Boltzmann constant

= 5.67×10^{-8} watts/(m² − K⁴)

A = Skin surface area, m² (see Chapter 10)

f_{eff} = Effective skin radiation area factor

(.725 for standing, .696 for sitting)

(Fanger et al., 1970)

f_{cir} = Increase in radiant area due to clothing

= $1 + .155 I_{clo}$

I_{clo} = Insulation value of clothing, clo

(see Table 23.1)

F_{cir} = Multiplier to radiant heat transfer

coefficient to adjust for clothing barrier

= $1/(1 + .155 (5.2) I_{clo})$ where 5.2 W/(m²
− C) is the reference radiant heat

transfer coefficient

e = Emissivity (nonvisible radiation: .95 for

skin, clothing; visible radiation: .8 for

black skin and dark clothing, .65 for

pastel clothing, .6 for white skin, .5 for

white clothing)

T_{skin} = Temperature in K of the skin, K =

C + 273

T_{mrt} = Mean radiant temperature in K of

environment, K = C + 273

The driving force is the difference between the two temperatures—each raised to the 4th power. The important number is 35 C—skin temperature in the heat. For T_{mrt} above 35 C, the body gains heat; below 35 C it loses. For a typical task with T_{mrt} = 43 C, R = 60 watts; with T_{mrt} = 37.8 C, R = 5.

Reduce radiant load by working in the shade since heat radiation behaves much as light radiation. Working in the sun can add 170 watts to a clothed person and 220 to an unclothed person.

Clothing (a mobile shield) is the first line of defense, so use hats and long-sleeved shirts. As a first approximation, clothing temperature is halfway between skin temperature and environmental temperature. For visible radiation (such as the sun), use light-colored clothing; clothing color does not matter for nonvisible radiation. Clothing material, density, and thickness do matter.

A fixed shield between the person and the source is a second line of defense. Use with ovens, welding torches and arcs, and molten glass. Reflecting heat shields tend to be more effective and economical than absorbing shields or water-cooled shields. The air on the source side of the shield becomes heated and rises—giving a welcome current of supply (makeup) air to the worker. See Figure 23.9. Aluminum is a very good shield since it has high reflectivity and doesn't corrode. If the operator must see the source, use a screen of chains or coated glass (glass, although more effective, tends to get broken or dirty). Cover an oven conveyor entrance and exit with a screen of hanging chains.

3.3.5   Convection   Convective heat transfer is:

$$C = b_c A f_{clo}(T_{air} - t_{skin})$$

where C = Convection gain (+) or loss (−), watts

b_c = Convective heat transfer coefficient,

watts/(m² − C)

= 4.5 W/(m² − C) for standing adults with

velocity .05 to 2 m/s

= $8.3 V^{.6}$ for seated adults

A = Skin surface area, m² (see Chapter 10)

f_{clo} = Multiplier to b_c for clothing

= $1/(1 + .155 (2.9) I_{clo})$

where 2.9 W/(m² − C) is b_c in still air

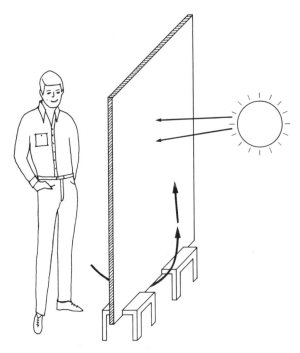

FIGURE 23.9   Radiant heat shields should have a gap at the bottom to aid convection air flow.

(.15 m/s)

I_{clo} = Insulation value of clothing, clo
(see Table 23.1)

V = Air velocity, m/s

t_{air} = Air temperature, C

t_{skin} = Skin temperature, C

The driving force is the difference between the two temperatures. As with radiation, keep temperature of the environment below 35 C. Second, increase air velocity on the skin. Figure 23.10 shows that air velocity above 2 m/s has little additional benefit. It is important to note that air velocity drops very, very rapidly with distance. The primary effect of clothing is not obvious in the equation—it drops air velocity next to the skin to effectively zero. Therefore, to maximize convective loss, wear little clothing if the temperature is below 35 C and insects, radiation, and social mores are not a problem. If clothing is worn, don't restrict air circulation at the neck, waist, wrist, or ankles.

Convective cooling is effective if air temperature is sufficiently below 35 C, perhaps at 15 C. The problem is to get the cool air. One solution is to pass air through water, which will normally be about 10–15 C. The air then is 15 C but humidity is 100%, say 12 torr. The worker then is cooled by convection. The technique works best with relatively high air velocities (.5 m/s).

3.3.6 Evaporation   Evaporative heat transfer is:

$$E = b_c A W F_{pcl}(VP_a - VP_s)$$

where E = Evaporative gain (+) or loss (−), watts

b_c = Evaporative heat transfer coefficient = 2.2 b_c

A = Skin surface area, m²

W = E/E_{max} Mislabeled the proportion of the skin that is wet, it actually is the proportion of actual sweat to maximum possible sweat. Values of .7 are a reasonable maximum and .5 are much more common.

F_{pcl} = Decrease in evaporative efficiency for permeable clothing

= $1/(1 + .143 (2.9) I_{clo})$ where 2.9 W/(m² − C) is b_c in still air (.15 m/s) for a sedentary person.

I_{clo} = Insulation value of clothing, clo (see Table 23.1)

VP_a = Vapor pressure of water in air, torr

VP_s = Vapor pressure of water on skin (45 torr if t_{skin} = 35 C)

Evaporation can be limited by the body or the environment.

Body limitations   The capacity of the body to sweat is quite large, especially after acclimatization. Each kg of sweat gives about 580 kcal/kg if evaporated from the skin. An unacclimatized man can sweat 1.5 L/h; within 10 days of acclimatization this can reach 3.0 L/h (Guyton, 1971). Skin lotions can restrict sweat evaporation (Spaul et al., 1985). Acclimatization to heat requires exercise in the heat. People living in a hot climate who do not exercise do not get the benefits of acclimatization. Vitamin C deficiency slows the rate of acclimatization; vitamin C aids sweating (Strydom, 1976). Use of 250 mg of vitamin C/day permitted acclimatization in an average of 5.2 days instead of 8.7 (Strydom et al., 1976). Acclimatized people are able to eliminate heat by sweating rather than vasodilation. The reduced skin blood flow results in heart rates 30–40 beats/min less for acclimatized people when exercising. Unfortunately, acclimatization can be lost in as little as 10 days, so workers returning from vacation must be careful of heat stress. For each day of nonexercise in the heat (such as vacations), the equivalent of .5 day of acclimatization is lost (Givoni and Goldman, 1973). As long as water is replaced, sweating can continue without health problems, but **dehydration**

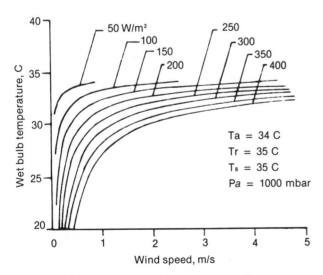

FIGURE 23.10 Equal cooling power lines show the effect of air velocity and metabolic rate when air and radiant temperature and vapor pressure are held constant. The person is assumed to be nude (Mitchell and Whillier, 1971). Note that there is little benefit of increasing air velocity on the skin beyond about 2 m/s.

of 3% causes physiological performance changes, 5% gives evidence of heat exhaustion, and at 7% hallucinations occur. Losses totaling 10% are extremely hazardous and lead to heat stroke; if not treated immediately, death will result. Dehydration potential is greater when under forced ventilation (fans, vortex tube cooling, air jets). However, the thirst drive is not sufficient to replace water loss during heavy sweating. Supervisors must insist that workers drink water; frequent, small amounts are better than occasional large amounts. Drink from a container because volume drunk from a water fountain tends to be small (30–60 ml/drink) if in a comfortable temperature environment (Coetzee and Bennett, 1978). An insulated mug keeps the water cool, which increases palatability. Cool water is more important than flavorings (Konz et al., 1983; Johnson and Strowman, 1987). "Tank up" before heat exposure.

Sweat is .2% to .4% salt (.4% at high sweat rates) whether you are acclimatized or not, as long as you are in positive salt balance. If salt balance becomes negative due to lack of salt intake or heavy sweating, the kidneys start to decrease urine salt content within 30 min. Sweat salt content declines after several days (Collins, 1963). Salt tablets rarely are desirable. People take them excessively and get stomach problems and high blood pressure. Salt on food usually is sufficient. Whenever more than 4 liters of water/day are required to replace sweat loss, provide extra salt—2 g of salt for each liter over 4. A person doing heavy work under hot conditions may need an extra 7 g/day (Committee on Nutritional Misinformation, 1974). The normal adult diet in the United States has 4 g of sodium and 10 g of salt/day; requirements are 2 of sodium and 5 of salt. A typical well-salted meal has 3–4 g of sodium; 1 cup of beef or chicken broth, consomme, or bouillon has about 1 g of sodium. If additional salt is absolutely necessary during work, add salt to a lime drink to reduce its concentration and increase palatability.

Environment limitations Evaporation also can be limited by the environment. "It isn't the heat, it's the humidity." When water vapor pressure is over 32 torr, a 1 C increase in wet bulb is equal to a 10 C increase in dry bulb. Reduce environmentally limited evaporation by (1) increasing air velocity or (2) decreasing water vapor pressure.

Figure 23.10 showed that beyond 2 m/s air velocity had little benefit. However, achieving 2 m/s at the skin is difficult. Figure 23.11 shows a plan view of air velocity for a typical fan. Figure 23.12 also shows that velocity drops off very rapidly with departure from the fan's axis. A rule of thumb is zero velocity beyond a distance from the fan of 30D, where D = fan diameter.

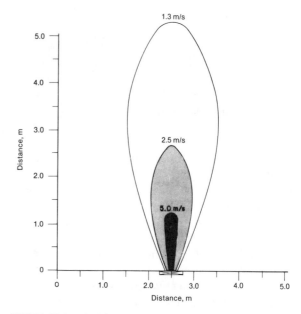

FIGURE 23.11 Fan velocity drops rapidly with distance, along the axis and especially with misaim. As a rule of thumb, assume no velocity beyond 30D, where D = fan diameter.

The second approach to improve evaporation is to decrease water vapor pressure in the air. Dehumidification, although expensive, makes a pleasant environment since sweat evaporates rapidly (discomfort is related to skin wetness).

3.3.7 Conduction
Astronauts have metabolic heat to be removed. The early spacesuits used

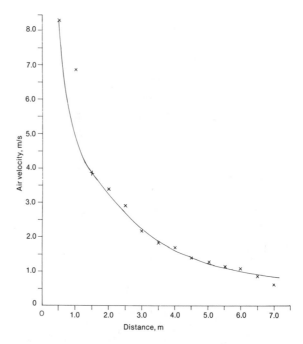

FIGURE 23.12 Fan placement is extremely critical because air velocity drops so rapidly with distance.

evaporative cooling to remove the metabolic heat, but blowers took considerable power and there was no place to dump the sweat-laden air. The solution was a network of small tubes on the torso through which small (1 L/min) volumes of cool (10 C) water were passed. Heat removal was excellent and there were no physiological problems from removing heat from only a localized area of the body. For earth-bound applications, the problems with this personal cooling technique are the high capital cost/unit (about $1,500) and the weight of the associated pump, compressor, and battery. If the weight is not kept on the back, the worker must be connected with a lifeline to the source of water and power (Konz, 1984).

Another technique is cooling with dry ice. Granulated dry ice is placed in pockets which are held next to the skin (Konz, 1984). This technique was used in the South African gold mines.

Figure 23.13 shows cooling with water ice. The key here is that the ice is in replaceable canisters. Thus, the person can continue being cooled without removing the nonpermeable protective garment usually worn which causes the heat stress.

FIGURE 23.13   Water cooling with ice is done two ways. In the model shown, the ice is in replaceable canisters and a pump circulates water by the ice and the warm skin. In another design, the ice is frozen in plastic packets and placed in pockets in a garment. There is no circulating liquid, pump, or battery but the ice is not replaceable while the operator is working (Konz, 1984).

4 COLD STRESS

4.1 Criteria of Stress

At low levels of cold stress, we experience discomfort. Additional cold gives loss of mental performance and manual dexterity. Then comes pain and potential loss of extremities such as fingers, ears, and feet. Finally, death. (At very low body temperatures (20 C), vital signs disappear, but the brain oxygen supply may still be sufficient; therefore, death from accidental hypothermia should be defined as "failure to revive upon rewarming.")

Use the formulas in the section on comfort and Figure 23.1 to evaluate whether people will consider themselves slightly cool, cool, or cold.

If the core is warm, dexterity does not drop until hand skin temperature reaches about 15 to 20 C (Poulton, 1970). Fox (1967), reviewing 68 references, says hand skin temperature (HST) is critical. Tactile sensitivity declines when HST is below 8 C and manual dexterity when HST is below 12-16 C (Fourt and Hollies, 1970). Poulton (1970, p. 150) reported less vigilance on ships when oral temperature dropped to 36.5 C. At a body temperature of 35 C, dexterity is reduced to the point where you cannot open a jackknife or light a match. When body temperature drops below 35 C, the mind becomes confused; at about 32 C there is loss of consciousness. At deep body temperature of 26 C, death occurs from heart failure. Local tissue, such as fingers and ears, freezes at −1 C instead of 0 due to osmotic pressure.

4.2 Environmental Limits

Just as people want one number to combine all factors for heat and use WBGT, they want one number for cold and use the **wind chill index,** I.

$$I, \text{kcal}/(\text{m}^2 - \text{h}) = h(t_{skin} + t_{air})$$

where h, kcal/(m² − h − C) = 10.45 − V + (100 V)$^{.5}$

$$V = \text{Air velocity, m/s}$$
$$t_{skin} = 33 \text{ C}$$
$$t_{air} = \text{Dry bulb air}$$
$$\text{temperature, C}$$

Table 23.2 gives the sensation for various values of I. The concept of wind chill also can be presented in the format of Table 23.3. Values give the 1.8 m/s air temperature that has the same value of I as the moving air. Air velocity can be estimated as:

   .8 m/s—paper on a desk moves

  1.5 m/s—a noticeable breeze

  2.5 m/s—light flag moves

    5 m/s—light flag fully extended

  7.5 m/s—raises newspaper sheet

  10 m/s—blowing and drifting snow

TABLE 23.2   Sensations for various values of I (Newburg, 1968, p. 423).

I kcal/(m² - h)	Sensation
50	Hot
100	Warm
200	Pleasant
400	Cool
600	Very cool
800	Cold
1000	Very cold
1200	Bitterly cold
1400	Exposed flesh freezes

Be cautious about rigorous use of the wind chill index as it is based on cooling a liter container of 33 C water, not a clothed human with a metabolism. A container of water does not have a metabolism, has no clothes, and has a much greater surface area/volume than a person. Wind chill can be used for livestock and as a crude index for precautions about frostbitten hands and ears.

4.3   Protection Against Cold Stress

Clothing is the primary defense in a heat-hungry environment. As clothing thickness increases, the insulation value increases to a point at which the increase is overcome by the increase in surface area. Thus, thin gloves may increase heat loss from a child's fingers! (If dexterity is not a problem, use mittens with liners rather than gloves.) The best insulator is air, so the Eskimos use two layers of fur. The Russian army since the Napoleonic War has issued winter boots one size too large—the soldiers stuff the extra space with straw or newspapers. The German army in World War II issued boots which fitted exactly and soldiers suffered many frostbitten feet.

Floors conduct heat (and cold) to the feet. When wearing normal shoes, optimal comfort for floors is 23 C (73.5 F) for standing or walking people and 25 C (77 F) for sedentary people (ASHRAE, 1993). Because of vasoconstriction, foot skin temperature usually is the lowest body skin temperature. (Normal skin foot temperature = 33.3 C for males but 31.2 for females (Oleson and Fanger, 1973).) Thus, cold feet may be due to a cold environment rather than a cold floor. Under-desk heaters may be requested when the real problem is the chair design (which cuts off circulation to the legs). Better chairs (lower and with rounded front edges) and foot rests may eliminate the desire for heaters.

Obviously, contact with cold metal should be avoided. Metal seats or metal handles for tools or controls should therefore be avoided. Wood, due to its lower conductivity, is better than plastic.

Provide windbreaks near the work. For clothing, a number of wind-proof layers tends to be more efficient than a single thick layer. It is especially important to protect the head since it does not vasodilate or constrict. Froese and Burton (1957) estimated:

$$H = 285 - 7.55\, T_{air}$$

where H = Head heat loss, kcal/(h − m² of head)

(the head was approximately .12 m²)

T_{air} = Air temperature, C

Therefore at −4 C, heat loss from the head may be 50% of resting metabolism. Adding 2.4 clo units over the head gives the same result as increasing the insulation over the 1.7 m² of body by 4 clo units. A stocking cap gives good protection and is easily

TABLE 23.3   Wind-chill "equivalent temperatures" predict the effect of air velocity at various temperatures. The number in the table gives the temperature at 1.8 m/s which has the same wind-chill as the dry bulb temperature at an air velocity. Thus, −12 @ 2 m/s, −6 @ 6 m/s, and 0 @ 8 m/s are approximately equivalent.

Dry Bulb Temperature, C	Air Velocity, m/s							
	2	4	6	8	10	12	14	16
+ 4	+ 3.3	+ 1.8	- 5.0	- 7.4	- 9.1	-10.5	-11.5	-12.3
+ 2	+ 1.3	- 4.2	- 7.6	-10.2	-12.0	-13.5	-14.6	-15.4
0	- 0.8	- 6.5	-10.3	-12.9	-15.0	-16.5	-17.6	-18.5
- 2	- 2.8	- 8.9	-12.9	-15.7	-17.9	-19.5	-20.7	-21.6
- 4	- 4.9	-11.3	-15.5	-18.5	-20.8	-22.5	-23.8	-24.8
- 6	- 6.9	-13.7	-18.1	-21.3	-23.7	-25.5	-26.9	-27.9
- 8	- 9.0	-16.1	-20.0	-24.1	-26.6	-28.5	-29.9	-31.0
-10	-11.0	-18.5	-23.4	-26.9	-29.5	-31.5	-33.0	-34.1
-12	-13.1	-20.9	-26.0	-29.7	-32.4	-34.5	-36.1	-37.2
-14	-15.1	-23.3	-28.6	-32.4	-35.3	-37.5	-39.1	-40.4
-16	-17.2	-25.7	-31.3	-35.2	-38.2	-40.5	-42.2	-43.5
-18	-19.2	-28.1	-33.9	-38.0	-41.1	-43.5	-45.3	-46.6
-20	-21.2	-30.5	-36.5	-40.8	-44.0	-46.5	-48.3	-49.7

removed and stored when not in use. In very severe cold, use a face mask.

Another technique is to warm the hands by putting on a jacket. The jacket reduces heat loss on the torso and thus lets warm blood rather than cold blood flow to the hands (Sundheim & Konz, 1990). See Box 18.1 for comments on gloves. Special precautions need to be taken if a worker is handling evaporative liquids (gasoline, alcohol, or cleaning fluids) because of the danger of cold injury due to evaporative cooling.

One serious problem with clothing is exercise, because the exerciser becomes a tropical person in arctic clothing. Exercise causes sweating; the sweat may accumulate in the clothing and freeze. Eskimos avoid this in two ways; they constantly pull off their outer fur parka so that much of the time their skin is cool. More important, the clothing is designed with many ventilation areas, gaps, and drawstrings. Eskimos bending over while wearing a parka show the bare skin of their backs. Another exercise clothing tip is that wool is superior to cotton: (1) It doesn't absorb water (wet clothes lose 90% of their clo value), and (2) the weave permits sweat to move from the skin to the outside of the garment (evaporation on the skin removes more heat from the body than evaporation from clothes).

Relative humidity makes little difference since a 10% rh may give an absolute vapor pressure of 1 torr vs. a 100% rh giving 7 torr. Skin vapor pressure is about 44 torr so $44 - 1 = 43$ is not much more driving force than $44 - 7 = 37$. A cold rain, however, puts cold water on the skin. Then raising each kg takes 1 kcal/C. Thus, raising 1 kg of water from 5 C to the 33 C of skin temperature would require 28 kcal; if the water then is evaporated, it takes 80 more kcal/kg from the body.

Respiratory heat loss is one reason you can eat more calories in the cold weather. For 0 C air, dry loss is 3 kcal/h and wet loss is 10 kcal/h, so at the end of the day you can have an extra 8 spoonfuls of sugar, or 1 beer. In the cold, urine production increases to about three times normal—leading to dehydration.

Counter with warm drinks such as soup or hot chocolate; avoid coffee with caffeine (which causes vasodilation).

Trauma sustained in freezing weather requires special attention because an injured worker is predisposed to secondary cold injury.

There is acclimatization to cold just as there is acclimatization to heat. The exact physiology is relatively unknown but probably involves improved nonshivering thermogenesis (increased heat generation without shivering). The location of the nonshivering thermogenesis in humans is debated but is probably the liver, the muscles, and fat. Metabolism in the muscles increases up to 50% before the physical movement of shivering begins. The cold-acclimatized person retains the ability to use shivering also.

REVIEW QUESTIONS

1. For a sedentary (i.e., desk) task, how much more ventilation is needed in the office if smoking is permitted?

2. Sketch the psychrometric chart showing: dry bulb temperature, absolute humidity, relative humidity, wet bulb temperature, effective temperature, and new effective temperature.

3. What is the clo value of the clothing you are presently wearing?

4. Sketch the core body temperature vs. environmental temperature for low, medium, and high metabolic rates.

5. What is the difference between wet bulb and natural wet bulb temperature? How is each measured?

6. Why should a radiant heat shield have a gap at the bottom?

7. How does heat acclimatization reduce heat stress?

8. What is the primary defense against heat stress? What is the primary defense against cold stress?

9. Why is it so important to protect the head in cold weather?

REFERENCES

ACGIH. *1993-1994 Threshold Limit Values and Biological Exposure Limits.* Cincinnati: American Congress of Governmental Industrial Hygiene, 1993.

ASHRAE. *Handbook of Fundamentals*, Parsons, R. (ed.). Atlanta: American Society of Heating, Refrig-

eration and Air Conditioning Engineers, 1993.

ASHRAE. *HVAC Applications.* Atlanta: American Society of Heating, Refrigeration and Air Conditioning, 1991.

ASHRAE. *HVAC Systems and Applications.* Atlanta:

American Society of Heating, Refrigeration and Air Conditioning Engineers, 1988.

ASHRAE. *Ventilation for Acceptable Indoor Air Quality* (Std. 62-1989). Atlanta: American Society of Heating, Refrigeration and Air Conditioning Engineers, 1989.

Berenson, P. and Robertson, W. Temperature. In *Bioastronautics Data Book*. Washington, D. C.: National Aeronautics and Space Administration, 1973.

Beshir, M. A comprehensive comparison between WBGT and Botsball. *American Industrial Hygiene Association J.,* Vol. 42 [2], 81-87, 1981.

Coetzee, J. and Bennett, C. The efficiency of a drinking fountain. *Applied Ergonomics,* Vol. 9 [2], 97-100, 1978.

Cohen, J. Heading off heat stress. *EPRI Journal,* 22-28, July–August 1988.

Collins, K. Endocrine control of salt and water in hot conditions. *Federation Proceedings,* Vol. 22, 716-20, 1963.

Committee on Nutritional Misinformation. *Water Deprivation and Performance of Athletes.* Washington, D. C.: National Academy of Sciences, 1974; p. 2.

Fanger, P. Assessment of man's thermal comfort in practice. *British J. of Industrial Medicine,* Vol. 30, 323-24, 1973b.

Fanger, P. Conditions for thermal comfort — A review. *Thermal Comfort and Moderate Heat Stress.* London: Her Majesty's Stationery Office, 1973a.

Fanger, P. Hidden olfs in sick buildings. *ASHRAE Journal,* Vol. 30 [11], 40-43, November 1988b.

Fanger, P. The olf and decipol. *ASHRAE Journal,* Vol. 30 [10], 35-38, October 1988a.

Fanger, P. *Thermal Comfort.* New York: McGraw-Hill, 1972.

Fanger, P., Angelius, O. and Kjeruf-Jensen, P. Radiation data for the human body. *ASHRAE Transactions,* paper 2168, Part 1, 1970.

Fanger, P. and Christensen, Perception of draught in ventilated spaces. *Ergonomics,* Vol. 29 [2], 215-35, 1986.

Fanger, P., Hojbjerre, J., and Thomsen, J. Thermal comfort conditions in the morning and in the evening. *International J. of Biometeorology,* Vol. 18 [1], 16-22, 1974.

Fourt, L. and Hollies, N. *Clothing Comfort and Function.* New York: Marcel Dekker, 1970.

Fox, W. Human performance in the cold. *Human Factors,* Vol. 9 [3], 203-20, 1967.

Froese, G. and Burton, A. Heat losses from the human head. *J of Applied Physiology,* Vol. 10 [2], 235-41, 1957.

Gagge, A. and Nishi, Y. Physical indices of the thermal environment. *ASHRAE Journal,* Vol. 18 [1], 47-51, 1976.

Givoni, B. and Goldman, R. Predicting effects of heat acclimatization on heart rate and rectal temperature. *J. of Applied Physiology,* Vol. 35 [6], 875-79, 1973.

Green, G. The effect of indoor relative humidity on absenteeism and colds in schools. *ASHRAE Transactions,* Vol. 80, Part 2, 1974.

Grether, W. Human performance at elevated environmental temperatures. *Aerospace Medicine,* Vol. 44 [7], 747-55, 1973.

Guyton, A. *Textbook of Medical Physiology,* 4th ed. Philadelphia: W.B. Saunders, 1971.

Hancock, P. The limitation of human performance in extreme heat conditions. *Proceedings of the Human Factors Society,* 74-78, 1981.

Hancock, P. and Vercruyssen, M. Limits of behavioral efficiency for workers in heat stress. *International J. of Industrial Ergonomics,* Vol. 3, 149-58, 1988.

Haslam, R. and Parson, K. Using computer-based models for predicting human thermal responses to hot and cold environments. *Ergonomics,* Vol. 37 [3], 399-416, 1994.

Johnson, R. and Strowman, S. Effects of cooling and flavoring drinking water on psychological performance in a hot environment. *Proceedings of Human Factors Society,* 825-29, 1987.

Konz, S. Personal cooling garments: A review. *ASHRAE Transactions,* 499-518 [4], 1984.

Konz, S., Rohles, F., and McCullough, E. Male responses to intermittent heat. *ASHRAE Transactions,* 79-100, Part 1B, 1983.

Konz, S., Rohles, F., Zuti, W., and Skipton, D. Physiological responses of 262 seated sedentary subjects in neutral and heat stress environments. *ASHRAE Transactions,* Part 2, 1980.

Lind, A. A physiological criterion for setting thermal environmental limits for everyday work. *J. of Applied Physiology,* Vol. 18 [1], 51-56, 1963.

Meese, G., Kok, R., Lewis, J., and Wyon, D. A laboratory study of the effects of moderate thermal stress on the factory workers. *Ergonomics,* Vol. 27 [1], 19-43, 1984.

Minard, D. and Copman, L. Elevation of body temperature in health. In *Temperature—Its Measurement and Control,* Hertzfield, C. (ed.). New York: Reinhold, 1963.

Mitchell, D. and Whillier, A. Cooling power of underground environments. *J. of South Africa Institute of Mining and Metallurgy,* 93-99, October 1971.

National Academy of Sciences. *Recommended Dietary Allowances.* Washington, D. C.: National Academy of Sciences, 1974; p. 90.

Nevins, R. Energy conservation strategies and human comfort. *ASHRAE Journal,* Vol. 17 [4], 33-37, 1975.

Nevins, R. and Gorton, R. Thermal comfort conditions. *ASHRAE Journal,* Vol. 16, 90-93, January 1974.

Newburg, L. (ed.). *The Physiology of Heat Regulation and the Science of Clothing.* New York: Hafner Publishing, 1968.

Nielsen, B. Heat stress and acclimation. *Ergonomics,* Vol. 37 [1], 49-58, 1994.

Nielsen, M. Die Regulation der Korpertemperaturbei Muskelarbeit. *Skandinavian Archives Physiologue,* Vol. 79, 193-230, 1938.

Oborne, D. Cognitive effects of passive smoking. *Ergonomics,* Vol. 26 [12], 1163-78, 1983.

O'Brian, J., Bernard, T., and Kenney, W. Personal monitor to protect workers from heat stress. *Proceed-*

ings of IEEE Fourth Conference on Human Factors and Power Plants, June 1988.

Oleson, B. and Fanger, P. The skin temperature distribution for resting man in comfort. *Archives des Sciences Physiologigues,* Vol. 27 [4], A385-93, 1973.

Pirnay, F., Petit, J., and Deroanne, R. A comparative study of the evolution of heart rate and body temperature during physical effort at high temperature (in French). *Internationale Zeitschrift fur Angewandte Physiologie Einschliesslich Arbeitsphysiologie,* Vol. 28, 23-30, December 1969.

Poulton, E. *Environment and Human Efficiency.* Springfield, Ill.: C. T. Thomas, 1970.

Ramsey, J. Heat stress standard. *National Safety News,* 89-95, June 1975.

Ramsey, J., Burford, C., Beshir, M., and Jensen, R. Effects of workplace thermal conditions on safe work behavior. *J. of Safety Research,* Vol. 14, 105-14, 1983.

Ramsey, J. and Kwon, Y. Simplified decision rules for predicting performance loss in the heat. In *Proceedings Seminar on Heat Stress and Indices,* 337-72. Luxembourg: Commission of the European Communities, 1988.

Rohles, F., Hayter, R., and Milliken, G. Effective temperature *(ET*)* as a predictor of thermal comfort. *ASHRAE Transactions,* Vol. 81, Part 2, 148-56, 1975.

Rohles, F., Konz, S., and Munson, D. Estimating occupant satisfaction from effective temperature *(ET*)*. *Proceedings of the Human Factors Society,* 223-27, 1980.

Rohles, F. and Nevins, R. Thermal comfort: New directions and standards. *Aerospace Medicine,* Vol. 44, 730-48, July 1973.

Rosen, E. Comfort and cooling with box fans. M. S. thesis, Kansas State University, 1982.

Spaul, W., Boatman, J., Emling, S., Dirks, H., Flohr, S., Crocker, W., and Glazeski, M. Reduced tolerance for heat stress environments caused by protective lotions. *American Industrial Hygiene Association J.,* Vol. 46 [8], 460-62, 1985.

Strydom, N. Effect of ascorbic acid on rate of heat acclimatization. *J. of Applied Physiology,* Vol. 141 [2], 202-205, 1976.

Strydom, N., van der Walt, W., Jooste, P., and Kotze, H. Note: A revised method of heat acclimatization. *J. South Africa Institute of Mining and Metallurgy,* Vol. 76, 448-52, 1976.

Sundheim, N. and Konz, S. Keeping bare hands warm with extra clothing on the body. *Advances in Industrial Ergonomics and Safety II,* B. Das (ed.). London: Taylor and Francis, 1990.

Wing, J. Upper thermal tolerance limits for unimpaired-mental performance. *Aerospace Medicine,* Vol. 36, 960-64, October 1965.

Wing, J. and Touchstone, R. A Bibliography of the Effects of Temperature on Human Performance. Technical Report AMRL-TDR-63-13. Ohio: Wright Patterson AFB, February 1963.

Wyndham, C. and Strydom, N. The effect of environmental heat on comfort, productivity and health of workmen. *The South African Mechanical Engineer,* 208-21, May 1965.

24 | TOXICOLOGY

OVERVIEW

Toxicology deals with the long-term effects of foreign chemicals on the body—it considers health. Although dermatitis is a common problem, toxins generally enter the body through breathing. Threshold limit values give the recommendations for permitted exposures. Toxicology controls are divided into engineering controls, administrative controls, and personal protective equipment.

CHAPTER CONTENTS

1 Poisons
2 Poison Routes
3 Poison Targets
4 Poison Elimination
5 Threshold Limit Values
6 Controls

KEY CONCEPTS

biological monitoring
biotransformation
dermatitis
excursions/peaks
leaky bucket
Material Safety Data Sheets
    (MSDSs)

permissible exposure limits
    (PELs)
poison targets
protective clothing
recommended exposure limits
    (RELs)

short-term exposure limits
    (STELs)
teratogens
threshold limit values (TLVs)
toxicology/cumulative trauma
work environment exposure limits
    (WEELs)

1 POISONS

Toxicology deals with long-term effects of foreign chemicals upon the body—it considers health. A toxicology problem might be the effect of 10 years' exposure to a solvent. Safety deals with the short-term effects of physical agents upon the body—it considers accidents. A safety problem might be cutting off a finger in a punch press.

However, the real world is not so discrete. For example, some chemicals can injure the body within hours rather than years. There also is considerable attention paid to repetitive strain or cumulative trauma. A cumulative trauma problem might be a sore shoulder due to repetitive reaching for an object over your head. Cumulative trauma is discussed in more detail in Chapters 15, 16, and 17.

Toxicology is a very complex subject and this chapter touches the subject only briefly. Toxicology problems have been present for many years. For example, French hatters in the 17th century used mercuric nitrate to aid fur felting and the resultant chronic mercury poisoning led to the expression "mad as a hatter." Chemicals affect the body with doses producing a response.

1.1 Effect

An effect could be a permanent physical change such as death or could be reversible, such as change in dark adaptation of the eye. The **threshold limit values (TLVs)** given later in this chapter primarily are based on nonreversible functional changes in an organ (usually the liver and kidneys) and define maximum acceptable exposure to various materials. A severe practical problem is that there may a 20-year lag between dose and response. For example, exposure to asbestos during the 1940s caused an increase in certain types of cancer in the 1970s and 1980s. In the 1940s organizations did not realize that use of asbestos would impair the health of the employees many years later. Another problem is that when there is such a long lag between exposure and the consequences (which are not certain, but just increased odds of poor health), many workers tend to ignore the hazard. A political problem is that the financial benefits accrue to organizations but the health costs to individuals.

1.2 Body

One technique of determining which chemicals are dangerous is to give the chemicals to humans and see what happens. This may be hard on the people! Thus, most TLVs are developed from studies on animals rather than humans. But there are marked species differences. For example, **teratogens** are substances which cause defects in fetal development. In rats and mice, thalidomide had no effect at a dose of 4,000 mg/kg of body weight

but a dose of .5 mg/kg in humans was teratogenic (Mastromatteo, 1981). The TLV approach has been to consider that humans respond as the most sensitive animal species and then use a safety margin between the dose which produces the effect and the TLV. But you can see that the TLVs are not as precise as they may seem. In addition, exposure to humans may indicate that a reduction in the TLV is needed. For example, some workers in polyvinyl chloride factories came down with a rare cancer many years later and so the TLV was reduced. The point here is that if it had been a common cancer the danger probably never would have been noticed.

1.3 Dose/Response

With present technology, it often is possible to detect chemicals in a concentration of one part per billion. This is roughly equivalent to measuring 1 second in a period of 33 years! Mere detection of a chemical is not enough. What needs to be considered is the dose in relation to the response. Water in excess will kill as will salt, sugar, alcohol, sulfanilamide, or arsenic. The problem is to define "in excess."

Think of the human as a **leaky bucket** with components on shelves at various levels. See Figure 24.1. The problem is whether the liquid will rise high enough to cause corrosion. As in the analogy, poisoning depends upon the rate of poison input (liquid input), the kind of liquid, the body size (bucket size), target organ susceptibility (shelf level), and the poison removal capability (hole size).

Since the same amount of poison rises much higher in a small bucket (small person), the same dose is more dangerous for small people. For simplicity, the TLVs are given for a 70-kg adult rather than

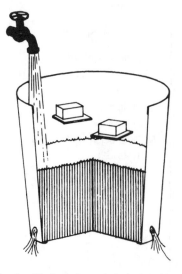

FIGURE 24.1 Consider the body as a leaky bucket with components on shelves at various heights. Will the liquid rise high enough to cause corrosion?

per kg of body weight. There also is a large individual range (5–10 fold or more) of susceptibility to toxins, just as there is a need for vitamins, the good chemicals. The hole size depends primarily on the ability of the kidneys and liver to transform the poison into a less toxic compound and to eliminate it from the body. Examples of people with less liver and kidney capacity are older people, people with hepatitis, and people who drink alcohol to excess.

2   POISON ROUTES

Figure 24.2 gives a schematic view of the body, making the point that entrance to the mouth or lungs is not entrance to the body. To enter the body, a poison must enter the blood. Thus, from a toxicology viewpoint, if your little brother swallows a penny and it goes through the stomach and is eliminated with the feces, it has not been in the "body." However, if he swallows or breathes an object or chemical which is broken down in the intestinal tract or is absorbed by the blood in the lungs, he has greater potential to be poisoned. Thus, an important characteristic of potential poisons is their ability to penetrate the body's perimeter. In general, inorganic materials are slow to penetrate the barrier, polar organic materials penetrate more quickly, and nonpolar organic materials are absorbed most quickly.

The most important potential entrance points are skin, the mouth, and the lungs.

2.1   Skin
The skin is a superb barrier; Webb and Annis (1967) showed that it could resist even the vacuum of space. Common experience has shown that most compounds run off the skin rather than penetrate the barrier. Therefore, people can insert their hands into compounds which, if they would penetrate the skin, would kill them. The TLV table (Table 24.1) has the word *skin* next to compounds which penetrate the intact skin. Clothing or shoes wetted with a toxic agent increase contact time with the skin and thus increase danger. Cuts and abrasions to the skin permit toxins (as well as germs) to enter the body. However, in general, poisons entering the body through the skin are not a serious toxicology problem.

2.2   Mouth
The second entrance route is the mouth, by eating or drinking the poison. Although children often consume a poison by itself, the most common problem for adults in industry is toxic compounds in food or drink. For example, in a pesticide factory, dust in the air from the process may fall on sandwiches, in open coffee cups, or on cigarettes. A famous example of inserting items into the mouth was the women who painted radium dials on watches. They occasionally pointed the tips of their brushes—by licking them! The best precaution seems to be to forbid eating, drinking, or smoking in work areas at any time. This protects the product from contamination as well as the workers. On the other hand, clean, convenient areas must be provided for eating and drinking. However, in general, poison entering through the mouth is not much of a problem in most industries.

2.3   Lungs
The biggest problem in poison absorption is the third route, the lungs.

The lungs are a problem due to both the physical nature of the poison (a gas or finely dispersed aerosol or mist, often invisible) and the design of the lungs (very good at moving molecules from the gas to the lungs).

The most important characteristic of particles in regard to inhalation is their size. Figure 24.3 shows a typical log-normal distribution of airborne dust and

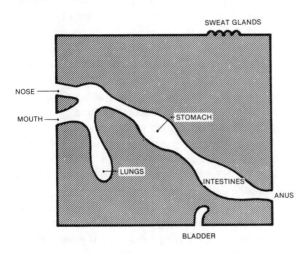

FIGURE 24.2   Lungs, stomach, and intestines are not in the body (from a toxicology viewpoint).

TABLE 24.1   TLVs for chemical substances for 1993–1994 (ACGIH, 1993). Only the first five substances are listed below.

Substance	TWA		STEL	
	ppm[a]	mg/m³[a]	ppm[a]	mg/m³[b]
Acetaldehyde	100	180	150	270
Acetic acid	10	25	15	37
Acetic anhydride	C 5	C 20		
Acetone	750	1780	1000	2375
Acetonitrile-skin	40	70	60	105

[a]Parts of vapor or gas per million parts of contaminated air by volume at 25 C and 760 torr.

[b]Milligrams of substance per cubic meter of air. When entry is in this column only, the value is exact; when listed with a ppm entry, it is approximate.

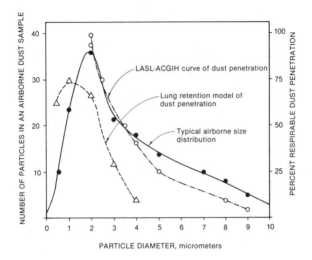

FIGURE 24.3 Airborne dust (solid line) has a log-normal distribution (Frazer, 1973) with a peak about 2 μm. Particle penetration (dashed circle line) shows how the larger particles are filtered out (the line is lower than the solid line) (Carson, 1974). Retention (dashed triangle line) is even less (Carson, 1974).

respirable dust. The peak is about 2 μm. The next question is what sizes are retained in the lungs.

Wright (1973) points out that, when lungs of 50- to 60-year-olds are dissected, about 50% of all particles are .5 μm or less in diameter. Effectively 100% are 5 μm or less; less than .002% are over 10 μm. (1 micrometer = 1 μm = 1 micron = .000 001 m. A human hair has a 100 μm diameter.) Particles in the air larger than 10 μm are removed completely in the nose and upper airways; particles between 5 and 10 μm or less are likely to be retained in the lung depths. (Particles from .1 to 1 μm significantly reduce visibility since their diameters are comparable to the wavelength of light in the visible region.)

A fiber will be defined as having a length more than three times the diameter. Almost all fibers will have a length of less than 50 μm. Since fibers tend to orient their long axis to the airstream, straight fibers and short fibers penetrate deeper than curved, U-shaped, or long fibers.

(To better understand the physical dimensions, bring the objects to human scale by using 1 μm = 25 mm. Then a .1 μm particle is a BB, a 1 μm particle is a golf ball, and a 10 μm particle is a basketball. The trachea (actually 25 mm) becomes 625 m in diameter with the cilia (actually 10 μm long) becoming slightly longer than a pencil. The terminal bronchioles (actually 300 μm) and the alveolar sacs (actually 75 to 300 μm diameter) become as the walls of a small classroom. The three-layer membrane separating the gas from the blood (actually .2 μm) becomes the thickness of a book cover.)

3   POISON TARGETS

As discussed in the leaky bucket analogy, poison targets various organs. For discussion here, **poison targets** will be divided into interior and exterior (skin).

3.1   Interior   The first target is the respiratory system. One response might be an increase in airflow resistance (asthmatic response); this response usually ends when the mucociliary escalator removes the irritant. A second response (chronic bronchitis) might occur if there is continuing irritation to the mucosa due to high levels of irritant gas or particles. There is an increase in coughing and sputum. A third response (acute bronchitis or pneumonia) may occur if the cilia movements are paralyzed—thus leading to inadequate lung cleaning and possible improved colonization of bacteria. A fourth response (chronic interstitial lung disease) may occur from dusts containing micro-organisms or animal proteins. In various forms, it is called bagassosis (baggasse is the fibrous material of sugar cane), mushroom picker's disease, wheat thresher's lung, pigeon breeder's disease, feather plucker's disease, malt worker's lung, and paprika slicer's disease (Buechner, 1975).

Particles which penetrate the lung are engulfed by macrophages when they settle on the lung surface. The macrophages, having the power of independent motion, draw the particles through the tissues of the lung wall and (a) directly into the blood, or (b) to surrounding bronchioles for ciliary removal to the lymph system, or (c) simply remain permanently attached to the wall. Inert particles present a problem. Free silica repeatedly kills the macrophages (silicosis); this leads to scar tissue and a loss of lung surface area (in effect, slow suffocation). Long asbestos fibers (over 5 μm) stick out of the macrophages and also cause scar tissue and loss of surface area. Cancer of the lungs is another possibility—especially from particles of chromium, nickel, uranium, and asbestos.

Finally, a compound may pass into the blood and attack other organs. Lead primarily affects nervous tissue, cadmium affects the kidneys, carbon monoxide and cyanide affect hemoglobin, and sulfur dioxide and hydrogen sulfide affect the lungs (cough). Hydrogen sulfide also causes pulmonary edema.

A fetus is especially endangered since the baby functions as a sponge to toxins; that is, it does not have a good ability to eliminate poisons. Substances which cause defects in fetal development are called teratogens. Thus TLV standards which protect the normal adult are not strict enough for pregnant women. The most critical period is the first three months of pregnancy. Zenz (1984) discusses the following teratogens: lead and its compounds, benzene, carbon monoxide, DBCP, chlordecone, chloroprene (2-chlorobutadiene), epichlorohydrin, ethylene dibromide, ethylene oxide, mercury, vinyl chloride, anesthetic gases, ionizing radiation, and PCBs.

In 1975, General Motors was sued by the United Auto Workers for discrimination in a battery factory. General Motors had prohibited women of childbearing age from working there although sterile women and women past childbearing age could work there. The lead concentrations varied from 1/2 to 3/4 of the TLV. The judge agreed with GM and said of the two competing social priorities of protection of health and equal opportunity employment, that protection of health should take priority.

3.2 Skin

Dermatitis accounts for 35 to 40% of reported industrial disease. (Dermatosis denotes disease of the skin; **dermatitis** is more limited, referring to inflammation of the skin.) Birmingham (1973) reported that 1% of the working population suffers occupational skin disease during a year. Birmingham (1975) reported that 41% of occupational illnesses were skin ailments causing 25% of lost working days.

The skin is the largest organ system of the body. Structurally the skin is composed of two layers—the epidermis and the dermis. The epidermis has an outer layer of dead cells (horny layer, keratin layer, or stratum corneum), which is a fair protection against chemicals (except alkalis and solvents). Next is the Malpighian layer of living cells. The dermis or true skin has connective tissues, nerves, hair follicles, oil and sweat glands, and blood and lymph vessels.

Some people have a genetic predisposition to atopic dermatitis (AD). AD tends to flare up when a person is exposed to triggering factors:

- dry skin
- low humidity
- skin infections
- heat, humidity, and sweating
- emotional stress
- irritants and allergens

Irritants include solvents, industrial chemicals, detergents, some soaps and fragrances, fumes and tobacco smoke, paints, bleach, woolens, acidic foods, and astringents and other alcohol-containing skin-care products. Allergens are usually proteins from food, pollens, or pets (*All About Atopic Dermatitis,* 1989).

The two main categories of contact dermatitis are irritant contact dermatitis (75% of cases) and allergic contact dermatitis (25% of cases). With allergic contact dermatitis it takes 14 to 21 days for an allergy to develop following initial contact. However, once sensitized, response becomes obvious within 12 h of exposure; it may not even occur at the contact site. Medications that are notorious for sensitizing individuals include topical anesthetics containing benzocaine, topical antibiotics containing neomycin, topical antihistamine creams, fungi creams, and the skin disinfectant thimerosal (merthiolate) (Hogan, 1986).

Causes of occupational dermatoses are:

- mechanical and physical—abrasions or wounds, sunlight for outdoor workers, fiberglass, and asbestos
- chemical—subdivided into strong irritants such as chromic acid and sodium hydroxide and marginal irritants such as soluble cutting fluids and acetone, which require prolonged contact over time.
- plant poisons—woods such as West Indian mahogany, silver fir, and spruce (when being sandpapered or polished)
- biological agents—anthrax contracted by handlers of skins or hides from infected animals; grain or straw itch contracted by food and grain handlers from handling produce infected with mites, etc.

In addition, some metals can sensitize the skin; nickel-plated earrings or stainless steel (which contains nickel) may be a problem.

Protective clothing helps—especially aprons and gloves. Aprons protect the chest and legs from liquids which might soak into the clothing and thus maintain contact with the skin for a long time. There are many types of gloves. See Box 18.1. Gloves for chemicals naturally must not be permeable to that specific chemical. Chemicals can penetrate the gloves (1) if the material degrades (change in physical property), (2) through penetration on a nonmolecular level (seams, holes, and even pores of materials such as leather), and (3) through permeation (chemical penetration on a molecular level). In addition, chemical protection gloves should have long sleeves (gauntlets) with a turn-back cuff to reduce chemicals dripping into the glove. Rinse gloves before removal; wash and air-dry gloves between wearings. A cotton liner may reduce skin irritation due to sweating. Gloves also can protect the skin against dry irritants and against cuts and abrasions. Machinists tend to cut their hands by wiping them with rags contaminated with metal shavings; they need to keep rags for wiping up separate from rags for hands—perhaps by using two colors of rags.

Exposure can be reduced by good housekeeping around the workstation, by designing machines with splashguards, and by educating the workers concerning the danger of various compounds (so they don't expose themselves unnecessarily).

Barrier creams, a substitute for gloves, are designed to prevent dermatitis, not to treat it. The two types are water-repellent and oil-repellent. Sunscreens can reduce the risk of nonmelanoma skin cancer by 90% (Hogan, 1986).

Personal cleanliness is the most important measure in preventing occupational skin disease (Birmingham, 1975). Younger workers and males develop more occupational dermatoses than older workers and females due to their lack of care in handling injurious materials and lower personal cleanliness. Of course, engineers can help by providing adequate washing facilities near the work area.

When cleaning the hands, the cleaner needs to be specific for the substance to be removed. Powdered cleaners may be too abrasive; inorganic scrubbers (borax, silica) are more harsh than organic scrubbers (corn meal, rice hulls). Liquid soaps with a neutral pH are good when the soil is light and alkaline. Avoid soaps with perfumes. Waterless cleaners are better than a raw solvent but tend to dry the skin when used repeatedly. After using them, wipe the hands clean with a clean dry towel, then wash and rinse the hands with mild soap and water, and then add a skin moisturizer.

Personal clothing worn on the job, including undergarments, should be washed thoroughly before reuse. Cohen and Positano (1986) found that unwashed shirts resuspended two to eight times more particles than washed shirts. "Bystanders" (primarily the person who washes the clothing but also possibly the children in the home) can be affected by the toxins (Grandjean and Bach, 1986). Two solutions are a double locker system with intervening shower and having the clothing cleaned professionally rather than at home. Zirschky and Witherell (1987) discuss the problems of cleanup of mercury contamination of thermometer workers' homes.

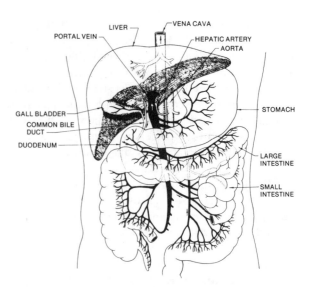

FIGURE 24.4   Liver cells are the primary source of biotransformations of poisons entering the body. About 30% of the liver's blood enters from the general circulation (through the hepatic artery) and 70% from the stomach and intestines (portal vein). Blood from the stomach and intestines must pass through the liver before entering the general circulation. Transformed materials leave the liver in bile (bile duct to small intestine) or in the blood to the general circulation (hepatic vein to vena cava).

the original compound not yet transformed) goes to the various organs, including the kidney. The kidney, which receives 30% of cardiac output, takes the transformed compound and puts it into the urine, with which the transformed poison leaves the body. Not all the blood goes to the liver and not all of the compound is transformed each time; in some cases the liver even metabolizes a compound into something more toxic than the parent compound. This permits toxins to reach various target organs and cause ill effects before the liver eventually detoxifies the body. The various metabolic reactions are very complex and interactions become very important.

4   POISON ELIMINATION

Although the lungs eliminate carbon dioxide and sweat glands eliminate some salts, the primary organs for poison elimination are the liver and the kidneys.

The liver biotransforms the toxins in the blood through oxidation, reduction, hydrolysis, and conjugation; it converts fat-soluble compounds to water-soluble compounds. Figures 24.4 and 24.5 give schematics of the liver and kidney. After the compound has been transformed in the liver, it reenters the blood. Some transformed compounds are carried by the bile to the intestines for excretion. The blood with the transformed compound (as well as some of

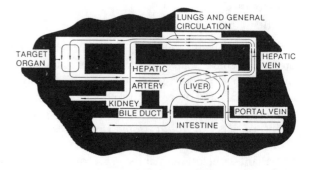

FIGURE 24.5   **Biotransformation** is shown schematically. The poison can enter the general circulation directly from the lungs or by penetration of the skin. If it comes from the intestines it must pass through the liver. On each passage through the liver, a proportion of the poison is transformed. The transformed compounds in the blood are water-soluble and are transferred in the kidney to the urinary tract for elimination.

For example, two compounds (such as alcohol and barbiturates, or cigarette smoke and cotton dust or asbestos) may both compete for the same liver enzyme (and thus slow down biotransformations) or both act on the same or different target organs (e.g., lungs, brain). Johns-Manville Corporation, citing a study reporting that smokers who are occupationally exposed to asbestos have a 92 times greater chance of developing lung cancer than the general population, has banned smoking in its asbestos facilities. It expects the policy to give asbestos workers the same chance of lung cancer as the general nonsmoking population. Alcohol and its products disturb liver metabolism and damage liver cells. Heavy metals, such as lead and methyl mercury, inhibit enzymes.

5 THRESHOLD LIMIT VALUES

The American Conference of Governmental Industrial Hygienists (ACGIH, 6500 Glenway Ave., Building D-7, Cincinnati, OH 45211; phone (513) 661-7881) issues updated TLV recommendations for chemical substances, dusts, and physical agents each year.

Note: Although the ACGIH uses TLV-TWA, the U. S. government calls its limits **permissible exposure limits (PELs),** the National Institute for Occupational Safety and Health (NIOSH) calls its limits **recommended exposure limits (RELs),** and the American Industrial Hygiene Association (AIHA) calls its limits **work environment exposure limits (WEELs).**

Table 24.1 gives the TLVs for the first five values listed for chemical substances. In addition to chemical substances, values are given for respirable dusts (fibrogenic dusts which cause scar tissue in the lungs) and nuisance dusts (which cause insignificant scar tissue). Three different types of threshold limit values are given: a time-weighted average (TLV-TWA), a **short-term exposure limit (TLV-STEL),** and a ceiling (TLV-C).

The TLV-TWA is the concentration, for a normal 8-h workday and 40-h work week, to which nearly all workers may be repeatedly exposed, day after day, without adverse effect. It primarily recognizes chronic (long-term) effects.

The TLV-STEL is concerned with acute (short-term) effects of (1) irritation, (2) chronic or irreversible tissue damage, and (3) narcosis of sufficient degree to increase the likelihood of accidental injury, to impair self-rescue, or to reduce work efficiency. The STEL is a 15-min time-weighted exposure which should not be exceeded any time during the day even if the TLV-TWA is met. There should not be more than 4 STEL exposures of 15 min per day and

there should be at least 60 min between each STEL exposure.

The TLV-C is the concentration which should not be exceeded during any part of the working day.

Excursions (peaks) are the recorded values of the toxin. They are permitted above the TWA and STEL but, of course, not above the C. For example, acetic acid has a TWA of 10 ppm. An exposure of 12 for 4 h can be balanced by an exposure of 8 for the remaining 4 h. You could even have an exposure of 80 for 1 h balanced by exposure of 0 for the remaining 7 h. This, however, is where the STEL comes in. The STEL for acetic acid is 15. Thus the exposure of 80 for 1 h is too high. An exposure of 80 for 1 min and exposure of 0 for 14 min would give an exposure of 6 for 15 min, which would be within the STEL. Note that, for administrative simplicity, the TLVs assume that concentration × time = a constant. That is, you would get the same effect from 8 aspirins taken once in 8 h as from one aspirin taken each hour for 8 h. This is unlikely to be true so don't push the excursions too far. For TWAs with no STEL listed, ACGIH (1993–94) recommends "STEL should exceed 3 times the TWA for no more than 30 min/day; under no circumstances should STEL exceed 5 times TWA, provided that TWA is not exceeded."

The TLVs in Table 24.1 are given both by volume and by weight. To convert (at 25 C):

$$ppm = \frac{24.45 \ (mg/m^3)}{Molecular \ weight}$$

The TLV is based on an 8-h exposure. The value needs to be adjusted if (a) concentration varies during the day, (b) the day is not 8 h, or (c) exposure is to more than one substance.

First, assume a worker was exposed to acetone for 4 h at 500 ppm, 2 h at 750 ppm, and 2 h at 1,500 ppm. The equivalent exposure is:

$$TWA = \frac{C_a t_a + C_b t_b \ldots C_n t_n}{8}$$

where TWA = Time-weighted average (equivalent 8-h exposure)

C = Concentration of $a, b, c \ldots$, ppm or mg/m³

t = Time of exposure to concentration $a, b, c \ldots$, h

For the above example, $TWA = [500(4) + 750(2) + 1,500(2)]/8 = 6,500/8 = 812$. Since 812 is more than the TLV-TWA of 750, the exposure is not acceptable.

Next, assume a worker was exposed to 1,250 ppm of acetone during the entire working day but only worked 6 h/day. Then $TWA = [1,250(6) + 0(2)]/8 = 7,500/8 = 940$. Since 940 is more than

750, the exposure is not acceptable. Assume the worker was exposed to 900 ppm of acetone for a 10-h shift. Then $TWA = 900(10)/8 = 9,000/8 = 1,125$. Since 1,125 is over 750, the exposure is not acceptable.

The worker also might be exposed to a mixture of substances. The assumption is that the effects are additive, not independent. Assume exposure for 8 h to acetone at 500 ppm, 2-butanone of 45 ppm, and toluene of 40 ppm. The TLV-TWAs are 750, 200, and 100 ppm. Then

$$TWA_{mixture} = \frac{C_1}{TLV_1} + \frac{C_2}{TLV_2} + \frac{C_3}{TLV_3}$$

where $TWA_{mixture}$ = Equivalent TWA mixture exposure (maximum of 1 permitted)

$C_{1,2,3}$ = Concentration (8 h) for a specific substance

$TLV_{1,2,3}$ = TLV for a particular substance

For the numbers above, $TWA_{mixture} = 500/750 + 45/200 + 40/100 = .667 + .225 + .4 = 1.292$. Since 1.292 is over 1, the exposure is not acceptable.

The above calculations have been performed mathematically. The engineer should remember, however, the accuracy of the measuring equipment used in establishing the TLVs and the unknowns faced by the scientists. Thus, treat the TLV values as guides rather than precise numbers. In general, the TLV standards are getting lower as more evidence is obtained.

In addition there are measurement errors in obtaining concentrations at the job. For example, exercise makes a difference; Soderlund (1975) reported that the air in lungs contained twice as much toluene during light work as during rest. People with sedentary occupations inhale about 5 m³ of air/8 h while those working very hard may inhale 20/8 h (see Chapter 11). Measurement of the contaminant concentration should be representative of the worker's breathing zone. The standard technique is to use a battery-powered pump with the sampling device in the breathing zone. Use of a direct readout unit (where sampling and analysis occur in one step) has the advantage that the second-by-second readout can be related to specific worker actions. This is much more useful than just knowing the total exposure over an 8-h period.

Although the TLVs are designed to protect the normal person during exposure for 40 h/week for the entire working life, they are not designed for hypersensitive people. Some examples of hypersensitive people are those with kidney and liver problems and the fetuses of pregnant women (especially in the first three months).

6   CONTROLS

Controls will be divided into engineering, administrative, and personal protective equipment.

Often the management staff tolerate working conditions for the employees that they would not tolerate if they had to do the work themselves. A key question is, "Would you let your own child work there?" If the answer is no, then the employees shouldn't either.

6.1 Engineering Controls   Engineering controls are the first priority for prevention. See Table 24.2.

6.1.1 Substitute a less harmful material
Use water-base cleaning compounds instead of organic-base. Use solvents with higher TLVs (methyl chloroform with TLV = 350 instead of carbon tetrachloride with TLV = 5; toluene with TLV = 100 instead of benzene with TLV = 10). In addition, some substances are much more volatile than others. Table 24.3 gives the ratio of vapor pressure, ppm/TLV-TWA, ppm. Use substances lower in the table rather than higher.

TABLE 24.2   Controls for respiratory hazards can be subdivided into engineering controls and administrative controls (Revoir, 1973). Engineering controls are more desirable than administrative controls.

Engineering Controls	Administrative Controls
1. Substitute a less harmful material.	1. Screen potential employees.
2. Change the material or process.	2. Periodically examine employees (biological monitoring).
3. Enclose (isolate) the process.	3. Train engineers, supervisors and workers.
4. Use wet methods.	4. Reduce exposure time.
5. Provide local ventilation.	
6. Provide general ventilation.	
7. Use good housekeeping.	
8. Control waste disposal.	

TABLE 24.3 Comparison of toxicity (TLV) and vapor pressure (P_0) vs. vapor hazard ratio (VHR) for selected industrial solvents, rank order (Popendorf, 1984).

Compound	1987 TLV ppm	P_0 at 25°C mmHg	VHR ppm/ppm
Allyl chloride	1	365	480 300
Carbon disulfide	10 skin	361	47 500
Dichloropropene	1 skin	36	47 400
Carbon tetrachloride	5 skin	113	29 700
Chloroform	10	195	25 700
Benzene	10	95	12 540
1,2-Dichloroethane	10	79	10 400
1,1,2,2-Tetrachloroethane	1 skin	6	7850
Triethylamine	10	68	8910
Methylene chloride	100	431	5670
Hexane	50	151	3975
1,1,2-Trichloroethane	10 skin	24	3200
Trichlorethylene	50	74	1955
Ethyl ether	400	534	1760
1,1-Dichloroethane	200	227	1490
Cellosolve	5 skin	5.3	1395
Methanol	200 skin	122	800
Methyl ethyl ketone	200	95	625
Methyl isobutyl ketone	50	19	500
Perchloroethylene	50	18	484
1,1,1-Trichloroethane	350	121	455
Aniline	2 skin	0.67	440
Acetone	750	230	400
Toluene	100 skin	28	374
Styrene	50	6.1	160
Phenol	5 skin	0.35	92
o-Xylene	100 skin	6.6	87
Ethanol	1000	59	78
m- and p-Xylenes	100 skin	3.6	48
Cresol	5 skin	0.17	45
Diazinon	0.008[A] skin	2×10^4	33
Nonane	200	4.3	28
Parathion	0.008[A] skin	6×10^5	10
Dibrom	0.19[A] skin	3×10^4	2.1
Trichlorofluoromethane (Freon 11)	1000	0.8	1.1
Malathion	0.74 skin	4×10^5	0.07

[A]Value listed in 1982 TLV publication only as mg/m³.

6.1.2 Change the material or process Reduce carbon monoxide by using electric-powered fork trucks instead of gasoline-powered. Use safety cans instead of glass bottles (which can break). Reduce dust by using low-speed oscillating sanders instead of high-speed rotary sanders. Paint with a brush or roller instead of spray painting. Remove grinding particles or solder fumes with a vacuum cleaner instead of blowing them into the room with compressed air. When hand grinding steel castings, workers are exposed to silica (sand) burned into the casting surface; you can reduce the risk of silicosis by coating mold surfaces to reduce sand burn-in, by shotblasting the castings, and by using high-velocity downdraft exhaust ducts.

6.1.3 Enclose (isolate) the process Usually it is less expensive to capture substances and vapors before they get out than once they have dispersed. As a first approximation, cost of air handling is proportional to the volume moved. Two guidelines are (1) *physically enclose the process or equipment* and (2) *remove air from the enclosure (hood) fast enough so that air movement at all openings is into the enclosure (i.e., negative pressure)*. A mechanical supply of air (a fan) usually is superior to depending on air infiltration. The operating cost is twofold: (1) the direct cost of electrical power for the fans and (2) the hidden cost of replacing the conditioned air (heated or cooled and humidified and purified to desired values).

The plastic strips used for strip doors also can be used to enclose machines or processes (such as solvent tanks) which require passage of product on conveyors. Enclosure also reduces the number of workers exposed. Isolate pumps which could leak toxic compounds. A sealer coat on concrete floors reduces dust; less dust means fewer air changes/h and thus lower heating costs. Maintenance workers need protection too. For example, an automotive plant found that maintenance workers were exposed to beryllium dust when repairing copper alloy welding tools.

Factories are not the only location of toxic compounds. Embalmers may be exposed to too much formaldehyde (Plunkett and Barbela, 1977). Asbestos fiber concentrations are high in some offices. Asbestos has been used as a fire retardant on steel beams. Over time, vibrations shake fibers loose to be circulated by the ventilation system. The problem is the worst when the area between the dropped ceiling and the floor above is used as a ventilating plenum.

6.1.4 Use wet methods
Water alone may not be enough, so use a wetting agent and dispose of the wetted particulate before it dries. Wet floors before sweeping. In rock drilling, use hollow drills through which water is passed. Steam cotton. Use moistened flint in potteries. Use a high-pressure water jet instead of abrasive blasting to clean castings.

6.1.5 Provide local ventilation
The general air-flow order should be input air, worker, contaminant, exhaust air. That is, the worker is upwind of the contaminant. Local ducts are much more efficient if they have a flange (flat plate perpendicular to the duct axis) at the entrance.

Dumping the exhaust "out the window" is not satisfactory. Clean the air with filters, cyclones, vapor traps, precipitators, and so forth. (Recycle the heat with rotary wheels, fixed plates, heat pipes, and runaround coils.) In some cases, this trapped waste product can be sold for a profit. One example is spraying alfalfa dust in a cyclone with liquid lard from a rendering company. The resulting mixture is sold at a profit as cattle feed. Paper mills formerly dumped sulfite liquor into the river. When forbidden to do this, they found they could sell it at a profit as a dust suppressor on roads.

When analyzing exposures, make a videotape of the operation. A first approximation is that toxin exposure is proportional to time exposed to the toxin. This estimate can be refined if a direct reading instrument can be used to estimate real-time exposure. For example, in a nonferrous foundry, lead exposure primarily came from moving an unvented ladle. It took 10% of the job time but gave 33% of the lead exposure (Edmonds et al., 1993). Edmonds et al. did not actually measure lead exposure but used an instrument which measured respirable aerosols. It was assumed that the lead concentration of the aerosol did not vary. The data were entered on a spreadsheet so average concentration, cumulative time, and cumulative exposure (product of average exposure and cumulative time) could be calculated.

6.1.6 Provide general (dilution) ventilation
Use general ventilation when the contaminant is released from nonpoint sources; for point sources, local exhaust is more efficient. Forced ventilation (fans, blowers) is preferable to natural ventilation (open doors, windows) since air direction, volume, and velocity can be controlled. Inadvertent recirculation of exhausted air is a problem. Discharge exhaust air so that it escapes from the "cavity" which forms as a result of wind movement around buildings. Also, take precautions when buildings are fogged with insecticides. The application should be done on Friday evening and the building should be well ventilated over the weekend. Before fogging, remove or store coffee cups, stationery, and clothing (Currie et al., 1990).

6.1.7 Use good housekeeping
Remove dust from floor and ledges to prevent dust movement by traffic, vibration, and air currents. Eliminate piles of open containers of chemicals. Fix leaking containers. Immediately clean up spills of volatile chemicals.

6.1.8 Control waste disposal
Each disposal problem should be considered separately and specific procedures should be established for safe disposal of unused dangerous substances, toxic residues, contaminated wastes, material containers which are no longer needed, and containers with missing labels. Be sure the policies are followed under strict supervision. In one GM plant, management had sewer covers tack-welded shut to prevent employees from dumping wastes down the sewer. Make provision for leaking and broken containers. That is, the storage area should be fail-safe so that when a drum leaks, the contents are contained. Note also that drain systems become chemical storage systems. Be sure you know what is going to mix in your drains.

6.2 Administrative Controls
See Table 24.2.

6.2.1 Screen potential employees
The goal is to not use workers who are hypersensitive to the substance. Examples are people with allergic contact dermatitis, pregnant women who would be exposed to teratogens, or people with impaired livers or kidneys (to be screened out just as people

with bad backs have lifting restrictions). Another example is screening out cigarette smokers from exposure to asbestos or cotton plant bracts.

6.2.2 Periodically examine employees
So far, the concept has been to monitor the environmental air and compare the concentration of a substance to its TLV. Another approach is to monitor the body of an individual (**biological monitoring**) through measurement of a substance concentration in blood, urine, hair, fingernails, or expired air. Table 24.4 gives biological exposure indices (BEIs) for mercury and carbon monoxide (ACGIH, 1993).

6.2.3 Train the supervisors, engineers, and workers
In 1983, the U. S. government published the Hazardous Communication Standard (29 CFR 191.1200), which is commonly called the "Worker's Right to Know." Manufacturers and distributors of hazardous chemicals must provide **Material Safety Data Sheets** (commonly called **MSDSs**) that identify the physical and health hazards of their products. Employers are responsible to inform their employees about the chemicals and to train them in safe use of the chemicals.

To meet training needs, give supervisors overviews with emphasis on costs and legal aspects; give technical principles and details to engineers and workers. Give the workers specific information rather than glittering generalities, for example, "Change your respirator filter once every four hours" rather than "Change filter when needed."

For comments on warnings, see Box 25.3.

6.2.4 Reduce exposure time
Reduced exposure time increases the recovery time/exposure time ratio. For example, if a worker is exposed 8 h/day, there is 16-h recovery or a recovery ratio of $16/8 = 2$. If two workers each are exposed for 4 h, the ratio goes to $20/4 = 5$. For more on work/recovery, see Box 16.1. In addition to job sharing, scheduling can help. For example, mines schedule blasting at the end of the shift so the dust can settle for 16 h before the next shift arrives. Schedule maintenance during off hours so the fewest possible people are exposed.

6.3 Personal Protective Equipment
Personal protective equipment is the last line of defense. Often it fits poorly, workers abuse it, it is not maintained, and workers are not taught how to use it. In addition, workers can get a false sense of security from the equipment.

Figure 24.6 shows some respirators (which cover the nose and mouth and provide filtered air). Figure 24.7 shows another approach—cover the entire head with a helmet and provide clean external air (which also can be cooled).

As pointed out previously, personal clothing after contamination is a risk for the spouse and family if it is brought home. Have the clothing cleaned professionally by people who know how to remove the toxins; have the worker shower before going home, especially if dealing with beryllium, asbestos, lead PCBs, or chlorinated hydrocarbons (Bellin, 1981).

If cleaned at home (e.g., cleaning farming clothes contaminated with pesticides), do the following (*Farmsafe 2000*, 1994):

- If the clothing is saturated, throw it away.
- Wash clothing as soon as possible.
- Wash separately from other clothing.
- Wear rubber gloves.
- Prespot the toxin.
- Use hot water and 1.5 (normal amount) of heavy-duty detergent.
- Air dry on an outside line.
- Run an empty cycle with hot water and detergent before washer is used again.

TABLE 24.4   Biological exposure indices (BEIs) from ACGIH (1993). Only two are given as an example.

Indices	Timing	BEI	Notation
CARBON MONOXIDE			
Carboxyhemoglobin in blood	End of shift	3.5% of hemoglobin	B*, Ns+
Carbon monoxide in end-exhaled air	End of shift	20 ppm	B*, Ns+
MERCURY			
Total inorganic mercury in urine	Preshift	35 µg/g creatinine	B*
Total inorganic mercury in blood	End of shift at end of work week	15 µg/L	

* B indicates significant background levels usually are present in people not occupationally exposed.
+Ns indicates determinant is nonspecific. Nonspecific tests are preferred.

Source: ACGIH, *1993-1994 Threshold Limit Values and Biological Exposure Limits.* Copyright © 1993 by American Conference of Governmental Industrial Hygienists, Cincinnati, OH. Reprinted with permission.

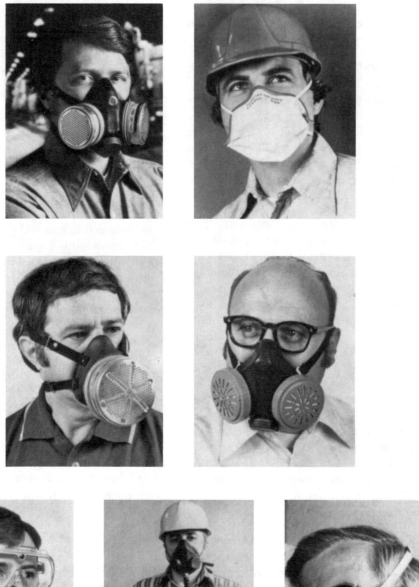

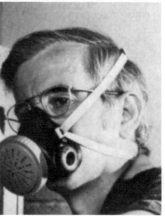

FIGURE 24.6   Respirators put a filter at the last line of defense—the nose and mouth. The OSHA respirator standard is: atmospheric oxygen content, maximum of 5 mg of condensed hydrocarbons per cubic m of gas, maximum of 5,000 ppm of CO_2 and 20 ppm of CO.

FIGURE 24.7   Helmets with clean, cool external air are an alternative to filtering ambient air. Keep the pressure positive to minimize infiltration.

REVIEW QUESTIONS

1. Differentiate toxicology, cumulative trauma, and safety.

2. Discuss the body as a "leaky bucket."

3. Is the primary toxin entry point the skin, lungs, or mouth?

4. What do macrophages do?

5. How does silicosis kill?

6. What is a teratogen? List three.

7. Briefly describe the epidermis and the dermis.

8. What is a gauntlet?

9. Briefly describe how people in the home can be protected against toxins in work clothing.

10. What is the difference between respirable dusts and nuisance dusts?

11. What is the difference between TLV-TWA and TLV-STEL?

12. Assume the TLV-TWA for acetone is 750 ppm. What exposure is permitted for 2 h? Show calculations.

13. List four engineering controls. Give an example for each.

14. Should the worker be upwind or downwind of the contaminant?

15. Briefly discuss the concept of biological monitoring.

16. Show with an example how reducing exposure time improves the recovery/exposure ratio.

REFERENCES

ACGIH. *1993–1994 Threshold Limit Values and Biological Exposure Limits.* Cincinnati: American Conference of Governmental Industrial Hygienists, 1993.

All About Atopic Dermatitis. Eczema Association for Science and Education, 1221 Yamhill, Portland, OR 97205, 1989.

Bellin, J. Don't take your work home with you. *Occupa-*

tional Health and Safety, Vol. 5 [6], 39-42, 1981.

Birmingham, D. Occupational dermatoses: Their recognition, control and prevention. In *The Industrial Environment,* Chapter 34. Washington, D. C.: Supt. of Documents, 1973.

Birmingham, D. *The Prevention of Occupational Skin Disease.* New York: Soap and Detergent Association, 1975.

Buechner, H. Organic dust: Critical emerging health hazard. *Health and Safety,* 22-24, January 1975.

Carson, G. Sampling particulates in the industrial environment. *ASHRAE Journal,* Vol. 16 [5], 45-49, May 1974.

Cohen, B. and Positano, R. Resuspension of dust from work clothing as a source of inhalation exposure. *Am. Industrial Hygiene Assoc. J.,* Vol. 47 [5], 255-58, 1986.

Currie, K., McDonald, E., Chung, L., and Higgs, A. Concentrations of diazinon, chlorpyrifos, and bendiocarb after application in offices. *Am. Industrial Hygiene Association J.,* Vol. 51 [1], 23-27.

Edmonds, M., Gressel, M., O'Brien, D., and Clark, N. Reducing exposures during the pouring operations of a brass foundry. *Am. Industrial Hygiene Association J.,* Vol. 54 [5], 260-66, 1993.

Farmsafe 2000. Vol. 3 [1], NIOSH, 4676 Columbia Parkway, Cincinnati: OH 45226, 1994.

Frazer, D. Sizing methodology. In *The Industrial Environment,* Chapter 14. Washington, D. C.: Supt. of Documents, 1973.

Grandjean, P. and Bach, E. Indirect exposures: The significance of bystanders at work and at home. *Am.*

Industrial Hygiene Assoc. J., Vol. 47 [12], 819-24, 1986.

Hogan, D. Skin disorders are high on the list of occupational health hazards. *Occupational Health and Safety,* 42-45, October 1986.

Mastromatteo, E. On the concept of threshold. *American Industrial Hygiene Association J.,* Vol. 42 [11], 763-70, 1981.

Plunkett, E. and Barbela, T. Are embalmers at risk? *American Industrial Hygiene Association J.,* Vol. 38 [1], 61-62, 1977.

Popendorf, W. Vapor pressure and solvent vapor hazards. *American Industrial Hygiene Association J.,* Vol. 45 [10], 719-26, 1984.

Revoir, W. Control of respiratory hazards. In *Safety Sentinel.* Southbridge, Mass.: American Optical Co., 1973.

Soderlund, S. Exertion adds to solvent inhalation danger. *Health and Safety,* 42-43, January 1975.

Webb, P. and Annis, J. The principle of the space activity suit. *NASA Contractor Report,* NASA CR-973, December 1967.

Wright, G. The influence of industrial contaminants on the respiratory system, Chapter 33. In *The Industrial Environment.* Washington, DC: Supt. of Documents, 1973.

Zenz, C. Reproductive risks in the workplace. *National Safety News,* 38-46, September 1984.

Zirschky, J. and Witherell, L. Cleanup of mercury contamination of thermometer workers' homes. *American Industrial Hygiene Association J.,* Vol. 48 [1], 81-84, 1987.

25 | SAFETY

OVERVIEW

Safety deals with accidents and short-term injuries (in contrast to cumulative trauma). Accident reduction should focus on reduction of unsafe conditions rather than unsafe acts. Reduce equipment failure; use distance and guards; and design the correct control, display, and environment.

CHAPTER CONTENTS

1 Introduction
2 Reduction of Unsafe Conditions
3 Unsafe Acts
4 Medical Management

KEY CONCEPTS

AND gates/OR gates
fail-safe
fault trees
ground fault circuit interrupter
layered defense
let go current

lockout/tagout
machine guards
mean time between failures
   (MTBF)
open manhole analogy
OSHA recordable accidents

parallel/standby
Pareto analysis
unsafe acts/unsafe conditions
warning

1   INTRODUCTION

As pointed out in Chapter 24, toxicology deals with health, while repetitive strain or cumulative trauma deals with injuries which occur from repeated insults to the body. For ways of reducing cumulative trauma, see Chapter 16. Safety deals with accidents and injuries. (Of course in a specific industrial organization, they may all be the responsibility of the safety department or the industrial hygiene department or some other title or may be split between departments.)

1.1   Which Problem
The safety goal is to (1) reduce accident frequency, (2) reduce the proportion of accidents that become injuries (i.e., injury frequency), and (3) reduce lost days/injury (i.e., injury severity). Because injuries are relatively rare events, it is easiest to work on the accidents. However, even accidents occur relatively rarely (they are the tip of the iceberg). Carter and Menckel (1985) recommend expanding the database to consider near accidents, hazardous disturbances, and even nondangerous disturbances. The worker should record, each day, all incidents that, at worst, could have resulted in personal injury; this should be done for at least 10 days. Laugherty and Vaubel (1993) confirm that there is a great deal of similarity between the circumstances of minor and major (**OSHA recordable**) **accidents.** Therefore, minor accidents are a useful database in prevention of major accidents.

To decide which problems to work on, consider using a **Pareto analysis** (insignificant many, mighty few). For example, if your organization over the last five years had X dollars of accident expense in the shipping department, 1.5 X dollars in the packaging department, and 9 X dollars in the machining department, focus your safety program in the machining department. Pareto analysis also can be done by type of injury, for example, electrical, cuts, slips and falls, and so forth. In addition, nonroutine tasks (construction, maintenance) have higher risk. For example, Helander (1991) pointed out construction has 6 times as many fatalities/h and 2 times as many disabling injuries as manufacturing work.

High risk also is associated with use of high-energy sources. Focus on energy sources (electrical, chemical, biological, physical) and (1) eliminate the source, (2) substitute for the source, (3) isolate the source, and (4) reduce exposure to the source.

1.2   Management Approach
As with any other activity, results occur where emphasis is placed. Thus, safety needs to be emphasized by top management. The degree of emphasis can be observed by the following: Is safety discussed at production meetings? Is safety listed first on meeting agendas? Do supervisory bonuses and raises depend upon safety performance of supervised employees? How large is the safety budget?

Accidents can be categorized as caused by **unsafe conditions** (equipment failure) and by **unsafe acts** (human failure). Supervisors tend to blame an accident on unsafe acts rather than unsafe conditions. Unsafe conditions reflect on them; unsafe acts can be blamed on "irresponsible or stupid" workers. The ergonomic approach is to consider all accidents as due to unsafe conditions and thus to focus on reducing unsafe conditions. Commercial aircraft safety is a good example of what can be done in an inherently unsafe situation.

More attention is focused on safety when workers' compensation costs, medical costs, and the like are charged directly to the responsible department rather than to plant or division overhead.

Employee attention on the safety program can be focused a number of ways. For example, a large green light at the plant entrance can change to red when there is a lost-time accident or workers can play "Safety Bingo," in which one number is drawn each day until there is a winner or a lost-time accident. The prize might be 4 h off with pay.

The **open manhole analogy** is an easy way to remember which approach to use. The warning approach to safety is to put up signs that state "don't step into open manholes." The guarding approach is to have a guard around the open manhole. The engineering approach is to put a cover on the hole. The engineering approach is best because it works best and because it is permanent rather than temporary. Thus, the sequence is (1) design out the problem, (2) guard against the problem, and (3) warn about the problem.

2   REDUCTION OF UNSAFE CONDITIONS

There are four ways to deal with unsafe conditions: (1) reduce equipment failure, (2) design the proper control, display, and environment, (3) use distance, and (4) use guards.

2.1   Reduce Equipment Failure
For more on error reduction, see Chapter 12.

2.1.1   Failure rate
Reduce the failure rate (that is, increase the **mean time between failures or MTBF**) through failure locations, safety factors, redundant equipment, and maintenance policies.

Failures occur at specific locations—that is, locations act as a "series circuit." Therefore, reduce the number of failure locations. For example, leaks on pipes usually occur at joints and valves so reduce joints and valves. For highway safety, reduce the number of railroad–highway grade crossings. In a work or home environment, reduce the number of sharp edges and protruding objects.

Safety factors (ratio of strength/stress) involve designing so that the unit can take a greater stress (load) than the anticipated stress (load). Examples of more strength are extra-thick insulation, a larger than required motor, or a stronger than required brace. Examples of less stress are reducing heat on electronic equipment or running a motor at less than full load. Derating is using a component with a design life greater than the equipment life (such as an auto transmission which would last 50,000 miles beyond the expected vehicle life).

Redundant equipment can be in **parallel** (e.g., two batteries in a car, either of which will start it) or in **standby** (diesel-powered electrical generator in a hospital). The parallel system has the characteristic that the redundant unit is in service and is wearing out (although the stress on the unit may be low due to the use of two units, thus giving a long MTBF for each unit). In the standby design, the unit isn't wearing out from use (although some units deteriorate without use); the disadvantage is that the standby unit must be switched in and out of the system (giving time delays plus the possibility of switching failures). Humans using the buddy system can use either a parallel (both work) or standby (one works and one is "lifeguard") mode. Decisions also can be made redundant. For example, there are voting circuits in some computers (the problem is solved in three independent circuits and if the answer is not the same, the output of the circuits that agree is used). Another example is a second opinion from a physician concerning an operation.

Maintenance can replace a component before failure. This preventive maintenance can be open loop (such as changing an engine based on hours of use) or can be closed loop. The closed loop utilizes feedback from built-in signals (noise from metal in brake pads when pads are worn) or failure signals observed by the operator ("It's running hot today" or "It seems to shake more than usual"). A trained operator usually is required to take the signal and convert it into an action.

2.1.2 Hazard

Even if equipment fails, hazard can be eliminated or reduced.

An example of an eliminated electrical hazard is use of a **ground fault circuit interrupter** (GFCI).

Even if a grounded person contacts the line, the GFCI breaks the circuit before injury occurs. See Box 25.1. Another example is replacing the glass in storm doors with plastic, which even if shattered, has no sharp edges. Auto safety glass is another example. **Fail-safe** design (fuse in electrical circuit, deadman throttle on locomotive or lift truck) is another example of eliminating the hazard.

Examples of reducing hazards include a battery-powered electric drill instead of a 110 V drill, a compressed air drill instead of an electrical drill, a less caustic chemical in place of a more caustic chemical, or a small amount of a dangerous material stored instead of a large amount. Radial tires and front-wheel drive give better vehicle control than bias tires and rear-wheel drive.

2.2 Design the Proper Control, Display, and Environment

As shown in Figure 19.1, a person communicates to a machine through controls, receives information through displays, and does the task in an environment.

2.2.1 Controls

See Chapter 19 for a discussion of controls and especially Guideline 3, "Prevent unintended activation of controls."

2.2.2 Displays

See Chapters 20 and 32 for a discussion of visual displays of words. See Box 25.3 and Table 32.9 for comments on warnings. See Chapter 20 for instrument displays.

In transportation environments, a locomotive or emergency vehicle may signal its presence with a siren or horn. However, this horn (which typically has an intensity of 118 dbA at 10 m) may not be heard (Miller and Beaton, 1994; Seshagiri and

---

BOX 25.1 *Electrical safety*

Table 25.1 gives the effect of various levels of current and Table 25.2 gives some electrical safety tips.

Ohm's Law says:

$$I = E/R$$

where

I = Current, amps
E = Voltage, volts
R = Resistance, ohms

Although the principles for electrical safety are well known, there are still many deaths and injuries.

TABLE 25.1 Effects of 60-cycle AC current (Hammer, 1989). DC currents, for the same effects, are 3 to 5 times the AC value. Frequencies of 20–200 Hz are especially dangerous because they cause ventricular fibrillation. See also section 6.2 of Chapter 18.

MILLIAMPERES	EFFECT
1	Perceptible shock
5–25	Lose muscle control. For 60 Hz, **let go current** (the current at which people can still let go) depends on weight. Typical values are 6 for women and 9 for men.
25–75	Very painful and injurious. Death if paralysis lasts over 3 min.
75–300	Death if for over 1/4 s (due to ventricular fibrillation)
2,500	Clamps (stops) heart. Burns to skin and internal organs. Immediately applied resuscitation may succeed.

TABLE 25.2 Electrical safety tips (Hammer, 1989).

De-energize the circuit. Don't forget to discharge capacitance-stored charge.
Use ground fault circuit interrupters (GFCIs). A GFCI monitors the circuit. If it senses imbalance in the current, it breaks the circuit. It does not work for a line-to-line contact, only line-to-ground.
Insulate with distance (isolate). Put distance between people and current. Barriers can replace physical distance.
Insulate the person. Provide insulating material to stand and/or sit on, such as nonconductive shoes, rubber gloves.
Warn people. Active warning = lights, sounds; passive = signs, colored backgrounds.

FIGURE 25.1 Sound pressure level of an emergency siren (assumed at 118 dbA at 10 m) at the driver's ear. A vehicle attenuation of 20 dbA is assumed. For more or less attenuation, move the entire curve vertically.

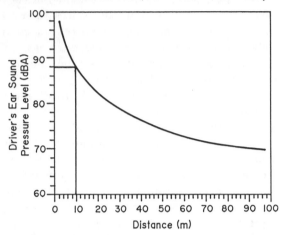

Stewart, 1992). Why not? Figure 25.1 shows the sound pressure at the operator's ear as a function of distance. In addition to the distance effect, the sound will be attenuated by the vehicle (20 dbA is assumed in Figure 25.1). There also may be interior noise (radio, conversation), and for detection of an auditory signal, its sound pressure level should exceed the background noise by 8–12 dbA. Since a vehicle with windows up and radio off has a typical noise of 70 dbA, the siren level at the driver's ear needs to be 78–82 dbA. Assuming 78 dbA, from Figure 25.1, the driver would hear the warning only when the distance from the noise source is about 32 m. Thus, auditory alarms should be complemented by other methods (e.g., flashing lights, crossing gates, etc.).

2.2.3 Environment/task

Housekeeping is a potential problem. Examples are grease and oil on the floor, projecting nails from scrap lumber and pallets, and sharp edges on machining chips. Poor housekeeping can cause fires. The key to good housekeeping is prevention, not cleanup. Solutions include "a place for everything and everything in its place" and organized scrap and waste disposal. Reduce congestion with shelving and drawers (use cube of space). Drains and sloped floors reduce standing liquids and thus falls. Overwaxed floors can be a problem, especially at entrances where people can track in snow and rain. Use mats at entrances for people to dry their feet. Saari and Nasanen (1989) reported that an improved housekeeping program in a shipyard not only decreased accidents associated with housekeeping but also decreased other accidents substantially.

In the office, appliance power cords (radio, coffeepot, etc.) are a common trip hazard.

The operator also can be "overloaded" either from the job or the environment. A job overload can be due to an incentive workpace or just working at the top of the specific operator's capability (i.e., there is no margin). The overload also could come from the environment with a continuous overload (e.g., heat stress, cold stress, blowing dust, glare, noise) or intermittent overload (a visual or auditory distraction).

2.3 Use Distance

Distance (isolation) is a powerful protective technique. Separating people and equipment reduces the chance of injury. Some examples are putting a machine which throws sparks or chips away from the aisle (or at least changing its orientation) so the objects don't hit people; putting a dike around all chemical tanks and a drain under all valves (sooner or later all tanks and valves leak); not permitting people to walk under cranes which are carrying a load; separating people from a process by automation (e.g., automated warehouses, machine-loading of a press, remote manipulators for radio-active materials); and using a push–pull jig for loading machines and saws. Also essential is to provide sufficient vertical clearance under beams, pipes, and conveyors (to avoid head contact); sufficient vertical clearance between vehicles and obstacles (use tie downs for antennas to prevent contact with power lines); and sufficient horizontal clearance between vehicles and obstacles (e.g., people on sides of railroad cars squeezed by insufficient space between car and a building).

A fence (barrier, wall) can increase the effective distance without changing actual distance. A fence is a type of guard.

2.4 Use Guards

Guards are not acceptable if (1) the guard can be defeated easily and (2) the guard defeat or failure is not easily detectable by the user *and* the user organization. For example, if a press is designed to be operated by two separate buttons to keep the hands out of the die, but one button is tied down, this defeat should be obvious to any observer. Another example is an electrical guard (such as insulation) which might fail and so pass a lethal current. A third example is a fire extinguisher which appears functional but is not. A poor guard may be a hazard in that it gives an impression of protection which doesn't exist.

Guards will be divided into machine guards and people guards.

2.4.1 Machine guards

Machine guards are attached to equipment to prevent people from dangerous contact. Some machine guard examples are a barrier to prevent reaching into run-in or nip points (rotating machinery, shears) or entering an area (electrical substation, robot workspaces), guards on sharp objects (saws), guards to prevent electrical shocks, guards (e.g., insulation) to prevent burns, guards to catch falling objects under overhead conveyors or protect lift truck drivers, guards to intercept flying material (chips, sparks, splashes), and guards to prevent escape of liquid or vapor (enclosures). A building example is guards and handrails on stairs, ramps, and docks to prevent falls to a lower level.

Note that guards and barriers are not perfect. People drive around railroad grade crossing barriers; children go through gates to swimming pools; maintenance people often remove machine guards to do maintenance; and so on. Consider guards and barriers as part of a layered defense.

An example of a **layered defense** is the Occupational Safety and Health Administration (OSHA) confined space regulation. (As a side note, in government documents, the word *shall* means required and *should* means recommended.) A confined space is dangerous due to oxygen deficiency, combustibility, or toxicity. Defenses include (1) employer evaluation of such spaces, (2) written permits required to enter such spaces, (3) establishment of safe procedures for safe entry, (4) attendants outside the space while it is occupied, and (5) trained rescue workers.

Lockout/tagout procedures require machine controls to be locked so the machine cannot be turned on by someone else while maintenance is being performed. See "Prevent Unintended Activation" in Chapter 19.

Purchase machine guards from the equipment manufacturer when the machine is purchased; locally manufactured guards tend to be poorly designed and constructed. (Tip: Have someone from purchasing on the safety committee.) Guards should not impair machine function (operator vision, maintenance) or they will tend to be removed. For example, if an opaque guard interferes with vision, make the guard transparent. Well-designed guards will be sufficiently rugged to withstand predictable events (impact from lift trucks, bursting of abrasive wheels, stock kickback).

Thompson (1989) gives the safe object distance from seven heights of barriers for 99% of British males. He emphasizes the need to test barriers with people instead of manikins.

In addition to their function of protecting the body, guards also reduce or amplify noise. See Chapter 22 for noise reduction comments.

2.4.2 People guards

If you don't use machine guards, there remains, as a last line of defense, to put guards on people—that is, protective clothing. Examples are safety shoes for the feet, aprons and leggings for the legs and torso (including clothing to protect against heat, cold, chemicals, welding, radiation), gloves and gauntlets (nonpermeable for chemicals; tough for abrasion resistance) for the hands, respirators to protect the lungs, earmuffs and earplugs to protect the ears against noise, helmets to protect the skull, and hairnets to protect the hair from rotating machinery. Special precautions need to be taken for loose clothing (such as neckties, sleeves, gloves) near rotating machinery. In addition, workers should remove jewelry such as rings, watches, and earrings.

For comments on shoes and standing, see Chapter 15. For clothing and climate, see Chapter 23. For gloves, see Box 18.1. For earmuffs and earplugs, see Chapter 22.

There are two problems with protective clothing as the last line of defense. First, it is the last line; if it fails, injury results. Second, much protective clothing decreases the comfort or performance of the person and thus causes a temptation not to use it. Comfort of the protective device (helmet, safety glass, glove, etc.) needs to be emphasized when it is purchased as there are many examples of nonuse of uncomfortable safety equipment. To encourage compliance, safety-conscious managements impose severe penalties on employees not using safety equipment (e.g., one day off without pay on the first offense, one week off on the second offense, and dismissal on the third offense). Unions strongly support such policies (as long as they are fairly administered), as unions are very safety-conscious.

Since workers rarely have the technical knowledge to select safety equipment properly and are tempted to purchase inferior protection to save money, organizations should purchase protective clothing and give it to the employees at no cost. Organizations should also control its maintenance.

3   UNSAFE ACTS

As discussed previously, for accident-prevention purposes, treat all accidents as unsafe conditions. Unsafe acts will be divided into lack of knowledge and deliberate risk.

3.1 Lack of Knowledge   Generally, there is someone somewhere in the organization who knows what to do; the problem is to get this knowledge to the person at risk. The person at risk usually is the operator but can be a maintainer or even someone passing by.

To minimize risk, make a fault analysis of all possible failures. See Box 25.2. Exactly what should be done by whom to return the system to normal? This fault analysis should consider not only operator errors but also equipment failure. For example, in the famous Three Mile Island nuclear accident, the system had a failure. A backup system was bringing the system back to normal but a display did not present the proper information to the operator and the operator turned off the backup system, which thus allowed the primary system to fail catastrophically. What should be done if a valve or tank ruptures? A decision structure table (see Chapters 7 and 32) is a good tool for the safety analyst to use to record what should be done.

Next, this information needs to be communicated to the people at risk, either through training (i.e., memorization) or a job aid (i.e., information always available); see Chapter 32. One of the problems with training is that it is not permanent. Not only do operators change jobs, but people also forget. For error reduction, see Chapter 12; for training, see Chapter 32. Thus, practice makes perfect (fire drills, aircraft simulators). See Box 25.3 on warnings. Job aids (such as instruction books) may be helpful in some circumstances but instruction manuals never seem to be around when you are looking for them and, in an emergency, you may not have time to read the manual.

3.2 Deliberate Risk   Workers may take a deliberate risk because the risk is low, cost of compliance is high, and the rewards are large and immediate. People also tend to be poor at estimating how risky something is. Verhaegen et al. (1985) report a

---

BOX 25.2   *Fault trees*

Table 25.3 shows fault tree symbols and Figure 25.2 shows a **fault tree.** The concept, originally developed at Bell Laboratories, shows how a "top event" (a potential accident) can occur. A fault tree shows the complexity of a situation in a graphical form which improves communication among those investigating the problem. It includes not only equipment failures but also personal factors.

Although fault trees are usually used without mathematics, it is possible to quantify the probabilities and calculate probabilities of the various events. The key problem is determination of the probabilities. For example, what is the probability

that an operator will drop a part during an 8-h shift? Is it .01, .001, .0001, or what?

However, the nonmath tree still is very useful. Remembering that **AND gates** attenuate probabilities and **OR gates** multiply probabilities, try to redesign the situation to increase the number of AND gates and decrease OR gates. Another technique is to look for "single-point failures"—that is, a situation where failure at a single event will cause the top event. Usually this is through a series of OR gates. Another way a single-point failure could occur is at an AND gate if the other input is a normal event.

BOX 25.3   *Warnings*

A **warning** is information about a possible negative consequence (Ayres et al., 1989). There is little evidence that product warnings work—that is, change people's behavior (Lehto and Miller, 1986; Horst et al., 1986). Any cigarette smoker illustrates this point. There have been many warnings about the dangers of smoking and yet some people continue to smoke. Although useful change probably should focus on the individual (user knowledge and knowledge requirements), there has been considerable effort to improve the warning itself. The key point of this effort is the belief that an effective warning *changes behavior.*

The acronym PRUMAE from Lehto and Miller (1986), points out some of the problems.

P (present). The information must be present. If the warning is in an instruction manual which is never seen, the warning can't work. If the warning label is no longer on the machine, it can't be read.

R (read). The next problem is to get the warning read. People find many excuses not to read material. We are all subject to information overload, and people learn to filter out extraneous information. (Although an organization's lawyer may prefer to have detailed warnings about every possible problem placed on products, the resulting information overload would probably decrease reading of the warnings (Ayres et al., 1989).) Since most warnings concern rare events, there is little penalty from not reading the warning.

The person needs to be "information seeking." It may help if the warning stands out from the background.

U (understand, comprehend). Language problems can occur when the reader does not understand the warning language (e.g., native Spanish-speaker reading English). Even for a native English reader, problems can occur with long or complicated words. See Table 32.2. Pictographs are an attempt to reduce this problem; unfortunately some of them are as intelligible as hieroglyphics. See comments on pictographs in Chapter 20.

Understandability can be improved with sentence construction and layout of the message. Make the signal stand out from the noise. Some people recommend the warning be divided into four statements: signal word, hazard, consequence, and

instruction. For example, DANGER, HIGH VOLTAGE WIRES, CAN KILL, STAY AWAY or WARNING, CONTAMINATED WATER, ILLNESS MAY RESULT, DO NOT DRINK. Wogalter et al. (1987) reported the hazard statement is the most important; the other statements may be redundant information for informed users. See Figure 20.9. In general, "danger" indicates potential death, "warning" indicates serious harm or injury, and "caution" indicates minor harm or injury.

If there are multiple warnings in the message, then it is more difficult for the user to remember them all.

M (memory). Once motivated to input the information to the brain, the person must commit it to long-term storage and then, when needed, recall it. The decision to store the information may depend upon the credibility of the information—its "believability."

A (act, comply). Upon retrieval of the information from the brain, the person must be motivated to translate this into action. An important point is the cost of compliance. Reducing the cost of compliance should improve the probability of the person complying with the warning. For example, complying with the warning "Don't use broken door" was 94% if another door was adjacent, 6% when another door was 50 ft away, and 0% when another door was 200 ft away (Wogalter et al., 1987). Cost of compliance is another way of discussing benefit/cost. Athletes may consider the benefits of steroids greater than the costs; cigarette smokers may consider the benefits of smoking greater than the costs. People often have poor estimates of the true probabilities of rare events. In addition, they often have the opinion that rare events won't happen to them.

E (effective). The person needs the necessary ability to do the desired behavior. Then the person needs the skill and training to do the behavior effectively.

For a warning to work, all of the above conditions must be met. Thus, warnings should be considered a last line of defense. If a warning message is important, field test it on representative users under representative conditions to determine whether it modifies behavior.

TABLE 25.3   Fault tree symbols.

Symbol	Title	Comments
▭	EVENT	Event resulting from an AND or OR gate. An event is a dynamic change of state that occurs in a system element.
⬠	NORMAL EVENT	"House" symbol represents the event which "normally" occurs.
◇	EVENT	The diamond (sometimes a circle) is an event which will not be analyzed further but is included for completeness.
△	CONNECTOR (TRANSFER)	The triangle shows remote connections in the diagram (for ease of drawing).
OUTPUT ⌒ INPUT	AND gate	Output occurs if **all** inputs occur. $F_0 = f_1 f_2 \dots f_n$. Thus for three inputs with probability of .01 each, output probability = (.01)(.01)(.01) = .000 001. The *AND gate attenuates* probabilities.
OUTPUT ⬡—⬭ INPUT	AND gate with INHIBITING CONDITIONS	Hexagon depicts special case of AND gate. Ovals show **inhibiting** inputs (conditions).
OUTPUT ⌓ INPUT	OR gate	Ouput occurs if **any** inputs occur. $F_0 = 1 - (1 - f_1)(1 - f_2) \dots (1 - f_n)$. Thus for 3 inputs with probability of .1 each, output probability = $1 - (1 - .1)(1 - .1)(1 - .1) = 1 - (.9)(.9)(.9)$ = 1 - .729 = .271. The *OR gate multiplies* probabilities.

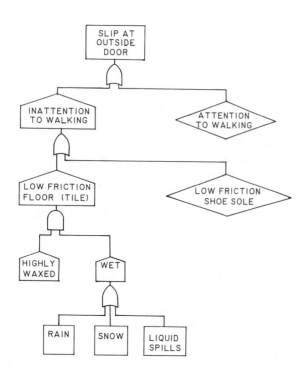

FIGURE 25.2   Fault trees show which events lead to the top event. Note that, even with the AND gate, a single-point failure could occur with a low friction floor since highly waxed tile is a normal event.

relation between accidents and absenteeism. People who have more accidents are absent more. Verhaegen et al. attribute the accidents to greater risk taking due to a more negative attitude toward the firm and its rules. Low-probability events (accidents) are ineffective in controlling present behavior. The worker may even consider the accident as due to bad luck, instead of bad behavior. Management must reinforce safe behavior with "carrots" and punish unsafe behavior with "sticks." Some carrots are praise, public recognition of a group's behavior, free T-shirts or hats with safety slogans, and so forth. Sticks include verbal reprimand by supervision, disciplinary action, and social pressure by colleagues. A good attention-getting device is a "safety traffic light" at the plant entrance. It stays green as long as there is no accident, yellow whenever there is an accident but no lost time, and flashes red whenever there is a lost-time accident.

4   MEDICAL MANAGEMENT

Although accident prevention is the first goal, provide good medical management to reduce the effect of an accident.

The first level is emergency care. This includes a plant nurse, first aid training for employees, first aid equipment, provision of emergency phone numbers, arrangements with local hospitals, and so forth.

The second level is rehabilitation and return to work. The rehabilitation work is done by specialists such as physicians, occupational therapists, physical therapists, and others. The consensus is that injured workers should return to work as soon as possible (often in light-duty jobs) rather than stay at home.

REVIEW QUESTIONS

1. State the open manhole analogy and explain which approach to safety is best and why.
2. What is a ground fault circuit interrupter?
3. Give four examples of using distance to reduce accidents.
4. Give three examples of machine guards.
5. Give four examples of people guards.
6. Why should accidents be considered as due to unsafe conditions rather than unsafe acts?
7. Briefly discuss the acronym PRUMAE.
8. Show a fault tree for slipping on the floor.
9. Do AND gates attenuate probabilities? Do OR gates multiply probabilities?

REFERENCES

Ayres, T., Gross, M., Wood, C., Horst, D., Beyer, R., and Robinson, J. What is a warning and when will it work? *Proceedings of the Human Factors Society,* 426–30, 1989.

Carter, N. and Menckel, E. Near-accident reporting: A review of Swedish research. *J. of Occupational Accidents,* Vol. 7, 41–64, 1985.

Hammer, W. *Occupational Safety Management and Engineering,* 4th ed. Englewood Cliffs, N. J.: Prentice-Hall, 1989.

Helander, M. Safety hazards and motivation for safe work in the construction industry. *Int. J. of Industrial Ergonomics,* Vol. 8, 205–23, 1991.

Horst, D., McCarthy, G., Robinson, J., McCarthy, R., and Krumm-Scott, S. Safety information presentation: Factors influencing the potential for changing behavior. *Proceedings of the Human Factors Society,* 111–15, 1986.

Laugherty, K. and Vaubel, K. Major and minor injuries at work: Are the circumstances similar or different? *Int. J. of Industrial Ergonomics,* Vol. 12, 273–79, 1993.

Lehto, M. and Miller, J. *Warnings: Vol. 1 Fundamentals, Design and Evaluation Methodologies.* Ann Arbor,

Mich.: Fuller Technical Publications, 1986. See also Miller, J. and Lehto, M. *Warnings: Vol. 2 Annotated Bibliography.* Ann Arbor, Mich.: Fuller Technical Publications, 1987.

Miller, M. and Beaton, R. The alarming sounds of silence. *Ergonomics in Design,* 21–23, January 1994.

Saari, J. and Nasanen, M. The effect of positive feedback on industrial housekeeping and accidents: A long-term study at a shipyard. *Int. J. of Industrial Ergonomics,* Vol. 4, 201–11, 1989.

Seshagiri, B. and Stewart, B. Investigation of the audibility of locomotive horns. *Am. Industrial Hygiene Association J.,* Vol. 53 [11], 726–35, 1992.

Thompson, D. Reach distance and safety standards. *Ergonomics,* Vol. 32 [9], 1061–76, 1989.

Verhaegen, P., Strubbe, J., Vonck, R., and van den Abeele, J. Absenteeism, accidents, and risk-taking. *J. of Occupational Accidents,* Vol. 7, 177–86, 1985.

Wogalter, M., Godfrey, S., Fontenelle, G., Desaulniers, D., Rothstein, P., and Laughery, K. Effectiveness of warnings. *Human Factors,* Vol. 29 [5], 599–612, 1987.

CHAPTER

26 | DETERMINING TIME/JOB

OVERVIEW

After a good method is determined, there are a number of reasons to determine how long the job takes. As with any measurement technique, the measurement can be "quick and dirty" (Type II, nonengineered standards) or more precise (Type I, engineered standards). The two techniques used to set engineered standards are covered in more detail in other chapters: time study in Chapter 27 and standard data in Chapters 29 and 30.

CHAPTER CONTENTS

1 Why Determine Time/Job
2 Establishing Time Standards
3 Documenting, Using, and Maintaining Standards

KEY CONCEPTS

acceptable day's work
audit
discipline level
earned hours
engineered standards
high-task/low-task
labor efficiency

measured daywork
nonengineered standards
occurrence sampling
performance ratios
restricted output
should take/did take standards
standard

standard data
stopwatch time study
time logs
unmeasured hours
work measurement

1   WHY DETERMINE TIME/JOB

What you cannot measure,
your knowledge is of a
meagre and unsatisfactory kind.

Lord Kelvin

So far, the emphasis in this book has been on job design. However, the six chapters in Part 6 discuss determining how long a job takes, commonly called **work measurement.** (Time measurement is not really "work" measurement since work is not measured in either the physiological or mechanical sense.) Hopefully the people in the industrial engineering department will allocate most of their effort to job design as that is where productivity improvements are made. But it must be recognized that knowing the time/unit is useful for many parts of the organization and someone must obtain time/unit.

As an aside, with increasing computerization in industry, the direct labor contribution to total cost has been dropping. Increasing emphasis has been put on indirect departments (especially white-collar departments). Table 26.1 gives some **performance ratios** for other than labor. The goal is to measure output and connect employee actions to outputs.

TABLE 26.1   Performance ratios other than direct labor (Engwall, 1987).

Engineering
- Number of software instructions/number of software engineers
- Cost of repair drawings/number of drawings produced
- Number of engineering change orders/number of engineers

Quality
- Cost of quality/cost of sales
- Total receiving inspection hours/lots received
- Production earned hours/quality engineered support hours

Procurement
- Purchasing department budget/number of purchase orders placed
- Lots received on time/total lots received
- Purchase order errors/purchase orders audited

Finance
- Number of pricing proposals/number of pricing people
- Operations budget/finance department budget
- Receivables over 60 days/total receivable

Information Systems
- Hardware uptime/total hardware time
- Out-of-service terminals/total number of terminals
- User complaints/hours of usage

Production
- Wait time/direct labor hours
- Indirect hours/direct labor hours
- Setup hours/earned hours

When the goal is to influence behavior, a simpler index is better than a complicated one.

Table 26.2 gives an estimate of the number of direct labor standards that can be maintained by one person.

Time/unit is useful for five groups of applications:

1. cost allocation
2. production and inventory control
3. evaluation of alternatives
4. acceptable day's work
5. incentive pay

1.1   Cost Allocation   Without time/unit, you can't determine costs. In Table 4.4, the cost of making a fan was discussed. Among other costs was direct labor of $1.00 with the notation "Time to stamp out blades, paint blades, assemble knob to shaft." In practice, there would be a computer listing for the fan including the following partial list:

Item/Operation	Make/Buy	Number Required	Total Material, $	Total Labor, h
Components				
123 Motor	B	1	1.00	
242 Bearing	B	2	.20	
430 Blade	M	1	.50	.12
Assemblies				
500 Assemble	M	1		.05
520 Attach guard	M	1		.01

Periodically the standard cost of the product will be calculated from (a) the current cost of purchased components and raw materials, (b) multiplying direct labor time by the current labor efficiency and rate/h, and (c) adding the other overhead or burden costs. Burden costs typically are determined by using the direct labor cost and multiplying by a standard rate (say 2.5). That is, if direct labor cost/unit is $2, then burden is $5.

The key point is that an accurate estimate of direct labor time per unit is needed or there will not

TABLE 26.2   A large farm equipment manufacturing facility had a plantwide incentive plan. Standards are individual rates except group rates on assembly and finishing lines. Rates are set by time study or elemental standard data based on time studies (Koop, 1982).

Shop Area	Number of Active Standards/Technician
Press and shear	5000
Machine shop	2500
Welding	800
Assembly, finishing	300

be a reasonable estimate of the cost of the product. The difference between the cost of the product and the price of the product is the profit of the product. If you make only a single product you could estimate your annual profit by counting the amount of money left over at the end of the year—assuming all income and expenses came in within exactly the year and that you had no capital expenses (such as machinery). In practice, payment periods overlap, there are capital expenses, and, most important, firms make more than one product. Thus, without accurate costs, which require accurate time standards, you can't determine your net income on any product.

1.2 Production and Inventory Control

Without time/unit, you can't schedule or staff. How many machines should be assigned to a job? When can the units be delivered? What workers should do which tasks on the assembly line? When should components be made? Production and inventory control is impossible without time/unit.

1.3 Evaluation of Alternatives

Without time/ unit, you can't compare alternatives. Should a bushing be made on a turret lathe or an engine lathe? Should a building be built of brick, steel, or wood? Should a mechanic repair a part or replace it with a new one? The decisions can be at a simple level (cost of cleaning a hypodermic needle in a hospital is $1 vs. cost of $3 for a new needle); the decisions can be at a complex level (programs into which operation research people feed a_{ij} values into the jaws of their computers). In all cases, the decision models are based on accurate input values; incorrect times will result in incorrect decisions.

1.4 Acceptable Day's Work

Without time/ unit, supervisors can't judge the performance of people working for them—an **acceptable day's work.** George made 29 widgets today—is that good or bad? Betty Jo sewed 500 sleeves today—good or bad? Sam picked 1,600 cases from the warehouse today—good or bad? A standard time is necessary so actual times can be compared to the standard, which leads to more effective management. Many applications of standards to repetitive work have shown improvement of output by 30% or more when measured daywork systems are installed in place of nonengineered standards. Output improves approximately 10% more when a group incentive payment plan is used and 20% more when an individual incentive is used (Kopelman, 1987). See also Table 26.3. (As a side comment, this higher productivity means less need for employees. Thus, during the period of the time study, have people work overtime rather than hiring new employees. Then when the standards are installed, the improved productivity will be partly reflected in reduced overtime rather than layoffs.)

TABLE 26.3 Productivity (%) as a function of work measurement and supervision (Sellie, 1992). The data are based on over 1,000 productivity audits and work measurement installations.

PERFORMANCE MEASUREMENT	SUPERVISION		
	Poor	Average	Good
Measured	60–80	70–90	80–95
Unmeasured	30–70	50–75	60–85

Source: C. Sellie, "Predetermined Motion-Time Systems and the Development and Use of Standard Data," in *Handbook of Industrial Engineering*, 2nd ed., G. Salvendy (Ed.). Copyright © 1992 by John Wiley & Sons. Reprinted by permission of John Wiley & Sons, Inc.

1.5 Incentive Pay

Without time/unit, you can't fairly pay people based on the number of units they produce. Only a minority of firms use the pay-by-results (the carrot) approach; the majority use the acceptable day's work (the stick) approach. For example, a textile firm might conclude that a typical worker could sew 450 sleeves/h. If a typical wage in that locality was $9.00/h, then the firm might decide to pay $.02/sleeve instead of a flat $9.00/h. If Betty Jo sewed 500 sleeve/h, then her pay would be $10/h because for every 1% increase in output she would receive a 1% increase in pay. What if she produced less than 100%? Then the policy depends upon the firm but a common plan is 100% pay for all output less than 100%—unless a discipline level is reached (see Box 26.1). Thus a worker has everything to gain and nothing to lose. Why would a firm implement such a policy when direct labor costs increase as fast as output? Because overhead costs do not increase proportionally. For example, if Betty Jo produces 450 sleeves/h, her labor cost would be $.02/unit, the supervisor's salary might add another $.001/unit, the machine depreciation add another $.0001/unit, etc. If she produces 20% more, the other costs do not increase so the burden cost/unit decreases and the total cost/unit decreases. See Box 26.2.

Box 26.3 briefly describes incentives for nonincentive employees.

2 ESTABLISHING TIME STANDARDS

Having discussed why there is a need for time/unit, the next question is which technique should be used. The cost of inaccurate information (inaccurate time standard) can be great. For example, a product

BOX 26.1   *Consequence of not making standard*

Organizations use different concepts of **standards.** See Figure 26.1. The curves show the potential long-range performance of the industrial population. Thus, for a low-task standard, over 99% of the industrial population would be able to reach standard.

Although it is tempting for the personnel department to attempt to prescreen those who "won't make it anyway," this approach has many possible legal problems. In the long run, having the individual try it and quit or try it and fail to make standard and thus not retain the job is the best policy.

The next question, however, is not what workers are potentially able to do but what they actually do. What if the standard is picking 100 cases/h and Joe picks only 50? It depends. See Figure 26.2.

If Joe is learning the job, he would be expected to achieve a certain percent of standard each week. For example, 50% the first week, 80% the second week, 90% the third week, 95% the fourth week, 98% the fifth week, and 100% the sixth week. Previously qualified workers returning to a job might be assumed to start part-way through a learner's standard. See Chapter 28 for more on learning. If Joe was assigned temporarily to the job, he could either be exempt from the **discipline level** or put in the learning schedule.

If Joe is a permanent, experienced worker the performance could be considered "excused" or "nonexcused." (Performance usually is considered for a period of a week rather than a day so that random fluctuations average out.) Excused failures to meet standard are for temporary situations—bad parts from the supplier, back injuries, pregnancy for females, and so forth. For example, "employees returning to work from workers' compensation due to a loss-of-time accident in excess of 30 days will be given consideration based upon the medical circumstances of each individual case."

Nonexcused are those for which the worker is considered capable of making standard but did not achieve it. Generally the first assumption is that training was not adequate. But if performance stays low, then penalties begin to be imposed. See Table 26.4. Most organizations have a "forget" feature; for example, one month of acceptable performance drops you down a step. The use of an established discipline procedure allows workers to self-select themselves on a job. The organization can have only minimal pre-employment screening and thus not be as subject to discrimination charges.

Standards are based on 8 h/day but people sometimes work less than or more than 8 h/day. Examples of longer shifts are overtime and working 4 shifts of 10 h instead of five of 8 h. Because of fatigue, in theory, people can produce more/hour when working shorter hours and less/hour when working longer hours. In practice, in today's society most standards are not very tight. People can pace themselves and the output/hour tends to be constant over the shift. In addition, it is awkward to change the standard when people work longer or shorter hours because then it is not a standard. If an adjustment is to be made, it is more practical to modify the discipline level.

Any change in discipline level for fatigue should be a function of the amount of fatigue allowance. That is, if a task has a 20% fatigue allowance, the task should be more subject to fatigue than a task with a 5% fatigue allowance. Fatigue also should be greater for longer shifts (e.g., 12 h vs. 10 h vs. 8 h) and longer periods of overtime (e.g., 4 weeks vs. 1 week). Rodgers (1986) reports, for one study at Kodak, that performance for overtime was 5–10% below the performance for 8 h. This implies that the discipline level for hours over 8 h might be 5–10% below the discipline level for 8 h. However, discipline levels should be based on longer time periods than just hours 8 and 9 of a day. Thus, if 95% discipline is used for 8 h and 90% for hours 8 and 9, then the daily discipline level would be $(8(95) + 2(90))/10 = 94\%$. However, remember that performance should be compared to the standard on a weekly basis, not a daily basis.

The level at which discipline takes place is negotiable between the organization and the union. For example, it may be 95% of standard. That is, as long as workers perform above 95% of standard, they are considered satisfactory. However, this tends to get overall performance from the group running around 98% of standard. (As soon as the workers get above the discipline level they take more leisure time.) The best long-range strategy probably is to set discipline at 100% of standard. Anything less will give a long-run loss in production—especially if a measured daywork system is used instead of incentives.

There are several strategies that organizations can take to improve the group performance and

(continued)

BOX 26.1 *Continued*

reduce output restrictions. Basically they allow the employees as well as the organization to benefit from output over 100%.

The primary technique is to give money for output over 100%. A 1% increase in pay for a 1% increase in output is the prevalent system (Panico, 1992).

Another alternative is to give the employee time off for output over 100%. For example, allow individuals to "bank" weekly hours earned over 100%. These banked hours then can be used to compensate for weeks when their individual performance is less than the discipline level. Most people will run up a positive balance to use as "insurance." This can be combined with a plan in which all hours in the bank over (say) 20 h are given as scheduled paid time off. With the use of paid time off (taken in minimum amounts of 8 h), absenteeism tends to drop as employees can use the paid time for personal business.

Another possibility is to let the discipline level for individuals decrease when average performance of the group is over the discipline level. For example, if the group's performance is 98% and required = 95%, then the discipline level for individuals that week is 95 − 3 = 92%. This allows the group to "carry" poor performers. Of course, management gets higher performance for many people while only 1 or 2 people are carried.

When discipline measures are taken, the standards themselves naturally begin to be questioned. These challenges typically can be resolved at the local level if the firm has been open and above-board in conducting the standards setting and has detailed records showing how each standard was set. If conflicts cannot be resolved, the standard may go to arbitration (grievance procedure). Arbitration typically has several levels with the lowest level being the plant, then division, and then outside-the-organization arbitration.

---

may be priced too high or too low. A method and machine may be retained or replaced incorrectly. In the very common **measured daywork** system, people tend to work only at 100% of standard but no more. Thus, if a standard is set at .05 h/unit when it should have been .048 h/unit, the workers will take

.05 h/unit instead of .048 h/unit. This is a loss of .002 h/unit or 4%. If the worker's annual cost is $20,000, this is a loss of $800/year.

The techniques to use depend upon the cost of obtaining the information and the benefits of using the information.

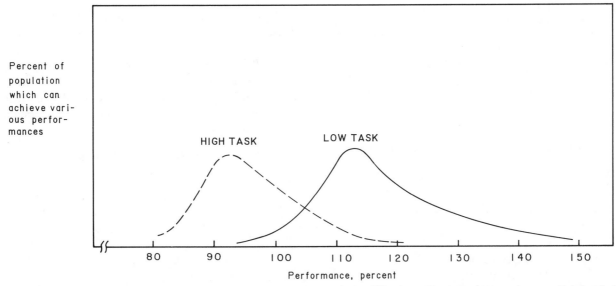

FIGURE 26.1 Normal pace (100% pace) is not the same from organization to organization. Different concepts of normal pace can be grouped into **low-task** paces and **high-task** paces. While it is difficult for many to maintain a 100% pace in the high-task concept (i.e., there is considerable worker selection), in the low-task pace, almost everyone can achieve it. Karger and Hancock (1982) report "almost everyone" is 95% for MTM and time studies using MTM's definition of 100%. When allowances are added to the normal time, over 99% of the industrial population is capable of achieving standard. However, performance may not match capability.

2.1 Nonengineered (Type II) Estimates

"Quick and dirty" information can be obtained with low cost; it is useful in some situations. Note that **nonengineered standards** are not preceded by methods or quality analysis and thus are **did take standards,** not **should take standards.** Nonengineered estimates have many subjective and few objective characteristics.

Often the time study group is under pressure to give standards on many jobs immediately and they don't have the resources to do everything immediately. If type II standards are used in such situations, they should be identified as temporary and should expire automatically after a specified date, say 60 days after issue. At the time of issue the IE department also should provide a schedule for upgrading standards from type II to type I.

2.1.1 Historical records

Historical records may be used for very "dirty" estimates. For example, how many cases can be picked/h in a warehouse? You might count how many cases were shipped in January, February, and March and find out how many employees were employed in each of those months, assume a monthly h/employee, and calculate h/case. Assume product mix does not change. Assume delays and idle time are "reasonable." Note that this is a "did take" time, not a "should take" time.

2.1.2 Ask expert

Another approach is to ask a knowledgeable person how long a job will take. For example, ask the supervisor of the maintenance department how long it will take to paint a room or install a conveyor system. Ask the sales manager how many customers can be contacted in a week. Ask the traffic supervisor how many miles a driver can cover each day.

A problem is the estimate may not be good and it is difficult to cross-check. Estimates also may vary depending on how hungry the group is for work.

2.1.3 Time logs

It may be that the time/job is being used for accounting purposes and need not be known before the job is done. Then **time logs** can be used. An engineer, for example, might write down:

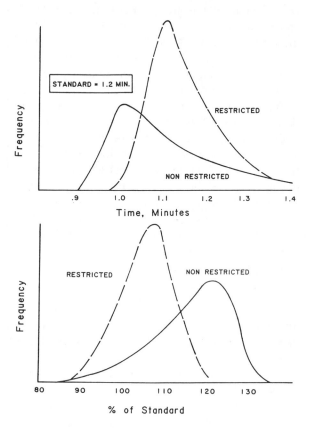

FIGURE 26.2 **Restricted output** has the characteristics of small variance and negative skewness (as a percent of standard). The key is a small amount of variation between or within individuals. The consistent achievement of the target over the reporting period (e.g., 1 week) may conceal wide fluctuations by period (days) but with the characteristic that, on the last day, cumulative actual performance for the period slightly exceeds cumulative required performance. For example, if the weekly goal is 100%, then performance may be 95% on Monday, 90% on Tuesday, 90% on Wednesday, 100% on Thursday, and 125% on Friday. Often this pattern repeats as it becomes a game to the worker.

TABLE 26.4 Example discipline levels for not producing enough. A published "set of rules" ensures that everyone is treated fairly. Most organizations have a similar set of rules for tardiness and absenteeism.

Step	Description
0	Normal operator, acceptable performance
1	Oral warning
2	Oral warning; detailed review of method with supervisor or trainer
3	Written warning; additional training
4	Written warning; some loss of pay
5	Written warning; larger loss of pay
6	Discharge from job

Day	Project	Hours
Monday	A	4.0
	B	2.5
	C	2.0
Tuesday	C	3.5
	A	.5
	B	4.0

Naturally, this approach does not give any information on should take times, just on did take times. In addition, even the did take times tend to be

BOX 26.2 *Incentive applications*

At one time, individual incentive wages were common. Kopelman (1987, p. 37) reports that wage incentives were used by 75% of industrial organizations in 1935, 52% in 1939, 27% in 1958, and 20% in 1982. Wage incentives are becoming relatively rare as it is more and more difficult to find jobs which (1) are highly repetitive, (2) are employee paced, not machine paced, (3) are performed relatively independently of other workers, and (4) have outputs which can be tested for quality and, if found acceptable, counted. The shift has been from individual incentives to group incentives.

In a group incentive, the group of workers may be told to produce 250 shirts every hour. For every percent that the team is over standard, they get a percent increase in pay. Another approach is to reward the entire factory based on standard costs of the entire product line; if standard costs go down X%, then wages go up by Y% (where X may not equal Y). An even more indirect incentive is an annual bonus that depends upon the firm's profit. (The U. S. auto industry has such a plan.) In this case the worker bonus depends not only on individual performance but also on management performance, product design and marketing, how tough the competition is, and other factors. Management has an advantage in this kind of plan since the worker income is not a fixed cost. That is, if profits are good, everyone can get a bonus but if profits are bad, labor costs stay low. The workers benefit because management is more willing to give bonuses when times are good but strongly resists making them a fixed cost to be paid regardless of profits. (Of course, many people feel they are "just workers" and none of their income should be "at risk.")

BOX 26.3 *Financial incentives for nonincentive employees*

By definition, nonincentive employees' wages do not vary with their output. However, it still is possible to use incentives to cut costs. (In addition, there are plans to increase revenue. Examples would be employees on commission, dealers, and distributors. See *Everything You Ever Wanted to Know.*)

Example cost-reduction goals are to reduce standard maintenance downtime on company vehicles, reduce inservice accidents involving company vehicles, reduce tardiness, reduce absenteeism.

Each goal should have an objective and a performance level specified. For example:

Goal: Reduce tardiness
Objective: Reduce tardiness 30%
Performance level: No unexcused lateness in each week.

The rules for the game should be simple and specific (have written rules), attainable (every participant should have a chance of winning), and measurable.

---

inaccurate. Possible reasons for inaccuracies are as follows:

- Totals will be at least 8 h/day even if the individual did not work 8 h. That is, no idle time will be reported.
- Small jobs tend to be omitted. For example, a 10-min phone call concerning job P might be lumped with job A.
- The person probably will try to remember at the end of the day how much time was spent on each project rather than recording as the day goes by.

- The time log may be deliberately biased. For example, assume an engineer has a time budget of 500 h on project A and no more hours left for project B, which was supposedly completed last week. Yet 2 hours have to be spent Tuesday completing some work on project B. It is very easy to not make waves and just charge the project B time to project A.

A time log is a form of traveler. A folder accompanies the job and each person working on the job writes the time used on each day and initials it.

2.1.4 Occurrence sampling

Occurrence sampling is described in detail in Chapter 8. The observer intermittently observes operations many times over a given period. For example, a study might establish that a secretary did letters 40% of the workday. If the study were over a 3-week period and the work week was 40 h, this means that .4(3) (40) = 48 h was spent doing letters. If 72 letters were done during this period, then it took 48/72 = .67 h/letter. Note that the method is not questioned or recorded since the operation is not studied beforehand. Note also that the worker was not rated (i.e., pace was assumed to equal 100%). Note also that the resulting time includes allowances. Although occurrence sampling can give approximate times for intermittent jobs, note that the workers can bias the results by slowing down during the observation period or by not using good methods.

2.2 Engineered (Type I) Estimates

For more accurate estimates, use **engineered estimates**. Engineered estimates must be preceded by a methods and quality analysis so they give a should take time, not a did take time. Engineering estimates have few subjective and many objective characteristics.

If a plant or area is going to have type I standards and formerly had no standards or type II standards, use the following steps:

1. Select a logical organizational unit and decide to set type I standards for all touch labor in the area. It is not recommended that you set standards for only a few people in each department.
2. Orient supervisors and workers.
3. Improve the work methods.
4. Break the tasks into elements and record them and the work conditions.
5. Determine the time, either through time study or standard data.
6. Calculate standards, including allowances.
7. Check and debug standards with check studies and audits.
8. Implement for a trial period, for example, one month.
9. Go to full implementation, including discipline and incentives.

There very well may be employee resistance to the new standards, and full acceptance may take several years.

MIL-STD-1567A, Military Standards: Work Measurement, requires for all individual type I standards (assuming the basic standards system is in place):

- documentation that the method was analyzed before time was determined
- a record of the method or standard practice followed when the time standard was developed
- a record of rating (if time study was used)
- a record of the observed times (if time study) or predetermined time values used
- a record of the computations of standard time, including allowances

There are two basic ways of determining time/job: stopwatch time study and standard data.

2.2.1 Stopwatch time study

Stopwatch time study is described in detail in Chapter 27. When discussing time/job, time is reported at three levels:

1. *Observed time:* The "raw" (unadjusted) time taken by the worker. It does not have any rating or allowance adjustment.
2. *Normal time:* Normal time = (Observed time) (Rating). The observer estimates the pace of the worker in relation to normal pace. See Chapter 27 for more details.
3. *Standard time:* Standard time = Normal time / (1 − Allowances) (This formula is based on allowances expressed as a percent of shift time. See Chapter 31 for an alternative formula for allowances expressed as a percent of work time.) The normal time is increased to account for personal, fatigue, and delay allowances.

The observer should use the following steps:

1. Make sure the operator is following a good method. After all, the goal is a should take time, not a did take time. The method should be analyzed and recorded in detail, including quality requirements. The improved performance typically accompanying engineered work standards comes not only from improved methods themselves but also the attention focused on delays.
2. Break the job into elements to facilitate methods recording and analysis and for standard data input. Record the observed time; determine proper rating.
3. Calculate normal time for each element.
4. Increase normal time with allowances to get standard time.
5. Decide how often each element of the job occurs and multiply standard time by occurrence/unit.

Stopwatch time study requires an operator doing the operation; thus, it cannot be done ahead of production. In general, it requires the operator to do the operation over and over, rather than doing many different tasks intermittently (such as might be done in office or maintenance work).

In repetitive work, where detailed methods analysis is desired, a videotape of the task can be made and the analyst can study the tape rather than the live person. See Box 16.2.

2.2.2 Standard data

Standard data are described in detail in Chapter 29 (micro level) and Chapter 30 (macro level). In this case, the analyst visualizes what the job entails. A danger is that the analyst may not think of some of the steps (elements) needed to complete the job. After recording the method, the analyst uses a table or formula for the amount of time for each element. The database elements are expressed in normal time (i.e., rating is included), so no additional rating is required. Then normal time is increased with allowances to obtain standard time.

Compared with time study, the standard data method has three advantages. (1) Cost of determining a standard is low (assuming you have a database with element times). (2) Consistency is high as everyone using the database should get the same times. (3) Ahead-of-production standards are helpful in many planning activities. One disadvantage is you may not have the money to build the database. (The databases are built from stopwatch studies and predetermined times.) Another disadvantage is that the analyst must imagine the work method; even experienced analysts may overlook important details or low-frequency elements.

3 DOCUMENTING, USING, AND MAINTAINING STANDARDS

3.1 Documenting Standards

Standards are part of a goal-setting system and control is essential to any goal-setting system. The attainment of goals must be monitored. Therefore, firms set up many documents that compare performance to the plan.

3.1.1 Quantification and identification of output

For maximum control, each output step of each component or assembly would be recorded. For example, operator 24 completed operation 7 on one unit of part 25 at 9:00, on a second unit at 9:05, on a third unit at 9:09, and so on. With bar coding of individual parts and computer input terminals at specific workstations, this technique is not only feasible but becoming more economical every day. Generally, however, the operator would report completion only of the lot. For example, operator 24 completed

operation 7 on 250 items at 11:36. Noncomputerized systems may just report that all operations were completed on 250 units of part 25 on July 17th. The cruder the data, the less useful the analysis. (Note Figure 5.6 and the river with an eddy analogy. Much rework (eddies) is not reported in the reporting system.) Be sure that quality is recorded as well as quantity. Firms tend to develop quite elaborate codes for various events such as direct labor, downtime, machine breakdown, defective components, setup, material handling, union activities, and so forth.

As mentioned before, nonengineered estimates can be used for temporary situations; engineered estimates, however, are considered permanent. In theory, permanent means forever, eternity. Yet you would not expect a time standard set in 1895 to be valid in 1995, because things are different. This is the key point. Establishing the present work method is not difficult—you can just go look. But what was the method when the standard was set in 1980? Any engineered standard therefore must be accompanied by a detailed method description so you can *prove* how things changed.

3.1.2 Audits

MIL-STD-1567A requires an annual audit. The **audit** should determine the validity of the prescribed coverage, the percentage of type I and II coverage, use of labor standards, accuracy of reporting, attainment of goals, and results of corrective actions regarding variance analysis. That is, the audit encourages keeping the work measurement systems up to date, accurate, and useful (Jorcyk and Castle, 1987).

A question is which standards to audit. Audits are taken to see if the present method is sufficiently different from the method used when the standard was installed, so that the time has changed significantly. "Significantly" often is defined as a 5% change in time.

One possibility is to audit jobs when there is a major change in the work method. This tends to be difficult to implement as many changes are very small by themselves. However, the cumulative effect of many small changes can be quite large.

Auditing only jobs for which the operators report performance over "X" percent of standard (e.g., 120%) doesn't work. As soon as the operators learn that 120 is the magic number, the never report performances over 120% again. They just take more breaks when given that job.

Audits should be periodic, that is, on a standard schedule. Auditing all standards in a specific department at the same time (e.g., every March) means that some jobs would be audited after a year and some would be audited only a few days after they were set. A better procedure is to set an expiration date (12 or

24 months) on each standard at the time it is set. Then when the standard expires, and if it still is an active job, an audit is made. If it is not active, the standard would be converted from permanent to temporary. Then if the job is resumed, the temporary can be used for a short period of time (e.g., 30 days) until a new permanent standard can be established. An advantage of a known expiration date is that if a standard is audited (and perhaps tightened), the operator will not feel "picked on."

If the resources available for auditing are not sufficient for doing all the audits required, then use the Pareto principle and audit the "mighty few" and don't audit the "insignificant many." However, the standards on the "insignificant many" that have passed the expiration date should be converted from permanent to temporary standards until they can be audited. Note also that temporary standards are type II standards, with all the problems of type II standards.

When auditing time standards, keep track of the delay time over all the audits. This is an excellent check on whether the former delay allowance is still valid. For example, assume 40 jobs were audited in time studies lasting a total of 60 hours. During the 60 hours, unavoidable delay time was 3 hours. Then the delay percent was 3/60 = 5%. This can be checked against the delay allowance presently in use.

3.2 Using Standards

Note that information on standard performance vs. actual performance (variances) is fed back not only to management but also to workers. Feedback to the workers provides knowledge, motivation, and reinforcement. They know the effects of their actions, are motivated to continue their actions, and receive reinforcement to increase the frequency of desired behavior and reduce the frequency of undesired behavior (Smith, 1987). See Table 26.5 for an example weekly form. (MIL-STD-1567A requires a labor performance report for each work center at least weekly.) Daily reports are useful to highlight delays and production problems; monthly reports smooth the fluctuations and help show long-run trends. Consistent low production by a specific worker often indicates a need for more training. Naturally these data are useless if the input data have been incorrectly entered or falsified.

Downtime is an area in which employees can make entries to benefit themselves; unfortunately, some managements consider all downtime as evidence of employee "deviousness" and forget that most downtime is due to management error.

The following list defines terms from Table 26.5.

- Type I. The hours, either actual (reported) or earned (standard) set with type I (engineered) standards.
- Type II. The hours, either actual (reported) or earned (standard) set with type II (nonengineered) standards.
- Actual hours is the time reported by an operator for a task covered by either a type I (engineered) or type II (nonengineered) standard. It normally includes rework/repair/scrap due to operator error.
- **Earned hours** is the time from the time standard.
- **Unmeasured hours** is the time for tasks not on standard.
- Delay is the clocked-out time due to delays beyond operator control (e.g., wait for material, downtime, rework due to engineering or vendor problems); it also is called unavoidable delay. Delays less than 5 min generally are not clocked out and are covered by a delay allowance in the standard.
- **Labor efficiency** or efficiency is (earned hours/actual hours) × 100. In Table 26.5 on 7-13, this is $(14.7 + 2.2)/(16.7 + 4.0) = 82\%$.
- Variance, which is not tabulated in Table 26.5, is actual time on standard minus earned time on standard.
- Delay percent is delay hours/total actual hours. In Table 26.5 on 7-13, it is $3.3/(16.7 + 4.0 + 3.3) = 14\%$.
- Percent coverage is the percent of standard hours that are type I hours. In Table 26.5 on 7.13, this is $16.7/(16.7 + 4.0) = 81\%$. MIL-STD 1567A requires 80% coverage of touch labor hours.
- Realization is the ratio of total actual hours/ earned hours. Total actual hours includes hours on standard + unmeasured + delay. In Table 26.5 on 7-13, this is $(16.7 + 4.0 + 3.3)/(14.7 + 2.2) = 1.42$. While labor efficiency measures the operator, realization measures the operator + organization. Some realization factors are:

1. learning (in using the method)
2. technical (engineering changes, tooling errors, instruction errors, scrap/rework)
3. logistics (incorrect hardware, part shortages, waiting for other operators)

TABLE 26.5 Example supervisor's weekly labor report (Jorcyk and Castle, 1987).

Dept. *175*   Supervisor: *Sullivan, G.*   Shift *1*   Date: *July 19, 1986*

| | Actual Time Dist., h | | | | Earned Hours | | Performance | | | |
Date	Type I	Type II	Unmeasured	Delay	Type I	Type II	% Eff.	% Delay	% Coverage	Realization
7-13-86	16.7	4.0		3.3	14.7	2.2	82	14	81	1.63
7-14-86	18.1	3.0		2.9	14.9	1.0	75	12	86	1.61
7-15-86	28.0		6.8	4.0	20.1		72	10	80	1.93
7-16-86	27.5			4.5	20.2		73	14	100	1.58
7-17-86	28.1	1.5		2.4	20.9	0.8	73	8	95	1.53
7-18-86										
7-19-86										
Totals	118.4	8.5	6.8	17.1	90.8	4.0	75	13	89	1.66
Goals							85	10	85	1.30

TRENDS

% Eff.

```
100%
95
90
85
80
75
70
     Week
     1  2  3  4  5  6  7  8
```

% Delay

```
20%
18
16
14
12
10
8
6
    Week
    1  2  3  4  5  6  7  8
```

Realization Factor

```
2.4
2.2
2.0
1.8
1.6
1.4
1.2
    Week
    1  2  3  4  5  6  7  8
```

REVIEW QUESTIONS

1. List the 5 groups of applications for time/unit.

2. Briefly discuss the concept of a discipline level.

3. List 4 different ways of determining nonengineered (type II) time estimates.

4. List the 2 ways of setting engineered (type I) time estimates.

5. Define observed time, normal time, and standard time.

6. Give 3 advantages and 2 disadvantages of standard data vs. time study.

7. Give the 5 requirements of MIL-STD-1567A for individual type I standards.

8. Define and contrast labor efficiency and realization.

REFERENCES

Engwall, R. Work measurement in the organization of the future. *1987 IIE Integrated Systems Conference*, November 1987.

Everything you ever wanted to know about running an incentive program. Bulova Watch Co., 75–20 Astoria Blvd., Jackson Heights, NY 11370.

Jorcyk, G. and Castle, D. Work measurement and MIL-STD-1567A. *IE News: Work Measurement and Methods Engineering*, Vol. 22 [1], Summer 1987.

Karger, D. and Hancock, W. *Advanced Work Measurement.* New York: Industrial Press, 1982.

Koop, J. What we were told. *IE Work Measurement and Methods Newsletter,* Autumn 1982.

Kopelman, R. *Managing Productivity in Organizations.* New York: McGraw-Hill, 1987.

Panico, J. Work standards: Establishment, documentation, usage, and maintenance. In *Handbook of Industrial Engineering,* 2nd ed. Salvendy, G. (ed.), Chapter 59. New York: Wiley, 1992.

Rodgers, S. (ed.). *Ergonomics Design for People at Work, Vol. 2.* New York: Van Nostrand Reinhold, 1986; p. 284.

Sellie, C. Predetermined motion–time systems and the development and use of standard data. In *Handbook of Industrial Engineering,* 2nd ed. Salvendy, G. (ed.), Chapter 63. New York: Wiley, 1992.

Smith, G. Work measurement under attack: The IE's response. *Proceedings of 1987 IIE Integrated Systems Conference,* 112–17, 1987.

OVERVIEW

Stopwatch time study should be preceded by a methods analysis so the resulting standard is a "should take" time, not a "did take" time. The time study is a sample from which the population is predicted. Better results are obtained if the sample is representative of the population, accurate measurements are taken, and data are carefully gathered. The observed time is multiplied by the rating to obtain normal time.

CHAPTER CONTENTS

1 Overview
2 Preparation
3 Timing
4 Rating

KEY CONCEPTS

allowances	irregular elements	present world/future world
continuous/snapback	machine time/manual time	rating
elements	micromotion	sample/population
expectancy	normal pace	standard time
flat ratings	normal time	
foreign elements	observed time	

1 OVERVIEW

Stopwatch time study is one of the two methods to establish a type I time standard. (For standard data, the other method, see Chapter 30.)

After determining the proper method, the analyst will observe one or more operators continuously and record the time taken. This time is called **observed time.** Then this observed time is adjusted (given a **rating**) to obtain the time that a typical experienced operator would take.

(Observed time) (Rating) = **Normal time**

(In England, normal time is called basic time.) However, normal time is not representative of the time the experienced worker would take working all day. Additional time must be given for personal, fatigue, and delay **allowances.** See Chapter 31 for more on allowances. The resulting time is called **standard time.** There are two possible formulas:

Standard time = (Normal time) / (1 − Shift
allowances)

and

Standard time = (Normal time) (1 + Work time
allowances)

For example, if observed time = .01 h/unit and the rating was 120, then the normal time = .012 h/unit. Assuming allowances were given as a percent of the shift (time at work, not working time) and were 10%, the standard time = .012/.9 = .0133 h/unit. If the operator had an 8-h shift, the operator should produce 8/.0133 = 601 units/shift.

The basic time study procedure is to take a small sample of the times from the **present world** (now) and predict the times of the **future world** (then). There are potential inaccuracies in going from a **sample** of the present world to the **population** of the present world (i.e., from the person or persons studied to all people who have that job). There are also potential inaccuracies in going from the present world to the future world.

The procedures in this chapter attempt to reduce inaccuracies from the sample present world estimates of the population of present world times. In addition, as the months pass, the present world changes and the relevance of the existing standard becomes more and more in question. General changes can be predicted from individual and organization learning. See Chapter 28. Specific changes should be observed through audits. See Chapter 26.

2 PREPARATION

Before the study, the analyst should do a methods analysis and select the operator to be studied.

2.1 Methods Analysis
The goal of a type I standard is a "should take" time, not the "did take" time of a type II standard. If a time study is done without a preceding methods analysis, the result is a type II standard.

Design of the job has been covered in the previous chapters of this book. Most of the book is devoted to job design since that is where productivity changes are made and worker health safeguarded. A poorly designed job can easily result in low productivity and poor quality. Having an accurate time standard for a low-productivity job doesn't give as much benefit to the organization and the workers as a type II standard on a well-designed job. Thus, from a productivity viewpoint, supervisors of industrial engineering departments should devote most of the department's resources to job design and productivity and only a small portion to time standards. However, from a service viewpoint, many other groups need accurate time standard information. Thus, their needs for information must be balanced against the natural desire of the industrial engineering department to improve jobs.

The primary reason for doing a methods analysis before doing a time study is to establish a safe, productive job. A secondary, but still important, reason is to leave a permanent record of the method so that future audits will have a basis for comparison. Typical items that should be recorded are date of observation, person observed, person observing, machine used, tools/fixtures used, feeds, speeds, handtools, part number processed, and the like.

After a good method is established (say by using the checklists of Chapter 5 or the guidelines given in Chapters 12 through 25), the job should be broken into **elements.**

For ease of understanding, consider the task to be polishing a pair of shoes. Rather than timing the entire task as a whole, the job might be divided into (1) get the shoes, (2) polish the shoes, and (3) put the shoes away.

There are 4 reasons to break a job into elements:

1. Elements make it possible to reuse the data. Assume the next job required the operator to get 4 shoes at a time instead of 2, or that someone else brought the shoes. Then a new time study would be required for element 1 only, not the entire task. In some tasks, elements may be used in different

sequences, so having the job split allows the elements to be used for standard data. Think of the elements as bricks used to build a structure of times. We will reuse bricks instead of using new bricks.

2. Elements permit different ratings for different elements. The rating can be 90% for element 1, 100% for element 2, and 105% for element 3—but only if the job is broken into elements. Without elements, only one overall rating can be given. Elements should not combine machine time (which is always rated 100%) with manual time (which can have any rating). **Machine time** is time in which a machine can operate unattended; **manual time** is time when an operator is required. Rating by element is especially important if the element will become a standard datum.

3. Elements permit consistency checks, within the study and between studies. For example, if getting the first pair of shoes took 1.2 min, the second took 1.3, and the third took 2.1, the long time of 2.1 stands out. When only the overall time is recorded, chance fluctuations in other elements may make the total time for all 3 elements be 4.1, 4.2, and 4.5. Then the long time in element 1 would not be noticed. Elements in this study can be compared with similar elements in other studies. If the overall task had a different combination of elements, the overall times could not be compared.

4. Elements improve methods descriptions. A constant problem of audits is determining the method and quality of the original study so it can be compared to the present situation. Breaking the job into elements improves the methods description.

After the task is broken into elements, write the element description on the time study form with a description of its end point (EP). (It is also called termination point, TP.) Select a very definite, easily defined point as the EP; an EP that makes a sound usually is best. For example:

Element	Description	EP
1	Get 1 pair of shoes from bedroom closet.	Release shoes
2	Polish shoes with dauber, brush, paste polish, and rag.	RL rag
3	Put shoes away in closet.	RL shoes

2.2 Operator Selection

Once the proper method has been developed, the next question is who should be studied. Under no circumstances

should the timing be done without the knowledge of the worker and the supervisor.

Remembering that this is a sample from a population, try to make the sample as representative of the population as possible. In many cases there is no choice as there is only one worker for the job.

If there is a choice (multiple shifts for the same job, multiple people doing the task on the same shift), select experienced rather than inexperienced workers. The standard will apply to experienced workers. Experienced means not only experienced in a specific type of work but experienced doing this specific operation on this specific part or assembly—that is, reasonably far along the learning curve. Work methods of novices have an unusually high number of delays, fumbles, hesitations, and slow decisions; it is quite difficult for the time study technician to determine the precise rating correction for these difficulties. Thus, the standard from studying inexperienced operators probably will be loose. If a standard is required now but the worker is at the start of the learning curve, my recommendation is to establish a type II (i.e., temporary) standard until the worker has more practice.

If there are multiple experienced operators, it may be possible to do a time study on several of them. If possible, select average or typical workers rather than someone who is unusually slow or fast. (For methods analysis, study a fast worker, because the speed is likely to be due to a good method rather than a fast pace; for the time recording, however, it is better to use an average worker.)

There are two reasons to use an average worker for time study: rating accuracy and worker acceptance.

Rating (discussed in more detail later in this chapter) requires the time study technician to normalize the observed time. For example, assume the problem is to find the standard time for college males to run 1,500 m. If timing a world-class distance runner, the time might be 3.95 min; if timing Joe Student, the time might be 6.5 min; if timing Professor Konz, the time might be 12.0 min. When timing the expert, the analyst knows the runner is fast but the question is how fast. The rating might be 160%. The resulting normal time then would be $3.95 \times 1.60 = 6.32$ min. The rating on Joe Student might be 100%, so normal time $= 6.5 \times 1.00 = 6.50$ min. The rating on Konz might be 50%, so the normal time $= 12 \times .50 = 6.00$ min.

Rating accuracy is greater if the performance is close to 100%. As performance gets farther and farther from 100%, the absolute error of rating increases. (Your firm may wish to set a policy of not using any time studies in which the operator performs at lower than 70% or higher than 130%.) Thus, to improve rating accuracy, study an average operator if possible.

The second reason to study an average operator is to improve acceptance of the standard. Workers really don't trust ratings. The more the observed time is adjusted the less they trust it. As pointed out in the previous paragraph, their mistrust of extreme adjustments is well founded.

To make the sample more representative, the time study should occur at different times of the day and week. That is, not all studies should be done on Monday or in the early morning or on the first shift.

3 TIMING

The normal technique is to time a live performance. However, some people like to make a videotape of the operation and then time the videotape. The primary advantage of the videotape is the permanent record of the method. (Remember, however, to record feeds, speeds, operator's name, observer's name, and other relevant details, as is done with a conventional time study; that is, the videotape supplements the time study form; it does not replace it.) When timing the videotape, note that using just the tape counter does not give an accurate time. This is due to tape slippage. See Box 16.2 for comments on videotaping.

After discussing timing techniques, we will discuss the number of observations to make and what to do with irregular and foreign observations.

3.1 Timing Techniques There are 4 alternatives: 1 watch with continuous hand movement, 1 watch with snapback, 3 watches with snapback, and an electronic digital watch with a hold circuit. In the **continuous** technique, the watch runs without stopping and the completion time of each element is noted. In the **snapback** technique, the watch is reset to zero after noting the time for each element.

3.1.1 One watch, continuous First consider using one stopwatch, starting the watch at the start of the study (e.g., 1:02) and letting it run until the end of the study (e.g., 1:32). The elapsed time was 30 min or .5 h. As a check, the sum of all the elements should total .5 h. During the study, the eye of the observer was aligned with the task and the watch. When an element was completed (the observer noticed an end point), the observer noted the position of the watch hand and then wrote the time down on the form. (In some cases element end points can be detected by sound.) The watch hand continues moving. Figure 27.1 shows the resulting record of the data. Rather than writing the entire number down (.95, 1.02, 1.21, 1.87, 2.06), the data are simplified by omitting decimal points and omitting redundant first numbers (95, 102, 21, 87, 206). Then later, in the office, the numbers are subtracted, written down using ink or a contrasting color, and the analyst learns what the times for each element are. Figure 27.2 shows how the form looks after the subtraction. The advantage of continuous recording is that the clock never stops, so no time is omitted; workers like that. The disadvantages are that the observer does not know, at the job, how individual elements vary and the observer is trying to read a moving target.

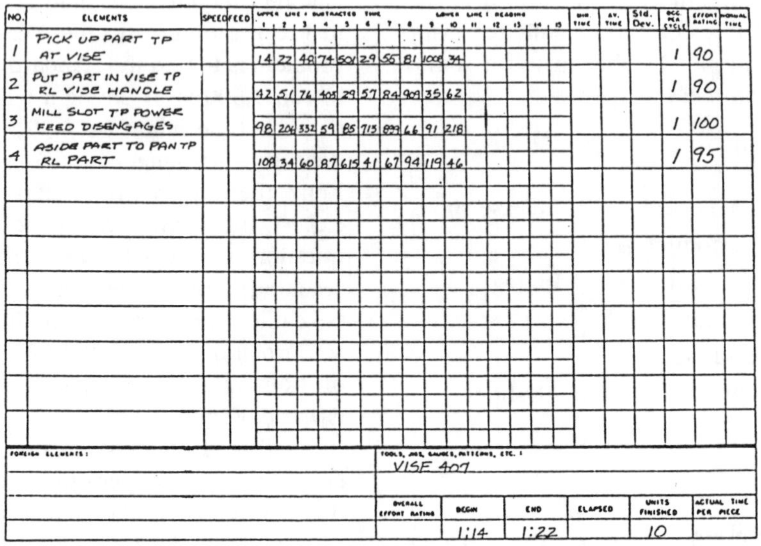

NO.	ELEMENTS	SPEED	FEED	1	2	3	4	5	6	7	8	9	10	11	12	13	14	15	MIN. TIME	AV. TIME	Std. Dev.	OCC. PER CYCLE	EFFORT RATING	NORMAL TIME
1	PICK UP PART TP AT VISE			14	22	48	74	501	29	55	81	1008	34									1	90	
2	PUT PART IN VISE TP RL VISE HANDLE			42	51	76	405	29	57	84	909	35	62									1	90	
3	MILL SLOT TP POWER FEED DISENGAGES			98	204	332	59	85	713	899	66	91	218									1	100	
4	ASIDE PART TO PAN TP RL PART			108	34	60	87	615	41	67	94	119	46									1	95	

UPPER LINE: SUBTRACTED TIME; LOWER LINE: READING

FOREIGN ELEMENTS:

TOOLS, JIGS, GAUGES, PATTERNS, ETC.: VISE 407

OVERALL EFFORT RATING	BEGIN	END	ELAPSED	UNITS FINISHED	ACTUAL TIME PER PIECE
	1:14	1:22		10	

FIGURE 27.1 Continuous analog one-watch form before subtraction. The reverse side of the form would show a workstation sketch. Times usually are coded, that is 14, instead of .0014.

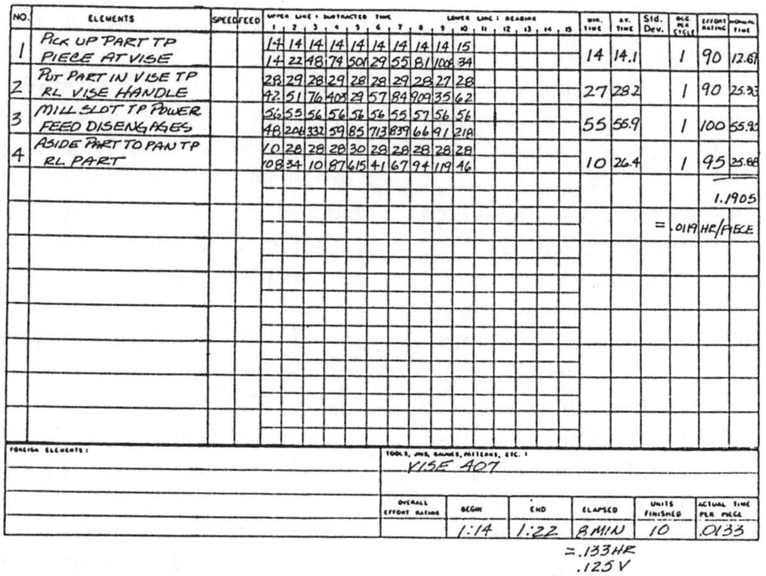

NO.	ELEMENTS	SPEED	FEED	1	2	3	4	5	6	7	8	9	10	MIN. TIME	AV. TIME	Std. Dev.	OCC. PER CYCLE	EFFORT RATING	NORMAL TIME
1	PICK UP PART TP PIECE AT VISE			14 / 14	14 / 22	14 / 48	14 / 74	14 / 501	14 / 29	14 / 55	14 / 81	14 / 1008	15 / 34	14	14.1		1	90	12.69
2	PUT PART IN VISE TP RL VISE HANDLE			28 / 42	29 / 51	28 / 76	29 / 403	28 / 29	28 / 57	29 / 84	28 / 909	27 / 35	28 / 62	27	28.2		1	90	25.38
3	MILL SLOT TP POWER FEED DISENGAGES			56 / 48	55 / 206	56 / 332	56 / 59	56 / 85	56 / 713	55 / 839	57 / 66	56 / 91	56 / 218	55	55.9		1	100	55.90
4	ASIDE PART TO PAN TP RL PART			10 / 108	28 / 34	28 / 10	28 / 87	30 / 615	28 / 41	28 / 67	28 / 94	28 / 119	28 / 46	10	26.4		1	95	25.08
																			1.1905
																	=	.0119	HR/PIECE

(Upper line: subtracted time; lower line: reading.)

TOOLS, JIGS, GAUGES, PATTERNS, ETC.: VISE 407

OVERALL EFFORT RATING	BEGIN	END	ELAPSED	UNITS FINISHED	ACTUAL TIME PER PIECE
	1:14	1:22	8 MIN	10	.0133

= .133 HR
.125 V

FIGURE 27.2 Continuous one-watch form after subtraction later in the office. Use a contrasting color (such as red) or ink for the subtracted times. The times then are totaled, averaged, and multiplied by the rating to get the normal time/element of .0119 h/unit. As a check, the elapsed time on the wristwatch from the start to stop of the study (in lower right of form) was 8 min; 8 min/10 units studied = .8 min/unit = .0133 h/unit. The sum of the element times of 14.1 + 28.2 + 55.9 + 26.4 = 124.6 × .001 = .0124 h/unit.

3.1.2 One watch, snapback

As before, the observer aligns the eye, the watch, and the task. However, when the element ends, the time is read while simultaneously snapping the hand back to zero where it starts moving forward again. Then the element time is written on the form. See Figure 27.3. The advantage is that the subtraction is eliminated and the observer can see the pattern of element times while recording the times. Disadvantages are the need to read the moving target (watch hand) and remember where it was, the difficulty of reading 2 short (say 3 s) elements in a row, the fact that the moving hand takes a short time to come to zero and start again (about .2 s), which shortens every observed time, and the possibility that the observer may not record each and every cycle since observers tend to judge foreign and irregular data (discussed later in this chapter) as delays and stop the watch when it should be recording. There also is a tendency of the observer to stop the watch when confused (as when an operator changes element sequence). The lack of complete accounting of all

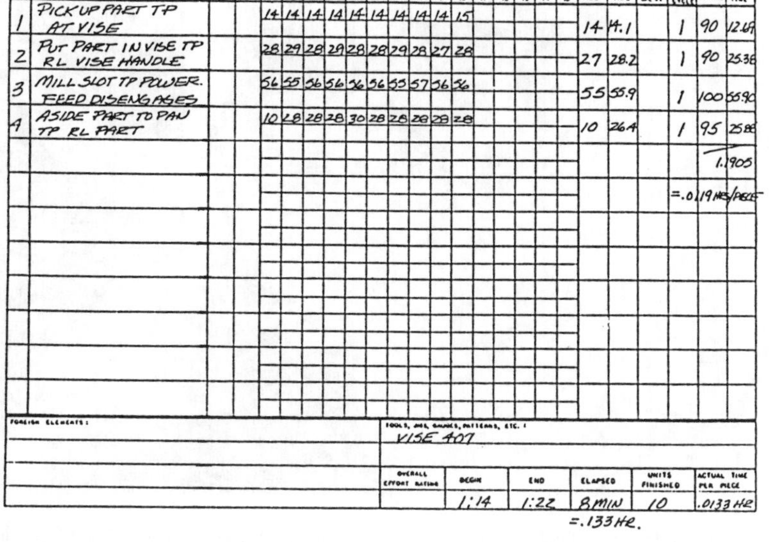

NO.	ELEMENTS	SPEED	FEED	1	2	3	4	5	6	7	8	9	10	MIN. TIME	AV. TIME	Std. Dev.	OCC. PER CYCLE	EFFORT RATING	NORMAL TIME
1	PICK UP PART TP AT VISE			14	14	14	14	14	14	14	14	14	15	14	14.1		1	90	12.69
2	PUT PART IN VISE TP RL VISE HANDLE			28	29	28	29	28	28	29	28	27	28	27	28.2		1	90	25.38
3	MILL SLOT TP POWER. FEED DISENGAGES			56	55	56	56	56	56	55	57	56	56	55	55.9		1	100	55.90
4	ASIDE PART TO PAN TP RL PART			10	28	28	28	30	28	28	28	28	28	10	26.4		1	95	25.08
																			1.1905
																	=.0119	HRS/PIECE	

TOOLS, JIGS, GAUGES, PATTERNS, ETC.: VISE 407

OVERALL EFFORT RATING	BEGIN	END	ELAPSED	UNITS FINISHED	ACTUAL TIME PER PIECE
	1:14	1:22	8 MIN	10	.0133 HR

= .133 HR.

FIGURE 27.3 Snapback with analog watch has the advantage of no subtraction later, although it has many other disadvantages.

the time makes it quite difficult to sell the resulting time standard to the worker. If a time study is to be made with only one conventional watch (an obsolete technique), my recommendation is to use the continuous system rather than the snapback system.

3.1.3 *Three-watch system*

A better alternative is the three-watch system. See Figure 27.4. All three watches are controlled by the same lever. Initially the hands of the first watch are moving, the second's are stopped at some time value, and the third's are stopped at zero. At the end of element A, the observer depresses the lever. Watch 1 moves to zero, watch 2 starts recording, and watch 3 stops at a time. The observer now writes down the time for element A from the stationary hand of watch 3. At the end of element B, the observer depresses the lever again. Now watch 1 starts, watch 2 stops at a time, and watch 3 goes to zero. The observer writes down the time for element B from the stationary hand of watch

2. The data form will be similar to that of Figure 27.3. The only disadvantage of the 3-watch system is that it requires 3 watches instead of 1 and the watch must be actuated by the crown (about 10% more expensive). However, the watch cost (amortized over watch life) is trivial compared to the extra clerical cost of the 1-watch conventional system. The three-watch system is also less susceptible to the errors of the snapback system.

3.1.4 *Electronic watches*

These watches can be used in either the snapback or the continuous mode. The key feature is that, when the user depresses the button, the display time is frozen while the clock continues timing. It thereby eliminates the moving target and the problems with watch hand movement time to zero position. It also has the advantage of being a digital display instead of analog, so it is easier to write the digital number down on the form instead of translating from the analog display.

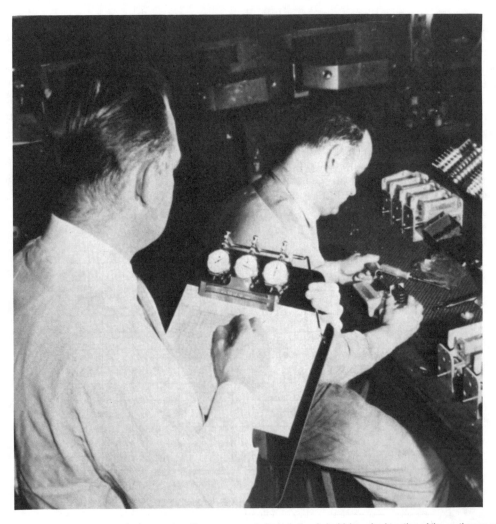

FIGURE 27.4   Three-watch analog systems give improved reading accuracy and eliminate the clerical labor of subtraction of the continuous analog watch system.

3.2 Number of Observations

There are two approaches to determining the number of observations: (1) statistical and (2) importance of decision.

3.2.1 Statistical

A time study of sequential observations is a sample from a population of times. The goal is to estimate the population mean from the sample mean. Sample means are not precise (unbiased) estimators of the population mean. That is, if the mean time from a time study (a sample) is .1 h/unit, we cannot say that the population of times (time/unit over the weeks, months, and years) will have a mean of exactly .1 h/unit. To minimize the difference between the sample and population means, increase the sample size; that is, make more observations. Unfortunately, increasing observations increases the cost of taking the sample. From statistics, the number of observations to record depends on (1) accuracy desired, (2) confidence desired, and (3) data variability. (The times vary primarily due to operator variations in grasp, position, and fumbles rather than moving or reaching faster.)

Accuracy desired   Accuracy can be given in relative or absolute terms. For a .2-h element, a ±5% relative accuracy is the same as an absolute accuracy of .2(.05) = .01 h. For a .02-h element, a ±5% relative accuracy is the same as an absolute accuracy of .02(.05) = .001 h. The limits of precision for the first case would be .19 to .21 h (a target width of .02 h). Both have a ±5% relative accuracy but the absolute accuracy is 10 times different! The size of the target determines the number of "shots" needed to hit it. It is more difficult to hit a smaller target, so more shots are needed. In a similar manner, more observations are needed for a small target (.002 h) than a big one (.02 h).

Unfortunately people often are not precise in their statements. They may just say "5% accuracy" without specifying whether it is absolute or relative accuracy.

Confidence desired   Continuing with the target analogy, shots may not always hit the target. For 90% confidence, you want shots to hit the target 90% of the time; for 95% confidence, you want shots to hit the target 95% of the time. In terms of a time study, 90% confidence with ±5% accuracy for a .02-h element means that if there were 100 different time studies, then in 90 studies the time study mean would be between .019 and .021 h, assuming the population mean was .02 h. In actuality, only 1 time study is made, so the confidence is really the "long-run" confidence. Any individual time study could be wrong.

Data variability   The more variability there is in the data, the more observations are required to hit the target with confidence; that is, fewer shots are required with a good rifle, a good shooter, and no wind gusts than with a poor instrument, unskilled technician, and random variations in the task. To the time study technician, the data variability is unknown when the study begins. Therefore, it must be estimated.

Estimate the population variability by taking a subsample (e.g., 10 cycles), calculating the subsample's variability, and, from the subsample's variability, estimating the population variability. See Box 27.1.

3.2.2 Importance of decision

The discussion in Box 27.1 assumes that the number of cycles should be determined solely from probability. However, in real life you can drown in a river that is only one meter deep on the average. Telling your boss you will be correct most of the time doesn't help when you have made a mistake on an important decision. The same problem was faced in statistical sampling in quality control. The Military Standard 105 tables recommend larger sample sizes for larger lot sizes—that is, larger samples for more important decisions. (From probability, required sample size does not depend upon lot size.) That is, cost is not only the observer cost but the cost of bad decisions from inaccurate information.

Tables 27.2 and 27.3 give three approaches to the concept of larger samples for more important decisions. Note that the tables do not agree. For example, for time/piece of .017 h (1 min), the Westinghouse recommendations are 20, 25, or 50 cycles (depending on quantity); GE recommends 40 (regardless of quantity); and Niebel recommends 25 to 40 cycles (depending on quantity). Thus, the recommendations should not be expected to apply exactly to all situations.

If accuracy of the time standard is especially important (as on a job with many hours/year), consider making two or three shorter studies rather than one long study. You will be estimating the population characteristics from several samples instead of one.

Generally, there is a fixed cost in making a time study (contacting the supervisor, analyzing the job, recording observations, making calculations, and writing it up). Therefore, it is probably worthwhile to spend at least 15 minutes recording times, regardless of the statistical calculations of Box 27.1. This seems to be the general approach of Tables 27.2 and 27.3. To make the benefits of the time study exceed its cost, increase sample size when (1) cycle time is short, (2) activity/year is large, and (3) cost of an inaccurate standard is high.

BOX 27.1 *Statistical formula for number of observations*

Accuracy desired, confidence desired, and data variability are related by the following equation:

$$A = z\sigma'_{\bar{x}}$$

where A = Amplitude (absolute accuracy) of the target, that is, precision = $S\,(\Sigma X/N)$
S = Relative accuracy desired, decimal
z = Number of standard deviations corresponding to the confidence desired
$\sigma'_{\bar{x}}$ = Standard deviation of the population times

The standard deviation of the population times and subsample times are related by:

$$\sigma'_{\bar{x}} = \frac{\sigma'_x}{N'}$$

where N' = Number of observations required to meet the criteria of precision and accuracy
σ'_x = Standard deviation of the subsample
$$= \frac{1}{N}\sqrt{N\Sigma X^2 - (\Sigma X)^2}$$
N = Number of times (observations in subsample)

Substitution and solving for N' gives:

$$N' = \left[\frac{\sigma'_x}{\sigma'_{\bar{x}}}\right]^2 = \left[\frac{Z}{A}\sigma'_x\right]^2 =$$

$$\left[\frac{Z\frac{1}{N}\sqrt{N\Sigma X^2 - (\Sigma X)^2}}{A}\right]^2$$

Using the subsample size = N as an estimate of population variability gives a "biased" estimate. For an "unbiased" estimate, use $N - 1$ instead of N. (Think of the unbiased estimate as a factor of safety, since the formula using $N - 1$ makes the variance look larger and thus the user will be more cautious in using the data.) Then

$$N' = \left[Z\frac{\sqrt{\dfrac{\Sigma X^2 - (\Sigma X)^2/N}{N - 1}}}{A}\right]^2$$

The equation can be simplified in a number of ways. First, precision level, A, can be expressed in terms of X. Then 5% relative accuracy becomes $.05\,(\Sigma X/N)$; 10% relative accuracy becomes $.1\,(\Sigma X/N)$. Second, the normal distribution can be assumed and 95% confidence rounded to 2 standard deviations instead of 1.96. Thus, for 5% precision and 2σ confidence:

$$N' = \left[\frac{40\,N\sqrt{\dfrac{\Sigma X^2 - (\Sigma X)^2/N}{N - 1}}}{\Sigma X}\right]^2$$

For 10% precision and 2σ confidence:

$$N' = \left[\frac{20\,N\sqrt{\dfrac{\Sigma X^2 - (\Sigma X)^2/N}{N - 1}}}{\Sigma X}\right]^2$$

Third, the population variability can be estimated from the range of the sample rather than the standard deviation of the sample. Although the range is not as efficient an estimator of the variability as the standard deviation, it is simpler to calculate.

$$R = d_2\sigma_x$$
so
$$\sigma_x = R/d_2$$
where R = Mean range of a subgroup of specified size
d_2 = Number of standard deviations that the mean range includes for a specified subgroup sample size
σ_x = Standard deviation of individual times

For example, for a subgroup of 10 sample times, $d_2 = 3.078$ from Table 27.1. That is, the range of a sample of 10 will, on the average, include 3.078 standard deviations.

As given before:

$$A = z\sigma'_{\bar{x}}$$
$$= \frac{z\bar{R}}{d_2 N'} \quad \text{so } N' = \left[\frac{z\bar{R}}{d_2 A}\right]^2$$

For 5% precision, 2σ confidence, and a subsample $N = 10$:

$$N' = \left[\frac{2\bar{R}}{.05 d_2 \bar{X}}\right]^2 = \left[\frac{40\bar{R}}{d_2 \bar{X}}\right]^2 = \left[\frac{40\bar{R}}{3.078\bar{X}}\right]^2 = \left[\frac{13\bar{R}}{\bar{X}}\right]^2$$

For 10% precision, 2σ confidence, and a subsample $N = 10$:

$$N' = \left[\frac{2\bar{R}}{.10 d_2 \bar{X}}\right]^2 = \left[\frac{6.5\bar{R}}{\bar{X}}\right]^2$$

In practice, the range of the sample is used instead of $\bar{R}$.

Assume 10 times were obtained in a time subsample: 20, 22, 20, 22, 20, 18, 18, 20, 19, and 21.

(continued)

BOX 27.1   *Continued*

The mean = 20. For 5% precision and 2σ confidence, $N' = 8$. That is, the subsample is sufficient and no more observations are needed. Using the range method, the subsample range is 4 giving $N' = 7$. In general, however, the standard deviation method gives a smaller N' than the range method (Hicks and Young, 1962).

Calculate N' for each element in the time study. If N' is equal to or less than the subsample size already taken, enough observations have been taken and the study can stop. If, for example, N' is 18 for element 3, then element 3 needs 8 more observations. In practice it would be best to get 8 more observations on all elements and thus improve their accuracy and confidence also rather than just studying 8 more cycles of element 3. Note also that the total of all the elements will have greater precision than individual elements.

TABLE 27.1   The range of a sample includes d_2 standard deviations, on the average.

Subgroup Number, N	Number of Standard Deviations in Range
5	2.326
6	2.534
7	2.704
8	2.847
9	2.970
10	3.078
15	3.472
20	3.735
25	3.931

3.3   Irregular and Foreign Observations   This section will be divided into concept and recording technique.

3.3.1   Concept   After initially analyzing the job, the observer will have recorded the elements in the order they should occur. But during the study, the operator may perform unexpected activities.

One possibility is that this is a rare or **irregular element**—at least the observer didn't anticipate it. It needs to be included like any other element. The observer must determine how often the element should be allowed per unit produced.

Another possibility is that the data are not normal work but are **foreign elements.** (Foreign elements are not allowed directly as part of the

TABLE 27.2   Minimum number of cycles to study.

IF	AND Westinghouse Electric Value (Westinghouse, 1953)			OR General Electric Values (Shaw, 1978)
Time/Piece or Cycle is Over	Activity/yr is under 1,000	Activity/yr is from 1,000 to 10,000	Activity/yr is over 10,000	Activity/yr is any value
.002 h (under)	60	80	140	200
.002	50	60	120	175
.003	40	50	100	125
.004	35	45	90	100
.005	30	40	80	85
.008	25	30	60	60
.012	20	25	50	40
.020	15	20	40	30
.035	12	15	30	18
.050	10	12	25	15
.080	8	10	20	13
.120	6	8	15	10
.200	5	6	12	9
.300	4	5	10	8
.500	3	4	8	5
.800	2	3	6	3
1.000	2	3	5	3
2.000	1	2	4	3

TABLE 27.3　Minimum number of cycles to study (Niebel, 1992).

IF CYCLE TIME, MIN (h)	AND ACTIVITY IS LESS THAN			
	1,000/yr	1,000 to 5,000/yr	5,000 to 10,000/yr	OVER 10,000/yr
<1 (.017)	40	45	50	60
1 to 2 (.017 to .033)	25	30	35	40
2 to 5 (.033 to .083)	18	20	22	25
5 to 10 (.083 to .167)	15	16	18	20
10 to 20 (.167 to .333)	9	10	11	12
20 to 40 (.333 to .667)	7	8	9	10
40 to 60 (.667 to 1.0)	5	6	7	8
>60 (1.0)	3	4	5	6

Source B. Niebel, "Time Study," in *Handbook of Industrial Engineering*, 2nd ed., G. Salvendy (Ed.). Copyright © 1992 by John Wiley & Sons. Reprinted by permission of John Wiley & Sons, Inc.

recorded times, but the time for this delay may be allowed through an allowance.) A delay can be avoidable (meaning the time will not be included in the standard) or unavoidable (meaning the time will be included in the standard). Examples of avoidable delay might be: stopped working to talk to friend, blowing nose, drinking a cup of coffee. Some time is allowed for these types of activities but under the classification of personal allowances and fatigue allowances rather than work time. Examples of unavoidable delay might be: talk to supervisor about work, idle because of lack of supplies, breaking a tool. Time is allowed for these types of activities under the classification of delay allowances.

Personal and fatigue allowances (see Chapter 31) are set from standard tables. The job gets a certain

allowance and the operator time spent in this type of activity during the study does not affect the allowance given. Delay allowances, however, should be determined from the delays occurring on the job. The delays occurring during a time study (especially longer studies lasting an hour or so) can give an estimate of the percentage of delays to allow for the standard. That is, if during the time study there was 5% unavoidable delay, then 5% may be a reasonable number to use for the delay allowance.

3.3.2　Recording technique

At the time the unusual event begins, the observer does not know whether it is an irregular element, an unavoidable delay, or an avoidable delay. The data must be recorded as they occur; later the decision can be made whether to include them in the standard.

Figure 27.5 gives examples of the following problems.

Missed reading　The worker may be going through cycles in a standard manner and the observer just misses the end of an element. Put an M in the spot where the time should have been. See column 2 in Figure 27.5. Do not guess at the time as this will bias not only the time of the missed element but also the following element.

Omitted element　The worker may omit an element. Put a dash in the spot where the time should have been. See column 4 in Figure 27.5. An omitted element may indicate an inexperienced operator. More likely is that the operator is trying to confuse the observer or that the element is not needed 100%

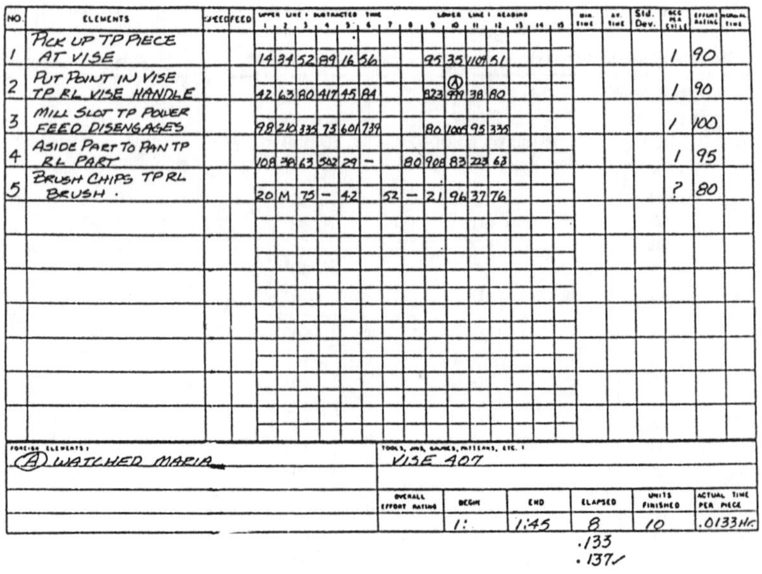

FIGURE 27.5　Missed times are shown by an M as in column 2. When the worker omits an element, put down a dash as in column 4. When an element is done out of order, as in columns 6–8, put a dash down for the omitted element, record the next element over one column, and when the omitted element occurs, move over another column. If an unexpected element occurs, as in column 10, record the time and next to it a letter. Then, elsewhere on the form, write a short note describing the event.

of the time. Before a time study, workers may describe in detail many different things necessary to do the job. Then, during the time study, they may become engrossed in doing the job and do it the way they usually do it and forget to add the embellishments for the observer.

Element out of order   Operators may do an element out of the order the observer has on the form (say element 5, then 4 instead of 4, then 5) either because the sequence isn't critical or because the operator is trying to confuse the observer. See columns 6–8 of Figure 27.5. The simplest technique is to put a dash down for the omitted element (4), put the time down for the next element (5) one column over, and then, when the skipped element (4) occurs, move over another column on the form and record it. Then element 5 is marked with a dash and the normal sequence resumes.

Unexpected element   As pointed out above, this may be an irregular element or avoidable or unavoidable delay. Record such elements as they occur; do not stop the watch. The decision of what to do with the time can be made later. See column 10 of Figure 27.5. As the events occur, move over a column and code them A, B, C, etc. Then elsewhere on the form write a short note (1–3 words) defining the meaning of the A, B, C.

4 RATING

4.1 Normal Pace   During the time study, the observer will rate the worker, that is, determine the adjustment to convert the observed time to the time that a normal, experienced worker would take.

Before this a **normal pace** must be defined. This is not a trivial problem. Unless normal is defined, any time observation becomes a type II standard (i.e., a did take time instead of a should take time). Without a definition of normal pace, recording time is analogous to recording temperature without specifying whether the Fahrenheit or Celsius scale is being used.

Fein (1972a, 1972b) recommends first defining the motivated productivity level (MPL). Then, from this, a specified distance away (called **expectancy**) is acceptable productivity level (APL), also called normal. See Figure 27.6.

$$APL = MPL - Expectancy$$

Motivated productivity level (MPL): The work pace of a motivated worker possessing sufficient skill and effort to do the job, physically fit to do the job after adjustment to it, and working at an incentive pace

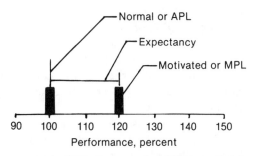

FIGURE 27.6   Fein (1972a, b) recommends establishing motivated productivity level (MPL) first as the "anchor." Then, a negotiated distance away (Incentive Expectancy) determines acceptable productivity level (APL), also called normal. Fein (1972b) states Work-Factor is at MPL, MTM is APL, and Expectancy is 25%. Boepple (1977) says Work-Factor and British Standard 3138 are equal, both are at MPL, and there is a 20% difference between them and MTM, which is at APL. However, *Introduction to Work Study* (1978) gives expectancy for British Standard 3138 as 33%, as incentive rate is defined as walking 4 miles/h. Work-Factor defines walking at 3.7 miles/h as MPL. MTM defines 3.57 miles/h as APL. The ratio of 3.57/3.7 is 96.5% so expectancy is only 3.5% on walking; thus, MTM probably has an expectancy of about 30% on arm motions to counterbalance the 3.5% on walking—assuming the overall is 20–25%. Many firms define 3 miles/h as 100% and don't concern themselves with MPL.

that can be maintained day after day without harmful effect.

Normal or acceptable productivity level (APL): The work pace established by management or jointly by management and labor at a level that is considered satisfactory; it is established at a given relationship to motivated productivity level.

Expectancy = Incentive expectancy, percent

MPL is a function of human work capacity. It is the maximum at which average workers can be expected to work. APL, on the other hand, is a "discounted" value—it is how hard you expect workers to work. (If MPL were defined as 100% the discount would be clearer but, by convention, APL is defined as 100% instead of MPL.) A smaller value of expectancy means workers are working closer to their limits; a larger expectancy means a looser standard. Fein (1972b) says typical expectancy values were 10 to 15% during the 1930s and 20 to 25% during the 1940s. During World War II, the U. S. War Labor Board ruled that 30% was fair and equitable. Rice (1977), surveying 1,500 U. S. firms, reported target for daywork averaged 101% while target for incentive was 119%—a difference of 18%. Actual performance for daywork averaged 93% and for incentive averaged 123%—a difference of 30%. British Standard 3138 defines incentive as walking 4 miles/h and daywork as walking 3 miles/h—an expectancy of 33% (*Introduction to Work Study,* 1978). Fein (1972b) says MTM (which has APL times) has an expectancy vs.

Work-Factor (which has MPL times) of 25%. However, Boepple (1977) of Work-Factor says MTM has a 20% expectancy vs. Work-Factor.

Expectancy also is related to allowances. If allowances are low, the expectancy discount tends to be high. If allowances are liberal, the expectancy discount tends to be low.

Thus, expectancy is really a political rather than a scientific decision. That is, whether a worker works at 20, 25, or 30% less than capacity really depends on the relative strength of management and labor, the work ethic of the country and firm, and so forth.

A related question is the discipline level vs. standard. For example, if a firm, using an APL standard, has a standard of .1 h/unit, what happens if the worker does not produce at 100% of standard but produces at 99%, 95%, 90%, 75%, and so on? If standard of 100% is considered as a minimum and cause for dismissal if not achieved, this is quite different from 100% as a goal and cause for praise if achieved. See Box 26.1 for a more extensive discussion of discipline levels.

Once the decision has been made by management (with the formal or informal agreement of labor) on the definition of APL (say that walking 3 miles/h = 100 = APL), then engineers can rate performance of specific individuals vs. the defined standard.

Since most jobs involve more than walking and MTM is widely known, a practical strategy is to define an individual firm's APL in terms of MTM. That is, the firm's APL = MTMs or 90% of MTMs or 110% of MTMs. Then, on a specific job, the firm's standard could be compared to a standard set using MTM times.

4.2 Rating Techniques

Devotta (1988) surveyed the literature on rating. The section will be divided into problem, proposed solutions, and improving rating accuracy.

4.2.1 Problem

Tasks can be broken down into **micromotions.** Using MTM nomenclature (see Chapter 29), there are moves, reaches, grasps, positions, and so forth. The essential problem of rating is that each of these micromotions changes its proportion of the total task as the pace changes (Sakuma, 1975).

For example, at a 100% pace, the total task may take 1.0 min, composed of .1 min (10%) for reach, .2 min (20%) for grasp, .3 min (30%) for move, and .4 min (40%) for position. However, assume the worker speeds up so that the overall task takes only .8 min (i.e., overall pace is 1.0/.8 = 125%). The reach may have taken only .067 min (150% pace, 8.4% of total); grasp and move may take .16 and .24 min (125% pace, keeping at 20 and 30%); while position may take .333 min (120% pace, 42% of total).

The rater's problem is which micromotion to watch. The low-skill (difficulty) micromotions such as move and reach generally change more than the overall task; the high-skill (position, grasp) micromotions change less than the overall task. In addition, the proportion of low-skill to high-skill micromotions varies from element to element, so the rater cannot rate just one element and assume the rating is valid for the entire task (Jinich and Niebel, 1970).

Table 27.4 shows levels of methods detail. Level 1, methods controlled by management, gives design details such as tools and equipment, fixtures and containers, workplace layout, and the material flow. Level 2, which management tries to control, gives the general motion pattern. Level 3, controlled by the skill and training of the worker, gives hand–hand coordination, unnecessary motions, fumbles and hesitations, and eye–hand coordination. The fine points of these level-3 micromethods are not detected by the usual rater even though they are critical to the time taken by the worker (Gershoni, 1969).

TABLE 27.4   Divide methods into level 1, management-controlled; level 2, which management tries to control; and level 3, operator-controlled. Although level 3 greatly influences the operator's proficiency, most time study personnel have difficulty in detecting changes in level 3.

Level	Detail of Methods Description
1. Management-controlled	Tools and equipment Fixtures and parts containers Workplace layout Material flow
2. Management attempts to control	General motion pattern
3. Operator-controlled (influenced by training)	Specific motion pattern, including hand coordination Unnecessary motions Types of motions (undesirable or highly refined) Amount of fumbling Eye-hand coordination Delay intervals between motions

4.2.2 Proposed solutions
Two approaches are pace rating and objective rating.

Pace rating Pace rating (also called speed rating and tempo rating) is the simplest system. The observer estimates the pace, that is, primarily concentrates on the dynamic micromotions such as reach and move rather than the stationary micromotions such as position and grasp. This single-factor technique is the most common approach.

Objective rating There are 3 steps. First, the observer rates the speed. Second, the observer estimates task difficulty, also called effort (in effect, estimates what proportion of the total time is composed of easy, average, and difficult micromotions). Third, the observer multiplies the speed factor by the difficulty factor to get the overall factor, pace. Objective rating is a descendant of the original Westinghouse system, first used in 1925. The original Westinghouse system had 4 components: skill (proficiency at following a given method), effort (will to work), conditions, and consistency (low consistency went with low pace).

The following example shows how objective rating distinguishes speed from difficulty.

> An unburdened man walked 8 m across a level, unobstructed floor in 6 s; his pace is rated as 110%. After picking up a heavy load, he carries it 8 m up an inclined, obstructed ramp in 6 s. When walking up the ramp the speed does not change but difficulty does, so the pace increases. If, instead of picking up the load, he had returned the 8 m in 5 s instead of 6 s, the speed would increase and the difficulty would be constant, so pace increases.

4.3 Improving Rating Accuracy
After background comments, remarks will be given for rating procedures and rater training.

4.3.1 Background
There have been a number of studies of pace rating. Investigators usually have a large number of raters (perhaps 100) from a number of different companies rate a number of different operations. When they calculate the variability of the total group, approximately 50% of the variability is within-company error and 50% is between-company error (Moores, 1972). That is, raters have been trained for their companies' definitions of APL and the APL varies among companies. Excluding the between-company variability, Fein (1972b) reports that 1,200 U. S. raters had a mean standard deviation of 6%; Moores reports 100 British raters had a mean standard deviation of 7%.

4.3.2 Rating procedures
Perhaps the first question is whether to rate on individual cycles, on each element, or give the same rating on all elements. Rating on individual cycles is not practical unless the cycles are fairly long (e.g., 1 min). On the other hand, giving the same rating to all elements can lead to problems. Consider an order-picking job in a warehouse which consisted of some mental elements (such as read list, decide how to stack different size boxes on pallet truck, etc.) and some manual elements (such as obtain box from rack, stack boxes on pallet truck, etc.). One worker might be good at the mental work and another at the physical work. Rating just overall performance would give biased estimates of the individual elements—a particular problem if the elements were used for standard data. If the observer cannot rate each element, the observer should try to divide element rating into categories of mental, fine manipulative (finger work), and muscular (arm) work. Naturally, machine-time elements should be kept separate from operator-controlled elements.

A rating technique which should improve rating accuracy is to rate each element twice—once before starting the watch and once when the study is complete. Then use the average for the rating. For long elements, rating might even be done more often.

As mentioned previously, it is difficult to rate extreme pace. Thus, a firm may have a policy that if a time study has a rating of under 70% or over 130%, it will not be used.

Rating of inexperienced operators is difficult due to their many fumbles, delays, and hesitations. Although the rater may attempt to compensate, the resulting standard is likely to be loose. Then with learning the operator has a loose standard. Rather than exceed the informal production ceiling, the operator will take increased leisure time. Taking a time study on an inexperienced operator may therefore permanently restrict productivity on that operation.

Operators vary from cycle to cycle (see Section 3.2 of this chapter) (Murrell, 1974). That is, the observer is tracking a moving target. Thus, enough cycles need to be studied to estimate the mean pace. When observing movies of operations, Andrews and Barnes (1967) found no appreciable difference in rating accuracy whether the observation was for 5, 10, or 15 s. As a practical matter, the observer should be conscious of the pace throughout the time study although perhaps recording it only at the end. In a longer study (e.g., 30 min total), the rating can be done several times during the study (every 15 min) and then averaged.

As mentioned before, the important differences in performances are due to skill in the stationary motions of position and grasp rather than the dynamic motions of reach and move. Gershoni (1969) found that raters did not detect very important differences in micromethods. His recommended solution was to use a predetermined time system of notation in the element description, assuming the system forces the rater to pay more attention to the exact motion pattern. Too much attention on reach and move results in "flat" rating (Moores, 1972; Wygant, 1984). See Figure 27.7.

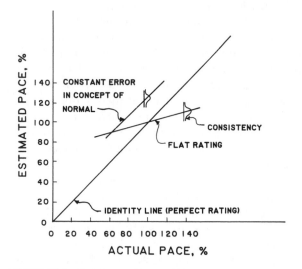

FIGURE 27.7   Perfect pace ratings would have each estimate fall on the identity line. If raters had no bias but had variability, values would fall randomly above and below the identity line. If they have a constant systematic bias (e.g., rating loose), their values would fall around a line parallel to the identity line. If they pay too much attention to reaches and moves instead of positions and grasps, they will give **flat ratings**—being tight when the pace is fast and loose when the pace is slow.

Rating tasks requiring movement of heavy loads is especially difficult. The worker will move quickly so as to minimize the time with the load, probably more quickly than with a light load. Although the effect of the weight should be accounted for through the allowances, the rater needs to be careful not to rate the pace high and thus negate the allowance.

Jobs that require high skill, application of pressure, and inspection are difficult to rate. The analyst has trouble determining exactly what is being done and whether the operator is doing what is required or is embellishing it for the time study.

In general, observers will do a better job rating if they are familiar with the job than if they are novices to it. Although it is not feasible in every job, it would be good practice for observers actually to do the jobs they study. In addition to teaching humility, it would give them a better understanding of the operator's problems.

Gambrell (1959) said raters trained to estimate to the nearest percent (97%, 109%) had less variability than those who rounded to the nearest 5% (95%, 110%).

4.3.3   Rating training

Rating is a skill. As with any other skill, those with more experience and training do better than novices. Rating also is a subjective judgment in which the rater compares performance with a remembered standard. Therefore, the rater must know the standard and be calibrated with it.

The standard training procedure is to show the

TABLE 27.5   The Society for Advancement of Management (SAM) has 18 time study rating films that are subdivided into three series. For each operation a worker (who actually did the task) is shown in an introductory scene plus five scenes at various paces. For copies of videotapes, contact Tampa Mfg. Institute, 6300 Flotilla Dr., Holmes Beach, FL 33510; phone (813) 778-4722.

Series	Reel	Operations
5201	1	Deal cards, toss blocks, transport marbles
5201	2	Dink tile squares, fold gauze, pack gaskets
5201	3	Countersink, kick press, shear rubber tile
5201	4	Cut cork tiles, deburr, form rug cups
5201	5	Feed rolling mill, shovel sand, stack cartons
5201	6	Pack cans, seal cartons, tape boxes
5201	7	Bolt pipe flange, check tires, fill radiator
5201	8	Collate papers, staple papers, tear bills
6302	1	Collate papers, comptometer, typing-electric
6302	2	Ditto, key punch, posting
6302	3	Calculator-multiplication, filing, staple papers
6302	4	Card sorter, mail sorting, tear bills
6503	1	Filing, mail sorting, typing-nonelectric
6503	2	Check writing, comptometer, key punch
6503	3	Drill press, punch press, turret lathe
6503	4	Engraving, form rug cups, mechanical assembly
6503	5	Sweep floors, wash windows, wet-mop floors
6503	6	Pack cans, seal cartons, stack cartons

rater a movie or videotape of a reference operation. Films of reference operations are available. See Table 27.5. Firms also can videotape 15 to 20 of their own operations to use as standards. (When studying different paces, have the worker speed up and slow down rather than speeding up or slowing down the projection speed, since the micromotion composition changes with pace.) Das (1965) has studied the training procedure in detail. Most films mix operations (a scene of Form Rug Cups, then Cut Cork Tube, then Deburr, etc., each at a different pace). Das says this is good for testing but, for training, a number of paces should be shown for Form Rug Cups, then a number of paces for Cut Cork Tubes, and so on. After each scene, the rater should be given the true pace immediately after putting down the estimate; if the estimate is in error by more than 5%, back up and show the scene over. In training terminology, apply the principles of knowledge of results and immediate feedback.

REVIEW QUESTIONS

1. If observed time = .1 h, rating = 110%, and shift allowances = 20%, what is standard time?

2. Is the goal of a type I standard a "should take" time or a "did take" time?

3. Give the 4 reasons to break a job into elements.

4. Does rating accuracy depend on the deviation from a 100% pace or is it relatively constant for all paces?

5. Give the 4 timing techniques. Which would you use and why?

6. If a .2 h element is to have a relative accuracy of ±10%, what is the absolute accuracy?

7. From a statistics viewpoint, the number of observations depends upon what three factors?

8. Assume that, during a time study, the supervisor came by and gave some instructions to the worker. How would this time be incorporated into the standard time?

9. Give the formula relating MPL, APL, and expectancy. What is defined as 100%? What is a typical value of expectancy?

10. As a person improves performance, would you expect the proportions of time devoted to stationary micromotions to be increased or decreased? Why?

11. In pace rating, does the observer concentrate on the stationary or dynamic micromotions?

12. Justify why rating should be for each element rather than one rate for the entire task.

13. Using actual pace as the X axis and estimated pace as the Y axis, show (1) the identity line, (2) consistent loose rating, and (3) flat rating.

REFERENCES

Andrews, R. and Barnes, R. The influence of the duration of observation time on performance rating. *J. of Industrial Engineering*, Vol. 18 [4], 243–47, 1967.

Boepple, E., coordinator of research and development of Wofac Co. Personal communication, May 17, 1977.

Das, B. Applying programmed learning concepts to instruct in performance rating. *J. of Industrial Engineering*, Vol. 16 [2], 94–100, 1965.

Devotta, A. A survey of performance rating research in work measurement. M. S. thesis, Manhattan, Kan.: Kansas State University, 1988.

Fein, M. Work measurement today. *Industrial Engineering*, Vol. 4 [8], 14–20, 1972a.

Fein, M. Work measurement: Concepts of normal pace. *Industrial Engineering*, Vol. 4 [9], 34–39, 1972b.

Gambrell, C. The independence of pace rating vs. weight handled. *J. of Industrial Engineering*, Vol. 10 [4], 318–22, 1959.

Gershoni, H. An analysis of time study based on studies made in the United Kingdom and Israel. *Am. Institute of Industrial Engineers Transactions*, Vol. 1 [3], 244–52, 1969.

Hicks, C. and Young, H. A comparison of several methods for determining the number of readings in a time study. *Journal of Industrial Engineering*, Vol. 13 [2], 93–96, 1962.

Introduction to Work Study, 3rd ed. Geneva, Switzerland: Int. Labour Office, 1978.

Jinich, C. and Niebel, B. Synthetic leveling—How valid? *Industrial Engineering*, Vol. 2 [5], 34–37, 1970.

Moores, B. Variability in concept of standard in the performance rating process. *Int. J. of Production Research*, Vol. 10 [2], 167–73, 1972.

Murrell, H. Performance rating as a subjective judgment.

Applied Ergonomics, Vol. 5 [4], 201–08, 1974.

Niebel, B. Time study. In *Handbook of Industrial Engineering,* 2nd ed., Salvendy, G. (ed.), Chapter 62. New York: Wiley, 1992.

Rice, R. Survey of work measurement and wage incentives. *Industrial Engineering,* Vol. 9 [7], 18–31, 1977.

Sakuma, A. New insight into pace rating. *Industrial Engineering,* Vol. 7 [7], 32–39, 1975.

Shaw, A. *Time Study Manual of the Erie Works of the General Electric Co.* Erie, Penn., 1978.

Westinghouse Electric Corp. *Work Measurement Techniques and Application,* R-131-REV. Pittsburgh: Westinghouse Electric, 1953, p. 57.

Wygant, R. An analysis of performance rating. *Proceedings of Annual Ind. Engineering Conference,* 632–36, 1984.

28 | PROGRESS CURVES

OVERVIEW

Practice makes perfect. Both organizations and individuals improve with practice. Learning has been found in many different industries in a variety of countries. The amount of this learning can be predicted. Time standards that make the assumption that time/unit is constant vs. cumulative units produced can cause major errors in scheduling, evaluation of alternatives, acceptable day's pay, and incentive pay.

CHAPTER CONTENTS

1 Two Locations for Learning
2 Quantifying Improvement
3 Applications

KEY CONCEPTS

doubled quantities	log–log scale
going slowly less often	manufacturing progress
learning	manufacturing vs. construction
learning effect on time standards	unit/average cost/unit

1 TWO LOCATIONS FOR LEARNING

Learning (improvement in performance) usually is described by reduced time, although reduced errors sometimes is used. Learning occurs in the individual and the organization.

1.1 Individual Learning

Individual learning is improvement with a constant product design and constant tools and equipment.

Consider Joe playing golf at the Riverview course. He has a driver, a 5 iron, a 9 iron, a putter, and some slightly used golf balls. For his first round of 18 holes he shoots 137. His next round is 131. Then 127, 115, 118, 112, 114, 110. Eventually he levels out at a score of 100.

All during the time that his score was improving he played the same course; his required output was the same. He used the same clubs, tees, and balls; his tools and equipment were the same. He modified the same work method by reducing jerks, getting his feet set just right, and holding the club slightly tighter.

In the same way, Joe might go to his job and turn out more and more widgits/h as he gained more experience even though the design of the widgits did not change and his tools and equipment did not change. The improvement is due to better eye–hand coordination, fewer mistakes, reduced decision time, and so forth.

1.2 Organization Learning (Manufacturing Progress)

This is improvement with changing product design, changing tools and equipment, and changing work methods. It is individual learning plus organization learning; often it is called **manufacturing progress.**

Using the golf example, Joe might switch golf courses to the Sunnyside course. There he can average 96 since there are not as many sand traps. This is similar to a change in product design; less is required to complete a unit.

He could modify his tools and equipment. One step might be to get a complete set of clubs; another might be to get new golf balls.

He could modify his work method further. Perhaps he could use a different putting stance. He could improve his technique for estimating distance.

For an occupational example, consider the server Maureen serving coffee and doughnuts. During the individual learning period, Maureen learned where the coffee cups and the coffeepot are, the prices of each product, and so forth. The amount of time it took her to serve a customer declined to a plateau. Now let's consider some organizational

changes which might occur. Management might set a policy to serve coffee in cups without saucers and to furnish cream in sealed one-serving containers so that the container need not be carried upright. These changes in product design (learning by the organization) reduce time for Maureen and all other servers. Tools and equipment also can be modified. Put one coffeepot at each end of the counter. Redesign the sales slip to speed computations and reduce errors. Buy a coffeepot with a better handle so less care is required to prevent burns. These changes in tools and equipment also help all servers reduce time per customer. The organization may even decide to have the servers leave the bill with the customer when the last food item is served so as to further reduce time per customer. Organization progress includes all the changes—not just individual learning.

For an industrial product, such as an airplane, manufacturing progress includes the increased skill of the assembly operator plus the reduced supervision needed from the supervisor as more and more planes are built; the reduced time needed by the fork truck driver since the driver knows exactly when to bring the materials; the reduced repairs required due to incorrect drawings as the initial errors on drawings are eliminated; the reduced need for the operators to look up information because by the 50th airplane they remember that the red part goes above the green part instead of having to look it up; and so on. It includes better material usage as parts are trimmed with less scrap and better materials are used.

Organization progress comes from three factors: (1) operator learning with existing technology, (2) influence of new technology, and (3) economies of scale.

Point 1 was just discussed. Examples of new technology are the subsurface bulblike nose on the front of tankers (a nose that increased tanker speed at very low cost), improved turbine blade design which increased turbine efficiency, and solid-state electronics which improved the performance of TV sets and computers.

Economy of scale occurs when equipment with twice the capacity costs less than twice as much; then capital cost/unit is reduced. Operator costs are reduced as fewer work hours are needed/unit of output.

2 QUANTIFYING IMPROVEMENT

2.1 Log–Log Concept

Practice makes perfect has been known for a long time. Wright (1936) took a key step when he published manufacturing progress curves for the aircraft industry. Wright made two major contributions. First, he quantified the amount of manufacturing progress for a specific product. It's not too useful to executives to tell them that time to assemble an airplane is going down. They want to know how much it is going down. Wright told them. See Figure 28.1.

This was quite a big step, but giving a manager the complex equation $y = ax^b$ was too much in the 1930s; in fact, engineers should be careful even today in presenting formulas to managers. Managers like simplicity. So Wright took the curve out of the data and presented the data in a straight line. The managers hardly noticed that he had put the curve in the axes. Figure 28.2 has the same data plotted on a **log–log scale** (both axes are a log scale). See Figure 28.3 for Cartesian coordinates, semilog coordinates, and log-log coordinates.

On Figure 28.2 look at the "cumulative number of airplanes" axis. Call the physical distance between the first and second airplane "Δx". It is equal to 1 airplane. Mark off this same physical distance from 2; it will reach to 4. Mark off the same physical distance from 4; it will each to 8. The same physical distance on a log scale is represented by a series of 1, 2, 4, 8, 16, ... We also can say that every time we move the same distance to the right we double the quantity (1, 2, 4, 8, 16, . . .).

Next consider the straight line fitted to the data. The rate of decline of the line, the slope, is $\Delta y/\Delta x$. If we pick the Δx as the distance between **doubled quantities,** then Δx will be the same no matter whether we are dealing with 8 airplanes to 16 airplanes, 16 airplanes to 32 airplanes, or 25 airplanes to 50 airplanes.

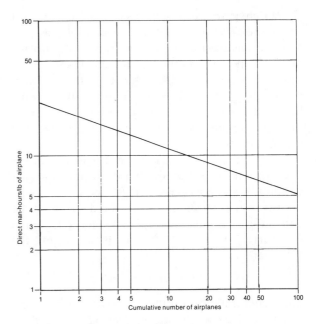

FIGURE 28.2 Supervisors like straight lines. Plotting $y = ax^b$ on a log–log paper gives a straight line. The key piece of information desired by supervisors is the rate of improvement—the slope of the line. The convention is to refer to reduction with doubled quantities. If quantity $x_1 = 8$, then quantity $x_2 = 16$. Then if cost at x_1 is $y_1 = 100$ and cost at x_2 is $y_2 = 80$, this is an "80% curve."

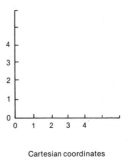

Cartesian coordinates

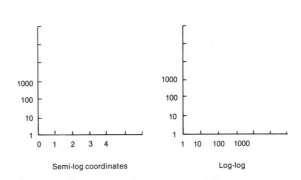

Semi-log coordinates   Log-log

FIGURE 28.3 Cartesian coordinates have equal distances for equal numerical differences; that is, the linear distance from 1 to 3 is the same as from 8 to 10. On a log scale, the same distance represents a constant *ratio;* that is, the distance from 2 to 4 is the same as 30 to 60 or 1,000 to 2,000. Semi-log paper has one axis cartesian and one axis log. Log–log (double log) paper has a log scale on both axes.

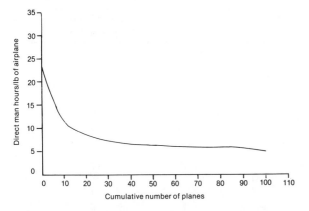

FIGURE 28.1 Practice makes perfect. As more and more units are produced the fixed cost is divided over more units, so fixed cost/unit declines. In addition, however, variable cost/unit declines as fewer mistakes are made, less time is spent looking up instructions, better tooling is used, and so on. The variable cost data usually can be fitted with an equation of the form $y = ax^b$.

Wright gave learning curves in terms of new cost when the quantity doubled. For example, assume that the manufacturing cost/lb of the 8th airplane was 12 h and cost of the 16th airplane was 9.4. Then this is considered to be "a 78% curve"; that is, for $x_2 = 2 x_1$, $y_2 = .78 y_1$. If cost at unit 16 was 10, it would be 10/12 or an 83% curve. (If you wish to determine the b in the equation $y = ax^b$, use the conversion from Table 28.1).

The exponent $b = \log \Phi / \log 2$, where Φ is the slope parameter of the learning curve function. For example, for an 80% curve (log .8) (log 2) $= -.322$.

Given the equation $Y = ax^b$, and assuming the time for cycle a (i.e., cycle 1) = 10 min and there is a 90% curve (i.e., $b = -.152$), what is the time for the 50th unit?

$$Y = 10(50)^{-.152} = 10/(50)^{.152} = 5.52 \text{ min}$$

2.2 Determining a Curve

The formula just recommended uses the straight line $\log y = \log a + b (\log x)$. Scholarly articles often advocate fitting more complicated formulas to learning data. (A statistician has been defined as one who draws a mathematically precise line from an unwarranted assumption to a foregone conclusion.) Carlson (1973) recommended adding two more terms to the polynomial to get $\log y = \log a + b(\log x) + c(\log x)^2 + d(\log x)^3$. Others recommend other curves. The trade-off is accuracy of estimate vs. difficulty of obtaining the estimate and ease of use. In brief, for $y = ax^b$, use log x and log y; for $y = ae^{bx}$, use x and log y; for $y = a + bx^n$, where n is known, use a plot of x^n vs. y; for $y = x/(a + bx)$, plot x vs. x/y or $1/x$ vs. $1/y$. See Chapter 30. The simple formula also facilitates comparisons among references. The sheer simplicity of log-log plots has another advantage; they are done by hand. Since computers are available, people no longer plot the data and fit a curve by eye; instead they key some numbers and dump them into a computer curve-fitting routine. The unfortunate result is that keying errors, transcription errors, accounting errors, and the like are no longer detected, because "garbage in becomes Bible out" and the computer prints all numbers to 6 decimal points.

(Conversely, you might have some very ordinary data that are not very convincing alone. Make them more convincing by having the data printed by a computer; even more convincing to the uninitiated is the same data plotted by a computer.) The consensus has been that the simple straight line on log-log paper is best.

Table 28.2 shows how these complicated equations are obtained. During the month of March various people wrote down on charge slips a total of 410 hours against this project's charge number. The average work hours/unit during March then becomes 29.3. The average x coordinate is (1 + 14)/2 = 7.5. Because the curve shape is changing so rapidly in the early lots, some authors recommend plotting the first lot at the 1/3 point [(1 + 14)/3] and points for all subsequent lots at the midpoint. What they are really saying is that one straight line for the early lots and a

---

TABLE 28.1   Factors for various improvement curves. The multipliers in columns 3 and 4 are large quantity approximations. For example, for a 90% curve, the table value in column 4 is 1.18. A more precise value at a quantity of 10 = 1.07, at 50 = 1.13, and at 100 = 1.17. A more precise value for an 85% curve at a quantity of 100 = 1.29; a more precise value for a 95% curve at a quantity of 100 = 1.077.

Improvement Curve, % Between Doubled Quantities	Learning Factor, b, for Curve $y = ax^b$	Multiplier to Determine Unit Cost if Average Cost is Known	Multiplier to Determine Average Cost if Unit Cost is Known
70	-.515	.485	2.06
72	-.474	.524	1.91
74	-.434	.565	1.77
76	-.396	.606	1.65
78	-.358	.641	1.56
80	-.322	.676	1.48
82	-.286	.709	1.41
84	-.252	.746	1.34
85	-.234	.763	1.31
86	-.218	.781	1.28
88	-.184	.813	1.23
90	-.152	.847	1.18
92	-.120	.877	1.14
94	-.089	.909	1.10
95	-.074	.926	1.08
96	-.059	.943	1.06
98	-.029	.971	1.03

TABLE 28.2 Time and completed units as they might be reported for a product.

Month	Units Completed (Pass Final Inspection)	Month's Direct Labor Hours Charged to Project	Cumulative Units Completed	Cumulative Work Hours Charged to Project	Average Work h/Unit
March	14	410	14	410	29.3
April	9	191	23	601	26.1
May	16	244	39	845	21.7
June	21	284	60	1129	18.8
July	24	238	84	1367	16.3
August	43	401	127	1708	13.4

second, less steep, line for the later lots would fit the data more accurately, but it is easier to fit one "adjusted" line to the total data.

During April, 9 units passed final inspection and 191 hours were charged against the project. Cumulative hours of 601 divided by cumulative completed output of 23 gives average hours/unit of 26.1. The 26.1 is plotted at (15 + 23)/2 = 19 in Figure 28.4 using log–log paper. Complicated curve-fitting routines can be used but the accuracy of the data rarely justifies anything beyond a line fitted by eye. The purpose of using progress curves is to predict the future. Calculating the future to 6 decimal points is foolish. Fitting by eye has an advantage in that it requires a plot of the data, and outliers and abnormal points are obvious. Computer routines will fit a straight line (or whatever curve you specify) whether the straight line is valid or not. For these data, the percent is about 79. A simple way to calculate it is to read the cost from the line at a quantity, the quantity doubled, and quantity quadrupled. For example, 24.9 for 20, 19.7 for 40, and 15.5 for 80. Then 15.5/24.9 = .6225 and the square root of .6225 = .79. If errors in reading are not so important, just use one doubling. For example, 19.7/24.9 = .79. If someone reports a curve as 78.898%, have that person gather the data next time.

Although average cost/unit is what is usually used, you may wish to calculate cost at a specific unit. Conversely, the data may be for specific units and you want average cost. Table 28.1 gives the multiplier for various slopes. The multiplier for a 79%

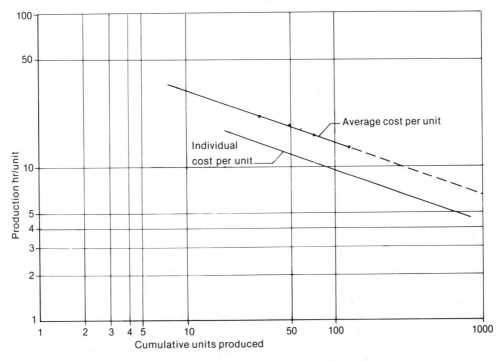

FIGURE 28.4 Average cost/unit from Table 28.1 gives a 79% curve. Cost/unit is the cost of the nth unit; **average cost/unit** is the sum of the unit costs/n. Cost/unit can be estimated by multiplying average cost/unit by the factor from Table 28.1. The average cost of the first 20 units is estimated as 25.9 from the fitted line; the cost of the 20th unit is 25.9 (.658) = 17.0 h.

curve is estimated as (.641 + .676)/2 = .658. Thus, if we wished to estimate the cost of the 20th unit, it would be (24.9 h) (.658) = 16.4 h. This multiplier is based on the fact that the average cost curve and the unit cost curve are parallel after an initial transient. "Initial transient" usually is 20 units and sometimes as few as 3.

Cost/unit is especially useful in scheduling. For example, if 50 units are scheduled for August, then work-h/unit (79% curve) at unit 127 = (13.4) (.656) = 8.8 and at unit 177 = 7.8. Therefore, between 390 and 440 hours should be scheduled.

Price/unit often is used for between-company comparisons since price/unit is available while cost/unit probably isn't. If cost/unit is calculated from price/unit, the additional assumption must be made that profit margins do not change.

Looking at the curve in Figure 28.4, you can see how the extrapolated line would predict average cost/unit at 200 to be 11.4 h, at 500 to be 8.3, and at 1,000 to be 6.6. If we add more cycles on the paper the line eventually reaches a cost of zero at cumulative production of 200,000 units. Can cost go to zero? Can a tree grow to the sky? No.

The log–log paper has increased understanding of improvement, but it also deceives. Note that cost for unit 20 was 24.9 h. When output was doubled to 40 units, cost dropped to 19.7; doubling to 80 dropped cost to 15.5; doubling to 160 dropped cost to 12.1; doubling to 320 dropped cost to 9.6; doubling to 640 dropped cost to 7.6. Now consider the improvement for each doubling. For the first doubling from 20 units to 40 units, cost dropped 5.20 h or .260 h/unit of extra experience. For the next doubling from 40 to 80, cost dropped 4.2 h or .105 h/unit of extra experience. For the doubling from 320 to 640, cost dropped 2.00 h or .006 h/unit of extra experience. In sum, the more experience, the more difficult it is to show additional improvement.

Yet the figure would predict zero cost at 200,000 units and products just aren't made in zero time. One explanation is that total output for the product, in its present design, is stopped before 200,000 units are produced. In other words, if we no longer produce Model Ts and start to produce Model As, then we start on a new improvement curve at zero experience. A second explanation is that the effect of improvement in hours is masked by changes in labor wages/h. For example, in 1910, when 12,300 Model T Fords had been built, the price was $950. When it went out of production in 1926 after a cumulative output of 15,000,000 the price was $270: $200 in constant prices plus inflation of $70.

The third explanation is that straight lines on log–log paper are not perfect fits over large ranges of cycles. If output is going to go to 1,000,000 cumulative units over a 10-year period, you really shouldn't expect to predict the cost of the 1,000,000th unit (which will be built 10 years from the start) from the data of the first 6 months. There is too much change in economic conditions, managers, unions, material availability, and other factors. Anyone who expects the future to be perfectly predicted by a formula has not lost money in the stock market.

2.3 Typical Values for Organization Progress

What are typical values for various products?

Tables 28.3 and 28.4 give values of organization progress curves reported in the literature. These include all progress: better designs, more efficient factory layouts, economies of scale, individual learning, improved technology from supplier industries, and others. Those values that represent sales prices also include the decline in cost/unit of initial investment in engineering, buildings, machinery, and the like.

In summary, the rate of improvement depends on the amount that can be learned. The more that *can* be learned, the more that *will* be learned. The amount that can be learned depends on two factors: amount of previous experience with the product and the degree of mechanization.

Titleman (1957) gives an explanation, shown with Figure 28.5, which he says predicts the learning curve ratio "with less than 2.5% error." To estimate learning rate from Figure 28.5, divide total time for an operation into two categories: manual time and mechanical time. Then calculate the manual ratio, which is manual time over total time. A job requiring 10 min manual time and 20 min mechanical (machine) time would have a manual ratio of 33% and an estimated learning curve of 95%.

Percent of Task Time		Manufacturing Progress, %
Manual	Machine	
25	75	90
50	50	85
75	25	80

Thus, if you were dealing with a product that required 1,000 h/unit, that was scheduled for total production of 70 units, and that required about 75% manual time, you might estimate an 80% curve. If this same product was scheduled for total production of 2,000 units, you would plan to purchase more machines, set up a better production line, select raw material sizes more carefully, and so on, and less learning would occur during this production—85% might be appropriate. Then if cumulative production over a 15-yr period reached 10,000 units, a rate of 87,

TABLE 28.3   Manufacturing progress rates reported in the literature with their references.

Rate, %	Number of Cycles	Examples
60		Production work hours/cumulative units of steel produced since 1867 (Hirshman, 1964)
68	1000	Test and adjust time for product B (Conway and Schultz, 1959)
70	10,000,000 to 3,000,000,000	Average revenue/unit for silicon transistors industry accumulated volume (Conley, 1970)
	3,000,000	Price/unit for integrated circuits for industry total accumulated volume (1964-1968) (Conley, 1970)
	300,000,000 to 3,000,000,000	Price/lb of polyvinyl chloride (constant dollars) for industry accumulated volume; from 30,000,000 to 300,000,000 it was 95% (Conley, 1970)
	50,000,000 to 80,000,000	Average price/unit (constant dollars) for free-standing gas ranges for cumulative industry volume (1952-1967) (Conley, 1970)
		Overhead cost/airplane (Wright, 1936)
72	1,000,000 to 2,000,000,000	Price/unit of integrated circuits (1963-1972) (Noyce, 1977)
73	11,000	Electronic assembly of product A (Conway and Schultz, 1959)
75	8,000	Electro-mechanical assembly of product B (Conway and Schultz, 1959)
		$/ton for U.S. electric utility coal 1948 to 1971 (in 1970 prices), 1.5 to 5.5 billion tons. Sharp increase in price upon passage of Clean Air Act of 1970 (Fisher, 1974)
		Cents/kwh for electricity price in USA from 1926 to 1970 in 1970 dollars. 0.4 to 22 kwh $\times$ 10^{12}. Price rose above trend during 1930s. Discontinuity in curve in 1971 due to effect of Clean Air Act of 1970 (Fisher, 1974)
76		Work hours/barrel of petroleum refined in United States since 1888 (Hirschman, 1964)
		Assembly labor/unit on 15 different models of machine tools (Hirsch, 1952)
		Maintenance work hours/shutdown in a General Electric plant
78-84		Production hours/unit for Liberty ships, Victory ships, tankers, and standard cargo vessels in United States during World War II (Hirsch, 1952)
79	1,000	Service time—IBM electronic machine (Kneip, 1965)
80		Cost/barrel of catalytic cracking unit capacity (inflation has changed actual cost to 94%) (Hirschman, 1964)
		Work hours/airframe in United States during World War II (Hirsch, 1952)
		Labor cost/airplane (Wright, 1936)
		$/barrel of retail gasoline processing cost for United States 1919 to 1969. 1970 dollars. 2 to 37 C units where C = 10^{16} BTU (Fisher, 1974)
81	20,000	Service calls/150 machines (IBM electro-mechanical) (Kneip, 1965)
82		Total labor/unit on 16 different models of machine tools (Hirsch, 1952)
84	8,000	Final assembly labor/unit on product B (Conway and Schultz, 1959)
85	2,000	Labor hours/gun barrel on boring machine (Andress, 1954)
86	12,300 to 15,000,000	Cost/unit for Ford Model T. Price in 1910 when 12,300 had been built was $950. In 1926, when 15,000,000 had been built, price was $270 ($200 in constant prices) (Hirschman, 1964)
	1,000	Service time—IBM electronic machine (Kneip, 1965)
88	20,000	Machining labor/unit on 15 different models of machine tools (Hirsch, 1952)
		Cost of purchased subassemblies on airplanes (Wright, 1936)
90		Output of a fluid catalytic cracking unit
95		Lbs of raw material/airplane (Wright, 1936)
		$/barrel of price for crude oil at the well in United States 1869 to 1971. 1970 prices. .01 to 80 C units where C = 10^{16} BTU (Fisher, 1974)
96	11,000 to 120,000	Hours/unit for electronic assembly of product A (Conway and Schultz, 1959)

90, or even 93% might be appropriate for this last cycle (1,000 to 10,000 units) of the log–log paper.

Allemang (1977) recommends a procedure using product design stability, a product characteristics table (complexity, accessibility, close tolerances, test specifications, and delicate parts), parts shortage, and operator learning to calculate an allowance. Advantages cited are reduced judgement and more documented estimates (especially important for government auditors).

2.4 Typical Values for Learning   Table 28.5

gives some learning values reported in the literature.

Smyth (1943) used a different approach, a job evaluation-type plan. In place of Titleman's two factors, he used six: length of cycle, education required, rules and regulations that needed to be known, rhythm and dexterity, mental dexterity, and physical demand. For a spin and tin wire operation in a radio factory, he assigned (for the factors in order) "over 240 units/h," "equivalent to 8th grade," "rules and regulations at minimum," "medium amount of rhythm and dexterity," "minimum," and "light physical effort." The factors had a total of 69 points, which

TABLE 28.4   Japanese organization progress curves (Morooka and Nakai, 1971). Konz has estimated cumulative units, cumulative work time, early times, and late times from the Japanese figures and tables to give the reader a "feel" for each situation.

Industry	Firm & Product	Operation	Learning %	Learning % Based on Cumulative Units	Prod. Period, Days	"Early" Time/ Unit	"Late" Time/ Unit
Airline	V1	Maintenance/flying hr	91	396,000	2,000	4.45	2.44
Automobile	B1	Total	80		800	200.	45.
	B2		78		1,100	160.	95.
	J	Assembly	84				
	J	Lot work	87				
	J	Mechanical milling	93				
	J	Drilling	85				
	J	Lathe	87				
	Q	Assembly	83	3,892	300	1.52	0.24
	U1	Total, workers/unit	84	900,000		2.3	0.7
	U2	Engine	82		70	700.	150.
Chemical	D1	Paper (days/10,000 tons)	74	350,000 tons		800.	70.
	D2	Paper (days/10,000 tons)	75	1,300,000 tons		800.	40.
	F	Tire	89	1,146	21	276.	150.
Construction	E	Aluminum welding touchup, h	86	153	30	53.1	17.7
Electrical and Electronical	C1	Assembly line, min	86		400	0.400	0.135
	C2	Assembly line, min	88		300	0.090	0.037
	C3	Assembly line, min	89		280	0.080	0.037
	C7	Assembly line, min	82		28	0.15	0.055
	C8	Automatic lead insert machine, min	86		100	0.1	
	C9	Automatic lead insert machine, min	87		100	0.1	
	C4	Defect percent	81		100	0.65	0.17
	C5	Defect percent	90		100	0.45	0.18
	C6	Defect percent	70		70		0.15
	01	Total	83	26,200	220	9.2	4.4
	01	Total	83	26,200 to 106,800	180	4.4	3.1
	P	Tape recorder total	86	652	300	84.	16.
	Y1	Inspection	89	22,500	210	288.	164.
	Y2	Inspection	88	5,600	210	198.	117.
	Y3	Inspection	83	1,921	105	48.	30.
	Y4	Inspection	85	787	105	43.	33.
Food Processing Machine	T	Total	93	2,500	22	1.37	0.96
	A	Lathe	84				
	A	Drilling	90				
	A	Milling	87				
	H1	Grinding for casting	94	up to 8,000			
	H1	Grinding for casting	83	8,000 to 16,000	250	2.27	1.61
	H2	Grinding for casting	96	up to 8,000			

(continued)

TABLE 28.4   (Continued)

	H2	Grinding for casting	87	8,000 to 16,000	250	2.25	1.74
	L1	Mechanical cutting	93				
	L2	Assembly	87				
	N1	Total for cooling products	88	370			
	N2	Total for cooling products	91	350	505.		
	N4	Total for cooling products	91	400			
	N5	Total for cooling products	90	390			
	N6	Total for cooling products	85	400			
	N7	Total for cooling products	87	470			
	N8	Total for cooling products	91	450			
	N9	Total for cooling products	92	450			
	N10	Total for cooling products	89	650	970.		
	N11	Total for cooling products	93	500			
	Z	Machine assembly	87	700	350.	150.	
Metal	X	Stainless Steel	87		800	15.	11.
	W	Metal grinding	97	15,200		0.72	0.53
Precision	I2	Daily loss time, min (job 2)	76	110		250.	50.
	I2	Daily loss time, min (job 20)	83	20		11.	6.
	I2	Daily loss time, min (job 79)	79	80		200.	70.
	I2	Daily loss time, min (job 40)	79	80		38.	10.
	I3	Assembly (cash register A), h	96	40		190.	150.
	I3	Assembly (cash register B), h	93	35		170.	85.
	I3	Assembly (cash register C), h	92	90		50.	30.
	K1	Nut making	80	1,000		1.	
	K2	Nut making	81	1,000		1.	
	K3	Nut making	82	1,000		1.	
	K4	Nut making	86	1,000		1.	
	K5	Nut making	90	1,000		1.	
	K6	Nut making	95	1,000			
	K7	Nut making	98	1,000			
	S2	Thread	83	10,000		0.9	0.4
	S4	Thread	90	1,700		1.8	1.
Shipping	G	Total (small ships), h/ship	76	48,200 (53 ships)		3,320.	667.
	R1	Machining, %	86	460 (7 ships)		100.	66.
	R2	Boiler, %	88	480 (7 ships)		100.	69.
	R3	Tank and pipe, %	84	430 (7 ships)		100.	61.
	R4	Stringer, %	84	423 (7 ships)		100.	60.

FIGURE 28.5   Predict learning percent from the manual ratio for an operation (Titleman, 1957). For example, a job with a 10 min manual time and 5 min mechanical time has a 67% manual ratio and learning percent is predicted as 88%.

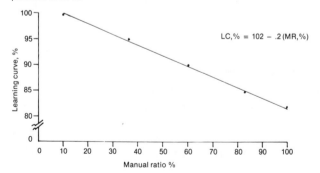

LC,% = 102 − .2 (MR,%)

Source: Reprinted with permission of *Product Engineering Magazine* (a Morgan-Gampian Production).

was equivalent to about 40 hours of learning time. The most complicated job was allowed 384 hours of learning time. Hancock and Bayha (1982) point out that age of the operator affects the learning rate. They say rate of learning declines after age 35, that the rate is slower for those who have not been required to learn in the past, and that "nervous" capabilities decline more with age than "physical" capabilities. They emphasize the importance of good training programs. Turban (1968) reports that inexperienced workers (either new workers or veteran workers learning new skills) take twice as long to learn as experienced workers.

The improvement takes place through reduction and elimination of fumbles and delays rather than greater movement speed (Salvendy and Seymour, 1973). Stationary motions such as position and grasp improve the most while move and reach improve little (Rohmert and Schlaich, 1966). It is reduced "information processing" time rather than faster hand speed that makes the reduction. Salvendy and

TABLE 28.5   Learning rates reported in the literature with their references.

% Rate	Number of Cycles	Examples
68		Truck body assembly (Glover, 1966)
70	12	Bag-molded aircraft cowls (Sheffler, 1957)
72	50	Complex, 300 h/unit assembly (Conway and Schultz, 1959)
74		Machining and fitting of small castings (Glover, 1966)
78	212	Complex cored large radrome (Sheffler, 1957)
80	400,000	Keyboard entry on business machines (Killbridge, 1959)
		Precision bench assembly (Nadler and Smith, 1963)
	6,000	Press molded housings (Sheffler, 1957)
82		Burring, sanding and hand forming (McCambell and McQueen, 1956)
		Shearing plates (Titleman, 1957)
		Grinding (Nadler and Smith, 1963)
83		Fitting (Titleman, 1957)
		Power sawing (Nadler and Smith, 1963)
	40	Sorting cards into compartments (Crossman, 1959)
	4,000	Substituting letters for symbols (Crossman, 1959)
		Radio tube assembly (Glover, 1966)
		Servicing automatic transfer machines (Glover, 1966)
84	2,000,000	Cigar making (90% for cycles 2,000,000 to 10,000,000) (Crossman, 1959)
		Lathes (Nadler and Smith, 1963)
85		Gas cutting, thin plates — machines (Titleman, 1957)
		Work attendance hours — washing machines; effective hours were 88%, wasted hours rate was 80% (Glover, 1966)
87	1,000	Drill, ream, and tap (Konz, 1960)
		Drilling (Nissley, 1949)
88		Welding — manual (Titleman, 1957)
89	10,000	Punch press (Conway and Schultz, 1959)
		4 2 cycle assembly (Barnes et al., 1940)
		Washers on pegs (initial cycle time = 20 s) (Daniels, 1966)
		Milling (Nadler and Smith, 1963)
90		Punch press (McCambell and McQueen, 1956)
		Bench inspection (Nadler and Smith, 1963)
91	8,000	Adding pairs of digits (Crossman, 1959)
92		Assembly with jig (Titleman, 1957)
	14,000	Assembly (70% from 14,000 to 25,000; 95% from 25,000 to 600,000 when put on piecework) (Youde, 1947)
		Gas cutting, thick plates — machine (Titleman, 1957)
		Welding (Nadler and Smith, 1963)
		Deburring, cleaning (Nadler and Smith, 1963)
94	300	Pegs in pegboard (Youde, 1947)
		Welding, submerged arc (Titleman, 1957)
95	450	Countersink (Konz, 1960)
	10,000	Punch press — average of 5 operators on 5 operations (range 89 to 98)
	450	Screwdriver work (85% for cycles 450 to 3500) (Barnes and Amrine, 1942)
		Reduction of running speed (m/s) vs. distance (km) for world-class male runner (.94 for females) (Riegel, 1981)
96.5		Milling — no jig (Titleman, 1957)
98.5		Grinding — manual; chipping — pneumatic, blast cleaning; milling — with jig; assembly — no jig (Titleman, 1957)

Pilitsis (1974) reported that kcal/cycle declined with experience as well as time/cycle.

The range of times and the minimum time of elements show little change with practice. The reduction in the mean performance times of elements does not arise from an overall lowering of these times; rather it is due to a shift in their distribution, whereby the shorter times are achieved more and more frequently and the slower times less frequently. Higher rates of performance come not from going more quickly but from **going slowly less often** (Salvendy and Seymour, 1973).

Carlson and Rowe (1976) report that interruptions in work cause forgetting, with the amount of forgetting depending on the length of time between "batches."

Work-Factor learning factors (see Table 28.6) can be approximated by two learning rates—around an 80% curve for cycles 1–50 and around 90% for cycles 50–500 (Work-Factor, 1969). The table permits estimation of the mean time/unit required for quantities less than 500. For example, if the Work-Factor time was 1.5 min/unit and 100 units were to be made in this release to the shop, then the estimated mean time for 100 units would be 1.5 (1.3) = 1.95 min/unit; total time would be 1.95 (100) = 195 min. This estimate does not include allowance time, so if allowances were 15%, the total time would be

TABLE 28.6   Work-Factor learning allowances for individual operators (Work-Factor, 1969).

Quantity of Pieces to be Produced	Work-Factor Time Cycle			
	-3.0 Min.	-6.0 Min.	-12.0 Min.	>12.0 Min.
	Quantity Class Multipliers (Learning Allowances Included)			
1	5.0	5.0	5.0	5.0
2	4.1	4.2	4.3	4.4
3	3.6	3.7	3.9	4.0
4	3.3	3.4	3.6	3.8
5	3.0	3.2	3.4	3.6
6	2.9	3.0	3.2	3.4
7	2.7	2.9	3.1	3.3
8	2.6	2.8	3.0	3.2
9	2.5	2.6	2.9	3.1
10	2.4	2.6	2.8	3.0
15	2.1	2.2	2.5	2.7
20	1.9	2.0	2.3	2.5
30	1.7	1.8	2.0	2.2
40	1.6	1.7	1.9	2.0
50	1.5	1.6	1.7	1.9
60	1.4	1.5	1.7	1.8
70	1.4	1.5	1.6	1.7
80	1.3	1.4	1.5	1.7
90	1.3	1.4	1.5	1.6
100	1.3	1.4	1.5	1.6
110	1.2	1.3	1.4	1.5
120	1.2	1.3	1.4	1.5
130	1.2	1.3	1.4	1.5
140	1.2	1.2	1.4	1.5
150	1.2	1.2	1.3	1.4
160	1.2	1.2	1.3	1.4
170	1.2	1.2	1.3	1.4
180	1.1	1.2	1.3	1.4
190	1.1	1.2	1.3	1.4
200	1.1	1.2	1.3	1.3
250	1.1	1.1	1.2	1.3
300	1.1	1.1	1.2	1.2
350	1.1	1.1	1.1	1.2
400	1.1	1.1	1.1	1.2
450	1.06	1.1	1.1	1.2
500	1.05	1.1	1.1	1.1

195/.85 = 229 min.

The effect of similar work can be considered:

$$ESTIME = WFTIME \ (SMFACT \ [QMULT - 1] + 1)$$

Where *ESTIME* = Estimated time/unit including learning allowances

WFTIME = Work-Factor time/unit

SMFACT = Similarity factor

= 1.0 for 0 to 30% similar

= 0.5 for 31 to 60% similar

= 0.25 for 61 to 90% similar

= 0 for over 90% similar

QMULT = Quantity multiplier from Table 28.6

For example, if *WFTIME* = 1.5, the quantity released is 100, and 80% of the operations were fully familiar for this operator, then *SMFACT* = .25 and *QMULT* = 1.3. Then *ESTIME* = 1.5 (1.075) = 1.61 min/unit.

Table 28.6 also allows estimation of the specific unit at which 100% of Work-Factor time should be achieved. For a 90% curve, the unit curve is about 85% of the mean curve (see Table 28.1). A line on log–log paper parallel to the *QMULT* values and at 85% of their values cuts the axis between 200 and 400 units (depending on the cycle time and the similarity). Thus, a worker is expected to achieve Work-Factor standard after 200 to 400 cycles. Remember, however, that Work-Factor assumes the worker is experienced at the start and only familiarity with a specific unit is needed.

3   APPLICATIONS

As discussed in Chapter 26, use of time/unit has 4 applications in all organizations: cost allocation, scheduling, acceptable day's work, and evaluation of alternative methods. In some organizations there is a fifth application since wages are based on output. For more examples, see Nanda and Adler (1977).

The fact that labor h/unit declines as output increases makes computations for all the applications more complex than if it could be assumed constant. Ah, for the simple life!

3.1   Cost Allocation

For cost allocation, many authors have pointed out the necessity of using the projected lower costs in bidding on new contracts or make-buy decisions. For example, assume average cost of the first 100 units is 250 h/unit, you have an 80% manufacturing progress curve, and you must estimate manufacturing costs for a reorder of 100 units. Then you should estimate average cost of .8 (250) = 200 h/unit for the cumulative output of 200. Thus the first 200 will take 40,000 hours; the first 100 took 25,000 hours so this reorder of 100 will require 15,000 hours. It is not necessary to charge a lower price due to the improvement—you may wish to have higher profits—but it certainly is desirable to know your true costs when making your bid.

Shared costs are important. That is, if a component is used on more than one final product (standardization), it can progress much faster on the curve because its sales come from multiple sources. In addition, the costs of all the final products are pulled down. Remember, however, that low costs do not guarantee success if the product is obsolete. That is, products have life cycles of birth, growth, decay, and death. For more on cost estimating, see Ostwald (1991).

3.2   Scheduling

For scheduling, changing labor h/unit is important. If you wish to keep units/week constant, then you should expect to decrease your labor force for the project as time passes. If you wish to keep your labor force constant, then your output/week should increase week by week. Consider also the effect of different lot sizes and labor turnover. For example, consider the effect on economic lot size for spare part components. If a unit has a standard manufacturing cost of $11.33 based on a lot size of 100 units, a setup cost of 3 h, run cost of 1 h/unit, labor cost of $10/h (including overhead), and material cost of $1/unit, then standard cost is:

Material	$1.00
Labor	
Setup (3/100) ($10)	0.33
Run 1 ($10)	10.00
	$11.33

Assume that due to poor scheduling or a customer emergency it is decided to run a special lot of 6. Assume a learning curve of 95%. Dividing 100/2 = 50; 50/2 = 25; 25/2 = 12.5; 12.5/2 = 6.25 (which is close to 6 units). This is four doublings. Then run time = $1.0/(.95)^4$ = 1.23 h/unit and total cost becomes:

Material	$1.00
Labor	
Setup (3/6) ($10)	5.00
Run 1.23 ($10)	12.30
	$18.30

3.3 Evaluation of Alternatives

Evaluation of alternative methods is simpler when labor h/unit are assumed to be constant; we want accurate predictions, however, so use a changing cost/unit. People often are disappointed by the high cost of a new method or model early in its curve when comparing with the old method or model, which had years of experience.

For example, consider a decision whether to manufacture an item in a factory (**manufacturing**) or in the field (**construction**). Think of the construction at the site as a little factory. The real factory produces 10,000 units/yr and has an 80% progress curve. The decision is whether to make a 200-unit order in the factory or in the field.

The production ratio is 10,000/200 = 50. Unit production cost in the factory = 1. Then, for the .8 curve, cost is changed by 1/.8 = 1.25 for each doubling. There are 5.65 doublings in 50 ($2^{5.65} = 50$), so cost is $(1.25)^{5.65} = 3.53$ times higher to manufacture in the field than in the factory. Fisher (1974) used this example to explain the high cost of nuclear power plants. Due to economies of scale for heat recovery, it is desirable to build very large nuclear power plants, but because the plants are so large, components are built on the site rather than in centralized factories, and therefore costs are very high.

Although the example is simplified, the calculations help explain the advantages of manufacturing over construction.

When evaluating performance in laboratory experiments, the typical assumption is that performance with a small amount of practice (e.g., 50–100 cycles) is representative of long-term performance. However, the average time and errors should decrease with practice. Another problem is that learning may not be at the same rate for different alternatives. That is, alternative A may be best after 50 cycles of practice, but alternative B may be best after 1,000 cycles of practice. But, based on the initial trials, the decision may be made to use alternative A. Engineers should be very cautious about initial trials and base their recommendations on experienced operators with more trials.

3.4 Acceptable Day's Work

For an acceptable day's work, units produced/standard time should increase with experience. If a new operator's output is plotted against the typical learning curve for that job, it is possible to predict quite early whether an operator will ever make standard. See Knowles and Bell (1950) for examples of learning curves used to select operators; see Nissley (1949) for a form used to give a daily target for operators as they progress along the curve. Glover (1967) says good predictions can be made after two weeks.

Assume a time standard, y, is set at experience level x. Assume further, for ease of understanding, that $y = 1.0$ min and x is 100 units. That is, a time study technician, Bill, made a time study on the first 100 units produced by Roger, calculated the average time, and got 1.0 min/unit. What happens after Bill leaves and Roger continues working? See Table 28.7. Bill started the study at 7:30 A.M. and studied 100 units, getting an average time of 1.0 min. He stopped the study when Roger started his coffee break. Let's assume that 95% is the appropriate rate for this job for the first 5,000 units. Roger completes his 200th piece shortly before lunch and his 400th before the end of the day. His average time/unit for the first day is about .90. He has performed at 111% of the standard time of 1.0 min/unit. The second day, Roger completes his 800th unit early in the afternoon; he will complete his 3,200th unit within a week of the time study. His average time/unit of .77 min/unit during the week was 129% of standard!

No time standard is accurate unless accompanied by a run quantity.

There are a number of points which can be made from the example.

First, assume that this unit is produced in lots of varying sizes depending on sales. Roger had a

TABLE 28.7   Demonstration of the **learning effect on time standards**.

Learning Curve, %	Time/Unit at			Percent of Standard at		
	2X	**4X**	**32X**	**2X**	**4X**	**32X**
98	.98	.96	.90	102	104	111
95	.95	.90	.77	105	111	129
90	.90	.81	.59	111	123	169
85	.85	.72	.44	118	138	225

Note: Even a small learning rate can have a major effect on performance. X = experience level of the operator when time study was taken, for example, 50, 100, or 500 cycles. The table gives time/unit based on a time standard of 1.0 min/unit; therefore, if actual time standard was 5.0 min/unit, then time/unit at 98% and 2X would be .98 (5.0) = 4.9.

requirement of 3,200 units, and, at .77/min, the 3,200 units took 41.1 hours. He turned in a reported time for the lot of 44 hours, and, since standard was 53.3 hours, his efficiency was 53.3/44 = 121%. The supervisor was happy and Roger was happy. Four months later, an order for 400 units is released to Bill O'Dell. Bill, who is trying to become supervisor, worked very hard and produced them in .87 min. He reported 400 × .87/60 or 5.8 hours for the job. The foreman compared 5.8 with 6.67 standard hours and calculated his efficiency as 6.67/5.8 = 115% and said to himself, "Good but not as good as Roger." For accuracy the foreman should have used a standard time of .77 min/unit for Roger and a standard time of .90 for Bill. Roger's real efficiency was 41.1/(.77 × 3,200/60) = .94 while Bill's was 5.8/(.90 × 400/60) = .97. Thus, unless learning curves are considered, supervisors will have an incorrect idea of whether a specific operator is doing an acceptable day's work.

Second, the magnitude of the learning effect should be emphasized. Criticism is made of the accuracy of the rating procedure used in time study.

Considerable effort and training are used to reduce errors to ±5%; that is, a person's pace might be estimated as 95% when in actuality it might be between 90% and 100%. However, this 10% range is minor compared to the errors which can occur if a standard time is used regardless of the run quantity.

The major cause of inaccurate time standards is failure to consider learning. If none of the operators in your plant ever turn in a time that improves as they gain experience, does that mean you have obtuse operators? Might it be stupid supervisors?

3.5 Incentives

If you pay by results, failure to recognize that labor time changes with output causes a number of problems. One is dissatisfaction of the workers as some get easy jobs (N is large) while others get tough jobs (N is small). Another effect is restriction of output. Rather than report output of 200% of standard (resulting in complaints of high wages from fellow workers and possible changing of the standard by organization), the employees just increase their leisure time.

REVIEW QUESTIONS

1. How did Wright make a curve into a straight line?
2. What is meant by "garbage in, Bible out?"
3. Assume a Work-Factor time was .5 min/unit and 200 units were to be made on this release. If allowances are 20%, what is the estimated standard time?
4. Assume a time standard was set as .05 h/unit after observing the first 50 cycles. If the task has a 98% learning curve, what would be the average time/unit at 100 cycles, 200 cycles, 400 cycles, 800 cycles, and 1,600 cycles? What percent of standard could be expected at each number if the operator worked at full capacity?
5. Is learning or inaccurate rating a greater potential source of incorrect time standards? Explain your answer.

REFERENCES

Allemang, R. New technique could replace learning curves. *Industrial Engineering,* Vol. 9 [8], 22–25, 1977.

Andress, F. The learning curve as a prediction tool. *Harvard Business Review,* Vol. 32, 87–97, January–February 1954.

Barnes, R. and Amrine, H. The effect of practice on various elements used in screwdriver work. *J. of Applied Psychology,* 197–209, 1942.

Barnes, R., Perkins, J., and Juran, J. A study of the effect of practice on the elements of a factory operation. *University of Iowa Studies in Engineering Bulletin* 22, No. 387, 1940.

Carlson, J. Cubic learning curves: Precision tool for labor estimating. *Manufacturing Engineering and Management,* 22–25, November, 1973.

Carlson, J. and Rowe, A. How much does forgetting cost? *Industrial Engineering,* Vol. 8 [9], 40–47, 1976.

Conley, P. Experience curves as a planning tool. *IEEE Spectrum,* 63–68, June 1970.

Conway, R. and Schultz, A. The manufacturing progress function. *J. of Industrial Engineering,* Vol. 10 [1], 39–54, 1959.

Crossman, H. A theory of the acquisition of speed skill. *Ergonomics,* 153–65, 1959.

Daniels, R. Factors affecting industrial learning on interrupted production schedules. M. S. thesis, Kansas State University, 1966.

Fisher, J. *Energy Crisis in Perspective.* New York: Wiley-Interscience, 1974.

Glover, J. Manufacturing progress functions II. Selection of trainees and control of their progress. *International J. of Production Research,* Vol. 5 [1], 43–59, 1966.

Glover, J. Manufacturing progress functions III. Production control of new products. *International J. of Production Research,* Vol. 6 [1], 15–24, 1967.

Hancock, W. and Bayha, F. The learning curve. In *Handbook of Industrial Engineering,* Salvendy, G. (ed.), Chapter 4.3. New York: Wiley & Sons, 1982.

Hirsch, W. Manufacturing progress functions. *The Review of Economics and Statistics,* Vol. 34, 143-55, May 1952.

Hirschman, W. Profit from the learning curve. *Harvard Business Review,* 125-39, January-February 1964.

Kilbridge, M. Predetermined learning curves for clerical operations. *J. of Industrial Engineering,* 203-209, 1959.

Kneip, J. The maintenance progress function. *J. of Industrial Engineering,* Vol. 16 [6], 398-400, 1965.

Knowles, A. and Bell, L. Learning curves will tell you who's worth training and who isn't. *Factory Management & Maintenance,* 114-15, June 1950.

Konz, S. Learning curves for drill press operations. M. S. thesis, University of Iowa, 1960.

McCambell, E. and McQueen, C. Cost estimating from the learning curve. *Aero Digest,* 36-39, October 1956.

Morooka, K. and Nakai, S. *Learning Investigation.* Report by Committee of Learning Investigation to Japan Society of Mechanical Engineers, February 1, 1971.

Nadler, G. and Smith, W. Manufacturing progress functions for types of processes. *International J. of Production Research,* Vol. 2[2] 115-35, 1963.

Nanda, R. and Adler, G. (eds.). *Learning Curves Theory and Application,* Monograph 6. Norcross, GA: Institute of Industrial Engineers, 1977.

Nissley, H. The importance of learning curves in setting job shop standards. *Mill and Factory,* Vol. 44, 119-22, May 1949.

Noyce, R. Microelectronics. *Scientific American,* Vol. 237 [3], 63-69, 1977.

Ostwald, P. *Engineering Cost Estimating,* 3rd ed. Englewood Cliffs, NJ: Prentice-Hall, 1991.

Riegel, P. Athletic records and human endurance. *American Scientist,* Vol. 69, 285-89, May-June, 1981.

Rohmert, W. and Schlaich, K. Learning of complex manual tasks. *International J. of Production Research,* Vol. 5 [2], 137-45, 1966.

Salvendy, G. and Pilitsis, J. Improvement in physiological performance as a function of practice. *International J. of Production Research,* Vol. 12 [4], 519-31, 1974.

Salvendy G. and Seymour, W. *Prediction and Development of Industrial Work Performance.* New York: Wiley & Sons, 1973; p. 17.

Scheffler, F. Estimating for reinforced plastics. *Modern Plastics,* 135-50, 243, May 1957.

Smyth, R. How to figure learning time. *Factory Management and Maintenance,* 94-96, March 1943.

Titleman, M. Learning curves—Key to better labor estimates. *Product Engineering,* 36-38, November 18, 1957.

Turban, E. Incentives during learning. *J. of Industrial Engineering,* Vol. 19, 600-607, 1968.

Work-Factor learning time allowances. Ref. 1.1.2, Moorestown, N. J.: Wofac Company, 1969.

Wright, T. Factors affecting the cost of airplanes. *J. of Aeronautical Sciences,* Vol. 3, 122-28, February 1936.

Youde, L. A study of the training time for two repetitive operations. M. S. thesis, State University of Iowa, 1947.

29 PREDETERMINED TIME SYSTEMS

OVERVIEW

PTS developed from therbligs; they offer detailed methods analysis as well as times. The two most popular systems are Methods-Time Measurement (MTM) and Work-Factor (WF). Each system has several levels of detail, allowing a trade-off between analysis detail vs. analysis time. The chapter describes MTM-1, MTM-2, and MTM-3 as well as Ready Work-Factor.

CHAPTER CONTENTS

1 History and Development
2 Methods-Time Measurement
3 Work-Factor
4 Comments on Predetermined Time Systems

KEY CONCEPTS

apply pressure	incentive work pace	preposition
assemble	independence and additivity	reach
body motion	limited motions	Ready Work-Factor
combined motions	mental processes	release
disassemble	move	therbligs
disengage	MTM-1	transport
efficient work method	MTM-2 and MTM-3	turn
eye motion	position	use
grasp	predetermined time systems (PTS)	

Predetermined time systems (PTS) are an excellent technique to analyze work methods in addition to furnishing times. For additional information on workstation design, see Chapter 15. For information on standard data (i.e., times for macro elements), see Chapter 30.

1   HISTORY AND DEVELOPMENT

Frederic Taylor, developer of scientific management, first applied the scientific method to the mundane world of work rather than just to laboratory experiments. Frank and Lillian Gilbreth took a more detailed look at work than Taylor and broke work into 17 micro elements **(therbligs)**. Through micromotion analysis of bricklaying, a task which had been done for 2,000 years before Gilbreth, Frank improved bricklaying productivity 300%!

Step 1 is to determine an efficient work method. Step 2 is to determine the time/unit for the efficient work method. PTS can be used for step 2 as well as for step 1.

Table 29.1 gives the 17 therbligs, as refined by Barnes (1980).

Therbligs were the first step in that they broke work down into elements. The second step was to assign time values to each of the elements. Then, to determine the total time for a task, the times for each of the elements are totaled.

In ordinary words the concept is similar to constructing a building. A building is composed of elements—doors, walls, beams, bricks, plumbing. The structure is the sum of the elements. Likewise, a job is also considered to consist of elements, the total of which is the sum of the elements. In formal words, the assumption is that each job element is independent and additive; that is, each element does not affect what happens before or after it **(independence and additivity).**

A number of critics of PTS have shown that these independence and additivity assumptions are not perfectly valid for all situations. That is, time for an element is affected by the preceding and following element. The PTS do compensate for these assumptions—at least to some extent—since most elements occur in standard sequences. However, at the present state of knowledge, there is no realistic alternative to the PTS. They should be used with caution since their time values, especially in the hands of inexperienced users, often can depart from actual times by 25–50%. Trained users do better. Furthermore, actual times also vary, depending on worker and work pace on a specific day.

The concept of basic, universal units of work with accompanying standard amounts of time is an attractive idea. Many felt called to develop a system; few were chosen. The most popular systems (as of the present) are Methods-Time Measurement and Work-Factor; they are described in this chapter. Some other systems (which may or may not be more accurate, easier to apply, more consistent, more relevant to a specific industry, etc.) are described by Sellie (1992).

2   METHODS-TIME MEASUREMENT

2.1   Basic Concept   Methods-Time Measurement was developed by Maynard, Stegemerten, and Schwab (1948) in 1946 from motion pictures of sensitive drill press operations at Westinghouse. The MTM Association, which publishes news of research and applications in the MTM journal, has emphasized improving ease of application and development of simplified versions of MTM. MTM is probably the most widely used PTS in the world. For a detailed (503-page) explanation of MTM, see Karger and Bayha (1987).

This section describes the basic MTM system (MTM-1) as well as two simplified systems (MTM-2 and MTM-3). The MTM material was reviewed by Karl Eady, Director of Training Development. The MTM tables are reprinted with the permission of the MTM Association, 1411 Peterson Avenue, Park Ridge, IL, 60068; phone (708) 823-7120. People who wish to actually use MTM (in contrast to students) should take a training course approved by the MTM Association (approximately 24 to 80 classroom hours) so they can use the procedures accurately and consistently.

2.2   MTM-1   In the most detailed system, **MTM-1**, motions are broken down into 10 categories: Reach, Move, Turn, Apply Pressure, Grasp, Position, Release, Disengage, Body (leg–foot, horizontal, and vertical) Motions, and Eye Motions. Times for each of these are given in time measurement units (TMUs), which is a fancy name for .000 010 h. Thus 1 TMU = .000 010 h = .000 600 min = .036 s. Conversely, 1 s = 27.78 TMU; 1 min = 1,667 TMU; and 1 h = 100,000 TMU.

The times are for an experienced operator working at a normal pace (100%). (See Chapter 27 for a definition of normal.) No allowances are included in the times. See Chapter 31 for allowances.

Another question is how many cycles of practice it takes before a worker can achieve MTM standard.

TABLE 29.1 Therbligs and their present descendants.

Therblig	Code	Definition
Transport empty	TE	Reach for an object with an empty hand. Became REACH or TURN in MTM and ARM MOTION in Work-Factor.
Search	Sh	Search begins when the eyes or hands begin to hunt for an object and ends when the object has been found. Incorporated into the various cases of REACH and GRASP of MTM and into the "work factors" of ARM MOTIONS and GRASP.
Select	Se	Select one object from several. Often combined with search. Incorporated into the cases of GRASP in both MTM and Work-Factor.
Grasp	G	Grasp begins when the hands or fingers first make contact with an object and ends when control is obtained. Grasp is a key element in both MTM and Work-Factor.
Transport loaded	TL	Transport loaded is moving an object (or an empty hand vs. resistance). Became MOVE or TURN in MTM and an ARM MOTION with work factors in Work-Factor.
Preposition	PP	Preposition for the next operation is locating an object in a predetermined place or position. Becomes either a REGRASP or POSITION in MTM but is called PREPOSITION in Work-Factor.
Position	P	Position begins when the hand begins to turn or orient an object during a move and ends when the object has been placed in the desired location. Became POSITION in MTM and part of ASSEMBLE in Work-Factor.
Assemble	A	Assemble is placing one object on or in another object. Became MOVES, REGRASPS, and POSITION in MTM and ASSEMBLE in Work-Factor.
Disassemble	DA	Disassemble is separating one object from another object of which it is an integral part. Not an element in MTM except when resistance to separation is present (DISENGAGE). Became DISASSEMBLE in Work-Factor.
Release load	RL	Release load is letting go of an object. Became RELEASE in MTM and RELEASE in Work-Factor.
Use	U	Use begins when a hand begins to manipulate a tool or device and ends when the hand ceases the application. Described by individual elements in MTM. Became USE in Work-Factor.
Hold	H	Hold is retention of an object after grasp with no movement taking place. Not an element in MTM. One of two types of Balance Delay in Work-Factor (the other type is wait).
Inspect	I	Inspect is examining an object (with sight, hearing, touch, odor, or taste) to compare vs. a standard. Covered very briefly by MTM with VISUAL ACTION; covered in much more detail by Work-Factor with MENTAL PROCESS.
Avoidable delay	AD	Avoidable delay is a delay which operators may avoid if they wish. Not an element in either MTM or Work-Factor.
Unavoidable delay	UD	Unavoidable delay is a delay beyond the operator's control. If it occurs due to an interruption in the process the time should be considered by adding an allowance or by having the operator "clock out." A second possibility is a delay of one part of the body while another part is busy. This is considered by the SIMO Table in MTM; it is called Balance Delay in Work-Factor.
Plan	Pn	Plan is a mental reaction preceding the physical movement. Covered by learning allowances in both MTM and Work-Factor as both systems assume an "experienced" operator.
Rest to overcome fatigue	R	Rest is a fatigue or delay factor to permit the operator to recover from fatigue incurred by the work. Not given in either the MTM or Work-Factor systems as both systems assume a percentage will be added to the times for "personal, fatigue, and delay time."

See Chapter 28 for a discussion of learning curves. Rivett (1972) developed learning curves for 4 tasks; MTM standard was achieved at 900, 135, 3,100, and 3,300 cycles. Karger and Bayha (1987) give a complicated technique to apply learning curves to MTM. In their two examples, the worker achieved MTM standard after 1,900 cycles for method 1 and 2,300 cycles for method 2. For Rivett's 4 examples, at 200 cycles a worker would be expected to produce at 82%, 108%, 71%, and 74% of standard. Thus, complaints about inaccurate MTM times may be primarily due to failure to apply learning allowances. An important point is that the MTM standard will not be met for occasional work (for anything less than, say, 2,000 cycles).

2.2.1 Reach

Reach is usually movement with an empty hand or finger while Move is generally movement with an object in the hand. Reach is subdivided into 5 cases (see Table 29.2), which can be modified in some instances by removing the effect of acceleration and/or deceleration.

TABLE 29.2   MTM-1 Reach times (TMU) are for five cases at various distances. Case A and B Reaches also can omit an acceleration or deceleration. A typical code is R6B.

Distance Moved, Inches	Time, TMU				Hand in Motion		Case and Description
	A	B	C or D	E	A	B	
¾ or less	2.0	2.0	2.0	2.0	1.6	1.6	A Reach to object in fixed location, or to object in other hand or on which other hand rests.
1	2.5	2.5	3.6	2.4	2.3	2.3	
2	4.0	4.0	5.9	3.8	3.5	2.7	
3	5.3	5.3	7.3	5.3	4.5	3.6	B Reach to single object in location which may vary slightly from cycle to cycle.
4	6.1	6.4	8.4	6.8	4.9	4.3	
5	6.5	7.8	9.4	7.4	5.3	5.0	
6	7.0	8.6	10.1	8.0	5.7	5.7	
7	7.4	9.3	10.8	8.7	6.1	6.5	
8	7.9	10.1	11.5	9.3	6.5	7.2	C Reach to object jumbled with other objects in a group so that search and select occur.
9	8.3	10.8	12.2	9.9	6.9	7.9	
10	8.7	11.5	12.9	10.5	7.3	8.6	
12	9.6	12.9	14.2	11.8	8.1	10.1	
14	10.5	14.4	15.6	13.0	8.9	11.5	
16	11.4	15.8	17.0	14.2	9.7	12.9	D Reach a very small object or where accurate grasp is required.
18	12.3	17.2	18.4	15.5	10.5	14.4	
20	13.1	18.6	19.8	16.7	11.3	15.8	
22	14.0	20.1	21.2	18.0	12.1	17.3	
24	14.9	21.5	22.5	19.2	12.9	18.8	E Reach to indefinite location to get hand in position for body balance or next motion or out of way.
26	15.8	22.9	23.9	20.4	13.7	20.2	
28	16.7	24.4	25.3	21.7	14.5	21.7	
30	17.5	25.8	26.7	22.9	15.3	23.2	
Additional	0.4	0.7	0.7	0.6			TMU per inch over 30 inches.

The 5 cases of Reach are:

A. Reach to an object in a fixed location, or to an object in the other hand, or on which the other hand rests. (The concept is of minimum eye control and emphasis on proprioceptive feedback.)

B. Reach to a single object whose general location is known. Location may vary slightly from cycle to cycle. (The concept is that some visual control is necessary and that the following grasping motion will be simple and not slow down the reach.)

C. Reach to object jumbled with other objects in a group so that search and select occur. (The concept is that considerable visual, muscular, and mental control is necessary and that the following grasp motion will be complex, so the hand will slow down during the terminal portion of the reach to prepare for the complex grasp. It is the most difficult reach.)

D. Reach to a very small object or where accurate grasp is required. (The concept is that considerable visual control is necessary and that the following location grasp motion will be precise, so the hand will slow down during the terminal portion of the reach to prepare for the careful grasp.)

E. Reach to an indefinite location to get the hand in position for body balance or next motion or out of the way. (The concept is of minimum mental control. The movement often is "limited out" as other motions are done simultaneously.)

Table 29.2 gives the TMU for each case of Reach for various distances. The distances are for the motion path of the hand knuckle or fingertip rather than the straight-line distance between two points, although this fine point is often overlooked by users not intensively trained in MTM. Novices also tend to overlook the reduction in Reach distance due to shoulder movement and pivoting—called *body assist*.

If a movement is not given in the table, the user can interpolate (time for R15A is average of time for R14A and R16A) or just use the next higher value. The MTM Association recommends interpolation when the tables are given in inches but use of the next higher value when the tables are given in even cm or multiples of 5 cm.

In addition to breaking Reach down into 5 cases and a number of different distances, the effect of acceleration and deceleration also can be considered. The normal motion has the hand stopped at the beginning and end of the motion. Hands also could be in motion either at the beginning or end of the cycle, so either an acceleration or deceleration time can be omitted. The shortest time would occur in a situation in which the hand was in motion both at the beginning and end of the cycle and both acceleration and deceleration can be omitted. When one acceleration or deceleration can be omitted, time is decreased as is shown by the hand-in-motion columns; when both acceleration and deceleration can be omitted, double this decrease is subtracted.

2.2.2 Move

Move differs from Reach in that in Move the hand is usually holding something; occasionally the hand is pushing or dragging an object.

Move is subdivided into 3 cases (see Table 29.3). As with Reach, each Move can be refined to consider the effect of acceleration or deceleration or both. There is also a refinement for object weight or resistance to movement.

The three cases of Move are:

A. Move object to the other hand or against stop. (The concept is that there is little need to control the last portion of the move, except perhaps to prevent damage to the object.)

TABLE 29.3 MTM-1 Move times (TMU) are for three cases at various distances. Case B Moves can omit acceleration and/or deceleration. The effect of weight is calculated by multiplying effective net weight/hand by the dynamic factor and then adding the static constant. A typical code without weight is M8A; a typical code with a weight of 7 lb is M8A7.

Distance Moved Inches	Time TMU				Weight Allowance			Case and Description
	A	B	C	Hand In Motion B	Wt. (lb) Up to	Dynamic Factor	Static Constant TMU	
¾ or less	2.0	2.0	2.0	1.7				
1	2.5	2.9	3.4	2.3	2.5	1.00	0	
2	3.6	4.6	5.2	2.9				
3	4.9	5.7	6.7	3.6	7.5	1.06	2.2	A Move object to other hand or against stop
4	6.1	6.9	8.0	4.3				
5	7.3	8.0	9.2	5.0	12.5	1.11	3.9	
6	8.1	8.9	10.3	5.7				
7	8.9	9.7	11.1	6.5	17.5	1.17	5.6	
8	9.7	10.6	11.8	7.2				
9	10.5	11.5	12.7	7.9	22.5	1.22	7.4	
10	11.3	12.2	13.5	8.6				B Move object to approximate or indefinite location
12	12.9	13.4	15.2	10.0	27.5	1.28	9.1	
14	14.4	14.6	16.9	11.4				
16	16.0	15.8	18.7	12.8	32.5	1.33	10.8	
18	17.6	17.0	20.4	14.2				
20	19.2	18.2	22.1	15.6	37.5	1.39	12.5	
22	20.8	19.4	23.8	17.0				
24	22.4	20.6	25.5	18.4	42.5	1.44	14.3	
26	24.0	21.8	27.3	19.8				C Move object to exact location
28	25.5	23.1	29.0	21.2	47.5	1.50	16.0	
30	27.1	24.3	30.7	22.7				
Add'l.	0.8	0.6	0.85		TMU per inch over 30 inches			

EFFECTIVE NET WEIGHT

Effective Net Weight (ENW)	No. of Hands	Spatial	Sliding
	1	W	$W \times F_c$
	2	W/2	$W/2 \times F_c$

W = Weight in pounds
F_c = Coefficient of Friction

B. Move object to an approximate or indefinite location. (The concept is a move in which some control is needed at the end of the move but not a great deal of control.)

C. Move object to exact location. (The concept is a move with considerable control needed at the end of the move. A case C Move very often is followed by a position.)

The three cases of Move thus differ (as did the five cases of Reach) by the nature of their destination. Thus, MTM, in effect, says that all elements are not purely independent but that both Moves and Reaches are influenced by the motions preceding or following them.

The shorthand for a 5-inch case C Move is M5C.

As with Reach, the hand acceleration or deceleration may not be needed for some Moves. If omitted, it is indicated by "m" before or after the code, for example, mM6B, or M6Bm.

The last adjustment to Move is the effect of the weight of the object moved. Weight up to 2.5 lb/hand is included in the time shown in the Move table. Thus, if both hands move a 5-lb object, no extra time is allocated. If an object is slid rather than lifted, then take the object weight times the coefficient of friction (0.4 for wood–wood and wood–metal, 0.3 for metal–metal).

Next, all weights between 2.5 and 7.5 lb are considered alike. First, the TMU value is multiplied by a factor and then a constant is added. For example, an

M6B = 8.9. For a 5-lb weight, time would be M6B5 = 8.9 (1.06) + 2.2 = 11.8. For an 18-lb weight moved 12 inches with both hands, time for an M12C would be M12C9 = 15.2 (1.11) + 3.9 = 20.8.

MTM also has a more precise method of allowing for weights. Additional time for weight has a static component for obtaining control and a dynamic component for additional travel time. The static time component in TMU = .975 + .345 (weight, lbs). The dynamic time component is 1.1%/lb, at any given distance.

Table 29.4 gives the times for a variation of moving: *cranking*. For continuous cranking for 5 revolutions against a 1-lb load with a 6-inch diameter crank, the time would be 5 × 12.7 = 63.5 TMU plus 5.2 for start and stop = 68.7 TMU. Then the resistance is considered by multiplying by 1.11 (from Move table) and adding 3.9, giving a total of 68.7 × 1.11 = 76.3 + 3.9 = 80.2 TMU. The code is 5C6-10.

2.2.3 Turn
The third type of motion with the hand is turn (see Table 29.5). **Turn** is a movement that rotates the hand, wrist, and forearm about the long axis of the forearm. The amount of time depends on the number of degrees turned as well as on the weight of the object or the resistance against which the turn is made.

2.2.4 Apply pressure
Apply Pressure (See Table 29.6) is the application of force without resultant movement. An APA is the basic element; an APB is an APA plus a Regrasp (G2).

TABLE 29.4   MTM-1 cranking motion times (TMU) are for light resistance. A typical code is 6C5. With 10-lb resistance: 6C5-10. See also Table 19.4.

Diameter of Cranking (inches)	TMU (T) Per Revolution	Diameter of Cranking (inches)	TMU (T) Per Revolution
1	8.5	9	14.0
2	9.7	10	14.4
3	10.6	11	14.7
4	11.4	12	15.0
5	12.1	14	15.5
6	12.7	16	16.0
7	13.2	18	16.4
8	13.6	20	16.7

FORMULAS:
A. Continuous Cranking (Start at beginning and stop at end of cycle only)
$$TMU = [(N \times T) + 5.2] \cdot F + C$$
B. Intermittent Cranking (Start at beginning and stop at end of each revolution)
$$TMU = [(T + 5.2) F + C] \cdot N$$

C = Static component TMU weight allowance constant from move table
F = Dynamic component weight allowance factor from move table
N = Number of revolutions
T = TMU per revolution (type III Motion)
5.2 = TMU for start and stop

TABLE 29.5 MTM-1 Turn (forearm swivel) times (TMU) vary with angle turned and weight in the hand. Typical codes are T30S and T45M.

Weight	Time, TMU for Degrees Turned										
	30°	45°	60°	75°	90°	105°	120°	135°	150°	165°	180°
Small-0 to 2 pounds	2.8	3.5	4.1	4.8	5.4	6.1	6.8	7.4	8.1	8.7	9.4
Medium-2.1 to 10 pounds	4.4	5.5	6.5	7.5	8.5	9.6	10.6	11.6	12.7	13.7	14.8
Large-10.1 to 35 pounds	8.4	10.5	12.3	14.4	16.2	18.3	20.4	22.2	24.3	26.1	28.2

TABLE 29.6 Apply Pressure times (TMU) have two categories. An APB is an APA plus a regrasp.

Symbol	TMU	Full Cycle Description	Symbol	TMU	Components Description
APA	10.6	AF + DM + RLF	AF	3.4	Apply Force
			DM	4.2	Dwell, Minimum
APB	16.2	APA + G2	RLF	3.0	Release Force

2.2.5 Grasp

The next 4 motions—Grasp, Position, Disengage, and Release—are the skill motions. Improvement in performance times usually is a result of reductions in times for these motions rather than increased speed of movement for Move, Reach, or Turn.

Grasp is the motion used when the purpose is to gain control of an object or objects; it almost always is followed by Move. Grasp is divided into five categories; Pickup and Select are subdivided further (see Table 29.7).

1. *Type 1:* Pickup Grasp. Usually follows an A or B reach.

   - Case 1A—small, medium, or large object by itself, easily grasped
   - Case 1B—very small object or object lying close against a flat surface
   - Case 1C1—interference with Grasp on bottom and one side of nearly cylindrical object; diameter greater than .50 inch
   - Case 1C2—interference with Grasp on bottom and one side of nearly cylindrical object; diameter .25 to .50 inch

TABLE 29.7 Grasp times (TMU) are given for five types of grasp; pickup and select are subdivided further. A typical code is G1A.

Type of Grasp	Case	Time, TMU	Description	
PICK-UP	1A	2.0	Any size object by itself, easily grasped	
	1B	3.5	Object very small or lying close against a flat surface	
	1C1	7.3	Diameter larger than ½″	Interference with Grasp on bottom and one side of nearly cylindrical object.
	1C2	8.7	Diameter ¼″ to ½″	
	1C3	10.8	Diameter less than ¼″	
REGRASP	2	5.6	Change grasp without relinquishing control	
TRANSFER	3	5.6	Control transferred from one hand to the other	
SELECT	4A	7.3	Larger than 1″ × 1″ × 1″	Object jumbled with other objects so that search and select occur.
	4B	9.1	¼″ × ¼″ × ⅛″ to 1″ × 1″ × 1″	
	4C	12.9	Smaller than ¼″ × ¼″ × ⅛″	
CONTACT	5	0	Contact, Sliding, or Hook Grasp.	

- Case 1C3—interference with Grasp on bottom and one side of nearly cylindrical object; diameter less than .25 inch

2. *Type 2:* Regrasp. This Grasp is used to change or improve control of an object which had previously been grasped. It often is performed during a Move and is "limited out."

3. *Type 3:* Transfer Grasp. This Grasp is used to transfer control of an object from one hand to the other.

4. *Type 4:* Jumbled Grasp. Follows a C Reach.
   - Case 4A—object jumbled with other objects, so search and select occur; larger than 1 inch × 1 inch × 1 inch
   - Case 4B—object jumbled with other objects, so search and select occur; .25 inch × .25 inch × .12 inch to 1 × 1 × 1 inch
   - Case 4C—object jumbled with other objects, so search and select occur; smaller than .25 inch × .25 inch × .12 inch

5. *Type 5:* Contact, sliding, or hook Grasp. This Grasp usually occurs between a Reach and Move. No time is required to make contact with an object.

2.2.6 Position

Original **position** is the collection of minor hand movements (distance moved to engage no more than 1 inch) for aligning, orienting, and engaging one object with another object. It usually follows a C Move. Align is orienting the longitudinal axes of the two items. Orient is rotation about the long axis to align mating features (key in lock). Engage is to move along the longitudinal axis to mate the parts. Disengage is the complement of the engage portion of position. It is assumed that the items are already aligned and oriented, so time for that is zero.

Position times (see Table 29.8) vary with amount of pressure needed to fit, with symmetry of the object, and with ease of handling.

There are three classes of fit:

1. *Loose:* no pressure required (gravity sufficient); code = 1

2. *Close:* light pressure required (1 APA); Most common fit; code = 2

3. *Exact:* heavy pressure required (3 APA + G2); code = 3

There also are three classes of symmetry:

1. *Symmetrical (Code = S):* This class is demonstrated by a round peg in a round hole. The concept is that, no matter in which orientation the part might happen to be, no rotation is necessary for assembly.

2. *Nonsymmetrical (Code = NS):* This class is demonstrated by a cylinder with a key. There is one and only one orientation in which the two parts will mate. A turn of 180° (either clockwise or counterclockwise) is the maximum rotation required. Because some preorientation is done during most moves, only 75° is considered to be typical (i.e., an extra 4.8 TMU is allowed).

3. *Semisymmetrical:* All Positions that are neither symmetrical nor nonsymmetrical are considered semisymmetrical. An average turn of 45° is considered typical.

There are two classes of ease of handling: easy and difficult. They differ by a Regrasp, that is, 5.6 TMU. All flexible materials are considered difficult. A complete Position code includes all three variables; e.g., P1SE, P2SSE, P1NSD, P3SD. Alignment to a point or line (without engagement) within .25 to .5 inch requires only an M_C Move, within .06 to .25, an M__C Move plus a P1SE or P1SD, and less than .06, an M__C Move plus a P2SE or P2SD.

The middle table of Table 29.8 gives an alternative description for Position while the lower table gives the times for Secondary Engage.

Secondary Engage is a component of Supplementary Position data which is not a variable in original Position (top table of Table 29.8). It is the time required to insert an object into another for varying depths of insertion. It is used in analyses as a separate code when the radial clearance of two mating objects changes during insertion as may be the case in chamfering or countersinking.

2.2.7 Disengage

Disengage is the breaking of contact between one object and another. It includes the involuntary movement resulting from the sudden end of resistance (see Table 29.9). The three factors of class of fit, ease of handling, and care in handling are combined in a 2-factor table. The three classes of fit are given as:

1. *Loose:* very slight effort, blends with subsequent Moves (recoil up to 1 inch)

2. *Close:* normal effort, slight recoil (over 1 inch to 5 inches)

3. *Tight:* considerable effort, hand recoils markedly (over 5 inches)

Again there are only two classes of ease of handling: easy and difficult. A complete disengage code includes both fit and ease of handling, for example, D1E, D1D, D2D.

2.2.8 Release

Release is the relinquishing of control of an object by the hand or fingers. Table 29.10 shows that there are only 2 categories. The

TABLE 29.8   MTM-1 Position (align, orient, and engage) times depend on the pressure required, the symmetry, and the ease of handling. Using the top table (Original Position), a typical code is P2SSE, denoting close fit, semisymmetrical orientation, and easy handling. Insertion up to one inch is implied in all original Position codes. The middle table (Supplementary Position) classifies fit in terms of radial clearance between the mating objects as 21, 22, or 23. In addition to align, orient, and primary engage time, a secondary engage time is added to account for varying insertion depth. Select the Position time from one of the four Depth of Insertion columns. For example, to align, orient, and engage a semisymmetrical object with a radial clearance of 0.25 inch between the mating parts to a depth of one inch, coded P21SS4, requires 14.6 TMU. The Align Only column is used for surface alignments having no engagement depths, as when sliding a ruler to a point. The lowest table gives Secondary Engage times (TMU) as a function of class of fit and depth of insertion. It is used as a separate code when the radial clearance changes after the initial engagement and before completing the insertion. A typical code is E22-2 followed by a P21 Position.

Class of Fit		Symmetry	Easy To Handle	Difficult To Handle
1—Loose	No pressure required	S	5.6	11.2
		SS	9.1	14.7
		NS	10.4	16.0
2—Close	Light pressure required	S	16.2	21.8
		SS	19.7	25.3
		NS	21.0	26.6
3—Exact	Heavy pressure required	S	43.0	48.6
		SS	46.5	52.1
		NS	47.8	53.4

SUPPLEMENTARY RULE FOR SURFACE ALIGNMENT

P1SE per alignment: $>\!\frac{1}{16}\!<\!\frac{1}{4}''$ P2SE per alignment: $<\!\frac{1}{16}''$

*Distance moved to engage—1″ or less.

Class of Fit And Clearance	Case of Symmetry†	Align Only	Depth of Insertion (per ¼″)			
			0 $>\!0\!\leq\!\frac{1}{8}''$	2 $>\!\frac{1}{8}\!\leq\!\frac{3}{4}''$	4 $>\!\frac{3}{4}\!\leq\!\frac{5}{4}''$	6 $>\!\frac{5}{4}\!\leq\!\frac{7}{4}''$
21 .150″ to .350″	S	3.0	3.4	6.6	7.7	8.8
	SS	3.0	10.3	13.5	14.6	15.7
	NS	4.8	15.5	18.7	19.8	20.9
22 .025″ to .149″	S	7.2	7.2	11.9	13.0	14.2
	SS	8.0	14.9	19.6	20.7	21.9
	NS	9.5	20.2	24.9	26.0	27.2
23* .005″ to .024″	S	9.5	9.5	16.3	18.7	21.0
	SS	10.4	17.3	24.1	26.5	28.8
	NS	12.2	22.9	29.7	32.1	34.4

*BINDING—Add observed number of Apply Pressures.
  DIFFICULT HANDLING—Add observed number of G2's.

†Determine symmetry by geometric properties, except use S case when object is oriented prior to preceding Move.

Class of Fit	Depth of Insertion (per ¼″)		
	2	4	6
21	3.2	4.3	5.4
22	4.7	5.8	7.0
23	6.8	9.2	11.5

TABLE 29.9 MTM-1 Disengage times (TMU) depend on the class of fit and ease of handling. The lower table aids in selecting the proper category. A typical code is D2E.

Supplementary		
Class of Fit	**Care in Handling**	**Binding**
1—LOOSE	Allow Class 2	
2—CLOSE	Allow Class 3	One G2 per Bind
3—TIGHT	Change Method	One APB per Bind

Class of Fit	Height of Recoil	Easy to Handle	Difficult to Handle
1—LOOSE—Very slight effort, blends with subsequent move	Up to 1″	4.0	5.7
2—CLOSE—Normal effort, slight recoil	Over 1″ to 5″	7.5	11.8
3—TIGHT—Considerable effort, hand recoils markedly	Over 5″ to 12″	22.9	34.7

TABLE 29.10 MTM-1 Release times (TMU) have only two subdivisions. The normal release is coded RL1.

		SUPPLEMENTARY
Case	**Time TMU**	**Description**
1	2.0	Normal release performed by opening fingers as independent motion.
2	0	Contact Release.

most common Release, a simple opening of the fingers, is given 2 TMU. In the contact Release, for which no time is allowed, the release begins and is completed at the instance the following Reach motion begins.

The two codes are RL1 and RL2.

2.2.9 Other Motions

Other motions include (1) body, leg, and foot motions; (2) eye motions; and (3) combined and limited motions.

Body, leg, and foot motions The previous tables gave motions of the hand and arm. Table 29.11 gives **body motions** (motions of the leg-foot, horizontal torso motions, and vertical torso motions). The descriptions of the motions in the tables are reasonably self-explanatory. A foot motion is hinged at the ankle, a leg motion is hinged at the knee or hip or both, and the body centerline doesn't move appreciably. In a sidestep, the time depends on the distance the centerline moves. A pace is considered to cover 2.8 feet or 34 inches at a relatively fast pace of 3.57 miles/h. For walking through obstructed areas,

use 17 TMU/pace. Use 15 TMU/pace for stairs. For carrying a load up to 34 lb, use a 30-inch pace and 15 TMU/pace. For more on walking, see Box 11.2. For vertical torso motions, going up takes more time than going down.

Eye motions In addition to motions of the limbs and torso, time is allowed, in certain cases, for activities of the eyes. Table 29.12 gives the two basic elements of **eye motions:** Eye Focus and Eye Travel, and for formula for Read. Eye Focus is the focusing of the eye once it has an object in its line of sight. Eye Travel is the movement of the eyes from one point to another. From geometry, when $T/D = 1$, the angle swept by the eyes is 45°; for $T/D = 2$, the angle is 90°; for $T/D = 3$, the angle is 135°. Thus, 15.2 TMU is allowed per 45° sweep (.33 TMU/degree) with the limitation, however, of a maximum allowable for Eye Travel of 20 TMU, because the eyes are generally restricted to a maximum movement of 70°. Reading 100 words takes 505 TMU (18.2 s).

Combined and limited motions So far the motions have been described as if the person performed

TABLE 29.11 MTM-1 times (TMU) for motions of the leg and torso have three major subdivisions. Typical codes are W5P and SS15C1.

	Type	Symbol	TMU	Distance	Description
LEG—FOOT		FM	8.5	To 4″	Hinged at ankle
		FMP	19.1	To 4″	With heavy pressure
			7.1	To 6″	
	MOTION	LM—	1.2	Ea. add'l inch	Hinged at knee or hip in any direction
HORIZONTAL MOTION	SIDE	SS—C1	17.0	<12″	Use Reach or Move time when less than 12″;
				12″	complete when leading leg contacts floor
			0.6	Ea. add'l inch	
	STEP	SS—C2	34.1	12″	Lagging leg must contact floor before next motion
			1.1	Ea. add'l inch	can be made
	TURN	TBC1	18.6	—	Complete when leading leg contacts floor
	BODY	TBC2	37.2	—	Lagging leg must contact floor before motion can be made
	WALK	W—FT	5.3	Per Foot	Unobstructed
		W—P	15.0	Per Pace	Unobstructed
		W—PO	17.0	Per Pace	When obstructed or with weight
	VERTICAL	SIT	34.7	—	From standing position
		STD	43.4	—	From sitting position
		B,S,KOK	29.0	—	Bend, Stoop, Kneel on one knee
		AB,AS,AKOK	31.9	—	Arise from Bend, Stoop, Kneel on one knee
	MOTION	KBK	69.4	—	Kneel on both knees
		AKBK	76.7	—	Arise from Kneel on both knees

TABLE 29.12 MTM-1 times (TMU) for the eye are Eye Focus, Eye Travel, and Read. Eye Focus assumes eye travel is completed. Eye Travel is the movement from one point to another at .33 TMU/degree of motion. Reading time, TMU, = 5.05 N, where N = number of words.

Eye Travel Time = 15.2 $\times$ T/D, with a maximum value of 20 TMU.
where T = the distance between points from and to which the eye travels
D = the perpendicular distance from the eye to the line of travel T

Eye Focus Time = 7.3 TMU.

SUPPLEMENTARY INFORMATION
—Area of Normal Vision = Circle 4″ in diameter 16″ from eyes
—Reading Formula = 5.05 N where N = the number of words

one at a time. Often we have combined motions or simultaneous motions.

Combined motions are those which occur when two or more motions are performed by the same body member at the same time (Turn a part of the hand while moving it; Regrasp during a Move). The time to allocate is the greater of the two times. Thus, a Regrasp during an M3A would be given 5.6 for time with a slash line through the M3A. A Regrasp during an M6A would be given 7.0 for time with a slash line through the G2. See Figure 29.1 for a combined motion.

Motions also occur simultaneously, such as right and left hand, hands and feet, or eyes and hands. If they are truly simultaneous, allow only the longer time. If they are only apparently simultaneous, allow both times. Use Table 29.13 to decide between truly

and apparently simultaneous. Truly or apparently depend on which combinations are considered, on whether they are in the area of normal vision (objects within 4 inches of each other at a distance of 16 inches from the eye), and the amount of practice (one common definition is 500 cycles, although others use 1,000 or 2,000).

Thus, for truly simultaneous motions such as an M8A with the left hand and an M10A with the right hand, allow only the 11.3 of the M10A. If apparently simultaneous but according to Table 29.13 not actually simultaneous (such as an M8C with the left hand and an M8C with the right hand), allow 11.8 + 11.8—a total of 23.6.

In many cases where truly simultaneous motions are not allowed by Table 29.13, one hand can drift toward the target while the other does the motion.

METHODS ANALYSIS CHART   REFERENCE No. _____

PART __ 30 Wooden pegs & (1) pegboard __ DATE 21 July 76 __ STUDY No. 1A
OPERATION __ Assemble 30 pegs to pegboard __ ANALYST SK __ SHEET No. 1 OF 1 SHEETS

DESCRIPTION — LEFT HAND	No.	LH	Time	RH	No.	DESCRIPTION — RIGHT HAND
① Reach to (1) peg in bin w/RH. Pick up and place nose FIRST in hole.						
See sketch on back for hole sequence. Repeat 30 times.						
			13.6	R11C	1	Reach to bin for (1) peg from table edge
			333.5	R6C	29	Reach to bin for (1) peg from pegboard center
			219.0	G4A	30	Grasp (1) peg
			354.0	M8C	30	Move (1) peg to hole
			—	G2	30	Regrasp during Move
			168.0	P1SE	30	Position (1) peg in hole
			60.0	RL1	30	Release peg
			11.2	R11E	1	Return hand to table edge
			1159.3			

o.	ELEMENT DESCRIPTION	ELEMENT TIME TMU	CONVERSION FACTOR / LEVELED TIME	% ALLOWANCE	ELEMENT TIME ALLOWED	OCCURRENCES PER PIECE OR CYCLE	TOTAL TIME ALLOWED
1	Get 1 peg at a time. Place in board. Repeat 30 times	1159.3					

FIGURE 29.1   Combined Motion (Regrasp during the Move to the hole) is shown with a slash through it. Note how repeated steps can be described with the number column. Time to assemble 30 pegs in the pegboard = 1159.3 TMU = .70 min.

For example, while the right hand does an M8C, the left hand does an M8B. The M8B can be done simultaneously with the M8C and is "limited out"; it is conventional to circle **limited motions** (see Figure 29.2 for an example). Then, when the M8C is completed, the left hand has only a small distance (an MfC where f stands for fractional) remaining. The time allowed is 11.8 for the right hand plus a 2.0 for the left—a total of 13.8.

2.3 MTM-2 and MTM-3
How much error can be tolerated in the time determined for a specific task: 1%, 5%, 10%, 30%? How much are you willing to pay for accuracy? The answers are not the same for all situations so the International MTM Directorate (an association of 12 national MTM organizations) developed two simplified systems called **MTM-2 and MTM-3.** (The basic system described previously is called MTM-1.)

In MTM-1 it takes about 250 times the cycle time to analyze the task while in MTM-2 it takes about 100 times, and in MTM-3 about 35 times (Sellie, 1992). Figure 29.3 shows expected error limits (95% confidence) as a function of the nonrepetitive manual time in a task and the level of analysis. For example, for a specific task with a 1,000-TMU cycle time, an MTM-1 analysis would take about 3.5 h and would be expected to be accurate within ±7%; an MTM-3 analysis would take about .5 h and would be accurate within ±20%. Magnusson (1972) found that the use of MTM-2 gave times 0.1% higher than MTM-1 while MTM-3 gave values 0.9% higher than MTM-1. (This "bias" is considered to be insignificant.) The remaining random error (after this systematic error) will decrease with increasing sample size or, in this case, with longer cycle times. A 10,000-TMU cycle with MTM-1 would therefore be expected to be accurate within ±2%. However, Knott and Sury (1986)

TABLE 29.13 If motions are easy to perform simultaneously, allow just the longer of the two times. If they can be done without practice, use judgement on whether to allow both times. If difficult, allow both times.

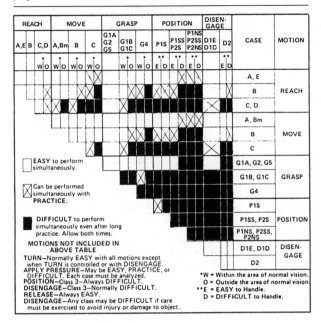

☐ EASY to perform simultaneously.

☒ Can be performed simultaneously with **PRACTICE**.

■ **DIFFICULT** to perform simultaneously even after long practice. Allow both times.

MOTIONS NOT INCLUDED IN ABOVE TABLE
TURN—Normally EASY with all motions except when TURN is controlled or with DISENGAGE.
APPLY PRESSURE—May be EASY, PRACTICE, or DIFFICULT. Each case must be analyzed.
POSITION—Class 3—Always DIFFICULT.
DISENGAGE—Class 3—Normally DIFFICULT.
RELEASE—Always EASY.
DISENGAGE—Any class may be DIFFICULT if care must be exercised to avoid injury or damage to object.

*W = Within the area of normal vision.
O = Outside the area of normal vision.
**E = EASY to Handle.
D = DIFFICULT to Handle.

reported that accuracy was the same for MTM-1, MTM-2, and MTM-3, so the only trade-off was the desired level of methods description with the increase time to calculate a time. MTM-2 and MTM-3 require considerable analyst judgment.

2.3.1 MTM-2

Table 29.14 shows the MTM-2 table; there are only 37 times in all.

The two key motion categories are GET (combining Reach, Grasp, and Release) and PUT (combining Move and Position). Which of the 15 values of GET or PUT to use depends on the case of GET or PUT, the distance, and the weight or resistance to motion.

Figure 29.4 shows the decision tree for GET to determine case. Then the user estimates (not measures) the distance and uses one of the five rows. Third, if necessary, add 1 TMU/2 lb moved if the object to be moved weighs 4 lb or more per hand.

Figure 29.5 shows the decision tree for PUT to determine case. Then the user estimates (not measures) the distance and uses one of the five rows.

Third, if necessary, add 1 TMU/10 lb moved if the object weighs 4 lb or more per hand.

There remain seven motions:

A. Apply Pressure—an action with the purpose of exerting muscular force on an object

R. Regrasp—a hand action performed with the purpose of changing the grasp on an object

E. Eye Action—an action with the purpose of either

recognizing a readily distinguishable characteristic of an object or shifting the aim of the axis of vision to a new viewing area

C. Crank—a motion with the purpose of moving an object in a circular path of more than half a revolution with the hand or fingers

S. Step—either a leg motion with the purpose of moving the body or a leg motion longer than 12 inches

F. Foot Motion—a short foot or leg motion when the purpose is not to move the body

B. Bend and Arise—a bend, stoop, or kneel on one knee, and the subsequent rise

See Figure 29.6 for an MTM-2 analysis of the two-hand assembly of the pegboard.

2.3.2 MTM-3

Table 29.15 shows the MTM-3 table. Now there are only 10 times.

The two key motion categories have been reduced to HANDLE (getting control over an object with the hand or fingers and placing the object in a new location) and TRANSPORT (placing an object in a new location with the hand or fingers). Which of the four values of HANDLE or TRANSPORT to use depends on the case and the distance.

Figure 29.7 shows the decision tree for both HANDLE and TRANSPORT. Distances are estimated as either equal to or less than 6 inches or over 6 inches. The 7 additional elements of MTM-2 have been reduced to 2; SF and B. SF combines the S and F categories of MTM-2 while B is the same as in MTM-2. Crank would now be a TRANSPORT while Apply Pressure, Regrasp, and Eye Motion are included in the HANDLE and TRANSPORT.

See Figure 29.8 for an MTM-3 analysis of the pegboard task assembly.

Additional MTM systems (Sellie, 1992) are MTM-GPD (General Purpose Data), MTM-C (Clerical), MTM-V (Machine Tool Users), and MTM-M (Microscope Assembly).

3 WORK-FACTOR

3.1 Basic Concept

The idea for the system was conceived at the Philco Radio Corporation in Philadelphia, Pennsylvania, between 1930 and 1934 because rate setting with stopwatch time study was unacceptable to union employees working under tightly controlled incentive pay systems. The original motion time data were developed between 1934 and 1938 by a group of experienced time study engineers directed by Joseph Quick, assisted by Samuel Benner, William Shea, and Robert Koehler. The purpose of

METHODS ANALYSIS CHART   REFERENCE No. _____

PART ___30 wooden pegs (1) pegboard___ DATE _21 July 76_ STUDY No. _1B_

OPERATION ___Assemble 30 pegs to pegboard___ ANALYST ___JK___ SHEET No. _1_ OF _1_ SHEETS

DESCRIPTION — LEFT HAND	No.	LH	Time	RH	No.	DESCRIPTION — RIGHT HAND
① Reach into bin w/BH simo. Get (1) peg in each hand. Move peg to board and beveled end first. See sketch on back for hole sequence.						
To bin from table edge	1	R11C	13.6	R11C	1	to bin from table edge
To bin from board center	14	R8C	161.0	(R8C	14	to bin from board center
Turn hand 60°	14	T60)		+60	14	Turn hand 60°
Grasp (1) peg	15	G4A	109.5			
			109.5	G4A	15	Grasp (1) peg
Move (1) peg toward board *	15	M8B	177.0	M8C	15	Move (1) peg to board
Regrasp during Move	15	G2		G2	15	Regrasp during Move
Turn hand 60°	15	+60°		+60°	15	Turn hand 60°
			84.0	P1SE	15	Position peg in hole
Move (1) peg to hole	15	M7C	78.0			
Position peg in hole	15	P1SE	84.0			
Release peg	15	RL1	30.0	RL1	15	Release peg
Hand to table edge	1	R11E	11.2	R11E	1	Hand to table edge
			809.8			
* Although two rows are within area of normal vision, lack of practice precludes simo motion.						

ELEMENT DESCRIPTION	ELEMENT TIME TMU	CONVERSION FACTOR / LEVELED TIME	% ALLOWANCE	ELEMENT TIME ALLOWED	OCCURRENCES PER PIECE OR CYCLE	TOTAL TIME ALLOWED
Get (1) peg in each hand. Place in board. Repeat 15 times	809.8					

FIGURE 29.2 Limited motions are circled. Note that working smart with two hands gives an assembly time of 809.8 TMU = .49 min or 70% of the one-hand method.

the 1934 research was to develop an objective work measurement technique to eliminate the stopwatch and the observer's performance rating judgments in establishing work standards and setting production output rates.

The motion times were applied worldwide for work measurement in actual factory operations at Radio Corporation of America (RCA) between 1938 and 1946. They first were published in 1945. James Duncan and James Malcolm, working with Quick, developed techniques to simplify their application. Quick, Duncan, and Malcolm (1962) described the Detailed, Ready, and Abbreviated systems in detail. In 1949 they began their research on human mental processes which led to the compilation of the Mento-Factor Manual of Detailed Work-Factor Mental Process Times in 1965.

Work-Factor time is based on the output level of the "average experienced operator working with good skill and good effort." Popularly, it is identified as an **incentive work pace.** If the MTM pace = 100%, then Work-Factor and British Standard 3138 = 120% (Boepple, 1977). That is, if the time for a specific task was 1.0 min/unit according to MTM, then Work-Factor time would be 1/1.2 = .833 min/unit. Naturally, this ratio is averaged over many tasks because, for specific tasks, details of the application rules might lead to slightly larger or smaller differences. No allowances are included in the Work-Factor time tables. Workers are expected to achieve standard time in 200 to 400 cycles (Work Factor, 1969). See Table 28.6.

This Work-Factor text was reviewed by Emerson Boepple, Coordinator of Research and Development

FIGURE 29.3 Analysis time and accuracy can be predicted for MTM-1, MTM-2, and MTM-3, given the nonrepetitive cycle time (Magnusson, 1972). Magnusson estimated MTM-1 took 350 times the cycle time, so a 2,000-TMU (.02 h) task would take 350 (.02) = 7 h to analyze by MTM-1; predicted accuracy (with 95% confidence) is ±4.9%.

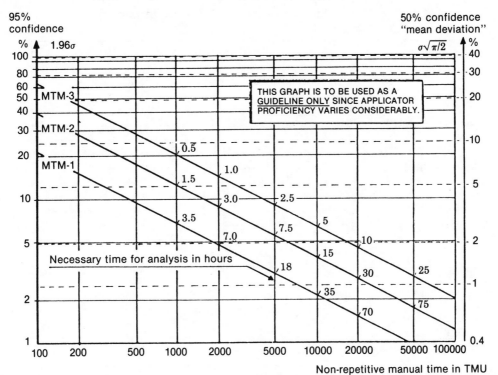

RANDOM DEVIATIONS AND TIME REQUIRED FOR ANALYSIS

TABLE 29.14 MTM-2 is a very simplified version of MTM-1; task analysis time should take about 150 times the cycle time. Times are in TMU.

				MTM-2			
RANGE	Code	GA	GB	GC	PA	PB	PC
Up to 2″	-2	3	7	14	3	10	21
Over 2″ to 6″	-6	6	10	19	6	15	26
Over 6″ to 12″	-12	9	14	23	11	19	36
Over 12″ to 18″	-18	13	18	27	15	24	36
Over 18″	-32	17	23	32	20	30	41

GW 1-per 2 lb						PW 1-per 10 lb.	
A		R	E	C	S	F	B
14		6	7	15	18	9	61

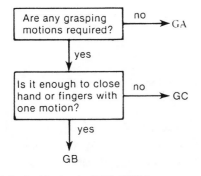

FIGURE 29.4 Decision tree for GET in MTM-2.

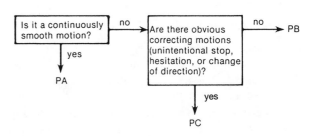

FIGURE 29.5 Decision tree for PUT in MTM-2.

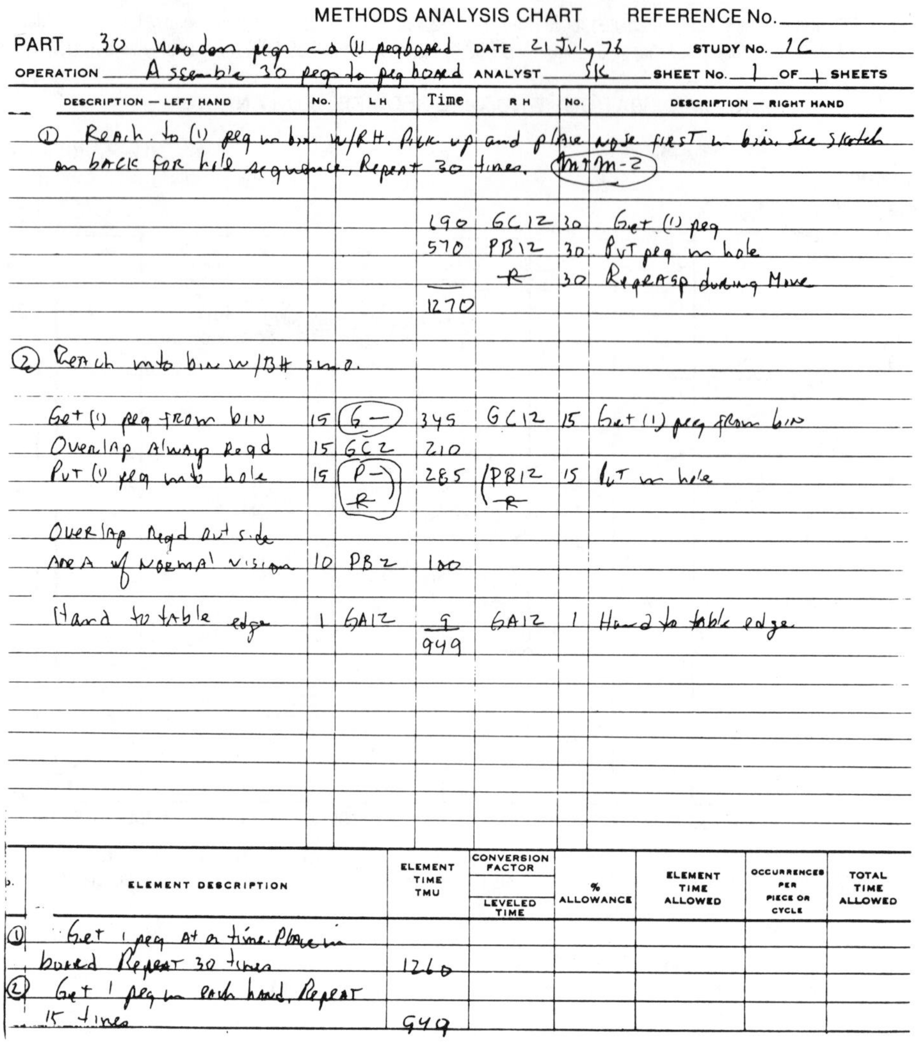

METHODS ANALYSIS CHART REFERENCE No.

PART 30 wooden pegs and (1) pegboard DATE 21 July 76 STUDY No. 1C

OPERATION Assemble 30 pegs to pegboard ANALYST SK SHEET No. 1 OF 1 SHEETS

DESCRIPTION — LEFT HAND	No.	LH	Time	RH	No.	DESCRIPTION — RIGHT HAND
① Reach to (1) peg in bin w/RH. Pick up and place nose first in bin. See sketch on back for hole sequence. Repeat 30 times. (MTM-2)						
			190	GC12	30	Get (1) peg
			570	PB12	30	Put peg in hole
			—	R	30	Regrasp during Move
			1270			
② Reach into bin w/BH simo.						
Get (1) peg from bin	15	G —	345	GC12	15	Get (1) peg from bin
Overlap Always Reqd	15	GC2	210			
Put (1) peg into hole	15	P — R	285	PB12 R	15	Put in hole
Overlap Reqd outside area of normal vision	10	PB2	100			
Hand to table edge	1	GA12	9	GA12	1	Hand to table edge
			949			

ELEMENT DESCRIPTION	ELEMENT TIME TMU	CONVERSION FACTOR / LEVELED TIME	% ALLOWANCE	ELEMENT TIME ALLOWED	OCCURRENCES PER PIECE OR CYCLE	TOTAL TIME ALLOWED
① Get 1 peg at a time. Place in board. Repeat 30 times	1260					
② Get 1 peg in each hand. Repeat 15 times	949					

FIGURE 29.6 MTM-2 analysis of the one-hand and two-hand assembly of the pegboard previously analyzed in Figures 29.1 and 29.2, with MTM-1 showing total time = 1,270 TMU = .76 min and 949 TMU = .57 min.

TABLE 29.15 MTM-3 has only 10 times; analysis time should take about 50 times the cycle time. Times are in TMU.

MTM-3					
RANGE	CODE	HA	HB	TA	TB
Up to 6″	-6	18	34	7	21
Over 6″	-32	34	48	16	29
		SF	18	B	61

Source: Copyrighted by the MTM Association for Standards and Research. No reprint permission without written consent from the MTM Association, 1411 Peterson Avenue, Park Ridge, IL 60068.

of Wofac Company. The figures and tables are reprinted with the permission of Science Management Corp., 721 Route 202-206, Bridgewater, NJ 08807; phone (908) 722-0300.

There are three system levels—Detailed, Ready, and Brief—each of which is based on the original Detailed time data. Each system is self-sufficient, yet each is completely consistent and interchangeable if a mixture is desired.

3.1.1 Detailed Work-Factor® This first-level system provides elemental times for motion study and measurement of mass production (i.e., fingers, hands, arms, legs, feet, trunk, and head). Use

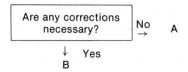

Are any corrections necessary?	No → A

↓ Yes
B

FIGURE 29.7   Decision tree for both HANDLE and TRANSPORT in MTM-3.

when exact work measurement is important. Detailed Work-Factor has 31 elemental descriptions and its motion time tables have 764 time classes.

3.1.2 Detailed Mento-Factors®
This first-level system provides elemental times for identifiable mental processes. Use when precise measurement is required for human mental functions occurring in

inspection (audio, visual, kinesthetic), reading, proof-reading, calculating, color matching, and the like. Its mental process times have 710 time classes.

3.1.3 Ready Work-Factor®
Ready Work-Factor is a second-level system that provides a simplified table of motion times which are averages of times for finger, hand, arm, foot, leg, and trunk motions. Time standards correspond closely to those set with Detailed Work-Factor but are applied more rapidly. Ready requires a shorter training period and thus is useful for training supervisors and employees in work simplification. Ready Work-Factor has 10 elemental descriptions and its motion time tables have 154 time classes.

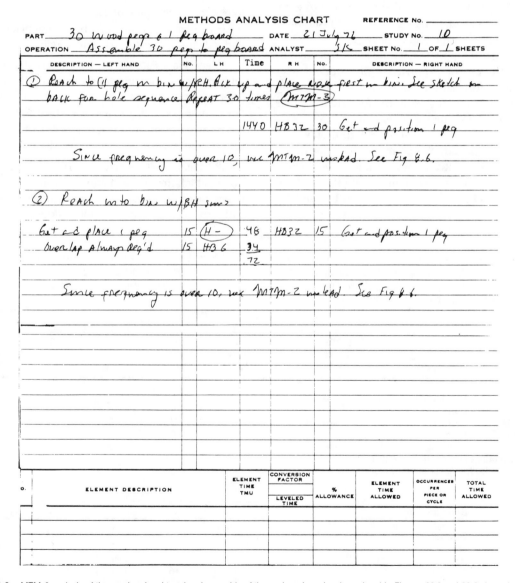

FIGURE 29.8   MTM-3 analysis of the one-hand and two-hand assembly of the pegboard previously analyzed in Figures 29.2 and 29.6 shows how you revert to MTM-2 if the frequency is over 10 (except for SF).

3.1.4 Brief Work-Factor®

This third-level system was introduced in 1975 to replace Abbreviated Work-Factor. It offers the simplest motion timetable, combining Reaches, Grasps, and Moves to form work segments called Pickups. It is applied to situations requiring much less detailed measurement, such as nonrepetitive operations with long cycle times occurring in maintenance, clerical, and indirect labor functions. Brief Work-Factor has 8 elemental descriptions, and its time tables have 32 classes.

3.1.5 WOCOM®

is a computerized version of all the Work-Factor Systems (except Brief, which is so simple to apply that the computer is not useful). WOCOM's modules include the Ready Work-Factor System, the MTM-1 System, Multiple Regression Analysis, Learning Time Allowances, Production Line Balancing, Standard Time Data Development, and Standard Time Data Revision. The concept is to eliminate routine calculations and thus substantially reduce engineering time and improve accuracy.

If you wish to apply the Work-Factor Systems (in contrast to the cursory knowledge of a student who receives an appreciation course), take a training course ranging from 10 class hours for Brief to 80 for Ready to 120 for Detailed Work-Factor.

The remainder of this chapter will deal only with the second level, Ready Work-Factor. Times for Ready Work-Factor are in Ready Work-Factor Time Units (RU); 1 RU = .060 s = .001 min = .000 017 h. Conversely, 1 h = 60,000 RU, 1 min = 1,000 RU, and 1 s = 16.7 RU.

3.2 Ready Work-Factor

There are eight standard elements: Transport, Grasp, Release, Preposition, Assemble, Use, Disassemble, and Mental Process. Walk is classified as a Transport.

3.2.1 Transport

Table 29.16 gives Transport (Reach and Move) times.

For the degree of difficulty, use 1 Work-Factor, if it occurs, for each of the following manual control factors involved in the **Transport** Motion (Definite Stop, Steer, Precaution, or Change Direction).

Definite Stop Work-Factor   This is used for motions which are stopped voluntarily by the operator at or near a specific location (> 2-inch diameter or circle). All reach motions followed by grasp have at least a Definite Stop. All move motions, other than tosses, motions to indefinite locations, or motions suddenly arrested by rigid objects, have at least a Definite Stop. Examples of Definite Stop are reach to a pack of cigarettes, move a bolt to a place on a die surface, and move the hand to touch a large button.

Steer Work-Factor   This is used for motions which steer a body member or object to a target. For tolerances ≤ 2 inches, use Steer plus Definite Stop. For tolerances ≤ ⅝ inches, use Move followed by an Assemble. Examples are move bolt to hole, move to touch small button, and move washer to end of bolt.

Precaution Work-Factor   This is used for motions which must be made carefully. Examples are move razor to face, reach near moving saw blade, and move open container full of liquid.

TABLE 29.16 Transport (Reach and Move) times (RU) for Ready Work-Factor. All distance and weight ranges are over the smaller value and up to and including the larger value. To determine Work-Factors due to weight or resistance (lb), enter lower table on the appropriate row, go right to approximate weight/limb, and then up to read the Work-Factor. −X means over the previous distance, up to and including X; e.g., 20 in the upper table means "more than 10 inches, up to and including 20 inches."

			WORK-FACTORS				
Code	Description	Distance, Inches*	0 Very Easy	1 Easy	2 Average	3 Difficult	4 Very Difficult
4	Very short**	-4	2	3	4	5	6
10	Short	-10	4	5	6	7	8
20	Medium	-20	5	7	9	11	13
30	Long	-30	7	9	11	13	15
40	Very long	-40	9	11	13	15	17

*For trunk motions multiply distance by 2.
**Use this row also for Forearm Swivel motions.

	WORK-FACTORS				
Body member	0	1	2	3	4
Finger-hand	-1	-2	-3	-5	Over 5
Arm	-2	-4	-6	-10	Over 10
Foot	-3	-8	Over 8	—	—
Leg	-5	-16	Over 16	—	—
Trunk	-7	-32	Over 32	—	—

Change Direction Work-Factor   This is used for motions which follow a curved path sharper than a circle. Examples are reach to 15-inch-distant object over 10-inch-high barrier and move casting 20 inches to fixture over the side of a 12-inch-high box side.

In addition to a maximum of 1 WF each for Definite Stop, Steer, Precaution, or Change Direction, Work-Factors can be given for weight. The lower portion of Table 29.16 indicates that a 1-lb weight moved by the hand would be 0 WF but a 1.5-lb weight moved by the hand would be 1 WF. A 5-lb weight moved by the arm would be 2 WF.

Table 29.17 has example Transport Motions and their codes. Typical codes would be 20-0, 30-1, 10-2, and 40-3; that is, the distance, a dash, and the number of Work-Factors.

Now that the degree of difficulty for the motion has been determined (0 to 4 WF), estimate the distance moved or reached. Measurement is not necessary since estimations are within the accuracy of the Ready Work-Factor system. Then read the time from Table 29.16. An "average" Move for a "medium" distance takes 9 RUs or, if you wish, a 2 WF transport for over 10 to including 20 inches takes an average of 9 RUs.

3.2.2 Grasp   Contact Grasps are allowed as part of the preceding Reach and have no time added from the Grasp Table. Table 29.18 gives the time for the other Grasps as well as Release and Preposition. In a Visual Grasp, the operators see their fingers while grasping; in a Blind Grasp, they don't. Single Motion Grasps require only 1 Finger Motion; Pinch Grasps require 0 WF; and Wrap-Around Grasps require 1 WF. Multiple Motion Grasps require more than 1 Finger Motion; that is, 2, 3, or 4

TABLE 29.17   Example Transport motions with their work-factors and times.

Description	Distance Code	Manual Control WF	Weight WF	Code	Time, RU
Push small switch button with finger.	4	0	0	4-0	2
Toss 5 lb bracket 18 inches to truck.	20	0	0	20-0	5
Move arm 15 inches aside to indefinite location after placing part in machine.	20	0	0	20-0	5
Return trunk 11 inches to normal relaxed position. (double distance for trunk)	30	0	0	30-0	4
Reach 24 inches to grasp handle of screwdriver.	30	Definite Stop	0	30-1	9
Reach 15 inches to grasp bolt from bin.	20	Definite Stop	0	20-1	7
Move .25 inch bolt 6 inches to hole in bracket.	10	Definite Stop; Steer	0	10-2	6
Move eraser 6 inches to specific location on typed letter.	10	Definite Stop; Steer	0	10-2	6
Reach 18 inches to grasp piece of wood near revolving circular saw blade.	20	Definite Stop; Precaution	0	20-2	9
Reach 20 inches over side of box 12 inches high to grasp casting.	40	Definite Stop; Change Direction	0	40-2	13
Move 3 lb bolt 36 inches to hole in I-beam.	40	Definite Stop; Steer; Weight	1	40-3	15
Pull machine lever 2 inches against stop (3 lb resistance)	4	0	1	4-1	3
Toss 3 lb object 24 inches into truck.	30	0	1	30-1	9
Toss 8 lb piece of scrap iron 24 inches to scrap pile.	30	0	3	30-3	13
Push heavy carton 15 inches along conveyor with one arm (15 lb resistance).	20	0	4	20-4	13
Using leg, depress pedal 5 inches against 15 lb resistance.	10	0	1	10-1	5

TABLE 29.18   Grasp, Release, and Preposition times (RU) for Ready Work-Factor.

	WORK FACTORS				
No. of finger motions	Single Motion (Simple)		Multiple Motion (manipulative)		
DESCRIPTION	0	1	2	3	4
Element[a]	Very Easy (Pinch)	Easy Wrap-Around Transfer	Average	Difficult	Very Difficult
Simple and Manipulative Grasp—Visual[c]	1[b]	2[b]	3	5	8
Simple and Manipulative Grasp—Blind[c]	1[b]	2[b]	4	6	8
Major Dimension, inches			-.25	-.25	-.25
Diameter, inches			.25	-.25	All
Thickness, inches			.05	-.05	All
Complex Grasp—Visual[d]	1	2	3	5	
Complex Grasp—Blind[d]	1	2	4	6	8
Release[a]	1	2			
Preposition[e]	4	5	6	7	8

[a]Contact grasp and release have zero time.
[b]Multiply time by 2 for weight >3 lb.
[c]Add 2 RU if SIMO manipulative; SIMO does not apply to simple grasp.
[d]Add 1 RU if entangled (e) or nested (n); add 1 RU if slippery (slp); add 2 RU if SIMO.
[e]Multiply RU by percent occurrence to closest 25%; round result to nearest RU.

WF depending on their complexity. Table 29.19 has example Grasps and their codes. The Grasp code is the number of Work-Factors (with x 2 if weight grasped is more than 3 lb), a dash, and symbols for the add-on times, if any.

3.2.3   Release (code = R1)
Contact Release (the opposite of Contact Grasp) requires no time. Single-motion releases require only 1 Finger Motion; Gravity Releases (the opposite of Pinch Grasp) get 1 WF; but Unwrap Releases (the opposite of Wrap-Around Grasp) get 2 WF. Multiple-motion releases require more than 1 Finger Motion. They occur only when the object clings or sticks to the fingers and are analyzed using Transport rules.

3.2.4   Preposition (code = PP)
A **preposition** (turning or orienting an object to a correct position for a subsequent element) that can be completed by a single arm motion or forearm swivel is allowed for in the move and has no time added from the preposition row in Table 29.18. One-hand prepositions are 2 WF if the object's longest dimension is ≤ 3/8 inch, 1 WF if > 10, and 0 WF if in-between. Two-hand prepositions are 3 WF if the longest dimension is ≤ 10 inches and 4 if it is > 10. A very easy

preposition (0 WF) would be a one-hand manipulation of a cylinder between 3/8 and 4 inches long. Easy (1 WF) is for over 4 to 10 inches long). Average (2 WF) would be one-hand manipulation of an object ≤ 3/8 inch long. Difficult (3 WF) would be two-hand manipulation of a cylinder 4 inches long. Very difficult (4 WF) would be two-hand manipulation of a solid over 10 inches long.

If the preposition does not occur every cycle, multiply the RU value by percent occurrence (rounded to nearest 25%); round resulting time to nearest RU.

For example, preposition (turn end for end) a 1-inch-long bolt following grasp in 0 WF. RU = 4. Since this preposition occurs only 50% of the time, allow 4 RUs. Preposition a 1/4-inch-long screw in the fingers following grasp in 2 WF. RU = 6. Since this preposition occurs only 50% of the time, allow 3 RUs. Preposition (turn over) a flat board 9 × 9 × 1/4 inch using both hands. RU = 7. If it occurs every cycle, allow 7 RUs.

The preposition code is the number of Work-Factors and the percent occurrence (nearest 25%).

3.2.5   Assembly-type motions
These are subdivided into assembly, use, disassemble, and SIMO-assemble.

TABLE 29.19 Example Grasp Motions with categories, Work-Factors, and times.

Description	Category	WF	Code	Time, RU
Place hand against carton to push along conveyor.	Contact grasp		Gr-C	0
Place fingers on piece of paper to hold it steady during writing.	Contact grasp		Gr-C	0
Grasp pencil from desk.	Single motion—pinch grasp	0	0	1
Transfer ruler from one hand to the other.	Single motion—transfer grasp	1	1	2
Grasp handle of suitcase.	Single motion—wrap-around grasp	1	1	2
Grasp 1 inch diameter piece of pipe.	Single motion—visual	1	1	2
Grasp 8 lb piece of pipe.	Single motion—wrap-around grasp	1	1-x2	4
Grasp bracket (major dimension .75 inch) from box of brackets.	Multiple motion—blind	2	2-B	4
Grasp flat punched card from table.	Multiple motion—visual	3	3	5
Grasp 2 bolts (.37 diameter, 2 long) SIMO from pile.	Complex blind—SIMO	2	2-Bs	6
Grasp paper clip (1 long, .03 thick) from pile.	Complex visual—entangled	3	3-B	6
Grasp ends of tape stuck to surface.	Multiple motion—visual	4	4	8
Grasp washer (thickness .015 inch; diameter of .22 inch).	Complex visual—oily	4	4-slp	9
Grasp dished, oily washer (thickness .12; diameter .25) from pile.	Complex visual—nested-oily	4	4-nslp	10

Assemble (code = Asy) **Assemble** follows Move and has the subelements of Align (bring plug and target opposite each other), Upright (put plug and target on the same axis), Index (rotate plug and target), Insert (move plug into target or vice versa), and Seat. Assemble has two classes: Mechanical Assemble and Surface Assemble. In Surface Assemble no mechanical features aid correct orientation. Examples are put a book on a desk top, a stamp on a letter, or a nail point on a mark.

Assemble times and adjustments are given in Table 29.20. The move immediately preceding Assemble always contains both Definite Stop and Steering Work-Factors.

Assemble is based on the plug (male-type object) vs. target (female-type object) concept. The plug can be put into the target or the target onto the plug.

For Mechanical Assemble, first determine whether the target is closed or open. Closed targets permit the plug or target to move along only one axis (such as a washer on a bolt or a hammer head onto a hammer handle). Open targets reduce the number of aligns necessary as movement is possible in two axes (such as a bolt into vise jaws, a cylinder into a slot). Second, determine target size, plug size, and plug/target ratio. If the target size varies (as with a beveled hole), use the outside diameter of the bevel. If the plug diameter varies, the mating dimension is which

of the following is greater: (1) actual diameter across the flat or (2) .33 times the plug shank diameter, if the plug shaft is rounded or pointed.

Adjustments are given as a multiplier of the align time. The basic value combines Align, Upright if required, and Insert.

Gripping distance adjustment occurs when the fingertips are more than two inches from the insert point. Between-targets adjustment occurs when two or more plugs and targets are being assembled simultaneously. A temporary blind adjustment occurs when the target or plug or both are in the operator's view during the move preceding assemble but not during the aligns that are a part of assemble. In a permanent blind assemble, the view is blocked during both the move and the assemble.

The assemble code gives, on one line, close target (CT) or open target (OT), the row used, and the column used (e.g., CT − < ⅜ − > .9). Align, Add-ons, Index, SIMO, and Seat are each on separate lines. Table 29.21 shows example calculations.

Use (code = Use) The act of using machines, equipment, instructions, tools, or devices is called **Use.** In fact, a body member itself may be a tool (mark with a finger).

Operator-controlled motions (tighten screw with screwdriver, drive nail, cut with scissors, paint with

TABLE 29.20   Assemble time (RU) and adjustments for Ready Work-Factor. The table gives base time (total assemble time) followed by the align time in parentheses. If there is an align time, align time may require adjustments. If appropriate, multiply align time by factor in adjustment table. If SIMO, add .5 (align time + align adjustment time). Then, if appropriate, add times for Index. Then, if appropriate, multiply total time by factor in weight table. Rounding is done at each step of the calculation as shown in Table 29.21.

		MECHANICAL ASSEMBLE						SURFACE ASSEMBLE	
		Open Target			Closed Target			Open Target	Closed Target
PLUG/TARGET RATIO		−.4	−.9	> .9	−.4	−.9	> .9		
Target dimension, inches	up to ⅛	6(4)	6(4)	10(4)	9(7)	9(7)	13(7)	8(5)	12(9)
	⅛ to ⅜	3(1)	4(2)	8(2)	5(3)	6(4)	10(4)	5(2)	6(3)
	Over ⅜	2(0)	3(1)	7(1)	2(0)	3(1)	7(1)	3(0)	3(0)

Align adjustment multiplier:

	Distance, inches						
	-1	-2	-3	-5	-7	-15	> 15
Gripping distance	—	—	.1	.2	.3	.3	.7
Distance between	—	.2	.3	.5	.7	*	*
Temporary blind	—	.2	.3	.5	.7	1.5	—
Permanent blind	.3	.5	.7	1.5	2.5	5.0	—

*Consider as 2 assemblies; > 15 inches requires MP of 5 RU between the 2 assemblies.
SIMO adjustment to align: .5 (Align time + Align adjustment time)
Index   Mechanical 3 RU   Surface 4 RU
Weight adjustment:

	Weight, lb				
	-2	-4	-6	-10	> 10
Weight multiplier	1.0	1.3	1.5	1.5	2.0

TABLE 29.21   Examples of Mechanical Assemble and Surface Assemble in Ready Work-Factor.

Description	Target Plug Ratio	Target	Base	Times, RU Align Adjustment	Index	Total
Mechanical Assemble						
Assemble .25 inch diameter wooden dowels to .5 inch hole in board, one at a time.	.5	Closed	3			3
Assemble .49 inch diameter steel shaft to bearing with .50 inch diameter.	.98	Closed	7			7
Assemble end of .06 inch diameter shaft into .25 inch diameter hole.	.24	Closed	5			5
Assemble washer with .5 inch diameter hole over end of .25 inch diameter bolt.	.50	Closed	3			3
Same as previous example except 2 assemblies, one in each hand, are made simultaneously, assemblies 6 inches apart.	.50	Closed	3	2		5*
Assemble ¼ × ¼ × ½ inch piece of steel with square cross section in open jaws (.31 open target) of vise.	.80	Open	4		3	7
Surface Assemble						
Move .5 inch diameter metal disc and place on flat surface within a .56 inch diameter circle (target = .06 inch).	—	Closed	12			12
Place 15 inch ruler at 2 points 6 inches apart, for drawing straight line.	—	Open	8	4	4	16**

*Distance between targets = .7 × 1 = .7 = 1. SIMO factor (see section on SIMO-Process "Times") = .5 × 2 = 1.
**For 6 inch distance between targets the factor is .7; 5 × .7 = 3.5 = 4.

brush) have their times determined from the appropriate tables of Transport, Grasp, or others.

Process-controlled motions (pour liquid, pull caulking-gun trigger) are determined from time study or mathematical calculation.

Machine or process time (tool cutting time) also is determined from time study or mathematical calculation.

Disassemble (code = Dsy)   Analyze **Disassemble** (take apart) from the Transport Table.

SIMO-Disassemble   Add SIMO time when assemble, preposition, or multiple-motion grasp occurs entirely or partially simultaneously—one in each hand. All other elements can be done without SIMO time. See Table 29.22.

For assemble, add 50% of align time. For preposition, add 50% of total preposition time. For multiple-motion grasp, add 2 RU.

3.2.6 Mental processes   These **mental processes** include Focus, Inspect, React, and Mento.

Focus (2 RU) (code = Fo)   Focus is given each time the eyes shift to a new location. At distances of 12–18 inches, 1 Focus is given for each "inspection unit" (a 3-inch square) inspected.

Inspect Interval (3 RU/Interval) (code = I)   This is given following Focus. One Inspect Interval occurs, at a specified viewing distance, and under specified lighting, for the following:

1. The presence and identity of or the absence of one or any group of visible and distinguishable

characters, objects, symbols, or characters (referred to as Work-Factor Inspection Characters) is determined. A visible and distinguishable Character subtends a visual angle of not less than 1 min. (At "normal fix distance," a character dimension subtending 1 min = .004 400 inches.) Under perfect visual conditions, 1 min of arc can be seen with 20/20 vision. For practical applications, however, minimum dimension characters are not sufficient (see Chapter 20 for additional discussion).

2. The presence or absence of exactly 1, 2, 3, or 4 Inspection Characters is determined.

3. Up to 3 digits of a number are recognized.

4. Words $\leq$ 6 letters are recognized, provided the word is common to the reader's language. For familiar words of 7 or 8 letters, give 2 Inspect Intervals. For familiar words of $\geq$ 9 letters or for unfamiliar words, give 1 Inspect Interval for each 3 letters.

5. Even if there are < 3 items in a group, give 1 Inspect Interval for each group of numbers, letters, or symbols.

6. When an Inspect Unit is inspected for the presence or absence of a specified Inspection Character, give 1 Inspect Interval for the initial inspection and 1 Inspect Interval for each Character (such as a scratch, spot, stain, or scar) that could be mistaken for the specified Character.

React   The two categories of React are Anticipated Simple React (2-choice decision in which the signal is anticipated) for which 0 time is allowed; Unanticipated Simple React (2-choice decision in which the signal is unanticipated) for which 2 RU are allowed; Anticipated Choice React for which 2 RU are allowed if there are 3 to 5 choices and 4 RU if there are more than 5 alternative choices; and Unanticipated Choice React (select from multiple alternatives) for which 2 RU is allowed (8 RU maximum) for each 2 alternate choices, from which the operator must select the correct one.

Mento (1 RU) (code = Mt)   This category is subdivided into Memorize, Compute, and Recall. Memorize (2 RU) is the Work-Factor term for the act of fixing a number (word, idea) in the mind so it is available for use by the operator a few seconds or minutes later. It is not comparable to memorizing a poem, speech, or table of numbers. Memorize occurs (1) immediately following Compute or (2) when memorized information must be retained in instant readiness to be used after an intervening Manual Motion or a Mental Process (other than an Inspect Interval). Compute (2 RU) is the Work-Factor term for adding, subtracting, multiplying, or

TABLE 29.22   SIMO-Factor, Mental Process, and Walk times (RU) for Ready Work-Factor.

Element	Times, RU
SIMO-Factor:	
Multiple-motion grasp	2
Assembly	.5 (Align time)
Preposition	.5 (Preposition time)
Mental Process:	
Focus	2
Inspect	3/Inspection
React	2
Mento	1
Walk:	
Normal	12 + 8 (Number of paces)
Restricted	12 + 10 (Number of paces)
Up and down steps:	
General	10 (Number of steps)
Restricted	13 (Number of steps)
Stand up	13
Sit down	9

dividing 1-digit numbers. For numbers with > 2 digits, Compute is performed in steps using 1 digit of each number per step. Recall (2 RU) is the Work-Factor term for recalling previously memorized information. Recall is given if a motion or Mental Process other than one Inspect Interval intervenes between Memorize and the use of the memorized information.

3.2.7 Walk   See Table 29.22. A pace is 30 inches. Multiply the number of paces by 8 RU/pace and add 12 RU for start plus stop; speed is 3.7 mph. For restricted travel (restricted space, slippery floors, heavy loads) give 10 RU/pace; speed is 2.96 mph. See Box 11.2.

Figure 29.9 gives a Ready Work-Factor analysis of the 1-hand pegboard assembly; Figure 29.10 gives a Ready Work-Factor analysis of the 2-hand pegboard assembly.

Note that the analyst worked smart instead of hard by writing "R to bin from table edge" instead of "Reach to bin from table edge" or "Reach to the bin from the edge of the table."

4 COMMENTS ON PREDETERMINED TIME SYSTEMS

In theory, PTS, when applied by a trained analyst, can accurately predict the amount of time for a task. However, there is extensive evidence that the theory and reality don't agree. If the reader is interested in the extensive literature on the subject, see Frederick (1960), Schmidtke and Stier (1961), Bailey (1961), Sellie (1961), Taggart (1961), Davidson (1962), Schmidtke and Stier (1963), and Sanfleber (1967). For data on angle of movement (a variable neglected by the PTS), see Konz, Jeans, and Rathore (1969); for data on weight allowances during moves see Konz and Rode (1972). An impartial observer must conclude that the PTS cannot always accurately predict the time a worker will take for a task.

So? Do you throw away your bowling ball because it gives strikes only 7 times in 10?

In this imperfect world we must use the available tools—even though not perfect—until better tools come along.

WORK-FACTOR TWO-HAND ANALYSIS FORM

Part Name Pegboard - 30 Pegs			Sheet No. 1 of 1	Company		Department	Part No.		Sub.	Oper. No.

Operation Name & Description Asy pegs to board - 1 hand

No	LEFT HAND Elemental Description	Analysis	Time Units	Cumulative Time	Time Units	RIGHT HAND Analysis	Elemental Description	No
1							R to 1st peg from edge of	1
2				7	7	20-1	table	2
3				10	3	2-	Gr 1st peg	3
4				12	2	0-50%	P P peg	4
5				18	6	10-2	M peg to hole	5
6							Asy beveled end of	6
7				23	5	CT-.4-3/8	peg to hole	7
8				24	1	0-	Rl peg	8
9				29	5	10-1	R to 2nd peg	9
10				46	17	E1 3-8	Gr and Asy 2nd peg	10
11				442	396	18 x EL 9-10	PU and Asy 18 more pegs	11
12				447	5	10-1	R to 21st peg	12
13				448	1	0-	Gr peg (isolated)	13
14							M peg to board -	14
15				454	6	10-2	PP internal	15
16				459	5	CT-.4-3/8	Asy peg to hole	16
17	↓			460	1	0-	Rl peg	17
18	Hold Board	BD	622	622	622	162	9 x EL 12 - 17	PU and Asy 9 more pegs	18
19								19
20								20
21								21
22								22
23								23
24								24

wofac	Date 1 July 76	Analyst E. Boepple	Total 622	Time in Minutes .622	Multiplier		13.5/2/ (69/1)

FIGURE 29.9   Ready Work-Factor analysis of one-hand pegboard assembly which was analyzed by MTM-1 in Figure 29.1. Total time = 622 RU = .62 min.

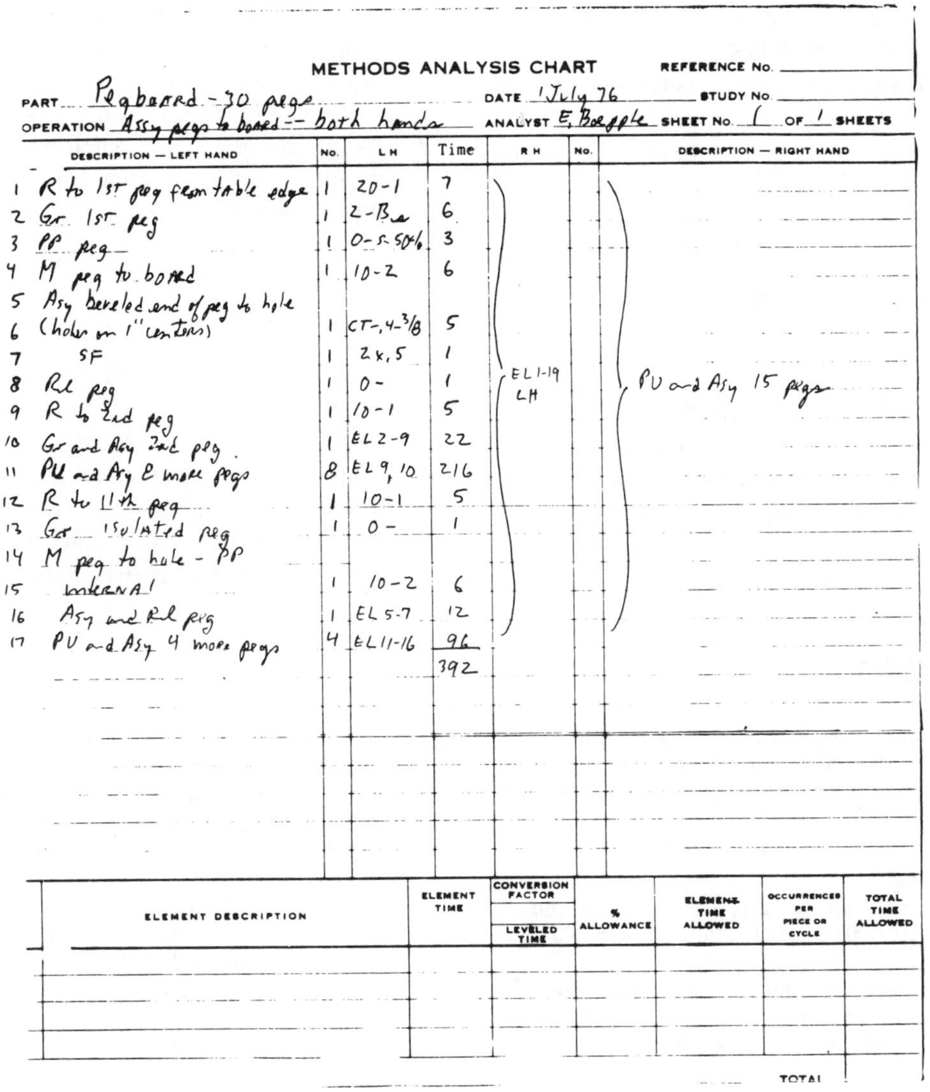

METHODS ANALYSIS CHART

PART Pegboard - 30 pegs   REFERENCE No.
DATE 1 July 76   STUDY No.
OPERATION Assy pegs to board -- both hands   ANALYST E. Boepple   SHEET No. 1 OF 1 SHEETS

	DESCRIPTION — LEFT HAND	No.	L H	Time	R H	No.	DESCRIPTION — RIGHT HAND
1	R to 1st peg from table edge	1	20-1	7			
2	Gr 1st peg	1	2-Ba	6			
3	PP peg	1	0-S 50%	3			
4	M peg to board	1	10-2	6			
5	Asy beveled end of peg to hole						
6	(holes on 1" centers)	1	CT-,4-³⁄₈	5			
7	SF	1	2×,5	1			
8	Rl peg	1	0-	1	EL 1-19		PU and Asy 15 pegs
9	R to 2nd peg	1	10-1	5	LH		
10	Gr and Asy 2nd peg	1	EL 2-9	22			
11	PU and Asy 8 more pegs	8	EL 9,10	216			
12	R to 11th peg	1	10-1	5			
13	Gr isolated peg	1	0-	1			
14	M peg to hole - PP						
15	internal	1	10-2	6			
16	Asy and Rl peg	1	EL 5-7	12			
17	PU and Asy 4 more pegs	4	EL 11-16	96			
				392			

ELEMENT DESCRIPTION	ELEMENT TIME	CONVERSION FACTOR / LEVELED TIME	% ALLOWANCE	ELEMENT TIME ALLOWED	OCCURRENCES PER PIECE OR CYCLE	TOTAL TIME ALLOWED
						TOTAL

FIGURE 29.10   Ready Work-Factor analysis of the two-hand assembly analyzed with MTM-1 in Figure 29.2, with MTM-2 in Figure 29.6, and with MTM-3 in Figure 29.8. Total time = 392 RU = .39 min.

One problem with the predetermined time systems is that they are not automatic, that is, analyst judgment is required. Different analysts get different times for the same job because of different interpretations of the various rules. Very detailed rules have been found to be feasible if applied by a computer, but the programming is very complex. At least 2 computer systems for applying MTM-1 and 1 for applying MTM-2 are in worldwide use.

The popularity of the quick and dirty systems such as MTM-2 and Ready Work-Factor suggest that most managements do not need a great deal of accuracy and are quite concerned with analysis cost. The point here is: "What is the purpose of making a PTS study?" The purpose is (1) to make a methods analysis of the job to determine an **efficient work method** and (2) to determine the amount of time necessary to do the job.

PTSs force the analyst to consider use of both hands, whether a grasp is simple or complex, whether the distance moved is 6 or 9 inches, and so on. If a PTS system analysis results in an efficient work method, then it has accomplished its most important task.

The second and less important task is to assign a time to the efficient work method. Time per unit is necessary for several purposes:

1. *Cost accounting:* How much should we charge for the unit?

2. *Scheduling:* If we want to be done by the 15th, when should we start?

3. *Evaluation of alternatives:* Should the job be done with one hand or two?

4. *Acceptable day's work:* How many should Joe do in a week?

5. *Pay-by-results:* What should Betty Jo's pay be today?

In most of these applications, a 10% deviation from the estimated time is not critical. This especially is true if the error is a consistent error. That is, if a PTS time system consistently estimates times that are 7–12% shorter than an organization achieves, the supervisor can adjust without much difficulty by just adding 10% to the standard time.

REVIEW QUESTIONS

1. What are the 2 purposes of making a PTS analysis?

2. What is the pace for MTM? for Work-Factor?

3. At which number of cycles is a worker predicted to achieve MTM time? Work-Factor time?

4. In MTM-1, what is the difference between a Reach and a Move?

5. In MTM some motions may or may not be simultaneous. How is the decision made whether the motion is simultaneous or not?

6. Designers now specify Phillips-head screws instead of slot-head screws for use with automatic screwdrivers. What positioning motions are eliminated with the Phillips-head screws?

REFERENCES

Bailey, G. Comments on an experimental evaluation of the validity of predetermined time systems. *J. of Industrial Engineering,* Vol. 12, 328–30, September–October 1961.

Barnes R. *Motion and Time Study.* New York: Wiley & Sons, 1980.

Boepple, E. Coordinator of Research and Development of Work-Factor; personal communication, May 17, 1977.

Davidson, J. On Balance—The validity of predetermined elemental time systems. *J. of Industrial Engineering,* Vol. 13, 162–65, May–June 1962.

Frederick, C. On obtaining consistency in application of predetermined time systems. *J. of Industrial Engineering,* Vol. 11 [1], 18–19, January–February 1960.

Karger, D. and Bayha, F. *Engineered Work Measurement,* 4th ed. New York: Industrial Press, 1987.

Knott, K. and Sury R. An investigation into the minimum cycle time restrictions of MTM-2 and MTM-3. *IIE Transactions,* 380–91, December 1986.

Konz, S., Jeans, C., and Rathore, R. Arm motions in the horizontal plane. *AIIE Transactions,* Vol. 1, 359–70, December 1969.

Konz, S. and Rode, V. The control effect of small weights on hand–arm movements in the horizontal plane. *AIIE Transactions,* Vol 4, 228–33, September 1972.

Magnusson, K. The development of MTM-2, MTM-V, MTM-3. *J. of Methods-Time Measurement,* Vol. 17, 11–23, February 1972.

Maynard, H., Stegemerten, G., and Schwab, J. *Methods-Time Measurement.* New York: McGraw-Hill, 1948.

(See also Maynard, H. (ed.). *Industrial Engineering Handbook,* McGraw-Hill.)

Quick, J., Duncan, J., and Malcolm, J. *Work-Factor Time Standards.* New York: McGraw-Hill, 1962. (See also Maynard, H. (ed.). *Industrial Engineering Handbook,* McGraw-Hill.)

Rivett, H. Learning curve prediction development using MTM and the computer. *MTM Journal,* Vol. 1, 32–42, February 1972.

Sanfleber, J. An investigation into some aspects of the accuracy of predetermined motion time systems. *International J. of Production Research,* Vol. 6 [1], 25–45, 1967.

Schmidtke, H. and Stier, F. An experimental evaluation of the validity of predetermined elemental time systems. *J. of Industrial Engineering,* Vol. 12, 192–204, May–June 1961.

Schmidtke, H. and Stier, F. Response to the comments. *J. of Industrial Engineering,* Vol. 14, 119–24, May–June 1963.

Sellie, C. Comments on an experimental evaluation of the validity of predetermined elemental time systems. *J. of Industrial Engineering,* Vol. 12, 330–33, September–October 1961.

Sellie, C. Predetermined motion-time systems and the development and use of standard data. In *Handbook of Industrial Engineering,* 2nd ed., Salvendy, G. (ed.). New York: Wiley, 1992.

Taggart, J. Comments on an experimental evaluation. *J. of Industrial Engineering,* Vol. 12, 422–27, November–December 1961.

Work-Factor. *Work-Factor Learning Allowances,* Ref. 1.1.2, January 1969.

OVERVIEW

Once the database of standard data has been developed, use it for cost savings and consistency and because it can be done ahead of production. Data elements can be either at the motion or element level of detail. Variable elements allocate time more accurately than constant elements. Curve fitting has been simplified by having a computer do the calculations, but the analyst must select the proper equation from those investigated.

CHAPTER CONTENTS

1 Reasons for Standard Data
2 Standard Data Structure
3 Curve Fitting

KEY CONCEPTS

advantages of standard data
coefficient of correlation
coefficient of determination
coefficient of variation
constant/variable elements
database
exponential curve

family of curves
hyperbola
imagine the task
least-squares equation
motion/element
parabola
polynomial

random error/constant error
residual
SAS
standard error
straight line

1 REASONS FOR STANDARD DATA

Rather than measure new times for each and every new operation, consider reuse of previous times—use of standard data. Standard data have three advantages and two disadvantages.

1.1 Advantages
The 3 **advantages of standard data** are cost, consistency, and ahead of production.

1.1.1 Cost
It is cheaper to look up a number in a table or solve an equation than to make a time study to determine the same number. For example, assume you wished to know the time to reach 300 mm to a washer in a bin, grasp it, and move it to assembly. You could use the MTM or Work-Factor tables (see Chapter 29). Or you could set up a workstation and operator to do the task, have the operator do it while being timed, analyze the data, and determine the time. Or assume you wished to know the amount of time required to load pallets into a truck which holds 24 pallets and you presently have available data for 16, 20, and 28 pallets/truck. You could solve an equation and estimate the time for 24 pallets. Or you could have operators load, for example, 15 trucks while you recorded the time. In both examples, it is much cheaper to do the calculations or look the figures up in a table than to obtain new data.

1.1.2 Consistency (fairness)
Time values obtained from standard data are more consistent than time values obtained from one study since they come from a bigger database. In statistical terms, $\sigma'_{\bar{x}} = \sigma'_{x}/N$. That is, the mean standard deviation of the average of multiple studies (represented by $\sigma'_{\bar{x}}$) is smaller than the mean of the standard deviation of N individual time studies (represented by the σ'_{x}). In any 1 time study, there is individual variability (hopefully corrected by rating) and thus possible error. Random errors will tend to cancel over many studies. Using standard data minimizes the random error due to individual operators or individual ratings.

Although it may seem to be heresy, it is more important in time standards to be consistent (small **random error**) than to be accurate (small **constant error**). (Error equals actual time minus standard time). It's nice of course to be both accurate and consistent. See Figure 30.1 for a sketch of random and constant errors. For example, assume actual time for both job A and job B is 1.0 min. A consistent system might set time for A at .90 min and B at .91, or time for A at 1.05 and B at 1.05; this system has low random error and high constant error. A random but accurate system might set time for A at .9 and B at 1.1 or B at .96 and A at 1.04; this system has a high random error and a low constant error. Best of all would be a system which sets time for A at 1.0 and time for B at 1.0; this system would have low random error and low constant error.

Constant errors are much easier to correct or handle administratively. For example, assume for the

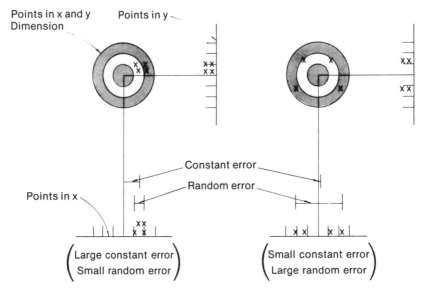

FIGURE 30.1   Random and constant errors are shown on a 2-dimensional bull's-eye (Error = Bull's-eye − Hit location). The lower figures on the side and bottom with the Xs show error in 1 dimension. Constant errors can be corrected by calibration. Random errors can be corrected by improved techniques, better devices, and by training the observer. Random error also can be compensated for by increasing sample size. That is, if you "fire enough shots," you can "hit the bull's-eye," since random error approaches zero as *n* approaches infinity.

whole department that the standard averaged .90 min and performance averaged 1.0 min. Then department efficiency is 90%. The supervisor can make simple adjustments of the standards when doing schedules and evaluating employees, the cost accounting department can easily estimate product cost, and so on. However, if the standards were in error randomly, scheduling would be difficult, costs would be inconsistent, and other problems would surface. An important consequence is that everyone would lose faith in the time standard system since it gives incorrect answers.

1.1.3   Ahead of production

The third advantage of standard data is that the standard can be determined ahead of production. Timing an operation with a stopwatch requires that the operation be observed. There must be a workstation with tools and equipment, parts to be worked on, and an experienced operator. But times often are needed for cost estimates on bids, for determining which method to use for scheduling of machines and operators—that is, before production.

1.2   Disadvantages

But among the three roses are two thorns: imagining the task and database cost.

1.2.1   Imagining the task

To visualize the task when doing a time study, the analyst simply goes to the workstation and watches the operator; the items worked can be "held in the hand." In addition, the operator can be consulted for information. In contrast, with standard data the analyst must **imagine the task,** including each and every element to be done. There is no operator to consult, there are no items to examine, or quality levels to compare, and there is no workstation to examine. Thus, the analyst must be quite familiar with the task to be done. Even experienced analysts occasionally forget to include elements that are done rarely (e.g., machine maintenance elements, bin replenishment elements).

1.2.2   Database cost

As pointed out above, it is better to reuse data than continually have to develop new values. The problem is that you may not have the **database.**

If we are dealing with a database at the motion level (that is, use of MTM or Work-Factor), the firm needs to invest money in having its staff trained in the specific predetermined time system.

If we are dealing with a database at the element level (that is, elements such as "Load milling machine, using jib crane = .05 h" or "Load truck with 24 pallets = .33 h"), the firm needs to invest money in developing the database (either from the micro systems or from time studies), training staff in use of the system, and maintaining the information (typically in a computer and supporting notebooks). Developing the database requires both money and time; if your firm didn't do it in the past, then you will not have a database to save the firm money in the future.

2   STANDARD DATA STRUCTURE

Two decisions that must be made about the database are whether to use motions or elements and whether to use constant or variable elements.

2.1   Motions vs. Elements

The database can be built at various levels of detail. Using an analogy, the goal is to build a "house." The components may be at the micro (**motion**) level (e.g., boards and nails) or at the macro (**element**) level (panels, doors, cabinets). See Table 30.1.

MTM and Work-Factor are examples of times at the motion level. Here each component time is very small—.5 or even .1 s.

In an element system, there is a collection of individual motions and the elements might be "Transfer box from conveyer to pallet" or "Load fixture and clamp." Here the component time is much larger—10, 100, or even 1,000 s. The elements can come from a buildup of motion times by an analyst, from time studies, from curve fitting, or from a combination. Using the house analogy, it is easier to build the house from larger components than from individual boards and nails. However, in the house you must have the proper size and shape component or it won't work. In the same way, an element system is fine if the element is appropriate to the job; if not you will either have to accept "sloppy construction" (errors in standard times compared to what worker actually takes) or build it from scratch (with time study or predetermined times).

2.2   Constant vs. Variable

The element in the database can be considered either a constant element or a variable element. See Table 30.1. **Constant elements** occur or don't occur; the time allowed is either "A" or 0. A constant element for an order picker in a warehouse might be "Drive truck from loading dock to dispatch station." If the operator did the element, a time of .01 h might be assigned. Constant elements are easy to apply but tend to have large random error. That is, they are reasonable, on the average, in the long run but are inaccurate for specific operators doing specific cycles.

TABLE 30.1   *Levels of detail for standard time systems.*

Motions (Typical time range from .1 to 1 s)

Element	Code	Time, TMU	
Reach	R10C	12.9	
Grasp	G4B	9.1	1 s = 27.8 TMU
Move	M10B	12.2	
Position	P1SE	5.6	1 TMU = .036 s
Release etc.	RL1	2.0	

Constant elements (Typical time range 1 s to 1000 s)

Element	Time, s
Assemble bracket to unit	8
Get equipment	90
Polish a shoe	130
Put equipment away	47
Load carton on pallet	15
Pack box	

Variable elements (Typical time range 1 s to 1000 s)

Assemble bracket	Time = f (number of holes, type of fastener)
Pack box	Time = f (type of component packed, number packed)
Walk	Time = f (distance traveled, load carried)
Repair TV set	Time = f (mfg. of set, type of defect)

The element "Drive truck from loading dock to dispatch station" also might be made a **variable element.** The time might vary depending upon the type of truck used and from which dock the driver is coming to dispatching, but the random error would be smaller. The constant error also would tend to be smaller as real applications (1) probably would not have exactly the same proportional use of truck J and truck K as the constant element assumes and (2) probably would not have the same average distance traveled, as the constant element assumes. However, time is more difficult to calculate.

Calculation difficulty will not be as large a problem as it was previously, however, if the calculations are done by computer. For example, the computer may look up in its memory that operator Joe is assigned truck J. The computer may also look up in its memory that the previous order picked was delivered to dock 7. Since it knows the location of dock 7, for the following order it obtains the distance traveled (say 90 m) and calculates an appropriate travel time (giving due allowance for acceleration and deceleration time).

When making a variable element, try to use variables that are easy to count (pages of drawings, meters of weld wire used, number of pieces made, tons melted). Try to use a variable that is known ahead of production. Consider dimensional categories (part, length, thickness), process categories (machining rev/min, distance from pan to machine), and material properties (tensile strength, difficulty).

Next you must see if the actual data vary with your variable. That is, you may assume walking time is a function of distance traveled and load carried.

You need to plot your data vs. the variable and determine the equation. See the following section on curve fitting. Naturally it is absolutely essential that the methods description on the time study include sufficient detail (length of cut, size of box, weight in box, etc.).

You may consider stepwise multiple regression. In this procedure, you give identifiers for all the "X" variables associated with a specific time. For example, a time of .1 h may be identified with a distance variable of 10 m, a weight variable of 5 kg, a temperature variable of 35 C, a container variable of sack, and so forth. You enter all data into the computer and it attempts to fit a straight line for each variable; it identifies the variables that are promising for you to analyze further (perhaps with a different curve shape or family of curves).

If specific data points are not well described by the equation, you must use judgement on whether the data point is an outlier and an aberration or whether your predictive model lacks some variables.

All formulas should be valid only within specified ranges since extrapolation can lead to absurd numbers.

2.3   Developing the Standard

A series of steps need to be taken to convert motions (micro level) or elements (macro level) into the task standard (the house in the analogy). Keep elements in normal time; add allowances only after all elements have been totaled into a task.

2.3.1   Plan the work

First, plan what the standard data system will cover. Warehouse, factory or both? All assembly? All machining operations? Just lathe work? In general, use standard data when jobs/tasks are related and repetitive. In general, set up standard data one area at a time. For example, set up the standard data system for order pickers in a dry grocery warehouse. Then add order pickers in a perishable section of the warehouse or add the fork truck drivers who replenish the racks. That is, build a step at a time.

2.3.2 Classify the data
Code the data with identifiers such as types of machines, types of product. Examples would be to identify elements as lathe work, packing work, grinding work; machining with steel, plastic, or brass; machine shop work, electronics work, assembly work, warehouse work. The problem is very similar to the coding problems of group technology. You are trying to find "relatives" so you can group them into "families." It is very important when making a time study to describe the element and environment in detail so the elements can be classified properly.

2.3.3 Group the elements
Now organize your data. It may be that data fit into only one category (e.g., machine shop) but many elements must be put in several categories (e.g., machine shop, maintenance, material handling). Being able to find a relevant element from several directions makes the database more useful.

2.3.4 Analyze the job
Make a preliminary breakdown of the job into elements. How many of these elements are in your database? You need not have 100% coverage of all the job elements from the standard data. If an element is missing, one possibility is to develop a new standard element for this job (either from a predetermined time or a time study) and then put it into the database after using it for this specific job. A second possibility is to give a blanket allowance for the nonspecified elements. For example, if time for elements 1, 2, and 4 (from the database) is .15 h, you might just add a blanket allowance (say 25% additional time) for elements 3 and 5. A third possibility is to have the entire job done "off standard." That is, there is no standard set for this job.

2.3.5 Develop the standard
Using the selected elements, do the required table lookups and formula calculations. Add allowances. Note that computers can have decision rules built into them in addition to the normal programs permitting easy computation. That is, the program can have diagnosis and error-checking capability. Although these programs are expensive to develop, a number of firms have developed them (Sellie, 1992). For example, the "Micro-Matic Methods and Measurement (4M)" system reduces the time to apply an MTM standard to 25% of manual time. In addition, it calculates a number of indices showing where methods might be improved:

- MAI—percent utilization of both hands
- RMB—percent of reach/move motions time in total cycle
- GRA—cycle time indicating grasp complexity
- POS—cycle time indicating position complexity
- PROC—ratio of waiting-for-process to total time

See Box 30.1 for a brief description of a warehouse standard data system.

3 CURVE FITTING

To analyze experimental data: (1) plot the data, (2) guess several appropriate curve shapes, (3) use a

---

BOX 30.1 *Order picking in dry-grocery warehouses*

Order pickers drive electric pallet jacks (each holding 2 pallets) along the aisles in a warehouse. The computerized picking list (with peel-off labels) indicates how many cases of each product to stack on a pallet. After the list is picked, the load is stretch-wrapped. The pallets then are transported to a truck. Then the picker goes to a dispatch station and gets a new pick list.

Since the standard data system is computerized, an individual time is calculated for each specific order.

After allowing time for obtaining the order, the program calculates the distance to the first pick location and assigns a time for the travel. Then a pick time is assigned depending on the weight of the specific item, type of container, and whether it is on the first- or second-level rack. A weight allowance is calculated from a regression equation, entering the item weight to the closest pound. Then the distance is calculated to the next pick, etc.

computer to determine the constants for the selected shapes, and (4) select which equation you want to use.

You can tell how well the **least-squares equation** (the sum of the Y deviations of the data points from the line squared) fits the data from two different indices: (1) the absolute error in Y (given by the standard error) and (2) the explained error in Y (given by the coefficient of determination, r^2). See Figure 30.2. That is, if you have plotted distance walked in meters as X and time in hours as Y, then the standard error (e.g., SE = .01 h) will give you the standard deviation of the points from the equation in hours. You can convert absolute error into relative error. Divide standard deviation by the mean value of Y; this ratio is called the **coefficient of variation.** In Figure 30.2, an SE of .01 h divided by a mean of .5 h gives a relative error of 6.7%.

The equation can explain anywhere from 0 to 100% of the variability of the data. If it explains 100%, then all the points will fall on the line. The ratio of explained variation/total variation, r^2, is called the **coefficient of determination.** (r is the **coefficient of correlation**.) An r^2 value of .60 means that the equation explained 60% of the total variability.

Although it is possible to fit equations to data by eye, the ease of using computers and preprogrammed calculators makes it worthwhile to use the least-squares programs. A strong advantage of the programs is that for a given set of data and equation form, everyone not only gets exactly the same answer (there is no judgment in the fit) but also they get the best answer. The only judgment required is the trade-off between equation complexity and goodness of fit (indicated by r^2 or SE). Note that with increasing number of terms, r^2 always increases (as the curve can make more bends). However, with increasing number of terms, SE will decrease for a while and then increase as the cost of dividing by a larger and larger N overcomes the benefit of a slightly lower unexplained error. If the equation will be stored in a computer and solved by computer, additional equation complexity is not the disadvantage it was with manual calculations. However, if the equation is so complex you can't put it on a graph, simplify it.

Note also that the program will give you the best fit for the curve shape you specified. If you specified a straight line and the data are best fitted by a parabola, the program will give you a best-fit straight line. As a general policy, have the program print out the actual value of Y, the predicted value of Y, and the difference (also called **residual**). If the residuals have a pattern, consider another equation form. For example, consider 10 points and a fitted straight line. If the deviations are randomly + and −, the line is satisfactory. But if the first 2 deviations and the last 2 deviations are + and the middle 6 are −, then a curve would give a better fit.

3.1   Curve Shapes

Three possible results will be discussed.

3.1.1   Y independent of X

See Figure 30.3. It may be that Y is not related to X. Then the estimate of $Y = A$; that is, Y is a constant. You can determine that Y is not related to X by the value of SE (if SE is too high) or r^2 (if r^2 is too low). From an engineering viewpoint, what is too high or too low will vary with the situation. However, from a statistical

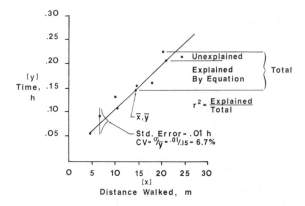

FIGURE 30.2   Goodness of fit of equations is determined by the spread of the points from the line or curve. The absolute error is given by the **standard error**; the relative error is the coefficient of variation; the ratio of explained variability/total variability is the coefficient of determination.

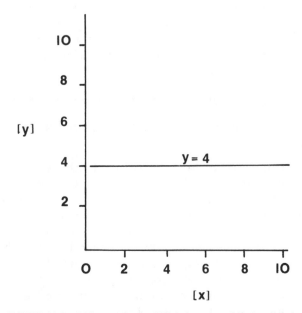

FIGURE 30.3   If Y is not related to X (is independent of X), then $Y = A$, where A is a constant.

viewpoint, the critical value of r^2 can be given for various risk levels. (Risk is defined as the chance that the equation predicting Y from X occurred by chance. That is, there is a 5% chance that the coefficients of X in the equation are really not significantly different from 0.)

3.1.2 Y depends on X, one variable   There are many possibilities.

1. The equation has the form $Y = A + BX$, that is, a straight line. See Figure 30.4. If the value of $A = 0$, the line goes through the origin.

2. The equation has the form $Y = AX^B$; that is, a geometric curve. See Figure 30.5. Geometric curves can be approximated by a straight line if both the X and Y axes are logarithmic. Generally you would do this by using log-log paper and plotting X and Y, but it also could be done by using regular graph paper and plotting the log X vs. log Y.

3. The equation has the form $Y = Ae^{BX}$, that is, an **exponential curve** where $e = 2.178\ldots$, the base of natural logarithms. See Figure 30.6. Exponential curves can be approximated by a straight line with a plot of x vs. log Y. (The equation also can be expressed as $Y = r^x$ where $x = e^B$.)

4. The equation has the form $Y = A + BX^n$. If n is known or suspected, a plot of X^n vs. Y approximates a straight line.

5. The equation has the form $Y = X/(A + BX)$ or $X/Y = A + BX$, that is, a hyperbola with asymptotes $X = -A/B$ and $Y = 1/B$. See Figure 30.7. The hyperbola can be approximated by a straight line with a plot of X vs. X/Y or of $1/X$ vs. $1/Y$.

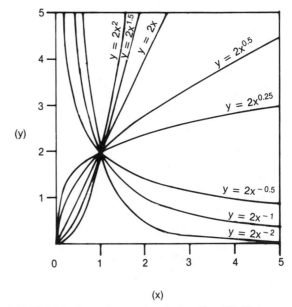

FIGURE 30.5 Geometric curves have the form $Y = AX^B$. All six curves have $A = 2$ but B has different values. The geometric curves with B positive all pass through the points (0, 0) and (1, A), and as one variable increases, so does the other. The curves with B negative all pass through the point (1, A); they have $X = 0$ and $Y = 0$ as asymptotes, and as one variable increases, the other decreases.

6. The data may be best described by three coefficients A, B, and C. To show the **polynomial** relationship:

$$Y = AX^0 = A(1) = A$$
$$Y = AX^0 + BX^1 = A + BX$$
$$Y = AX^0 + BX^1 + CX^2 = A + BX + CX^2$$

When the exponent 2 is positive, the equation is called a **parabola.** When the exponent 2 is negative,

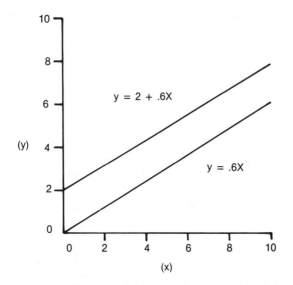

FIGURE 30.4 Straight lines have the form $Y = A + BX$. If $A = 0$, then the line goes through the origin. If B is positive, the line slopes upward; if B is negative, the line slopes downward.

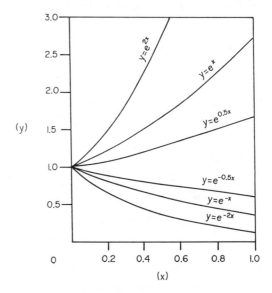

FIGURE 30.6 Exponential (logarithmic) curves have the form $Y = Ae^{BX}$. All six curves drawn have $A = 1$ but B has different values. The curves all pass through the point (0, A) and have $Y = 0$ for an asymptote.

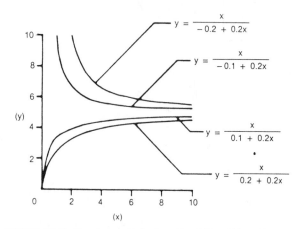

FIGURE 30.7 Hyperbolas with the form $Y = X/(A + BX)$ or $X/Y = A + BX$ have asymptotes $X = A/B$ and $Y = 1/B$. The four curves all have the value of $B = .2$ but A has different values.

the equation is called a **hyperbola.** See Figure 30.8. Each term in the polynomial beyond 2 gives 1 bend to the curve. That is, using coefficients A and B, the equation is a straight line; using coefficients A, B, and C, there is 1 bend to the curve; using coefficients A, B, C, and D, there are 2 bends to the curve; and so on.

In addition, the data may be fitted by 1 equation over part of the range of X and another equation over another part of X. Progress curves (see Chapter 28) often have one straight line (on log–log coordinates) for beginning experience and another straight line for mature experience.

3.1.3 Y depends on X, multiple variables

Another possibility is a **family of curves.** For example, the time of walking vs. distance may be one line for unburdened walking, another line for walking with a load in the hands, and another line for walking while pushing a cart. In mathematical terms: $Y = A + BX + CZ$ where X = distance walked and Z = type of walking. In statistical terms, this is known as multiple regression since there are multiple variables.

3.2 Example Application

Table 30.2 gives example data for walking without and with a load (called "walk" and "carry"). The problem is to fit an equation to both sets. Box 30.2 gives the Statistical Analysis System **(SAS)** program to determine the least-squares equations for a number of different alternatives.

After the selected data sets are read in, the program calls up the desired data set, does desired transformations, and then fits as many equations as desired.

In this example, the walk data set was fitted with the following equations:

$$\text{Walk time, h} = .0054 + .01 \text{ (Distance, m)}$$
$$r^2 = .986 \quad \sigma = .0073 \text{ h}$$

$$\text{Walk time, h} = -.01 + .014 \text{ (Distance, m)}$$
$$- .00013 \text{ (Distance, m)}^2$$
$$r^2 = .989 \quad \sigma = .0067 \text{ h}$$

$$\text{Walk time, h} = -.13 + .11 \text{ (log}_e \text{ Distance, m)}$$
$$r^2 = .966 \quad \sigma = .012 \text{ h}$$

$$1/\text{Walk time, h} = .24 - .96 \text{ (1/Distance, m)}$$
$$r^2 = .881 \quad \sigma = .021 \text{ h}^{-1}$$

The best fit is the parabola with $\sigma = .0067$ h, with the straight line a close second with $\sigma = .0073$ h. The user would have to decide whether the small gain in accuracy with the parabola is worth the additional equation complexity.

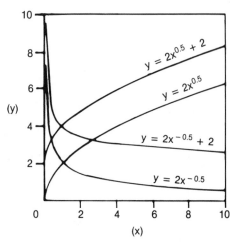

FIGURE 30.8 Parabolas or hyperbolas with a third constant have the form $Y = A + BX^n$. Two of the curves have $A = 0$ and two have $A = 2$. If n is positive, the curve is a parabola with a Y intercept $= A$. If n is negative, the curve is a hyperbola with asymptotes $X = 0$ and $Y = A$.

TABLE 30.2   Walk and carry normal times (min).

| Activity | Distance, m | | | |
	5	10	15	20
Walk	.0553	.1105	.1654	.2205
	.0590	.1170	.1751	.2205
	.0550	.1105	.1660	.2090
	.0521	.1045	.1680	.2200
	.0541	.1080	.1625	.2080
	.0595	.1200	.1800	.1980
Carry	.0691	.1116	.1571	.1984
	.0732	.1158	.1681	.1962
	.0693	.1094	.1660	.1881
	.0651	.1066	.1626	.1958
	.0676	.1080	.1511	.1872
	.0737	.1188	.1674	.1762

BOX 30.2   *SAS curve fitting*

You will need four types of entry lines:

1. job control lines
2. program lines
3. data lines
4. program lines

JCL Lines	Comment
Line 1 //*++REGION 1000 K	
Line 2 //EXEC SAS	
Line 3 //SYSIN DD *	

Program Lines

Line 4	DATA name;	Put your name in the location "name" (e.g., WALK).
Line 5	INPUT DIST TIME;	All program lines in SAS *must* end in a semicolon. The distance value will be entered first and then the time value with a space between entries.
Line 6	DISTSQ = DIST**2;	This creates a new variable, DISTSQ.
Line 7	DISTLOGE = LOG (OF DIST);	New variable is log base e of DIST.
Line 7a	DISTLTEN = LOG10 (OF DIST);	Alternative new variable is log base 10.
Line 8	IDIST = 1/(DIST);	Inverse of DISTANCE.
Line 9	ITIME = 1/(TIME);	Inverse of TIME.
Line 10	CARDS;	

Data Lines

One line per point; do *not* end with semicolon.

Line 11	5	.0553
Line 12	5	.0590
Line 35	20	.1980

Program Lines

Line 36	PROC PRINT DATA=name;	Prints out your data.
Line 37	PROC REG DATA=name;	Uses Regression procedure on your data set.
Line 38	MODEL TIME=DIST/P;	Fits st. line TIME = $A + B$ (DIST).
Line 39	TITLE 'ST.LINE OF name';	Labels output.
Line 40	PROC REG DATA=name;	Runs a new analysis of same data.
Line 41	MODEL TIME = DIST DISTSQ/P;	Fits Time = $A + B$ (DIST) $+ C$ (DIST)^2.
Line 42	TITLE 'PARABOLA OF name';	Labels output.
Line 43	PROC REG DATA=name;	Runs new analysis of same data.
Line 44	MODEL TIME = DISTLOGE/P;	Fits TIME = $A + B$ (log of DIST). You may wish to use alternative of line 7A.
Line 45	TITLE 'EXPONENTIAL OF name';	Labels output.
Line 46	PROC REG DATA = name;	Runs new analysis of same data.
Line 47	MODEL ITIME=IDIST/P;	Fits ITIME = $A + B$ (IDIST).
Line 48	TITLE 'HYPERBOLA OF name';	Labels output.
Line 49	DATA name;	Use CARRY this time.
50-56		Repeat of lines 5 to 10.
57-81		Entries for CARRY.
82-91		Repeat of 36 to 48 with CARRY for name.

The carry data set was fitted with the following equation: Carry time, h = A + B (Distance, m)

The results were:

Carry time, h = .030 + .0082 (Distance, m)
$$r^2 = .975 \quad \sigma = .0077 \text{ h}$$

Carry time, h = .013 + .012 (Distance, m)
$$-.00014 \text{ (Distance, m)}^2$$
$$r^2 = .981 \quad \sigma = .0069 \text{ h}$$

Carry time, h = −.078 + .087 ($\log_e$ Distance, m)
$$r^2 = .960 \quad \sigma = .0098 \text{ h}$$

1/Carry time, h = 2.36 + 60.8/Distance, m
$$r^2 = .981 \quad \sigma = .52 \text{ h}^{-1}$$

The best fits are the parabola and hyperbola. Again the user must decide between accuracy of fit vs. complexity.

If the curves for walk and carry are parallel, the following form could be used:

Time, h = A + B (Distance) + C

where   C = Difference in time between the curves.

For example, C = 0 for walk and C = C_1 for carry.

This also could be formulated as A + B (Distance) with one value of A for walk and another for carry.

REVIEW QUESTIONS

1. Give the 3 advantages and 2 disadvantages of standard data systems.
2. Assuming your true weight was 70 kg and you were weighing yourself on a scale, give an example of random error and systematic error.
3. Give an example of standard data at the motion level and at the element level.
4. For 1 element, give as a constant element and give as a variable element.
5. How is the standard error related to the coefficient of variation?
6. What is the definition of the coefficient of determination?
7. Give the equation (using a consistent format) for (a) Y unrelated to X, (b) Y related to X by a straight line, and (c) Y related to X by a parabola.

REFERENCE

Sellie, C. Predetermined motion-time systems and the development and use of standard data. In *Handbook of Industrial Engineering*, 2nd ed., Salvendy, G. (ed.), Chapter 63. New York: Wiley, 1992.

31 | ALLOWANCES

OVERVIEW

Allowances can be given as a percent of work time or as a percent of shift time. In addition, allowances are related to the definition of normal pace and the consequences of not achieving standard.

Allowances usually are given for personal needs, fatigue, and delay. The allowances given in this chapter are rational and consistent but cannot be considered scientific.

A large allowance indicates a large potential for job redesign.

CHAPTER CONTENTS

1 Definitions
2 Personal Allowances
3 Fatigue Allowances
4 Delay Allowances

KEY CONCEPTS

delay allowances	machine allowances	shift allowances/work allowances
dynamic load	mental fatigue	short cycle
environmental fatigue allowances	pace	standard time/normal time
fatigue allowances	personal allowances	static load
inside work	restrictive clothing	

1   DEFINITIONS

1.1   Shift vs. Work Allowances

To review, a time study analyst observes a person working and records the time taken for a task (observed time)—perhaps .1 h. The analyst also records the "pace" of the worker, using the concept of 100% as normal—perhaps 110%. Then the two are multiplied to determine **normal time,** the time required by a normal worker—.1(1.10) = .11 h.

However, this is the time taken during the direct observation. For an entire day the worker would need some additional time for personal needs (drink water, go to toilet), fatigue (if job has fatigue), and delays (if job has delays). This extra time is called allowances. **Standard time** is normal time after allowances.

Allowances can be added 2 ways because allowances can be defined 2 ways: (1) **work allowances**—allowances as a percent of work time and (2) **shift allowances**—allowances as a percent of shift time. For example, a person might have an allowance of 10% of time worked; then .11 h + .011 = .121/unit. On the other hand, a person might have an allowance of 10% of a shift of 8 h or .8 h. Then .11 h/(1 − .1) = .11/.9 = .122 h/unit.

The basic allowance relationship is as follows:

$$STTIME = NOTIME + ALTIME$$

where

$STTIME$ = Standard time for a task over an entire shift, h/unit

$ALTIME$ = Allowance time, h/unit

But the allowance can be given as a proportion of work or a proportion of the shift:

$$SHFTAL = ALTIME/STTIME$$
$$WORKAL = ALTIME/NOTIME$$

where

$SHFTAL$ = Shift allowance (allowance based on percent of shift time), %

$WORKAL$ = Work allowance (allowance based on percent of time worked), % (British standard 3138 specifies that allowances be given as work allowances.)

Thus:

$$STTIME = NOTIME/(1 − SHFTAL)$$

or

$$STTIME = NOTIME (1 + WORKAL)$$

The allowances given in this chapter are used by some firms as shift allowances and by other firms as work allowances. Note that if the same percent allowance is given as a shift allowance instead of a work allowance, the operator is given slightly more time. In the above example, with a 10% work allowance, standard time was .121 h/unit while with a 10% shift allowance, standard time was .122 h/unit.

Giving the illustration another way, if $SHFTAL$ = 10%, then for a 480-min shift, there would be 48 min of allowances and 480 − 48 = 432 min of work. To give the same amount of time (1 + $WORKAL$) (432) = 480 and $WORKAL$ = 480/432 = 11.1%.

1.2   Allowance vs. Pace

When comparing allowances among organizations, in addition to comparing whether the allowance is a work allowance or shift allowance, more important distinctions are whether an allowance is given for *all* jobs or only *some* jobs (that is, only for cause) and the amount of the allowance.

For example, assume 3 different firms all used shift allowances and all had the same packing task, which had the same measured time of .01 h/unit. The pace was rated as 120% in all 3 firms. But assume that firm A gives all workers 15% for personal allowances, firm B gives all workers 5% for personal allowances, and firm C gives all workers 0% for personal allowances. Assume that A, B, and C all give 10% allowance for task difficulty.

Then, as shown in Table 31.1, standard time would be .0160 h/unit in firm A, .0141 h/unit in firm B, and .0133 h/unit in firm C. In this case, all 3 firms had the same definition of normal **pace** but different allowances, so the standard varied.

The firms also could have the same definition of standard (.016 h) but different allowances (25, 15, 5). Then, as shown in the lower part of Table 31.1, the rating, for the same observed time, would have to be 120% in firm A, 136% in firm B, and 152% in firm C.

Thus, when comparing allowances between jobs and organizations, comparisons need to consider not only the allowances but also the definition of normal pace used.

1.3   Machine Time Allowances

In some situations, workers work with automatic or semiautomatic machines. As shown in Figure 31.1, it may be possible to take some of the personal, fatigue, or delay allowances during the machine time (**machine allowances**). These situations can become quite complicated (especially with multiple machines). Each firm needs to set consistent rules so that people working on manual jobs, semiautomatic machines, and automatic machines are all treated fairly.

TABLE 31.1 Standard time for a task depends on a combination of the definition of normal pace and the base allowance given every task.

	Organization		
	A	B	C
Measured time, h/unit	.0100	.0100	.0100
Rating (if all 3 use the same definition of normal pace), %	120	120	120
Normal time, h/unit	.0120	.0120	.0120
Base allowance, %	15	5	0
Allowance for the same difficult task, %	25	15	10
Standard time, h/unit	.0160	.0141	.0133
Measured time, h/unit	.0100	.0100	.0100
Rating (if all 3 use the same definition of standard time), %	120	136	152
Normal time, h/unit	.0120	.0136	.0152
Base allowance, %	15	5	0
Allowance for the same difficult task, %	25	15	5
Standard time, h/unit	.0160	.0160	.0160

1.4 Who Determines Allowances

The decisions are made at 2 levels: (1) setting up the basic tables and (2) interpreting tables for specific jobs.

Setting up the basic tables for an organization is a political rather than a scientific decision. The technical staff can give advice about the allowance procedures to be used and what type of allowances should be used, but the amounts of the allowances will affect the standard time. Note also that firms can have different concepts of a standard. That is, some firms consider 100% to be a minimum level of performance and a performance level to be achieved by everyone. Others consider 100% as what a typical worker will achieve. Other firms consider 100% as a goal and something that only a few workers will achieve. Thus, the amount of the allowance, the definition of a 100% pace, and the consequences of not making 100% all are political decisions to be worked out by management and labor.

Once these political decisions are made, the technical staff needs to set up a rational, consistent allowance procedure that treats everyone fairly. To do this, allowances usually are broken down into personal allowances, fatigue allowances, and delay allowances.

2 PERSONAL ALLOWANCES

Personal allowances are given for such things as blowing your nose, going to the toilet, getting a drink of water, smoking, and so forth. They do not vary with task but are the same for all tasks in the organization.

There is no scientific or engineering basis for the percent to give.

Lazarus (1968) reported that the mean of 23 industries (235 plants) was 5.6%; with the exception of 10% for the lumber and wood industry and 7.6% for primary metals, all the others were between 4.6% and 6.5%. The International Labor Office (1979) reports common numbers are 5% for males and 7% for females; however, different allowances for men and women no longer are legal in the United States or England.

In many organizations there are standardized break periods (coffee breaks)—for example, 10 min in the first part of the shift and the same in the second part. The question is whether to consider this time as part of the personal allowance time or in addition to it. The lunch period obviously is a time

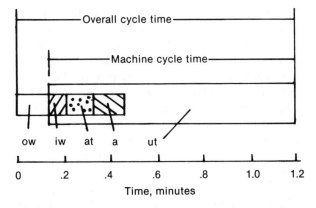

FIGURE 31.1 Machine time allowance terms are shown in the figure of 1 worker and 1 machine. OW = Outside Work = work which must be done outside the machine (process)-controlled time. IW = **Inside Work** = work which can be done within the machine time. UT = Unoccupied Time = operator not engaged in side work, attention time, or taking authorized rest. (UT usually has 0% allowance.) A = Allowances. Personal and environmental fatigue allowances must be taken away from the machine but physical and mental fatigue allowances, in some cases, may be taken at the machine. It is difficult to use incentive pay when the outside work time is a small proportion of the total time. Generally the goal is a 1% increase in pay for a 1% increase in output. But if the machine cycle dominates, even very good performance on the outside work has little effect on the total time.

during which the worker can attend to personal needs and recover from fatigue. Yet it usually is not considered part of the allowance—even if the lunch period is paid. When comparing allowances at different firms, therefore, consider whether coffee and lunch breaks are included in allowances.

Some firms grant an additional break if overtime if worked. For example, if more than 10 h are worked per shift, there is an additional break of 10 min after the 9th hour.

Note that if 20 min are given for breaks in an 8-h shift, this is an allowance of 20/480 = 4.2% (assuming you use shift-time allowances). If, however, the second shift works only 450 min instead of 480, then they get an allowance of 20/450 = 4.4%.

In addition, some firms give special allowances for certain jobs. For example, there may be a cleanup allowance (either to clean up the machine or the person) or a travel allowance. In minework, there may be portal-to-portal pay; this means that pay begins when the worker crosses the mine portal, even though the worker may not arrive at the working surface until some time later.

3   FATIGUE ALLOWANCES

3.1 Background
Fatigue allowances are given for tasks which cause fatigue; they are based on the task, not the operator. Because machines do not fatigue, fatigue allowances are given only for the manual portion of a cycle, not the machine portion. In addition, the operator can rest (at least specific muscle groups) during machine time.

Perhaps because there is little agreement on what fatigue is, there is little agreement on what factors to consider, much less how much allowance to give.

Authors don't consider the same factors. They divide the problem differently (e.g., temperature and humidity in one combined factor or two separate factors), they give the same factor different names (monotony and repetition of cycle), and allowances for the same degree of a factor differ drastically. Although fatigue allowances are supposed to be given only for factors which increase time to do the job, some allowances (such as monotony, odors, noise) just reflect an unpleasant job. My recommendation is to give fatigue allowances only if time/unit is affected. For unpleasant jobs, don't change time/unit but increase the wage rate/hour.

For additional comments on allowances, see the excellent article by Mital et al. (1991).

The four primary authors are Page (1964), Williams (1973), Cornman (1970), and ILO (1979).

Page developed his values while he was at the University of Michigan. A basic allowance of 3% is given for all jobs.

Williams' values are an amalgam of industrial values in England; no evidence is given for their validity although "the results are now in use for the Imperial Group." All his values assume a "standard fatigue allowance of 10%, which includes personal allowances, is given to everyone."

Cornman's values are from an American consulting firm. His procedure is to add allowances from all his factors and then subtract 25%. That is, if the value from the tables was 35%, then the fatigue allowances equal 10%.

The ILO values were supplied by a British consulting firm. Use of the ILO tables is a little complex. Remembering that fatigue allowances are given for work time only (not machine time), sum the applicable fatigue allowance points. Then, using Table 31.2, convert points to percent time. The percent includes a 5% personal time for coffee breaks.

I will group the factors from these authors into 3 categories: physical, mental, and environmental.

3.2 Physical
There are 4 divisions: dynamic load (material handling), short cycle, static load (posture), and restrictive clothing.

3.2.1 Dynamic load
Dynamic load allowances are primarily for manual material handling. See Chapter 17 for a discussion of material handling itself.

Page (see Table 31.3) varies the allowance depending upon both percent of time handled and amount of the load. When allowance is plotted vs. load for each percent of time, the plot gives a family of straight lines with steeper slopes for the greater percentages. Cornman's values (see Table 31.4) are 4-step functions. Williams (see Table 31.5) gives 2 straight lines, with a steeper line for durations over .1 min. The ILO has three curves (see Table 31.6) that are straight lines when plotted vs. percent allowances.

From an application viewpoint, the use of computers has eliminated the need for calculational simplicity implied by Cornman's steps at long intervals, and a weight allowance calculated to the nearest kg certainly is desirable—especially in the critical 0–20 kg range where almost all tasks will fall. It may be difficult to determine the percentage of time the force occurs and this may lead, in practice, to assuming it occurs 100% of the time. The ILO makes a distinction among carrying loads, lifting loads, and force applied, whereas the other systems group ignores these categories; this ILO distinction seems desirable.

TABLE 31.2 Conversion from points allowance to percent allowance for ILO. The second column (0) gives the 10s and the remaining columns give the units. Thus, 30 points (0 column) = 15%; 31 points (1 column) = 16%; 34 points = 17%. The percent allowance is for manual work time (not machine time) and includes 5% personal time for coffee breaks.

Points	0	1	2	3	4	5	6	7	8	9
0	10	10	10	10	10	10	10	11	11	11
10	11	11	11	11	11	12	12	12	12	12
20	13	13	13	13	14	14	14	14	15	15
30	15	16	16	16	17	17	17	18	18	18
40	19	19	20	20	21	21	22	22	23	23
50	24	24	25	26	26	27	27	28	28	29
60	30	30	31	32	32	33	34	34	35	36
70	37	37	38	39	40	40	41	42	43	44
80	45	46	47	48	48	49	50	51	52	53
90	54	55	56	57	58	59	60	61	62	63
100	64	65	66	68	69	70	71	72	73	74
110	75	77	78	79	80	82	83	84	85	87
120	88	89	91	92	93	95	96	97	99	100
130	101	103	105	106	107	109	110	112	113	115
140	116	118	119	121	122	123	125	126	128	130

TABLE 31.3 Fatigue allowances (Page, 1964). Page states "this table is empirical." All the allowances (if relevant) are additive. A personal allowance = 5% and delay (miscellaneous, unspecified, shop loss) = 2% is "reasonable." Every job gets a basic fatigue allowance = 3%. Additional allowances depend on the factor and the percent of time the factor applies.

Factor and Comment		Percent of Time Factor Applies				
		0-19	20-39	40-59	60-79	80-100
Position						
Standing		.5	1.0	1.5	2.0	2.5
Straining		1.0	2.0	3.0	4.0	5.0
Weight (or equivalent)						
lbs (4 lb or more)						
4.0 - 9.9		.5	1.0	1.5	2.0	2.5
10.0 - 19.9		1.0	2.0	3.0	4.0	5.0
20.0 - 39.9		2.0	4.0	6.0	8.0	10.0
40.0 - 59.9		3.0	6.0	9.0	12.0	15.0
60.0 - 79.9		4.0	8.0	12.0	16.0	20.0
80.0 - up		5.0	10.0	15.0	20.0	25.0
Conditions						
Exposure to hazard, vibration,	Poor	0	.5	1.0	1.5	2.0
extremes in temp, noise, etc.	Very poor	.5	1.0	2.0	3.0	4.0
Coordination (Hand and Eye)						
Close application		0	0	0	.5	1.0
Delicate or minute		.5	1.0	1.5	2.5	3.5

Freivalds and Goldberg (1988) comment that the ILO allowances really should consider lifting frequency and lifting height as well as load; the allowances also should not be for loads exceeding the NIOSH Recommended Weight Limit (RWL).

Note that the four allowance tables are trying to combine two different things: local muscle load and whole body (metabolic) load. For analyses treating them separately, see Freivalds and Goldberg (1988), Price (1990a, 1990b) and Mital et al. (1991).

TABLE 31.4 Physical demand allowance (%) (Cornman, 1970). Add 3% if work is done in a difficult work position.

Level of Effect, kg	Time Effort is Applied			
	Up to 15%	15% to 40%	40% to 70%	Over 70%
Up to 2.2	0	0	3	3
2.2 to 11	0	0	3	7
11 to 27	0	3	7	10
Over 27	3	7	10	13

Source: Reprinted with permission from *Industrial Engineering* magazine, Vol. 2 [4], 1970. Copyright, Institute of Industrial Engineers, 25 Technology Park/Atlanta, Norcross, GA 30092.

TABLE 31.5 Weight, force, or pressure allowance (%) (Williams, 1973.)

Weight, Force, or Pressure, kg		Duration	
		Up to .1 min	Over .1 min
0 to	2.5	0	0.3
2.6	5.0	1.1	2.0
5.1	7.5	2.2	3.6
7.6	10.0	3.3	5.7
10.1	12.5	5.2	7.9
12.6	15.0	6.8	10.1
15.1	17.5	8.5	12.3
17.6	20.0	10.1	14.9
20.1	22.5	11.8	17.6
22.6	25.0	13.4	20.4
25.1	30.0	15.9	24.5
30.1	40.0	20.8	31.8

Source: Reprinted with permission from *Industrial Engineering* magazine, December, 1973. Copyright, Institute of Industrial Engineers, 25 Technology Park/Atlanta, Norcross, GA 30092.

Murrell (1965) gave the following formula for metabolic load:

$$PREST = (MET - LTMET)/(MET - RSTMET)$$

where

$PREST$ = Percent rest time

MET = Metabolic rate of the task, W

$LTMET$ = Long-term metabolic rate, W (e.g., 350 W or 5 kcal for males and 4.2 kcal for females, assuming 1/3 of aerobic capacity for each)

$RSTMET$ = Resting metabolic rate, W (e.g., 100 W or 1.5 kcal)

This formula implies no metabolic allowance is needed if $MET < LTMET$. See Mital and Shell (1984) for additional analysis of this formula.

When lifting or carrying loads, observed times for very heavy loads may actually be shorter than for lighter loads because the operator tries to minimize the time under stress. Thus, allowances for very heavy loads need to be considerably higher since they will be applied to a shorter time base.

Note that the allowances say nothing concerning who is doing the task, since allowances are based on the task rather than the person. However, it is obvious that lifting a 5-kg weight is more difficult for a 50-kg person than a 100-kg person. Thus, a small person may feel the allowances are low while a large person may be satisfied.

3.2.2 Short cycle
Short cycle allowances can have a psychological or physiological basis. Table

TABLE 31.6 Carrying, lifting, and body force allowance (ILO, 1979). The ILO tables go to 64 kg. Push includes foot pedal push and carry on the back. Carry includes hand carry and swinging arm movements. Weight is averaged over time. A 15-kg load lifted for 33% of a cycle is 5 kg.

Weight or Force, kg	Push Points	Carry Points	Lift Points
1	0	0	0
2	5	5	10
3	8	9	15
4	10	13	18
5	12	15	21
6	14	17	23
7	15	20	26
8	17	21	29
9	19	24	32
10	20	26	34
11	21	29	37
12	23	31	40
13	25	33	44
14	26	34	46
15	27	36	50
16	28	39	50
17	30	40	53
18	32	42	56
19	33	44	58
20	34	46	60

31.7 gives the effects of discipline and monotony according to Williams. Table 31.8 gives the ILO table, which is justified on the basis of lack of time for muscles to recover.

3.2.3 Static load
Static load (also called body posture and body position) is considered in Table 31.3 (Page), Table 31.9 (Williams), Table 31.10

TABLE 31.7 Short-cycle allowances according to Williams (1973). Discipline considers the need for exact timing or conformance to a machine cycle. Cycle durations are for guidance; allowances need not be given for all cycles of less than .2 min. Monotony considers the absence of mental stimulation due to task tediousness, lack of variation in the work, absence of competitive spirit or companionship, and degree of repetition.

Discipline Demand	Cycle Length, min.	Allowance, %
Negligible	Over .20	0
Low	.16 to .20	0 to 2
Medium	.09 to .15	2 to 3
High	Below .09	3 to 5

Monotony, %	Cycle, min.	Allowance, %
Long cycle, repetitive	Over .10	0
Average repetitive or nonrepetitive		0 to 3
Highly repetitive	Less than .05	3 to 5

Source: Reprinted with permission from *Industrial Engineering* magazine, December, 1973. Copyright, Institute of Industrial Engineers, 25 Technology Park/Atlanta, Norcross, GA 30092.

TABLE 31.8 Short-cycle allowance (ILO, 1979).

Points	Cycle Time, min
1	.16 - .17
2	.15
3	.13 - .14
4	.12
5	.10 - .11
6	.08 - .09
7	.07
8	.06
9	.05
10	Less than .05

(Cornman) and Table 31.11 (ILO). Rodgers (1986) discusses static load in more detail; also see Guideline 1 of Chapter 15. For comments on the difficulty of setting allowances, see Mathiassen and Winkel (1992).

3.2.4 Restrictive clothing
Table 31.12 (Williams) and Table 31.13 (ILO) give allowances for **restrictive clothing**. For more on clothing, see Chapter 23.

3.3 Mental Fatigue Allowances
The concern in **mental fatigue** is with discipline, concentration, mental and visual demand, and mental and visual

TABLE 31.9 Posture and motion allowances (Williams, 1973).

Allowance, %	Category
0	Sedentary work, no muscular strain (light assembly, packing or inspection)
1	Sitting, slight muscular strain (operating low-pressure foot pedal)
2	Standing or walking, body erect and supported by both feet on ground
3	Standing on one foot (operating foot press control)
4	Unnatural postures (kneeling, bending, stretching, or lying down), light shoveling, holding unbalanced loads
5	Crouch, working with manual restraint (restricted tool movement in confined areas)
7	Awkward posture in conjunction with heavy work, bending or stooping in lifting heavy weights, carrying heavy loads on level ground or on slopes, one or both hands as convenient
10	Carrying or moving awkward and heavy loads over rising ground, climbing stairs or ladders with heavy loads, working with hands above shoulder height (painting ceilings)

Source: Reprinted from *Industrial Engineering* magazine, May 1973. Copyright (c) 1973, Institute of Industrial Engineers, 25 Technology Park/Atlanta, Norcross, Georgia 30092.

TABLE 31.10 Position allowances (Cornman, 1970).

Allowance, %	Category
2	Sit or combination of sit, stand, and walk where change of position is not more than 5 min apart; arm and head positions at normal working height
3	Stand or combination of standing and walking where sitting is allowed only during rest periods; also for situations where arms and head are out of normal working range for periods of less than 1 min
5	Workplace requires constant stooping or standing on toes; also for work requiring extension of arms or legs
7	Body is in cramped or extended positions for long time periods; also for where attention requires motionless body

Source: Reprinted with permission from *Industrial Engineering* magazine, Vol. 2, 4, 1970. Copyright, Institute of Industrial Engineers, 25 Technology Park/Atlanta, Norcross, GA 30092.

strain. Goldberg and Freivalds (1988) comment on mental strain, monotony, and tediousness.

Table 31.3 (Page) covers these as hand–eye coordination allowances. Table 31.14 (Williams) gives a discipline allowance. Table 31.15 (Cornman) divides mental allowances into mental (visual) demand,

TABLE 31.11 Posture allowance (ILO, 1979).

Points	Activity
0	Sitting easily
2	Sitting awkwardly or mixed sitting and standing
4	Standing or walking freely
5	Ascending or descending stairs, unladen
6	Standing with a load; walking with a load
8	Climbing up or down ladders; some bending, lifting, stretching, or throwing
10	Awkward lifting; shoveling ballast to container
12	Constant bending, lifting, stretching, or throwing
16	Coal mining with pickaxes; lying in a low seam

TABLE 31.12 Restrictive clothing allowance (Williams, 1973).

Allowance, %	Category
0	Clothing or apparatus not restrictive
1 to 5	Additional weight for protection against elements restricts movement, gloves affect handling, face mask affects breathing
6 to 15	Heavy breathing apparatus, cumbersome clothing (asbestos suit) restricting movement

Source: Reprinted from *Industrial Engineering* magazine, May 1973. Copyright © 1973, Institute of Industrial Engineers, 25 Technology Park/Atlanta, Norcross, Georgia 30092.

TABLE 31.13 Restrictive clothing allowances (ILO, 1979). Consider the clothing weight in relation to effort and movement. Also consider whether it affects ventilation and breathing.

Points	Clothing
1	Thin rubber (surgeon's) gloves
2	Household rubber gloves; rubber boots
3	Grinder's goggles
5	Industrial rubber or leather gloves
8	Face mask (e.g., for paint spraying)
15	Asbestos suit or tarpaulin coat
20	Restrictive protective clothing and respirator

TABLE 31.14 Discipline allowance (Williams, 1973). Discipline considers the need for exact timing or conformance to a machine cycle. Cycle durations are for guidance; allowances need not be given for all cycles of less than .2 min.

Allowance, %	Discipline Demand	Cycle Length, Min
0	Negligible	Over .20
0–2	Low	.16 to .20
2–3	Medium	.09 to .15
3–5	High	Below .09

Source: Reprinted from *Industrial Engineering* magazine, May 1973. Copyright © 1973, Institute of Industrial Engineers, 25 Technology Park/Atlanta, Norcross, Georgia 30092.

repetition of cycle, and duration. Tables 31.16 and 31.17 (ILO) give concentration/anxiety and monotony allowances.

My personal opinion is that allowances for boredom, monotony, lack of a feeling of accomplishment, and the like are quite questionable. These factors are unlikely to cause fatigue and thus increase time/cycle. These factors primarily reflect unpleasantness and thus should be reflected in the wage rate per hour rather than in the time/unit.

3.4 Environmental Fatigue Allowances Page
(Table 31.3) lumps all **environmental fatigue allowances** together but the following discussion is divided into climate, noise/vibration, and visual.

See Freivalds and Goldberg (1988) for comments on heat-stress and noise allowances.

3.4.1 Climate Table 31.18 (Williams) gives thermal and atmospheric allowances. Table 31.19 (Cornman) divides climate into temperature, humidity, and ventilation. Table 31.20 and 31.21 (ILO) assign temperature/humidity, wet, and ventilation to climate and have a separate table for dust, dirt, and fumes.

Hancock and Chaffin (1977) give a systematic procedure for heat-stress allowances. See Konz et al. (1983) for data on 12 heat-acclimated males who

TABLE 31.15 Mental allowances according to Cornman (1970). Mental (visual) demand considers the degree of mental and visual fatigue sustained through mind–eye concentration and coordination. Repetition of cycle considers the hypnotic, monotonous effect of short-cycle repetitive operations. Duration considers the amount of time to complete a job and obtain a feeling of accomplishment or of being finished.

Allowance, %	Mental (Visual) Demand
0	Only occasional mental or visual attention
2	Operation practically automatic or attention required only at long intervals
3	Frequent mental and visual attention; intermittent work or operation involves waiting for a machine or process to complete a cycle (some checking)
5	Continuous mental and visual attention for either safety or quality reasons; usually repetitive operations requiring constant alertness or activity
8	Concentration or intense mental and visual attention in layout or doing complex work to a very close accuracy or quality or coordinating a high degree of manual dexterity with close visual attention for sustained periods of time; also all purely inspection operations (checking quality is prime object)

Allowance, %	Repetition of Cycle
3	Operator varies operation pattern or schedules own work; operations vary from day to day or operations may not be done daily
7	Reasonably fixed operation pattern or where deadlines or pressure to complete are present; task is regular although operator can vary operations from cycle to cycle.
10	Periodic completion of operations scheduled; operations regular in occurrence; thought and motion patterns made at least 10 times/day
13	Completion of thought and motion patterns more than 10 times/day; machine-paced operations; most piece-rated operations; operators suffer boredom and lack of control

Allowance, %	Duration
3	Operation or suboperation completed in 1 min or less
7	Operation or suboperation completed in 15 min or less
10	Operation or suboperation completed in 60 min or less
13	Operation or suboperation completed in over 60 min

Source: Reprinted with permission from *Industrial Engineering* magazine, Vol. 2, 4, 1970. Copyright, Institute of Industrial Engineers, 25 Technology Park/Atlanta, Norcross, GA 30092.

each were exposed for 8 h/day in 24 conditions. The amount of rest required depended on both the environmental temperature and the amount of scheduled rest.

3.4.2 Noise/vibration Table 31.22 (Williams) combines climate and noise. Table 31.23 (Cornman)

TABLE 31.16 Concentration/anxiety allowance (ILO, 1979) considers what would happen if the operator relaxed attention, responsibility, need for exact timing, and accuracy or precision required.

Allowance Points	Degree
0	Routine, simple assembly; shoveling ballast
1	Routine packing, washing vehicles, wheeling trolley down clear gangway
2	Feed press tool (hand clear of press); topping up battery
3	Painting walls
4	Assembling small and simple batches (performed without much thinking); sewing machine work (automatically guided)
5	Assembling warehouse orders by trolley; simple inspection
6	Load/unload press tool; hand feed into machine; spray painting metalwork
7	Adding up figures; inspecting detailed components
8	Buffing and polishing
10	Guiding work by hand on sewing machine; packing assorted chocolates (memorizing patterns and selecting accordingly); assembly work too complex to become automatic; welding parts held in jig
15	Driving a bus in heavy traffic or fog; marking out in detail with high accuracy

TABLE 31.17 Monotony allowances (ILO, 1979) consider the degree of mental stimulation and if there is companionship, competitive spirit, music, etc.

Allowance Points	Degree
0	Two people on jobbing work
3	Cleaning own shoes for .5 h on one's own
5	Operator on repetitive work; operator working alone on nonrepetitive work
6	Routine inspection
8	Adding similar columns of figures
11	One operator working alone on highly repetitive work

and Table 31.24 (ILO) give two other approaches. See Chapter 22 for more on noise.

In general, studies have been unable to detect a difference in performance as a function of noise level (if auditory communication is not part of the job). Thus, noise allowances probably reflect unpleasantness rather than increased time/unit.

3.4.3 Visual
Table 31.25 (Williams), Table 31.26 (Cornman), and Table 31.27 (ILO) give three approaches to visual allowances. See Chapter 21 for more on eye and illumination.

Goldberg and Freivalds (1988) said Table 31.27 could be replaced with three values if the situation is compared with the IES values in Table 21.5. If the situation is in the same illuminance category (i.e., same row) as recommended in Table 21.5, give 0% visual allowance. If actual lighting was one category lower, give 2% allowance, and if actual lighting was two categories lower, give 5% allowance.

3.5 Overview of Fatigue Allowances
In general, most fatigue allowances seem to have an inadequate range. Cornman, for example, gives a 1% allowance for excellent lighting and 3% to working in the dark! In addition, remember that Cornman has "25% deductible."

Another problem is the use of allowances that give more time/unit as a substitute for unpleasant working conditions.

None of the authors specify the length of the day for which the fatigue allowance is applicable. Presumably it is 8 h/day, which was a typical workday in the 1960s and 1970s. The length of the work week presumably is 5 days, but again nothing is specified. If applied to work days of less than 8 h, presumably the standard is tighter. I personally do not recommend attempting to adjust allowances to length of the day (i.e., changing the time standard). If it is felt that some adjustment is necessary because the day or week is longer or shorter, I feel the adjustment should be in discipline level. See Box 26.1.

TABLE 31.18 Thermal and atmospheric allowances (%) (Williams, 1973) consider the demands despite presence of protective clothing or equipment.

Ventilation and Circulation and Humidity	Dry Bulb Temperature, C			
	Below −1	−1 to 13	13 to 24	24 to 38
Adequate ventilation and circulation; normal climatic humidity	10 to 20	1 to 10	0	1 to 10
Inadequate ventilation and circulation. Nonstandard climatic conditions, causing some discomfort	20 to 25	5 to 10	0 to 5	5 to 15
Very poor ventilation and circulation, fumes, dust, steam, causing irritation to eyes, skin, nose, throat	20 to 30	10 to 20	5 to 10	10 to 20

Source: Reprinted with permission from *Industrial Engineering* magazine, Vol. 2 [4], 1970. Copyright, Institute of Industrial Engineers, 25 Technology Park/Atlanta, Norcross, GA 30092.

TABLE 31.19 Temperature, humidity and ventilation allowances (Cornman, 1970). For temperature, consider average temperature in performing daily duties. For humidity, consider extra concentration due to perspiration, wiping of brow, pulling at clothing. For ventilation, consider oxygen availability or repulsion of human body to the surroundings.

Allowance, %	Category
1	Mechanical or electrical control of temperature for comfort (for inactive or office personnel, usually from 22-24 C; for normally active or plant, usually from 20-21 C)
2	Temperature controlled by job requirements (heat from machines, ovens, materials); for inside work, from 24-30 C; for outside work, from 27-32 C; normal circulation; temperatures controlled by job requirements (heat from machines, ovens, materials); for inactive or office personnel, below 18 C or above 27 C; for outdoor work or where normal air circulation, below 4 C or above 32 C
3	When normal air circulation is available, temperature below 2 C and above 35 C; when normal air circulation is not available, above 32 C

Allowance, %	Category
1	Normal, comfortable humidity supplied by air conditioning or heating systems; no sensation of dryness or humidity; for temperatures of 21-24 C, humidities of 40-55%
2	Unusually dry conditions (after 30 min have skin sensation or burning nostrils); less than 30% humidity or high humidity, noticeable upon entrance to area by clammy skin sensation (60-85% humidity) Unusually high humidity; clothing becomes damp quickly (humidity over 80%)
3	Humidity or wetting conditions such as steam rooms or outdoors in rain where special clothing must be worn

Allowance, %	Category
1	Normal operations outdoors or in air conditioned area where filter or washing of air supplies fresh, odor-free air
2	Normal nonair conditioned plant or office; occasional stuffiness; movement of air supplied by movement of personnel or from machines; no air filtration
3	Extremely small and enclosed surroundings; air movement nil or dusty conditions caused by the job (regardless of dust type); limited smoke (either foreign or operator-generated)
5	Extremely smoky, toxic, or dusty conditions; nauseating or mentally disturbing fumes (although not injurious to health); air movement or exhausting does not remove effects

Source: Reprinted with permission from *Industrial Engineering* magazine, Vol. 2 [4], 1970. Copyright, Institute of Industrial Engineers, 25 Technology Park/Atlanta, Norcross, GA 30092.

Mital and Shell (1984) present a model for allowances for energy-intensive physical activities (the person is working at maximum capacity). Randle (1988) points out that static work is an important part of most physical jobs and that oxygen consumption may not be a good index of fatigue.

4 DELAY ALLOWANCES

Delay allowances are meant to compensate the operator for short delays beyond the control of the operator. Examples are machine breakdowns, interrupted material flow, conversations with supervisors, and so forth. If the delay is long (e.g., 30 min), the operator clocks out (records the start and stop time of the delay on a form) and works on something else during the clocked-out time. Delay allowances should vary with the task but not with the operator.

Engineered delay allowances can be determined from occurrence sampling or from the delays recorded during time studies. For example, if there were 8 min of delays during 100 min of time study, then 8% could be used for the delay allowance.

Errors in delay allowances can occur from poor sampling or from changing conditions.

To obtain a valid sample of delays, the sample must represent the total shift, not the middle of the shift. That is, the delays must be observed starting at the start of the shift, not in the middle of the shift. Include the delays just before and after breaks and the delays just before and after meals and just before the end of the shift. Include delays on the second and third shifts as well as on the first shift.

Often a delay allowance for a task is set when the task is first performed. Then, over the years, as the organization learns, delays go down. If the delay allowance is not updated, the standard has become looser. A reasonable procedure is to give delay allowances an expiration date, for example, two years after being set. After two years, they must be redetermined.

Delays usually permit the operator to take some personal time and reduce fatigue; that is, they also serve as personal allowances and fatigue allowances.

TABLE 31.20 Climate allowances (ILO, 1979) consider temperature/humidity, wet, and ventilation. For temperature/humidity use the average environmental temperature. For wet, consider the cumulative effect over a long period. For ventilation, consider quality/freshness of air and its circulation by air conditioning or natural movement.

Points for Temperature/Humidity

Humidity, %	Temperature		
	Up to 24 C	24-32	Over 32
Up to 75	0	6-9	12-16
76-85	1-3	8-12	15-26
Over 85	4-6	12-17	20-36

Points	Wet
0	Normal factory operations
1	Outdoor workers (e.g., postman)
2	Working continuously in the damp
4	Rubbing down walls with wet pumice block
5	Continuous handling of wet articles
10	Laundry washhouse, wet work, steamy, floor running with water, hands wet

Points	Ventilation
0	Offices; factories with "office-type" conditions
1	Workshop with reasonable ventilation but some drafts
3	Drafty workshops
14	Working in sewer

TABLE 31.21 Dust, dirt, and fumes allowance (ILO, 1979). For dust, consider both volume and nature of the dust. The dirt allowance covers "washing time" where this is paid for (e.g., 3 min for washing). Do not allow both time and points. For fumes, consider the nature and concentration; whether toxic or injurious to the health; irritating to eyes, nose, throat, or skin; odor.

Points	Dust
0	Office, normal light assembly, press shop
1	Grinding or buffing with good extraction
2	Sawing wood
4	Emptying ashes
6	Finishing weld
10	Running coke from hoppers into skips or trucks
11	Unloading cement
12	Demolishing building

Points	Dirt
0	Office work, normal assembly operations
1	Office duplicators
2	Dustman (garbage collector)
4	Stripping internal combustion engine
5	Working under old motor vehicle
7	Unloading bags of cement
10	Coal miner; chimneysweep with brushes

Points	Fumes
0	Lathe tuning with coolants
1	Emulsion paint, gas cutting, soldering with resin
5	Motor vehicle exhaust in small commercial garage
6	Cellulose painting
10	Molder procuring metal and filling mold

TABLE 31.22 Physical environment allowances (Williams, 1973). The physical environment considers discomfort caused by dirt, oil, grease, water or other liquids, ice, chemicals, etc., and noise irritation (pitch, volume, irregularity). Vibration and instability includes physical movement because it affects work demands, precision, or safety.

Allowance, %	Physical Environment Category
0	Clean, bright, dry surroundings; normal machine and human noise
0 to 3	Dirty, wet, greasy, contaminated surroundings
0 to 4	Uncomfortable noise
0 to 8	Combination of several factors

Allowance, %	Vibration and Instability Category
0 to 1	Very low and steady (hand drill, saw)
2 to 3	Medium; significant but predictable vibration
4 to 7	High, distressing, difficult to control; floor vibration, pneumatic drilling, unevenly moving surface

TABLE 31.23 Noise allowances (Cornman, 1970) consider fatigue to the nervous system (changes in noise as well as loudness).

Allowance, %	Category
1	Normal noise level in average office or in industrial plant making lightweight products (30-60 dBA); intermittent music easily heard and enjoyed
2	Unusually quiet area where noise is almost absent (library; less than 30 dBA); also for constant loud noise (tin shop, knitting room, city street); 60-90 dBA of constant noise; music may not be heard with pleasure
3	Normally quiet surroundings with intermittent loud or annoying noise (nearby riveter, elevated train, punch press); noises of sharp nature and above 90 dBA; also nonintermittent noises above 100 dBA (boiler factory)
5	High-frequency or otherwise annoying noise, intermittent or constant

TABLE 31.24 Noise and vibration factors (ILO, 1979). Consider whether the noise affects concentration, is a steady hum or a background noise, is regular or occurs unexpectedly, is irritating or soothing. Consider the impact of the vibration on the body, limbs, or hands and the addition to mental effort due to it, or to a series of jars or shocks.

Points	Noise Category	Points	Vibration Category
0	Working in a quiet office, no distracting noise; light assembly work	1	Shoveling light materials
1	Work in a city office with continual traffic noise outside	2	Power sewing machine; power press or guillotine if operator is holding the material; cross-cut sawing
2	Light machine shop; office or assembly shop where noise is a distraction	4	Shoveling ballast; portable power drill operated by 1 hand
4	Woodworking machine shop	6	Pickaxing
5	Operating steam hammer in forge	8	Power drill (2 hands)
9	Riveting in a shipyard	15	Road drill on concrete
10	Road drilling		

TABLE 31.25 Visual allowances (%) (Williams, 1973) consider the closeness of the work as it affects visual attention and strain.

Category	Adequate Lighting	Inadequate or Disturbed Lighting
No special eye attention required	0	0
Close eye attention intermittently; or continuous eye attention with varying focus	0	1
Continuous eye attention with continuous focus	1 to 3	4 to 8

Source: Reprinted with permission from *Industrial Engineering* magazine, December, 1973. Copyright, Institute of Industrial Engineers, 25 Technology Park/Atlanta, Norcross, GA 30092.

TABLE 31.26 Light allowances (Cornman, 1970) primarily consider eye strain to focus unless light is so poor as to require extra body motions.

Allowances, %	Category
1	"Normal" lighting (200 to 500 lux in most industries; 500 to 1000 if offices and inspection); absence of glare is apparent
2	Occasional glare is inherent part of job or where substandard or special lighting is required. Continual glare is inherent part of job; also for work requiring constant change from lighted area to darkness (less than 50 lux); also work requiring "venetian blind" effect (shiny and dull surface in lathe turning)
3	Work in absence of light or where sight is obstructed (noticeable by feel of fingers or feet); eyes not used or are straining (photo darkroom, mechanic under machine)

Source: Reprinted with permission from *Industrial Engineering* magazine, Vol. 2 [4], 1970. Copyright, Institute of Industrial Engineers, 25 Technology Park/Atlanta, Norcross, GA 30092.

TABLE 31.27 Eye strain (ILO, 1979) considers the lighting conditions, glare, flicker, illumination, color, and closeness of work and for how long strain is endured.

Points	Category
0	Normal factory work
2	Inspection of easily visible faults; sorting distinctively colored articles by color; factory work in poor lighting
4	Intermittent inspection for detailed faults; grading apples
8	Reading a newspaper in a bus
10	Continuous visual inspection (cloth from a loom)
14	Engraving using an eyeglass

REVIEW QUESTIONS

1. What is the difference between a shift allowance and a work time allowance?

2. How are allowances related to the definition of normal pace?

3. Are fatigue allowances given for machine time?

4. Should allowances be given for unpleasant jobs? Explain your answer.

5. When carrying or lifting heavy loads, what happens to observed time?

6. Are allowances based on the job or the person?

7. How should an allowance be given for heat stress due to a hot summer?

REFERENCES

Cornman, G. Fatigue allowances—A systematic method. *Industrial Engineering,* Vol. 2 [4], 10-16, 1970.

Freivalds, A. and Goldberg, J. A methodology for assigning variable relaxation allowances: Manual work and environmental conditions. In *Trends in Ergonomics/Human Factors V,* Aghazadeh, F. (ed.), 457-64. New York: Elsevier, 1988.

Goldberg, J. and Freivalds, A. A methodology for assigning variable relaxation allowances: Visual strain, illumination and mental strain. In *Trends in Ergonomics/Human Factors V,* Aghazadeh, F. (ed.), 161-68. New York: Elsevier, 1988.

Hancock, W. and Chaffin, D. A practical method for industrial heat stress allowance determination. *AIIE Transactions,* Vol. 9 [2], 144-54, 1977.

International Labour Office. *Introduction to Work Study,* 3rd ed. Geneva, Switzerland: ILO, 1979.

Konz, S., Rohles, F., and McCullough, E. Male responses to intermittent heat. *ASHRAE Transactions,* Part 1B, 79-100, 1983.

Lazarus, I. Inaccurate allowances are crippling work measurement. *Factory,* 77-79, April 1968.

Mathiassen, S. and Winkel, J. Can occupational guidelines for work-rest schedules be based on endurance time data? *Ergonomics,* Vol. 35 [3], 253-59, 1992.

Mital, A., Bishu, R., and Manjunath, S. Review and evaluation of techniques for determining fatigue allowances. *Int. J. of Industrial Ergonomics,* Vol. 8, 165-78, 1991.

Mital, A. and Shell, R. Determination of rest allowances for repetitive physical activities that continue for extended hours. *Proceedings of Ind. Eng. Conference,* 637-45, 1984.

Murrell, K. *Human Performance in Industry.* New York: Reinhold, 1965.

Page, E. Determining fatigue allowances. *Industrial Management,* 1-3 and 14, February 1964.

Price, A. Calculating relaxation allowances for construction operatives—Part 1: Metabolic cost. *Applied Ergonomics,* Vol. 21 [4], 311-17, 1990a.

Price, A. Calculating relaxation allowances for construction operatives—Part 2: Local muscle fatigue. *Applied Ergonomics,* Vol. 21 [4], 318-24, 1990b.

Randle, I. Static work and recovery allowances. In *Contemporary Ergonomics 1988,* Megaw, E. (ed.). London: Taylor and Francis, 1988.

Rodgers, S. (ed.). *Ergonomic Design for People at Work: Vol. 2.* New York: Van Nostrand-Reinhold, 1986.

Williams, H. Developing a table of relaxation allowances. *Industrial Engineering,* Vol. 5 [12], 18-22, 1973.

32 JOB INSTRUCTION/ TRAINING

OVERVIEW

The primary resources of any group are the skills of its people. The primary instruction medium is visual; the message can be pictorial or text. Recommendations are given on how to improve messages. The worker can be trained or given a job aid. In either case, the training material is important.

CHAPTER CONTENTS

1 Problem
2 The Medium and the Message
3 Training (Memorization)
4 Job Aids

KEY CONCEPTS

active/passive
audio input
AVO
cross-training
decision structure tables
distributed practice/massed
    practice
expert systems
grade level

job knowledge requirements
kinesthetic input
limit aids
linear/branching mode
memorization/job aid
no stupid students, just terrible
    teachers
pictorial messages
pilot testing

plan/draft/revise
programmed learning
psychomotor information
routing sheets
self-instruction
simulation
sit by Nellie
text message
visual input

1 PROBLEM

1.1 Need   The primary resources of any company, community, or country are the skills (knowledge, education) of its people. On the micro level, this has been demonstrated repeatedly with the success of one restaurant vs. others in its locality, one store vs. another, one physician vs. other physicians, and so forth. At the intermediate level of the firm, some manufacturers do better than others, some universities do better than others. At the macro level of the nation, consider the success of resource-poor countries such as Switzerland, Holland, Japan, and Hong Kong.

Given the general need for people skills, the next question is what specific skill (knowledge) needs to be communicated to what target audience.

Give training programs the same attention you give to computer programming. When writing a computer program, the first step is to decide what you want the program to do. The same approach should be used for training people.

Assuming the target audience is your workers, what are their characteristics? Consider not only the mean of their characteristics but also the range. For example, the mean might be 12 years of schooling with a range of 8 to 19.

1.2 Matching Knowledge and Requirements

Instruction or training is needed when the worker's knowledge and the **job knowledge requirements** do not match. Four possibilities are:

1. static job and static worker
2. static job and changing worker
3. changing job and static worker
4. changing job and changing worker

1.2.1 Static job and static worker   Use training to improve the worker's job knowledge. For example, Table 32.1 shows a better way to close the top of a cardboard box to reduce repetitive strain problems on the operator's shoulder. Use Table 32.1 as a training aid. Alternatively, have a word processor operator learn how to use a new software program which will automatically check the spelling and punctuation of material typed. Office personnel could learn how to operate the new phone system, or an industrial engineer might learn about a technique called Total Quality Management.

1.2.2 Static job and changing worker   Even though the job does not change, individual employees quit, get promoted, get demoted, are fired, go on vacation, or are absent for the day. The replacement

TABLE 32.1   Closing the flaps on a cardboard box (see Figure 32.1) should be done close to the body to reduce repetitive strain on the shoulders. Use four playing cards or rectangular pieces of paper to visualize the pattern.

Bad: Tucking in corner away from body stresses shoulder.

Left Hand	Right Hand
Reach to 1B	Reach to 2R
Grasp 1B	Grasp 2R
	Fold 2 flat under 1
Fold 1 flat over 2	
Release 1	Release 2
Reach for 4L	Reach to 3
Grasp 4L	Grasp 3
Fold 4 flat over 1	
	Fold 3 flat over 2 and over 4
Release 4	Hold 3 flat
Reach for 2R	
Grasp 2R	
Pull 2R above 3T	
Release 2R	Release 3

Good: Tucking in corner close to the body has better leverage.

Left Hand	Right Hand
Reach to 2	Reach to 3B
Grasp 2	Grasp 3B
	Fold 3 flat under 2
Fold 2 flat over 3	Release 3
Release 2	
Reach to 1	Reach to 4
Grasp 1	Grasp 4
	Fold 4 flat
Fold 1 flat over 2 and under 4	Release 4
Release 1	Reach to 3B
Reach to 4	Grasp 3B
Hold 4 flat	Pull 3B above 4R
Release 4	Release 3

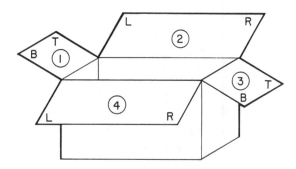

FIGURE 32.1   Seal the flaps on a cardboard box while the hands are close to the body to get the best lever arm.

worker needs to know how to do the job. Well-managed firms will not wait until the job is vacant before doing the training; they train before the vacancy. Pre-training employees is called **cross-training**—each worker is able to do several different jobs. For example, the word processor operator can do filing, the auto transmission specialist can work on the auto electrical system, the assembler for station 4 also can do the work of stations 3, 5, and 6. Cross-training not only fulfills desires for job enlargement and allows resting individual muscles, it also greatly simplifies the day-to-day administration of any group. Some firms also train potential employees, thus allowing better selection of employees from the applicants.

Even with constant technology, the number of workers may change as sales go up or down. Assume a firm decides to hire 10 more people to expand production. Naturally there is the training requirement for that 10. But, in addition, most firms fill positions through a "bidding" or "posting" procedure. That is, the job is posted and existing employees can bid on the new job. For example, a person who is now an assembler can bid on one of the new scheduling clerk jobs. But then the assembly job is open; it may be filled by an existing material handler. Then the material handler job is open; it may be filled by one of the new hires. Thus, for every new person hired, 3 or 4 people may have to be trained.

This movement of people also occurs with reductions in the workforce; the procedure is called bumping. In a layoff, a higher-seniority person can bump a lower-seniority person. For example, if a firm eliminates the second shift, a high-seniority press operator on the second shift may bump a low-seniority material handler on the first shift. The material handler may bump someone else. Thus, if 10 people are laid off, 30 people may change jobs. The number changing depends upon the rules of bumping at the firm. More will change if the seniority is plantwide instead of departmentwide and if required demonstrated proficiency requirements are low. For example, if a person with high seniority has 90 days to learn how to do another job, naturally the person will bid on the job because, at the minimum, the paycheck will continue for another 90 days. This can cause great quality problems during the 90 days. Previous cross-training can reduce these problems.

1.2.3 Changing job and static worker   The technology of the job may change. For example, the packing station may be changed so that the boxes are taken away by conveyor instead of on pallets; a typist may be given a word processor; an automatic loader might be installed on a machine to replace manual loading. Because of changing job requirements, it is a good idea to have a mechanical engineer learn about ergonomics so the engineer can become a better designer. In these examples, it is assumed that the same worker does the job as before; however, the worker needs to be trained in the new technology.

1.2.4 Changing job and changing worker
Here we are dealing with major changes in the job.

This can occur in an existing job (for example, replace manual spray painting with robotic spray painting) or a new job (opening a new factory or office and hiring new people to do the job). The job skills for the new job differ considerably from those for the old job. One of the characteristics of "lean production" (also called just-in-time) is use of a skilled, cross-trained worker. Another trend is the increase in worker involvement in many decisions (see Chapter 33). For this reason, manufacturers are becoming more selective in their hiring. A high school diploma has become a minimum as firms decrease the ratio of salaried to hourly workers. Chrysler, for example, had 25 hourly workers per each salaried worker in 1991 and 48 in 1994 and hopes to increase the ratio (Templin, 1994).

2 THE MEDIUM AND THE MESSAGE

Organize the information (message) to be presented through a medium.

2.1 Organization   Three parts of organizing information are (1) **planning**, (2) **drafting**, and (3) **revising.** (See Table 32.2.)

2.1.1 Planning   What is the user's ignorance? What needs to be communicated? What are your goals? What do the users know *before* your communication? What happens *during* your communication? What happens *after* your communication? How will your database be filled and organized?

Then decide how to communicate. In particular, how much of your message will be text and how much will be graphics? After this strategic decision, you will need to decide the tactics of specific types of text, graphs, tables, and so forth.

2.1.2 Drafting   Table 32.2 gives some comments on drafts. Perhaps the most important thing is that the first version is not the final version. Multiple drafts are the mark of good communicators.

2.1.3 Revising   The draft material needs **pilot testing** with the target audience; that is, the presentation needs a quality assurance test. Testing requires identifying the target audiences, finding a representative sample, and doing the test. Obtaining a sample of people who represent the target audience is important. In particular, they are unlikely to be similar to the people in the writer's office! A good test will evaluate at both the macro level (logic and structure of the presentation) as well as at the micro

TABLE 32.2 *Some guidelines for good instruction writing.*

PLAN

- Know the user's ignorance. This requires knowing the user.
- What will be communicated?
- Decide on strategy (medium; text vs. pictures).
- Decide on tactics (organization of message, types of tables, etc.).

DRAFT

- Plan on multiple drafts.
- Put the information in a logical order. Number steps.
- Have a specific objective for each instruction (who, what, where, why, when, how).
- Use both text and figures.
- Make the text at the appropriate reading level.
- Consider short sentences starting with action verbs. Instead of "The next step is to drill and chamfer the 50-mm hole," say:
  1. Drill the 50-mm hole.
  2. Chamfer the 50-mm hole.
- Use **active** statements, not **passive** or negative statements.
  Active: The large lever controls the depth of cut.
  Passive: The depth of cut is controlled by the large lever.
  Negative: The small lever does not control the depth of cut.

REVISE

- Identify the target audience (see PLAN).
- Select a sample population for testing.
- Run the test.
- Revise material.
- Continue the loop until satisfied.
- Allow sufficient time for testing and revision.

level (specific adjectives and nouns used, use of a bar graph as opposed to a line graph). After the test, the material is revised and then another test is administered to evaluate the revisions. Allow sufficient time for the testing and revisions.

The information (message) can be communicated to the trainee through many media.

2.2 Medium
The three primary input channels are kinesthetic, audio, and visual.

2.2.1 Kinesthetic
Learning by doing, primarily touch and **kinesthetic** (movement) **input,** has a long-established reputation of success. Hands-on experience results in learning with long retention. For example, most teachers feel that students learn more when they write notes rather than when they just underline or read with no physical action. Motor movements, such as riding a bicycle, are very resistant to extinction. Use kinesthetic input for highly repetitive psychomotor movements (skilled movement) so processing becomes automatic.

2.2.2 Audio
Auditory signals **(audio input)** that are mechanical tend to be quite simple (a horn

honking, a phone ringing). They present only one of a limited number of choices. In other words, they can be used for a signal but do not communicate much information. It is possible to get more elaborate in the mechanical signal through variations in pitch (a horn vs. a buzzer), duration (Morse code), and even volume, but the number of alternatives which can be indicated mechanically probably will be 10 or less for most situations.

The human voice, of course, can present a very wide range of alternatives (stop, go, turn right, slow down, call home, call Sally, drink Pepsi).

A problem of auditory messages is that they are transitory—they disappear with time. (An exception can be very simple audio messages such as buzzers, which can be continuous. However, an annoyance problem occurs within a few seconds.) What if someone tells you to call (913) 532-5606? Unless you write it down, the number probably will be forgotten within a few seconds. That is, there is no permanent reference. Complex information is difficult to communicate well by voice because people forget some of the details, which is why organizations have the rule **AVO**—Avoid Verbal Orders.

Another problem of the audio message is its pacing effect. You can hear the message only at the rate it is being presented. One example is Morse code—a series of auditory dots and dashes which are presented as short beeps and long beeps. You can receive the code only at the rate the sender is transmitting it to you. If it is too fast for you, you will miss it; if it is too slow, you get bored and impatient. Audio tapes have been used to talk workers through an assembly task (Dickey and Konz, 1969). The problem was the fixed rate. For the initial cycles of the task, the tape was too fast in relation to the worker's ability. After practice, the workers had to wait for the messages; that is, the rate of output was limited to the tape speed. If such audio tapes are used, they need to be available at a variety of speeds.

Computer prompt (help) information should follow the same principle, varying the information with the skill level of the user. That is, beginners want detailed information while experts want no information because they already know it. The user should be able to select the desired prompt level.

In most situations the instruction medium will be visual. Many choices remain.

2.2.3 Visual
The first choice when using the medium of **visual input** is whether the visual image will be text, a representation of the object, or a combination. All images are codes; codes must be decoded. Since many codes are easy to decipher (and for instructional purposes, that is the way it should be), the decoding (translation) problem often

is forgotten. A pictorial message requires the least translation. (A word equals .001 picture!) For example, show an operator the correct location of solder joints by using a picture of the assembly with circles about the solder joints; teach how to load a camera with pictures showing the loading sequence. Translation from the image to the task is simple and direct. Although pictures can communicate the *how*, text is very useful for communicating the *why*. The best approach seems to be not just a picture, not just text, but pictures and text (Dickey and Konz, 1969; Stern, 1984; Nugent, 1987).

Another choice is whether the information will be presented in a linear or branching mode. The traditional approach is the **linear mode.** An example is a movie. The information is presented in a standard sequence determined by the program designer. In a **branching mode,** the information branches, depending upon the needs of the learner. One branching example is a recitation class in which students ask questions of the instructor. What the instructor says depends on what questions were asked. Another example is computer help sequences. What information is given depends on what question is asked. Depending on the user, many different branches of information can be presented. (An advantage of videodisc over videotape is that a videodisc permits easy branching.) An advantage of computerized instruction is that branching is easy.

Another choice is whether the information will be communicated to individuals or to a group. Certain approaches lend themselves to group presentations (movies, slides and transparencies, lectures), some to small groups (TV, demonstrations), and some techniques lend themselves to individual instruction (books, printed photographs). Some approaches are flexible. Another determination is the importance of social interaction and involvement between the instructor and students and among students.

Another choice is ease of instruction modification. It is easy to replace a slide in a magazine or rearrange the slides; it is difficult to modify the images of a filmstrip or movie. It is difficult to modify a book or printed material; it is easy to modify typed material (especially with the use of word processors).

Is distributed or massed practice best? Evidence since the turn of the century indicates that, for the same total hours of instruction or practice, **distributed practice** (short training sessions) produces better learning than **massed practice** (long training sessions). Bradeley and Longman (1978), for example, emphasize the importance of the amount of training per day. They found that 1 session of 1 h/day was better than 1 session of 2 h/day or 2 sessions of 2 h/day. Bouzid and Crawshaw (1987) report that 2 30-min sessions for word processing training was better than 1 60-min session.

Another choice is the location of the instruction. Classroom instruction permits use of machines requiring power, projection screens, space for demonstrations, and the like. Field instruction may not allow use of powered equipment (unless it is battery-powered), seats for the participants, or space to set up instructional equipment. In some field situations, no equipment may be available and the instruction site may be dirty, vibrating, noisy, or otherwise distracting. Computerized instruction allows great flexibility in scheduling and location, especially when using remote terminals over telephone lines.

Do not overemphasize mechanical devices. Our society—and especially engineers—tends to overemphasize the machine (the medium) and neglect the message.

2.3 Message As mentioned previously, the message usually is visual. The primary divisions are pictorial messages and text messages.

2.3.1 Pictorial messages Pictures can be communicated many ways. **Pictorial messages** include a live demonstration, a recorded demonstration (film or video), still photographs (printed or projected), and line drawings (drawn, printed, projected, and even animated).

Moving or still? Still pictures, in addition to usually being cheaper, permit viewers to look at the pictures at their own individual paces; backing up or rearranging the pictures is easy. Movies and TV, on the other hand, can show relationships such as sequences of hand–arm motions. However, freezing the image or backing it up requires the designer to plan ahead or provide special equipment for the viewer. Even if the special equipment is available, a specific viewer may not wish to interrupt the group to see a scene over.

Color or black and white? Color film and color TV are so common that presenting the image in black and white probably will be more expensive than in color. However, color printing and photoduplication are expensive. In addition to being more interesting to view, a color image generally will be easier for the viewer to understand due to contrast, shading, and other features.

Image fidelity? A slide or printed photograph can be used in place of the physical object itself—if the representation gives all the relevant information. (Adequate representation is the basic principle of the simulator used for flight instruction.) Photographs may be out-of-focus, over- or underdeveloped, poorly lighted, and so forth; however, they do take little physical space (and thus are easy to store and transport), can be duplicated inexpensively, can be magnified, and can emphasize important features. Physical models (i.e., the real thing) can be manipulated by the operator for a better view, can be disassembled, touched, and so on. (If you use physical models of an assembly as a training aid, save workstation space by mounting the models on a turntable.)

Line drawings generally can communicate information better than photographs because they are carefully designed to emphasize important features and omit irrelevant details. Line drawings reproduce well and exploded views can be used. (See the right corner of Figure 23.1 for an exploded view.) If using a sketch, use isometric views because many people have difficulty understanding drafting conventions.

2.3.2 Text messages   The **text message** will be discussed in categories of words, style, and format.

Words   The vocabulary of most workers is not as advanced as the vocabulary of university graduates (who often write the instructions). Workers are not comfortable with multiple-syllable or uncommon words such as "subsequent, prior, chartreuse, incorporate, and simultaneously"; they like little substitutes such as "after, before, blue-green, put in, and at the same time." See Table 32.3. So eschew obfuscation! In general, use Anglo-Saxon derivative words (such as pig, end, begin, first, and go) rather than their Latin equivalents (swine, finish, commence, initial, and proceed).

In addition, remember that English may not be the native language of some of your workers.

Style   Consider the complexity of your writing. Complexity can be quantified with the two following formulas. The first formula was developed by the Gunning-Mueller Clear Writing Institute:

$$GL = .4 \, (A + P)$$

where GL = **Grade level,** years (Readers at that grade level would be expected to score 75% on a comprehension test.)

A = Average words per sentence. Treat independent clauses (they start with the words *and, or, but, nor, yet* or with semicolons) as separate sentences.

P = Percent (not proportion) of words with three or more syllables.

For P, omit:

a. Capitalized words (except the first word in a sentence).

b. Easy combinations such as screwdriver and guestworker.

c. Verbs that reach three syllables with the addition of -es or -ed.

A simplified Flesch reading ease formula is (Kincaid and Fishburne, 1977):

$$GL = .4 \, A + 12 \, S - 16$$

where S = Syllables/word. A computer algorithm (Coke and Koether, 1983) for estimating S is:

$$S = V - .343 \, W$$

where V = total vowels (aeiouy)

W = total words

TV Guide has a *GL* of 6 and the *Wall Street Journal, Time,* and *Newsweek* average 11. People prefer to read below their grade level. You may wish to have more than 75% comprehension.

Also consider the mechanics of your writing. Some computer programs can check spelling. In addition, many of these programs will count the words per sentence (although most don't recognize independent clauses). Some more complex programs, such as Writer's Workbench, will deliver detailed measures and comments about text readability, punctuation, word use, and abstraction. For example, *spellwwb* checks spelling, *punct* checks punctuation, *double* checks consecutive occurrences of the same word, *diction* searches for phrases classified as words or frequently misused, *splitinf* finds infinitives split by adverbs, and *style* gives 71 numbers including readability indices, distribution of sentence lengths, whether sentences are simple or complex, the percent of verbs that are passive, the

TABLE 32.3   Use simple words.

Bad	Better
Parameters	Values, variables
Scrutinize	Look at
Incorporate	Put in
Verification	Check
Precede	Before
Prior	Before
Facilitate	Help, make easy
Subsequent	After
Simultaneously	At the same time
Via	By
Inquire	Ask
Equivalent	Equal
Simulate	Pretend

percent of nouns that are nominalizations (nouns created from verbs, such as transformation, establishment, admittance), and the number of sentences that begin with expletives (it, there).

Finally, realize that messages can be negative as well as positive. That is, negative examples often are a good teaching technique. However, the consequences of the negative example (the punishment) should not be severe. For example, if you make a mistake on a computer input, it should not be difficult to recover.

Format   See Section 2 (Arrangement of Characters and Symbols) of Chapter 20 for comments on text, codes, abbreviations, formulas, menus, data tables, graphs, projected images, and symbolic messages.

Tabular formats are superior to narrative formats; flowcharts give fewer errors than do narrative formats (Kammann, 1975; Wright and Reid, 1973). Table 32.4 gives an improvement over the following text.

When time is limited, travel by Rocket, unless cost is also limited, in which case go by Space Ship. When only cost is limited, an Astrobus should be used for journeys of less than 10 orbs, and a Satellite for longer journeys. Cosmocars are recommended, when there are no constraints on time or cost, unless the distance to be traveled exceeds 10 orbs. For journeys longer than 10 orbs, when time and cost are not important, journeys should be made by Super Star.

Since paper is cheap, don't condense tables or force the user to make calculations. Use "white space" to show organization of information. Keep information in specific behavioral terms. See Table 32.5.

3 TRAINING (MEMORIZATION)

Assume the decision has been made to train the worker in specific knowledge. More specifically, the worker is to use **memorization** to "know it by heart." Section 4 of this chapter, Job Aids, briefly covers an alternative method. With a **job aid,** the worker does not memorize the information but looks it up each time.

Assume that the knowledge to be transferred has been developed. The challenge is to transfer it from its present location to the mind of the specific worker. This transfer requires:

- subject knowledge by the trainer
- knowledge of how to teach by the trainer
- training materials
- training time
- a trainee willing and able to learn

Assume that the trainee is not a problem. Challenges remain.

3.1 Who?   The first challenge is who will be the trainer. Four possibilities will be discussed.

One possibility the British call **sit by Nellie;** that is, learn by observing a fellow worker. But Nellie may not know the *best* way of doing a job, although she may know *a* way of doing the job. Nellie probably has not been trained how to teach. Nellie will have access to some physical demonstration aids such as the product and the workstation but will not have any theory materials. If Nellie's time is used for training, it will decrease production.

A second possibility is to let the supervisor do the training. The problems of subject knowledge and teaching materials remain; the supervisor probably

TABLE 32.4 **Decision structure tables** communicate with fewer errors than text or decision diagrams (boxes connected by lines). Use of the connective words (if, and, then) and verbs (is) aid communication. Single and double lines help also (Wright and Reid, 1973).

If TIME is	and COST is	and JOURNEY LENGTH is	then TRAVEL MODE is
limited	unlimited	any	Rocket
limited	limited	any	Spaceship
unlimited	unlimited	less than or equal to 10 orbs	Cosmocar
		more than 10 orbs	Superstar
unlimited	limited	less than or equal to 10 orbs	Astrobus
		more than 10 orbs	Satellite

will have better theory knowledge and poorer practical knowledge of the task. Teaching knowledge may be slightly better, but demands upon the supervisor's time are many, so time availability probably is worse.

A third possibility is to hire a teacher to teach the workers. Subject knowledge may be a problem because the teacher will have to know many subjects. Teaching knowledge should be adequate, but teaching materials are very important since the teacher will not be very familiar with each subject. Teaching time may be a problem because the teacher may be overworked one week and idle the next since demands for teaching fluctuate. The cost of the teacher, however, now has become a budget item and becomes visible. Thus, the organization may want the teacher to teach a group of workers rather than an individual so that the teaching time can be prorated over more people. Yet, because there are limited requirements for specialized knowledge, industrial class sizes tend to be small—5 to 10 students. If workers are released from a job to attend classes (and, most important, are paid to attend), this attracts additional attention to the training costs.

This has led to a fourth approach—eliminate the teacher. Self-service has been used in many fields, so why not education? In **self-instruction,** the trainer and trainee are the same person, but the trainer, of course, has no subject knowledge. Knowledge of how to teach is small although possibly somewhat offset by knowledge of what needs to be known. Scheduling problems are minimal. However, good training materials assume great importance since there will be no teacher to emphasize what is important, to answer questions, and to correct errors in the training materials. Self-instruction can benefit from the knowledge gained through research on programmed instruction (see Box 32.1) by Skinner.

3.2 Training Principles

How did Skinner train pigeons?

Consider teaching a pigeon to turn a clockwise circle. First, watch until it turns part of a clockwise circle, then give it a piece of grain. Continue giving grain for turning but gradually require more and more of the circle for the reward—that is, shape its behavior. As shown in the pigeon training by Skinner, there are five principles in **programmed learning:**

Principle	Pigeon Example
1. Define specifically the exact behavior you wish the trainee to do.	Turn a clockwise circle.
2. Present the information in small increments, steps, or modules.	Small arcs first.
3. Present information at a pace determined by the trainee.	Pigeon sets pace.
4. Give immediate feedback of the correct answer (knowledge of results) to the trainee.	Grain for correct movement.
5. Observe trainee behavior. If not satisfactory, modify program.	If pigeon doesn't learn, try another technique.

Research indicates that, when applied to humans, steps 2 through 4, although desirable, are not essential. The essential steps are 1 and 5.

Step 1, specific behavior, means we must specifically define which motor movements we want. Glittering generalities such as "know the job" or "understand the process" are replaced by "if the temperature on gauge 7 rises above 95 C, turn the red

---

BOX 32.1 *Programmed instruction*

During World War II, B. F. Skinner, a psychologist from Harvard, wanted to help the war effort. When they asked what he did he explained his research involved training rats to run through a maze. It turned out that the Air Force had a problem. When they dropped a bomb on a ship, the ship turned; thus the bomb missed. The Air Force thought it would be nice if the bomb could turn with the ship. Electronics was in its infancy so the "pilot" had to ride the bomb.

Although the Japanese used people as pilots, the Americans decided to use pigeons as pilots. The pigeon would look at a TV picture of the ship, sent by a camera in the bomb nose. Then the pigeon would peck at the picture of the ship, and by sensing where the screen was being pecked, the bomb could be steered. But who would train the pigeons? Enter Skinner!

The war ended before the project succeeded. Then one day in 1947 Skinner came home furious after visiting his daughter's kindergarten class. "I give my pigeons better training than my daughter gets." Skinner then began developing programmed learning for humans.

switch on column 7 to 'off' and call your supervisor." See Table 32.5. Give training programs the same attention as you give to computer programming. Poor teaching is often simply insufficient attention to detail.

Step 2, small steps, says that information is more digestible in small bites. Many studies have indicated better training when a large body of knowledge is divided so that training is in parts. The best size of the bite may depend on the appetite.

Step 3 is self-pacing. Let trainees receive the message at their own rates rather than at some rate appropriate for the average. If average is defined as those between the 40th and 60th percentile, then average includes only 20% of the group. Any presentation at a rate appropriate for the average is inappropriate for 80% of the group. Ideally, the level of difficulty can be varied as well as the rate of presentation so that instruction is personalized.

Step 4 is immediate feedback to the trainee. Delay of the correct answer for a previous question may mean that the trainee may be misled on much subsequent material. Even more important is the motivational value of quick confirmation of the correct response. Positive feedback tends to work better than negative feedback. Everyone likes praise.

Step 5, immediate feedback to the trainer, is important. Feedback distinguishes an efficient closed-loop system from an open-loop system in education just as it does in any other system; the feedback system reduces system error. Reduction of system error through program change can be summarized as "there are **no stupid students, just terrible teachers.**" It stems from Skinner's belief that he was smarter than any pigeon. If the pigeon did not learn, it was not the pigeon's fault; it was up to Skinner to develop a better teaching method. This concept, that poor learning is the teacher's fault, is hard for many teachers to accept. They reply that some students could not learn nuclear physics no matter how well they were taught; they have neither the capability nor the motivation. True. Therefore, use the "no stupid students, just terrible teachers" concept to keep teachers on their toes and doing the best possible job instead of looking for excuses.

3.3 Simulation

Most workers learn on the job. They produce product as they learn. However, there are some situations in which workers need a very high level of skill before they produce product. In that case, they can practice in a **simulation** of the environment. Some examples are the following:

- Actors and actresses have specific practices (rehearsals, dress rehearsals) before the paid performance; perhaps they will even have shows on the road before bringing it to the big city. Also, in movies and video, the final version is selected from many takes.
- Athletes practice specific skills and also have scrimmages, games against easy opponents, and so on.
- Soldiers spend most of their time practicing, at the individual level and at various levels of group activities.
- Pilots have classroom training but also training on machines called simulators. These simulators have considerable fidelity. The instruments respond, the simulator vibrates, and the visual and auditory environment of flight is simulated.

In general, use simulation for training when the cost of an error in real performance is high.

4 JOB AIDS

An alternative to memorization of information is to have the information constantly available for reference. Job aids (also known as job performance aids) can be divided into procedural and psychomotor.

4.1 Procedural

Present procedural (big picture) information in decision structure tables such as Tables 32.6 and 32.7 or a routing sheet such as Table 32.8. These tables give various decisions which have been made about processes. These decisions can be made in advance by experts rather than by inexperienced operators under time pressure, who lack all the relevant information. The tables also serve as a convenient storage location for tool numbers, standard costs, and so forth. They serve as a

TABLE 32.5 Keep information in specific behavioral terms.

Poor	Better
Unlock box.	With left hand, hold the latch on the left side. With the right hand, insert key 14 and turn clockwise.
Inspect nameplate to ensure that it has been properly installed.	Inspect nameplate. Place 75 mm from top and 100 mm from left side of cover. Nameplate should not be able to be pried up with fingernails.

TABLE 32.6   One explicit form of written standard instruction is a decision structure table. This specific example gives the check cashing policy of a grocery store.

If TYPE OF CHECK is	And AMOUNT OF CHECK is	And BANK OF CHECK is	And CUSTOMER ADDRESS is	Then CUSTOMER IDENTIFICATION REQUIRED is	And ON BAD CHECK list	Then DECISION IS
Two-Party	Any	Any	Any			Reject
Company	Up to $25	Any		0		Accept
	$25.01 to $200	Any	Any	1		Accept
	$200.01 and up					Reject
Personal	Up to $25	Local	Local	1	Yes No	Reject Accept
			Out of Town	1	Yes No	Reject Accept
		Out of Town	Local	1	Yes No	Reject Accept
			Out of Town	2	Yes No	Reject Accept
	Over $25	Any	Any			Reject

TABLE 32.7   An example of a decision structure table as applied to selecting either the drill size before tapping or the clearance drill size when you wish a hole large enough so that the drill will not touch the bolt threads.

National Special Thread Series			
If BOLT DIAMETER, INCHES is	And THREADS PER INCH is	Then DRILL SIZE FOR 75% THREAD is	Then CLEARANCE DRILL BIT SIZE is
1/16	64	3/64	51
5/64	60	1/16	45
3/32	48	49	40
7/64	48	43	32

good training aid. In computer-aided manufacturing (also called computer-integrated manufacturing), these tables, routing sheets, and the like are stored in a computer—a step toward the paperless office. The computer permits an easily updated central file, complex decision structure tables, and facilitates comparison with similar parts (group technology)—assuming the similar parts information also is in the database.

The tips in Table 32.9 were developed for emergency warning messages, but they apply to most job aids.

More complex versions of decision structure tables are called **expert systems.** The expert system may ask questions of a user and then suggest a solution (forward chaining), or it may take a solution suggested by the user and confirm or deny it (backward chaining). Development of the expert system often forces the experts to re-evaluate their knowledge and improves the experts' knowledge (Hartley and Rice, 1987).

Instruction manuals, whether on paper or computer, should be indexed. The indexing should be from the user's viewpoint, not from the author's viewpoint. Use organizing aids such as tab dividers, page headers, variable print size, and emphasis. Panel et al. (1993) point out that design of job aid documents should consider

1. *information readability* (typographic layout cues, including bold, italics, underlines, paragraphs, justification, etc.)

2. *information content* (appropriate content, balance between text and graphics)

TABLE 32.8 **Routing sheets** (also called operations charts) vary depending on the organization, but the table below is typical. Operation numbers are given in multiples of 10 so that operations originally omitted can be added without changing the numbers of the other operations. Some routing sheets also include bill of material information. The direct labor hours per unit should not include process time such as for drying.

Part name ___Punch___   Part number ___541-675___   Raw Material ___1040 10 mm Round___

Operations ___SK___   Date ___20 Jan 89___   Used on ___Model 80___

Operation Number	Operation Name	Machine	Tooling	Feed, mm/rev	Speed, rev/min	H/unit	Remarks
10	Turn 4 mm dia Turn 3 mm dia	J & L T. Lathe	#642 Box	.225	318		
	Cut off to length	J & L	#6 cutoff	Hand	318	.008	
20	Mill 5 mm radius	#1 Milwaukee	Tool 84	Hand		.004	
30	Heat-treat	#4 furnace				.006	
35	Degrease	Vapor degreaser					
40	Measure hardness	Rockwell tester				.002	
50	Store						

TABLE 32.9 Tips for emergency warning messages (Johnson, 1980).

- Use pictures, not just words
- Use 3-dimensional pictures, not 2-dimensional
- Use pictures rather than symbols (which require learning)
- Number sequential drawings
- Show both initial and final locations of controls
- Present what not to do as well as what to do
- Use drawings, not photographs (less irrevelant information, can emphasize important items, easier to change)
- Show time pressure with clockface; time changes in each picture
- Evaluate message on typical users (*not* equipment designer)

3. *information organization* (classifying and layering information into categories; giving information in the format of command verb, the action qualifier, and the objects)

4. *physical handling and environmental factors* (job aid must be rugged, lighting adequate, etc.)

4.2 Psychomotor

Present **psychomotor** ("fine-grain picture") **information** in pictorial instruction sheets (see Figure 32.2) or videotape (Booher, 1975). See Box 16.2. Emphasize grasp and position elements as these are the skill elements. Move and reach are less important. In general, using words only is fair, using pictures only is better, and words + pictures is best. The pictures explain the how; words explain the why. Words such as "element breakdowns" (as with MTM) may communicate to other engineers but don't give workers the necessary details. If using videotape, have operators add audio comments with a voiceover while viewing themselves. For photographs, instant photos work well because you can take repeated pictures until you get it just right. Then staple the picture to a sheet of paper and have the operator add supplementary words. Leave space for changes. See also Table 32.9.

Leave a copy of the instructions at the machine, as well as having a copy for the office. Avoid the "gold plating" of professional photographers, typed instructions, elegant mounting, and so forth. Put in a loose-leaf folder or laminate for protection. If an elegant copy is desirable, wait until instructions have been polished by the users; count on several drafts as most first drafts are poor. More polished instructions can be justified if the procedure is a standard manufacturing process and the instruction is widely distributed. If the instruction sheets are in the office only and not at the point of use, they lose most of their effectiveness.

Note that after workers obtain the job knowledge (e.g., how to fold a cardboard box), they will need many repetitions before knowledge becomes a skill.

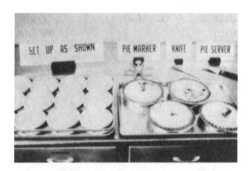

FIGURE 32.2a   Stacking enough plates to fill the tray is the first step of cutting a two-crust pie.

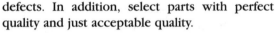

FIGURE 32.2b   Holding a marker by the edges makes it easier to center: (1) more accurate, (2) hands don't obscure the pie center.

FIGURE 32.2c   Use the edge of the pan as a guide to position the marker in the pie center.

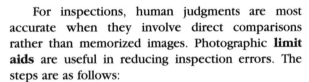

FIGURE 32.2d   Press the marker lightly on the crust. You need guide marks only.

For inspections, human judgments are most accurate when they involve direct comparisons rather than memorized images. Photographic **limit aids** are useful in reducing inspection errors. The steps are as follows:

1. Establish a panel of experts to define what is acceptable and not acceptable for each type of defect.

2. Select parts with defects. These defects should range from very bad defects to just rejectable defects. In addition, select parts with perfect quality and just acceptable quality.

3. Arrange the parts in a column with best on top, followed by descending quality. Put a gap between the just acceptable and just rejectable defect. Mount the parts on a display board or, if space is limited, as it usually is, photograph them.

4. Give each inspector and operator a set.

REVIEW QUESTIONS

1. What is the primary resource of any organization?

2. Give 3 advantages of cross-training.

3. Briefly discuss the transitory and pacing problems of audio messages.

4. Give an example of a linear presentation and of a branching presentation.

5. What are the Latin derivative words for the following Anglo-Saxon derivative words: *pig, end, begin, first, go?*

6. List the 6 things with which an independent clause begins.

7. Rewrite the sentence "The next step is to drill and chamfer the 50-mm hole," with 2 short sentences starting with an action verb.

8. What is the difference between training and a

job aid?

9. Using pigeon training for an illustration, list the 5 steps of programmed instruction.

10. Discuss the concept "There are no stupid students, just terrible teachers."

11. In training of psychomotor movements, is it more important to emphasize the Grasp and Position elements or the Move and Reach elements? Why?

12. Give the 4 steps of developing limit aids for inspection.

REFERENCES

Booher, J. Relative comprehensibility of pictorial information and printed words in proceduralized instructions. *Human Factors,* Vol. 17 [3], 266-77, 1975.

Bouzid, N. and Crawshaw, C. Massed versus distributed word processor training. *Applied Ergonomics,* Vol. 18 [3], 220-23, 1987.

Bradeley, A. and Longman, D. The influence of length and frequency of training sessions on the rate of learning to type. *Ergonomics,* Vol. 21 [8], 627-35, 1978.

Coke, U. and Koether, M. A study of the match between the stylistic difficulty of technical documents and the reading skills of technical personnel. *Bell System Technical Journal,* Vol. 62 [6], 1849-64, 1983.

Dickey, G. and Konz, S. Manufacturing assembly instructions: A summary. *Ergonomics,* Vol. 12 [3], 369-82, 1969.

Hartley, C. and Rice, J. A desktop expert system as a human factors work aid. *Proceedings of the Human Factors Society,* 1087-90, 1987.

Johnson, D. The design of effective safety information displays. *Proceedings of Human Factors and Industrial Design in Consumer Products,* Tufts University, 1980.

Kammann, R. The comprehensibility of printed instructions and the flowchart alternative. *Human Factors,* Vol. 17 [2], 183-91, 1975.

Kincaid, J. and Fishburne, R. Readability formulas for military training manuals. *Human Factors Society Bulletin,* July, 1977.

Nugent, W. A comparative assessment of computer-based media for presenting job task instructions. *Proceedings of the Human Factors Society,* 696-700, 1987.

Panel, S., Drawer, C., and Prabhu, P. Design and usability evaluation of work control documentation. *Proceedings of the Human Factors and Ergonomic Society,* 1156-60, 1993.

Stern, K. An evaluation of written, graphic, and voice messages in proceduralized instructions. *Proceedings of the Human Factors Society,* 314-18, 1984.

Templin, N. Auto plants, hiring again, are demanding higher-skilled labor. *Wall Street Journal,* March 11, 1994.

Wright, P. and Reid, F. Written information: Some alternatives to prose. *J. of Applied Psychology,* Vol. 57 [2], 160-66, 1973.

33 | RESISTANCE TO CHANGE

OVERVIEW

A technical proposal is useless unless it is accepted and implemented. Change involves the sociopolitical world as well as the world of facts. Small group techniques, which move decision making lower in the organization, not only improve fact gathering, but can also improve acceptance through resolution of emotional/attitudinal problems. Quality Circles are an example of a small group technique.

Recommendations are given on how to improve acceptance of proposals.

CHAPTER CONTENTS

1 Challenge of Change
2 Process of Change
3 Quality Circles
4 Implementing Change

KEY CONCEPTS

area of freedom	force diagram	project scope
client's servant	frequent consultation	Quality Circles
defense before offense	industrial democracy	quality vs. acceptance
emotions and attitudes	local experts	solution space
facilitator	Pareto diagram	Taylorism
fish diagram	passive approach/active approach	vertical barrier/horizontal barrier

1   CHALLENGE OF CHANGE

Most of the chapters of this text discuss obtaining a technically excellent design. Material is given concerning hand movement patterns, handtool design, workstation design, noise, toxicology, and so forth. However, when the engineer has decided upon a technically correct concept, the concept must be translated into practice. Translation of a concept into practice involves the sociopolitical world as well as the technical world. Engineering students, raised on a diet of mathematics and physical science, often have difficulty adjusting to this sociopolitical world, which treats their technical "solution" as merely a "proposal."

$$\left(\begin{array}{c}\text{Technical } \textbf{quality} \\ \text{of a proposal}\end{array}\right)\left(\begin{array}{c}\textbf{Acceptance} \\ \text{of a proposal}\end{array}\right) = \begin{array}{c}\text{Amount of} \\ \text{improvement}\end{array}$$

Leismann (1977), when reviewing this chapter in the first edition, commented:

> Without acceptance, nothing has happened. The technique or technology may be a thing of beauty, the analysis may be perfect and even foolproof, the proposal may be well written and well presented and overwhelmingly justified, with excellent payback. But it is nothing, zero, a total waste of time, unless it is acceptable to those who are impacted.
>
> Your skill at implementing change will determine how far and how fast you will be recognized by your supervision and your peers as having the potential for increased responsibility in your organization. This skill in total consists of 85% approach and 15% technical application.
>
> The bottom line of engineering is implementation—there is no other.

The statement that "where you stand depends upon where you sit" expresses the fact that whether you are for or against a change may depend upon which job you presently are holding. Staff units value change because that is how they prove their worth; line units value stability because change inconveniences them or reflects unfavorably on them. On the other hand, staff units are strongly committed to preserving control and rule systems (that's where they get their information and their power) while line units prefer flexible interpretation of control systems (we'll do the paperwork if we have time).

2   PROCESS OF CHANGE

2.1   Forces
Munson and Hancock (1969) break the change process down into five steps (see Table 33.1). The situation starts with stability, "unfreezes,"

Stage	Description
Frozen	stability
Unfreeze	rethink goals; consider alternatives
Move	try new methods; feelings of insecurity
Refreeze	"I like it;" "it really works"
Frozen	stability

"moves," "refreezes," and then is stable again. Only trivial technical changes arouse no resistance. The question, then, is how to overcome the resistance.

As an example, consider three workers (Joe, Pete, and Sam) rotating every two hours among jobs A, B, and C on the subassembly of product X. They are paid by the piece. The engineer proposed that Joe always do job A, Pete always do job B, and Sam always do job C. The engineer computes that their pay will increase 25% (due to increased specialization and thus more pieces/h) and the organization's cost/unit will drop (due to lower burden cost/unit even though labor costs/unit remains constant). However, the workers refuse to change.

Figure 33.1 gives a force diagram of this situation. The pointer can be pulled toward the present (the status quo) or toward what is proposed. Facts

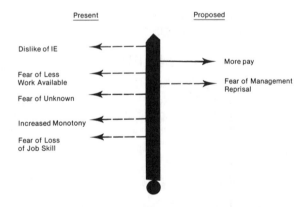

FIGURE 33.1   **Force diagram** shows the forces pulling a pointer toward and against change. Emotions and attitudes are very important factors in a potential change situation. Engineers tend to overlook the critical importance of the worker's social circle breakup, fears of the unknown, and so forth. One common technique of reducing financial fears is to have the organization guarantee no financial loss to the worker. If labor requirements are reduced, the worker is transferred to some other job (with no loss in pay) and the organization "eats" the cost of the surplus worker until attrition reduces the workforce. However, even with no financial loss, the workers may not approve of the social losses, the changes in power, and the disruptions and may therefore resist change.

are shown as solid arrows and emotions–attitudes are shown as dashed arrows. Note: (1) the many arrows (i.e., how complex even small problems can be) and (2) the preponderance of emotions–attitudes over facts.

The pointer can be made to go to the proposed side by increasing the forces toward the proposed (increase the $+$), by decreasing the forces toward the present (decrease the $-$), or by combining the two.

As a general strategy for dealing with **emotions and attitudes**, reduce the negative ones (pulling toward the present) rather than increase the positive ones (pulling toward change). People often resist the social aspects of change rather than the technical aspects. Increasing positive emotions and attitudes may be counterproductive. For example, increasing the employees' fears of management reprisal for not making the change may just set up counter emotions to resist the change ("it's them or us"; "solidarity forever").

Another example of counterproductive behavior can be bringing high levels of management into a problem when lower levels of management are "in error." The lower levels, then, in self-defense, become passionate defenders of the status quo. The change agent should try to avoid such defensive behavior by presenting the proposal in a way that avoids blaming individuals.

Breaking a large change into a series of smaller steps usually will aid implementation. Many people fear (with reason) giant steps. A common technique is the test market, where the change is implemented in one "market" and the results are studied before being implemented everywhere.

2.2 Area of freedom

Figure 33.2 shows how the **solution space** is reduced (limited, constricted, restrained) by various economic, legal, technical, or policy factors to the net **area of freedom.** For example, an engineer might consider a particular task satisfactory, but corporate policy may say no due to the risk of injury. (In addition to the moral responsibility to protect the workforce, there are severe economic penalties to a firm for injuries. For example, a back surgery may cost over $10,000. More and more corporations are charging the medical expenses of injuries to the plant, rather than retaining the charge at the corporate level. Another technique is to have corporate pay the actual expenses of an injury but charge a flat fee to the plant, such as $20,000 per permanent disability or $500/day for a lost time injury. This is corporate's way of getting the attention of plant-level management.)

2.3 Incomplete Communication

Resistance to change may be due to lack of knowledge of some of

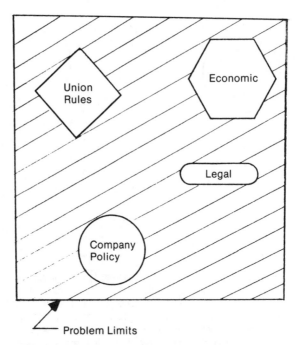

FIGURE 33.2   Decisions are limited by the area of freedom. Limitations might be economic (keep capital cost less than $10,000), union rules (all maintenance work must be done by in-plant workers), company policy (all reductions in the workforce must be by attrition), legal (noise level must be less than 90 dbA), and so forth.

the facts and emotions. For example, an IE might not know all the details of the union contract and may be proposing changes not permitted by the contract. Important facts to communicate include the reason why the change is being proposed and what the expected consequences are.

Figure 33.3 makes this point more explicitly. People concerned with a problem include technical experts or outsiders (such as IEs) and **local experts** or insiders (those doing the job). Note: (1) The technical experts tend to be "long on information and short on emotions," and (2) not all facts or emotions are known to each party. Thus, coordination among people should increase the knowledge available to everyone. This should improve not only solution quality but also solution acceptance.

2.4 Small Groups

If it is accepted that problems are complex, that there are emotions and attitudes as well as facts involved, and that no one individual knows everything, then the concept of having a group of people work on the problem seems logical.

Increasingly, this group of people includes not only staff and management personnel but also the workers themselves—the local experts. That is, the best combination, from a factual information viewpoint, is a combination of technical experts and local experts.

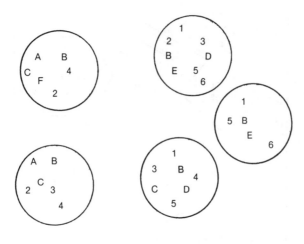

Technical experts   Local experts

FIGURE 33.3 Knowledge concerning a problem is spread through the members of both the local experts and the technical experts. Facts are depicted by letters (e.g., A = optimum work height is at the elbow, B = the machine has a bad bearing, C = capital cost of a roller conveyor is $2,000, etc.) Emotions and attitudes are depicted by numbers (e.g., 1 = we are underpaid, 2 = productivity in the department is unsatisfactory, 3 = that engineer is a stuck-up snob, etc.)

In some situations, an outside technical expert may become part of the team. This consultant may be useful because of special knowledge/ability, lack of time of the other technical experts, or because there is a lack of trust among the other participants.

Working with a diverse group of people with diverse backgrounds requires negotiation. The negotiation process has a number of advantages: (1) Negotiation takes time and thus permits people to change their concepts gradually, (2) clearly unacceptable designs will be eliminated (these bad designs generally occur because the technical experts didn't understand the complete problem or the complete consequences of their proposal), and (3) negotiation improves the communication between the groups. The communication gives the technical experts better facts for their proposal and reduces unnecessary fears among the inside experts. It helps the technical experts understand the (usually sound) reasons of inside experts for opposing changes; therefore, the necessary modifications can be made. The value of a dissenting minority is not so much in the correctness of minority members' position but in the attention and thought processes that result to refute their position. It very well may be that those resisting the change understand the situation better than those advocating the change! In particular, there often are differences in who pays the costs and who receives the benefits; there also may be a time gap between costs incurred by the workers and benefits received. The benefits of the group working on the problem

instead of a single staff person also tend to be long run, whereas the additional costs tend to occur quickly.

The negotiations may take place in 2 stages. The first stage is among members of the small group. The second stage is when the group presents its recommendation to management.

In addition to the factual benefits, since problems include emotions and attitudes, participation by the people directly affected by the change reduces emotional and/or attitudinal problems. Being involved in the change process may make people feel ownership of the idea and thus be more willing to implement it than with a change imposed from above. This can be considered **industrial democracy,** with many of the benefits that come from democracy in the political arena. However, many supervisors retain a desire for "aristocracy." Once you obtain power it is hard to give it up. The power sharing occurs mainly at the first-line supervision level and that tends to be where there is most resistance to industrial democracy.

Small groups of technical experts and local experts have been used on a limited scale in the United States for many years. The Hawthorne studies at Western Electric in the 1930s demonstrated the importance of worker involvement. Also see Coch and French (1948) and Chaney (1969). Other examples are the Scanlon plans (Frost et al., 1974), Evolutionary Operation of Processes (see Chapter 4), and the work simplification of Allan Mogensen. Safety committees commonly involve blue collar workers. See also the many articles on quality of work life (QWL).

The most popular form of these small groups is called **Quality Circles** (see the following section for more on Quality Circles). They also are called Productivity Circles, Quality Teams, and Effectiveness Teams. Data General calls its groups PRIDE Circles, where PRIDE stands for People Really Involved in Developing Excellence.

The Taylorism concept of workers blindly following the orders of their supervisors has been made obsolete by the increasing complexity of society and the increased education of workers (see Figure 33.4).

3 QUALITY CIRCLES

3.1 Historical Background   Quality Circles began in Japan in 1963, developing from the needs of Japanese industry and the characteristics of Japanese society and management.

The needs of Japanese industry concerned its long-range goals. Previously, Japan had competed on the basis of low labor costs. However, the Japanese

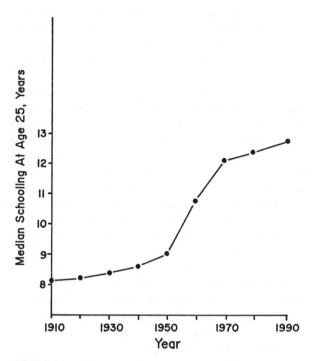

FIGURE 33.4 Education levels have increased since the time of Taylor. Workers had only a grade school education when Taylor formulated his policy of job design by experts without worker contribution.

wanted to improve their standard of living, and this required higher wages. In addition, there were millions of competitors in Asia willing to work for very low wages. The Japanese therefore decided to compete on the basis of quality and technology. To get that quality they decided to get "everyone" to help rather than just the elite group of engineers and managers—the idea being for quality (of product), use quantity (of people).

Japanese society and management style aided this concept of mass participation in decision making. Japanese management also emphasizes consensus by groups and a long-range goal.

The standard approach to job design in the developed societies since Frederick Taylor has been to have technical experts design the job and have the workers follow the instructions, a policy often called **Taylorism.** As pointed out in Chapter 3, it has been extraordinarily successful in improving productivity and the standard of living in many societies.

The Japanese, however, have a highly educated workforce—even more educated than the United States workforce shown in Figure 33.4. Japanese industry decided that the elite groups of engineers and managers would still work on the major problems. However, the masses would work on the many minor problems. The emphasis would be on quality problems.

Training would be necessary to give ordinary blue collar workers the ability to solve quality problems. This training would be at two levels: (1) specific technical techniques (how to plot a histogram, how to do a Pareto analysis) and (2) administrative skills (e.g., how to run a meeting, how to make a presentation).

It also would be necessary to give the workers an organizational mechanism through which to use the newly learned skills.

The training materials and procedures and the organizational forms were developed. A massive program was implemented.

The result is that Japanese products have a reputation for quality. Almost all of Japan's 54,000,000 workers have some quality training. Approximately 8,000,000 have had technical courses in quality from the Union of Japanese Scientists and Engineers. Membership in Quality Circles rose explosively from 10,000 (in 1,000 Circles) in 1964 to 80,000 (in 10,000 Circles) in 1966 to 6,000,000 (in 600,000 Circles) in 1980; 6,000,000/54,000,000 is about 12% of the workforce in Quality Circles (Arai, 1979; Berger and Shores, 1986).

By the late 1970s the Quality Circle concept had spread to Korea, Taiwan, Brazil, Europe, and the United States.

The technical group in the United States is the Association for Quality and Participation (AQP). In 1994, it had over 3,900 firms as members. No one has reported the number of Quality Circles in the United States, but one firm, Honda of America, has over 4,500 Circles. The AQP address is 801-B West 8th Street, Cincinnati, OH 45203; phone (513) 381-1959.

3.2 Circle Design   Circles generally are coordinated by a supervisory committee (board of directors; steering committee). The committee is responsible for establishing program policies, procedures, objectives, and resources. This overview responsibility includes publicity, training programs, coordination with management, and the like. Generally, a diverse committee works best—that is, people from line management, engineering, purchasing, perhaps a union member, and others—because a large responsibility of the committee is to maintain acceptance of Quality Circles throughout the organization.

The steering committee normally delegates its detailed supervision of Circles to a **facilitator.** The facilitator's job is to train the Circle members, set up Circles, perhaps run their meetings for the first few times, and, in general, to facilitate their operation.

Circles range from 3 to 25 members, but a typical size is 10. Membership is voluntary. Membership is primarily blue collar workers.

First of all, the Circle should have approximately five core operators from a specific group—these members should "touch the product." Next, the Circle composition should consider the problems of the horizontal and vertical organizational barriers.

The **horizontal barrier** is the barrier between hourly and salaried workers—between blue collar and white collar workers. A Circle should have one or two white collar workers. A common choice is the supervisor. This not only permits the Circle to break the horizontal barrier but also to allay the suspicions and mistrust supervisors might have if they are excluded from the group. In addition to bridging the horizontal barrier, another technique is to reduce it. For example, at a Cummings Engine plant in Columbus, Indiana, both production workers and managers wear company-issue khaki pants and dress shirts with no neckties.

The **vertical barrier** is the barrier between groups (Dept. A vs. Dept. B, operators vs. inspectors, operators vs. maintenance, operators vs. scheduling, etc.) The Circle should have members from these support groups.

Thus the Circle should have about five core blue collar workers and five support workers (two white collar, one being the supervisor).

Circles meet from once/week to once/month. The meeting can either be on company time or after work. In general, the members do not receive any additional pay for their work nor any financial reward for cost savings. However, there is a heavy emphasis on praise, publicity, and public thanks. For example, Ford Motor Co. has a policy by which it picks the best projects from its Quality Circles. Ford then flies the team members to Detroit for several days, tours them around various plants, and has them make their presentations to the chairman of the board in a fancy conference room. They all receive T-shirts saying "I made a presentation to the chairman of the board." They leave Detroit "feeling 10 feet tall."

Specific projects usually are selected by a Circle, with management and the steering committee having veto power. Projects primarily concern quality, although many Circles also include safety and housekeeping projects. In general, Quality Circles should stay away from cost reduction projects, especially those involving reducing labor. That type of project should remain with the technical staff and management. Circles should instead focus on the "insignificant many" problems and let the staff and management work on the "vital few."

3.3 Example Circle Project

The following example shows how Ms. F. Hashimoto's Quality Circle team at Matsushita Electric analyzed switches used for the volume control on stereos (Konz, 1979).

Step one was to select the project. Among the defects from the assembly line, the largest percentage was attributed to the volume switch. After consultation with management, the Circle decided to study switch defects.

Step two was to analyze present conditions. Figure 33.5 shows their **Pareto diagram** for defects over a 3-month period. The Y axis is percent defective. The X axis is a series of bars, arranged in descending magnitude, of various causes. They crosshatched the major cause for emphasis. Then they plotted cumulative defects (line with dots). (For additional discussion of Pareto diagrams, see Section 3 of Chapter 4.) Rotation caused 70% of the defects, so switch rotation was picked as the way to reduce switch defects.

Figure 33.6 shows an analysis of the types of rotation defects. 87% were uneven rotation. Then the Circle used the **fish diagram** (cause–effect diagram) given in Figure 33.7 to organize the problem and improve communication. With Figure 33.7 as a guide, they collected defect data from the in-process inspectors, sorted it by cause, and developed Figure 33.8.

The third step was to establish goals. At their next Circle meeting they established three goals:

1. Reduce rotation defect rate from 1.3% (as of January 1965) to 0.5% by December 1965, while introducing a more stable control system.

2. Develop an overall Circle implementation plan (Figure 33.9).

3. Selectively attack the problems for improvement using the Pareto chart order.

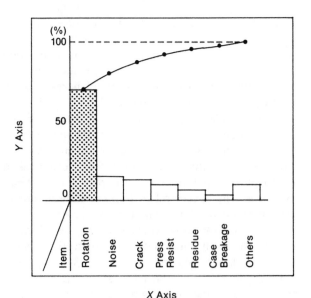

FIGURE 33.5   A Pareto analysis, using three-month data, showed that switch rotation accounted for 70% of the defects. Thus, it was selected for further analysis.

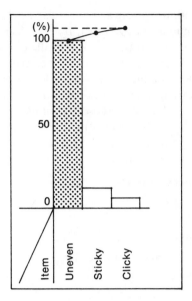

FIGURE 33.6   Another Pareto analysis, analyzing rotation defects by phenomena, showed that 87% of the problems were uneven rotation.

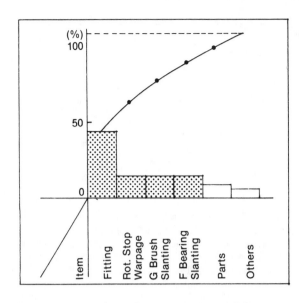

FIGURE 33.8   Using the fish diagram in consultation with the in-process inspectors, the Circle developed a new Pareto diagram of causes of uneven rotation.

The fourth step was to promote control activities. One activity was an *np* chart (Figure 33.10). For their production, a sample (n = 400) taken every hour was appropriate. When an out-of-control condition developed (point beyond control limits), the Circle had a meeting. Depending upon the nature of the problem, they either determined the problem cause or a measure to prevent recurrence of the problem. In order to control common deficiencies in previous operations, checklists were developed for critical control points. The operators used these checklists to check their own work. After improved procedures were put into practice and found workable, the standard procedure was revised to ensure continued use of the new method. Due to the smaller number of defects, the sample size was changed from 400 to 1,600.

As a result of these activities, by the end of a year the defect rate was reduced from 1.3% to 0.3% for an annual savings of 400,000 Y (about $1,000 at the exchange rate in 1965). In her paper, Ms. Hashimoto concluded that although they had achieved their goal, they had not completely eliminated the problems. They were determined to continue

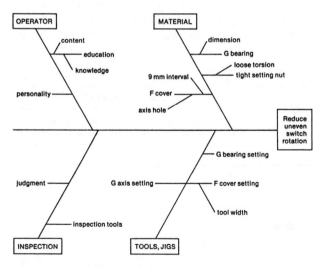

FIGURE 33.7   Uneven rotation errors were the target of a cause–effect diagram. The fish head is the goal, major bones are the major categories of the production process, and minor bones are the subdivisions of the major categories.

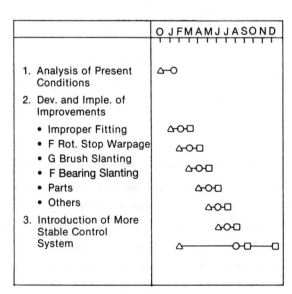

FIGURE 33.9   Goals and a schedule were set by the Circle at their next meeting.

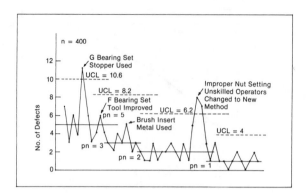

FIGURE 33.10   An *np* chart helped identify when defects occurred so their causes could be eliminated. The first control limits describe the original process. The second set of limits describes the results after the second and third Circle meetings. The third set describes results after the fourth, fifth, and sixth meetings, and the fourth set after the seventh meeting.

improvement to achieve a still better result. In addition, she commented that, as an inexperienced Circle leader, she had not worked enough with people from other departments.

A number of comments can be made about the example project. First, this was a small project with a $1,000 annual savings. Yet a considerable amount of work was necessary to achieve the savings. In the many examples of Japanese Quality Circles, I have not seen return on investment reported. Typically, the number reported is annual savings or percent defects. This implies that the engineering cost (that is, the cost of the time to make the analysis) and capital cost of making the change are not considered worth reporting. However, this analysis cost may be quite low as analysis is done during normal work time; meeting time probably is charged to training or general quality costs. Thus, a quality improvement was made that many United States firms, with a heavy emphasis on cost accounting, would not undertake.

Second, the Japanese blue collar Circle members used techniques in 1965 that, by American standards in 1994, are sophisticated. Pareto diagrams, fish diagrams, and *np* control charts are not even known to many American engineers—much less used. The key here is management's emphasis on training Circle members in the use of these techniques and encouraging their use. Gryna (1981), in discussing the need for training for Circles in the United States, gives the following training modules:

- introduction to Quality Circles
- brainstorming
- fish diagrams
- histograms
- checklists and data recording
- case study
- how to make graphs
- how to make a presentation

In Cummins' Columbus, Indiana, plant, new workers get 250 h of training, including 72 h of math and 36 h of statistical process control. (Note that this requires considerable selectivity in who is hired.) In addition, there is a need to train Circle leaders in interpersonal skills.

Third, quality problems were considered to be technical problems, not motivational problems. The improvements they implemented were modified handtools, modified assembly procedures, modified operator training, and use of checklists by the operators.

Fourth, many quality problems are communication problems. Fukuda (1978, 1981) summarized the results of 87 groups at Sumitomo (see Table 33.2). Figure 33.11 is a modified version of one of his figures showing that information must be known to both the operator and the technical staff in order to be

TABLE 33.2   Types of countermeasures found effective by Quality Circles at Sumitomo (Fukuda, 1978, 1981).

| Countermeasure type | Type of result (%) | | | |
| | Limited | | Considerable | |
	Slow	Quick	Slow	Quick
Warnings and suggestions put into easily readable visual form—defensive	41	25	57	68
Clearly defining standard operations—defensive	24	35	24	53
Developing tools enabling less effort yet more skill to be put forth—defensive	12	5	37	32
Making improvement in equipment—offensive	5	5	30	12
Making changes in manufacturing conditions—offensive	0	5	20	5

Quick = within 3 months
Slow = over 3 months

Limited = Reduction of defects of less than 40%
Considerable = Over 40%

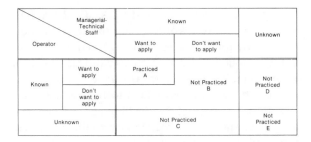

		Known		Unknown
	Managerial-Technical Staff			
Operator		Want to apply	Don't want to apply	
Known	Want to apply	Practiced A	Not Practiced B	Not Practiced D
	Don't want to apply			
Unknown		Not Practiced C		Not Practiced E

FIGURE 33.11 Many problems are communication problems. Area C, for example, shows that if information is known to the technical–managerial staff but not to the operator, the technique is not practiced.

practiced. Knowledge alone is not sufficient; note also that both groups must want to apply the knowledge. Fukuda summarized the figure in the saying **defense before offense.** In defense, try to move from categories B, C, and D to category A. Defense examples are to put warnings and suggestions into easily readable visible form, to clearly define standard operations, and to develop tools enabling less effort yet more skill to be put forth. Only when you are "farming as well as you know how" do you take the offense and make changes in equipment and manufacturing conditions.

One way to emphasize communication is to post the fish diagrams on the wall next to the machines and invite everyone to make comments and suggestions. You will have many detectives looking for clues rather than just one person. The goal is to get all workers to think beyond what they are told to do. The Circle then takes the best of these ideas for further investigation.

For investigation, you can use a **passive approach** and use a control chart such as that in Figure 33.10. When something on the chart changes, try to go back and find out what changed in the process. An alternative is to use an **active approach** and use EVOP (see Chapter 4). In the active approach, variables are selected for investigation and the chart shows the result of the selected change. The active approach is preferable since it produces more knowledge in a shorter time with less effort. Remember, however, as was pointed out in the EVOP discussion in Chapter 4, to make evolutionary changes rather than revolutionary changes.

4 IMPLEMENTING CHANGE

To accomplish change, the project needs to be managed. Then a proposal needs to be made to decision makers.

4.1 Project Management Five points will be emphasized (Konz, 1994).

4.1.1 Scope defined
The **project scope** should be agreed upon ahead of time: Who will do what by when? Who will help you for how many h/week? What resources (budget) will be available? What is the "work product" (i.e., report, computer analysis, etc.)? A tentative schedule for the various events (milestones) helps focus the problem. It is important that the project scope be put in writing. If your supervisor will not put the scope in writing, you should do it and submit it for approval. It is important to minimize potential misunderstandings as soon as possible.

4.1.2 Client's servant
Being your **client's servant** is an attitude. You are trying to help a client, a customer. You are not a master, giving commands. You are trying to satisfy client needs; the client is not trying to satisfy you.

4.1.3 Frequent consultation
Frequent consultation not only provides you with more facts but also improves the attitudes and emotions of clients toward you. Consult with the user, not just with staff. Information can be obtained from face-to-face consultation and tours (take notes) and from videotapes of existing operations. For videotapes of existing operations, be sure to take several minutes of tape with multiple views. Views of multiple operators are good because operators often use considerably different micromotions. It also may be possible to view similar operations in other facilities. Inspection of records tends to produce relatively little information, because recorded data generally are not complete and in convenient form. Written requests for information are fairly useless. No one wants to fill out forms, and if the forms are returned, you get "bare bones" answers.

The decision makers should be consulted at multiple approval points, not just at the end of the project. Consultation allows them to change their minds on various points early (rather than after you have done irrelevant work) and to change priorities. Early discussion of alternatives also tends to bring out potential problems that you may not have considered.

4.1.4 Use experts
Both line and staff people consider themselves experts in their jobs. Whether they are expert is not as important as their own concept of themselves as expert. You want them on your side, not against you.

Consult these experts during the project. For example, cost-accounting departments consider themselves expert in cost justifications. Even if you are able to do a cost justification by yourself, it is better to have the cost accountants go over your project during the project. You want to go into the final decision stages with them on your side, not with them unfamiliar with the project, or even worse, opposing you.

4.1.5 Give the buyer a choice

Every decision maker wants to make decisions. They do not like to be confronted with no alternatives and just rubber stamp something. The choices can be "yes" or "no" for the entire project, for pieces of the project, for the implementation schedule, and so forth, but if you make the alternative all or nothing it may be nothing.

Making a trade-off among multiple criteria may be easier for the decision makers if you present a summary such as Table 7.6. The table allows criteria with different dimensions (cost, quality, ease of supervision) to be totaled for various alternatives. Giving the best solution a value of 100% improves ease of comparison. Consult the decision makers ahead of time on how much weight each criterion has.

4.2 Written Proposal

Eventually the project will come to the official decision-making stage. A formal written proposal is needed. The goal of the proposal is acceptance and implementation of the proposal recommendation. Thus, the proposal needs to convey convincing information. The information itself is not sufficient. The proposal must convince the decision makers to accept the recommendation, and their decision will convert a recommendation into action, a deed.

Before writing the report, you need a plan—an outline. You need to answer the following: What is the purpose of the report? Who will read it? What do they not know that the report will tell them?

Some decision makers tend to avoid risk; others emphasize economic gain. You need to know your decision maker. Generally, decision makers will be interested primarily in the economic aspects of the proposal rather than the technical aspects. Since they probably will accept the proposal as technically sound, avoid technical overkill. Remember that the decision maker will be reviewing many proposals. Not all will be accepted. Why should yours be?

Decision makers are busy; thus, the report should present the proposal concisely. It also may present the proposal in depth, but the decision makers are unlikely to go through the detail. (They hired you to do the detail work!) They will want the detail to be available for their inspection, should they choose to inspect it. This leads to the concept of a short report, with the detail in appendices.

The following is a good general format with a logical structure obvious to both the reader and the writer. On the cover sheet, put the project title, the date, your name, and the word RECOMMENDATION followed by 50–100 words giving the recommendations. At the top of the second page, write the word PROBLEM followed by a 10- to 50-word statement of the problem. Next, write ANALYSIS followed by the analysis. This section may take 3 to 10 pages. Reduce the material by putting the detail in appendices. The appendices should be preceded by a table of appendices, and each appendix should be identified by a letter. Pages should be numbered in each appendix, (for example, B14 identifies page 14 in appendix B.) Finally, write CONCLUSIONS and give them briefly. Some decision makers also like an EXECUTIVE SUMMARY which summarizes the entire project, including recommendations, in about 400 words.

Determine the preferred writing style, and use it. Some people want an informal, active-voice style, such as "We measured background noise levels," or even "I measured background noise levels." Others want a formal, passive style such as "Background noise levels were measured."

Proposals should include the following economic information: annual cost of the present and proposed methods, capital cost of the proposed method (such as new equipment, installation costs, training costs), estimated project life, and expected savings. On simple proposals, an annual return on investment is sufficient, but expensive, long-range projects often require a cash flow analysis. State your assumptions specifically (product life = 3 more years at 8,000 units/yr, no change in product design, the change will not affect sales of spare parts, quality is not affected). Good proposals also have a schedule of "milestones" (1 May: proposal accepted; 15 May: change schedule OK by facility manager; 15 June: all machines moved; 20 June: production back to normal).

Proposals should go through several drafts. Be sure to schedule enough time for multiple drafts. Consider handwritten material as notes rather than a draft; too many errors of spelling, composition, and structure lurk in handwritten material to dignify it with the word "draft." Plan to have the report redone at least twice for a small project and more often for an important project. The report must not have any spelling, grammatical, or typographical errors if it is to have a reasonable chance of being accepted. Insist that the word processing operator use spelling checking programs. Many word processing programs now have grammar subroutines that check the use of active versus passive voice, agreement of subject and verb, trite sayings, and redundant words; they may also suggest alternative adjectives and so forth. Unless you are a skilled writer, you should use a grammar program routinely—whether you take the program's advice or not.

Chapter 20 has eight guidelines for good table design and discusses features of good and poor graphs.

The decision makers will generally be poorly prepared to evaluate the technical merits of your proposal. After all, you are the expert and they are decision makers. However, they can judge typing, spelling, and grammar; if these are poor, they will consider the technical material to be poor also.

4.3 Oral Presentation

Oral presentations are usually accompanied by visual aids. Speakers sometimes pay more attention to their clothes than to the quality of their visual aids, but they should remember that audiences focus on and remember the visual aids.

The first choice to make concerning visual aids is the medium. Transparencies or slides are used most often, but occasionally videos are the choice. See Table 33.3 for comments on good transparencies and slides.

Video is effective for showing existing operations. Use a remote control when presenting. For voiced videos, dub the voice after filming rather than while filming. Unless professionally made, videos should be used sparingly because audiences will compare them with other, more professional, videos they have seen.

Slides produce a better visual image than do overhead transparencies. Slides also have better options with color, for both photographs and text. Although the slide has the highest visual quality, it also takes the longest to prepare and costs the most. In addition, because slides are more formal, the speaker often takes more care with them. Transparencies are more likely to have poorly organized material presented with inferior images. If you do use transparencies, dark blue text on a clear background produces a good image with little loss in legibility. Multicolor transparencies can be made with computer graphics.

After the title slide (or transparency), there should be at least 1 slide (transparency)/minute. Using 2/minute is probably high for projections with considerable text, but pictorial scenes could be comprehended even at 4 or 5/minute. That is, a 5-min talk would have at least 6 projections; a 10-min talk would have at least 11. The problem usually is too few rather than too many.

An alternative to slides or transparencies is the flip chart, a large paper pad to use with a marker. Advantages are that flip charts can be used in any

---

TABLE 33.3 Effective overhead transparency and slide presentations. An important characteristic of slides and transparencies is that they force you to organize your presentation.

Overheads

Stay within 7½ x 9½ format, since projector platen is not 8½ x 11.

Use color to "outline" the talk as well as for figures and overlays.

Organize your presentation so it has a beginning, middle, and end. Tell what you will tell them, tell them, tell them what you told them.

Use high-contrast originals; be sure the transparency is easily readable from the farthest viewer position; check by projection, not hand-held viewing.

Use, for graphs, grids or graph paper under paper originals to get proper scales and relationships.

Keep projector on your left if you point with your left hand, on your right if you point with your right. Point to the transparency with a pointer or pen, not your finger. Face the audience, not the screen.

Have material prewritten on the film. Handwriting speed is about .4 words/s, speaking is 2 to 4 words/s, and reading is 3 to 9 words/s.

Mount frame borders on transparencies to serve as a notecard—to the audience you appear to be speaking without notes.

Slides

Don't put too much onto one slide. Three guidelines are: 1 slide/min of presentation, 20 words maximum per slide (6-7 words/line; 5 lines; 3 vertical columns), and maximum of 9 double-spaced lines high and 54 elite (45 pica) characters wide.

Use color, not black and white. Make color slides from black and white text by adding a yellow or light pastel overlay to the slide. Reverse slides (light letters on dark background) can be made in color from black and white text.

Keep information/slide to 1 idea; no more than 3 curves/graph.

Make material *easily* readable from the farthest viewing position. Graphs and tables that are satisfactory in print need to be simplified and lines emphasized for slides. Leave space—at least the height of a capital letter—between lines of text.

Use duplicate slides rather than backing up during a presentation.

Practice your talk. Do it *early* so you can make changes in the slides.

FIGURE 33.12   Oral presentation checksheet.

ORAL PRESENTATION CHECKSHEET

Student_____Team _____Time_____Grade_____

_____ Team attendance	_____ Transparency had frames	_____ Verbal speed
_____ Properly dressed	_____ Lettering readable	(slow, average, fast)
Title slide gave:	from rear	_____ Verbal delivery
_____ project name	_____ Transparency framed on	(varied, monotonous)
_____ date	screen	
_____ team members	_____ Transparency used color	Eye contact:
	(no, OK, excessive)	_____ decision maker
_____ Numbers too accurate	Pointed with:	_____ audience
	_____ finger	
_____ Had physical objects	_____ pen	Talked to:
	_____ pointer	_____ screen
_____ Had overlays	Talk organization:	_____ audience
	_____ intro/overview	
	_____ talk	
	_____ summary/conclusion	

room (e.g., a restaurant meeting room) and that the series of sheets are like frames that can be prepared in advance or even during the presentation. Two disadvantages are the lack of magnification and the difficulty of duplication.

Generally, oral presentations (briefings) should be relatively short, 15–30 min. Additional detail can be provided in answers to specific questions or through the written report. Do not assume the people present for the oral presentation have read the written report.

Visual aids and lighting are inadequate in many meeting rooms. If possible, practice your talk ahead of time in the room in which you will make the presentation. Check power cord lengths, light switch locations, projection distances (your material should be legible from the back of the room), sight angles, and so forth. For important presentations, rehearse not only the talk but also the question-and-answer period (as politicians do before appearing at a news conference).

The presentation itself should have a logical structure that the audience can follow. An effective sequence goes from *what* to *why* to *how*. (If the presentation is to a key executive, such as company president or division manager, find out what type of presentations that executive likes.) Describe the implementation plan, and conclude with a summary (e.g., table of cost savings, space requirements, etc.). If you don't know the answer to a question, say "I don't know, but I will find out." See Figure 33.12 for a checksheet.

After the presentation, you may wish to summarize the consensus or decision on a transparency, a flip chart, or the blackboard. Anyone who disagrees with your written interpretation can give immediate feedback.

The meeting should include those who can contribute, those with divergent views, those who can make the decisions, and those who will carry out the decisions.

REVIEW QUESTIONS

1. Give the formula relating quality, acceptance, and amount of improvement.
2. Discuss the statement "Where you stand depends upon where you sit."
3. Briefly describe a situation with resistance to change. Sketch a force diagram of the situation and discuss strategies to improve acceptance.
4. Give some examples of people who are considered to be local experts and technical experts.
5. Contrast the concepts of Taylorism and Quality Circles, relating them to education of the workforce.
6. Briefly discuss the 5 points of project management.

7. Give a sentence written in informal active voice. Give the same information written in formal passive voice.

8. When using a transparency projector, should you face the audience or the screen? Which hand should you use when pointing to the material on the projector?

9. In an oral technical presentation, about how much time should each transparency or slide be given?

10. List the 4 sections into which a written technical report should be divided.

REFERENCES

Arai, J. Japanese productivity: What's behind it? *Modern Machine Shop,* Vol. 52 [4], 117-25, 1979.

Berger, R. and Shores, D. (eds.). *Quality Circles: Selected Readings.* New York: Marcel Dekker, 1986.

Chaney, F. Employee participation in manufacturing job design. *Human Factors,* Vol. 11 [12], 101-106, 1969.

Coch, L. and French, J. Overcoming resistance to change. *Human Relations,* Vol. 1 [4], 512-32, 1948.

Frost, C., Wakely, J., and Ruh, R. *The Scanlon Plan for Organizational Development.* Michigan State University Press, 1974.

Fukuda, R. The reduction of quality defects by the application of a cause and effect diagram with the addition of cards. *International J. of Production Research,* Vol. 16 [6], 305-19, 1978.

Fukuda, R. Introduction to the CEDAC. *Quality Progress,* Vol. 14 [11], 14-19, 1981.

Gryna, F. *Quality Circles.* New York: American Management Association, 1981.

Konz, S. Quality Circles: An annotated bibliography. *Quality Progress,* Vol. 13 [4], 30-35, 1979.

Konz, S. *Facility Design: Manufacturing Engineering.* Scottsdale, Ariz.: Publishing Horizons, 1994.

Leisman, F. An engineer-in-charge of manufacturing development, General Motors Corp.; personal communication, June 2, 1977.

Munson, F. and Hancock, W. Problems of implementing change in two hospital settings. *American Institute of Industrial Engineers Transactions,* Vol. 1 [4-12], 166-76, 1969.

OVERVIEW

Job evaluation determines the relative worth of jobs. It is an analytical, consensus-building approach, not a procedure based on science. The first step is the job description. Then the jobs are arranged in sequence and a wage value is assigned. Labor grades simplify administration and permit paying for individual merit as well as the value of the job.

CHAPTER CONTENTS

1 Goal of Job Evaluation
2 Job Description
3 Arranging Jobs in Order

KEY CONCEPTS

differential pay
effort factors
entry level
job condition factors
job descriptions
labor grades

pay for knowledge
responsibility factors
skill factors
top of the bracket
two-tier
work to rule

1 GOAL OF JOB EVALUATION

1.1 Comparable Worth

Job evaluation is a method for determining the relative worth of jobs. It is concerned with providing hard data on which to make conclusions about the rate for the job. The "formula for fairness" offers system and stability if planned for and developed on the basis of consensus, cooperation, and concord. It is a rational consideration of the more important differences among the inputs of human work. The crux of the problem of different pay for different people (**differential pay)** is assessing and agreeing on the fairness of differential pay—the problems of equity and acceptability. The systems are not perfect. At root they are judgments, a thread in a total fabric, a skeleton framework for a system of equitable payment.

There are three goals: (1) make wages for jobs consistent within the organization, (2) make job wages comparable to job wages paid by competitors, (3) allow for individual merit.

For more information, see Bartley (1981), *Job Evaluation* (1986), and Milkovich et al. (1992).

1.2 Labor Grades

Wages should consider both the worth of the job and worth of the individual. See Figure 34.1. After determining the worth of the job (see Section 3 of this chapter), jobs similar in worth are grouped together into **labor grades.** For administrative purposes, all jobs in a labor grade are considered equal for pay purposes. Thus, the horizontal dimension of the labor grade considers job worth. The vertical dimension of a labor grade, on the other hand, considers individual merit.

One important decision to be made is the relative importance of job worth compared to individual merit. If merit has relatively little importance, the top and bottom of the bracket might be 5% from the middle; if merit is very important, the limits might be 50% either way. What number do you think is best?

Another design decision is the slope and shape of the curve connecting the labor grade means. How steep should the slope be? Should the shape be a straight line or a curve?

Another challenge is how to determine individual merit. In many firms it is seniority, or time-in-grade. In Japan, age often is the determining factor.

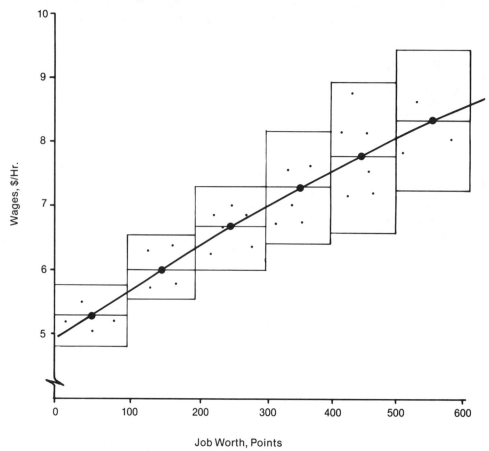

FIGURE 34.1   Labor grades (the boxes) summarize the equation of wage compared to worth. All points (jobs) within a box are considered of equal worth for administrative purposes. The vertical dimension of the box permits adjustment for different individual contributions to the same job; it should be greater for more complex jobs. The boxes generally have a vertical overlap so a person at the bottom of the bracket in a higher grade may earn as much as a top-of-the-bracket person in a lower grade. Both the overlap and the vertical spread from the mean wage for the grade should increase with complexity so individual jobs can be modified with minimal effect on an incumbent's pay and so there is more scope for merit rating.

In the United States, however, the job is the primary factor, although some allowance is made for the holder of the job. The individual allowance usually is some combination of merit and seniority.

Although in theory people are scattered vertically throughout a labor grade, in practice, after about five years most people are at the **top of the bracket.** This can lead to a very highly paid workforce and thus a very high-cost workforce. One of the advantages of a new plant is that its average labor costs are quite a bit lower because many people are at the bottom of the bracket instead of the top. In an effort to reduce this problem, some firms have instituted **two-tier** plans. In one form of the two-tier system, the top of the bracket is lower for new hires (**entry level**) than for existing employees. For example, the top of the bracket for new hires is $15/h while it is $18/h for existing employees.

In another form, the existing wage structure is maintained, but it takes longer for the new hire to reach the top. For example, present employees may have been able to make $18/h after 3 to 5 years but for new employees it will take 10 to 15 years.

Under both forms of two-tier plans, the new hires feel like second-class employees. Two-tier plans have been implemented in only a few firms.

2 JOB DESCRIPTION

Job evaluation has four steps: (1) describe the class of job (e.g., clerk-typist 2), (2) specify the minimal hiring requirements, (3) arrange the jobs in order of worth, and (4) describe a specific job at a specific location—a "position description" (clerk-typist for electrical engineering). See Figure 34.2.

2.1 Describe the Job Class   Describing the job class has two substeps: (1) securing the factual information and (2) writing the information in standardized format. The information to obtain includes:

- duties and the percent of time devoted to each
- responsibilities
- knowledge and skills needed
- performance standards to be met
- what is used
- working conditions, including hazards

Personnel staff generally secure the information using some combination of questionnaires, interviews, and observation of the jobs. Questionnaires, if used, need to be checked and supplemented by worker interviews for better worker acceptance and improved accuracy. Interviews with the workers are the most common approach. Use a structured interview. Observation is a good check but generally is not sufficient by itself.

Although the information gathering and write-up should be done by a trained person, the results should be checked for clarity and brevity by a committee of interested parties. For accuracy in the third step—ranking of the jobs—all job descriptions should be about the same length and contain about the same level of detail. Raters will be more consistent if the person writing the job description knows what types of information they want and includes them in the description.

Keep the number of job descriptions (and jobs) reasonably small. The most common problem is over-specification of jobs and resulting lack of flexibility as workers **work to rule** and refuse to do work that is not in their job descriptions. For example, a carpenter might refuse to turn on the power as that would be an electrician's work. All job descriptions should therefore have a miscellaneous duties section. In Figure 34.2 it is "Perform other duties as assigned."

Rather than the catch-all of a miscellaneous duties section, some firms have begun an aggressive program of reducing the number of job descriptions by broadening the language. Typically, the goal is 5 to 10 job descriptions for the entire blue collar workforce in a plant, replacing the former 50 to 100 descriptions. For example, all maintenance personnel might fall into 1 category instead of jobs of carpenters, electricians, plumbers, millwrights, and so forth. All inspectors might be called inspectors rather than electrical inspectors, mechanical inspectors, and so forth. However, this means that more emphasis is being placed on individual worth than on just the job. This, in turn, means more training for employees who now must be able to do it all. The big benefit for management is improved flexibility in shifting people among work assignments and thus reduced labor costs since there is less idle time. However, this policy of few labor grades and broadly trained workers (sometimes called **pay for knowledge**) is a sharp change from the conventional detailed job description approach and may be resisted by workers familiar with and comfortable with the concept of knowing just one job.

2.2 Specify Minimal Hiring Requirements
This step, job specification, has become a very difficult task with the increased emphasis on equal employment opportunity, nondiscrimination, and hiring the handicapped. In many situations, organizations are sued if they don't make a "reasonable effort" to modify the job to fit a handicap (such as a bad back). If job requirements are too rigorous, the organization may be challenged to prove that they

FIGURE 34.2   Job description for a clerk 2. **Job descriptions** should be accurate, brief, and clear statements of what the worker is expected to do. Begin sentences with verbs. Be specific. State duties rather than qualifications of the incumbent.

SECTION A: Position Purpose:
Explain concisely why the duties and responsibilities assigned to this position are essential to agency operations.

The person occupying this position functions as the office receptionist, typist, and clerk for the Electrical Engineering Department. Since this position is the only one of its type in the Department, it is obviously critical to the operation of the Department. This person is expected to meet students, staff, and the public, provide general information about the Department, and direct the visitor to the proper place or person for more information or help. Duties of this position include the typing of research manuscripts, daily correspondence, and other material submitted by faculty and staff.

SECTION B: Duties and Responsibilities:
Instructions: (1) Number each duty and indicate approximate percent of time spent on each major duty or group of duties. (2) Include specific data as to responsibility for direction of work of other employees; position numbers and class titles of employees supervised; degree of responsibility for funds or actions, decision making, and program and policy planning; nature, purpose, and level of contacts within and outside the agency. (3) Indicate how independently of supervision this position functions, or conversely, how closely and directly the position is supervised.

Duty No. and Percent of Time	Duties
1. 25%	Act as office receptionist, answer general questions about the Department, direct visitors, answer telephones, take messages. Person must have general knowledge about operation of the Department and University. Visitors include faculty, staff, students, and the general public.
2. 45%	Type research papers, proposals, correspondence, examinations, and forms. Most work is submitted to this person in the form of hand-written drafts. Work is checked by person submitting the material. Little supervision is required; work checked when completed.
3. 3%	Use duplicating machine and copying machine. Employee is responsible for meeting deadlines set by faculty, staff, and supervisor.
4. 25%	Maintain records of all students enrolled in Electrical Engineering. Related tasks include assigning advisors, transferring students, posting semester grades, recording drop-add slips, and filing all student-oriented transactions. Employee maintains line schedule, textbook lists, and course outlines. Little supervision is given. Results checked as files accessed.
5. 2%	Handle mail and maintain office supplies. Assist other staff members with clerical work. Perform other duties as assigned.

SECTION C: Minimum Qualifications (Education and Experience, Certificates, Licenses, Degrees, Skills Required)
Good oral and written abilities. Vo-Tech graduate with course emphasis on secretarial science, one-year experience.

are relevant in a court case. For example, can it prove in court that a high school diploma is necessary for job X? Desire to hire well-qualified employees is not proof! On the other hand, putting no requirements on a job doesn't help much either. Hiring people and then allowing them 30 to 90 days to prove they can do the job results in poor quality and may result in worker injury if the new hires aren't physically qualified. If an entry-level job is part of a job sequence (such as helper, apprentice, electrician), specify enough qualifications for the entry job or soon you will have people in the entry job who cannot complete the normal sequence.

2.3   Arrange Jobs in Order of Job Worth   This major step is discussed in detail in Section 3, "Arranging Jobs in Order."

2.4   Describe a Position   After the general categories of jobs are described (clerk 1, clerk 2, secretary 1, etc.) it is necessary to describe individual jobs at individual locations (the job in electrical engineering). Then it can be determined that this specific job should be a clerk 2. Sometimes the positions are described first and then the combinations are used to form a job description.

3 ARRANGING JOBS IN ORDER

3.1 Background

There are many different jobs: lion tamer, astronaut, cook, secretary, engineer, corporate executive, teacher, welder, truck driver, assembly operator, and so on. Fortunately, this variety can be reduced, because job evaluation has as a goal the arrangement of jobs within an organization, not between organizations. There still may be a great diversity, however. This diversity is reduced to a practical level by setting up multiple plans—for example, 1 plan for clerical/office, 1 for managerial/technical, and 1 for shop/maintenance. Different plans for the different groups are justified 2 ways:

1. Factors vary in importance by group. For example, effort and working conditions have relatively little importance in distinguishing among office jobs but considerable importance in shop jobs. Responsibility and initiative might vary considerably among professional/technical jobs and little among clerical jobs.

2. Wage curves are the goal. Wages for the work are, to some extent, determined by competition with other employers. The relative wages of shop jobs generally are higher than for clerical jobs since so many people are willing to exchange the white collar status of a clerical job for the higher wages of a shop job. In addition, professional/technical jobs have to compete on a national rather than a local basis. An engineer may look at wages all over the country, whereas shop people may compare wages only within the community; thus, the level of competition is higher for jobs in which people are more mobile, and organizations have to offer relatively higher wages to meet the competition. However, except for monopolies such as the government, organizations generally have a local wage curve (rather than a national curve) for both clerical and shop personnel.

There are 4 approaches to arranging jobs: (1) ranking, (2) classification (such as the Civil Service grades of the U. S. government), (3) factor comparison, and (4) point systems. This chapter will discuss only the most popular approach, point systems. For information on the others (and more detail on point systems), consult the references at the end of the chapter.

3.2 Point Plans

The general concept of a point system is to analyze the levels of the factors relative to a job. Each factor has a number of levels with points allocated for each level. The analyst just adds up the points to get job worth. Although simple in concept, judgment is required.

The first judgment step is in the selection of the factors and levels for the plan. As a general consensus over the last 50 years, the various factors have been grouped into categories of skill, effort, responsibility, and job conditions. Typical **skill factors** are "education," "experience," and "initiative and ingenuity." Another breakdown of skill is "pre-employment training," "employment training and experience," "mental skill," and "manual skill." Typical **effort factors** are "physical demand" and "mental or visual demand." Typical **responsibility factors** are for "equipment or process," "material or product," "safety of others," and "work of others." Another breakdown of responsibility is for "materials," "tools and equipment," "operations," and "safety of others." Typical **job condition factors** are "working conditions" and "hazards." Another breakdown is "surroundings" and "hazards."

Most plans use 10 to 15 different factors within each of these four categories. Since some factors are more important than others and wider ranges of points are given for some factors than others, a smaller number of factors probably would be sufficient from a technical viewpoint. However, from an administrative (political) viewpoint, the larger number of factors permits the administrator to say "our plan covers everything."

Table 34.1 gives the factors and points assigned for the Midwest Industrial Management Association plan for shop jobs; there is another plan for office jobs. Table 34.2 gives the levels of factor 10 (working conditions) and Table 34.3 gives the levels of factor 11 (hazards). If, for example, the job of a lathe operator were being evaluated, for working conditions it probably would be 2nd degree or 20 points; for hazards it probably would be 3rd degree or 10 points.

In the interests of accurate job evaluation and acceptance of the job evaluation, an evaluation of any specific job should never be made by a single individual. Each allocation of points should be checked by a committee of diverse backgrounds and interests and a consensus reached. Jobs should be reevaluated periodically (every 5 or 10 years), because jobs change considerably over time. The success of job evaluation plans has been founded on such analytical, detailed, consensus-building techniques to replace opinionated, general, whimsical approaches.

Many women feel that "typically female" jobs have not been given enough credit when points are assigned. The basic question is how much weight should be given to experience and initiative (often found in white collar jobs held by women) and how much weight should be given to physical demand

TABLE 34.1   Scoring system for the MIMA job evaluation plan for shop jobs.

Job Factors	Degrees				
	1st	2nd	3rd	4th	5th
SKILL					
1. Education or Trade Knowledge	14	28	42	56	70
2. Experience	22	44	66	88	110
3. Initiative and Ingenuity	14	28	42	56	70
EFFORT					
4. Physical Demand	10	20	30	40	50
5. Mental and/or Visual Demand	5	10	15	20	25
RESPONSIBILITY					
6. Equipment or Process	5	10	15	20	25
7. Material or Product	5	10	15	20	25
8. Safety of Others	5	10	15	20	25
9. Work of Others	5	10	15	20	25
JOB CONDITIONS					
10. Working Conditions	10	20	30	40	50
11. Hazards	5	10	15	20	25

Grade Ranges

Score Range	Grades	Score Range	Grades
139	12	250-271	6
140-161	11	272-293	5
162-183	10	294-315	4
184-205	9	316-337	3
206-227	8	338-359	2
228-249	7	360-381	1

Maximum Points 500

TABLE 34.2   Factor 10: Working conditions.

This factor measures the surroundings or physical conditions under which the job must be done and the extent to which those conditions make the job disagreeable. Consider the presence and relative amount of exposure to dust, dirt, heat, fumes, cold, noise, vibration, wetness, etc. When working conditions vary with specific work assignments, such as found in maintenance jobs, the degree selected must represent the weighted average of all the conditions encountered.

1st DEGREE
Excellent working conditions with absence of disagreeable conditions

2nd DEGREE
Good working conditions; may be slightly dirty or involve occasional exposure to some of the elements listed above

3rd DEGREE
Somewhat disagreeable working conditions due to exposure to 1 or more of the elements listed above to the extent of being objectionable; may be exposed to 1 element continuously or several elements occasionally, but usually not at the same time

4th DEGREE
Disagreeable working conditions where several of the above elements are continuously present to the extent of being objectionable

5th DEGREE
Continuous and intensive exposure to several extremely disagreeable elements; working conditions *particularly disagreeable*

TABLE 34.3   Factor 11: Hazards.

This factor measures the hazards, both accident and health, connected with or surrounding the job, considering safety clothing, devices, or equipment that have been installed or furnished the work location, the material being handled, the machines or tools used, the work position, and the probable extent of injury in case of accident.

1st DEGREE
Accident or health *hazards negligible;* remote probability of injury

2nd DEGREE
Accidents improbable, outside of minor injuries, such as abrasions, cuts, or bruises; health hazards negligible

3rd DEGREE
Exposure to lost-time accidents possible, such as severe injuries to hand or foot, loss of finger, or eye injury, etc.; some exposure to health hazards, not incapacitating in nature

4th DEGREE
Exposure to incapacitating accident or health hazards, such as loss of arm or leg, impairment of vision

5th DEGREE
Exposure to accidents or health hazards which may result in *total disability or death*

and working conditions (often found in blue collar jobs held by men). A point to remember is that the points reflect to some extent not only the requirements of the job but also the desire of people to do the job. As extreme examples, consider the jobs of selling women's clothing at retail and collecting garbage. The amount of initiative, responsibility, and the like is relatively high for the salesclerk but the supply of people who want to do this job is very high, so it is a minimum wage job. It is hard to find people who will collect the garbage, so the result is that they are relatively well paid.

REVIEW QUESTIONS

1. How is pay based both on the worth of the job and the worth of the person holding the job?
2. What are the three ways of obtaining information for a job description?
3. Why is it bad to have too many different job descriptions?
4. Why do organizations have different plans for different groups?
5. List the four major categories for the factors in the point systems.

REFERENCES

Bartley, D. *Job Evaluation.* Reading, Mass.: Addison-Wesley, 1981.

Job Evaluation. Geneva, Switzerland: Int. Labour Office, 1986.

Milkovich, G., Newman, J., and Brakefield, J. Job evaluation in organizations. In *Handbook of Industrial Engineering,* 2nd ed., Salvendy, G. (ed.). New York: Wiley, 1992.

INDEX